Wireless Networks

About the Authors

Clint Smith, P.E., is currently the CTO for Rivada Networks and has extensive experience in all aspects of fixed and mobile wireless telecommunications, including senior management, system design, project management, budgeting, implementation, operations, and sales. He is the author of several books published by McGraw-Hill Education, including *Wireless Network Performance Handbook* and *Wireless Telecommunications FAQs*. Mr. Smith has been awarded numerous patents related to wireless mobility and public safety. He also has specific design and operational experience with LTE, CDMA/1xRTT/EVDO, GSM/GPRS/EDGE, WCDMA/HSPA, TDMA, iDEN, AMPS, LMDS, WiMAX, WiFi, Microwave PtP, and public safety communications systems.

Daniel Collins, a 20-year veteran of the wireless communications industry, is currently Chief Technology Officer—Wireless at Jasper Wireless, Inc. He began his career at Ericsson, where he worked in product development and product management roles and helped develop some of the world's earliest GSM networks. Mr. Collins also served as an independent consultant for many years on 2.5G, 3G, and IP technologies to tier-one service providers around the world, and worked with Alcatel North America as Vice President of Fixed Mobile Convergence. He is the co-author of *Carrier-Grade VoIP,* published by McGraw-Hill Education.

Wireless Networks

Design and Integration for LTE, EVDO, HSPA, and WiMAX

Clint Smith, P.E.

Daniel Collins

Third Edition

McGraw Hill Education

New York Chicago San Francisco
Athens London Madrid
Mexico City Milan New Delhi
Singapore Sydney Toronto

McGraw-Hill Education books are available at special quantity discounts to use as premiums and sales promotions or for use in corporate training programs. To contact a representative, please visit the Contact Us page at www.mhprofessional.com.

Wireless Networks: Design and Integration for LTE, EVDO, HSPA, and WiMAX, Third Edition

1 2 3 4 5 6 7 8 9 0 DOC/DOC 1 2 0 9 8 7 6 5 4

ISBN 978-0-07-181983-1
MHID 0-07-181983-5

This book is printed on acid-free paper.

Sponsoring Editor
Michael McCabe

Editorial Supervisor
Stephen M. Smith

Production Supervisor
Pamela A. Pelton

Acquisitions Coordinator
Bridget L. Thoreson

Project Manager
Harleen Chopra,
Cenveo® Publisher Services

Copy Editor
Cenveo Publisher Services

Proofreader
Steven Moc

Indexer
Cenveo Publisher Services

Art Director, Cover
Jeff Weeks

Composition
Cenveo Publisher Services

This book is dedicated to my wife, Mary.

—Clint

Contents

Preface

Well, here we are again and there are no signs of this industry slowing down or becoming less complicated.

I was approached by McGraw-Hill Education to update and enhance the Second Edition of this book and I decided to take on the challenge. However, as all of us know, the devil is in the details and this task was daunting. A quick glance at the topics included in this new Third Edition will confirm that the amount of material has expanded greatly. Every attempt was made to make this book a very usable source of information by providing practical guidelines instead of the typical theory.

Presently, 4G implementation is in full swing and LTE seems to be a harmonizing technology for the wireless industry at the moment. However, as expected, discussions regarding 5G and beyond are appearing.

With that said, I have updated all the chapters to reflect various technology advances that have taken place since the publication of the Second Edition. I have also included several new chapters, including some topical chapters covering LTE, WiMAX, and backhaul.

As expected, the size of this book has grown, and some of the legacy network information, which I think is exceptionally important and required for migration strategies, needs to be available. Therefore, the 1G, 2G, and 2.5G technology chapters in the Second Edition have been reformatted as appendices and posted for download at www.mhprofessional.com/WirelessNetworks. I strongly recommend you take advantage of this material.

The exact form and function of the wireless industry are still evolving and will continue to do so as the quest for enhancing the user experience with an all-IP network with higher throughputs is continuously pursued. However, as we move forward it is the edge device and applications that are the true drivers of the wireless industry, and the network is just the enabler.

A brief summary of this book is in order to help bind the material that is contained within it.

- Chapter 1 is a general introduction to wireless communications and is essential for the remaining chapters.
- Chapter 2 discusses Third-Generation (3G) wireless systems.
- Chapter 3 covers UMTS, aka WCDMA, and also discusses HSDPA.
- Chapter 4 discusses CDMA2000, and EVDO is also discussed in depth.
- Chapter 5 covers TD-SCDMA.

- Chapter 6 covers TD-CDMA.
- Chapter 7 is a new chapter covering LTE.
- Chapter 8 is a new chapter covering WiMAX.
- Chapter 9 covers VoIP.
- Chapter 10 is a new chapter covering wireless broadband.
- Chapter 11 was revised to address numerous 3G RAN design issues.
- Chapter 12 was revised to address core network design issues related to 3G and 4G.
- Chapter 13 is a new chapter covering backhaul.
- Chapter 14 was updated regarding various antenna system considerations and design aspects.
- Chapter 15 addresses UMTS (WCDMA) system design guidelines.
- Chapter 16 addresses CDMA2000 with EVDO system design.
- Chapter 17 is about TD-SCDMA and TD-CDMA system design.
- Chapter 18 is a new chapter covering LTE system design.
- Chapter 19 is a new chapter covering WiMAX system design.
- Chapter 20 has many updates for communication-site designs, including in-building and colocation guidelines.
- Chapter 21 is a new chapter covering 5G and beyond.

As you can see, the breadth and depth of the material contained in this book will provide ample ammunition for design engineers either to use directly or to augment the existing knowledge base. Additionally, the breadth of material will enable design engineers and management to have a better understanding of competitive design and technology choices that have been or are being deployed. However, as with any book, it is a guide and not an absolute reference.

In short, I hope this edition will prove to be a valuable resource for you and help to improve the wireless engineering profession as it continues to evolve.

Acknowledgment

I wish to thank all of my colleagues who have helped with providing input and suggestions for the revision of this book, since none of this would be possible without them.

Clint Smith, P.E.

CHAPTER 1

Introduction

1.1 The Amazing Growth of Mobile Communications

Over recent years, telecommunications has been a fast-growing industry, with the focus shifting from voice-centric to one that is data-oriented, enabling through middleware the ability to provide a complete communications package instead of having multiple devices and technologies. This growth can be seen in the increasing revenues of major telecommunications carriers and the continued entry into the marketplace of new competitive carriers in addition to carrier consolidation. No segment of the telecom industry, however, has seen growth to match that experienced in mobile communications. From relatively humble beginnings, the last 25 years have seen an explosion in the number of mobile communications subscribers, and it appears that growth is likely to continue well into the future with an emphasis on broadband capability being delivered to end users no matter where they are located.

The growth in the number of mobile subscribers is expected to continue for some years, with the number of mobile subscribers surpassing the number of fixed network subscribers at some point in the near future. Although it may appear that such predictions are optimistic, it is worth pointing out that in the past, most predictions for the penetration of mobile communications have been far lower than what actually occurred. In fact, in several countries, the number of mobile subscribers already exceeds the number of fixed subscribers, which suggests that predictions of strong growth are well-founded. It is clear that the future is bright for mobile communications. For the next few years at least, that future means fourth-generation systems, the subject of this book.

Before delving into the details of fourth-generation systems, however, it is appropriate to review mobile communications in general, as well as first- and second-generation systems. As with most technologies, advances in wireless communications occur mainly through a process of steady evolution (although there is the occasional quantum leap forward). Therefore, a good understanding of third-generation systems requires an understanding of what has come before. In order to place everything in the correct perspective, the following sections of this chapter provide a history and a brief overview of mobile communications in general. Appendices A, B, and C provide some technical detail on first- and second-generation systems are available via download with the remaining chapters of the book dedicated to the technologies involved in third- and fourth-generation systems.

1.2 A Little History

Mobile telephony dates back to the 1920s, when several police departments in the United States began to use radiotelephony, albeit on an experimental basis. Although the technology at the time had had some success with maritime vessels, it was not particularly suited to on-land communication. The equipment was extremely bulky, and the radio technology did not deal very well with buildings and other obstacles found in cities. Therefore, the experiment remained just an experiment.

Further progress was made in the 1930s with the development of *Frequency Modulation* (FM), which helped in battlefield communications during World War II. These developments were carried over to peacetime, and limited mobile telephony service became available in the 1940s in some large cities. Such systems were of limited capacity, however, and it took many years for the mobile telephone to become a viable commercial product.

1.2.1 History of First-Generation Systems

Mobile communications as we know it today really started in the late 1970s with the implementation of a trial system in Chicago in 1978. The system used a technology known as *Advanced Mobile Phone Service* (AMPS), operating in the 800-MHz band. For numerous reasons, however, including the breakup of AT&T, it took a few years before a commercial system was launched in the United States. That launch occurred in Chicago in 1983, with other cities following rapidly.

Meanwhile, however, other countries were making progress, and a commercial AMPS system was launched in Japan in 1979. The Europeans also were active in mobile communications technology, and the first European system was launched in 1981 in Sweden, Norway, Denmark, and Finland. The European system used a technology known as *Nordic Mobile Telephony* (NMT), operating in the 450-MHz band. Later, a version of NMT was developed to operate in the 900-MHz band and was known (not surprisingly) as *NMT900*. Not to be left out, the British introduced yet another technology in 1985. This technology is known as the *Total Access Communications System* (TACS) and operates in the 900-MHz band. TACS is basically a modified version of AMPS.

Many other countries followed along, and soon mobile communications services spread across the globe. Although several other technologies were developed, particularly in Europe, AMPS, NMT (both variants), and TACS certainly were the most successful technologies. These are the main first-generation systems, and they are still in service today.

First-generation systems experienced success far greater than anyone had expected. In fact, this success exposed one of the weaknesses in the technologies—limited capacity. Of course, the systems were able to handle large numbers of subscribers, but when the subscribers started to number in the millions, cracks started to appear, particularly because subscribers tend to be densely clustered in metropolitan areas. Limited capacity was not the only problem, however, and other problems such as fraud became a major concern. Consequently, significant effort was dedicated to the development of second-generation systems.

1.2.2 History of Second-Generation Systems

Unlike first-generation systems, which are analog, second-generation systems are digital. The use of digital technology has a number of advantages, including increased capacity, greater security against fraud, and more advanced services.

Like first-generation systems, various types of second-generation technology have been developed. The three most successful variants of second-generation technology are *Interim Standard 136* (IS-136) *TDMA, IS-95 CDMA,* and the *Global System for Mobile* communications (GSM). Each of these came about in very different ways.

1.2.2.1 IS-54B and IS-136

IS-136 came about through a two-stage evolution from analog AMPS. As described in more detail later, AMPS is a *frequency-division multiple-access* (FDMA) system, with each channel occupying 30 kHz. Some of the channels, known as *control channels,* are dedicated to control signaling, and some, known as *voice channels,* are dedicated to carrying the actual voice conversation.

The first step in digitizing this system was the introduction of digital voice channels. This step involved the application of *time-division multiplexing* (TDM) to the voice channels such that each voice channel was divided into time slots, enabling up to three simultaneous conversations on the same *radiofrequency* (RF) channel. This stage in the evolution was known as *IS-54B* (also known as *Digital AMPS,* or *D-AMPS*), and it obviously gives a significant capacity boost compared with analog AMPS. IS-54B was introduced in 1990.

Note that IS-54B involves digital voice channels only and still uses analog control channels. Thus, although it may offer increased capacity and some other advantages, the fact that the control channel is analog does limit the number of services that can be offered. For this reason, among others, the next obvious step was to make the control channels also digital. This step took place in 1994 with the development of IS-136, a system that includes digital control channels and digital voice channels.

Today, AMPS, IS-54B, and IS-136 are all in service. AMPS and IS-54B operate only in the 800-MHz band, whereas IS-136 can be found both in the 800-MHz band and in the 1900-MHz band, at least in North America. The 1900-MHz band in North America is allocated to *Personal Communications Service* (PCS), which can be described as a family of second-generation mobile communications services.

1.2.2.2 GSM

Although NMT had been introduced in Europe as recently as 1981, the Europeans soon recognized the need for a pan-European digital system. There were many reasons for this, but a major reason was the fact that multiple incompatible analog systems were being deployed across Europe. It was understood that a single Europe-wide digital system could enable seamless roaming between countries as well as features and capabilities not possible with analog systems. Consequently, in 1982, the *Conference on European Posts and Telecommunications* (CEPT) embarked on developing such a system. The organization established a group called (in French) *Group Spéciale Mobile* (GSM). This group was assigned the necessary technical work involved in developing this new digital standard. Much work was done over several years before the newly created *European Telecommunications Standards Institute* (ETSI) took over the effort in 1989. Under ETSI, the first set of technical specifications was finalized, and the technology was given the same name as the group that had originally begun the work on its development—GSM.

The first GSM network was launched in 1991, with several more launched in 1992. International roaming between the various networks followed quickly. GSM was hugely successful, and soon, most countries in Europe had launched GSM service.

Furthermore, GSM began to spread outside Europe to countries as far away as Australia. It was clear that GSM was going to be more than just a European system; it was going to be global. Consequently, the letters *GSM* have taken on a new meaning—Global System for Mobile communications.

Initially, GSM was specified to operate only in the 900-MHz band, and most of the GSM networks in service use this band. There are, however, other frequency bands used by GSM technology. The first implementation of GSM at a different frequency happened in the United Kingdom in 1993. That service was known initially as *DCS1800* because it operates in the 1800-MHz band. These days, however, it is known as *GSM1800*. After all, it really is just GSM operating at 1800 MHz.

Subsequently, GSM was introduced to North America as one of the technologies to be used for PCS—that is, at 1900 MHz. In fact, the very first PCS network to be launched in North America used GSM technology.

1.2.2.3 IS-95 CDMA

Although they have significant differences, both IS-136 and GSM use TDMA. This means that individual radio channels are divided into time slots, enabling a number of users to share a single RF channel on a time-sharing basis. For several reasons, this technique offers an increase in capacity compared with an analog system, where each RF channel is dedicated to a single conversation. TDMA is not the only system that enables multiple users to share a given radiofrequency, however. A number of other options exist—most notably *Code-Division Multiple Access* (CDMA).

CDMA is a technique whereby all users share the same frequency at the same time. Obviously, since all users share the same frequency simultaneously, they all interfere with one another. The challenge is to pick out the signal of one user from all of the other signals on the same frequency. This can be done if the signal from each user is modulated with a unique code sequence, where the code bit rate is far higher than the bit rate of the information being sent. At the receiving end, knowledge of the code sequence being used for a given signal allows the signal to be extracted.

Although CDMA had been considered for commercial mobile communications services by several bodies, it was never considered a viable technology until 1989 when a CDMA system was demonstrated by Qualcomm in San Diego, California. At the time, great claims were made about the potential capacity improvement compared with AMPS, as well as the potentially improved voice quality and simplified system planning. Many people were impressed with these claims, and the Qualcomm CDMA system was standardized as IS-95 in 1993 by the U.S. *Telecommunications Industry Association* (TIA). Since then, many IS-95 CDMA systems have been deployed, particularly in North America and Korea. Although some of the initial claims regarding capacity improvements were perhaps a little overstated, IS-95 CDMA is certainly a significant improvement over AMPS and has had significant success. In North America, IS-95 CDMA has been deployed in the 800-MHz band, and a variation known as *J-STD-008* has been deployed in the 1900-MHz band.

CDMA is unique to wireless mobility in that it spreads the energy of the RF carrier as a direct function of the chip rate at which the system operates. The CDMA system using the Qualcomm technology has a chip rate of 1.228 MHz. The *chip rate* is the rate at which the initial data stream, the original information, is encoded and then modulated. The chip rate is the data rate output of the PN generator of the CDMA system. A chip is simply a portion of the initial data or message that is encoded through use of an XOR process.

The receiving system also must despread the signal using the exact same Pseudo-Random Number (PN) code sent through an XOR gate that the transmitter used in order to decode the initial signal properly. If the PN generator used by the receiver is different or is not in synchronization with the transmitter's PN generator, then the information being transmitted will never be received properly and will be unintelligible. Figure 1.1 represents a stream of data that is encoded, transmitted, and then decoded back to the original data stream for the receiver to use.

The chip rate also has a direct effect on the spreading of the CDMA signal. Figure 1.2 provides a brief summary of the effects that the chosen chip rate has on spreading the original signal. The heart of CDMA lies in the point that spreading of the initial information distributes the initial energy over a wide bandwidth. At the receiver, the signal is despread through reversing the initial spreading process, where the original signal is reconstructed for use. When the CDMA signal experiences interference in the band, the despreading process despreads the initial signal for use but at the same time spreads the interference so that it minimizes its negative impact on the received information.

The number of PN chips per data bit is referred to as the *processing gain* and is best represented by the equation illustrating processing gain in Figure 1.3. Another way of referencing processing gain is the amount of jamming, or interference, power that is reduced going through the despreading process. Processor gain is the improvement in the signal-to-noise ratio of a spread spectrum system and is depicted in Figure 1.3.

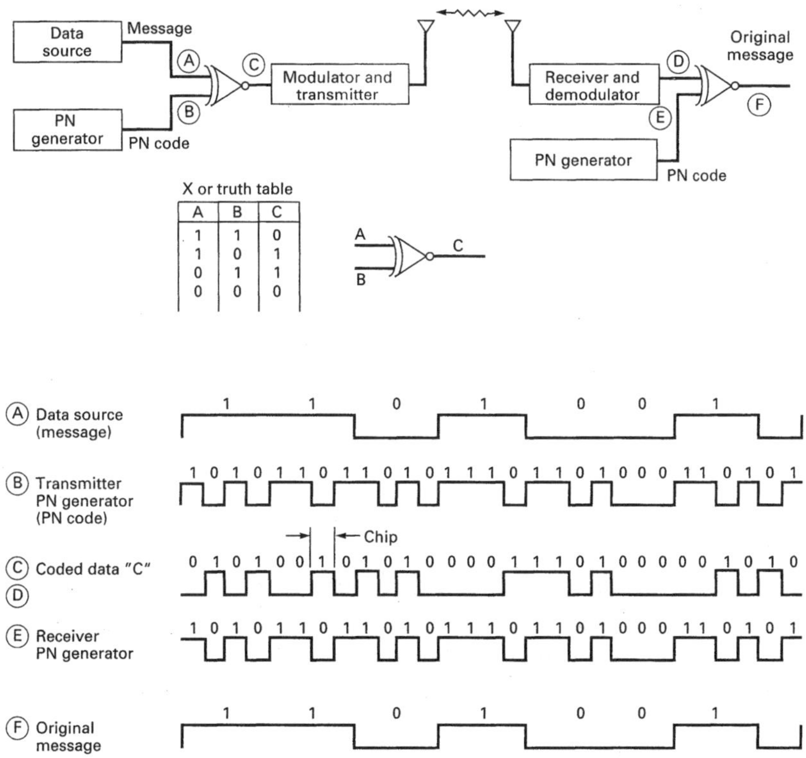

FIGURE 1.1 CDMA PN coding.

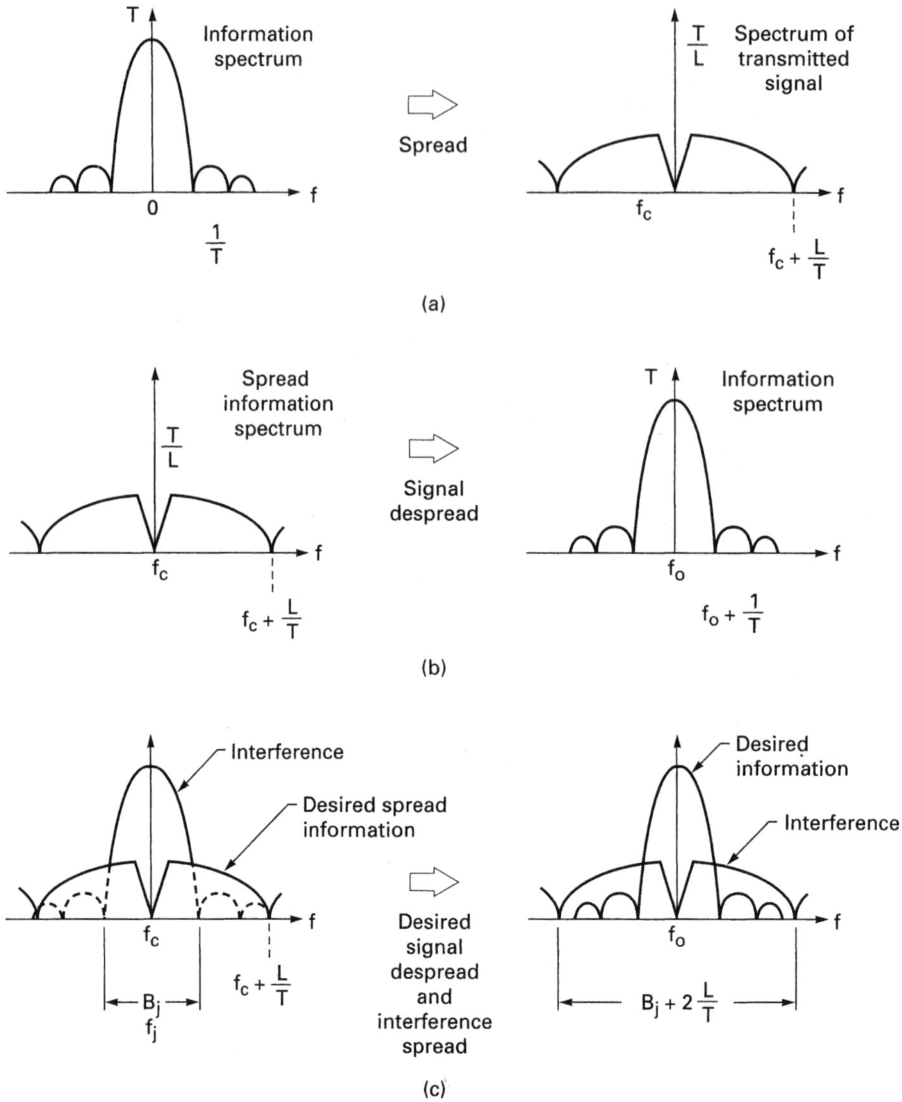

FIGURE 1.2 Summary of spread spectrum.

1.2.3 The Path to Third-Generation Technology

In many ways, second-generation systems have come about because of fundamental weaknesses in first-generation technologies. First-generation technologies have limited system capacity, they have very little protection against fraud, they are subject to easy eavesdropping, and they have little to offer in terms of advanced features. Second-generation systems are designed to address all these issues, and they have done a very successful job.

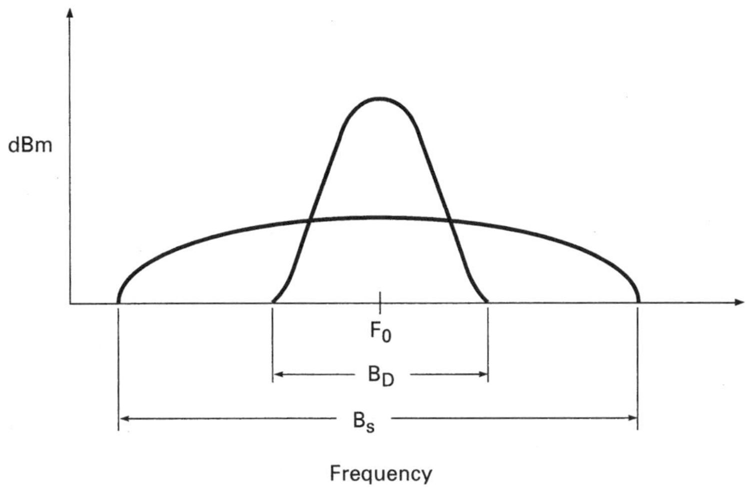

Figure 1.3 Processor gain.*
*Processing gain = B_s/B_d

Systems such as IS-95, GSM, and IS-136 are much more secure, and they also offer higher capacity and more calling features. They are, however, still optimized for voice service, and they are not well-suited to data communications. In the current environment of the Internet, electronic commerce, and multimedia communications, limited support for data communications is a serious drawback. Although subscribers want to talk as much as ever, they now want to communicate in a myriad of new ways, such as e-mail, instant messaging, the World Wide Web, and so on. Not only do subscribers want these services, but they also want mobility. To provide all these capabilities means that new advanced technology is required—third-generation technology.

The need for third-generation mobile communications technology was recognized on many different fronts, and various organizations began to address the issue as far back as the 1980s. The *International Telecommunications Union* (ITU) was heavily involved, and the work within the ITU was originally known as *Future Public Land Mobile Telecommunications Systems* (FPLMTS). Given the fact, however, that this acronym is difficult to pronounce, it was renamed subsequently *International Mobile Telecommunications—2000* (IMT-2000).

The IMT-2000 effort within the ITU has led to a number of recommendations. These recommendations address areas such as user bandwidth (144 kbps for mobile service and up to 2 Mbps for fixed service), richness of service offerings (multimedia services), and flexibility (networks that can support small or large numbers of subscribers). The recommendations also specify that IMT-2000 should operate in the 2-GHz band. In general, however, the ITU recommendations are mainly a set of requirements and do not specify the detailed technical solutions to meet the requirements. To address the technical solutions, the ITU has solicited technical proposals from interested organizations and then selected/approved some of those proposals. In 1998, numerous air-interface technical proposals were submitted. These were reviewed by the ITU, which in 1999 selected five technologies for terrestrial service (non-satellite-based). The five technologies are

- Wideband CDMA (WCDMA)
- CDMA2000 (an evolution of IS-95 CDMA)
- TDD-CDMA (Time Division–CDMA [TD-CDMA] and Time Division-Synchronous CDMA [TD-SCDMA])
- UWC-136 (an evolution of IS-136)
- DECT

These technologies represent the foundation for a suite of advanced mobile multi-media communications services and are starting to be deployed across the globe. TD-CDMA since has been defined as both TD-CDMA and TD-SCDMA. UWC-136 has not seen commercialization and has been abandoned as a viable technology alternative. This book deals with four of those technologies—WCDMA, CDMA2000, TD-CDMA, and TD-SCDMA.

1.2.4 4G and Beyond

There are numerous enhancements for the Third-Generation (3G) systems, each providing higher and higher data rates enabling ubiquitous data coverage and interoperability. However, the next advancement in wireless access technology is Fourth Generation (4G). 4G wireless access is defined under IMT-2000 Advanced and is usually referred to as *Long Term Evolution* (LTE); however, a later version of WiMAX and HSPA are also classified as 4G.

4G is an *Internet Protocol* (IP)–based solution and allows seamless mobility between 3G wireless networks and fixed wireless, allowing users to take advantage of the technology access method that best suits the environment in which they are located. The prevalence of IP ensures that this type of protocol will be in existence for many years to come.

Sometimes a brief chart showing the wireless access technology linkages helps bring the picture together. Figure 1.4 shows the relative relationship of 3G and 4G wireless access technologies as well as legacy protocols.

To further help in how 4G maps to 3G and future Gs, the following data are provided:

- 3G—IMT-2000 specification (CDMA, WCDMA, WiMAX, TD-SCDMA)
- 4G—IMT-2000 Advanced (LTE, WiMAX, HSPA)
- 5G—Not defined however LTE Advanced is referenced as 5G
- 6G and beyond not defined at this moment

1.2.5 Next-Generation Wireless

What exactly is a Next-Generation Wireless (NGW) system? An NGW system involves a fundamental change in architecture from that of circuit-switched mobility networks that are associated with 3G and other legacy wireless systems.

An NGW system is an all packet-based system enabling the wireless operator the ability to support services and applications desired by the end users instead of providing services and applications based on the underlying wireless transmission technologies.

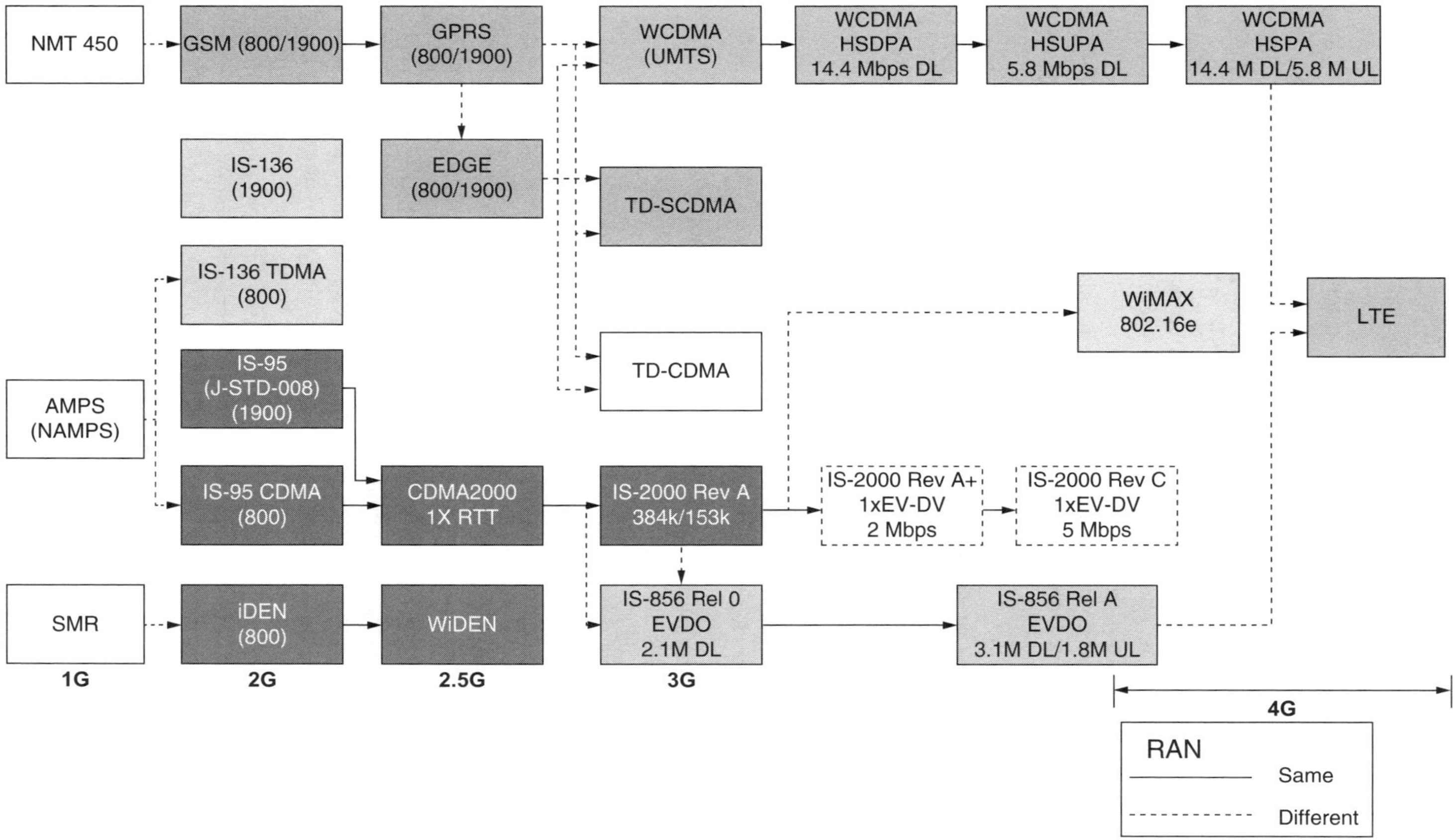

Figure 1.4 4G wireless access technology path.

In an NGW system the core network is designed to be independent of the radio access network. This is possible since the NGW architecture is designed with the goal of supporting packet-switched traffic with seamless mobility, Quality of Service (QoS), and minimal latency.

NGW systems are meant to provide seamless mobility for end users utilizing different radio access technologies and services. Unlike previously mobile wireless networks, they have been based on a circuit-switch design meant to support voice communications. Specifically, previously wireless networks were designed in a hierarchical approach where nodes like a *Base-Station Controller* (BSC) or *Radio Network Controller* (RNC) were an aggregation point in the network. In previous mobile wireless networks, packet networks were effectively overlaid onto the existing circuit switched–based wireless networks.

NGW systems have come about due to the prevalence of packet traffic resulting in the need to switch from a circuit to a packet-based network architecture and is needed to properly transport the increasing packet data that is being sent over wireless networks.

NGW fosters a packet network architecture which is a flat network that removes the hierarchical approach toward network topology. Specifically, a NGW system is packet-based, transport technology agnostic, and utilizes a flat architecture. As NGW system is packet-based it is able to support

- Multimedia services
- Voice and data simultaneous transport
- Quality-of-service enabled services
- Independent of the RF and physical transport technologies
- Heterogeneous networks
- Mobility
- Ubiquitous provisioning of services to end users

1.3 Mobile Communications Fundamentals

Even though the term *cellular* is used often in North America to denote analog AMPS systems, most, though not all, mobile communications systems are cellular in nature. Cellular simply means that the network is divided into a number of cells, or geographic coverage areas, as shown in Figure 1.5. Within each cell is a base station, which contains the radio transmission and reception equipment. It is the base station that provides the radio communication for those mobile phones that happen to be within the cell. The coverage area of a given cell depends on a number of factors, such as the transmit power of the base station, the transmit power of mobile, the height of the base-station antennas, and the topology of the landscape. The coverage of a cell can range from as little as 100 yards to tens of miles.

Specific radiofrequencies are allocated within each cell in a manner that depends on the technology in question. In most systems, a number of individual frequencies are allocated to a given cell, and those same frequencies are reused in other cells that are sufficiently far away to avoid interference. With CDMA, EVDO, WCDMA, HSPA, and

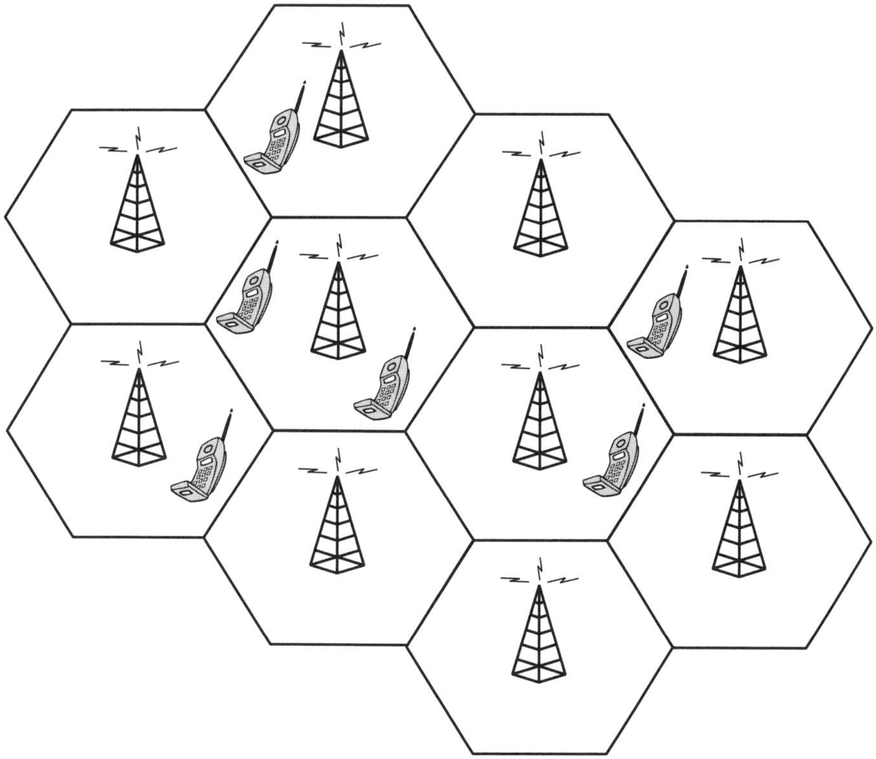

Figure 1.5 Cellular system.

LTE, however, the same frequency can be reused in every cell. Although the scheme shown in Figure 1.5 is certainly feasible and is sometimes implemented, it is common to sectorize the cells, as shown in Figure 1.6. In this approach, the base-station equipment for a number of cells is collocated at the edge of those cells, and directional antennas are used to provide coverage over the area of each cell (as opposed to omnidirectional antennas in the case where the base station is located at the center of a cell). Sectorized arrangements with up to six sectors are known, but the most common configuration is three sectors per base station in urban areas, with two sectors per base station along highways.

Of course, it is necessary that the base stations be connected to a switching network and for that network to be connected to other networks, such as the *Public Switched Telephone Network* (PSTN), in order for calls to be made to and from mobile subscribers. Furthermore, it is necessary for information about the mobile subscribers to be stored in a particular place on the network. Given that different subscribers may have different services and features, the network must know which services and features apply to each subscriber in order to handle calls appropriately. For example, a given subscriber may be prohibited from making international calls. Should the subscriber attempt to make an international call, the network must disallow that call based on the subscriber's service profile.

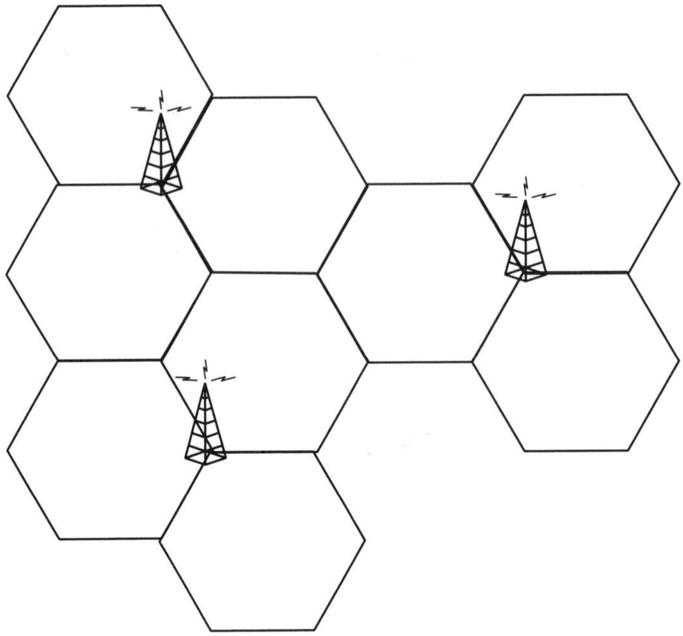

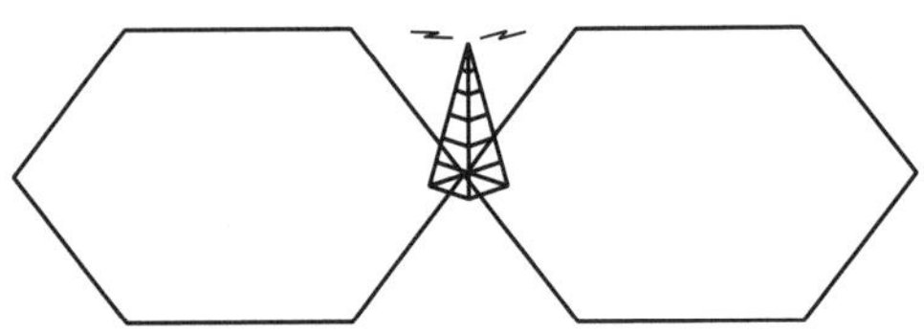

FIGURE 1.6 Typical sectorized cell sites.

1.3.1 Basic Network Architecture

Figure 1.7 shows a typical (although very basic) mobile communications network. A number of base stations, cell sites, are connected to a BSC. The BSC contains logic to control each of the base stations. Among other tasks, the BSC manages the handoff of calls from one base station to another as subscribers move from cell to cell. Note that in certain implementations, the BSC may be physically and logically combined with the mobile switching center. Additionally, the term used to describe a cell site and other network elements is different based on the wireless technology used which is described in more detail later in subsequent chapters. However, the fundamental concepts and layout are similar for all wireless mobility systems.

Connected to the BSC is the *Mobile Switching Center* (MSC). The MSC, also known in some circles as the *Mobile Telephone Switching Office* (MTSO), is the switch that manages

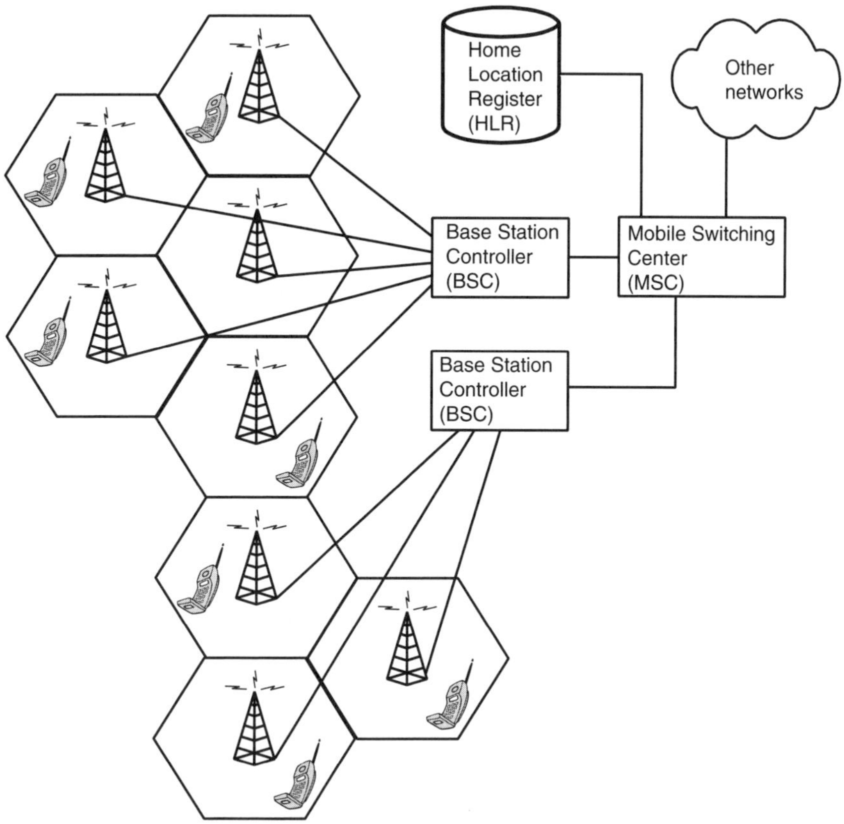

FIGURE 1.7 Basic network architecture.

the setup and teardown of calls to and from mobile subscribers. The MSC contains many of the features and functions found in a standard PSTN switch. It also contains, however, a number of functions that are specific to mobile communications. For example, the BSC functionality may be contained within the MSC in certain systems, particularly in first-generation (1G) systems. Even if the BSC functionality is not contained within the MSC, the MSC still must interact with a number of BSCs over an interface that is not found in other types of networks. Furthermore, the MSC must contain a logic of its own to deal with the fact that the subscribers are mobile. Part of this logic involves an interface to one or more *Home Location Registers* (HLRs), where subscriber-specific data are held.

The HLR contains subscription information related to a number of subscribers. It is effectively a subscriber database and usually is depicted in diagrams as a database. However, the HLR does do more than just hold subscriber data; it also plays a critical role in mobility management—that is, the tracking of a subscriber as he or she moves around the network. In particular, as a subscriber moves from one MSC to another, each MSC in turn notifies the HLR. When a call is received from the PSTN, the MSC that receives the call queries the HLR for the latest information regarding the subscriber's location so that the call can be routed correctly to the subscriber. Note that in some implementations, HLR functionality is incorporated within the MSC, which leads to the concept of a *home MSC* for a given subscriber.

The network depicted in Figure 1.7 can be considered to represent the bare minimum needed to provide mobile telephony service. These days, a range of different features' services are offered in addition to just the capability to make and receive calls. Therefore, most of today's mobile communications networks are much more sophisticated than the network depicted in Figure 1.7. As we progress through this book, we will introduce many other network elements and interfaces as we build from the fundamentals to the sophisticated technologies of 3G and 4G networks as the industry moves away from a circuit-switched system toward a system that is IP-based.

1.3.2 Air-Interface Access Techniques

Radio spectrum is a precious and finite resource. Unlike other transmission media such as copper or fiber facilities, it is not possible simply to add radio spectrum when needed. Only a certain amount of spectrum is available, and it is critical that it be used efficiently and be reused as much as possible. Such requirements are at the heart of the radio access techniques used in mobile communications.

1.3.2.1 FDD

The air interface consists of an access method coupled with a protocol method. The access method involves the use of either *Frequency-Division Duplex* (FDD) or *Time-Division Duplex* (TDD), with each having its own advantages and/or disadvantages. The protocol method uses FDMA, TDMA, and/or CDMA.

FDD uses two separate radio channels for communicating between the base station and the host terminal, also known as the CPE. One of the radio channels, f 1, is for the downlink from the base station to the host terminal. The other radio channel, f 2, is for the uplink from the host terminal to the base station, as shown in Figure 1.8.

The channels f 1 and f 2 normally are spaced a distance apart for isolation purposes. The FDD system uses dedicated channels for uplink and downlink communication. Figure 1.9 illustrates how the uplink and downlink channels are paired. The specific frequency band, channel size, and technology depend on the particular system.

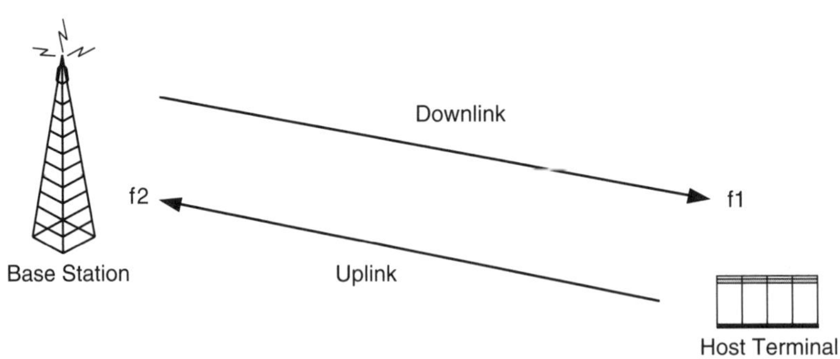

FIGURE 1.8 FDD system.

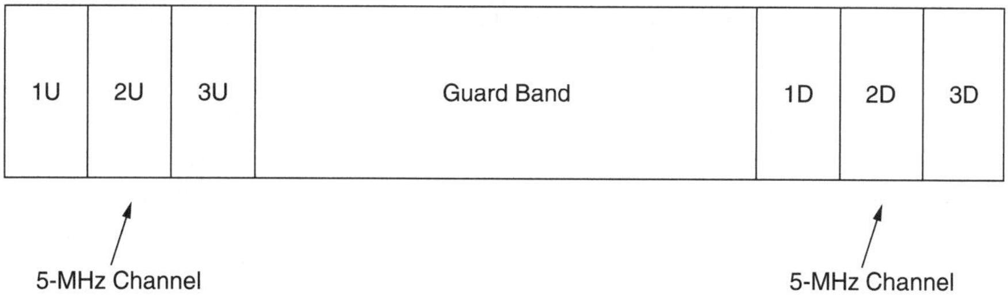

FIGURE 1.9 FDD spectrum-allocation example.

1.3.2.2 TDD

TDD systems use one radio channel for communication between the base station and the host terminal. The duplexing that is done is based on time and not frequency, as is done typically with an FDD system. The same channel, f1, is used for both uplink and downlink communication between the base station and the host terminal, as shown in Figure 1.10, and Figure 1.11 illustrates how the uplink and downlink channels are arranged.

TDD by its nature is designed to be more spectral efficient than an FDD system when involving data communication that is nonsymmetric, such as Internet traffic. If the traffic is nonsymmetric, as is IP Internet traffic, where the downlink accounts for,

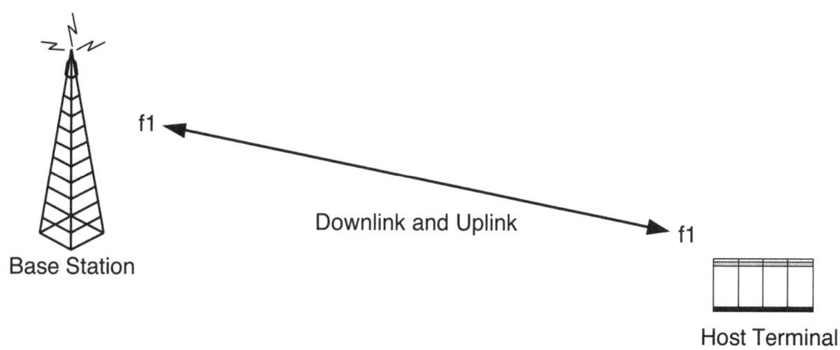

FIGURE 1.10 TDD system.

FIGURE 1.11 TDD spectrum-allocation example.

say, 75 percent of the traffic, then four (4) channels are used for TDD, which is the same as two (2) channels for FDD.

The complexity with a TDD system versus an FDD system lies in the interference issues and the need to ensure synchronization. TDD, however, can coexist with FDD in a mixed-technology market, that is, a TD-SCDMA and GSM system. Mixed-technology markets require spectrum segmentation and/or use of guard zones to coordinate the TDD system for both the base station and the host terminals with the FDD system because interference now comes from several point sources instead of one.

1.3.2.3 Frequency-Division Multiple Access (FDMA)

Of the common multiple-access techniques used in mobile communications systems, FDMA is the simplest. With FDMA, the available spectrum is divided into a number of radio channels of a specified bandwidth, and a selection of these channels is used within a given cell. In analog AMPS, for example, the available spectrum is divided into blocks of 30 kHz. A number of 30-kHz channels are allocated to each cell depending on the expected traffic load for that cell. When a subscriber wants to place a call, one of the 30-kHz channels is allocated exclusively to the subscriber for that call.

In most FDMA systems, separate channels are used in each direction—from network to subscriber (downlink) and from subscriber to network (uplink). For example, in analog AMPS, when we talk about 30-kHz channels, we are actually talking about two 30-kHz channels, one in each direction. Such an approach is known as *Frequency-Division Duplex* (FDD), and normally a fixed separation exists between the frequency used in the uplink and that used in the downlink. This fixed separation is known as the *duplex distance.* For example, in many systems in North America, the duplex distance is 45 MHz. Thus, in such a system, channel 1 corresponds to two channels (uplink and downlink) with a separation of 45 MHz between them. An FDD-FDMA technique can be represented as shown in Figure 1.12.

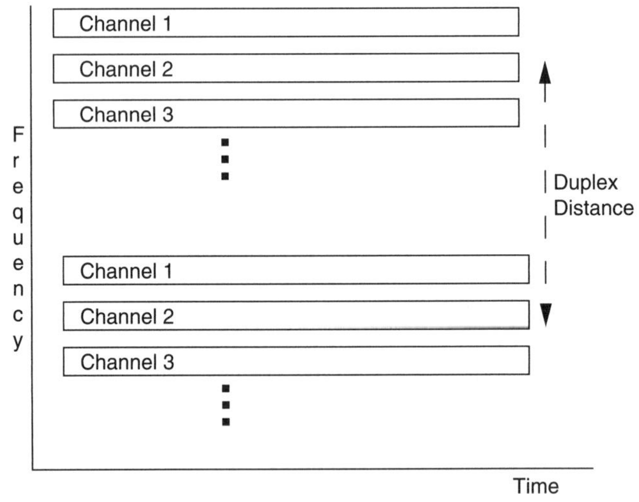

FIGURE 1.12 FDMA.

FDD is not the only duplexing scheme, however. Another technique, known as *Time-Division Duplex* (TDD), is also used. In such a system, only one channel is used for both uplink and downlink transmissions. With TDD, the channel is used very briefly for uplink, then very briefly for downlink, and then very briefly again for uplink, and so on. TDD is not very common in North America, but it is used widely in systems deployed in Asia.

1.3.2.4 Time-Division Multiple Access (TDMA)

With TDMA, radio channels are divided into a number of time slots, with each user assigned a given time slot. For example, on a given radiofrequency, user A might be assigned time slot number 1, and user B might be assigned time slot number 3. The allocation is performed by the network as part of the call-establishment procedure. Thus, the user's device knows exactly which time slot to use for the remainder of the call, and the device times its transmissions exactly to correspond with the allocated time slot. This technique is depicted in Figure 1.13.

Typically, a TDMA system is also an FDD system, as shown in Figure 1.13, although TDD is used in some implementations. Furthermore, TDMA systems normally also use FDMA. Thus, the available bandwidth is divided into a number of smaller channels as in FDMA, and it is these channels that are divided into time slots. The difference between a pure FDMA system and a TDMA system that also uses FDMA is that with the TDMA system, a given user does not have exclusive access to the radio channel.

Implementing a TDMA system can be done in many ways. For example, different TDMA systems may have different numbers of time slots per radio channel and/or different time-slot durations and/or different radio-channel bandwidths. Although in the

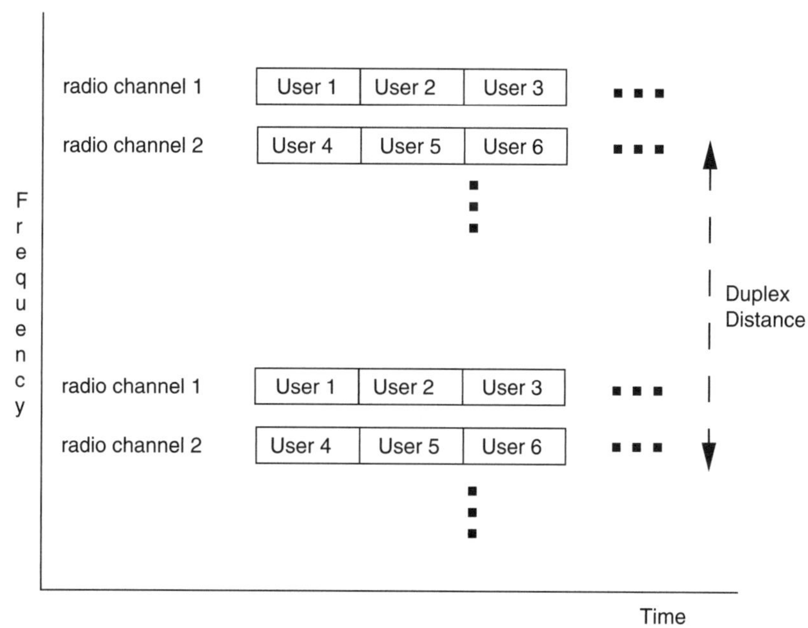

Figure 1.13 TDMA.

United States the term *TDMA* is used often to refer to IS-136, such a usage of the term is incorrect because IS-136 is just one example of a TDMA system. In fact, GSM is also a TDMA system.

1.3.2.5 Code-Division Multiple Access (CDMA)

With CDMA, neither the time domain nor the frequency domain are subdivided. Rather, all users share the same radiofrequency at the same time. This approach obviously means that all users interfere with one another. Such interference would be intolerable if the RF bandwidth were limited to just the bandwidth that would be needed to support a single user. To overcome this difficulty, CDMA systems use a technique called *spread spectrum,* which involves spreading the signal over a wide bandwidth. Each user is allocated a code or sequence, and the bit rate of the sequence is much greater than the bit rate of the information being transmitted by the user. The information signal from the user is modulated with the sequence assigned to the user, and at the far end, the receiver looks for the sequence in question. Having isolated the sequence from all the other signals (which appear as noise), the original user's signal can be extracted.

TDMA systems have a very well-defined capacity limit. A set number of channels and a set number of time slots exist per channel. Once all time slots are occupied, the system has reached capacity. CDMA is somewhat different. With CDMA, the capacity is limited by the amount of noise in the system. As each additional user is added, the total interference increases, and it becomes harder and harder to extract a given user's unique sequence from the sequences of all the other users. Eventually, the noise floor reaches a level where the inclusion of additional users significantly impedes the system's capability to filter out the transmission of each user. At this point, the system has reached capacity. Although it is possible to model this capacity limit mathematically, exact modeling can prove a little difficult because the noise in the system depends on such factors as the transmission power of each individual mobile device, thermal noise, and the use of discontinuous transmission (only transmitting when something is being said). By making certain reasonable assumptions in the design phase, however, it is possible to design a CDMA system that provides relatively high capacity without significant quality degradation.

CDMA2000, WCDMA, and TD-CDMA are widely deployed CDMA systems for mobile communications, with TD-SCDMA soon to be available commercially. All four protocols use CDMA, with CDMA2000 and WCDMA being FDD-based and TD-CDMA and TD-SCDMA being TDD-based.

For example, CDMA2000 uses a channel bandwidth of 1.23 MHz and is an FDD system. The fact that the bandwidth is 1.23 MHz means that the total system bandwidth (typically 10, 20, or 30 MHz) can accommodate several CDMA RF channels. Therefore, like TDMA, IS-95 CDMA also uses FDMA to some degree. In other words, within a given cell, more than one RF channel may be available to system users. More detail on the other CDMA protocols will follow in later chapters.

A significant advantage of any CDMA-based system is the fact that it practically eliminates frequency planning. Other systems are very sensitive to interference, meaning that a given frequency can be reused only in another cell that is sufficiently far away to avoid interference. In a commercial mobile communications network, cells are constantly being added, or capacity is being added to existing cells, and each such change must be done without causing undue interference. If interference is likely to be introduced, then retuning part of the network is required. Such retuning is needed frequently and can be an expensive effort. CDMA, however, is designed to deal with interference, and in fact, it allows a given RF carrier to be reused in every cell. Therefore, there is no need to worry about retuning the network when a new cell is added.

1.3.2.6 Orthogonal Frequency Division Multiplexing (OFDM)

OFDM modulation is widely used in communications systems and is sometimes called multicarrier or discrete multitone modulation.

OFDM is a spread-spectrum scheme and is used in wired and wireless networks and was initially designed to counter the effects of multipath interference and was actually introduced in the 1960s. However, OFDM was later exploited for wideband data communications, both wired and wireless.

The following are some of those communication systems that use OFDM.

- WiFi
- ADSL
- WiMAX
- HiperMAN
- LTE

Typical radio communication systems use a single channel, however. OFDM, however, uses multiple channels which are based on the concept of *Frequency-Division Multiplexing* (FDM). OFDM is exceptionally diverse and is used over many different mediums.

- Radio spectrum
- Coax cable
- Twisted pair
- Fiberoptic cable

In essence, OFDM divides the data stream into smaller parallel streams of data, as shown in Figure 1.14. With OFDM, the smaller parallel data streams are each sent on a different channel and different frequency.

Like all modulation schemes, each has positive and negative attributes that are used to determine if this modulation scheme meets the design requirements.

Positive attributes

- Improved spectrum efficiency bits per second per hertz, or bps/Hz
- Multipath and fading resiliency
- Improved signal-to-noise ratio
- Dynamic bandwidth allocation
- Data rates can be changed base
  - RF environment
  - Services required
  - By subscriber

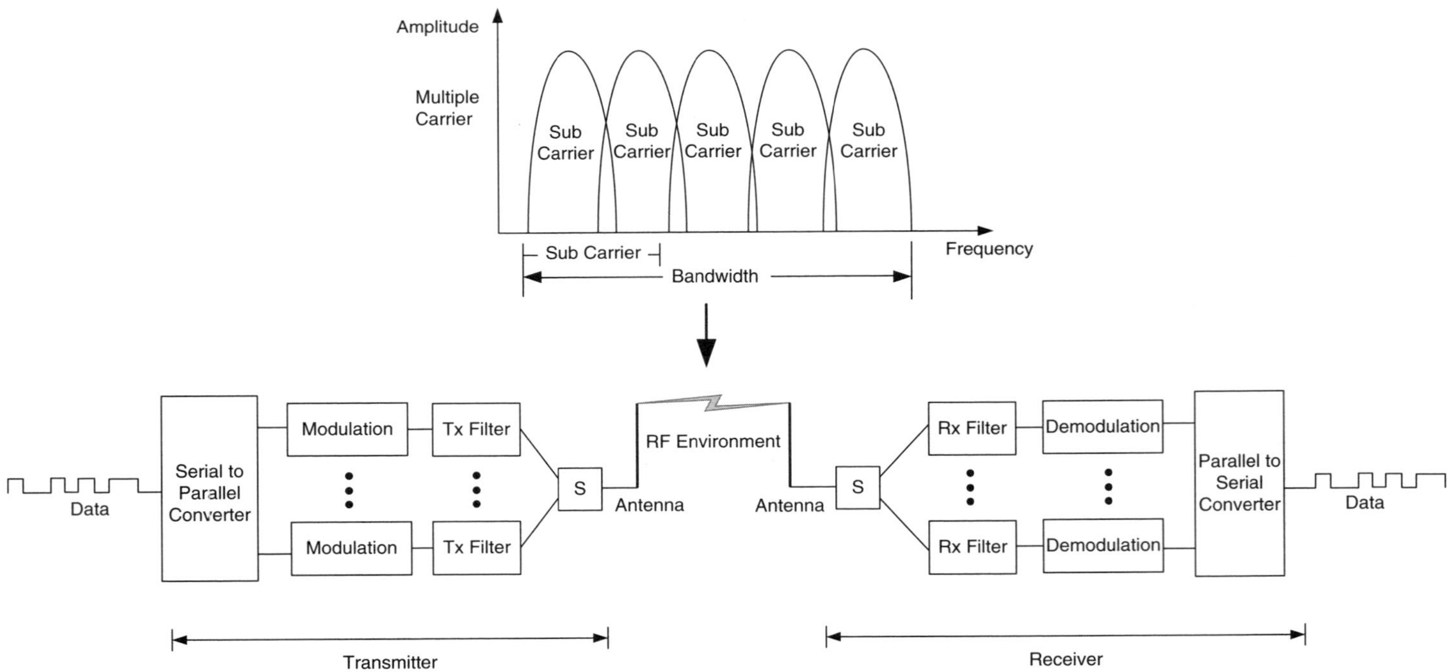

FIGURE 1.14 OFDM.

- Spread spectrum
  - OFDM has the ability to reject impulse noise.
  - OFDM spreads energy of an impulse noise over transmitted burst.

Negative attributes

- High Peak-to-Average Power Ratio (PAPR)
- Transmitter amplifier complexity
- Battery power consumption by mobile device

An OFDM channel is divided into subchannels or subcarriers where a subchannel is

- Less than the channel bandwidth
- Experiences flat fading
- ISI is reduced

To make OFDM work, all the subcarriers need to be orthogonal to each other and allow subcarriers to overlap without increasing ICI.

1.3.3 Roaming

The discussion so far has focused largely on the methods used to access the network over the air interface. The air-interface access is, of course, extremely important. Other aspects, however, are necessary to make a wireless communications network a mobile communications network.

Mobility implies that subscribers are able to move freely around the network and from one network to another. This requires that the network tracks the location of a subscriber to a certain accuracy so that calls destined for the subscriber may be delivered. Furthermore, a subscriber should be able to do so while engaged in a call.

The basic approach is as follows: First, when a subscriber initially switches on his or her mobile phone, the device itself sends a registration message to the local MSC. This message includes a unique identification for the subscriber. Based on this identification, the MSC is able to identify the HLR to which the subscriber belongs, and the MSC sends a registration message to the HLR to inform the HLR of the MSC that now serves the subscriber. The HLR then sends a registration cancellation message to the MSC that previously served the subscriber (if any) and then sends a confirmation to the new serving MSC.

When mobile communications networks were introduced, only the air-interface specification was standardized. The exact protocol used between the visited MSC and the HLR (or home MSC) was vendor-specific. The immediate drawback was that the home system and visited system had to be from the same vendor if roaming was to be supported. Therefore, a given network operator needed to have a complete network from only one vendor. Moreover, roaming between networks worked only if the two networks used equipment from the same vendor. These limitations severely curtailed roaming.

This problem was addressed in different ways on either side of the Atlantic. In North America, the problem was recognized fairly early, and an effort was undertaken to establish a standard protocol between home and visited systems. The result of that effort was a standard known as *IS-41*. This standard has been enhanced significantly over the years, and the current revision of the standard is revision D. IS-41 is used for roaming in AMPS systems, IS-136 systems, and IS-95 systems.

Meanwhile, in Europe, nothing was done to address the roaming issue for 1G systems, but a major effort was applied to ensuring that the problem was addressed in 2G technology—specifically GSM. Consequently, when GSM specifications were created, they addressed far more than just the air interface. In fact, most aspects of the network were specified in great detail, including the signaling interface between home and visited systems. The protocol specified for GSM is known as the GSM *Mobile Application Part* (MAP). Like IS-41, GSM MAP also has been enhanced over the years.

Strictly speaking, the term *MAP* is not specific to GSM. In fact, the term refers to any mobility-specific protocol that operates at layer 7 of the *Open Systems Interconnection* (OSI) seven-layer stack. Given that IS-41 also operates at layer 7, the term *MAP* is also applicable to IS-41.

1.3.4 Handoff/Handover

Handoff (also known as *handover*) is the ability of a subscriber to maintain a call while moving within a network. The term *handoff* typically is used with AMPS, IS-136, and IS-95, whereas *handover* is used in GSM. The two terms are synonymous.

Handoff usually means that a subscriber travels from one cell to another while engaged in a call and that the call is maintained during the transition (ideally without the subscriber noticing any change). In general, handoff means that the subscriber is transitioned from one radio channel (and/or time slot) to another. Depending on the two cells in question, the handoff can be between two sectors on the same base station, between two BSCs, between two MSCs belonging to the same operator, or even between two networks. (Note that internetwork handoff is not supported in some systems, often mainly for billing reasons.)

It is also possible to hand off a call between two channels in the same cell. This could occur when a given channel in a cell is experiencing interference that is affecting the communications quality. In such a case, the subscriber would be moved to another frequency that is subject to less interference. A handoff scenario is depicted in Figure 1.15.

How does the system determine that a handoff needs to occur? Basically, two main approaches are used. In 1G technologies, a handoff generally is controlled by the network. The network measures the signal strength from a mobile device as received at the serving cell. If it begins to fall below a certain threshold, then nearby cells are requested to perform signal-strength measurements. If a nearby cell records a better signal strength, then it is highly likely that the subscriber has moved to the coverage of that cell. The new cell is instructed by the BSC or MSC (typically just the MSC because 1G systems do not have BSCs) to allocate a channel for the subscriber. Once that allocation is performed, the network instructs the mobile device to swap to the new channel. This is known as a *network-controlled handoff* because the network determines when and how a handoff is to occur.

In more recent technologies, a technique known as *mobile-assisted handover* (MAHO) is most common. In this approach, the network provides the mobile device with a list of base-station frequencies (those of nearby base stations). The mobile device makes periodic measurements of the signals received from those base stations (as well as the serving base station), including signal strength and signal quality (usually determined from bit-error rates), and it sends the corresponding measurement reports to the network. The network analyzes the reports and makes a determination of if and how a

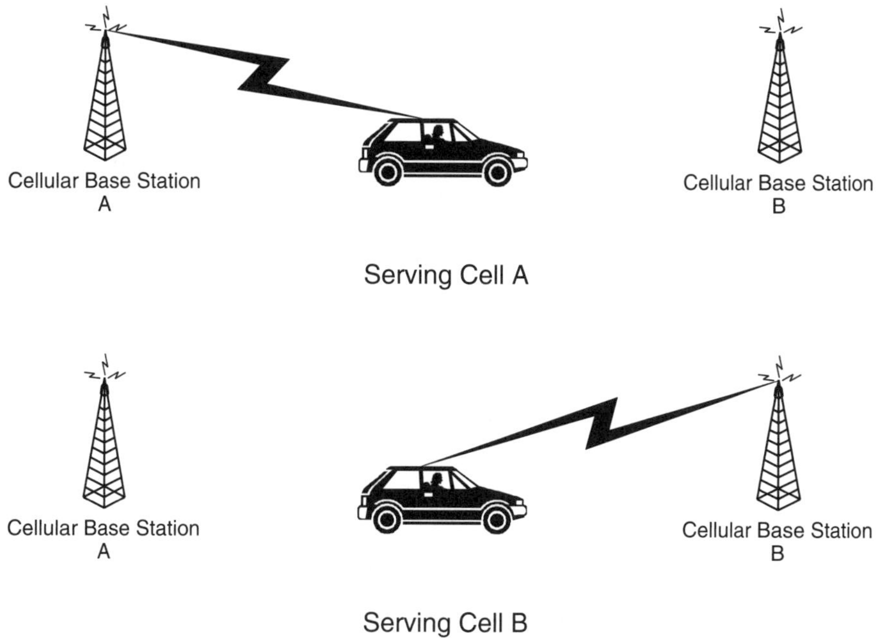

FIGURE 1.15 Handoff.

handoff should occur. Assuming that a handoff is required, then the network reserves a channel on the new cell and sends an instruction to the mobile to move to that channel, which it does.

1.4 Mobile Data

Mobile data are one of the logical futures for wireless and are the thrust of this book. Mobile data have different meanings and requirements depending on what application or demand is being served.

Mobile data have been used for decades, but the need to extend the corporate *local-area network* (LAN), surf the Internet for information, use *Voice over Internet Protocol* (VoIP) or perform an IP-to-IP call (PTT) is just beginning to be felt. The hope for wireless data has been heralded for many years as being just around the corner. The truth is, though, that wireless data are unknown for mobility in terms of user acceptance and real network benefits from the operators' aspect, besides the customer.

The central concept that needs to be kept in the forefront when addressing mobile data is that data are vertical—vertical in that every user's specific data requirements are different. For instance, your individual data requirements may be casual Internet browsing and *information management* (IM) capability. However, another user's requirements, may be order management and *File Transfer Protocol* (FTP) access. The point I am trying to address is that whereas mentioning wireless data on the surface appears rather simplistic, the actuality of it is profound and complicated.

Cellular data packet data (CDPD) is one of the examples that the wireless industry needs to always remember. CDPD was an excellent packet network that rivals many of the current wireless data offerings and was available close to a decade prior to 2.5G/3G. Applications that are specific to the user are what are needed; otherwise, wireless operators will be only pipe providers falling into constant price erosion owing to the fact that bandwidth will be treated as a commodity. CDPD is also being decommissioned or is in various last stages of its service life.

With this said, there are several key elements for mobile data that are relevant for this book. There are many mobile data technologies, but focus requires exclusion and wireless mobility for cellular/PCS is the focus. The key platforms used for 2G, 2.5G, and 3G wireless mobility are listed in Table 1.1.

In Table 1.1, WCDMA enhancements are referred to as *High Speed Downlink Packet Access* (HSDPA) and *High Speed Uplink Packet Access* (HSUPA). Both HSDPA an HSUPA are enchancements to WCDMA facilitating greater throughput. CDMA2000 reference to EVDO and EVDV also is an enchancement to the original CDMA2000 data rates. Both WCDMA and CDMA2000 and its subsequent variants are included in the

2G Technology	**Data Capability**	**Spectrum Required**	**Comment**
GSM	9.6 or 14.4 kbps	200 kHz	Circuit-switched data
IS-136	9.6 kbps	30 kHz	Circuit-switched data
IDEN	9.6 kbps	25 kHz	Circuit-switched data
CDMA	9.6/14.4 kbps	1.25 MHz	Circuit-switched data
(IS-95A/J-STD-008)	64 bps (IS-95B)		
2.5G Technology	**Data Capability**	**Spectrum Required**	**Comment**
HSCSD	28.8/56 kbps	200 kHz	Circuit/packet data
GPRS	128 kbps	200 kHz	Circuit/packet data
Edge	384 kbps	200 kHz	Circuit/packet data
CDMA2000-1XRTT	144 kbps	1.25 MHz	Circuit/packet data
3G Technology	**Data Capability**	**Spectrum Required**	**Comment**
WCDMA	144 kbps vehicular 384 kbps outdoors 2 Mbps indoors	5 MHz	Packet data
CDMA2000-EVDO/EVDV	144 kbps vehicular 384 kbps outdoors 2 Mbps indoors	1.25 MHz	Packet data
TD-CDMA	144 kbps vehicular 384 kbps outdoors 2 Mbps indoors	5 MHz	Packet data
TD-SCDMA	144 kbps vehicular 384 kbps outdoors 2 Mbps indoors	1.6 MHz	Packet data

Table 1.1 Wireless Mobility Data

later chapters. In addition to the previous topics mentioned, the most obvious missing component to the data offering is WiFi, WiMAX, WiMAN, cable, and *Digital Subscriber Line* (DSL) systems, which comprise the fixed-data-offering capability that can complement the mobile data portfolio when the subscriber is more stationary. Some of the possible configurations discussed in later chapters include harmonization of IMT-2000 technologies with WiMAX and WiFi all through the use of IP.

It is important to note that CDMA, GSM, EVDO, HSPA, and LTE are 3GPP and 3GPP2 standards while WiFi and WiMAX fall under the IEEE 802 standards.

1.5 WiFi

Wireless fidelity (WiFi) is a wireless LAN protocol based on the IEEE 802.11 standards and is a standard feature in all laptops and wireless handsets for mobility. WiFi operates in both licensed and unlicensed spectrums. However, the licensed spectrum is set aside for public safety. The unlicensed spectrum for WiFi involves the ISM (2.4 GHz) and UNII (5.8 GHz) bands.

There are multiple protocols for WiFi that are in play right now and they are

- 802.11 b/g (ISM)
- 802.11 a (UNII)
- 802.11 n (ISM/UNII)
- 802.11 g (ISM/UNII)

However, WiFi is not classified as a *Wireless Mobile Broadband Network* and is meant for small coverage areas.

WiFi has many advantages and disadvantages. Some of the advantages include the ability to establish a wireless LAN that eliminates cabling requirements. WiFi is an established standard and is universally accepted. What is important for wireless mobility networks is that WiFi is an excellent broadband traffic offload platform. However, some of the more notable disadvantages with WiFi are its limited coverage from access point, security, and the use of unlicensed spectrum.

Table 1.2 provides a high-level overview of some of WiFi attributes.

Function	802.11b	802.11g	802.11a	802.11n
Maximum Data Rate	11 Mbps	54 Mbps	54 Mbps	540 Mbps
Number of non-overlapping channels	3	3	11	11/3
Frequency Allocation Band	2.4 GHz ISM	2.4 GHz ISM	5 GHz UNII 2.4 GHz ISM	5 GHz UNII 2.4 GHz ISM
Modulation/Coding	CCK, HR/DSSS HR/DSS/short HRDSSS/PBCC	OFDM	OFDM	OFDM

Table 1.2 WiFi Attributes

1.6 Bluetooth

Bluetooth is a wireless protocol that operates in the 2.4-GHz ISM band, allowing wireless connectivity between mobile phones, Personal Digital Assistants (PDAs), and other similar devices for the purpose of exchanging information between them. Bluetooth is meant to replace the infrared telemetry portion on mobile phones and PDAs, enabling extended range and flexibility in addition to enhanced services.

Because Bluetooth systems use a radio link in the ISM band, there are several key advantages that this transport protocol can exploit. Bluetooth can effectively operate as an extension of a LAN or a peer-to-peer LAN and provide connectivity between a mobile device and the following other device types:

- Printers
- PDAs
- Mobile phones
- LCD projectors
- Wireless LAN devices
- Notebooks and desktop PCs

One of the key attributes that Bluetooth offers is the range over which the system or connection can operate. Since Bluetooth operates in the 2.4-GHz ISM band, (2.4000–2.4835 GHz), it has an effective range going from 10 to close to 100 m. The protocol does not require *Line-of-Sight* (LOS) for establishing communication. Its pattern is omnidirectional, thereby eliminating orientation issues, and can support both ISO- and asynchronous services, paving the way for effective use of TCP/IP communication.

Bluetooth comes under the IEEE specification 802.15 and is different from WiFi IEEE 802.11. Bluetooth is different from WiFi because it is meant to be a LAN extension providing connectivity with end-user devices not requiring high bandwidth. Bluetooth is also able to coexist with WiFi in the same frequency band.

Most of the wireless handsets currently used now have a Bluetooth radio built into it. However, there are currently multiple versions of Bluetooth ranging from 1.0, 1.2, 2.0, 2.1, 3.0, 4.0, and 4.1. Version 1 has several variants and all are considered basic and legacy version. Later versions of Bluetooth increase throughput and functionality and they are all backward-compatible.

Bluetooth, which operates in the 2.4-GHz ISM band, utilizes frequency hopping. There are several modulation schemes used by Bluetooth depending on the version of Bluetooth being used. However, all Bluetooth versions are capable of *Gaussian Frequency Shift Keying* (GFSK) enabling 1 Mbps. For Bluetooth Versions 2 and 3 *Phase Shift Keying* (PSK) is used where p/4 DQPSK provides 2 Mbps and 8 DPSK enables speeds up to 3 Mbps.

Figure 1.16 represents the Bluetooth-available channels in the 2.4-GHz ISM band. There are a total of 79 Bluetooth channels possible, however not all are allowed in every country. The channel spacing or size for Bluetooth is 1 MHz besides the implementation of guard bands.

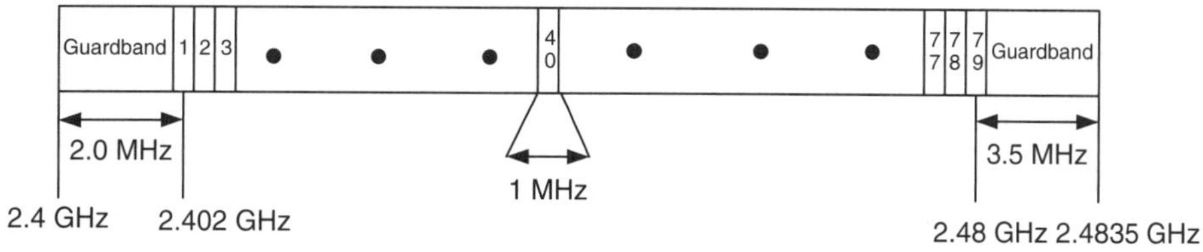

Figure 1.16 Bluetooth.

1.7 Cable Systems

The proliferation of cable modems, primarily in the United States, has brought broadband to many end users who previously were relying on dial-up IP. Cable operators have a unique advantage for delivering broadband services, as does the PSTN regarding the residential market, in that it has presence in many residential homes. Data services are delivered via a cable modem that meets the *Data Over Cable Service Interface Specification* (DOCSIS).

DOCSIS is an interface specification for cable modems enabling broadband to be delivered over a cable television network. The DOCSIS interface specification enables a cable television system to offer high-speed IP data between the subscriber location and the cable operator's head end.

The cable modem typically is connected to a two-way cable RF path over a low-split *hybrid fiber/coax* (HFC) cable system that uses fixed-wire facilities, unlike the radio counterpart. Because it uses a fixed medium, downlink or downstream data rates of between 27 and 36 Mbps are possible using a radio channel that is 50-MHz wide and operating around 750 MHz. The uplink or upstream data rate is between 320 kbps and 10 Mbps over an uplink radio channel that is between 5 and 42 MHz. For further information, www.cablemodem.com has a wealth of material and specifications regarding DOCSIS.

The common issue facing all broadband providers is the quality of their underlying transport layer. The quality of the cable plant itself dictates the services that effectively can be offered. The issue of quality of the cable plant is driven primarily by the number of drops that occur on any cable leg, which has a direct impact on the ingress noise problem that limits the ability of the cable plant to provide high-speed two-way communication. Since most of the information flow is from the head end to the subscriber, the system does not have to support symmetric bandwidth.

An HFC network has the capability of providing two-way communication for both voice and data besides the video service offering. The primary access method is physical medium, where the connection made to the subscriber at the end of the line is via coaxial cable. For increased distance and performance enhancements, fiberoptic cables often can be and are part of the cable network topology.

1.8 Wireless Migration Options

In previous sections of this chapter, some of the various technology platforms were briefly discussed. The existing wireless operators today, regardless of the frequency band or existing technology deployed, have or are making very fundamental decisions as to which direction they will take in the 3G/4G evolution. The decision on fundamental technology and how it will interface with legacy systems will define a company's position in the marketplace for years to come.

Some existing operators and new entrants are letting the technology platform be defined by the local regulator, thereby eliminating the platform decision. However, the majority of operators need to determine which platform they will use. Since the platforms to pick from use different access technologies, they are by default not directly compatible. The use of different access technologies for the realization of 3G also introduces several interesting issues related to the migration from 2G to 3G. The migration path from 2G to 3G is referred to as *2.5G* and involves an interim position for data services that are more advanced than 2G but not as robust as the 3G data services. Additionally, the decision for which 4G technology to implement and the associated legacy network migration or interoperability. The business plan as well as the spectrum portfolio of the wireless operator directly influence the migration options that are truely available.

Some of the migration strategies for an existing operator involve

- Overlay
- Spectrum segmentation

The 3G overlay approach typically involves implementing the 2.5G technology over the existing 2G system and then implementing 3G as either an overlay or in a separate part of the RF spectrum they are allocated, that is, spectrum segmentation. The 4G overlay approach involves deploying 4G in a separate frequency band, then legacy systems, and then migrating legacy systems.

The choice of whether to use an overlay or spectrum segmentation naturally depends on the technology platform that is currently being used, that is, 2G or 3G, the spectrum available, the existing capacity constraints, and marketing. Marketing is involved with the decision because of the impact on the existing subscriber base and services that are envisioned to be offered. More specifically with the plethora of technical options available the need to have a strong business plan is essential and this should provide the fundamental input to determine just what and how will implement 4G or any wireless technology with the legacy network.

Some of the decisions are rather straightforward, involving upgrading portions of the existing technology platforms that are currently deployed. Other operators have to make a decision as to which technology to use because they either are building a new system or have not migrated to a 4G platform, using only 2G or 3G networks.

In later chapters, various migration strategies are discussed relative to the underlying technology platform that exists.

1.9 Harmonization Process

Harmonization refers to the vision and objective of the IMT-2000 and IMT Advanced specifications that enable the various technology platforms that are defined in that specification to interact with each other. True harmonization relative to the capability of a wireless system is based on having subscriber units that operate in a radio access technology agnostic mode, commonly referred to as *software-definable radios*. Operating in a radio access technology agnostic mode would enable subscribers to achieve connectivity anywhere there is RF coverage. However, the access infrastructure that is able to support this is a goal, but not one that will exist in the near future.

1.10 Overview of Following Chapters

This chapter has served as a brief introduction to mobile communications systems. The brief overview that has been given, however, is certainly not a sufficient background to enable a good understanding of 3G and especially 4G technology. Therefore, before tackling the details of 3G systems, it is necessary to better describe 1G and 2G systems. Appendix A addresses 1G technology, and Appendix B delves into the 2G systems, and Appendix C addresses 2.5G systems. The remaining chapters focus on both 3G and 4G systems as well as covering unlicensed broadband and backhaul. Some possible migration paths are also covered for a mobile operator either to achieve IMT-2000 or IMT-2000 Advanced visions for a wireless packet data network.

References

AT&T, *Engineering and Operations in the Bell System,* 2d ed., AT&T Bell Laboratories, Murray Hill, NJ, 1983.

Barron, Tim, "Wireless Links for PCS and Cellular Networks," *Cellular Integration,* September 1995, pp. 20–23.

Brewster, R.L., *Telecommunications Technology,* Wiley, New York, 1986.

Brodsky, Ira, "3G Business Model," *Wireless Review,* June 15, 1999, p. 42.

Daniels, Guy, "A Brief History of 3G," *Mobile Communications International,* 65, October 1999, p. 106.

Gull, Dennis, "Spread-Spectrum Fool's Gold?" *Wireless Review,* January 1, 1999, p. 37.

Homa, Harri, and Antti Toskala, *WCDMA for UMTS,* Wiley, New York, 2000.

Smith, Clint, *Practical Cellular and PCS Design,* McGraw-Hill, New York, 1997.

Smith, Gervelis, *Cellular System Design and Optimization,* McGraw-Hill, New York, 1996.

Smith, Meyer, *3G Wireless with WiFi and WiMAX,* McGraw-Hill, New York, 2004.

www.3gpp.org.

www.ieee.org.

CHAPTER 2

Third-Generation (3G) Overview

2.1 Introduction

The rapid increase in the demand for data services, primarily *Internet Protocol* (IP), has been thrust on the wireless industry. Over the years, there has been much anticipation of the onslaught of data services, but the radio access platforms have been the inhibitor from making this a reality. *Third-Generation* (3G) is a term that has received and continues to receive much attention as the enabler for high-speed data for the wireless mobility market. 3G and all it is meant to be are defined in the *International Telecommunications Union* (ITU) specification *International Mobile Telecommunications 2000* (IMT-2000). IMT-2000 is a radio and network access specification defining several methods or technology platforms that meet the overall goals of the specification. The IMT-2000 specification is meant to be a unifying specification, enabling mobile and some fixed high-speed data services to use one or several radio channels with fixed network platforms for delivering the services envisioned:

- Global standard
- Compatibility of service within IMT-2000 and other fixed networks
- High quality
- Worldwide common frequency band
- Small terminals for worldwide use
- Worldwide roaming capability
- Multimedia application services and terminals
- Improved spectrum efficiency
- Flexibility for evolution to the next generation of wireless systems
- High-speed packet-data rates
    - 2 Mbps for fixed environment
    - 384 kbps for pedestrian
    - 144 kbps for vehicular traffic

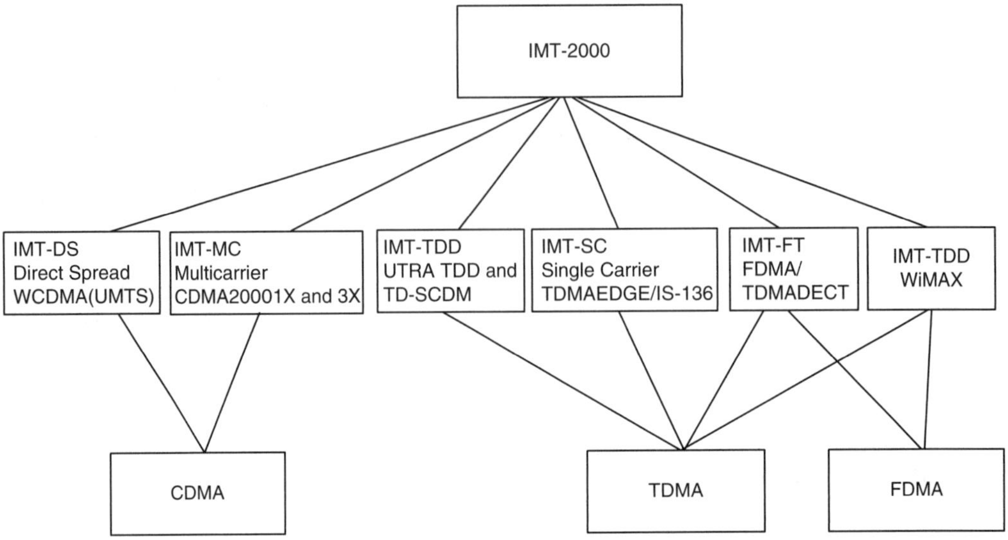

Figure 2.1 IMT-2000.

Figure 2.1 shows the linkage between the various platforms that make up the IMT-2000 specification group.

The definition of what exactly 3G encompasses usually is clouded in marketing terms, with the technical reader desiring a straightforward answer. The reason 3G is hard to pin down is primarily due to the fact that it involves radio access and network platforms that have not been fully realized yet. The standard that everyone is striving for is IMT-2000, and it incorporates several competing radio access platforms that will not achieve harmonization, if ever, until *Fourth-Generation* (4G) or beyond. The radio access platforms that make up the IMT-2000 specification are all different, and it should be no wonder that it is difficult to obtain a simple answer when asked to describe what a 3G system will look like.

- IMT-2000/3G can be described as being used to reference a multitude of technologies covering many frequency bands, channel bandwidths, and of course, modulation formats.
- No single 3G infrastructure platform, technology, or application exists.
- 3G is applied to mobile and stationary wireless applications involving high-speed data. IMT-2000 mandates data speeds of 144 kbps at driving speeds, 384 kbps for outside stationary use or walking speeds, and 2 Mbps for indoors.

Along with the different platforms that make up the IMT-2000 standard is the fact that existing *First-Generation/Second-Generation* (1G/2G) platforms need to transition into the 3G arena. The transition method that an operator must select and spend money on is, of course, a difficult decision and will determine how successful the wireless operator will be in the future. The interim platform that bridges the 2G systems into a 3G

Wireless Generation	Systems	General Service	Comments
First (1G)	AMPS, TACS, NMT	Voice	Traditional analog cellular depolyment scheme
Second (2G)	GSM,TDMA, CDMA (IS-95)	Primarily voice with SMS	• Digital modulation scheme implemented • Deployment in 800-, 900-, 1800-, and 1900-MHz bands • Spectrum clearing required for 1900 MHz in United States • Spectrum refarming required for existing 1G operators to implement 2G systems
Transition (2.5G)	CDMA (IS-95), GPRS/EDGE	Primarily voice with packet-data services being introduced	• Overlay approach used except in new spectrum • Packet-data enhancements to existing 2G operators
Third (3G)	CDMA2000/WCDMA, ULTRA TDD, TD-SCDMA WiMAX	Packet-data and voice services designed for high-speed multimedia data and voice	• Defined by IMT-2000 • Europe (UMTS—WCDMA/ULTRA TDD) • America (UMTS/CDMA2000/ULTRA TDD and WiMAX) • Asia (UMTS/CDMA2000/ TD-SCDMA/ULTRA TDD and WiMAX) • Overlay approach for existing operators of 2/2.5G networks

Table 2.1 3G

environment is referred to as *2.5G.* Table 2.1 attempts to group some of the major technology platforms by wireless generation.

What follows is a brief visualization of the interaction between the major 1G, 2G, 2.5G, and 3G platforms. Obviously, if an operator chooses to implement more than one technology platform for marketing and strategic reasons, then the lines of transition become more complicated than those shown in Figure 2.2.

Both WCDMA and CDMA2000 depicted in Figure 2.2 include various enhancements that are discussed in later chapters. Of specific interest for the enhancements includes HSDPA and HSUPA for WCDMA and Rev 0, A and B for EVDO. WiMAX is also included as a 3G technology and is discussed more in later chapters.

3G is a mobile radio and network access scheme that enables high-speed data to be used, allowing for true multimedia capabilities in a mobile wireless system. Presently, voice has been the primary wireless application, with the use of the *Short Message Service* (SMS) being the largest packet-data service.

Today's wireless cellular and *Personal Communications Services* (PCS) systems have the same radio bandwidth allocated for both voice and data. Some of the 2.5G transition

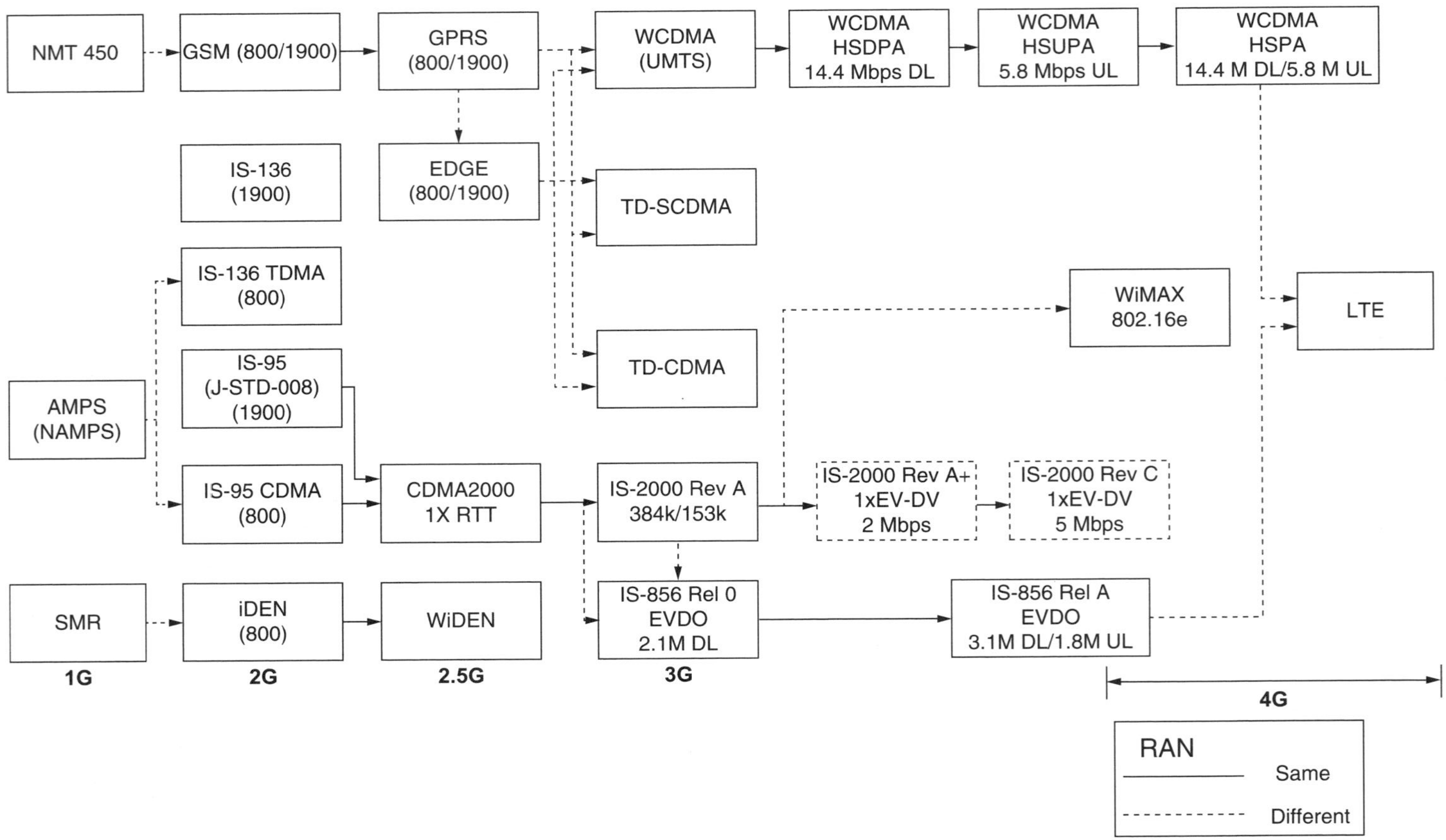

Figure 2.2 Migration path.

or migration plans call for the use of a dedicated spectrum just for data applications. IMT-2000 specifies that data speeds of 144 kbps for vehicular, 384 kbps for pedestrian, and 2 Mbps for indoor applications are the desired goals and have been built into the specifications.

Table 2.2 is a brief grouping of the various major technology platforms and the data speeds that are associated with each.

In examining Table 2.2, it is apparent that not all of the IMT-2000 platform standards are included, and that is on purpose. The platforms that are listed in both are *Wideband Code Division Multiple Access* (WCDMA), CDMA2000, TD-CDMA, and TD-SCDMA; these are the 3G platforms that will be discussed in some level of detail for the remainder of this textbook. WCDMA and CDMA2000 will receive more attention and the reason for the two-platform focus lies in the primary issue that a vast majority of wireless operators use one of these two standards, which are part of the IMT-2000 specification.

2G Technology	Data Capability	Spectrum Required	Comment
GSM	9.6 or 14.4 kbps	200 kHz	Circuit-switched data
IS-136	9.6 kbps	30 kHz	Circuit-switched data
iDEN	9.6 kbps	25 kHz	Circuit-switched data
CDMA (IS-95A/ J-STD-008)	9.6 bps/14.4 kbps, 64 bps (IS-95B)	1.25 MHz	Circuit-switched data
2.5G Technology	**Data Capability**	**Spectrum Required**	**Comment**
HSCSD	28.8/56 kbps	200 kHz	Circuit/packet data
GPRS	128 kbps	200 kHz	Circuit/packet data
Edge	384 kbps	200 kHz	Circuit/packet data
CDMA2000-1XRTT	153 kbps	1.25 MHz	Circuit/packet data
3G Technology	**Data Capability**	**Spectrum Required**	**Comment**
WCDMA	144 kbps vehicular 384 kbps outdoors 2 Mbps indoors	5 MHz	Packet data
CDMA2000-DO/EV	144 kbps vehicular 384 kbps outdoors 2 Mbps indoors	1.25 MHz	Packet data
TD-CDMA (Ultra TDD)	144 kbps vehicular 384 kbps outdoors 2 Mbps indoors	5 MHz	Packet data
TD-SCDMA	144 kbps vehicular 384 kbps outdoors 2 Mbps indoors	1.6 MHz	Packet data

Note: TD-SCDMA and TD-CDMA are TDD and are unpaired.

Table 2.2 2G, 2.5G, and 3G Comparison

CDMA2000 and WCDMA have received more acceptance right now than TD-CDMA and TD-SCDMA. TD-CDMA has experienced some level of rollout but not the same level that CDMA2000 and WCDMA have. In addition, TD-SCDMA implementation and adoption have localized to China.

In referencing Table 2.2 the data rates listed for 3G are those for compliance with IMT-2000. In practice each of the platforms listed have several enhancements that enable the technology to exceed the IMT-2000 requirements. The specific enhancements are discussed in later chapters devoted to each of the technology platforms.

2.2 Universal Mobile Telecommunications Service (UMTS)

When the ITU solicited solutions to meet the requirements laid down for IMT-2000, a number of technologies were proposed by various standards groups. These included both *Time-Division Multiple Access* (TDMA) solutions and *Code-Division Multiple Access* (CDMA) solutions. They also included both *Frequency-Division Duplex* (FDD) and *Time-Division Duplex* (TDD) solutions.

The *European Telecommunications Standards Institute* (ETSI) agreed on a WCDMA solution using FDD. In Japan, a WCDMA solution also was proposed, with both TDD and FDD options. In Korea, two different types of CDMA solutions were proposed—one similar to the European and Japanese proposals and one similar to a CDMA proposal being considered in North America (CDMA2000, which is an evolution of IS-95 CDMA), and lastly, China is pursuing its own standard, TD-SCDMA.

It was clear that a number of groups were working on very similar technologies, and it was fairly obvious that the most effective way forward was to pool resources. This led to the creation of two groups—the *Third Generation Partnership Project* (3GPP) and 3GPP2. 3GPP works on UMTS, which is based on WCDMA, and 3GPP2 works on CDMA2000 and the IEEE handles WiMAX. The following discussion provides a brief overview of UMTS.

2.2.1 Migration Path to UMTS and the Third Generation Partnership Project (3GPP)

The radio access for UMTS is known as *Universal Terrestrial Radio Access* (UTRA). This is a WCDMA-based radio solution that includes both FDD and TDD modes. The *radio access network* (RAN) is known as *UTRAN*. It takes more than an air-interface or an access network to make a complete system, however. The core network also must be considered. Because of the widespread deployment and success of *Global System for Mobile* (GSM) communications, it is appropriate to base the UMTS core network on an evolution of the GSM core network. In fact, as we shall see, the initial release of UMTS (3GPP Release 1999) makes use of the same core network architecture as defined for GSM/GPRS, albeit with some enhancements. Moreover, the core network is required to support both UMTS and GSM radio access networks (i.e., both UTRAN and the GSM BSS).

Evolution of the GSM base-station subsystem (BSS) has not stopped, however. As we shall see, enhancements such as the *Enhanced Data Rates for Global Evolution* (EDGE) have been made. With requirements for the continued evolution of GSM and for the GSM to meet UMTS requirements, it makes sense for the continued maintenance and evolution of GSM specifications to be undertaken by 3GPP. Consequently, 3GPP, rather than ETSI, is now responsible for GSM specifications as well as UMTS-specific specifications.

For several years, various enhancements to GSM have been developed according to yearly releases. Thus, for a given GSM specification, versions have been related to Release 1996, Release 1997, and Release 1998. Initially, the 3GPP determined to continue with this approach. Therefore, the first release of specifications from the 3GPP is known as *3GPP Release 1999.* The release includes not only new specifications for the support of a UTRAN access but also enhanced versions of existing GSM specifications (such as for the support of EDGE). The 3GPP Release 1999 specifications were completed in March, 2000. These, of course, will be subject to some revisions and corrections as errors and inconsistencies are discovered during test and deployment.

The next release of 3GPP specifications was originally termed *3GPP Release 2000.* This included major changes to the core network. The changes were so significant, however, that they could not all be handled in a single step. Thus Release 2000 was divided into two releases: Release 4 and Release 5. Going forward, the concept of yearly releases will no longer apply, and releases will be structured and timed according to defined functionality.

For the most part (although not exclusively), 3GPP Release 1999 focuses mainly on the access network (including a totally new air interface) and the changes needed to the core network to support that access network. Release 4 focuses more on changes to the architecture of the core network. Release 5 introduces a new call model, which means changes to user terminals, to the core network, and to the access network (although the fundamentals of the air interface remain the same). Given that the air interface is new in Release 1999 and that it does not change drastically in later releases, it is best to begin our description of UMTS technology with the WCDMA air interface. The primary focus of this book will be on the FDD mode of operation, with less emphasis on TDD. First, however, we need a few words about the types of services that UMTS can offer.

2.3 UMTS Services

Of course, the most notable capability promised by UMTS is a high data rate—up to 2 Mbps. There is, however, more to a given service than just the data rate that the service demands. Depending on what the end user is trying to do, various considerations must be made, of which data rate is only one.

UMTS specifications define four service classes, where the services within a given class have a common set of characteristics. The service classes are as follows:

- *Conversational.* This is characterized by low delay tolerance, low jitter (delay variation), and low error tolerance. The data rate requirement may be high or low but generally is symmetric. In other words, the data rate in one direction will be similar to that in the other direction. Voice, which is highly delay-sensitive, is a typical conversational application, one that does not require very high data rates. Video conferencing is also a conversational application. It has similar delay requirements to voice but is less error-tolerant and generally requires a higher data rate.
- *Interactive.* This consists typically of request/response-type transactions. Interactive traffic is characterized by low tolerance for errors but with a larger tolerance for delays than conversational services. Jitter (delay variation) is not a

major impediment to interactive services, provided that the overall delay does not become excessive. Interactive services may require low or high data rates depending on the service in question, but the data rate is generally significant only in one direction at a time.

- *Streaming.* This concerns one-way services, using low to high bit rates. Streaming services have a low error tolerance but generally have a high tolerance for delay and jitter. This is so because the receiving application usually buffers data so that they can be played to the user in a synchronized manner. Streaming audio and streaming video are typical streaming applications.
- *Background.* This is characterized by little, if any, delay constraint. Examples include server-to-server e-mail delivery (as opposed to user retrieval of e-mail), SMS, and performance/measurement reporting. Background applications require error-free delivery.

2.3.1 UMTS Speech Service

Although UMTS will be used for a variety of data services, speech may well remain the most widely used service. Speech has certain requirements in terms of data rate, delay, jitter, and error-free delivery, all of which are derived from human perceptions and expectations. Moreover, speech quality in UMTS needs to be comparable with that in fixed telephony networks and certainly no worse than that experienced in 2G wireless networks.

UMTS uses the *adaptive multirate* (AMR) speech coder. This is actually several coders in one and provides coding rates of 12.2, 10.2, 7.95, 7.40, 6.70, 5.90, 5.15, and 4.75 kbps. The 12.2-kbps rate is the same coding scheme as used in the GSM *Enhanced Full-Rate* (EFR) coding scheme. The 7.4-kbps rate is the same coding scheme as used in IS-136 TDMA networks. The reuse of existing coders means that the voice-coding scheme of UMTS should at least offer the same levels of quality as experienced in existing 2G networks.

The AMR coder allows for the speech bit rate to change dynamically during a call. As we shall describe later, the higher the bit rate of any service, the smaller is the effective footprint of a cell. Thus a user at the edge of a cell could change from a high speech-coding rate to a lower speech-coding rate to effectively extend the coverage for speech service. Each AMR speech frame is 20 ms in duration, and it is possible to change the speech-coding rate from one speech frame to the next. Thus the coding rate could change as often as every 20 ms, although this is unlikely to ever happen in reality.

The AMR coder also supports *voice activity detection* (VAD) and *discontinuous transmission* (DTX), with comfort noise generation. The net effect is that little or nothing is sent over the air interface when nothing is being said. Given that typical speech involves one person speaking, followed by the other, it is possible to reduce the amount of transmission over the air interface by as much as 50 percent. Of course, VAD and DTX are supported by most modern wireless technologies.

Many of the services supported by UMTS are packet-switched data services. Speech, on the other hand, at least in 3GPP Release 1999 and 3GPP Release 4, is a circuit-switched service. This means that a user in a speech call has access to dedicated resources throughout the call. In effect, a dedicated pipe is used between the two parties in a speech conversation. This is similar to the way speech is handled in a GSM/GPRS network, where a speech call uses a dedicated time slot on the air interface and uses dedicated transport and switching in the core network. Although the concept of time slots does not map well to WCDMA radio access, the assignment of dedicated resources still applies.

2.3.1.1 UMTS Data Service

Packet data are a key attribute. WCDMA is designed to offer great flexibility in transmission of user data across the air interface. For example, data rates can change on a frame-by-frame basis (every 10 ms). Moreover, it is possible to support a mix and match of different types of services. For example, a subscriber may be sending and receiving packet data while also involved in a voice call. When sending information over the air interface, physical control channels are used in combination with physical data channels. While the physical data channels carry the user information, the physical control channels carry information to support the correct interpretation of the data carried on the corresponding *Dedicated Physical Data Channel* (DPDCH) frame plus power control commands and feedback indicators.

Packet-data services in the 3GPP Release 1999 architecture use largely the same mechanisms as used for GPRS/EDGE data, the big difference being the user data rates that can be supported. Packet-data services are established in UMTS through the activation of a PDP context with an *Access Point Name* (APN), *Quality-of-Service* (QoS) criteria, etc.

From an air-interface perspective, UMTS provides greater flexibility than GPRS/EDGE in terms of how resources are allocated for packet-data traffic. Not only does UMTS offer a greater range of speeds, the WCDMA air interface has a selection of different channel types that can be used for packet data.

2.4 The UMTS Air Interface

The UMTS air interface is a *Direct-Sequence CDMA* (DS-CDMA) system. Given that this is a radical departure from the TDMA techniques of GSM, GPRS, and EDGE, it is worth briefly describing the concepts involved. In fact, all 3G platforms—WCDMA, CDMA2000, TD-CDMA, and TD-SCDMA—use variants of CDMA.

2.4.1 WCDMA Basics

DS-CDMA means that user data are spread over a much wider bandwidth through multiplication by a sequence of pseudorandom bits called *chips*. Figure 2.3 provides a conceptual depiction of this spreading. One can see that the user data, at a relatively low rate compared with the rate of the spreading code, are spread over a signal that has a higher bit rate. We also can see that the signal that is transmitted has pseudorandom characteristics. When transmitted over a radio interface, the spread signal looks like noise.

If multiple users transmit simultaneously on the same frequency, then the stream of data from each user needs to be spread according to a different pseudorandom sequence. In other words, each user data stream needs to be spread according to a different spreading code. At the receiving end, the stream of data from a given user is recovered by despreading the set of received signals with the appropriate spreading code. Of course, what is being despread is the complete set of signals received from all users that are transmitting.

Imagine, for example, two users (*A* and *B*) that are transmitting on the same frequency but with two different spreading codes. If, at the receiving end, the received

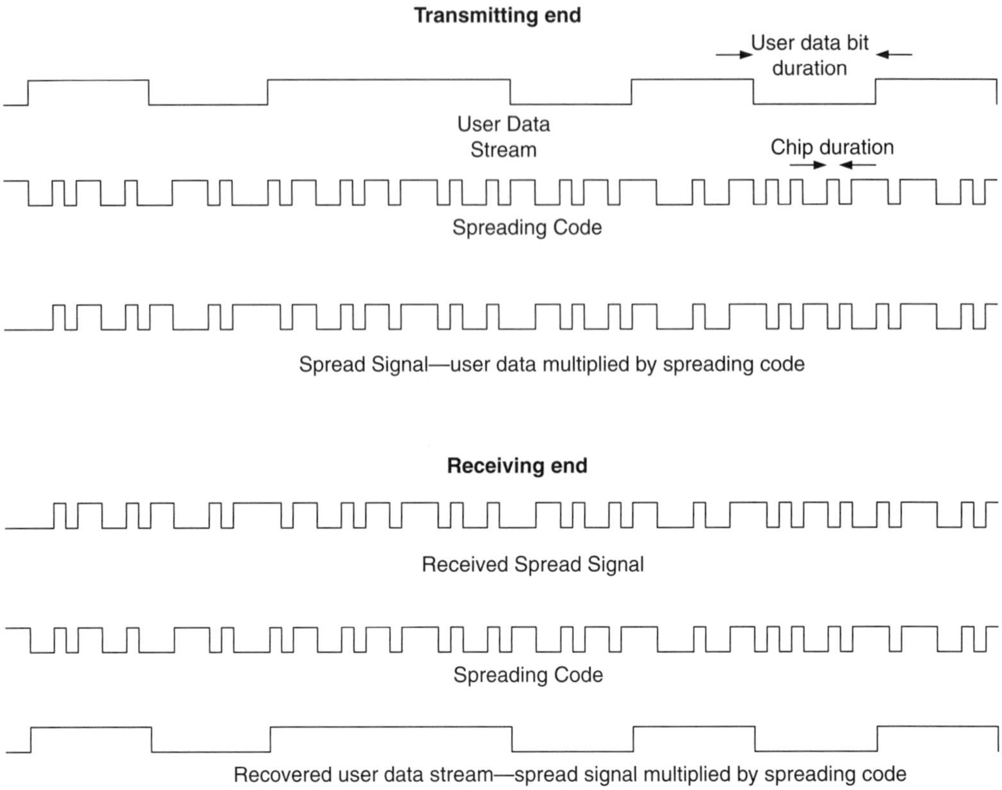

FIGURE 2.3 CDMA basic concept.

signal is despread with the spreading code applicable to user *A*, then the original data stream from user *A* is recovered. The data stream that is recovered does have some noise created by the fact that the received signal also contains user data from user *B*. The noise, however, is small.

Similarly, if the received signal is despread according the spreading code used by user *B*, then the original data stream from user *B* is recovered, with a little noise generated by the presence of user *A*'s data within the spread signal. Provided that the rate of the spreading signal (the chip rate) is far larger than the user data rate, then the noise (i.e., the interference) generated by the presence of other users will be sufficiently small not to inhibit recovery of the data stream from a given user. Of course, as the number of simultaneous users increases, so does the interference, and it eventually becomes impossible to recover a specific user's data with any confidence.

In other words, for a given bit of recovered user data, the signal-to-noise ratio must be sufficiently high. In CDMA, we refer to E_b/N_o, where E_b is the power density per bit of recovered user data, and N_o is the noise power density. Provided that E_b/N_o is sufficiently large, then the user data can be recovered.

The ratio of the chip rate to the user data symbol rate is known as the *spreading factor*. The ability to recover a given user's signal is directly influenced by the spreading factor. The higher the spreading factor, the greater is the ability to recover a given user's signal. In terms of transmission and reception, a higher spreading factor has an equivalent effect as transmitting at a higher power. Thus, the magnitude of the spreading factor can be considered a type of gain and is known as the *processing gain.* In decibels, the processing gain is given by 10 × 10 $\log_{10}$ (spreading rate/user rate). In some cases, this can be quite a large number and can help to overcome the effect of interference generated by the presence of other users.

If, for example, the processing gain for a given CDMA service were 20 dB, and if an E_b/N_o value of 5 dB were needed, then for a given user, the signal-to-interference ratio can be as low as –15 dB, and the user's signal still can be recovered. This is so because the despreading benefits from the processing gain of 20 dB. Note that for a given chip rate, the processing gain for low-bit-rate user applications is greater than for high-bit-rate applications, which often means that lower-bit-rate applications can tolerate more interference than high-bit-rate applications.

The WCDMA air interface of UMTS (hereafter simply WCDMA) has a nominal bandwidth of 5 MHz. While 5 MHz is the nominal carrier spacing, it is possible to have a carrier spacing of 4.4 to 5 MHz in steps of 200 kHz. This enables spacing that might be needed to avoid interference, particularly if the next 5-MHz block is allocated to another carrier.

The chip rate in WCDMA is 3.84×10^6 chips per second (3.84 Mcps). In theory, for a speech service at 12.2 kbps (and, for now, assuming no extra bandwidth for error correction), the spreading factor would be $3.84 \times 10^6/12.2 \times 10^3 = 314.75$. This would equate to a processing gain of 25 dB. In reality, however, WCDMA does include extra coding for error correction. Consequently, a spreading factor as high as 314.75 is not supported, at least not in the uplink. The supported uplink spreading factors are 4, 8, 16, 32, 64, 128, and 256. The highest spreading factor (256) is used mostly by the various control channels that require low data throughput. Some control channels also can use lower spreading factors, whereas user services generally use lower spreading factors, keeping in mind that the lower spreading rate allows for higher data throughput. Table 2.3 provides a summary of the spreading factors and the corresponding data rates on the uplink.

At first glance, it appears that the lowest spreading factor (4) provides a gross rate of only 960 kbps and a usable rate of only 480 kbps. This does not meet the requirements of IMT-2000, which states that a user should be able to achieve speeds of 2 Mbps.

Spreading Factor	Gross Data Rate (kbps)	User Data Rate (kbps) (Assuming Half-Rate Coding for Error Correction)
256	15	7.5
128	30	15
64	60	30
32	120	60
16	240	120
8	480	240
4	960	480

Table 2.3 Uplink Spreading Factors and Data Rates

In order to meet that requirement, UMTS supports the capability for a given user to transmit up to six simultaneous data channels. Thus, if a user wants to transmit user data at a user rate greater than 480 kbps, then multiple channels are used, each with a spreading factor of 4. With six parallel channels, each with a spreading factor of 4, a single user can obtain speeds of over 2 Mbps. This requires 6 × 5 MHz (30 MHz × 30 MHz) of FDD spectrum.

In the downlink, the same spreading factors are available, with a spreading factor of 512 also possible. One difference between the uplink and downlink, however, is the number of bits per symbol. As will be described in Chapter 3, the uplink effectively uses 1 bit per user symbol, whereas the downlink effectively uses 2 bits per user symbol. Consequently, for a given spreading factor, the user bit rate in the downlink is greater than the corresponding bit rate in the uplink. The user rate in the downlink is not quite twice that in the uplink, however, owing to differences in the way that control channels and traffic channels are multiplexed on the air interface. The details of uplink and downlink transmissions are provided in Chapter 3. Table 2.4 provides a summary of the spreading factors and the corresponding data rates on the downlink.

As is the case for the uplink, WCDMA supports multiple simultaneous user data channels in the downlink so that a single user can achieve rates of over 2 Mbps. It should be noted, however, that Table 2.4 does not tell the whole story of possible data rates on the downlink. WCDMA supports a concept known as *compressed mode,* whereby gaps exist in downlink transmission so that the terminal can take measurements on other frequencies. When compressed mode is used, a reduction will take place in the data rate compared with that shown in Table 2.4.

An important capability of WCDMA is that user data rates do not need to be fixed. In WCDMA, channels are transmitted with a 10-ms frame structure. It is possible to change the spreading factor on a frame-by-frame basis. Thus, within one frame, the user data rate is fixed, but the user data rate can change from frame to frame. This capability means that WCDMA can offer bandwidth on demand. Note that rate changes every 10 ms do not apply to AMR speech because each speech packet is 20 ms in duration, so the speech rate can change every 20 ms if needed, but not every 10 ms.

Spreading Factor	Gross Air-Interface Bit Rate (kbps)	User Data Rate (kbps) (Including Coding for Error Correction)	Approximate Net User Data Rate (kbps) (Assuming Half-Rate Coding)
512	15	3–6	1–3
256	30	12–24	6–12
128	60	42–512	21–25
64	120	90	45
32	240	210	105
16	480	432	216
8	960	912	456
4	1920	1872	936

TABLE 2.4 Downlink Spreading Factors and Data Rates

2.4.2 Spectrum Allocation

With the WCDMA FDD option, the paired 5-MHz carriers in the uplink and downlink outside North America are as follows: uplink—1920 to 1980 MHz; downlink—2110 to 2170 MHz. Thus, for the FDD mode of operation, a separation of 190 MHz exists between uplink and downlink. In the United States, the separation is 80 MHz and with the uplink—1850 to 1915 MHz; downlink—1930 to 1995 MHz. Although 5 MHz is the nominal carrier spacing, it is possible to have a carrier spacing of 4.4 to 5 MHz in steps of 200 kHz. This enables spacing that might be needed to avoid interference, particularly if the next 5-MHz block is allocated to another carrier. Additionally WCDMA will have other uplink and downlink separations to take advantage of the 800 and AWS frequency bands in the United States.

For the TDD option, a number of frequencies have been defined, including 1900 to 1920 MHz and 2010 to 2025 MHz. Of course, with TDD, a given carrier is used in both the uplink and the downlink, so no separation exists.

Of course, there is no reason why WCDMA could not be deployed at other frequencies. In fact, the use of other frequencies may well be necessary in some countries. You may have noticed that the frequency bands defined previously overlap significantly with frequencies used for PCS in North America. Therefore, in North America, it will be necessary to move some existing users from the PCS band and/or acquire a new spectrum in some other band. The movement of existing PCS users will happen only when a given carrier that wants to implement UMTS already has an existing PCS system and uses some of the spectrum for UMTS. The net result for such an operator, of course, will be limited spectrum for both PCS and UMTS.

UMTS is an overlay for an existing wireless operator using a 2G/2.5G RAN. The implementation of UMTS requires new spectrum or spectrum segmentation. In some instances, PCS operators may not have sufficient spectrum to offer WCDMA because they have only a single 5 × 5 MHz license that has existing GSM customers on it.

2.5 Overview of the 3GPP Release 1999 Network Architecture

Figure 2.4 shows the network architecture for 3GPP Release 1999, the first set of specifications for UMTS. Working our way from the top left, we see that a user device is termed the *User Equipment* (UE). Strictly speaking, the UE contains the *Mobile Equipment* (ME) and the *UMTS Subscriber Identity Module* (USIM). The USIM is a chip that contains some subscription-related information plus security keys. It is similar to the *Subscriber Identity Module* (SIM) in GSM.

The interface between the UE and the network is termed the *Uu interface.* This is the WCDMA air interface described previously. Strictly speaking, the WCDMA interface, at least at the physical layer, is between the UE and the *Base Transceiver Station* (BTS). In 3GPP specifications, the base station is known as *node B.* This was originally a temporary name that somehow stuck.

A node B is connected to a single *Radio Network Controller* (RNC). The RNC controls the radio resources of the node Bs that are connected to it. The RNC is analogous to a *Base-Station Controller* (BSC) in GSM. Combined, an RNC and the node Bs that are connected to it are known as a *Radio Network Subsystem* (RNS). The interface between a node B and an RNC is the *Iub interface.* Unlike the equivalent Abis interface in GSM, the Iub interface is fully standardized and open. It is possible to connect a node B to an RNC of a different vendor.

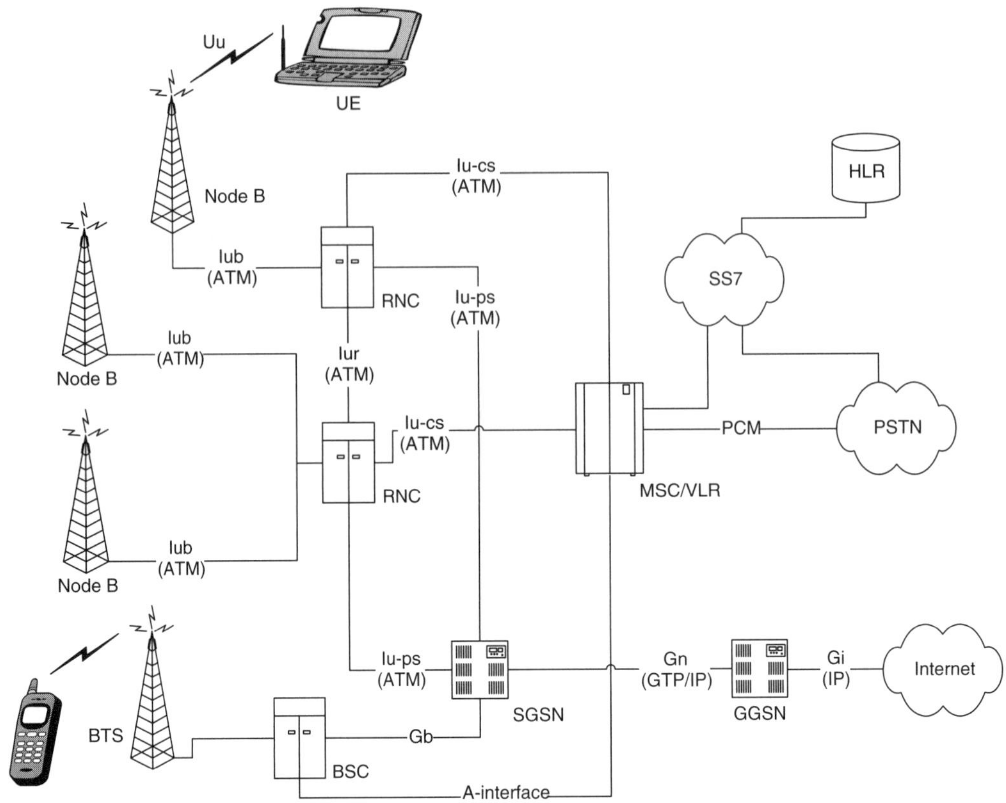

FIGURE 2.4 3GPP Release 1999 network architecture.

Unlike in GSM, where BSCs are not connected to each other, in the UMTS RAN (officially, the *UMTS Terrestrial Radio Access Network,* or UTRAN), an interface exists between the RNCs. This interface is termed *Iur.* The primary purpose of this interface is to support inter-RNC mobility and soft handover between node Bs connected to different RNCs. The Iur signaling in support of soft handoff is described in more detail in Chapter 3.

The UTRAN is connected to the core network via the Iu interface. The Iu interface, however, has two different components. The connection from UTRAN to the circuit-switched part of the core network is via the *Iu-CS interface,* which connects an RNC to a single *Mobile Switching Center* (MSC)/*Visitor Location Register* (VLR). The connection from UTRAN to the packet-switched part of the core network is termed *Iu-PS.* This connection is from an RNC to a Serving GPRS Support Node (SGSN).

It can be seen from Figure 2.4 that all the interfaces in the UTRAN of 3GPP Release 1999 are based on *Asynchronous Transfer Mode* (ATM). ATM was chosen because of its ability to support a range of different service types (such as a variable bit rate for packet-based services and a constant bit rate for circuit-switched services).

One also can see from Figure 2.4 that the core network uses the same basic architecture as GSM/GPRS. This was done purposely so that the new radio access technology could be supported by an established, robust core network technology. It should be possible for an existing core network to be upgraded to support UTRAN so that a given MSC, for example, could connect to both a UTRAN RNC and a GSM BSC.

In fact, UMTS specifications include support for a hard handover from UMTS to GSM and vice versa. This is an important requirement because the widespread rollout of UMTS coverage will take time to complete, and if holes exist in UMTS coverage, it is desirable that a UMTS subscriber should receive service from the more ubiquitous GSM coverage. If UTRAN and the GSM BSS are supported by different MSCs, then an inter-system handover could be achieved through an inter-MSC handover. Given that many of the functions of the MSC/VLR are similar for UMTS and GSM, however, it makes sense for a given MSC to be able to support both types of access simultaneously. Similar logic suggests that a given SGSN should be able to support an Iu-PS connection to an RNC and a Gb interface to a GPRS BSC simultaneously.

In most vendor implementations, many of the network elements are being upgraded to support GSM/GPRS/EDGE and UMTS simultaneously. Such network elements include the MSC/VLR, the *Home Location Register* (HLR), the SGSN, and the GGSN. For some vendors, the base stations deployed for GSM/GPRS/EDGE have been designed so that they can be upgraded to support both GSM and UMTS simultaneously. This is a major consideration for network operators who want to deploy a UMTS network in parallel with an existing GSM network. For some vendors, the BSC is being upgraded to act as both a GSM BSC and a UMTS RNC. This configuration is rare, however. The different interfaces and functions (such as a soft handover) required of a UMTS RNC mean that its technology is quite different from that of a GSM BSC. Consequently, it is normal to find separate UMTS RNCs and GSM BSCs.

2.6 Overview of the 3GPP Release 4 Network Architecture

Figure 2.5 shows the basic network architecture for 3GPP Release 4. The main difference between the Release 1999 architecture and the Release 4 architecture is that the core network becomes a distributed network. Rather than having traditional circuit-switched MSCs, as has been the case in previous network architectures, a distributed switch architecture is introduced, which is also referred to as IP Multimedia Services *(IMS)*.

In addition, in Release 4, *High-Speed Downlink Packet Access* (HSDPA) is introduced, enabling a more robust data offering in the downlink. The uplink remains the same as in Release 1999. For HSDPA, there are both hardware and software upgrades that need to be implemented, and these are discussed in Chapter 3.

Basically, the MSC is divided into an MSC server and a *Media Gateway* (MGW). The MSC server contains all the mobility management and call-control logic that would be contained in a standard MSC. It does not, however, contain a switching matrix. The switching matrix is contained within the MGW, which is controlled by the MSC server and can be placed remotely from the MSC.

Control signaling for circuit-switched calls is between the RNC and the MSC server. The media path for circuit-switched calls is between the RNC and the MG. Typically, an MG will take calls from the RNC and routes those calls toward their destinations over a packet backbone. In many cases, that packet backbone will use the *Real-Time Transport Protocol* (RTP) over the *Internet Protocol* (IP). As can be seen from Figure 2.5, packet-data traffic from the RNC is passed to the SGSN and from the SGSN to the GGSN over an IP backbone. Given that data and voice both can use IP transport within the core network,

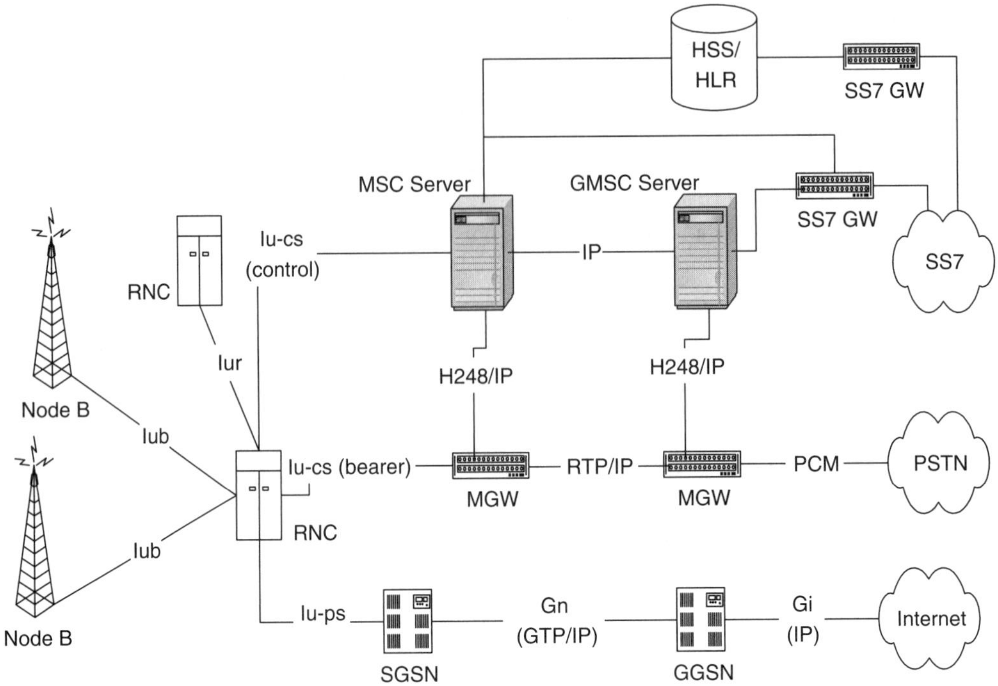

FIGURE 2.5 3GPP Release 4 distributed network architecture.

a single backbone can be constructed to support both types of service. This can mean significant savings of capital and operating expenses compared with construction and operation of separate packet- and circuit-switched backbone networks.

At the remote end, where a call needs to be handed off to another network, such as the *Public Switched Telephone Network* (PSTN), another MGW is controlled by a *Gateway MSC server* (GMSC server). This MGW will convert the packetized voice to standard *Pulse Code Modulation* (PCM) for delivery to the PSTN. It is only at this point that transcoding needs to take place. Assuming, for example, that speech over the air interface is carried at 12.2 kbps, then the voice does not need to be converted up to 64 kbps until it reaches the MGW that interfaces with the PSTN. This packetized transport can mean significant bandwidth savings on the backbone network, particularly if the two MGWs are some significant distance apart.

The control protocol between the MSC server or GMSC server and the MGW is the ITU H.248 protocol. This protocol was developed jointly by the ITU and the *Internet Engineering Task Force* (IETF). It also goes by the name *Media Gateway Control* (MEGACO). The call-control protocol between the MSC server and the GMSC server can be any suitable call-control protocol. The 3GPP standards suggest but do not mandate the *Bearer Independent Call Control* (BICC) protocol, which is based on ITU-T recommendation Q.1902.

In many cases, an MSC server also will support the functions of a GMSC server. Moreover, one MGW may have the ability to interface both with the RAN and with the

PSTN. In this case, calls to or from the PSTN can be handed off locally. This can represent another major savings.

Consider, for example, a scenario where an RNC is located in one city (city *A*) and is controlled by an MSC in another city (city *B*). Let's assume that a subscriber in city *A* makes a local phone call. Without a distributed architecture, the call needs to travel from city *A* to city *B* (where the MSC is), only to be connected back to a local PSTN number in city *A*. With a distributed architecture, the call can be controlled by an MSC server in city *B*, but the actual media path can remain within city *A*, thereby reducing transmission requirements and network operations costs.

One also will notice in Figure 2.5 that the HLR also may be known as a *Home Subscriber Server* (HSS). The HSS and HLR are functionally equivalent, with the exception that interfaces to an HSS will use packet-based transports such as IP, whereas an HLR is likely to use standard *Signaling System 7* (SS7)–based interfaces. Although not shown, a logical interface exists between the SGSN and HLR/HSS and between the GSN and HLR/HSS.

Many of the protocols used within the core network are packet-based, using either IP or ATM. However, the network must interface with traditional networks—through the use of media gateways. Moreover, the network also must interface with standard SS7 or CC7 networks. This interface is achieved through the use of an *SS7 Gateway* (SS7 GW). This is a gateway that on one side supports the transport of an SS7 message over a standard SS7 transport. On the other side, it transports SS7 application messages over a packet network such as IP. Entities such as the MSC server, the GMSC server, and the HSS communicate with the SS7 gateway using a set of transport protocols specially designed for carrying SS7 messages on an IP network. This suite of protocols is known as *SIGTRAN*. Many of the protocols mentioned in this brief discussion (e.g., RTP, H.248, and SIGTRAN) are described in greater detail in Chapter 6.

2.7 Overview of the 3GPP Release 5 All-IP Network Architecture

The next step in the UMTS evolution is the introduction of an all-IP multimedia network architecture (see Figure 2.6). This step in the evolution represents a change in the overall call model. Specifically, both voice and data are handled largely in the same manner all the way from the user terminal to the ultimate destination. This architecture can be considered the ultimate convergence of voice and data.

In Release 5, not only is the core enhanced, but the uplink data rate also is improved through the introduction of *High-Speed Uplink Packet Access* (HSUPA). This is discussed in Chapter 3.

As we can see from Figure 2.6, voice and data no longer need separate interfaces; just a single Iu interface can carry all the media. Within the core network, that interface terminates at the SGSN—there is no separate media gateway.

We also find a number of new network elements, notably the *Call State Control Function* (CSCF), the *Multimedia Resource Function* (MRF), the *Media Gateway Control Function* (MGCF), the *Transport Signaling Gateway* (T-SGW), and the *Roaming Signaling Gateway* (R-SGW).

An important aspect of the all-IP architecture is the fact that the user equipment is greatly enhanced. Significant logic is placed within the UE. In fact, the UE supports the *Session Initiation Protocol* (SIP), which is described in Chapter 9. The UE effectively becomes an SIP user agent. As such, the UE has far greater control of services than previously.

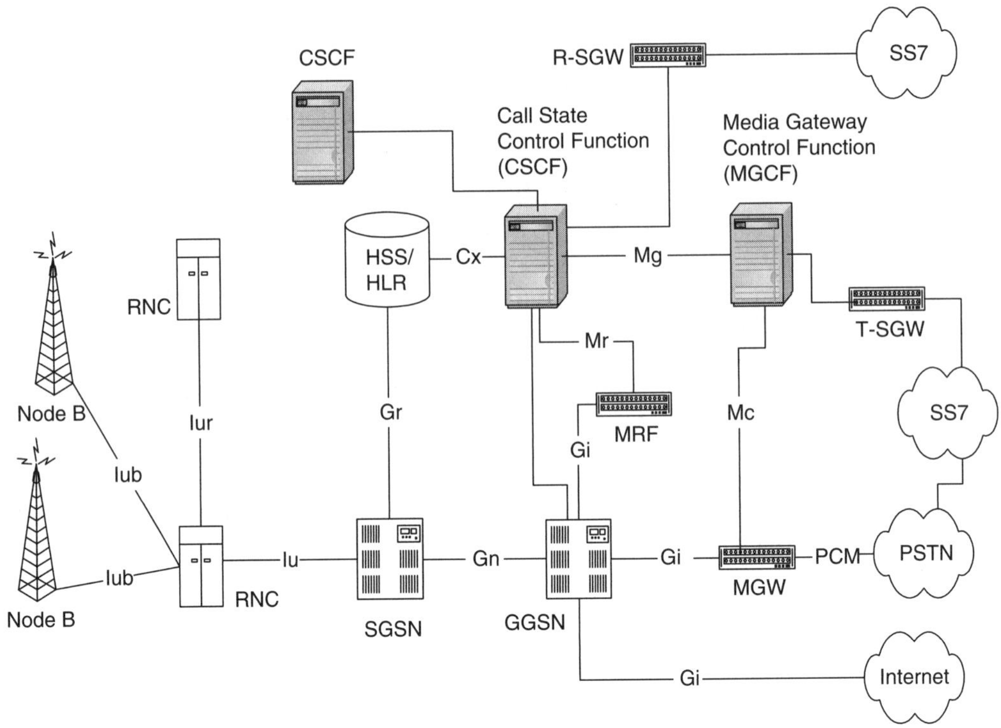

FIGURE 2.6 3GPP IP multimedia network architecture.

The CSCF manages the establishment, maintenance, and release of multimedia sessions to and from user devices. This includes functions such as translation and routing. The CSCF acts like a proxy server/registrar, as defined in the SIP architecture described in Chapter 9.

The SGSN and GGSN are enhanced versions of the same nodes used in GPRS and UMTS Releases 1999 and 4. The difference is that these nodes, in addition to data services, now support services that traditionally have been circuit-switched—such as voice. Consequently, appropriate QoS capabilities need to be supported either within the SGSN and the GGSN or, at a minimum, in the routers immediately connected to them.

The *Multimedia Resource Function* (MRF) is a conference bridging function used to support features such as multiparty calling and meet-me conference service.

The *Transport Signaling Gateway* (T-SGW) is an SS7 gateway that provides SS7 interworking with standard external networks such as the PSTN. The T-SGW will support SIGTRAN protocols. The *Roaming Signaling Gateway* (R-SGW) is a node that provides signaling interworking with legacy mobile networks that use standard SS7. In many cases, the T-SGW and R-SGW will exist within the same platform.

The MGW performs interworking with external networks at the media path level. The MGW in the 3GPP Release 5 network architecture is the same as the equivalent function within the 3GPP Release 4 architecture. The MGW is controlled by a *Media Gateway Control Function* (MGCF). The control protocol between these entities is ITU-T

H.248. The MGCF also communicates with the CSCF. The protocol of choice for this interface is SIP.

It should be noted that the Release 5 all-IP architecture is an enhancement to an existing Release 1999 or Release 4 network. It is effectively the addition of a new domain in the core network—the *IP-Multimedia* (IM) domain. This new domain, which enables both voice and data to be carried over IP all the way from the handset, uses the services of the *Packet Switched* (PS) domain for transport purposes. That is, it uses the SGSN, GGSN, Gn, Gi, etc., nodes and interfaces that belong to the PS domain.

2.8 Overview CDMA2000

CDMA2000 is a wireless platform that is part of the IMT-2000 specification and is an extension of the CDMAOne wireless platforms using the IS-95A/B and J-STD-008 standards. CDMA2000, being an IMT-2000 standard, is geared toward the transport and treatment of 3G wireless services supporting multimedia applications for fixed as well as mobile situations.

In the existing 2G platforms that are operational today for both cellular and PCS, the same radio bandwidth is allocated for voice and data. The data services are, of course, really circuit-switched services, without the capability to overbook the data service and thus increase the capacity of the wireless system through the appropriate use of data services.

However, with Evolution Data Only EVDO (IS-856), a separate carrier is allocated for data-only traffic. This enables current systems to operate circuit-switched services on a 1xRTT system and packet services on (EVDO). As shown in Figure 2.7, the divergence of packet- and circuit-switched services occurs with EVDO's introduction. Some level of convergence takes place with Evolution Data and Voice (EVDV), but the final vision for convergence of IS-2000 and IS-856 has not evolved at this moment.

2.8.1 Migration Path

The migration path that a wireless operator must take to realize CDMA2000 as envisioned for 3G is usually thought of as a staged approach for implementation. The concept behind the phased approach is to enable wireless operators using IS-95 platforms to migrate toward 3G without having to either forklift their existing platforms or acquire a new spectrum. However, upgrading to 1xRTT and/or EVDO will require hardware and software changes. An important concept is that CDMA2000 is backward-compatible with existing 2G CDMA systems, thereby speeding time to market.

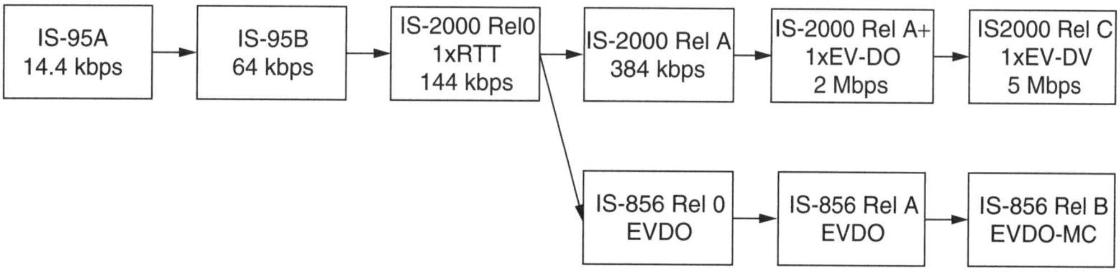

FIGURE 2.7 CDMA2000 evolution process.

From an operator's point of view, the migration from 2G to 3G and the realization of 3G must include the following:

- It needs to be cost-effective based on the capital infrastructure already in place.
- It needs increased capacity and throughput in both voice and data services that use existing spectrum allocations.
- It needs standard systems that enable both backward and forward compatibility with other network and data platforms.
- It needs the flexibility to meet ever-changing market conditions.

CDMA2000 phase 1 is an interim step between IS-95B and full realization of the IMT-2000 specification. The multicarrier approach for CDMA2000 has been abandoned at this time in favor of a more robust single-carrier solution. However, EVDO Rev B is planned to utilize multiple carriers but this specification has not been finalized at this writing. CDMA2000 can be and has been deployed in an existing IS-95 channel or system and will exhibit the numerous enhancements, some of which are included here:

- 1.25-MHz channel support
- 144-kbps packet-data rates
- 2× increase in voice capacity
- 2× increase in standby time
- Improved handoff

IS-95 and CDMA2000 1xRTT, EVDO, and EVDV can and will coexist in the same market and possibly at the same cell site. Obviously, one can take numerous approaches in the course of implementing any technology platform, and CDMA2000 is by no means unique to this situation. However, several common migration paths are being pursued for implementing CDMA2000. The migration path, of course, depends on whether the operator is currently using IS-95A/B, J-STD-008, or CDMA2000 (1xRTT) and upgrading to CDMA2000 EVDO/EVDV or is in the process of either installing a new system or segmenting existing spectrum to facilitate introduction of CDMA2000 EVDO/EVDV into the network (Table 2.5). Please keep in mind that there are several releases for CDMA2000 and EVDO that are driving toward harmonization in the future.

The following are three possible migration paths an operator may pursue:

1. CDMAOne (IS-95A)—CDMA2000 (1xRTT)—CDMA2000 (EVDO)
2. CDMAOne (IS-95A)—CDMAOne (IS-95B)—CDMA2000 (1xRTT)—CDMA2000 (EVDV)
3. CDMA2000 (1xRTT)—EVDV and EVDO

To complicate matters a little more for migration issues, several interim steps within the CDMA2000 implementation process bear mentioning relative to the single-carrier (1×) aspects. The expected migration path or, rather, the options for possible deployment of a CDMA2000-1x system are shown in Figure 2.7.

For instance there are several migration steps for EVDO starting with EVDO Rev 0 and then progressing to Rev A and then to Rev B. Many enhancements and also hardware and software changes are required to realize this migration, which is discussed in Chapter 4 in more detail.

Standard	Salient Issues
IS-95A	9600 bps or 14.4 kbps
IS-95B	Primarily voice, data on forward link, improved handoff, and data speeds of 64/56 kbps
CDMA2000	SR1 (1.2288 Mcps), voice and data (packet data via separate channel) 128 Walsh codes (256 future) 2× voice capacity over IS-95 144 kbps using 1xRTT with SR1
IS-856 (EVDO)	SR1 (1.2288 Mcps) Packet data only Higher data rate 3 Mbps DL 2 Mbps UL

Table 2.5 CDMA Path

2.8.2 System Architecture

The system architecture that will make up a CDMA2000 network is a logical extension of an existing CDMAOne network, with the fundamental difference being the introduction of packet-data services. The implementation of a CDMA2000 system is meant to involve upgrades to the BTS and BSC for the purpose of handling the packet-data services. Additionally, the use of packet-data services also necessitates the introduction of a packet server complex that may exist already to support services such as *Cellular Data Packet Data* (CDPD).

However, it is recommended that the existing packet-data network should not by default be considered for inclusion into the CDMA2000 network architecture. The system architecture for a CDMA2000 network, owing to packet-data services, can be either centralized or distributed. The decision whether the system uses a distributed or centralized system depends on the design requirements as well as operational issues. Figure 2.8 is an example of a stand-alone CDMA2000 system that has the inclusion of a Packet Data Serving Node (PDSN) for handling packet-data services.

2.8.3 Spectrum

The spectrum requirements for a CDMA2000 system have their roots in IS-95, but some differences exist. A comparison for spectrum requirements between IS-95, CDMA2000-1xRTT, and EVDO carriers is shown in Figure 2.9.

The channel depicted in the figure indicates that for whatever version of CDMA2000-1x the operator decides to deploy, it can be overlaid onto the existing IS-95 channel through a 1:1 or *N*:1 upgrade. Concurrently, an EVDO system can be overlaid onto an existing 1xRTT network in a 1:1 or *N*:1 method as well. CDMA2000-1x introduces several enhancements, including a reverse pilot channel, as well as improved Walsh codes, and these are covered in more detail in Chapter 4.

When the decision is made to migrate to an EVDO system, the operator allocates its own specific spectrum for EVDO and thus operates on its own channel, which is effectively an overlay system. Depending on the plan, the CDMA2000 channel locations will need to be thought through in advance, and this issue also is covered in Chapter 4.

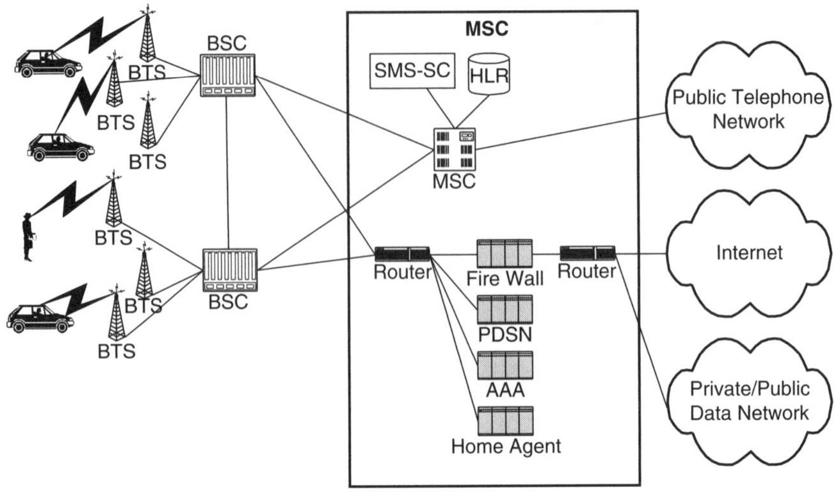

FIGURE 2.8 Generic CDMA2000 system architecture.

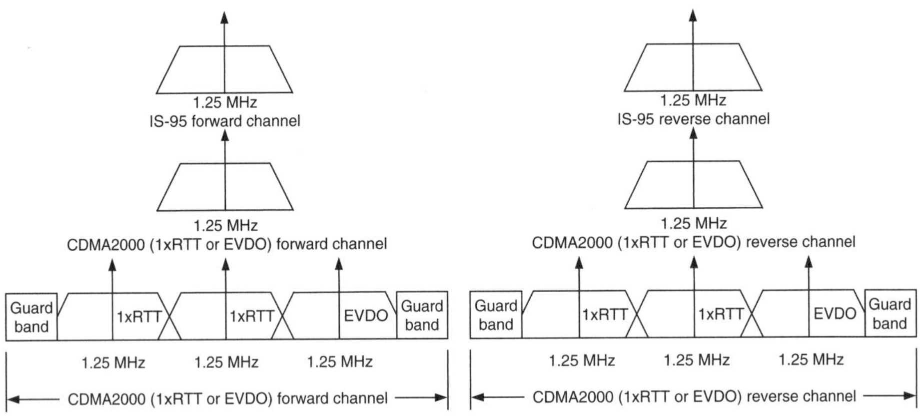

FIGURE 2.9 1xRTT and EVDO.

2.9 TD-CDMA

TD-CDMA, also referred to as *HCR TDD* and *UMTS TDD,* is an IMT-2000–compliant air-interface specification using packet data. The TD-CDMA system interfaces with a GSM/UMTS core network, which allows it to be overlaid on top of the existing network or built as a new system.

The TD-CDMA RAN uses a combination of the three basic multiple-access schemes—FDMA, TDMA, and CDMA—requiring a total of 5 MHz of frequency as compared with 10 MHz (5 × 5) for WCDMA.

The use of a TDD access method with TD-CDMA is designed to support asymmetric traffic such as IP. TD-CDMA systems use unpaired spectrum, enabling them to be more spectrum efficient than a WCDMA network when supporting IP traffic, unlike circuit-switched voice traffic.

The TDD air interface consists of 15 time slots that can be assigned by the operator for uplink or downlink use. Of the 15 total slots, 12 typically are allocated for traffic and 3 for signaling with the subscriber units. Each of the time slots that is assigned to carry traffic also uses 16 unique spreading codes, one for each user or segment of bandwidth. The TD-CDMA system also can assign a different modulation and coding scheme to each time slot in order to maximize the throughput for each of the subscribers in that sector in support of different services being supported at the same time.

TD-CDMA has many similar characteristics to TD-SCDMA, except that it uses a chip rate of 3.84 Mcps, whereas TD-SCDMA uses 1.28 Mcps. TD-CDMA's use of 3.84 Mcps harmonizes it with WCDMA for a symbiotic deployment.

TD-CDMA systems offer peak speeds of 7 Mbps per sector with a single carrier and up to 16 Mbps with two carriers. Higher rates are expected in future releases, and enhancements are anticipated with TD-CDMA with expected rates of 14 Mbps for a single carrier and 31.5 Mbps for two carriers. The dual carrier is important because 3G spectrum typically is allocated in an FDD manner.

Another variant of the TD-CDMA system is FDD-TDCDMA, which has increased bandwidth (10 MHz instead of 5 MHz). This is achieved by pairing channels for increased throughput by allowing more time slots to be concatenated.

2.9.1 Generic TD-CDMA Architecture

A TD-CDMA network architecture is very similar to other IMT-2000 networks. In fact, the RAN interface is the primary difference, where the core network uses Releases 99, 4, and 5. Figure 2.10 shows a sample TD-CDMA network that interfaces with an SGSN, enabling it to be used as a data-solution offering for a GSM/GPRS/EDGE, WCDMA, or even TD-SCDMA systems.

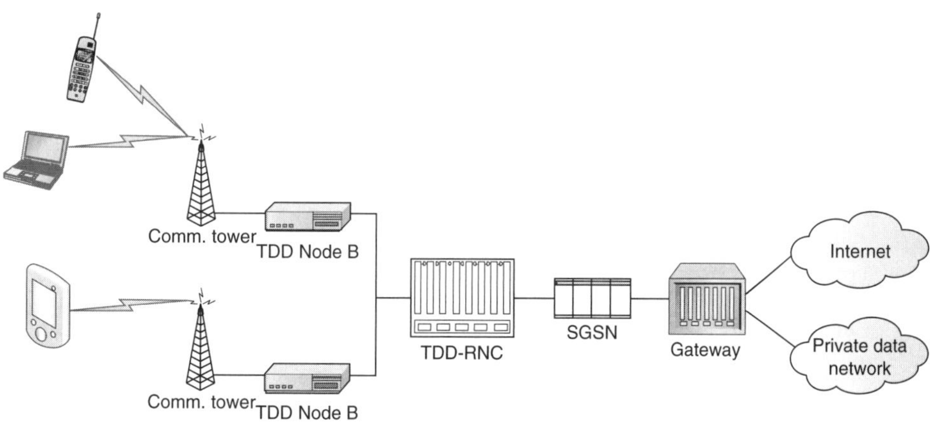

FIGURE 2.10 TD-CDMA system architecture.

2.9.2 Radio Network

TD-CDMA supports both circuit and packet services and is designed primarily to support the asymmetric characteristics of IP data, enabling broadband to the edge of the network. TD-CDMA uses TDD, and therefore, uplink and downlink traffic share the same physical radio channel. However, each radio channel has 15 time slots. Additionally, in each time slot, up to 16 separate codes, which are orthogonal to each other, can be transmitted, where each user can be allocated individual codes or groups of codes for higher data throughput.

Uplink and downlink channel allocation can be done by the system or the operator. A minimum of one uplink and one downlink always needs to be allocated per TD-CDMA carrier.

In order to support both packet and circuit services, TD-CDMA uses both shared channels and dedicated channels. This is in comparison with an FDD system such as WCDMA, which supports only dedicated channels on the uplink. TD-CDMA, however, supports shared channels in both directions, enabling it to be more efficient with bursty traffic such as IP traffic.

2.9.3 RAN

TD-CDMA requires 5 MHz of radio bandwidth to operate a single channel. The 5 MHz of bandwidth as well as the 3.84 Mcps is meant to facilitate coexistence with WCDMA networks, enabling operators to deploy both technologies in the same market. TD-CDMA can operate on all the IMT-2000 frequency bands.

The channel structure for the TD-CDMA radio access uses TDMA, FDMA, and CDMA. In addition, TD-CDMA uses the same chip rate, modulation, and bandwidth (nonduplex) that are used by WCDMA.

The heart of the TD-CDMA radio access is the TDD access method. Each radio carrier is divided into 15 time slots, each containing 16 separate and unique codes; this is depicted in Figure 2.11.

The specific uplink and downlink transmissions alternate on the same *radio-frequency* (RF) channel by allocating time slots to either the uplink or downlink, with the ratio determining the relative uplink and downlink bandwidth.

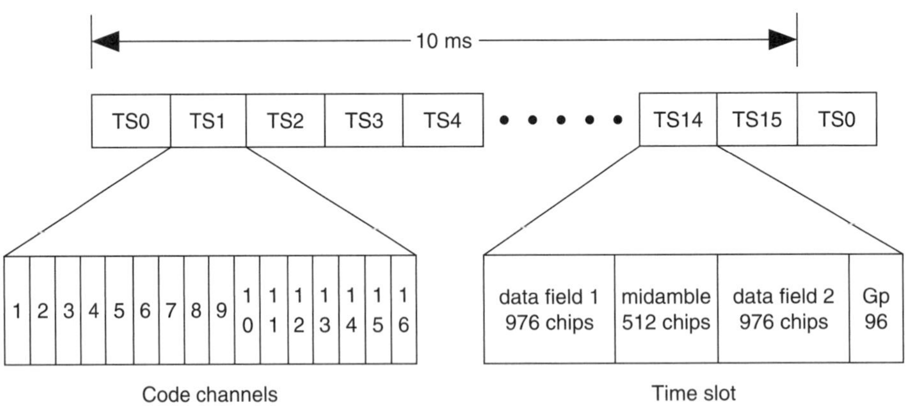

Figure 2.11 TD-CDMA channel structure.

An important aspect is that TD-CDMA supports both packet data (IP) and circuit-switched voice, and when it is supporting circuit-switched traffic, it allocates a symmetric uplink and downlink channel.

Several coding schemes are used with TD-CDMA. In TD-CDMA spreading, scrambling, and channelization codes are used. Spreading codes are an essential element in a TD-CDMA network. There are a total of 16 spreading codes, and all are orthogonal to each other and, like all CDMA codes, are part of a tree. In TDD, the power control in TD-CDMA is done via a closed loop for the downlink, and for the uplink, an open loop method is used.

2.9.4 Handover

For a TD-CDMA system, the parameters needed in the cell-selection monitoring set may include

- SIR
- Path loss
- Interference power
- Received power level on BCH, etc.

The handover process is implemented in the mobile unit and the RNS. Measurement of serving radio connection downlink performance and candidate cell received signal strengths and quality is made in the UE. The RNS measures the uplink performance, as well as the position information for the UE being served, and uses these measurements in conjunction with defined thresholds and handover strategy to make a handover decision.

2.9.5 Implementation

Multiple scenarios are possible for implementing a TD-CDMA system. However, all these implementation methods fall into one of two categories: new spectrum or overlay. The specific methodology for implementing a TD-CDMA system largely depends on two elements: your business plan relating to the services and the coverage area you want to service and the spectrum that you have to use. There are, of course, other considerations, but without spectrum and a valid business plan defining what you want to provide, it is difficult to establish a proper implementation plan.

2.10 TD-SCDMA

TD-SCDMA is one of the 3G wireless mobile services standards that makes up the IMT-2000 set of specifications discussed in Chapter 1. TD-SCDMA differs from CDMA2000 and WCDMA in that it is a TDD system as opposed to FDD. The use of TDD as the radio access method has several advantages over traditional FDD networks.

- TDD has no need for paired frequencies, using the same frequency for uplink and downlink transmission.
- TDD is suitable for asymmetric uplink and downlink transmission rates, especially for IP-type data services.

- The TDD system's major attribute is its spectral efficiency with asymmetric traffic such as IP. More specifically, TD-SCDMA adapts the uplink/downlink ratio according to the data load and provides data rates ranging from 1.2 kbps to 2 Mbps.

2.10.1 System Architecture

The fundamental architecture for a TD-SCDMA wireless system has a similar set of network components as all 3G networks. TD-SCDMA uses a 3GPP core network while at the RAN and therefore can be connected to an existing GSM, WCDMA, or TD-CDMA network. However, the primary implication is to interface with an existing GSM network, providing a migration path to 3G. The RAN portion of the TD-SCDMA, node B, network employs a combination of FDMA, TDMA, and CDMA protocols.

Figure 2.12 is a general depiction of a TD-SCDMA network that interfaces with a 3GPP core network. Only one TD-SCDMA cell is shown; other radio access systems are not shown, but they could be included easily, provided that they interface at the MSC/SGSN.

TD-SCDMA has an IP-based network following Release 4 and Release 5 for UMTS. The use of an IP-based core network facilitates many services and addresses the issue of interfacing with numerous appliances and services in the future.

TD-SCDMA uses TDD as the access method, allowing for both asynchronous and synchronous operation. Because TD-SCDMA is a TDD network, it needs to reduce interference through a combination of interference-mitigation techniques that include the use of "smart" antennas and joint detection (mobile rake receiver) and the previously mentioned uplink synchronization. All these techniques help to reduce interference, thereby increasing pole capacity.

The RAN interface only requires the use of 1.6 MHz of spectrum, enabling it to have multiple carriers in the same bandwidth as a WCDMA, CDMA2000, or TD-CDMA carrier. There are two variations to TD-SCDMA: TSM and LCR. LCR is *low chip rate* and requires that Release 4 be implemented in the core network. A TSM – TD-SCDMA system for mobile communication involves leveraging a GSM core platform.

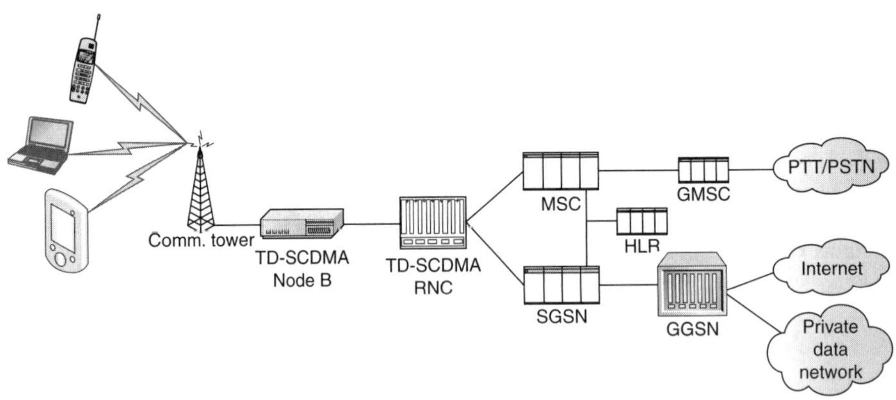

Figure 2.12 TD-SCDMA system architecture.

TD-SCDMA is designed currently to operate in the 2010- to 2025-MHz frequency band, but since it is IMT-2000-compliant, it is possible, via the specification, to operate in all IMT-2000 bands.

An important aspect of the all-IP architecture is the fact that the user equipment is greatly enhanced. Significant logic is placed within the UE. In fact, the UE supports the *Session Initiation Protocol* (SIP). The UE effectively becomes an SIP user agent. As such, the UE has far greater control of services than previously.

TD-SCDMA supports circuit-switched services in addition to packet (IP) services. Circuit-switched rates are defined as 12.2, 64, 144.4, 384, and 2048 kbps. Packet-data rates are defined as 9.6, 64, 144.4, 384, and 2048 kbps.

The radio spectrum required for a TD-SCDMA network can begin with as simple as 1.6 MHz of contiguous spectrum for an individual channel, with a possible use of a guard band of 100 to 200 kHz. The guard-band requirements will need to be specified when differing technologies are spectrally adjacent to each other. In many countries around the world, TDD spectrum has been allocated for potential use for mobility. In China, where TD-SCDMA is being sponsored heavily, a total of 155 MHz of frequency bandwidth has been allocated for TDD. Simple math indicates that with a 1.6-MHz carrier, a total of 96 unique TD-SCDMA carriers are possible.

2.10.2 Channel Structure

The unique frame structure of the TD-SCDMA radio channel enables it to be more adaptive to network evolution. TD-SCDMA network construction follows a layered approach to design and implementation. The TD-SCDMA radio resources for the uplink and downlink are allocated separately, even though the uplink and downlink are in the same carrier.

Figure 2.13 shows the channel structure for a TD-SCDMA carrier. There are a total of seven time slots for each TD-SCDMA carrier. Each of the carriers has a radio frame that is 10 ms in length, as shown in the figure. The radio carrier is made up of two separate subframes, and each subframe is made up of seven time slots.

A total of 16 spreading codes are used with TD-SCDMA. The codes are all orthogonal to each other and, like all CDMA codes, are part of a tree. The tree for the code sequence dictates the potential data throughput that can occur.

2.10.3 Interference-Mitigation Techniques

TD-SCDMA, because it uses a TDD method for access, has unique self-interference problems that need to be addressed in order to take advantage of system capabilities.

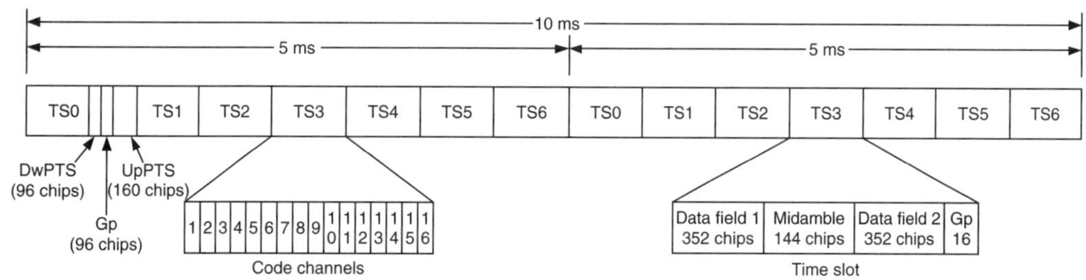

FIGURE 2.13 TD-SCDMA channel structure.

To help mitigate interference, TD-SCDMA uses a combination of interference-mitigation techniques that include the use of "smart" antennas and joint detection, terminal (uplink) synchronization, and dynamic channel allocation. All these techniques help to reduce interference, thereby increasing pole capacity:

- Smart antennas
- Joint detection
- Terminal synchronization

2.10.4 Handover

For a TD-SCDMA system, the parameters needed in the cell-selection monitoring set may include

- SIR
- Path loss
- Interference power
- Received power level on BCH, etc.

The handover process is implemented in the mobile unit and the RNS. Measurements of serving radio connection downlink performance and candidate cell received signal strengths and quality are made in the UE. The RNS measures the uplink performance as well as the position information for the UE being served and uses these measurements in conjunction with defined thresholds and handover strategy to make a hand-over decision.

2.11 Commonality among WCDMA, CDMA2000, TD-CDMA, and TD-SCDMA

WCDMA, CDMA2000, TD-CDMA, and TD-SCDMA share several commonalities that are part of the IMT-2000 platform specification. All four platforms use CDMA technology and require, in their final versions, a total of either 1.25, 1.6, or 5 MHz of spectrum. All four systems will be able to interoperate with one another, and it is possible for a wireless operator to deploy several or all of the 3G platforms in the same network, provided that the requisite spectrum is available.

One possible scenario would be CDMA2000 and TD-SCDMA being deployed jointly. WCDMA, TD-CDMA, and TD-SCDMA also use a GSM 2G network as a logical starting point, as does CDMA2000 with IS-95. Any level of perturbations is possible, but to make this a reality, it requires introduction of a multimode handset that is able to operate on any of the four technology platform plus legacy systems. However, at this time, no such handset is available, but it is possible with a *software-defined radio* (SDR).

All systems have a migration path from existing 2G platforms to 3G. However, the path both systems take is different and is driven by the imbedded infrastructure the existing operator already has deployed, which is driven by the business plans, capital, and operating costs. Since the end game is to offer high-speed packet-data services to the end user, the real issue between both these standards within the IMT-2000 specification is the methodology for how they realize the desired speed.

All four platforms are designed to operate in multiple frequency bands and can operate in the same frequency bands, provided that the spectrum is available. Therefore, the commonalities can be summed up in the following brief points that were introduced at the beginning of this chapter:

- Global standard
- Compatibility of service within IMT-2000 and other fixed networks
- High quality
- Worldwide common frequency band
- Small terminals for worldwide use
- Worldwide roaming capability
- Multimedia application services and terminals
- Improved spectrum efficiency
- Flexibility for evolution to the next generation of wireless systems
- High-speed packet-data rates
- 2 Mbps for fixed environment
- 384 kbps for pedestrian
- 144 kbps for vehicular traffic

References

Barron, Tim, "Wireless Links for PCS and Cellular Networks," *Cellular Integration*, September 1995, pp. 20–23.

Bates, Gregory, *Voice and Data Communications Handbook*, McGraw-Hill, New York, 1998.

Brewster, R.L., *Telecommunications Technology*, Wiley, New York, 1986.

Brodsky, Ira, "3G Business Model," *Wireless Review*, June 15, 1999, p. 42.

Daniels, Guy, "A Brief History of 3G," *Mobile Communications International* 65:106, October 1999.

DeRose, James E., *The Wireless Data Handbook*, Quantum Publishing, Mendocino, CA, 1994.

Dixon, Robert C., *Spread Spectrum Systems*, 2d ed., Wiley, New York, 1984.

Gull, Dennis, "Spread-Spectrum Fool's Gold?" *Wireless Review*, January 1, 1999, p. 37.

Harte, Hoenig, and Kikta McLaughlin, *CDMA IS-95 for Cellular and PCS*, McGraw-Hill, New York, 1996.

Harter, Betsy, "Putting the C in TDMA?" *Wireless Review*, January 2001, pp. 29–34.

Held, Gil, *Voice & Data Interworking*, 2d ed., McGraw-Hill, New York, 2000.

Homa, Harri, and Antti Toskala, *WCDMA for UMTS*, Wiley, New York, 2000.

IETF RFC 2543: Session Initiation Protocol (SIP).

IETF RFC 768: User Datagram Protocol (STD 6).

IETF RFC 791: Internet Protocol (STD 5).

IETF RFC 793: Transmission Control Protocol (STD 7).

ITU-T H.248: Media Gateway Control Protocol.

LaForge, Perry M., "cdmaOne Evolution to Third Generation: Rapid, Cost-Effective Introduction of Advanced Services," *CDMA World*, June 1999.

Louis, P. J., *M-Commerce Crash Course*, McGraw-Hill, New York, 2001.

McClelland, Stephen, "Europe's Wireless Futures," *Microwave Journal*, September 1999, pp. 78–107.

McDysan, Spohn, *ATM Theory and Applications*, McGraw-Hill, New York, 1999.

Molisch, Andreas F., *Wideband Wireless Digital Communications*, Prentice-Hall, Englewood Cliffs, NJ, 2001.

Mouly, Pautet, *The GSM System for Mobile Communications*, 1992.

Muratore, Flavio, *UMTS Mobile Communications for the Future*, Wiley, Sussex, England, 2000.

Newton, Harry, *Newton's Telcom Dictionary*, 14th ed., Flatiron Publishing, New York, 1998.

Oba, Junichi, "W-CDMA Systems Provide Multimedia Opportunities," *Wireless System Design*, July 1998, p. 20.

Ramjee, Prasad, Werner Mohr, and Walter Konhauser, *Third Generation Mobile Communication Systems*, Artech House, Boston, 2000.

Rusch, Roger, "The Market and Proposed Systems for Satellite Communication," *Applied Microwave & Wireless*, Fall 1995, pp. 10–34.

Salter, Avril, "W-CDMA Trial & Error," *Wireless Review*, November 1, 1999, p. 58.

Shank, Keith, "A Time to Converge," *Wireless Review*, August 1, 1999, p. 26.

Smith, Clint, *LMDS*, McGraw-Hill, New York, 2000.

Smith, Clint, *Wireless Telecom FAQ*, McGraw-Hill, New York, 2000.

3GPP TS 23.002: Network Architecture (Release 1999).

3GPP TS 23.002: Network Architecture (Release 4).

3GPP TS 23.002: Network Architecture (Release 5).

3GPP TS 25.101: UE Radio Transmission and Reception (FDD).

3GPP TS 25.104: UTRA (BS) FDD; Radio Transmission and Reception.

3GPP TS 25.211: Physical Channels and Mapping of Transport Channels onto Physical Channels (FDD).

3GPP TS 25.212: Multiplexing and Channel Coding (FDD).

3GPP TS 25.213: Spreading and Modulation (FDD).

3GPP TS 25.214: Physical Layer Procedures (FDD).

3GPP TS 25.301: Radio Interface Protocol Architecture.

3GPP TS 25.302: Services Provided by the Physical Layer.

3GPP TS 25.401: UTRAN Overall Description.

3GPP TS 26.090: AMR Speech Codec; Transcoding Functions.

Webb, William, "CDMA for WLL," *Mobile Communications International*, January 1999, p. 61.

Webb, William, *Introduction to Wireless Local Loop*, Vol. 2: *Broadband and Narrowband Systems*, Artech House, Boston, 2000.

Wesley, Clarence, "Wireless Gone Astray," *Telecommunications*, November 1999, p. 41.

Willenegger, Serge, "CDMA2000 Physical Layer: An Overview," Qualcomm, San Diego, CA.

William, C. Y., *Lee's Essentials of Wireless Communications*, McGraw-Hill, New York, 2001.

www.tdscdma-forum.org.

www.umtsworld.com.

CHAPTER 3

Universal Mobile Telecommunications Service (UMTS)

3.1 Introduction

Universal Mobile Telecommunications Service (UMTS) represents an evolution of *Global System for Mobile* (GSM) communications to support *Third-Generation* (3G) capabilities. UMTS has evolved from WCDMA to HSPA and is migrating to *Long Term Evolution* (LTE). In this chapter we examine the details of the UMTS, including the air interface and network architecture. The initial focus is on the air interface and *Radio Access Network* (RAN) because these represent the greatest change from the technologies of GSM, the *General Packet Radio Service* (GPRS), and the *Enhanced Data Rates for Global Evolution* (EDGE). Subsequently, we delve into the specifics of the core network and examine the evolution of UMTS to HSPA and in later chapters how this fits into LTE. First, however, we present some general information on UMTS technology.

3.2 UMTS Basics

As described briefly in Chapter 2, UMTS includes two of the air-interface proposals submitted to the *International Telecommunications Union* (ITU) as proposed solutions to meet the requirements laid down for *International Mobile Telephony 2000* (IMT-2000). These both use *Direct Sequence Wideband Code-Division Multiple Access* (DS-WCDMA). One solution uses *Frequency-Division Duplex* (FDD), and the other uses *Time-Division Duplex* (TDD). The FDD solution is likely to see the greatest deployment—particularly in Europe and the Americas. The TDD solution is likely to see deployment primarily in Asia. In this chapter we focus mainly on the FDD option.

In the FDD option, paired 5-MHz carriers are used in the uplink and downlink as follows: uplink—1920 to 1980 MHz; downlink—2110 to 2170 MHz. Thus, for the FDD mode of operation, a separation of 190 MHz is used between the uplink and downlink in Europe, China, and many other parts of the world. In the United States, UMTS can operate in the 850- and 1900-MHz and AWS bands and possibly the 700-MHz band as well, leading to different *transmission* (Tx) and *receiving* (Rx) spacing based on the specific band of operation.

Additionally, although 5 MHz is the nominal carrier spacing, it is possible to have a carrier spacing of 4.4 to 5 MHz in steps of 200 kHz. This enables spacing that might be needed to avoid interference, particularly if the next 5-MHz block is allocated to another carrier.

For the TDD option, a number of frequencies have been defined, including 1900 to 1920 MHz and 2010 to 2025 MHz. Of course, with TDD, a given carrier is used in both the uplink and the downlink, so no separation exists. TDD systems, because of their unique qualities, are covered in Chapters 5 and 6.

In any CDMA system, user data are spread to a far greater bandwidth than the user rate through application of a spreading code, which is a higher-bandwidth pseudo-random sequence of bits known as *chips*. The transmission from each user is spread by a different spreading code, and all users transmit at the same frequency at the same time. At the receiving end, the signal from one user is separated from those of other users by despreading the set of received signals with the spreading code applicable to the user in question. The result of the despreading operation is retrieval of the user data in question plus some noise generated as a result of the transmissions from other users.

The ratio of the spreading rate (the number of chips per second) to the user data rate (the number of user data symbols per second) is known as the *spreading factor*. The greater the spreading factor, the greater is the ability to extract a given user's signal from those of all other users. In other words, for a given user data rate, the higher the chip rate, the more users can be supported. Alternatively, for a set number of users, the higher the chip rate, the higher the data rates that can be supported for each user. Thus, the spreading rate is of major significance. Of course, one gets nothing for nothing—the higher the chip rate, the greater is the occupied spectrum. The chip rate in WCDMA is 3.84×10^6 chips per second (3.84 Mcps), which leads to a carrier bandwidth of between 4.4 and 5 MHz.

From a network architecture perspective, UMTS borrows heavily from the established network architecture of GSM. In fact, many of the network elements used in GSM are reused (with some enhancements) in UMTS. This commonality means that a given *Mobile Switching Center* (MSC), *Home Location Register* (HLR), *Serving GPRS Support Node* (SGSN), or *Gateway GPRS Support Node* (GGSN) can be upgraded to support UMTS and GSM simultaneously.

The radio access, however, is significantly different from that of GSM, GPRS, and EDGE. In UMTS, the *Radio Access Network* (RAN) is known as the *UMTS Terrestrial Radio Access Network* (UTRAN). The components that make up the UTRAN are significantly different from the corresponding elements in the GSM architecture. Therefore, the reuse of existing GSM base stations and *Base-Station Controllers* (BSCs) is limited.

For some vendors, GSM base stations were planned in advance to be upgradable to support WCDMA as well as GSM. Thus, for some vendors, it is possible to remove some number of GSM transceivers from a base station and replace them with some number of UMTS transceivers. For other vendors, a completely new base station is needed. A similar situation applies to BSCs. For most vendors, the technology of a UMTS *Radio Network Controller* (RNC) is so different from that of a GSM BSC that the BSC cannot be upgraded to act simultaneously as a GSM BSC and a UMTS RNC. Cases will occur, however, where a BSC can be upgraded to support both GSM and UMTS simultaneously, but that situation is less common.

3.3 The WCDMA Air Interface

As mentioned earlier, WCDMA uses a chip rate of 3.84 Mcps. As also mentioned earlier, CDMA technology in general uses a spreading code to separate one user's transmissions from those of another. In reality, however, there will be multiple simultaneous data streams from multiple users and multiple simultaneous data streams from a single base station. Therefore, not only is it necessary to separate the transmissions of one user or base station from those of another, but it is also necessary to separate the various transmissions that a single user might generate. In other words, if a single user (user *A*) is transmitting both user data and control information, the base station first must separate the set of transmissions from user *A* from the transmissions of all other users. Then it must separate the control information from the user data.

In order to support this requirement, WCDMA takes a two-step approach to the transmission from a single user, as shown in Figure 3.1. First, each individual data stream is spread to the chip rate by the application of a spreading code, also known as

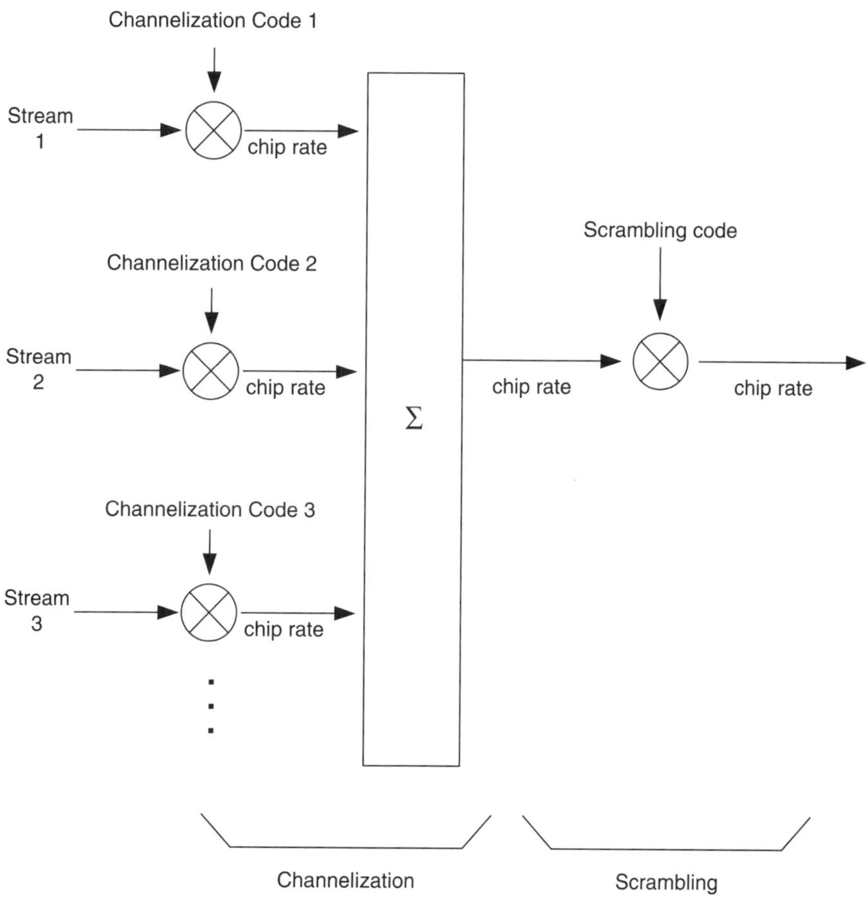

FIGURE 3.1 Spreading and scrambling, basic concept.

a *channelization code*, that operates at the chip rate of 3.84 Mcps. Then the combined set of spread signals is scrambled by the application of a scrambling code, which also operates at the chip rate. The channelization spreads the individual data streams and hence increases the required bandwidth. Since the scrambling code also operates at the chip rate, however, it does not increase the required bandwidth further. At the receiving end, the combined signal is first descrambled by application of the appropriate scrambling code. The individual user data streams then are recovered through application of the appropriate channelization codes. Clearly, it is important that different users employ different scrambling codes. Multiple users, however, can use the same channelization codes, provided, however, that no two transmissions from the same user employ the exact same channelization code.

3.3.1 Uplink Spreading, Scrambling, and Modulation

A *physical channel* is what carries the actual user data or control information over the air interface. A physical channel can be considered a combination of frequency, scrambling code, and channelization code, and in the uplink, as we shall describe later, the relative phase is also significant. For example, if a given user is transmitting user data and control information, then the user data stream will be carried on one physical channel, and the control information will be carried on a different physical channel.

A number of different physical channels are used in the uplink, with a given type of channel selected according to what the *User Equipment* (UE) is attempting to do—such as simply request access to the network, send a single burst of data, or send a stream of data. These channels are described in further detail later in this chapter. For now, let us focus on the situation where a user is transmitting a stream of data, which would happen in a voice conversation. In such a situation, the terminal normally will use at least two physical channels—a *Dedicated Physical Data Channel* (DPDCH) and a *Dedicated Physical Control Channel* (DPCCH). The DPDCH carries the user data, and the DPCCH carries control information. Depending on the amount of data to be sent, a single user can use just a single DPDCH, which will support up to 480 kbps of user data or as many as six DPDCHs, which will support up to 2.3 Mbps of user data.

A DPDCH can have a variable spreading factor. This simply means that the user bit rate does not have to be fixed to a specific value. The spreading factor for a DPDCH can be 4, 8, 16, 32, 64, 128, or 256. These correspond to DPDCH bit rates of 15 kbps ($3.84 \times 10^6/256 = 15 \times 10^3$) and up to 960 kbps ($3.84 \times 10^6/4 = 960 \times 10^3$). Of course, these are not the actual user data rates because a significant amount of coding overhead is included in the DPDCH to support forward error correction. In general, the user data rate is approximately half (or less) of the DPDCH rate. Thus, for example, a DPDCH operating at a spreading rate of 4 will carry data at a rate of 960 kbps. Of this, however, only about 480 kbps will correspond to usable data. The rest is consumed by additional coding required for error correction. If a single user wants to transmit user data at a rate greater than 480 kbps, then multiple DPDCHs can be used (up to a maximum of 6).

Figure 3.2 shows how multiple DPDCHs are handled. Also shown is the DPCCH, which is also sent whenever one or more DPDCHs are sent. The channelization codes ($C_{d,1}$–$C_{d,6}$) represent the channelization codes applied to each of the six DPDCHs. The channelization code applied to the DPCCH is represented as C_c. Each of the DPDCHs is spread to the chip rate by a channelization code. DPDCHs 1, 3, and 5 are channelized and weighted by a gain factor b_d. These DPDCHs are on the so-called I (in-phase) branch. DPDCHs 2, 4, and 6 plus the DPCCH are on the so-called Q (quadrature) branch.

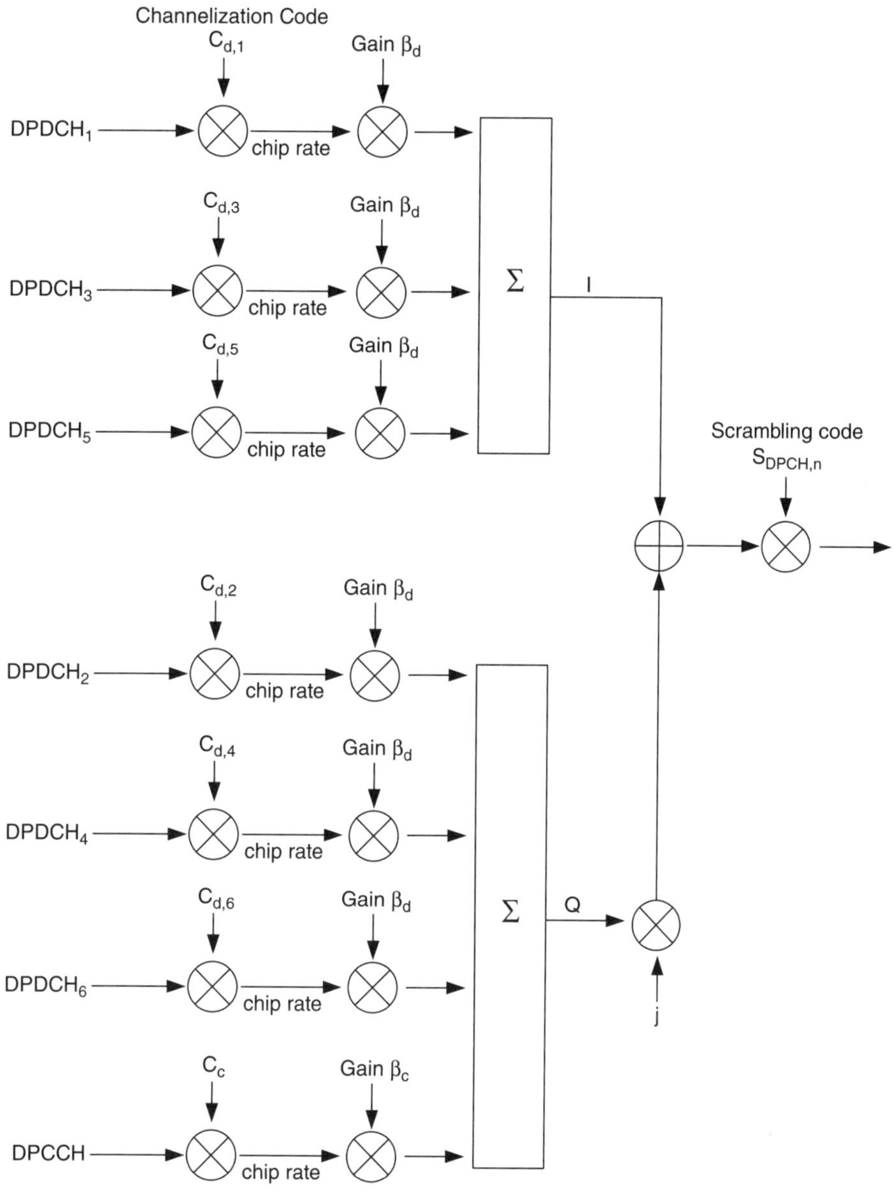

FIGURE 3.2 Uplink channelization and scrambling.

These are also channelized. These spread DPDCHs are also weighted by the gain factor b_d, whereas the spread DPCCH is weighted by the gain factor b_c. The two gain factors are specified as 4-bit words that represent steps from 0 to 1. Thus 0000 = off, 0001 = 1/15, 0010 = 2/15, and 1111 = 15/15 = 1. At any given instant, one of the two gain factors has the value of 1 (binary 1111 = 15/15 = 1).

Mathematically, the spread signals on the Q branch are treated as a stream of imaginary bits. These are summed with the stream of real bits on the I branch to provide a stream of complex-valued chips at the chip rate. This stream of complex-valued chips then is subjected to a complex-valued scrambling code that is aligned with the beginning of a radio frame.

3.3.1.1 Channelization Codes

As mentioned earlier, the channelization codes are used to separate multiple streams of data from a given user, whereas the scrambling codes are used to separate transmissions from different users. The channelization codes are known as *Orthogonal Variable Spreading Factor* (OVSF) codes. They are taken from the code tree shown in Figure 3.3. The generation of channelization codes is given by the following equations:

$$C_{ch,1,0} = (1)$$
$$[C_{ch,2,0}] = [C_{ch,1,0}\ C_{ch,1,0}] = (1,1)$$
$$[C_{ch,2,1}] = [C_{ch,1,0} - C_{ch,1,0}] = (1,0)$$
$$[C_{ch,2^{(n+1)},0}] = [C_{ch,2^{(n)},0}\ C_{ch,2^{(n)},0}]$$
$$[C_{ch,2^{(n+1)},1}] = [C_{ch,2^{(n)},0} - C_{ch,2^{(n)},0}]$$
$$[C_{ch,2^{(n+1)},2}] = [C_{ch,2^{(n)},1}\ C_{ch,2^{(n)},1}]$$
$$[C_{ch,2^{(n+1)},3}] = [C_{ch,2^{(n)},1} - C_{ch,2^{(n)},1}], \text{ etc.}$$

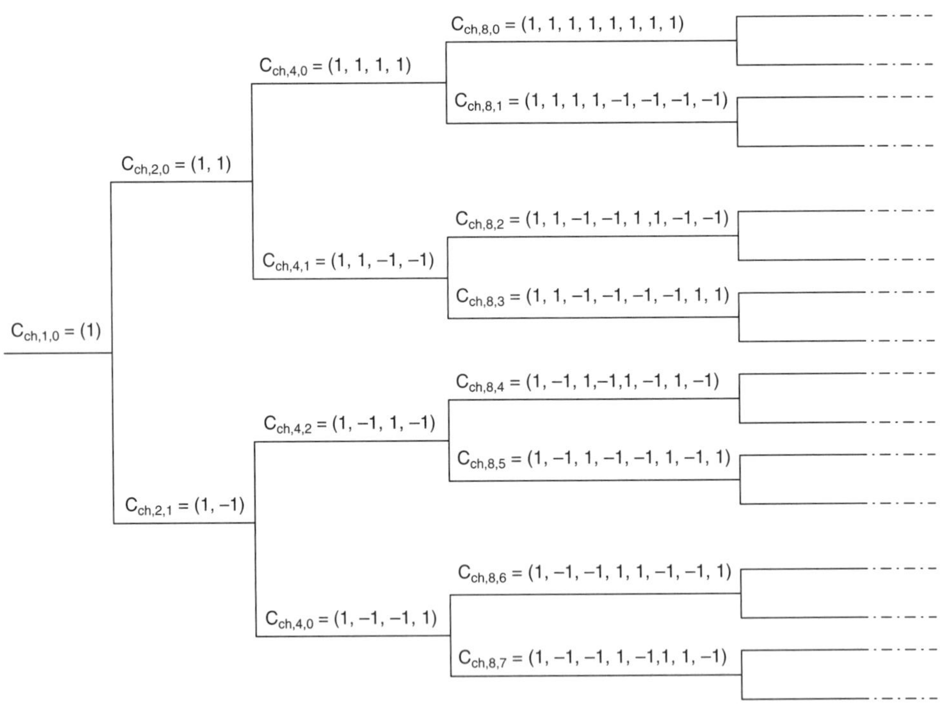

FIGURE 3.3 Channelization code tree.

In general, a given physical channel uses a channelization code that is related to the spreading factor being used for the channel. When only one DPDCH is to be transmitted, then the channelization code is $C_{ch,SF,k}$, where SF is the spreading factor, and $k = SF/4$. Therefore, if the spreading factor is 128 (as determined by the user data rate plus coding overhead), then the code to be used shall be $C_{ch,128,32}$. The spreading factor for the DPCCH is always 256, and the channelization code is $C_{ch,256,0}$.

When more than one DPDCH is to be transmitted (greater than 960 kbps of combined user data and coding overhead), then each DPDCH will have a spreading factor of 4, and the channelization code for each DPDCH will be $C_{ch,4,k}$. $k = 1$ for $DPDCH_1$ and $DPDCH_2$, $k = 2$ for $DPDCH_3$ and $DPDCH_4$, and $k = 3$ for $DPDCH_5$ and $DPDCH_6$. For example, $DPDCH_3$ and $DPDCH_4$ both would use the channelization code $C_{ch,4,2} = (1, 21, 1, 21)$. Given that channelization codes are used to separate different transmissions from a single user, the fact that certain codes can be used simultaneously on two channels is, at first glance, troubling. The fact, however, that those two channels always will be on separate I and Q branches means that they still can be separated.

Some important restrictions apply to the use of channelization codes. This is so because, in the case where more than one channel is being transmitted, the chosen channelization codes must be orthogonal. For example, consider the channelization code $C_{ch,4,0}$. This code is simply the sequence 1, 1, 1, 1 repeated over and over, with each sequence of 4 bits repeated 960,000 times per second. Consider the channelization code $C_{ch,8,0}$. This is simply the sequence 1, 1, 1, 1, 1, 1, 1, 1 repeated over and over, with each sequence of 8 bits repeated 480,000 times per second. Clearly, if one data stream from a given user is spread with the code $C_{ch,4,0}$ and a second data stream from the same user is spread with the code $C_{ch,8,0}$, the net effect is that they are spread in the same way and cannot be distinguished at the receiver. Consequently, channelization codes must be selected in a manner that ensures that each channel is spread differently.

3.3.1.2 Scrambling Codes

Once the different channels have been spread with appropriate channelization codes, they are combined as shown in Figure 3.2 and then scrambled by a particular scrambling code. Two types of scrambling codes exist—long and short scrambling codes, with 2^{24} possibilities for each type. The choice of a particular code is determined by the type of physical channel in question (as we shall see, other physical channels are available besides the DPDCH and DPCCH) and by the higher layer that requires the use of a channel in the first place. Depending on higher-layer requirements, a DPDCH or DPCCH can use either a short scrambling code or a long scrambling code.

Clearly, the channelization codes are far from random, and they do not need to have pseudorandom properties. Scrambling codes, however, must appear to be random and thus must have pseudorandom characteristics. The easiest way to generate a pseudorandom sequence is through the use of a linear feedback shift register such as that shown in Figure 3.4. This is basically a set of flip-flops that are clocked at a particular rate, and the output of the last flip-flop is copied back into one or more of the other flip-flops, possibly after an addition. In Figure 3.4., each of the gain values (g_n) is simply a 1 or a 0. Depending on the values of each g_n (i.e., exactly where the output is fed back), a different output pattern can be achieved. This output pattern can be described by a polynomial and is known as a *generator polynomial.*

It is possible to produce maximum-length sequences, known as *m-sequences.* This means that if a register has m elements, then it can produce a sequence of length $2^m - 1$.

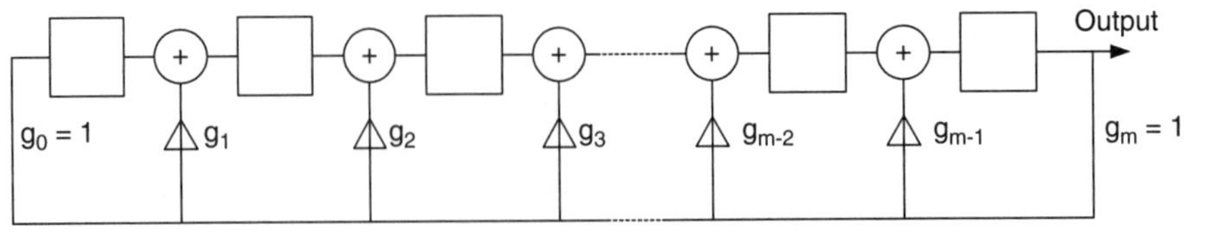

Figure 3.4 Linear feedback shift register.

For example, if a shift register has 10 elements, then it can produce a sequence of length $2^{10} - 1$ (i.e., 1023). This is a pattern that repeats after every 1023 bits. An m-sequence has a number of properties, including the property that, over the period of the sequence, there will be exactly 2^{m-1} ones and $2^{m-1} - 1$ zeros.

The long scrambling codes used in WCDMA are known as *gold codes* and are constructed from the modulo 2 addition of portions of two binary m-sequences. The portions used are segments of length 38,400. This is due to the fact, as will be explained later in this chapter, that the frame length in WCDMA is 10 ms, which corresponds to 38,400 chips. Because the long scrambling codes are generated from m-sequences, they have pseudorandom characteristics. The short scrambling codes also have pseudo- random characteristics. These, however, are much shorter, at a length of 256 chips. Long scrambling codes are used in the case where the base station uses a rake receiver. Short scrambling codes can be used when the base station uses advanced multiuser detection techniques such as a *Parallel Interference Cancellation* (PIC) receiver.

3.3.1.3 Uplink Modulation

WCDMA uses *Quadrature Phase Shift Keying* (QPSK) modulation in the uplink. This technique is depicted in Figure 3.5. The stream of spread and scrambled signals, such as the output shown in Figure 3.2, forms the complex-valued input stream of chips. The real and imaginary parts are separated, with the real part of a given complex chip forming the in-phase (I) branch and the imaginary part forming the quadrature-phase (Q) branch in the modulator.

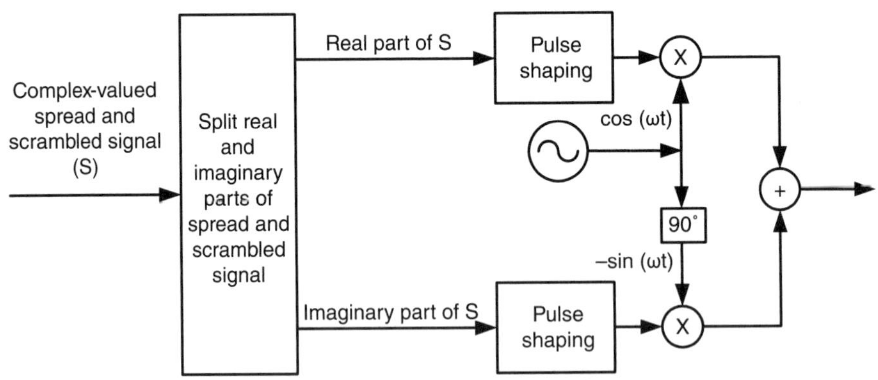

Figure 3.5 Uplink modulation (QPSK).

3.3.2 Downlink Spreading, Scrambling, and Modulation

As is the case for the uplink, a number of channels are used in the downlink. In fact, more channels are defined for the downlink than for the uplink. This is so because the downlink includes pilot channels, synchronization channels, channels used for the broadcast of system information, channels used for the paging of subscribers, and so on.

3.3.2.1 Downlink Spreading

With the exception of the *Synchronization Channels* (SCHs), the downlink channels are spread to the chip rate and scrambled, as shown in Figure 3.6. Each channel to be spread is split into two streams—the I branch and the Q branch. The even symbols are mapped to the I branch, and the odd symbols are mapped to the Q branch. The I branch is treated as a stream of real-valued bits, whereas the Q branch is treated as a stream of imaginary bits. Each of the two streams is spread by the same channelization code. The spreading code/channelization code to be used is taken from the same code tree as used in the uplink—that is, OVSF codes that are chosen to maintain the orthogonality between different channels transmitted from the same base station. The spreading rate for a given channel depends on the channel in question.

The I and Q streams then are combined such that each I and Q pair of chips is treated as a single complex value, and the result of combining them is a stream of complex-valued chips. This stream of chips then is subjected to a complex downlink scrambling code, identified as $S_{dl,n}$ in Figure 3.6.

One important difference occurs between spreading in the uplink and downlink. In the uplink, the data for a given physical channel (such as a single DPDCH) are directed either to the I branch or to the Q branch, as shown in Figure 3.2. Thus, on the uplink, no serial-to-parallel conversion takes place. Therefore, for a spreading factor of, say, 8, the data rate of the physical channel is simply 3,840,000/8 = 480 kbps.

On the downlink, however, each channel (with the exception of the synchronization channel) is subjected to a serial-to-parallel conversion, as shown in Figure 3.6. For a given spreading factor, the serial-to-parallel conversion effectively doubles the

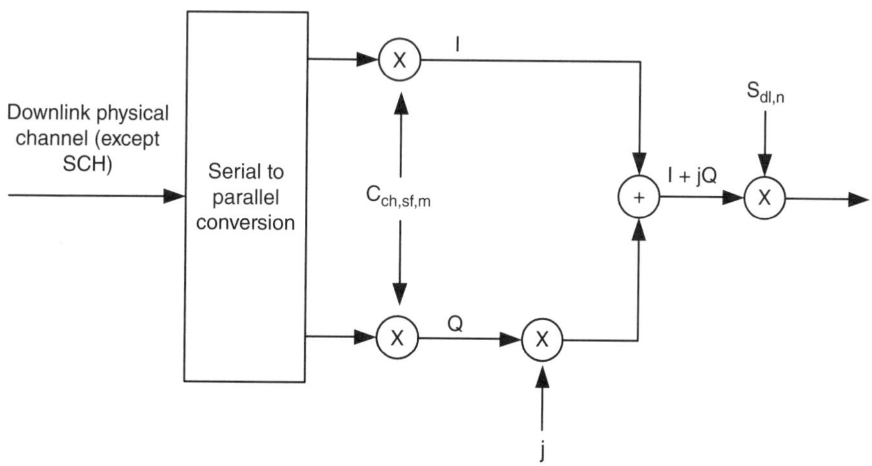

Figure 3.6 Downlink scrambling.

data rate of the physical channel. In other words, half the channel's data is carried on the I branch and half on the Q branch, and both of these are spread with the same spreading factor.

If, for example, we have a spreading factor of 8, then the data rate on the I channel is 480 kbps, and the data rate on the Q channel is also 480 kbps. The net data rate is 960 kbps—twice that achieved on the uplink for the same value of spreading factor. In reality, however, the data rate on the downlink is not quite twice that on the uplink. This is due to the fact, as explained later, that control information is time-multiplexed with a user data on the downlink. This reduces the net throughput for a given downlink data channel. Nonetheless, for a given spreading factor on the downlink, the effective throughput is significantly greater than the corresponding throughput on the uplink for the same spreading factor.

3.3.2.2 Downlink Scrambling

The downlink scrambling codes are used to separate the transmissions of one cell from those of another. The downlink scrambling codes are gold codes similar to the long scrambling codes used in the uplink. As is the case for the long codes used on the uplink, the codes used on the downlink are limited to a 10-ms duration. There is a total of $2^{18} - 1$ (262,143) downlink scrambling codes. Not all these codes are used, however. If all possible codes were to be usable, then one could find a situation where a terminal would have to check a received signal against all 262,143 codes. This could occur, for example, during cell selection. Clearly, checking against so many scrambling codes is impractical.

Therefore, the available downlink scrambling codes are separated into 512 groups. Each group contains one primary scrambling code and 15 secondary scrambling codes. Thus 512 primary scrambling codes and 7680 secondary scrambling codes exist, for a total of 8192 downlink scrambling codes. Table 3.1 shows the allocation of secondary downlink scrambling codes to primary downlink scrambling codes.

A cell is allocated one, and only one, primary scrambling code, which, of course, has 15 secondary scrambling codes associated with it. A given base station will use its primary scrambling code for the transmission of channels that need to be heard by all terminals in the cell. Thus paging messages need to be scrambled by the cell's primary scrambling code. For that matter, all transmissions from the base station simply can use the cell's primary scrambling code. After all, it is the scrambling code that identifies the cell, whereas the various channelization codes are used to separate the various transmissions (physical channels) within the cell.

Primary Scrambling Code Number	Secondary Scrambling Code Numbers
0	1–15
16	17–31
32	33–47
48	49–63
8176	8177–8191

Table 3.1 Allocation of Secondary Scrambling Codes to Primary Scrambling Codes

Primary Scrambling Group Number	Primary Scrambling Code Numbers
0	0, 16, 32, 48, 64, 80, 96, 112
1	128, 144, 160, 176, 192, 208, 228, 240
2	256, 272, 288, 304, 320, 336, 352, 368
63	8064, 8080, 8096, 8112, 8128, 8144, 8160, 8176

TABLE 3.2 Primary Scrambling Code Groups

A cell can, however, choose to use a secondary scrambling code for channels that are directed to a specific user and do not need to be decoded by other users. In general, it is a good idea for all transmissions from a cell to use the cell's primary scrambling code because this helps to minimize interference.

As described earlier, 512 primary scrambling codes are available. These are divided into 64 groups, each consisting of 8 scrambling codes, as shown in Table 3.2.

As mentioned earlier, downlink spreading and scrambling are applied to all downlink physical channels transmitted on a cell, with the exception of the SCH. This channel is added to the downlink stream, as shown in Figure 3.7. In fact, as explained later in this chapter, the SCH contains two subchannels—the primary SCH and the secondary SCH. The reason why these are transmitted without scrambling is the fact that they are the first channels decoded by a terminal. If they were scrambled, then the terminal would first have to know the scrambling code of the base station just to synchronize.

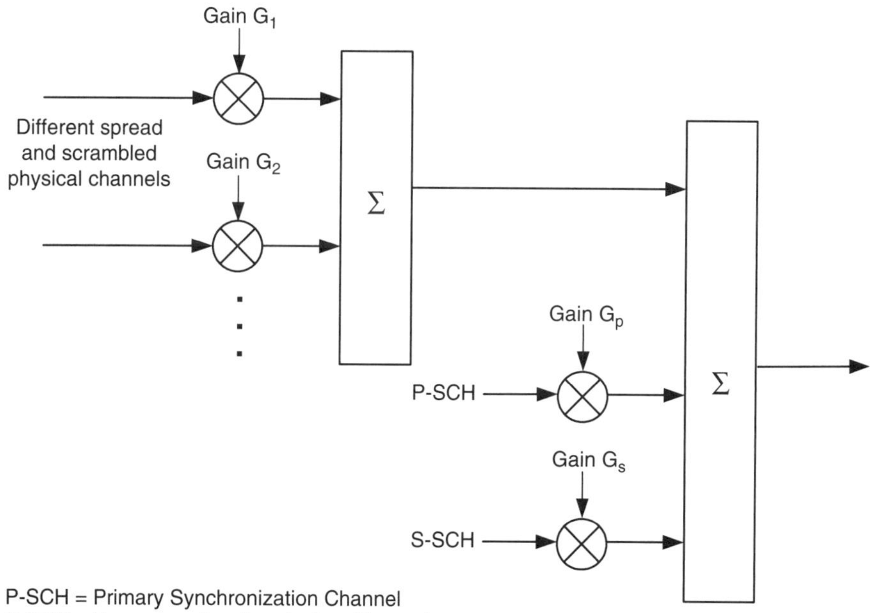

P-SCH = Primary Synchronization Channel
S-SCH = Secondary Synchronization Channel

FIGURE 3.7 Downlink multiplexing of the synchronization channel.

3.3.2.3 Downlink Modulation

As is the case for the uplink, the downlink uses QPSK modulation. The process in the downlink is the same as that shown in Figure 3.5 for the uplink. Each complex-valued chip is split into its constituent real and imaginary parts. The real part is sent on the I branch of the modulator, and the imaginary branch is sent on the Q branch of the modulator.

3.3.3 WCDMA Air-Interface Protocol Architecture

We have already mentioned some of the types of physical channels defined in WCDMA. In fact, many different channel types exist, and the various types of channels are defined in a logical hierarchy.

Figure 3.8 shows the overall logical structure of the WCDMA air interface. At the lowest level, we have the physical layer. The functions of the physical layer include *Radiofrequency* (RF) processing, spreading, scrambling and modulation, coding and decoding for support of forward error correction, power control, timing advance, and soft

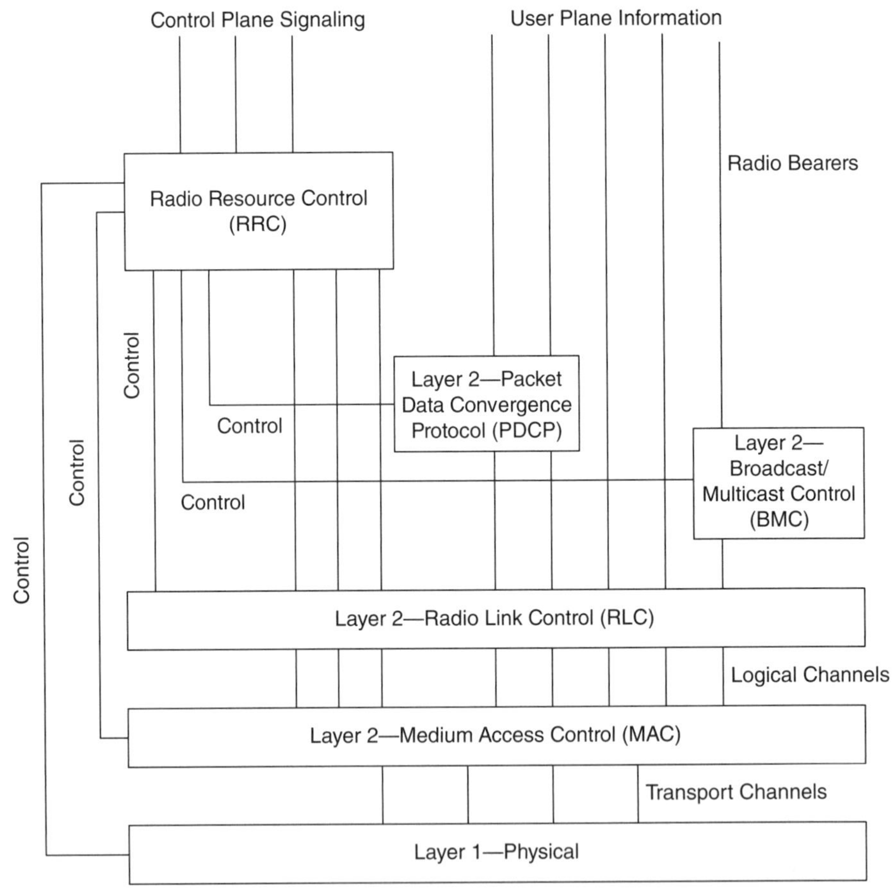

FIGURE 3.8 WCDMA air-interface protocol structure.

handover execution. Physical channels, such as those already mentioned, exist at the physical layer and are used for transmission across the RF interface. A given physical channel is defined by a combination of frequency, scrambling code, channelization code, and in the uplink, phase. Some physical channels exist solely for the correct operation of the physical layer. Other physical channels are used to carry information provided to or from higher layers.

Higher layers that want to transmit information across the RF interface pass information to the physical layer through the *Medium Access Control* (MAC) layer using a number of logical channels. The MAC maps these logical transport channels to channels. The physical layer maps transport channels to physical channels.

Above the MAC layer, we find the *Radio Link Control* (RLC) layer. Among the services provided by RLC are the following:

- *RLC connection establishment and release.* A given upper layer may request the use of a certain radio bearer. For each radio bearer, an RLC connection is established between the MS and the network.
- *Error detection.* RLC includes a sequence number check function that enables the detection of errors in received *protocol data units* (PDUs).
- *Ensuring error-free delivery through acknowledgments (if the upper-layer protocol has requested an acknowledged service).* RLC can request that the peer entity retransmit in the event that a PDU is received incorrectly, lost, or received out of sequence. Note that this type of error correction is different from the error correction that is achieved through coding schemes on the air interface.
- *In-sequence delivery.* This ensures that PDUs are passed to the upper layer in the correct order.
- *Unique delivery.* This ensures that a given PDU is passed to an upper layer only once, even if erroneously received twice at RLC.
- *Quality-of-Service (QoS) management.* Upper layers can request a certain QoS. It is RLC that ensures that the QoS is controlled.

RLC supports both acknowledged and transparent services. With transparent service, any errors in received PDUs will cause the PDU to be discarded, in which case it is up to the upper layer to recover from the loss according to its own capabilities. With acknowledged service, RLC recovers from errors in received data by requesting a retransmission by the peer entity (the UE or the network).

One of the protocols above the RLC layer is the *Packet Data Convergence Protocol* (PDCP). The main objective of PDCP is to enable the lower layers (RLC, MAC, and the physical layer) to be common regardless of the type or structure of the user data. For example, packet-data transfer from a UE could use either IPv4 or IPv6. One does not want the RLC and lower layers to be different depending on which of those two protocols a subscriber uses. Moreover, if new protocols are introduced, one would want them to be supported by the same radio interface. PDCP meets these objectives by maintaining a standard interface to RLC regardless of the type of user data. PDCP is similar to the *Subnetwork Dependent Convergence Protocol* (SNDCP) of GPRS.

In Figure 3.8 we also find the *Broadcast/Multicast Control* (BMC). This is a function that handles the broadcast of user messages across the cell. In other words, BMC supports the cell broadcast function, similar to cell broadcast as defined in GSM. This enables users in a cell to receive broadcast messages, such as traffic warnings and weather information. In GSM, cell broadcast also has been used as a means of informing users of the geographic zone that they are in as part of zone-based tariffing.

One of the most important components depicted in Figure 3.8 is the *Radio Resource Control* (RRC). RRC can be considered the overall manager of the air interface and, as such, is responsible for the management of radio resources, including determination of which radio resources will be allocated to a given user. As can be seen, all control signaling to or from users passes through RRC. This is necessary so that requests from a user or from the network can be analyzed and radio resources allocated as appropriate. Also, a control interface exists between RRC and each of the other layers. Among the functions performed or controlled by RRC are

- *The broadcast of system information.*
- *The establishment of initial signaling connections between the UE and the network.* When the user and network want to communicate, an RRC connection is first established. It is this RRC connection that is used for the transfer of signaling information between the UE and the network for the purpose of allocation and management of the radio resources to be used.
- *The allocation of radio bearers to a UE.* A given UE may be allocated multiple radio bearers for the transfer of user data.
- *Measurement reporting.* RRC determines what needs to be measured, when it should be measured, and how it should be reported.
- *Mobility management.* It is the RRC that determines when, for example, a call should be handed over. RRC also executes cell reselection and location area or routing area updates.
- *Quality-of-service (QoS) control.* The allocation of radio resources has a direct consequence for the QoS perceived by the user. Since RRC controls the allocation of radio resources, it has a direct influence on QoS. The resources allocated by RRC must be aligned with the QoS offered to the subscriber.

3.3.4 WCDMA Channel Types

At the physical layer, the UE and the network communicate via a number of physical channels. Many of these physical channels are used to carry information that is passed to the physical layer from higher layers. Specifically, information is passed to the physical layer from the MAC layer. The interface between the physical layer and the MAC layer consists of a number of transport channels that are mapped to physical channels. Moreover, information from the RLC layer to the MAC layer is passed in the form of logical channels. These logical channels are mapped to transport channels. The following subsections consider these various channels in a little more detail. We start with the transport channels.

3.3.4.1 Transport Channels

In general, two types of transport channels exist. These are common transport channels and dedicated transport channels. *Common transport channels* may be applicable either to all users in a cell or to one or more specific users. In the case when a common transport channel is used to transmit information to all users, no specific addressing information is required. When a specific user needs to be addressed by a common transport channel, then the user identification is included in-band (within the message being sent).

For example, the *Broadcast Channel* (BCH), which is a common transport channel, is used to transmit system information to all users in a cell and is not specific to any given user. On the other hand, the *Paging Channel* (PCH), which is also a common transport channel and is used to page a specific mobile unit, contains the identification of the user within the message being transmitted.

As we describe the various types of channels supported, we will make references to frames and slots. Basically, the various channels use a 10-ms frame structure, which corresponds to 38,400 chips. Each frame is divided into 15 slots, each with a length of 2560 chips, as shown in Figure 3.9. The content of each frame, and for that matter the content of each slot, depends on the type of channel in question.

The following common transport channels are defined:

- The *Random Access Channel* (RACH) is used in the uplink when a user wants to gain access to the network. It also may be used when a user wants to send a small amount of data to the network. The amount of data sent on the RACH is small—it lasts either 10 or 20 ms. This is in accordance with the fact that the RACH is used primarily for signaling related to initial system access. It must be possible for the RACH to be heard at the base station from any user in the cell coverage area—even from at the edge, at least when the RACH is used for initial access to the network. Because, as we shall see in Chapter 11, the effective coverage area of a cell decreases with the increasing bandwidth, it is necessary for the data rate on the RACH to be quite low. The RACH is available to all users in the cell. Consequently, the possibility of collision arises when multiple users attempt to access the RACH. UTRAN includes procedures at the physical layer for collision detection.
- The *Broadcast Channel* (BCH) is used in the downlink to transmit system information over the entire coverage area of a cell. For this reason, it is sent with

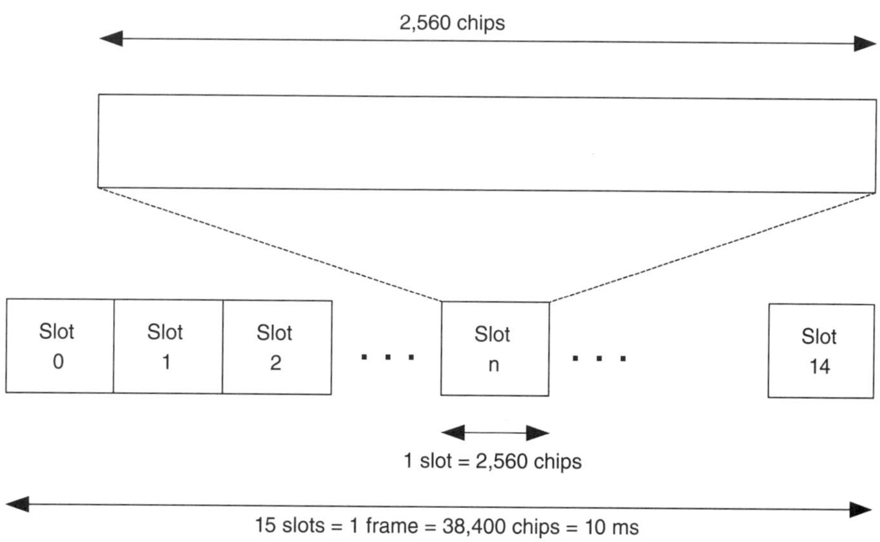

FIGURE 3.9 WCDMA channel framing structure.

a relatively high power level. Moreover, the data rate on the BCH is quite low compared with some other channels. When sent over the air interface, the BCH information is sent at 30 kbps, including coding overhead.

- The *Paging Channel* (PCH) is used in the downlink to page a given UE when the network wants to initiate communication with a user. A page for a given UE may be sent on a single cell or multiple cells depending on the location area/routing area configuration of the network. In a given cell, the PCH must be heard over the whole cell area.
- The *Forward Access Channel* (FACH) is used to send downlink control information to one or more users in a cell. If, for example, a user attempts to access the network on the RACH, then the response to the access request will be sent on the FACH. The FACH also can be used to send small amounts of packet data to a mobile unit. It is possible to have more than one FACH in a cell. At least one FACH, however, must have a sufficiently low data rate that all users in the cell can hear it.
- The uplink *Common Packet Channel* (CPCH) is similar to the RACH but can last for several frames. Thus, it enables a greater amount of data to be sent than is allowed by the RACH. It can be used, for example, when the terminal wants to send a single burst of data that cannot be accommodated on the RACH. The CPCH is available to all users in the cell. Consequently, the possibility of collision occurs when a user attempts to access the CPCH. The UTRAN includes procedures at the physical layer for minimizing the likelihood of collision and detecting collision when it does occur.
- The *Downlink Shared Channel* (DSCH) is used to carry dedicated user data or control signaling to one or more users in a cell. It is similar to the FACH but does not have to be transmitted over the entire cell area. Moreover, it supports higher data rates than the FACH, and the data rate on the DSCH can change on a frame-by-frame basis. The DSCH is always associated with one or more downlink-dedicated channels, described later.

Only a single *dedicated transport channel* type exists, known as the *Dedicated Channel* (DCH). This is a channel that carries user data and is specific to a single user. Although other channels can carry small amounts of bursty user data, they are not designed for large amounts of data or for extended data sessions. The DCH is used for those types of sessions.

For example, in a voice conversation, the coded voice uses the DCH. The DCH exists in the uplink and the downlink and is mapped to the physical channels DPDCH and DPCCH, described previously. In the uplink, at the physical layer, the combination of frequency, the scrambling code, the channelization code, and the phase is used to indicate a particular DPDCH or DPCCH. In the downlink, the DCH is mapped to a *Dedicated Physical Channel* (DPCH), which is identified in the downlink by a particular channelization code. The downlink DPDCH and DPCCH are time-multiplexed onto the downlink DPCH. The data rate on a DCH can vary on a frame-by-frame basis.

3.3.4.2 Physical Channels

As mentioned earlier, information from upper layers is passed to the physical layer through a number of transport channels. These transport channels are mapped to a number of physical channels on the air interface. In general, a physical channel is identified by a specific frequency, scrambling code, channelization code, duration, and in the uplink, phase.

In addition to the physical channels that are mapped to or from transport channels, a number of physical channels exist only for the correct operation of the physical layer. Such channels are not visible to higher layers. The following are the physical channels:

- The *Synchronization Channel* (SCH) is transmitted by the base station and is used by a UE during the cell search procedure. In order for a UE to retrieve broadcast information sent from the base station, it first must be synchronized properly with the base station. This synchronization is the primary purpose of the SCH. The SCH contains two subchannels—the primary SCH and the secondary SCH, as shown in Figure 3.7. The primary SCH contains a specific 256-chip codeword, known as the *Primary Synchronization Code* (PSC), that is identical in every cell. This specific codeword is created from a set of 16-bit chip sequences as follows:

  Let $a = (1, 1, 1, 1, 1, 1, -1, -1, 1, -1, 1, -1, 1, -1, -1, 1)$. Then the primary SCH contains a sequence of $(1 + j) \times (a, a, a, -a, -a, a, -a, -a, a, a, a, -a, a, -a, a, a)$.

  The secondary SCH is comprised of 16 codewords, each with a length of 256 chips. These 16 codewords are arranged into 64 different sequences of length 15. In other words, a *sequence* is a set of 15 codewords in a particular order, and there are 64 such sequences. The 64 available sequences are mapped to the 64 downlink primary scrambling code groups. Thus, when a terminal receives a particular secondary SCH sequence, it can identify the primary scrambling code group of the cell in question. Since only eight primary scrambling codes are in each primary scrambling code group, the UE then has relatively few primary scrambling codes to check before being able to decode transmissions from the base station. The SCH is transmitted in conjunction with the *Primary Common Control Physical Channel* (Primary CCPCH), described later.

- The *Common Pilot Channel* (CPICH) is a channel always transmitted by the base station and is scrambled with the cell-specific primary scrambling code. It uses a fixed spreading factor of 256, which equates to 30 kbps on the air interface.

  An important function of the CPICH is in measurements by the terminal for hand-over or cell reselection because the measurements made by the terminal are based on reception of the CPICH. Consequently, manipulation of the transmitted power on the CPICH can be used to steer terminals toward a given cell or away from a given cell.

  For example, if the CPICH power transmitted on a given cell is reduced, the effect is to make the CPICH reception from neighboring cells appear stronger, which may trigger a handover to a neighboring cell. This can be useful for load balancing in the RF network. It is possible to have more than one CPICH in a given cell. The primary CPICH is transmitted over the entire cell area. The secondary CPICH can be transmitted over the whole cell area or can be restricted by transmission on narrow-beam antennas to specific areas of the cell, such as areas of high traffic. The channelization code for the Primary CPICH is fixed to $C_{ch,256,0}$. An arbitrary channelization code of $SF = 256$ is used for the Secondary CPICH.

- The *Primary Common Control Physical Channel* (Primary CCPCH) is used on the downlink to carry the BCH transport channel. It operates at a spreading factor of 256, equivalent to 30 kbps on the air interface. In fact, the actual rate is reduced to 27 kbps on the air interface because of the fact that the Primary CCPCH is time-multiplexed with the SCH, as shown in Figure 3.10. For many

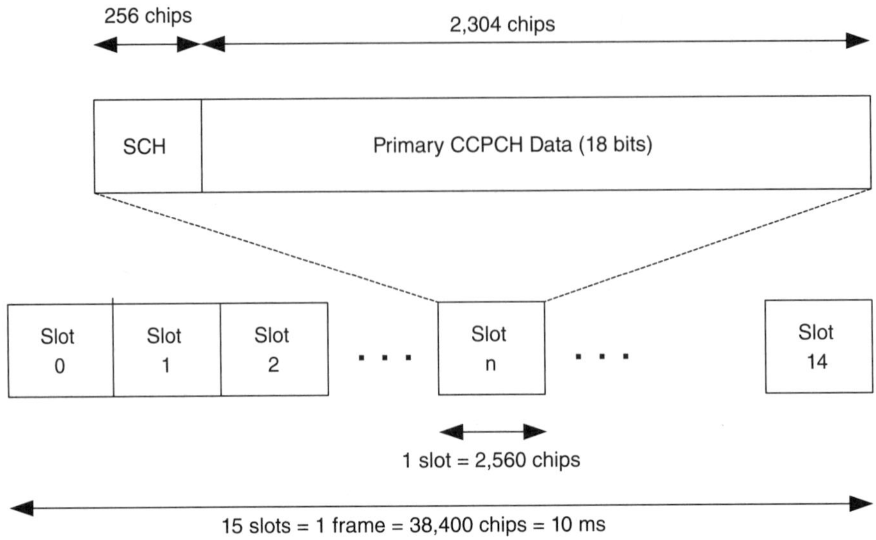

FIGURE 3.10 Multiplexing SCH and CCPCH.

of the channels on the air interface, all the chips in a slot are allocated to a particular physical channel. The Primary CCPCH is an exception in that it shares every slot with the SCH. The first 256 chips of each slot are used by the SCH. The remaining 2304 chips are used by the Primary CCPCH to carry the BCH transport channel. The 2304 chips allocated to the Primary CCPCH correspond to 18 bits of Primary CCPCH data. Moreover, the 18 bits include half-rate convolutional coding (to support forward error correction) so that the actual data rate is approximately 13.5 kbps.

- The *Secondary Common Control Physical Channel* (Secondary CCPCH) is used on the downlink to carry two common transport channels—the FACH and the PCH. The FACH and the PCH can share a single Secondary CCPCH, or each can have a Secondary CCPCH of its own. The Secondary CCPCH carrying the PCH must be transmitted over the whole cell area, which applies regardless of whether the physical channel carries just the PCH or both the PCH and the FACH. If a Secondary CCPCH is used just for the FACH, then it does not necessarily have to reach the whole cell coverage area.
- The *Physical Random Access Channel* (PRACH) is used in the uplink to carry the RACH transport channel. In the uplink, the PRACH has 15 access slots, each with a duration of 5120 chips. These access slots are arranged in different combinations, known as *RACH subchannels*, for which certain scrambling codes and signatures are available. A given UE may be allowed to use one or more RACH subchannels according to the class of the UE. The signatures and scrambling codes available for a particular RACH subchannel are broadcast on the BCH transport channel.

  The process for accessing the uplink begins with the transmission from the terminal of a specific preamble sent on a specific access slot. This preamble is

4096 chips long and consists of 256 repetitions of a 16-chip signature. The preamble is scrambled by one of 8192 long scrambling codes. These 8192 scrambling codes are grouped into 512 groups of 16 codes. A correspondence exists between a specific group of preamble scrambling codes and the primary downlink scrambling code used in the cell.

Once the base station detects the preamble, it uses the *Acquisition Indicator Channel* (AICH) to indicate to the UE that the preamble has been detected and that the UE either is or is not allowed uplink access. The AICH is also structured in slots, each of which is 5120 chips long. Each slot indicates a number of PRACH signatures and an indication for each as to whether the UE is allowed access to the uplink. The UE checks the AICH to see whether it has been granted access (as determined by checking for the signature it has just used). Assuming that the UE has been granted access, it transmits the actual RACH message (10- or 20-ms duration) on subsequent access slots.

- The *Physical Common Packet Channel* (PCPCH) is used in the uplink to carry the uplink CPCH transport channel. Given that the CPCH is somewhat similar to the RACH, the process for using the PCPCH is similar to that for using the PRACH. A preamble is first sent using a specific signature. The terminal then waits for a response from the base station on the *Access Preamble–Acquisition Indicator Channel* (AP-AICH), similar to what is done on the AICH for an access attempt on the PRACH.

  When the response is received on the AP-AICH, however, the terminal does not yet proceed to transmit the desired data. The reason is that the CPCH can support longer durations of data than the RACH. Thus, if there is a collision, a greater amount of data is lost. Therefore, the terminal next sends a specific *Collision-Detection* (CD) signature and waits for this to be echoed back from the base station on the *Collision Detection/Channel Assignment-Indication Channel* (CD/CA-ICH). At this point, the terminal can send the CPCH data on the PCPCH. The duration of the data transfer can last several 10-ms frames. The spreading factor can take any value from 4 to 256.

  As an option, the base station can support the CPCH Status Indication Channel, which is used to indicate the current state of affairs for any CPCH defined in the cell. By monitoring this channel, the terminal can determine, in advance, if resources are available to support the terminal's use of a CPCH. This avoids access attempts from the mobile unit that are doomed to fail.

- The *Physical Downlink Shared Channel* (PDSCH) is used in the downlink to carry the DSCH transport channel. Because the DSCH transport channel can be shared among several users, the PDSCH has a structure that enables it to be shared among users. A PDSCH has a root channelization code, and there may be multiple PDSCHs with channelization codes at or below the root channelization code. These various PDSCHs may be allocated to different UEs on a radio-frame-by-radio-frame basis. Within one radio frame, UTRAN may allocate different PDSCHs under the same PDSCH root channelization code to different UEs. Within the same radio frame, multiple parallel PDSCHs with the same spreading factor may be allocated to a single user. PDSCHs allocated to the same user on different radio frames may have different spreading factors.

- The indicator channels include the AICH, AP-AICH, and CD/CA-ICH, already mentioned. In addition, there is the *Paging Indicator Channel* (PICH). The purpose of the PICH is to let a given terminal know when it might expect a paging message on the PCH (carried on the Secondary PCPCH). When a user device registers with the network, it is assigned to a paging group. These paging groups are indicated through the use of paging indicators carried on the PICH. When a terminal is to be paged on the PCH, a paging indicator corresponding to the paging group in question is carried on the PICH. If a terminal decodes the PICH and finds that its paging group is indicated, then at least one terminal in its paging group is being paged, which means that the terminal must decode the PCH (carried on the Secondary PCPCH) to determine if it is being paged. If a given terminal's paging group is not indicated on the PICH, then the terminal need not decode the PCH.
- The DCH transport channel is mapped to the two physical channels—DPDCH and DPCCH, as mentioned previously. The DPDCH carries the actual user data and can have a variable spreading factor, whereas the DPCCH carries control information.

The mapping between the transport channels and the physical channel is shown in Figure 3.11.

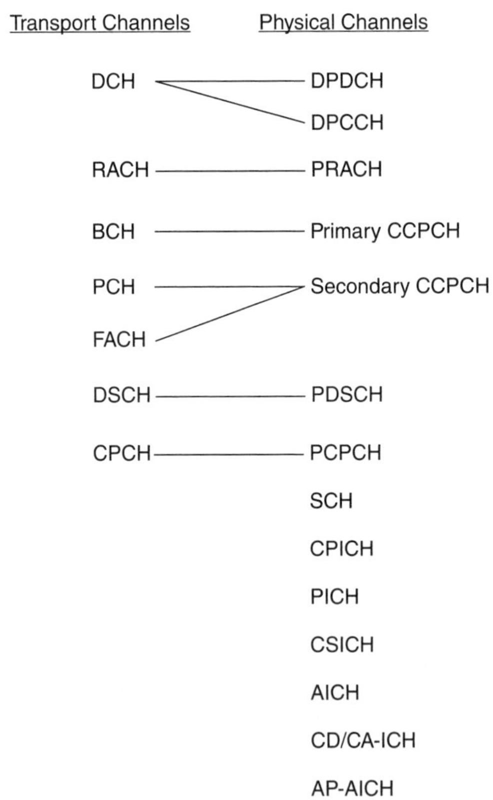

Figure 3.11 Mapping between transport channels and physical channels.

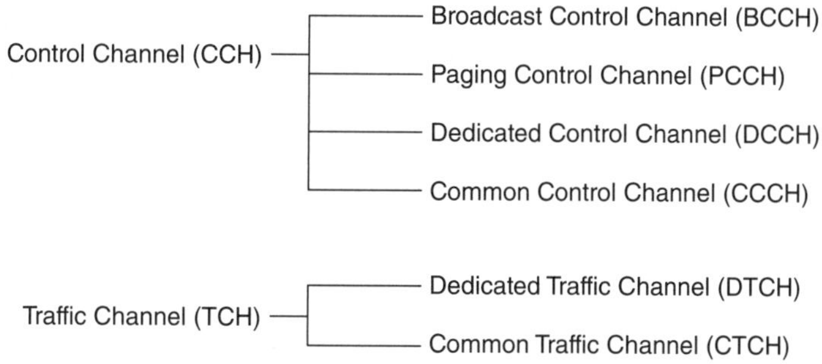

Figure 3.12 Types of logical channels.

3.3.4.3 Logical Channels

As shown in Figure 3.8, information is passed from the MAC layer to the physical layer in the form of transport channels. This information, however, can begin higher in the protocol stack, in which case it is passed from the RLC layer to the MAC layer in the form of logical channels. The logical channels are mapped to transport channels, which in turn, are mapped to physical channels.

As mentioned earlier, RLC interfaces with MAC through a number of logical channels. MAC maps these logical channels to the transport channels described previously. Logical channels relate to the information being transmitted, whereas transport channels relate largely to the manner in which the information is transmitted. Basically, two groups of logical channels exist—control channels and traffic channels. These are shown in Figure 3.12.

The *Broadcast Control Channel* (BCCH) is used for the downlink transmission of system information. The *Paging Control Channel* (PCCH) is used for the paging of an MS across one or more cells. The *Common Control Channel* (CCCH) is used in the uplink by terminals that want to access the network but do not already have any connection with the network. The CCCH can be used in the downlink to respond to such access attempts. The *Dedicated Control Channel* (DCCH) is a bidirectional point-to-point control channel between the MS and the network for sending control information. WCDMA also defines the *Shared Channel Control Channel* (SCCH), but this channel is used only in TDD mode.

Two types of logical traffic channels are available. The *Dedicated Traffic Channel* (DTCH) is a point-to-point channel dedicated to one UE for the transfer of user data. DTCHs apply to the uplink and to the downlink. The *Common Traffic Channel* (CTCH) is a point-to-multipoint unidirectional channel for the transfer of user data to all UEs or just to a single UE. The CTCH exists in the downlink only.

Numerous options are available for mapping between logical channels and transport channels. This mapping depends on a range of criteria, such as the type of information to be sent, whether it is to be sent to multiple UEs (in the downlink), and whether the UE already has an established connection with the network. The possible mapping between logical channels and transport channels for the FDD mode of operation is shown in Figure 3.13.

3.3.5 Power Control in WCDMA

In any CDMA system, power control is of critical importance. Because all users share the same frequency at the same time, it is important that no user transmit at such a high

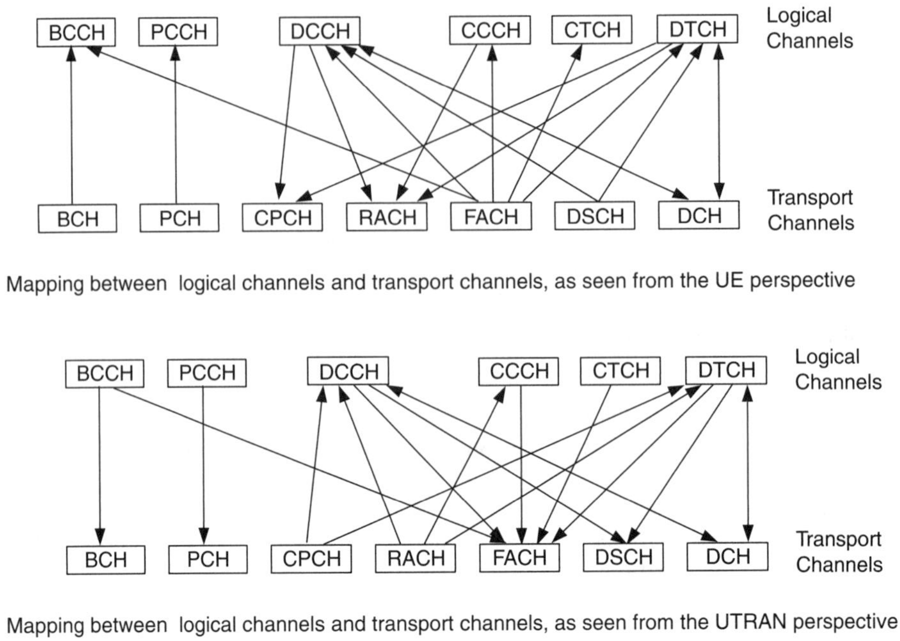

FIGURE 3.13 Mapping between logical channels and transport channels.

power that other users are drowned out. If, for example, a user near the base station were to transmit at the same power level as a user at the cell edge, then at the base station the signal from the nearby user would be so great that it would completely overpower the signal from the faraway user. The result is that the signal from the faraway user would be impossible to recover. This is known as the *near-far problem*.

To avoid this problem, mechanisms are required whereby the UE can be instructed to adjust its transmit power up or down so that all transmissions from all users in the cell arrive at the base station with the same power level. Not only is power control required to combat the near-far problem, but it is also required to combat the effects of Rayleigh fading, where the received signal can drop suddenly by many decibels as a result of multipath propagation, which results in multiple copies of a signal arriving at the receiver out of phase. Thus, power control is deployed both in the uplink and the downlink.

In general, power control in WCDMA uses two main techniques—open-loop power control and closed-loop power control. With *open-loop power control,* the terminal estimates the required transmission power based on the signal power received from the base station and information broadcast from the base station regarding the transmit power from the base station. Specifically, the base station broadcasts the transmit power used on the CPICH, and the terminal uses this information in conjunction with the received power level to determine the power that should be used on the uplink. In general, however, fading in the uplink and fading in the downlink are unrelated. Consequently, open-loop power control provides only a very rough estimate of the ideal power that the terminal should use. For this reason, open-loop power control is used only when the UE is making initial access on the PRACH or PCPCH. In all other situations, closed-loop power control is used.

Closed-loop power control means that the receiving entity (the base station or UE) measures the received *Signal-to-Interference Ratio* (SIR) and compares it with a target SIR value. The base station or UE then instructs the far end to increase the transmitted power if the SIR is too low or decrease the power if the SIR is too high. Closed-loop power control is also known as *fast power control* because it operates at a rate of 1500 Hz. In other words, power-control commands and changes happen at a rate of 1500 times per second. This rate is sufficiently fast to overcome path-loss changes and Rayleigh-fading effects for all situations except where the UE is traveling at high speed.

Closed-loop power-control commands are sent on physical control channels that are associated with physical data channels. Recall, for example, that in the uplink, the DPDCH has an associated DPCCH. Among other pieces of information, the DPCCH carries transmit power-control commands back to the base station. A power-control command is sent in every slot. Because 15 slots are available for each 10 ms, we have a rate of 1500 power-control commands per second. Each power-control command can instruct the sender to leave the transmitted power unchanged or to increase or decrease the transmitted power in steps of 1, 2, or 3 dB. Similarly, for the downlink DPDCH, an associated DPCCH sends power-control instructions to the UE, along with other functions.

There is also another form of power control known as *outer-loop power control,* with the primary objective being maintaining the service quality at an optimal level. In general, the objective of power control is to maintain the SIR at the receiver at the optimal level—not too high and not too low. The target SIR value, however, is a function of the required quality for the service to be supported. If we measure service quality in terms of *Frame Error Rate* (FER) on the air interface (as determined by a *Cyclic Redundancy Check* [CRC]), then the SIR can be considered a function of FER.

The acceptable FER can vary from service to service. Speech service using the *Adaptive Multirate* (AMR) coder at 12.2 kbps, for example, could support an FER of 1 percent without noticeable service degradation. A non-real-time data service could support much higher FER rates before retransmission, allowing retransmission to correct errors. The impact on such a service is greater delay and a lower overall throughput, but such an impact can be perfectly acceptable for a non-real-time service.

A real-time data service, however, may have a far more stringent FER requirement, perhaps 1×10^{23} or better. Consequently, depending on the service requirements, the FER may need to vary, which means that the required SIR may need to vary. This variation in the required SIR is known as *outer-loop power control.* It uses closed-loop power control to instruct the sender to vary the transmit power. With outer-loop power control, however, the reason for the change is the new SIR requirement.

3.3.6 User Data Transfer

WCDMA is designed to offer great flexibility in the transmission of user data across the air interface. For example, data rates can change on a frame-by-frame basis (every 10 ms). Moreover, it is possible to mix and match different types of service. For example, a subscriber may be sending and receiving packet data while also involved in a voice call. When sending information over the air interface, physical control channels are used in combination with physical data channels. Although the physical data channels carry the user information, the physical control channels carry information to support the correct

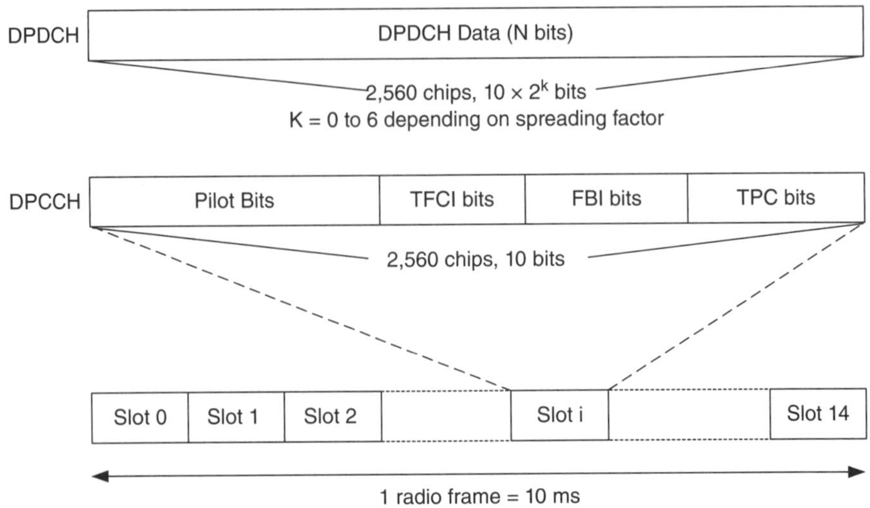

FIGURE 3.14 Uplink DPDCH and DPCCH frame and slot structure.

interpretation of the data carried on the corresponding DPDCH frame plus power-control commands and feedback indicators.

3.3.6.1 Uplink DPDCH and DPCCH

Figure 3.14 shows the structure of the uplink DPCCH as used with the uplink DPDCH. The DPCCH is transmitted in parallel with the DPDCH, and the information in a given DPCCH frame relates to the corresponding DPDCH frame.

The DPCCH always uses a spreading factor of 256. Thus each slot (2560 chips) corresponds to 10 bits of DPCCH information. These 10 bits are divided into pilot bits, *Transport Format Combination Indicator* (TFCI) bits, *Feedback Indicator* (FBI) bits, and *Transmit Power Control* (TPC) bits.

The pilot information bits are used for channel estimation purposes and include specific bit patterns for frame synchronization. The TFCI bits indicate the bit rate and channel coding for the DPDCH. A single DPDCH can carry multiple DCH transport channels.

If, for example, a user were invoking multiple simultaneous services, the associated DCH transport channels could be multiplexed together on a single DPDCH. In this case, the DPDCH is said to carry a *Coded Composite Transport Channel* (CCTrCH). The TFCI is used to indicate the format of each of the transport channels within the CCTrCH. The FBI bits are used in conjunction with transmit diversity at the base station. WCDMA supports downlink transmit diversity, whereby two antennas can be used for downlink transmission. When transmit diversity is used, it is possible for the power and/or phase on one transmit antenna to differ from that of the other. The FBI bits are used in the uplink to instruct the base station to change the power or phase difference associated with transmit diversity. Finally, the TPC bits are used to command the base station to change the transmit power when necessary.

The number of bits in each of the uplink DPCCH fields depends on the slot format for the DPCCH. A number of slot formats are possible, as shown in Table 3.3.

Slot Format	Transmitted Channel Bit Rate (kbps)	Channel Symbol Rate (ksps)	SF	Bits/ Frame	Bits/ Slot	Pilot Bits	TPC Bits	TFCI Bits	FBI Bits	Slots per Radio Frame
0	15	15	256	150	10	6	2	2	0	15
0A	15	15	256	150	10	5	2	3	0	10–14
0B	15	15	256	150	10	4	2	4	0	8–9
1	15	15	256	150	10	8	2	0	0	8–15
2	15	15	256	150	10	5	2	2	1	15
2A	15	15	256	150	10	4	2	3	1	10–14
2B	15	15	256	150	10	3	2	4	1	8–9
3	15	15	256	150	10	7	2	0	1	8–15
4	15	15	256	150	10	6	2	0	2	8–15
5	15	15	256	150	10	5	1	2	2	15
5A	15	15	256	150	10	4	1	3	2	10–14
5B	15	15	256	150	10	3	1	4	2	8–9

Table 3.3 Downlink DPCCH Slot Formats

As can be seen from Table 3.3, in some slot formats, the full 15 slots are not used in every radio frame. The reason for less than 15 slots per frame is due to the use of compressed mode. In compressed mode, gaps exist in both the uplink and downlink transmissions. These gaps are included to enable the UE to take measurements on other frequencies. By taking measurements on other frequencies and reporting those measurements, the UE enables the network to enable an interfrequency handover to another UMTS frequency or perhaps an intersystem handover to a GSM system.

Also, a number of different slot formats exist for the DPDCH, but these simply reflect the different spreading factors that can be applied to the DPDCH data. For example, a Spreading Factor (SF) of 256 for the uplink DPDCH means 10 bits per slot, whereas a spreading factor of 4 means 640 bits per slot.

3.3.6.2 Downlink DPDCH and DPCCH

Figure 3.15 shows the structure of the downlink DPDCH and DPCCH. The most notable characteristic is that the DPCCH is time- multiplexed with the DPDCH rather than being transmitted separately. In each slot on the downlink, two fields contain DPDCH user data, whereas three other fields maintain information on the pilot bits, the TFCI, and the TPC. As is the case for the uplink, a number of slot formats can be applied to the downlink DPDCH/DPCCH. Table 3.4 shows the various combinations.

As can be seen from Table 3.4, the actual DPDCH user throughput depends on the slot format used. Moreover, one can clearly see the effect of compressed mode, where less than 15 slots are used in the downlink.

3.4 The UTRAN Architecture

In most mobile communications networks, the network architecture can be split into two main parts—the access network and the core network. The *access network* is specific to the access technology being used, whereas the *core network* is shielded from

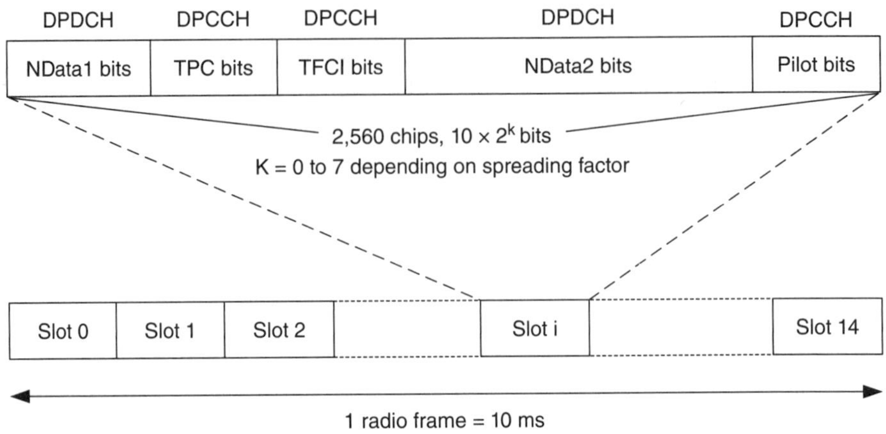

FIGURE 3.15 Downlink DPDCH and DPCCH frame and slot structure.

the vagaries of the access technology and ideally should be able to handle multiple access networks. This split applies quite well to UMTS, where the access network is known as the *UMTS Terrestrial Radio Access Network* (UTRAN). It is supported by a core network that is based on the core network used for GSM. In fact, the GSM core network can be upgraded to support both UTRAN and a GSM radio access network simultaneously.

The UTRAN architecture is shown in Figure 3.16 as it applies to the first release of the UMTS specification—3GPP Release 1999. The UTRAN consists of two types of nodes—the *Radio Network Controller* (RNC) and *Node B*, which is the base station. The RNC is analogous to the GSM *Base Station Controller* (BSC). The RNC is responsible for control of the radio resources within the network. It interfaces with one or more base stations, known as *Node Bs*. The interface between the RNC and the Node B is the *Iub interface*. Unlike the equivalent Abis interface in GSM, the Iub interface is open, which means that a network operator could acquire Node Bs from one vendor and RNCs from another vendor. Together, an RNC and the set of Node Bs that it supports are known as a *Radio Network Subsystem* (RNS).

Unlike in GSM, where BSCs are not connected to each other, UTRAN contains an interface between RNCs. This is known as the *Iur interface*. The primary purpose of the Iur interface is to support inter-RNC mobility and a soft handover between Node Bs connected to different RNCs.

The user device is the UE. It consists of the *Mobile Equipment* (ME) and the *UMTS Subscriber Identity Module* (USIM). UTRAN communicates with the UE over the *Uu interface*. The Uu interface is none other than the WCDMA air interface that we have already described in this chapter.

UTRAN communicates with the core network over the *Iu interface*. The Iu interface has two components—the *Iu-CS interface*, which supports circuit-switched services, and the *Iu-PS interface*, which supports packet-switched services. The Iu-CS interface connects the RNC to an MSC and is similar to the GSM A-interface. The Iu-PS interface connects the RNC to an SGSN and is analogous to the GPRS Gb interface.

Slot Format	Channel Bit Rate (kbps)	Channel Symbol Rate (ksps)	SF	Bits/ Slot	DPDCH Bits/Slot		DPCCH Bits/Slot			Transmitted Slots per Radio Frame
					NDATA1	NDATA2	TPC	TFCI	Pilot	
0	15	7.5	512	10	0	4	2	0	4	15
0A	15	7.5	512	10	0	4	2	0	4	8–14
0B	30	15	256	20	0	8	4	0	8	8–14
1	15	7.5	512	10	0	2	2	2	4	15
1B	30	15	256	20	0	4	4	4	8	8–14
2	30	15	256	20	2	14	2	0	2	15
2A	30	15	256	20	2	14	2	0	2	8–14
2B	60	30	128	40	4	28	4	0	4	8–14
3	30	15	256	20	2	12	2	2	2	15
3A	30	15	256	20	2	10	2	4	2	8–14
3B	60	30	128	40	4	24	4	4	4	8–14
4	30	15	256	20	2	12	2	0	4	15
4A	30	15	256	20	2	12	2	0	4	8–14
4B	60	30	128	40	4	24	4	0	8	8–14
5	30	15	256	20	2	10	2	2	4	15
5A	30	15	256	20	2	8	2	4	4	8–14
5B	60	30	128	40	4	20	4	4	8	8–14
6	30	15	256	20	2	8	2	0	8	15
6A	30	15	256	20	2	8	2	0	8	8–14
6B	60	30	128	40	4	16	4	0	16	8–14
7	30	15	256	20	2	6	2	2	8	15
7A	30	15	256	20	2	4	2	4	8	8–14
7B	60	30	128	40	4	12	4	4	16	8–14
8	60	30	128	40	6	28	2	0	4	15
8A	60	30	128	40	6	28	2	0	4	8–14
8B	120	60	64	80	12	56	4	0	8	8–14

Table 3.4 Downlink DPDCH/DPCCH Slot Formats

Slot Format	Channel Bit Rate (kbps)	Channel Symbol Rate (ksps)	SF	Bits/ Slot	DPDCH Bits/Slot		DPCCH Bits/Slot			Transmitted Slots per Radio Frame
					NDATA1	NDATA2	TPC	TFCI	Pilot	
9	60	30	128	40	6	26	2	2	4	15
9A	60	30	128	40	6	24	2	4	4	8–14
9B	120	60	64	80	12	52	4	4	8	8–14
10	60	30	128	40	6	24	2	0	8	15
10A	60	30	128	40	6	24	2	0	8	8–14
10B	120	60	64	80	12	48	4	0	16	8–14
11	60	30	128	40	6	22	2	2	8	15
11A	60	30	128	40	6	20	2	4	8	8–14
11B	120	60	64	80	12	44	4	4	16	8–14
12	120	60	64	80	12	48	4	8*	8	15
12A	120	60	64	80	12	40	4	16*	8	8–14
12B	240	120	32	160	24	96	8	16*	16	8–14
13	240	120	32	160	28	112	4	8*	8	15
13A	240	120	32	160	28	104	4	16*	8	8–14
13B	480	240	16	320	56	224	8	16*	16	8–14
14	480	240	16	320	56	232	8	8*	16	15
14A	480	240	16	320	56	224	8	16*	16	8–14
14B	960	480	8	640	112	464	16	16*	32	8–14
15	960	480	8	640	120	488	8	8*	16	15
15A	960	480	8	640	120	480	8	16*	16	8–14
15B	1920	960	4	1280	240	976	16	16*	32	8–14
16	1920	960	4	1280	248	1000	8	8*	16	15
16A	1920	960	4	1280	248	992	8	16*	16	8–14

*TFCI on the downlink is optional. If TFCI bits are not used, then discontinuous transmission is used in the TFCI field.

TABLE 3.4 Downlink DPDCH/DPCCH Slot Formats (*Continued*)

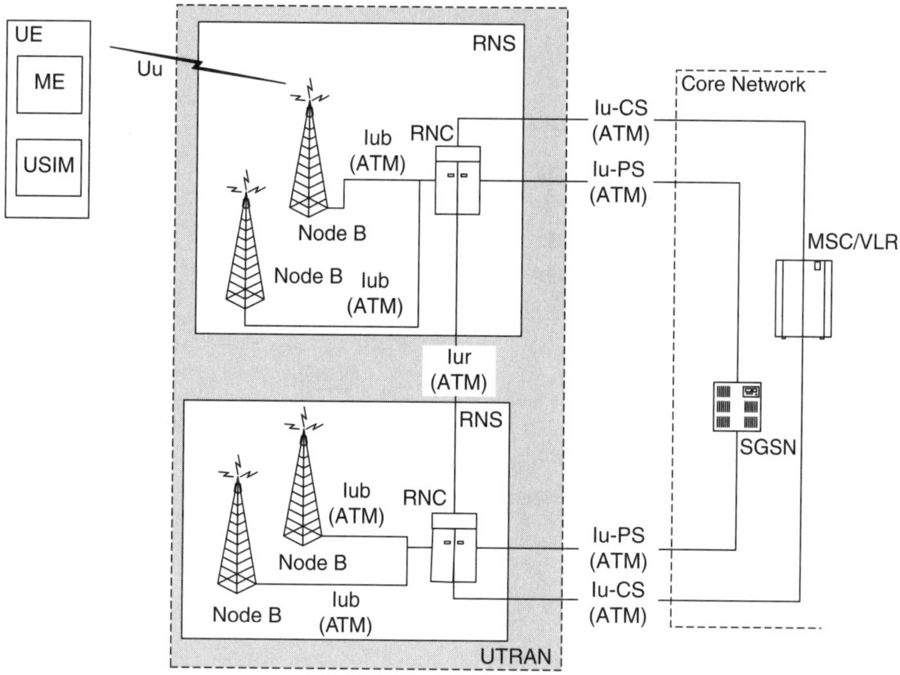

Figure 3.16 UTRAN architecture.

In 3GPP Release 1999, all the interfaces within UTRAN, as well as the interfaces between UTRAN and the core network, use *Asynchronous Transfer Mode* (ATM) as the transport mechanism.

3.4.1 Functional Roles of the RNC

The RNC that controls a given Node B is known as the *Controlling RNC* (CRNC). The CRNC is responsible for the management of radio resources available at a Node B that it supports.

For a given connection between the UE and the core network, one RNC is in control. This is called the *Serving RNC* (SRNC). For the user in question, the SRNC controls the radio resources that the UE is using. In addition, the SRNC terminates the Iu interface to or from the core network for the services being used by the UE. In many cases, though not all, the SRNC is also the CRNC for a Node B that is serving the user.

As depicted in Figure 3.17, UTRAN supports soft handovers, which may occur between Node Bs controlled by different RNCs. During and after a soft handover between RNCs, one may find a situation where a UE is communicating with a Node B that is controlled by an RNC that is not the SRNC. Such an RNC is termed a *Drift RNC* (DRNC). The DRNC does not perform any processing of user data (beyond what is required for correct operation of the physical layer). Rather, data to or from the UE are controlled by the SRNC and are passed transparently through the DRNC.

As a UE moves farther and farther away from any Node B controlled by the SRNC, it will become clear that it is no longer appropriate for the same RNC to continue to act as the SRNC. In this case, UTRAN may make the decision to hand control of the

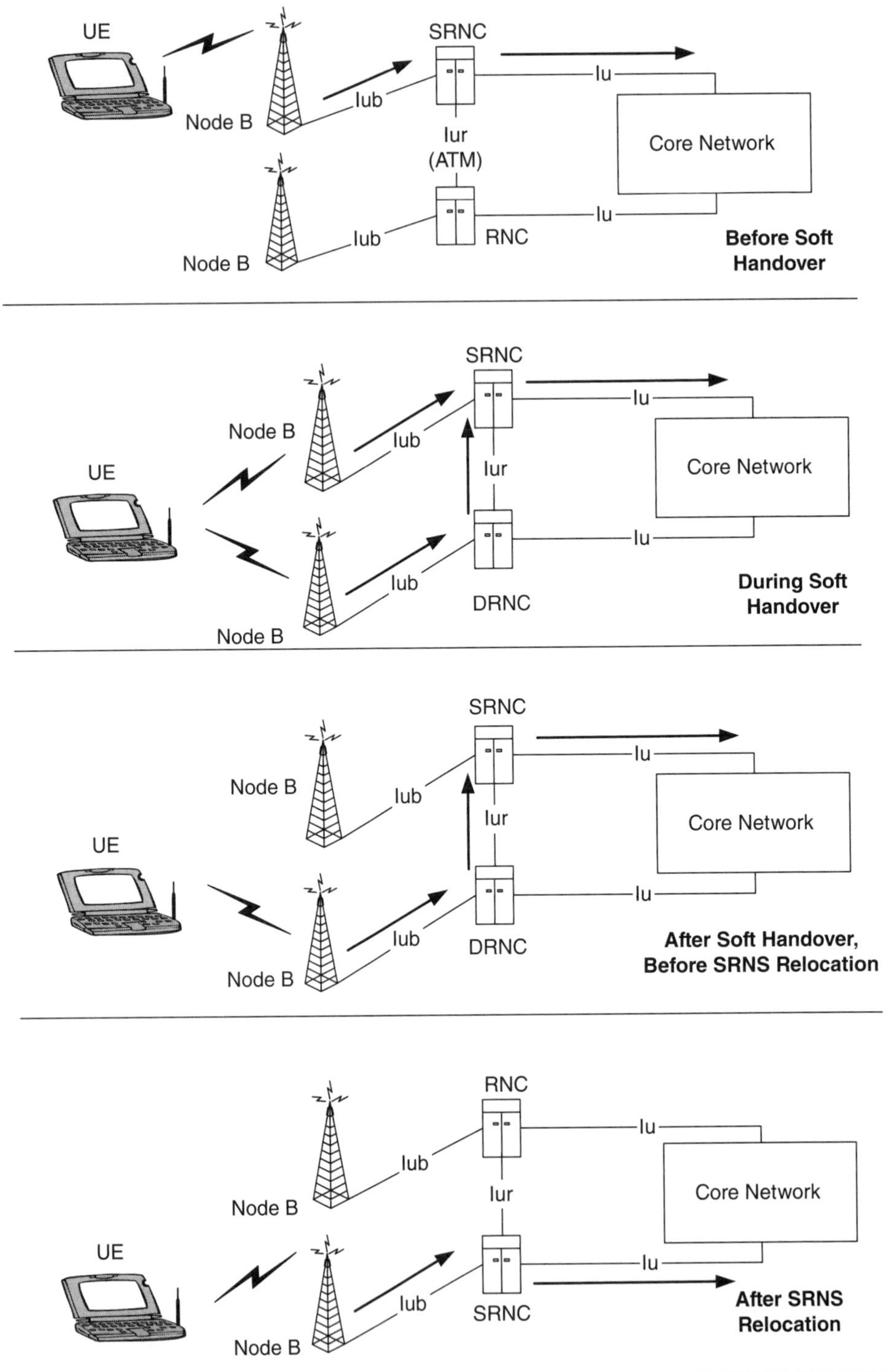

Figure 3.17 Soft handover and SRNS relocation.

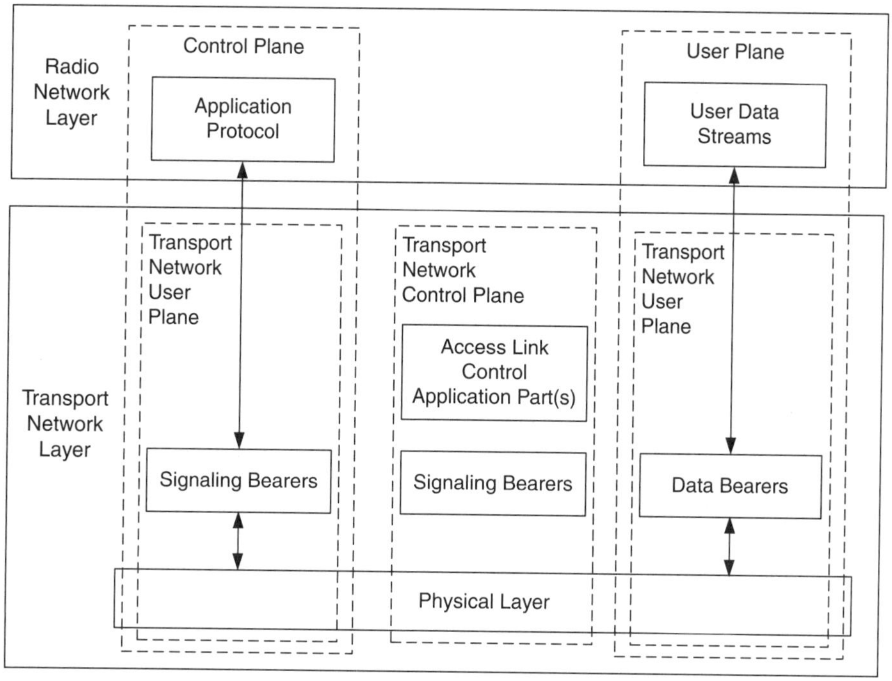

FIGURE 3.18 Generic model for UTRAN terrestrial interfaces.

connection over to another RNC. This is known as *Serving RNS* (SRNS) relocation. This action is invoked under the control of algorithms within the SRNC.

3.4.2 UTRAN Interfaces and Protocols

Figure 3.18 provides a generic model for the terrestrial interfaces used in UTRAN—the Iu-CS, Iu-PS, Iur, and Iub interfaces. Each interface has two main components—the radio network layer and the transport network layer. The radio network layer represents the application information to be carried—either user data or control information. This is the information that UTRAN actually cares about. The transport network layer represents the transport technology that the various interfaces use. In the case of 3GPP Release 1999, ATM transport is used, so the transport network layer represents an ATM-based transport. Another transport layer could be used instead. In such a case, the transport network layer would be different, but the radio network layer should not be.

Looking at Figure 3.18 in the vertical direction, we see three planes: the control plane, the user plane, and the transport network user plane. The control plane is used by UMTS-related control signaling. It includes the application protocol used on the interface in question. The control plane is responsible for establishment of the bearers that transport user data, but the user data are not carried on the control plane. As seen from the control plane, the user bearers established by the application protocol are generic bearers and are independent of the transport technology being used. If the application protocol were to view the bearers in terms of a specific transport technology, then it would not be possible to separate the radio network layer cleanly from the transport network layer. In other words, the application protocol would have to be designed to suit a particular transport technology. The signaling bearers that carry the application signaling are established by *Operations and Maintenance* (O&M) actions. These signaling bearers are analogous, for example, to the *Signaling System 7* (SS7) signaling links that are used between a BSC and an MSC in GSM.

The user plane is what carries the actual user data. These data could, for example, be data packets being sent or received by the UE as part of a data session. Each data stream carried in the user plane will have its own framing structure.

The transport network control plane contains functionality that is specific to the transport technology being used and is not visible to the radio network layer. If standard preconfigured bearers are to be used by the user plane and these are known to the control plane, then the transport network control plane is not needed. Otherwise, the transport network control plane is used. It involves the use of an *Access Link Control Application Part* (ALCAP). This is a generic term that describes a protocol or set of protocols used to set up a transport bearer. The ALCAP to be used depends on the user plane transport technology.

3.4.2.1 Iu-CS Interface

If we apply this generic structure to the Iu-CS interface (RNC to MSC), then it appears as shown in Figure 3.19. The application protocol in the control plane is the *Radio Access*

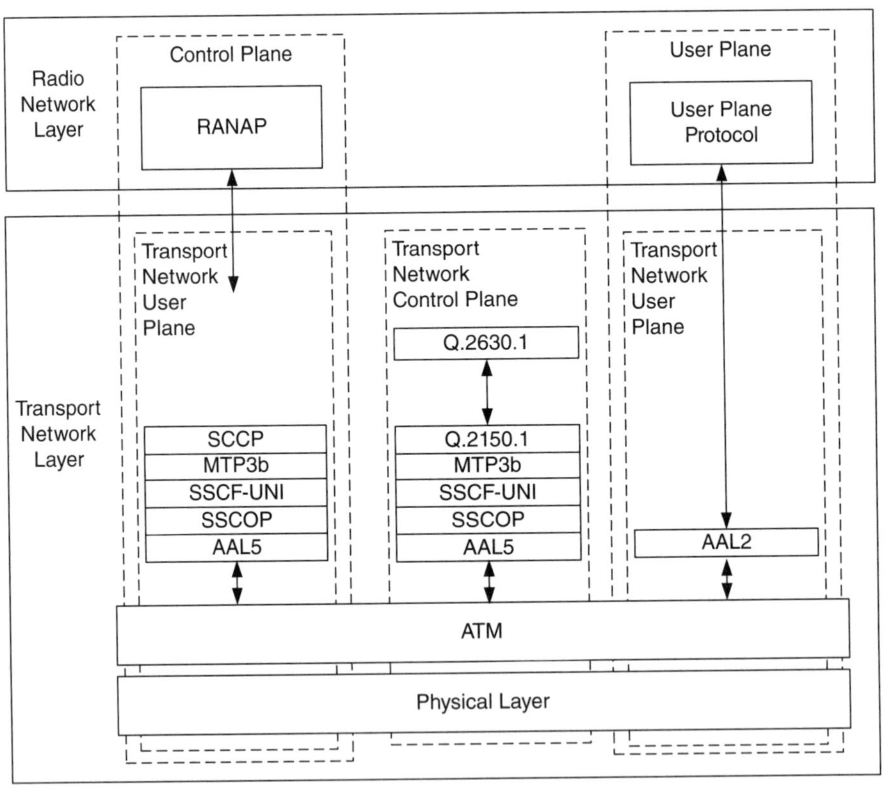

Figure 3.19 Iu-CS protocol structure.

Network Application Part (RANAP). This provides functionality similar to that provided by the BSSAP in GSM. Among the many functions supported by RANAP are the establishment of *Radio Access Bearers* (RABs), paging, the direct transfer of signaling messages between the UE and the core network, and SRNS relocation.

RANAP is carried over an ATM-based SS7 signaling bearer. This signaling bearer for the control plane consists of the *ATM Adaptation Layer 5* (AAL5), the *Service-Specific Connection-Oriented Protocol* (SSCOP), the *Service-Specific Coordination Function at the Network Node Interface* (SSCF-NNI), the *Layer 3 Broadband Message Transfer Part* (MTP3b), and the *Signaling Connection Control Part* (SCCP).

Different AAL layers may reside above the ATM layer depending on the type of service that needs to use ATM. In the case of signaling, it is normal to use AAL5, which supports variable-bit-rate services.

SSCOP provides mechanisms for the establishment and release of signaling connections. It also offers the reliable exchange of signaling information, including functions such as sequence integrity, error detection, message retransmission, and flow control. SSCF-NNI maps the requirements of the upper layer to the layer below. Together SSCOP and SSCF-NNI are known as the *Signaling ATM Adaptation Layer* (S-AAL).

MTP3b is similar to standard MTP3 as used in standard SS7 networks with some modifications to enable it to take advantage of the broadband transport that ATM can use. SCCP is the same SCCP as used in standard SS7 networks.

The same signaling stack is used for the transport network control plane—broadband SS7. Instead of SCCP, however, we find the Broadband ISDN ATM Adaptation Layer *Signaling Transport Converter* (STC) for the MTP3b (Q.2150.1). Above Q.2150.1, we have the ALCAP, which is AAL2 Signaling Protocol Capability Set 1 (Q.2630.1).

On the user side, things are much less complicated. We simply have the *ATM Adaptation Layer 2* (AAL2) as the user data bearer. This is an AAL specifically designed for the transport of short-length packets such as those we find with packetized voice. One advantage of AAL2 is that it enables multiple user packets to be multiplexed within one cell to minimize ATM overhead. At the radio network layer, we have the *User Plane Protocol* (UPP). This is a simple protocol that provides either transparent or supported service. In transparent mode, data simply are passed onward. In supported mode, the protocol takes care of such functions as data framing, time alignment, and rate control. Speech is an example of a service that would use supported mode.

3.4.2.2 Iu-PS Interface

The protocol architecture for the Iu-PS interface is shown in Figure 3.20. We first notice that no transport network control protocol is involved. It is not needed because of the protocol that is used in the user plane. Specifically, in the user plane, we find that the *GPRS Tunneling Protocol* (GTP) tunnel extends to the RNC. This is different from standard GPRS, where the tunnel ends at the SGSN, and a special Gb interface is used from SGSN to BSC. The fact that the tunnel extends to the RNC means that only a tunnel identifier and *Internet Protocol* (IP) addresses for each end are required for establishment of the bearer. These are included in the application messages used for establishment of the bearer, which means that no intermediate ALCAP is needed.

As mentioned earlier, the user plane employs the GTP (*GTP-U* indicates a GTP user plane). This protocol uses the *User Datagram Protocol* (UDP) over IP. AAL5 over ATM is used as the transport. For a packet-data transfer, the identification of individual user packets is supported within the GTP-U protocol. Consequently, it is not necessary to structure these user packets according to ATM cell boundaries. This means that

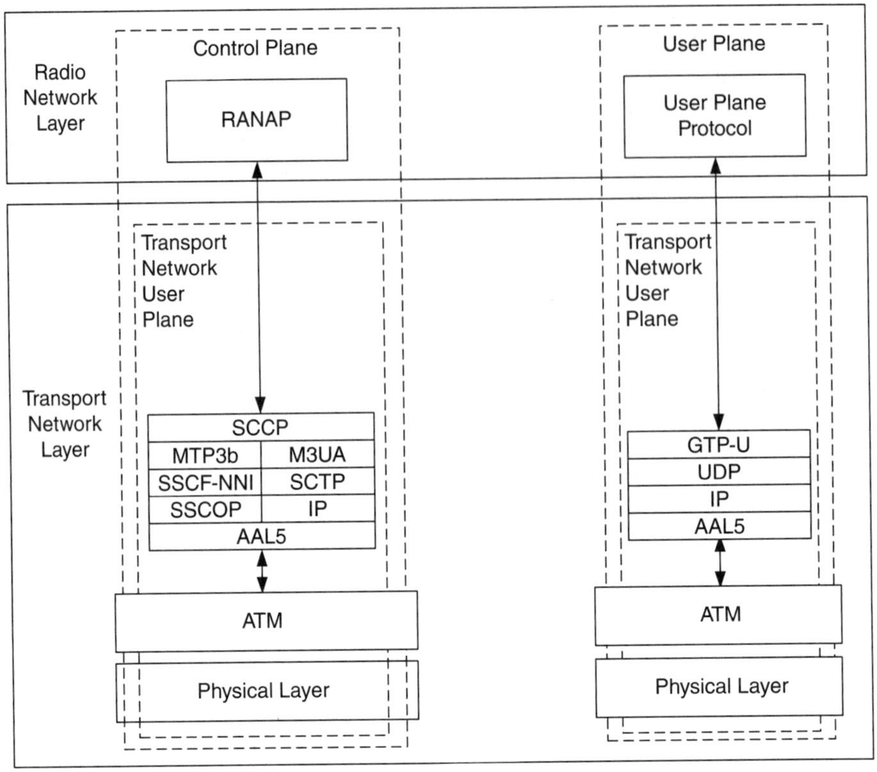

FIGURE 3.20 Iu-PS protocol structure.

multiple user packets can be multiplexed on a given ATM cell, thereby reducing ATM overhead.

In the control plane, we again find RANAP at the application layer. We have a choice of signaling bearer, however. One option is to use the standard ATM SS7 stack, as described previously, for the Iu-CS interface. Another option is to use SCCP over IP-based SS7 transport over ATM. For IP-based SS7 transport, we use the *MTP3 User Adaptation* (M3UA) protocol over the *Stream Control Transmission Protocol* (SCTP). Both these protocols are described in Chapter 9.

3.4.2.3 Iub Interface

The protocol architecture for the *Iub interface* is shown in Figure 3.21. This is the interface between an RNC and the Node B that it controls. In the protocol architecture, we again find the transport network control plane as was seen for the Iu-CS interface. In the control plane, we find the *Node B Application Part* (NBAP) as the application protocol. In the user plane, we find a number of frame protocols related to various types of transport channels described previously in this chapter. Basically, a specific framing protocol is applicable to each of the transport channels. Note that Figure 3.21 indicates the *Uplink Shared Channel* (USCH). This is a transport channel defined for TDD mode only.

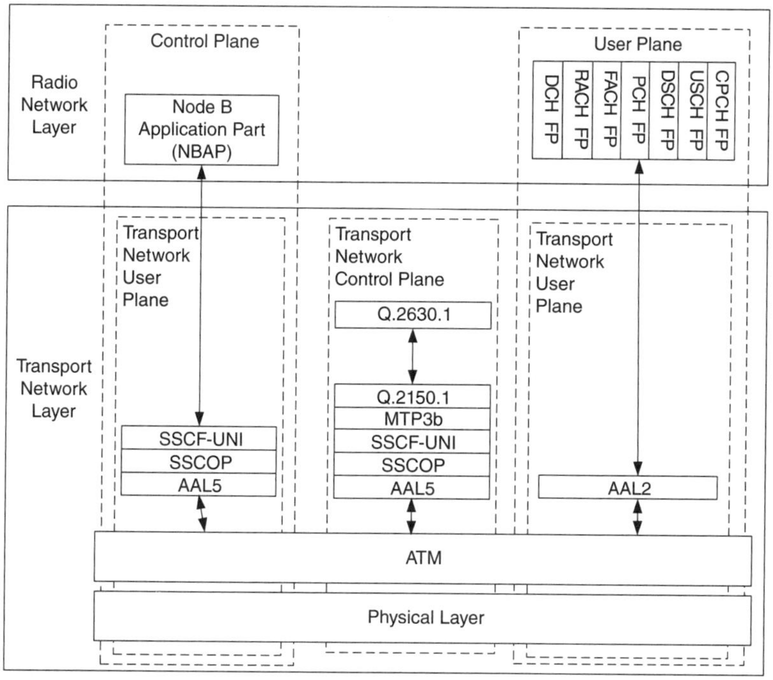

FIGURE 3.21 Iub interface protocol structure.

3.4.2.4 Iur Interface

The interface between RNCs is the *Iur interface.* The primary purpose of this interface is to support inter-RNC mobility (SRNS relocation) and a soft handover between Node Bs connected to different RNCs. The protocol architecture for the Iur interface is shown in Figure 3.22. The controlling application protocol is known as the *Radio Network System Application Part* (RNSAP). Signaling between RNCs is SS7-based, whereby RNSAP uses the services of SCCP. As is the case for the Iu-PS interface, the signaling can be transported on a standard ATM SS7 transport or can use an IP-based transport over ATM. The same applies for the transport network control plane.

The user plane contains two *Frame Protocols* (FPs), one related to dedicated transport channels, the *DCH FP,* and one related to common transport channels, the *CCH FP.* These user protocols carry the actual user data and signaling between the SRNC and the DRNC.

3.5 Establishment of a UMTS Speech Call

The procedure for the establishment of a basic speech call in UMTS is shown in Figure 3.23 (NBAP messaging has been omitted). The process begins with an access request from the UE. This access request is sent either on the RACH transport channel or the CPCH transport channel. The message sent is a request to establish an RRC connection, which must be done before signaling transactions or bearer establishment can take place. The RRC Connection Request includes an indication of the reason for the connection request.

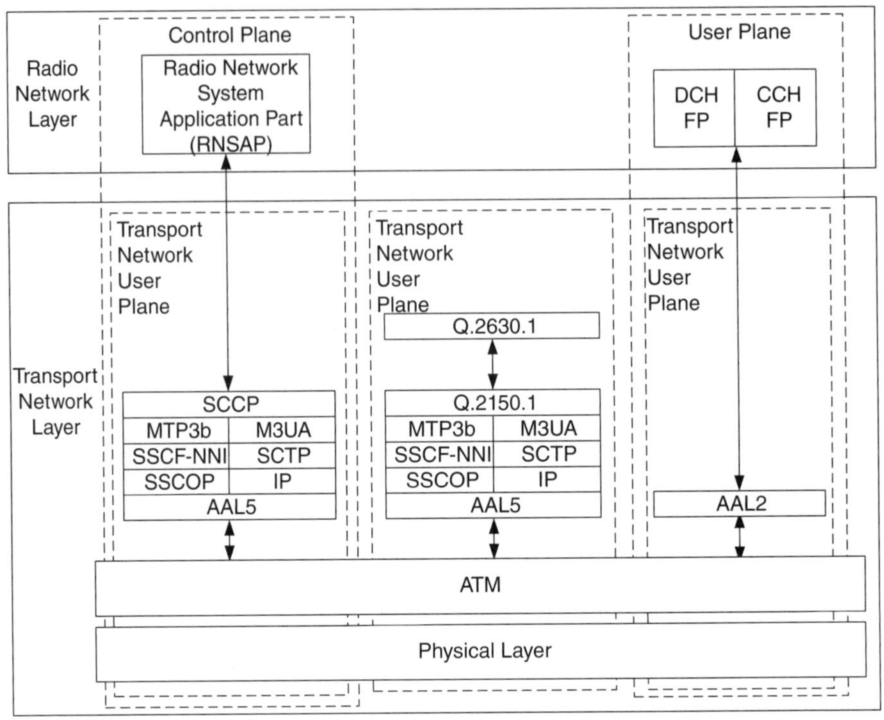

Figure 3.22 Iur interface protocol structure.

The RNC responds with an RRC Connection Setup message. This message will be sent on the CCCH logical channel (typically mapped to the FACH transport channel). At the discretion of the RNC, the RRC Connection Setup message may or may not allocate a DCH transport channel to the UE. If a DCH transport channel is allocated, then the RRC Connection Setup message indicates the scrambling code to be used by the UE in the uplink. The channelization code is determined by the UE and is indicated on the uplink itself. Recall, for example, that a DPCCH is associated with a DPDCH. The DPDCH contains the TFCI that contains spreading-factor information and enables the UTRAN to determine the channelization code for the DPDCH. If the RNC does not allocate a DCH, then further signaling is carried out on the FACH in the downlink and on the RACH or CPCH in the uplink.

The UE responds to the RNC with the message RRC Connection Setup Complete. This message is carried on the uplink DCCH logical channel, which is mapped to the RACH, CPCH, or DCH transport channel. Next, the UE issues a message destined for the core network. This is sent in an RRC Initial Direct Transfer message. The payload of a direct transfer message is passed directly between the UE and the core network. In the case that a signaling relationship has not been established between the UE and core network, then the RRC message Initial Direct Transfer is used. This indicates to the RNC, and subsequently to the core network, that a new signaling relationship needs to be established between the UE and the core.

The RNC maps the Initial Direct Transfer message to the RANAP Initial UE message and sends the message to the core network. In this case, the message is passed to

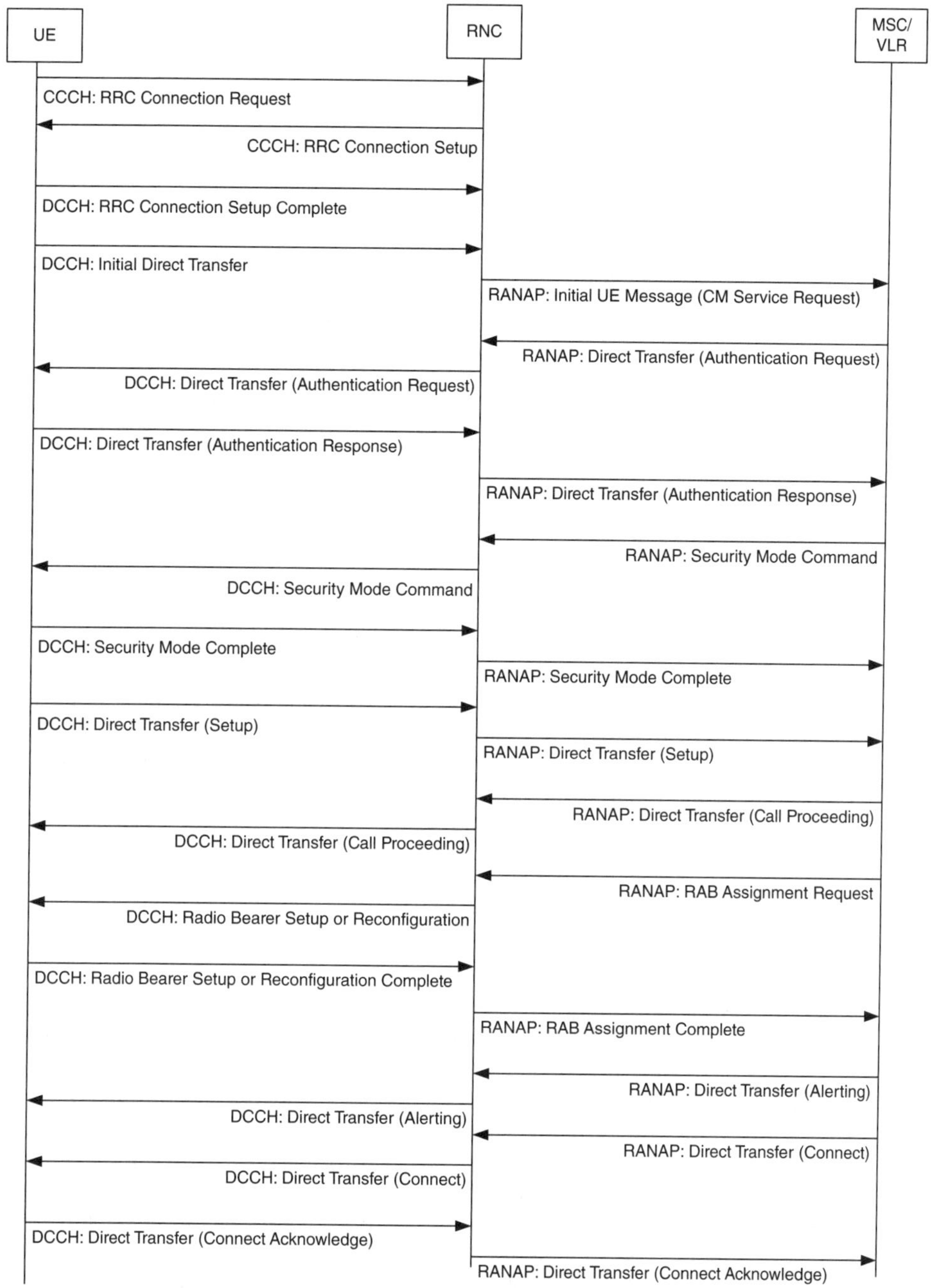

Figure 3.23 Establishing a speech call in UMTS.

the MSC. The choice of MSC or SGSN is made based on header information in the Initial Transfer message from the UE. The payload of the Initial Direct Transfer message is mapped to the payload of the RANAP Initial UE message to the MSC.

Next, the MSC will initiate security procedures. This begins with authentication, which uses a challenge-response mechanism similar to that used in GSM. One difference, however, is that the UE and network authenticate each other. Not only does the network send a random number to the UE to which a correct response must be received, but it also sends a *Network Authentication Token* (AUTN) that is calculated independently in the USIM and the HLR. The AUTN must match what the network is expecting. The authentication request is sent to the UE using the direct transfer messaging of RANAP and the RRC protocol.

Assuming that the AUTN is acceptable, the UE responds with an authentication response message that contains a response that the MSC checks. This message is also carried using the direct transfer capabilities of RANAP and RRC.

Next, the core network will instigate encryption (ciphering) and integrity procedures. This is similar to the ciphering that is performed in GSM, with the addition that integrity assurance is also enabled. This capability enables the network or UE to verify that signaling messages from the other entity have not been altered maliciously. Ciphering and integrity procedures are initiated by the core network but are executed between the UE and UTRAN. Therefore, the MSC sends the RANAP Security Mode Command message to the RNC. In turn, the RNC sends the RRC Security Mode Command message to the UE. The UE responds to the RNC with the RRC message Security Mode Complete, and the RNC responds to the MSC with the RANAP message Security Mode Complete.

At this point, the actual call-establishment information such as the called-party number is sent in a Setup message from the UE to the MSC using direct transfer signaling. Provided that the call attempt can be processed, MSC responds with the Call Proceeding message, much as is done in GSM. Next, it is necessary to establish a *Radio Access Bearer* (RAB) for transport of the actual voice stream from the user.

An RAB is a bearer between the UE and the core network for the transport of user data, either speech or packet data. It is mapped to one or more radio bearers on the air interface. Each RAB has its own identifier that is used in signaling between the UE and the network. An RAB establishment is requested by the core network through an RANAP RAB Assignment Request message.

Based on the information in the RAB Assignment Request, the RNC may set up a new radio bearer for the UE to use, or it may reconfigure any existing bearer that the UE has active. The RNC uses either the RRC message Radio Bearer Setup or the Radio Bearer Reconfiguration to instruct the UE to use the new or reconfigured radio bearers. The UE responds with either Radio Bearer Setup Complete or Radio Bearer Reconfiguration Complete. The RNC, in turn, responds to the MSC with the RANAP message RAB Assignment Complete. Now a bearer path exists from the UE through to the MSC. Note that the establishment of the bearer path also requires the establishment of a terrestrial facility between the Node B and the RNC and between the RNC and the MSC. The details of this establishment have not been shown in Figure 3.23. Suffice it to say that the transport bearer (using AAL2) will be established through the transport user control plane and the ALCAP described previously.

The remainder of the call establishment is quite similar to call establishment in GSM. It involves Alerting, Connect, and Connect Acknowledge messages carried over direct transfer signaling.

It should be noted that speech service in the 3GPP Release 1999 architecture is still a circuit-switched service. Although the speech actually is packetized for transfer over the air and also is packetized as it is carried over the Iub and Iu interfaces, a dedicated bearer is established for the duration of a call, even when discontinuous transmission is active and no speech packets are being sent.

3.6 UMTS Packet Data (R99)

Packet data with UMTS is evolving, as are the requirements for higher downlink and uplink data speeds. From a network perspective, packet-data services in the 3GPP Release 1999 architecture use largely the same mechanisms as used for GPRS data, the big difference being the user data rates that can be supported. One notable difference is that the Gb interface of GPRS (between the SGSN and the BSC) is replaced by the Iu-PS interface, which uses RANAP as the application protocol. This change includes the fact that IP over ATM is used between the SGSN and the RNC. Thus, an IP network is set up from GGSN to SGSN to RNC. Consequently, the GTP-U tunnel can be relayed from the GGSN through the SGSN to the RNC rather than terminating at the SGSN. The GTP-C tunnel, however, terminates at the SGSN because the application protocol between RNC and SGSN is RANAP rather than GTP. The establishment of the tunnel is still under the control of the SGSN. Figure 3.24 shows the Control Plane for packet-data services in UMTS, and Figure 3.25 shows the User Plane.

Packet data services are established in UMTS in largely the same manner as in GPRS—through the activation of a PDP context with an *Access Point Name* (APN), QoS criteria, and so on. One significant difference between UMTS and standard GPRS, however, involves SRNS relocation. Because of the fact that the GTP-U tunnel terminates at the RNC rather than at the SGSN, relocation of the UE to another RNC may require the buffering of packets at the first RNC and a subsequent relay of those packets to the second RNC once relocation has taken place. This relay occurs via the SGSN. In case the two RNCs are connected to two different SGSNs, then the path for buffered packets is from RNC1 to SGSN1 to SGSN2 to RNC2.

From an air-interface perspective, UMTS provides greater flexibility than GPRS in terms of how resources are allocated for packet-data traffic. Not only does UMTS offer a greater range of speeds, but the WCDMA air interface has a selection of different channel types that can be used for packet data. In the uplink, the RACH, CPCH, and DCH are available. In the downlink, the DCH, FACH, and DSCH are available. The choice of channel to be used is under the control of the RNC and is chosen depending on the characteristics of the session required by the user—for example, high-volume streaming versus low-volume bursty traffic.

For a non-bursty service such as video streaming or for large file transfers, the DCH might be the best option because it has the greatest throughput capability. It has the disadvantage, however, of taking time to establish. For small amounts of bursty traffic, the RACH or CPCH in the uplink is likely to be more suitable. These are faster to establish but cannot support rates as high as the DCH. In the case of the RACH, there is likely to be only one per cell (certainly no more than a few), whereas there can be many CPCH channels. Moreover, the CPCH can carry more data than the RACH.

In the downlink, the FACH is useful for small amounts of bursty user data. Like the RACH, however, the number of FACH channels is very limited. Another option in the

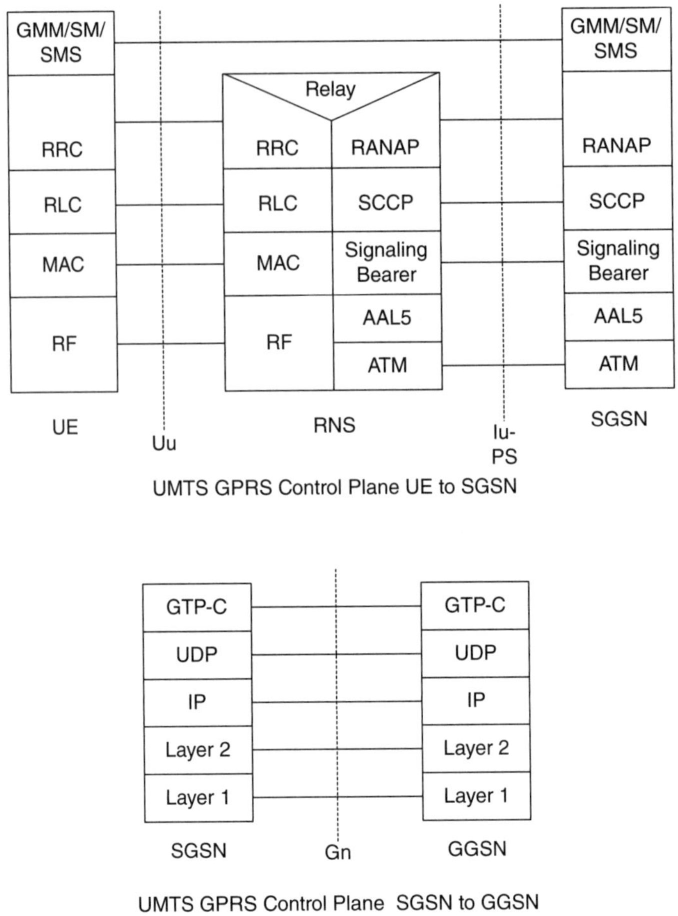

Figure 3.24 UMTS GPRS Control Plane protocol stack.

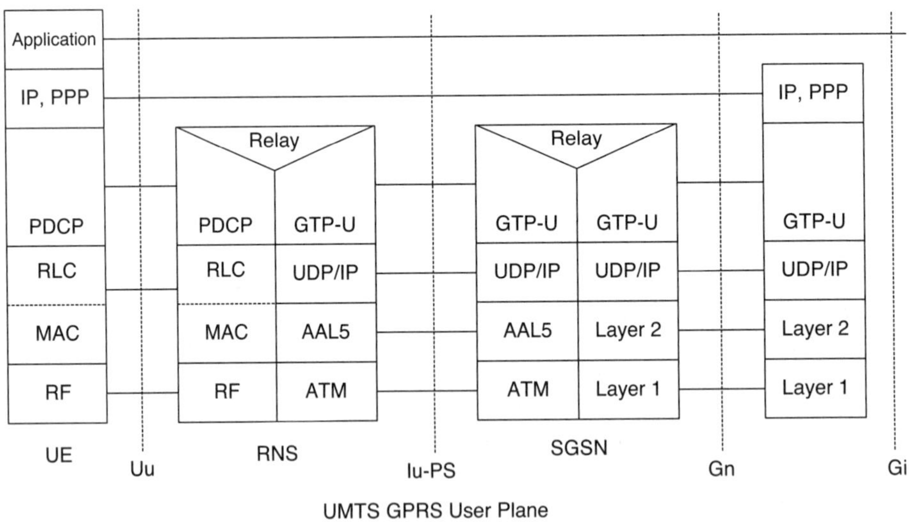

Figure 3.25 UMTS GPRS User Plane protocol stack.

downlink is the DSCH, which is a channel that is time-multiplexed among several users. It can support higher throughput than the FACH, although not as high as the DCH. It is, however, much more suited to bursty traffic than the DCH.

3.7 High-Speed Packet Data

Since the introduction of WCDMA, the need to improve the packet-data speeds was identified as urgent for both the downlink and uplink directions. Therefore, there are two primary improvements for WCDMA with data speeds: *High-Speed Downlink Packet Data* (HSDPA) and *High-Speed Uplink Packet Data* (HSUPA). HSDPA is associated with Release 5, whereas HSUPA is part of Release 6, which then leads to HSPA that combines both HSDPA and HSUPA features.

With WCDMA in Release 99, the architecture and mechanisms used were largely the same as exploited previously for GPRS/EDGE data. The primary differentiator, however, is the user data rates that can be supported. In UMTS Release 99, packet-data services are established through activation of a PDP context with an APN, QoS criteria, etc. Therefore, from an air-interface perspective, UMTS Release 99 provides greater flexibility than GPRS/EDGE in terms of how resources are allocated for packet-data traffic. Not only does UMTS offer a greater range of speeds, but also the WCDMA air interface has a selection of different channel types that can be used for packet data.

WCDMA was designed initially to offer great flexibility in transmission of user data across the air interface. For example, data rates can change on a frame-by-frame basis (every 10 ms). Moreover, it is possible to support a mix and match of different types of service. For example, a subscriber may be sending and receiving packet data while also involved in a voice call. When sending information over the air interface, physical control channels are used in combination with physical data channels. While the physical data channels carry the user information, the physical control channels carry information to support the correct interpretation of the data carried on the corresponding DPDCH frame plus power-control commands and feedback indicators.

With UMTS Release 99 the data rates are

- DL: 384 kbps
- UL: 64 kbps

A variation to Release 99 enables the UL to increase to 384 kbps. However, this does not make UMTS fully compliant with IMT-2000. Therefore, it became necessary to improve on the data rates that were possible for both uplink and downlink packet data with WCDMA. The solution for the data-rate improvements is handled in two different releases: HSDPA and HSUPA. Therefore, HSDPA and HSUPA are the next evolution for high-speed packet-data transport with UMTS/WCDMA. HSDPA is being developed for quick deployment to increase the data-handling capability of UMTS so that it is IMT-2000-compliant with Release 99. Figure 3.26 shows the high-level relationship between the various platforms and releases for UMTS and its migration toward higher-speed data at the edge of the network.

HSPA is a combination of both HSDPA and HSUPA (EUL). Unlike WCDMA, HSPA was designed to support multimedia services like streaming video, interactive e-mails and gaming, telemetrics, and, of course, faster downloads of high-resolution images and large files.

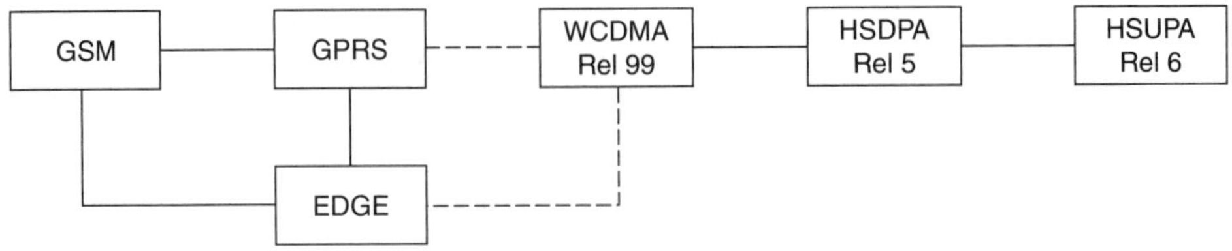

FIGURE 3.26 UMTS packet-data migration path.

HSPA also brought about better spectral efficiency with improved system performance supporting both Voice and Data services at the same time. HSPA with the improvements, however, still uses the same 5-MHz channel as in WCDMA and has a similar architecture as WCDMA (UMTS). HSPA is also backward-compatible to 3GPP Release 99.

HSPA benefits include

- Capacity gain over WCDMA 3GPP Release 99
  - 5 × downlink
  - 2 × uplink
- Peak data rates
  - Downlink: 14 Mbps
  - Uplink: 5.8 Mbps
- Reduced latency
- Greater system capacity

HSPA was effectively accomplished by moving many key scheduling and control processes from the RNC to the Node B, which is the precursor for LTE.

3.7.1 HSPA System Architecture

HSPA introduces some system architecture changes which are focused more on enhancements to Node B functionality. The enhancements with HSPA also paved the way for LTE with the Node B. The Node B enhancements largely introduced in HSDPA involve

- New Uplink Control Channel (UL DPCCH)
- CQI reporting
- DTX and DRX
- Enhanced cell FACH
- Single tunnel
  - Decreases latency
  - Direct tunnel between RNC and GGSN

The HSPA design architecture is moving toward a flat architecture as in later releases with LTE; however, HSPA is an interim step in that direction and an example of the HSPA architecture is shown in Figure 3.27.

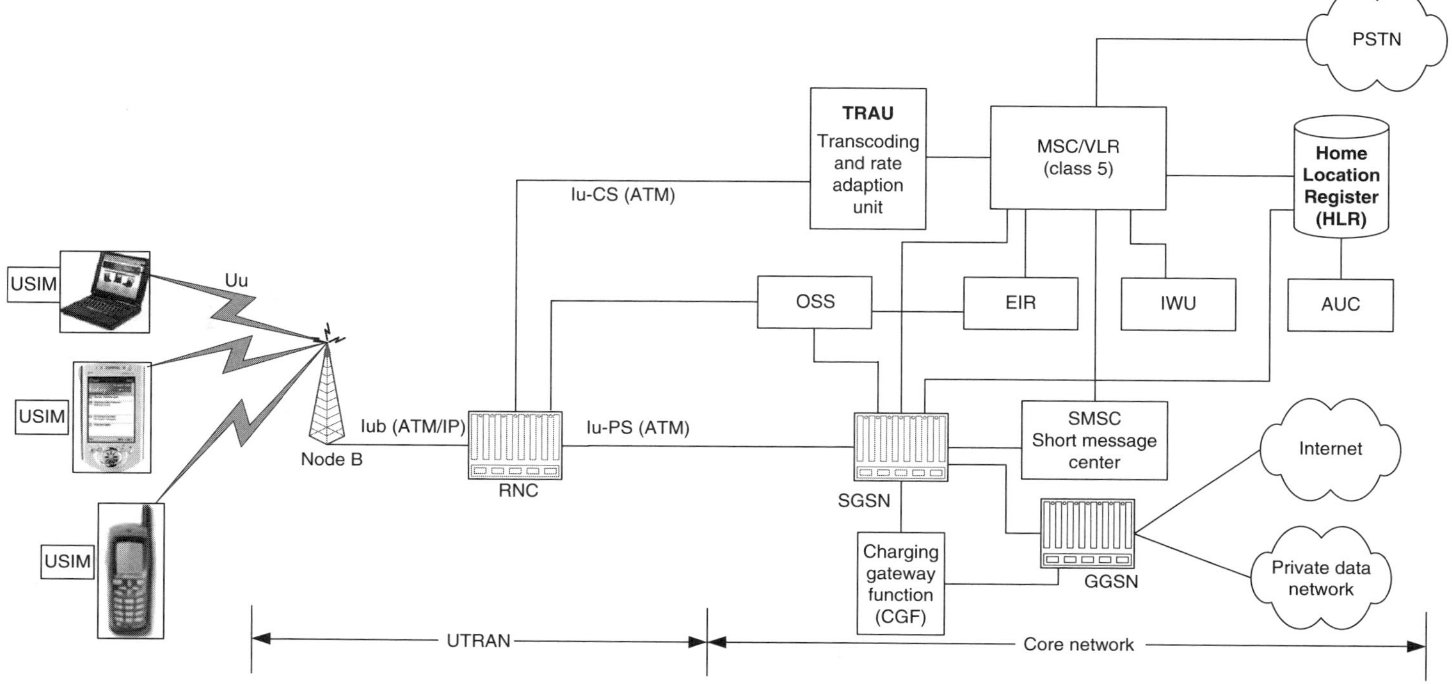

Figure 3.27 HSPA architecture.

The highlighted portions of the architecture shown in Figure 3.27 are the required changes associated with HSPA in a UMTS network. However, to describe HSPA it is best to discuss both HSDPA and HSUPA.

3.7.1.1 HSDPA

HSDPA is an enhanced modulation scheme that enhances WCDMA's throughput to a theoretical 14.4 Mbps in the downlink and uses an architecture and protocol that is designed to overlay onto an existing UMTS network for the purpose of supporting asymmetric data services, specifically multimedia, and is part of Release 5. HSDPA, as its name implies, is an improvement for the downlink path only, meaning that the uplink capability for HSDPA is the same as Release 99.
The primary features for HSDPA involve

- Shared channel transmission
- Adaptive Modulation and Coding (AMC)—used for power control
- Fast Hybrid Automatic Repeat Request (HARQ)
- Fast Cell Site Selection (FCSS)
- Short Transmission Time Interval (TTI)

To achieve this list of features, HSDPA involves modifications to the Node B, RNC, core network, and subscriber terminal to handle the new data rates and modulation format. In order to improve the data rate, enhancements to the modulation schemes used are needed, and HSDPA will use 16QAM in addition to QPSK. The modulation scheme used depends on the RF conditions. As in most communication systems, the RF environment determines if the higher modulation schemes are achievable, and this is coupled with the packet speed requested for service delivery. The overlay is not a physical overlay but rather an enhancement that can be integrated into the system in an N:1 configuration, enabling the operator to deploy HSDPA selectively. Table 3.5 shows the various throughputs available with HSDPA and the relationships between code schemes and modulation formats.

To achieve the throughputs referenced in the table, HSDPA has several improvements or rather enhancements over WCDMA. Some of the enhancements involve introduction of the *High-Speed Downlink Shared Channel* (HS-DSCH), which is a common high-speed downlink shared channel. HS-DSCH enables a downlink channel to be shared between users. Several new channel types also are introduced for access and

		Throughput (kbps)		
Modulation	**Coding Rate**	**5 Codes**	**10 Codes**	**15 Codes**
QPSK	1/4	600	1200	1800
QPSK	2/4	1200	2400	3600
QPSK	3/4	1800	3600	5400
16QAM	2/4	2400	4800	7200
16QAM	3/4	3600	7200	10700
16QAM	4/4	4800	9600	14400

Table 3.5 HSDPA Throughput (kbps)

control features. One of the new channels is the HS-DPCCH, which is used for uplink Channel Quality Indicator (CQI).

With a WCDMA system, the packet scheduling is done at RNC; however, with the introduction of HSDPA, all packet traffic that is associated with the HSDPA RF transmission is done at the Node B. Therefore, based on the CQI and power measurements, the RF channel dictates the actual transmission data rate. Naturally, as the user gets closer to the cell, a higher modulation and code rate becomes available.

Referring to Release 99, packet data in the downlink is handled by three different channels, namely the DCH, the DSCH, and the FACH. The DCH is the primary data channel and uses between 4 and 512 OVSF codes depending on the block error rate determined through both the inner and outer power-control loops. The DSCH offers the ability to time-multiplex different users. The FACH is used to send bursty traffic on a common channel.

HSDPA enhances the downlink capabilities introduced in Release 99 with DSCH, HS-DSCH, through the use of adaptive modulation, improved coding, HARQ, and Node B scheduling. This is important because this is also backward-compatible with Release 99 as well. Another important change with HSDPA is that it does not use a *Variable Spreading Factor* (VSF) and fast power control and relies on a fixed spreading factor. The use of a nonvariable spreading factor also allows for better power control.

Additionally, HSDPA allows for multiple codes to be assigned to a subscriber, enabling multiple services to be delivered simultaneously. The packet scheduling is performed at the Node B and not the RNC. The scheduling is based on the Channel Quality that is fed back to the Node B from the individual UE.

As indicated previously, HSDPA introduces several new channels for use. The high-level channel layout is displayed in Figure 3.28, where the new HSDPA channels are to the far right of the diagram and also displayed in bold.

Referring to the figure, the four new channels are

1 *HS-DSCH.* High-Speed Downlink Shared Channel (downlink transport channel)
2 *HS-PDSCH.* High-Speed Physical Downlink Shared Channel (physical channel)
3 *HS-SCCH.* High-Speed Shared Control Channel (downlink control channel)
4 *HS-DPCCH.* High-Speed Dedicated Physical Control Channel (uplink control channel)

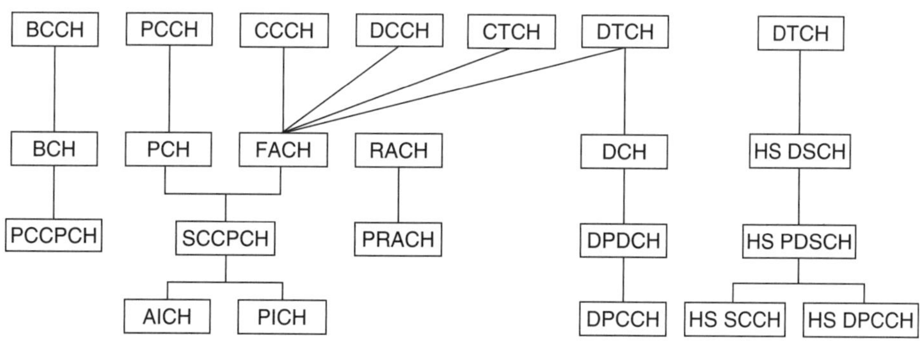

FIGURE 3.28 HSDPA channels.

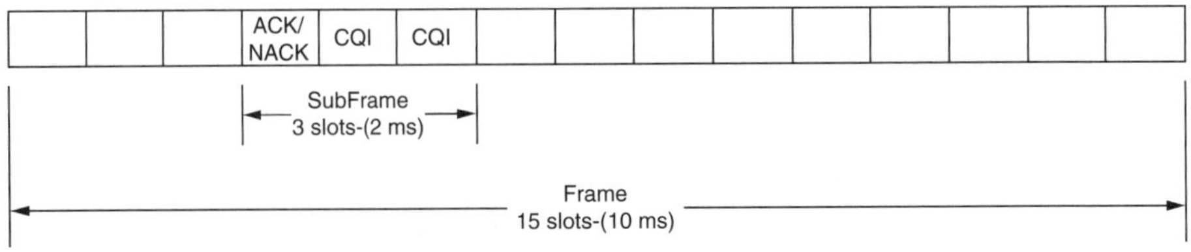

FIGURE 3.29 HSDPA frame.

HS-DSCH is the primary radio bearer, having a similar function to the DSCH and is downlink-only. An important aspect with HS-DSCH is that it allows for multiplexing. Multiplexing is possible because of the TTI, which consists of three slots taking a total time of 2 ms, as shown in Figure 3.29.

HS-DPCCH is used by the UE to send *Channel Quality Indicators* (CQIs) and ACK/NACKs.

Referring to Figure 3.29, the whole frame for HSDPA is 10 ms and is made up of 15 slots. Each frame consists of five subframes, where a subframe involves 1 TTI, and each TTI uses a constant SF of 16. There are a maximum of 15 codes, parallel, which can be assigned to any given user during a TTI. Alternatively, the 15 codes also can be assigned to multiple users within the same TTI, allowing for multiplexing.

However, to achieve this, the HS-DSCH requires more overhead and two new control channels have been introduced to support the HS-DSCH, namely the HS-SCCH and the HS-DPCCH. The HS-SCCH is a fixed-rate (60 kbps, SF = 128) channel used for delivering downlink signaling before the beginning of the next TTI for the UE. The information that the HS-SCCH contains includes identification of the UE HARQ. In addition, it is possible that the UE can be assigned up to four HS-SCCHs. The HS-SCCH structure is shown in Figure 3.30.

Referring to Figure 3.30, the HS-SCCH has three slots; one slot is called Part 1, and the remaining slots are called Part 2. For HS-SCCH, Part 1 provides the modulation information and OVSF code assignment used, whereas Part 2 provides the transport block size and HARQ parameters. In addition, the HS-SCCH used an SF of 128.

Continuing, the HS-DPCCH carries the ACK/NACK signaling indicating whether the packet transmission was decided successfully, in addition to the proper CQI for link adaptation, and uses an SF of 256. The CQI is based on the *Common Pilot Channel* (CPICH) and is used to estimate the transport block size, modulation type, and number of channelization codes that can be supported based on the downlink. *Adaptive Modulation and Coding* (AMC) is used in HSDPA to constantly optimize the code rate and the transmit power per code based on the CQI.

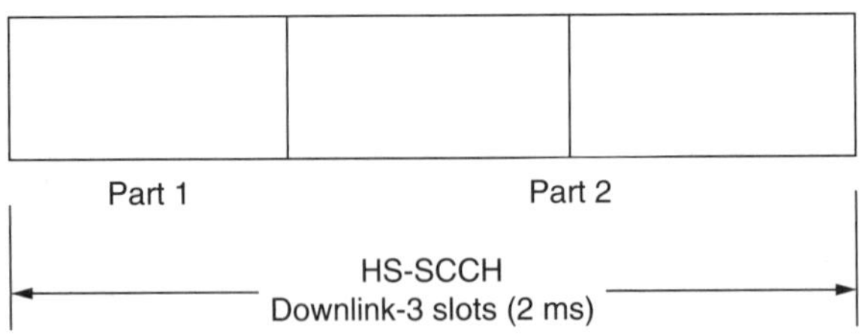

FIGURE 3.30 HS-SCCH.

The CQI is established through the UE measuring the downlink CPICH strength over three slots. This information, through its index number equating to CQI, then is sent to the Node B to establish the highest rate possible that can sustain an error rate of less than 10 percent. The CQI measurement slots are indicated in Figure 3.29.

Additionally, the ACK/NACK return is sent in both the uplink and downlink 5 ms after the block is received, equating to a minimum return rate of 10 ms. The UE can repeat up to four times the ACK/NACK.

All this enables the HSDPA system to deliver higher downlink data rates more efficiently. To also enable higher data rates, coding schemes are improved. The HS-DSCH-encoding scheme uses a 1/3 Turbo encoder, also used in Release 99. It improves the rate by puncturing the repetition to improve the overall granularity for the code to get a higher data rate. But this is combined with the higher modulation schemes of 16QAM and coupled with QPSK, allowing for different combinations of data rates based on the channel conditions.

For example, a rate of 119 kbps/code with QPSK is achievable with a 1/4 code rate, whereas 712 kbps/code can be achieved with 16QAM using a 3/4 code rate. HSDPA is capable of supporting 14.4 Mbps in the downlink (or 960 kbps/code) when 16QAM modulation is used with no coding (effective code rate of 1) and 15 multicodes. An HSDPA-capable UE can support 5, 10, or 15 multiple codes.

The retransmission mechanism selected for HSDPA is *Hybrid Automatic Repeat Request* (HARQ) with *Stop and Wait protocol* (SAW). Determines if *Receive* (rx) packet data has errors and informs the transmitter that errors are present. If errors are present the packets are then resent. HARQ allows the UE to rapidly request retransmission of erroneous transport blocks and is meant to help reduce delays in transmissions. HARQ combines ARQ and *Forward Error Correction* (FEC). The HARQ method is interesting in that with HARQ, the UE does not discard the energy from failed transmissions; instead, it saves it and later combines it with the retransmission(s) to increase the probability of successfully decoding the message.

HARQ uses two types of retransmissions, *Chase Combining* (CC), where the same bits are retransmitted when decoding errors, and *Incremental Redundancy* (IR). Additional parity bits are sent when decoding errors. The retransmissions are triggered by NACKs sent on the HS-DPCCH. This is a form of soft combining. HSDPA supports both Chase Combining (CC) and Incremental Redundancy (IR).

In HSDPA, the scheduler is essential and determines the overall behavior of the system and, to a certain extent, its performance or perception of performance for the customer. For each TTI, the scheduler determines which terminal the HS-DSCH should be transmitted to in conjunction with the data rate. Since the scheduler is now located at the Node B, it allows for better tracking of the UE channel condition, allowing for quicker adaptation of the data rate to the RF channel condition.

Scheduling tasks include

- Process received CQI, ACK/NACK
- Select HARQ process
- Priority queue
- Estimation of RF channel conditions
- Ranking of users
- Allocation of transport formats

The ranking criteria for the scheduler uses a ranking list which can be round robin, proportional faire, or max C/I. In the *Round Robin* the scheduler allocates an equal time for resources to each user. In the *Proportional Faire* the scheduler equates the channel rate and the throughput ratio by users, while in the *Max C/I* the user with the best channel condition is favored.

Node B scheduling is key to HSDPA and all of HSPA. The Node B makes the primary decisions and because it is closest to the UE it can respond much faster and hence the latency reduction.

Adaptive modulation and Coding (AMC):

- Rate control is done primarily through modulation and coding schemes
- HSDPA uses AMC to maximize data throughput in a changing RF environment
- CQI obtained from the UE also helps set the data rate
- AMC allows for a consistent power level to be achieved between Node B and UE
- AMC advantages are best effective with fast scheduling

Twelve new categories have been specified by Release 5 for HSDPA UE, as shown in Table 3.6. The maximum number of HS-DSCH multicodes that a UE can receive simultaneously is 5, 10, or 15, with a minimum inter-TTI time of two consecutive transmissions. All this means that the HS-DSCH and the inner TTI interval determine the UE peak data rate.

For example, the transport block size of the inner TTI interval for category 2 with a TTI of 7298 and three TTI (6 ms) leads to a data rate of 1.2 Mbps, as shown in Table 3.6.

Also, with HSDPA, soft handoffs are not used, as was the case in Release 99. The removal of soft handoff enables higher system and user throughput and also facilitates better power control of the UE itself. HSDPA instead uses a hard handover. The UE continuously monitors all the Node Bs in its active set and reports to UTRAN when a change in the best cell occurs.

Category	Code	Inter-TTI	TBS Size (min)	IP Buffer	Rate	Modulation
1	5	3	7298	19,200	1.2	QPSK/16QAM
2	5	3	7298	28,800	1.2	QPSK/16QAM
3	5	2	7298	28,800	1.8	QPSK/16QAM
4	5	2	7298	38,400	1.8	QPSK/16QAM
5	5	1	7298	57,600	3.6	QPSK/16QAM
6	5	1	7298	67,200	3.6	QPSK/16QAM
7	10	1	14,411	115,200	7.2	QPSK/16QAM
8	10	1	14,411	134,400	7.2	QPSK/16QAM
9	15	1	20,251	172,800	10.0	QPSK/16QAM
10	15	1	27,952	172,800	14	QPSK/16QAM
11	5	2	3630	14,400	0.9	QPSK
12	5	1	3630	28,800	1.8	QPSK

Table 3.6 HSDPA Categories

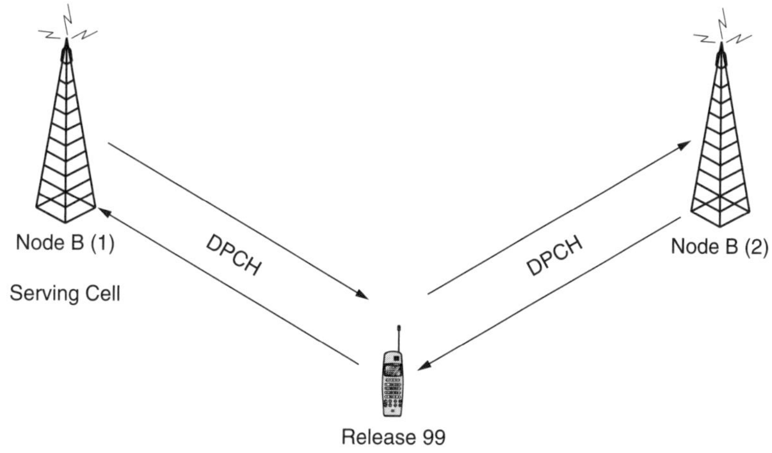

Figure 3.31 Release 99 soft handoff.

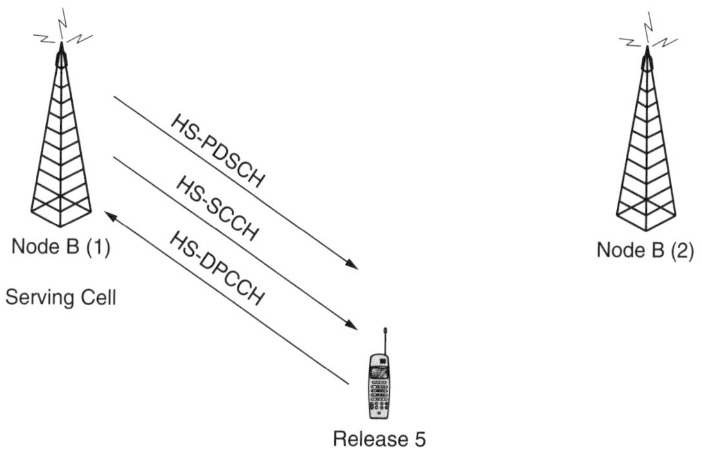

Figure 3.32 HDSPA—no soft handoff.

Figures 3.31 and 3.32 illustrate the fundamental differences between Release 99 and HSDPA signaling for handoffs. It is important to note that the UE monitors the neighbor cells for possible use based on the RF conditions through the use of the evaluation of the neighbor list. The nominal hysteresis value of 3 to 5 dB is used to stop unnecessary handoffs from occurring, that is, the pingpong effect.

To help put this all into perspective, the relationship between HS-DPCCH, HS-SCCH, and HS-PDSCH is shown in Figure 3.33, where each of the channels begins sending information prior to the completion of the other.

3.7.1.2 HSUPA

High-Speed Uplink Packet Access (HSUPA) is the uplink improvement that is following on the heels of HSDPA. HSUPA is officially referred to as *Enhanced Dedicated Channel* (E-DCH). The introduction of HSUPA enables uplink rates to achieve 1.4 Mbps, with

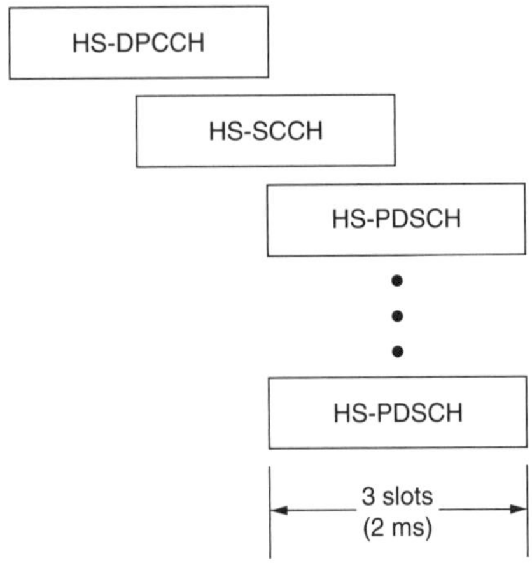

FIGURE 3.33 HSDPA channels.

later releases envisioned at 5.76 Mbps. The HSUPA specification at this moment is still under development and will be included in UMTS Release 6. This is Revision 5 and is a software and hardware upgrade for UMTS. However, this is backward-compatible with Release 99, Release 4, and Release 5 systems.

HSUPA will include several new channels both dedicated and shared and are meant to facilitate improved uplink data rates. Figure 3.34 illustrates the new channels including the E-DCHs. HSUPA also will employ link-adaptation methods similar to those in HSDPA, including shorter transmission intervals and HARQ.

The *Enhanced Absolute Grant Channel* (E-AGCH) is a shared channel which the rest are dedicated to the UE.

A packet scheduler also will be introduced for the uplink and will operate on a request-grant principle that will allow for many UEs to request resources at the same instant. As with HSDPA, scheduling is the key with HSUPA.

HSUPA does not utilize a shared channel for data transfer in the uplink and therefore relies on fast scheduling to reduce the overall transmission delay allowing for rapid changes in which UEs are transmitting and at what rate they are using. The transmission delay reduction is achieved through fast hybrid ARQ retransmissions, in a manner very similar to HSDPA.

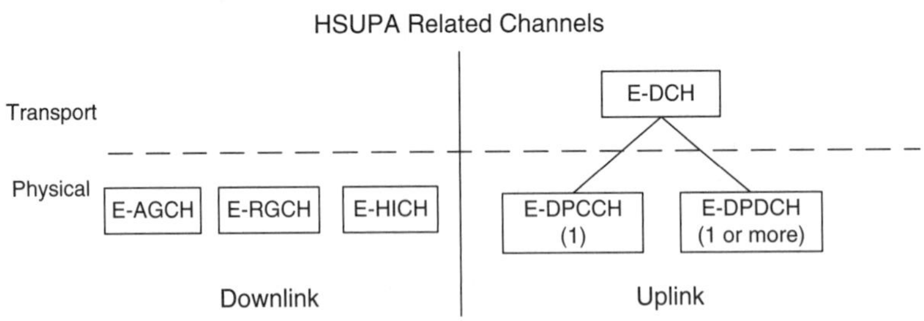

FIGURE 3.34 HSUPA-related channels.

With HSUPA, each UE already uses a unique scrambling code in the uplink so each UE already has a dedicated uplink connection to the network. The shared resource in the uplink is actually the interference level at the Node B, which the network manages through the fast closed-loop power control algorithm. As part of the interference level control, fast power control is used for the near-far problem and the UE needs to signal its buffer status to the scheduler.

With HSUPA, the scheduler in Node B controls when and what data rate UE is allowed to transmit. There are two methods for controlling the UE's transmit power on the E-DPDCH.

- Nonscheduled Grant
- Scheduled Grant

With Nonscheduled Grant, the UE is told the maximum block size that it can transmit on the E-DCH during a TTI call setup. The UE can then transmit a block of that size or less in each TTI until the call ends or the network modifies the Nonscheduled Grant via an RRC reconfiguration procedure. The block size deterministically maps to a power level, which is also configured by the network during call setup. Nonscheduling Grant is meant for delay-sensitive application such as Voice over IP (VoP).

Scheduled Grant involves the UE maintaining a Serving Grant that it updates based on information received from the network. The Serving Grant directly specifies the maximum power the UE can use on the E-DPDCH in the current TTI. UE can translate its Serving Grant to the maximum E-DCH block size it can use in a TTI (the mapping of power levels is determined by the E-TFCI Reference Power Offsets that are signaled at call setup).

There are several Serving Grant Methods, Absolute and Relative. The *Absolute Grant* is where the grant is transmitted on the shared E-AGCH downlink channel, which signals a specific, absolute number for the Serving Grant. The *Relative Grant* is transmitted using the downlink *Enhanced Relative Grant Channels* (E-RGCH) that incrementally adjust a UE's Serving Grant up or down from its current value. At any given point in time the UE will be listening to a single E-AGCH from its serving cell and to one or more E-RGCHs.

One major difference with the introduction of HSUPA is the inclusion of soft and softer handoffs with packet transmissions. In addition, power control will be different in that the main serving Node B will be able to issue power-up and power-down commands, but the other Node Bs will be able to issue only power-down commands.

3.7.2 HSPA

This section brings together both the HSDPA and HSUPA channels as shown in Figure 3.35.

Referring to Figure 3.35, the following channels are added in the uplink for HSPA as compared to UMTS:

- *E-DCH (Enhanced Dedicated Channel).* This is the uplink transport channel and an E-DCH can be mapped to more than one physical channel. The E-DCH is controlled by Node B and is made up of E-DPDCH and E-DPCCH.
- *E-DPDCH (Enhanced Dedicated Physical Data Channel).* This carries the user data bits to the network. This channel uses a variable spreading factor and is variable in number. The UE will select the most appropriate configuration based on its

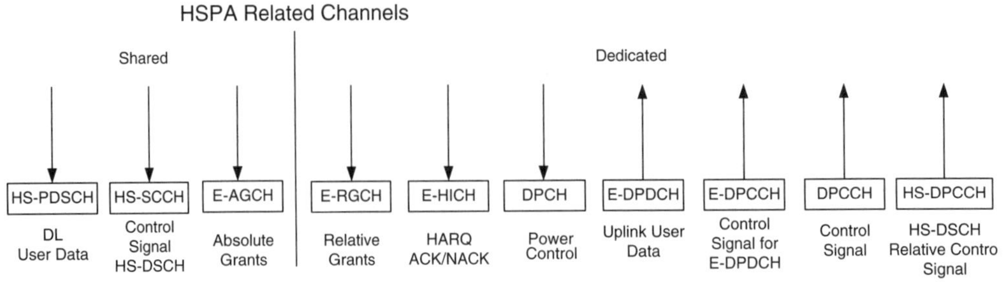

FIGURE 3.35 HSPA channels.

current Nonscheduled Grant or scheduled Serving Grant and the amount of data it has waiting to be sent.

- *E-DPCCH (Enhanced Dedicated Physical Control Channel).* This channel contains the transmission characteristics of the E-DPDCH.

In the downlink the following channels are added:

- *E-HICH (Enhanced Hybrid Indicator Channel).* This carries HARQ information to the UE like the ACKs and NACKs.
- *E-AGCH (Enhanced Absolute Grant Channel).* This is a shared channel that signals absolute values for the Serving Grant covering absolute power ratio allocation for uplink transmissions.
- *E-RGCH (Enhanced Relative Grant Channel).* This signals the incremental up/down/hold adjustments to the UE's Serving Grant.

HSPA also has several performance improvements over UMTS which involve combinations of improvements brought on by HSDPA and HSUPA. Specifically, the throughput is dependent upon

- UE device capability
- Number of on-site users competing for same resources
- RF Path conditions
- Backhaul and Core Network Capacity

HSPA Radio Resource Management is another performance improvement where the IP address is retained by the UE even when it is idle. The retainment of the IP address enables the UE to have sessions initiated from the network.

3.8 Handover

UMTS supports two main categories of handovers—soft handovers and hard handovers. However, with HSDPA, soft handovers are not supported; only hard handovers are supported. The main difference between a soft handover and a hard handover is the system resources involved with the maintenance of a call or data session. A soft handover is make-before-break, whereby communication exists between the UE and more than one cell for a period of time. A hard handover is break-before-make, whereby communication with the first cell is terminated before establishing communication with the second cell.

A soft handover has two variants—soft handover and softer handover. These two situations are depicted in Figure 3.36. A soft handover occurs between two cells or sectors that are supported by different base stations. The UE is transmitting to and receiving from

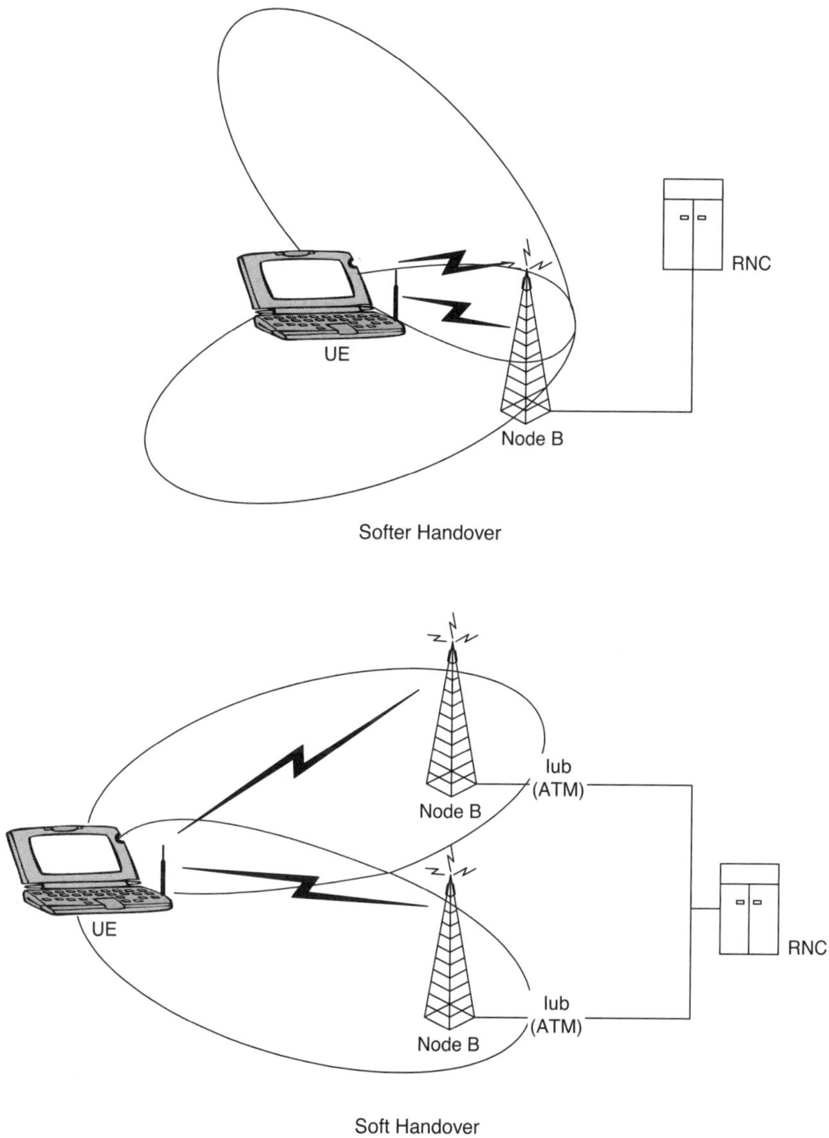

FIGURE 3.36 Soft handover and softer handover.

both base stations at the same time. The user information sent to the UE is sent from each base station simultaneously and is combined within the UE. In the uplink, the information sent from the UE is relayed from each base station to the RNC, where the combining takes place. In the case of a soft handover, each base station is sending power-control commands to the UE.

A softer handover occurs between two cells that are supported by the same base station. In this case, only one power-control loop is active and is controlled by the base station that serves both cells. Depending on RF coverage, both a soft handover and a softer handover may occur at the same time for a given UE.

A hard handover can occur in several situations, such as from one cell to another when the two cells are using different carrier frequencies or from one cell to another when the base stations are connected to different RNCs and no Iur interface exists between the RNCs. UMTS also supports a hard handover to and from GSM. This is a reasonable requirement because it takes time to roll out a UMTS network nationwide, and one would like UMTS subscribers to receive service from GSM in areas where holes occur in the UMTS coverage.

Regardless of the type of handover to take place, the decision when and how to invoke a handover is made at the serving RNC. This decision is based on measurements reported by the UE. The set of cells for which measurement reports are to be generated is broadcast from the network on the BCH or FACH. If a neighboring cell uses a different frequency and the RNC requires reports related to that cell, then the UE needs time periodically to tune to the frequency in question. This means that the UE and UTRAN must operate in compressed mode. This mode means that in a given radio frame, not all 15 slots are used. The unused slots correspond to durations where the UE can tune to another frequency to make the necessary measurements.

The fundamental differences between R99 and HSDPA signaling for handoffs is that the UE does not have a connection with the target Node B until the handover is completed.

A simple ladder diagram for a soft handover is shown in Figure 3.37 where the SRNC is the Serving RNC and the DRNC is called the *DRIFT RNC*. The DRNC is just another RNC that is not the serving RNC.

With HSUPA, the handoff process is slightly different than HSDPA since the E-RGCH is also shared by multiple UEs, but on this channel the UE is listening for a particular orthogonal signature rather than a higher layer identity. If it does not detect its signature in a given TTI, it interprets this as a "Hold" command, and thus makes no change to its Serving Grant.

In HSUPA, each cell that the UE is in soft handover with will transmit an E-RGCH serving cell responsible for scheduling operation. Cells in the serving RLS can give E-RGCH commands to raise, lower, or hold the current Serving Grant. Cells in the non-serving RLS can only give *hold* or *down* commands.

Figure 3.38 attempts to depict the channels used in HSUPA handovers.

3.9 HSPA Connection Establishment

Figure 3.39 is a simple ladder diagram of an HSPA connection. In HSPA the signaling for the connection is the same as HSUPA. In the ladder diagram of Figure 3.39, the specific channels used and the detailed signaling cannot be seen due to the clutter effect of having everything shown.

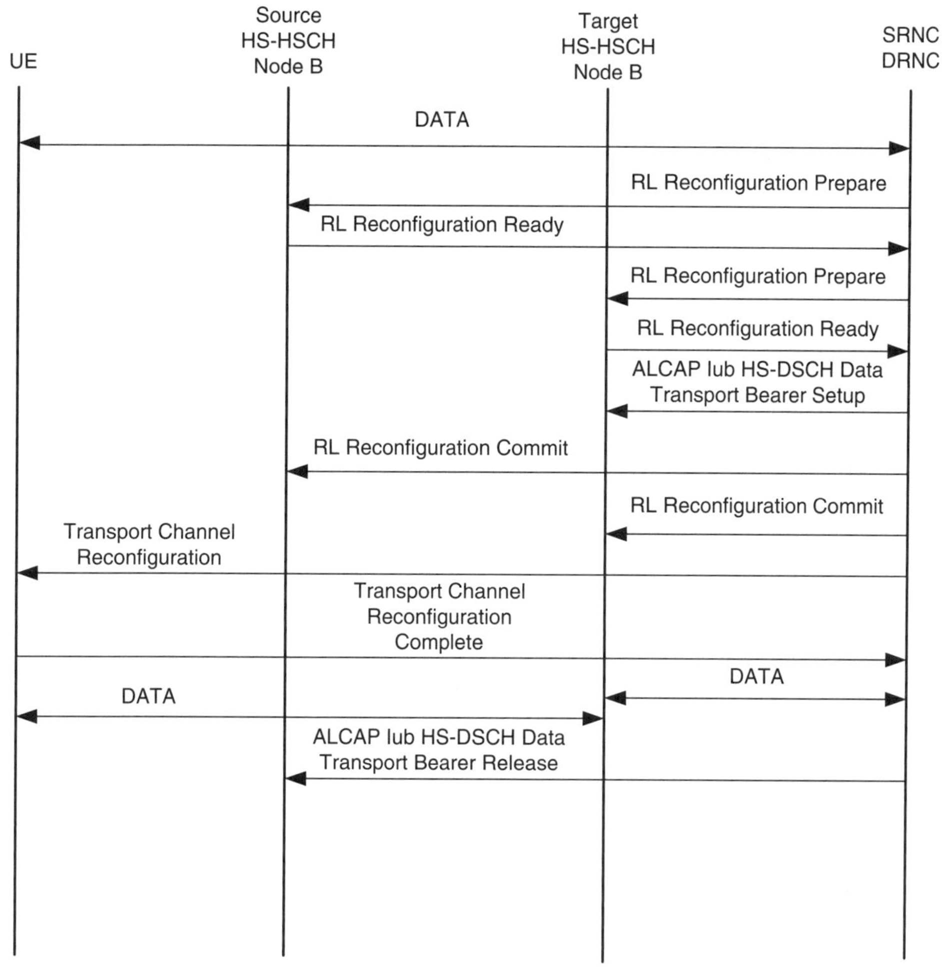

FIGURE 3.37 Soft handover ladder diagram.

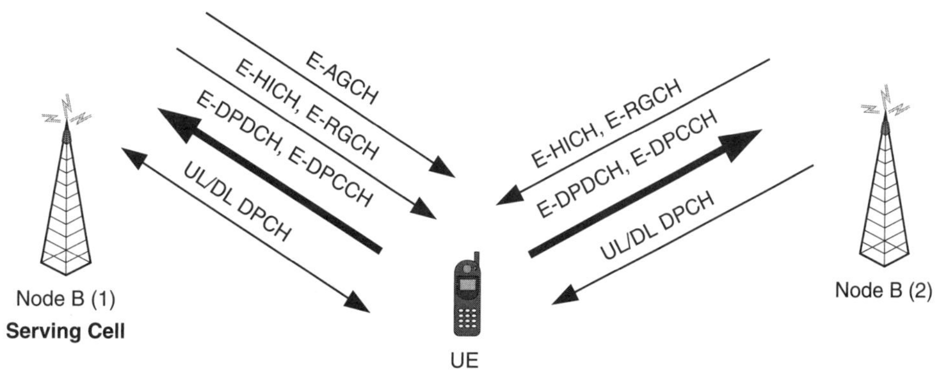

FIGURE 3.38 HSUPA handover.

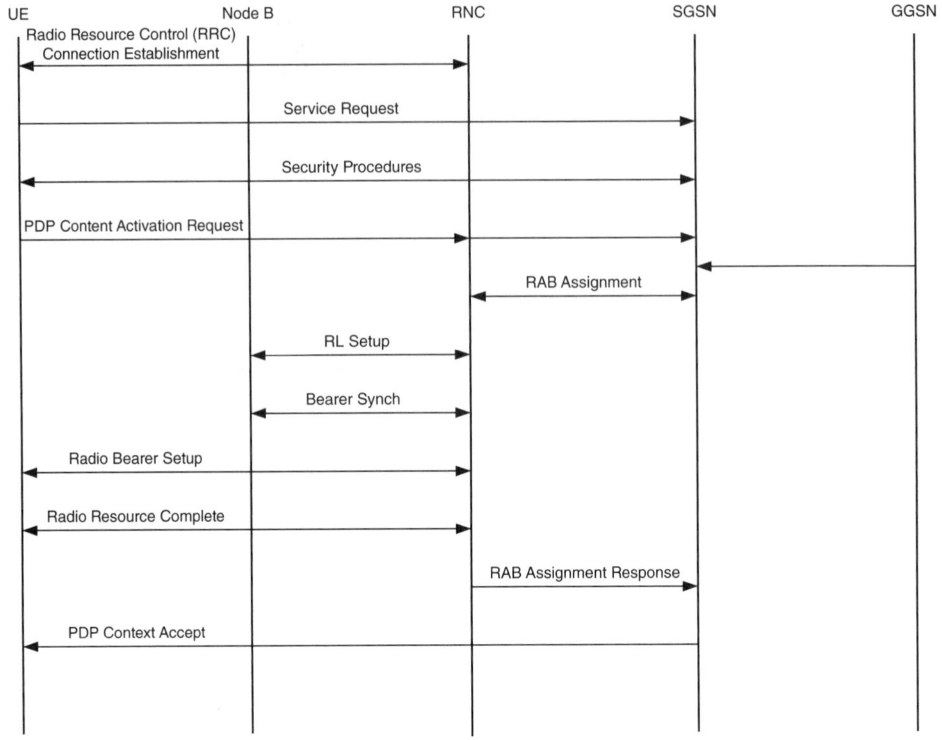

Figure 3.39 Connection establishment.

3.9.1 The 3GPP Release 4 Network Architecture

3GPP Release 4 introduces a significant enhancement to the core network architecture as it applies to the CS domain. Basically, the MSC is broken into constituent parts, and it is allowed to be deployed in a distributed manner, as shown in Figure 3.40. Specifically, the MSC is divided into an MSC server and a *Media Gateway* (MGW). The MSC server contains all the mobility management and call-control logic that would be contained in a standard MSC. It does not, however, reside in the media path. Rather, the media path is via one or more MGWs that establish, manipulate, and release media streams (voice streams) under the control of the MSC server.

Control signaling for circuit-switched calls is between the RNC and the MSC server. The media path for circuit-switched calls is between the RNC and the MGW. As far as the RNC is concerned, these two entities could be the same physical device, as would be the case when the RNC is communicating with a traditional MSC. Typically, an MGW takes calls from the RNC and routes those calls toward their destinations over a packet backbone. In many cases, the packet backbone will be IP-based, such that backbone traffic is VoIP, as described in more detail in Chapter 9. Given that the PS domain also uses an IP backbone, then only one backbone network is needed, which can mean significant cost savings for the network operator.

At the remote end, where a call needs to be handed off to another network, such as the *Public Switched Telephone Network* (PSTN), another MGW is controlled by a *Gateway*

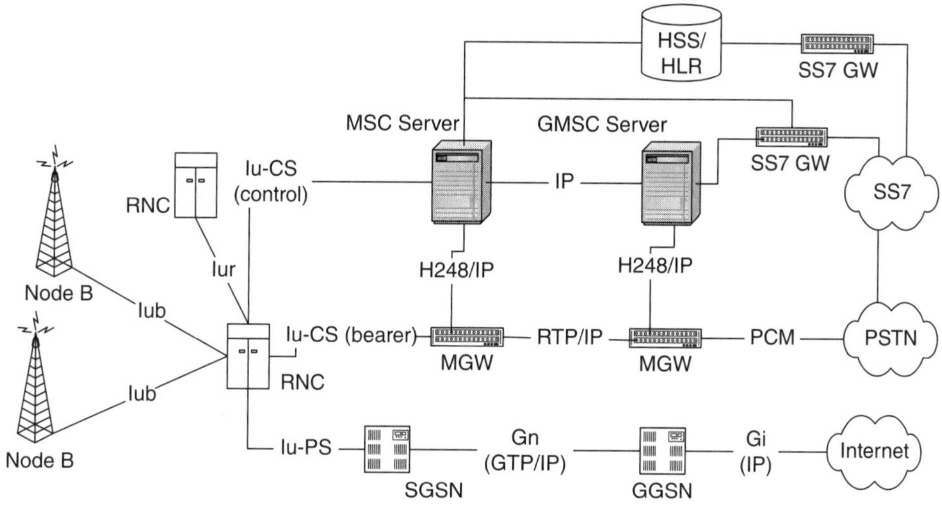

FIGURE 3.40 3GPP Release 4 distributed network architecture.

MSC server (GMSC server). This MGW converts the packetized voice to standard PCM for delivery to the PSTN. It is only at this point that transcoding needs to take place. Thus, voice can be carried through the backbone at a far lower rate than 64 kbps, with a step up to 64 kbps only at the last point. This represents a lower bandwidth requirement in the backbone network and therefore a lower cost.

The control protocol between the MSC server or GMSC server and the MGW is the *International Telecommunications Union* (ITU) H.248 protocol. This protocol is also known as *MEGACO*. The call-control protocol between the MSC server and the GMSC server can be any suitable call-control protocol. The 3GPP standards suggest, but do not mandate, the *Bearer Independent Call Control* (BICC) protocol, which is based on the ITU-T Recommendation Q.1902.

Figure 3.41 shows an example of a voice-call establishment using this architecture. For the sake of brevity, the messages are limited to those that relate to call establishment as seen from the core network. Thus, the RRC protocol messages between the UE and UTRAN have been omitted. These will be as shown in Figure 3.23. Figure 3.29 includes a number of H.248 messages. For details of the H.248 protocol, please refer to Chapter 9.

When the Setup message arrives from the UE, the MSC server performs a call-routing determination. It then responds with a Call Proceeding message to the UE. Based on the call-routing determination, the MSC server chooses an MGW to handle the call. It instructs (Add Request) the MGW to establish a new context and places a termination in that context. The termination in question (T1) will be on the network side of the MGW. Once the new context is established by the MGW, the MSC server requests the RAN to establish an RAB to handle the call. Once the RAB has been assigned, the MSC server is in a position to establish a media connection between the RNC and the MGW. Therefore, it requests the MGW to add a new termination to the context that has just been established. This new termination (T2) will face toward the RNC. Because the new termination is in the same context as termination T1, a path is created from one side of the MGW to the other.

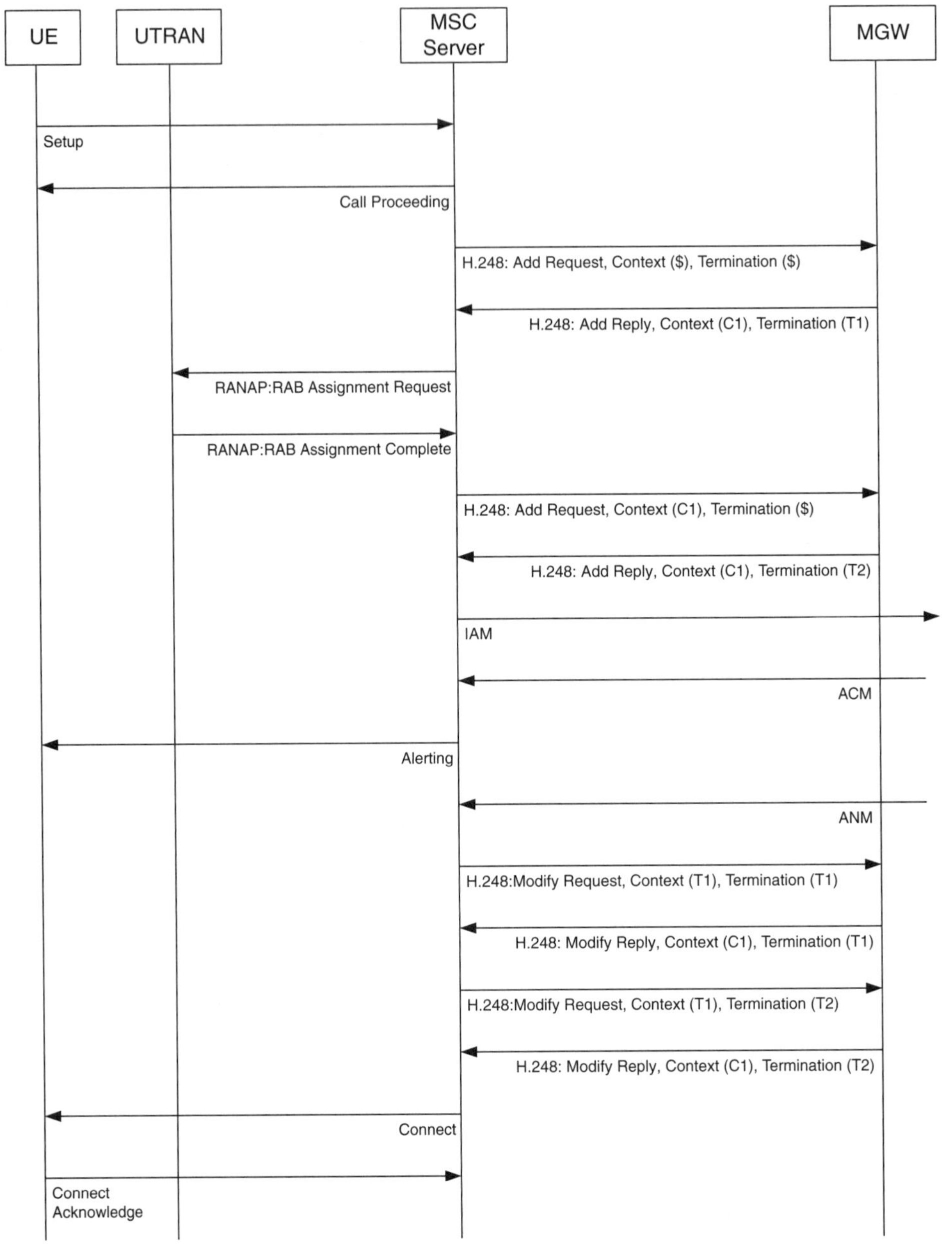

FIGURE 3.41 Mobile-originated call with 3GPP Release 4 distributed architecture.

The MSC server then sends the ISUP *Initial Address Message* (IAM) to the called network (such as the PSTN). On receipt of an *Address Complete Message* (ACM) from the far end, the MSC server sends an Alerting message to the UE. Typically, the called user will answer, which causes an ISUP *Answer Message* (ANM) to be received at the MSC server.

At this point, the MSC server optionally may modify the context established on the MGW. Specifically, when the terminations were established in the new context, they may have been configured to not provide a complete through-connection. For example, one or both of the terminations could have been configured to allow only a one-way media path (from the far end to UE for the purpose of receiving a ring-back tone). In such a case, the MSC server requests the MGW to modify the configuration so that a full two-way media path is established. Finally, the MSC server sends a Connect message to the UE, and the UE responds with a Connect acknowledge.

Note that the distributed architecture just described does not rely on WCDMA-based UTRAN access. The core network architecture just as easily could apply to standard GSM-based access, with BSCs instead of RNCs. In fact, as the distributed switching architecture is being deployed, it is likely that many deployments will support both UTRAN and GSM access networks simultaneously.

Note also that the signaling example of Figure 3.41 is just one possible sequence. Many different sequences are possible depending on the exact network configuration.

3.9.2 The 3GPP Release 5 IP Multimedia Domain

Figure 3.42 shows the network architecture for a new core network domain planned for 3GPP Release 5. This architecture is described already in Chapter 2. It is important to note that this architecture represents an addition to the core network rather than a change to the existing core. Instead, 3GPP Release 5 introduces a new core network domain in addition to the established CS and PS domains. This new domain is available for new user devices that have the capability and the call-model logic needed to take advantage of the new domain. Thus the UTRAN now can be connected to three different logical core network domains—the CS domain, the PS domain, and the *IP Multimedia* (IM) domain. When a terminal wants to use the services of the core network, it

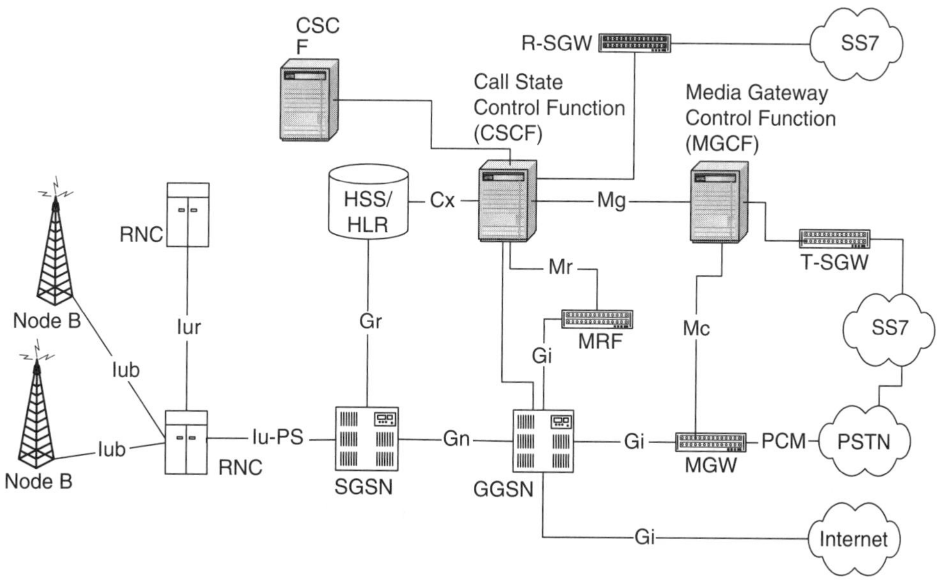

FIGURE 3.42 3GPP Release 5 IP multimedia network architecture.

indicates which domain it wants to use. Existing (pre-Release 5) terminals will continue to request the services of the CS or PS domain. New terminals also will be able to request the services of the IM domain.

Note that while the IM domain is a new domain, it uses the services of the PS domain. All IM traffic is packet-based and is transported using PS domain nodes such as the SGSN and the GGSN.

The IM domain is based on the *Session Initiation Protocol* (SIP), as described in Chapter 5. In fact, the *Call State Control Function* (CSCF) of Figure 3.42 is effectively an SIP proxy. The IM architecture enables voice and data calls to be handled in a uniform manner all the way from the UE to the destination. A complete convergence of voice and data takes place such that voice is simply a type of data with specific QoS requirements. This convergence enables a number of new advanced services. Moreover, the use of SIP means that a great deal of service control can be placed in the UE rather than in the network, making it easier for the subscriber to customize services to meet his or her particular needs.

References

Assaad, Mohamad, and Djamal Zeghlache, "On the Capacity of HSDPA," GLOBECOM 2003, IEEE, New York, August 2003.

Assaad, Mohamad, Badii Jouaber, and Djamal Zeghlache, "Effect of TCP on UMTS-HSDPA System Performance and Capacity," GLOBECOM 2004, IEEE, New York, May 2004.

Ghassane, Aniba, and Sonia Aýssa, "Resource Allocation in HSDPA Using Best-Users Selection Under Code Constraints," INRS-EMT University of Quebec Montreal (QC), Canada, September 2005.

ITU-T H.248: Media Gateway Control Protocol.

ITU-T Q.711: Functional Description of the Signaling Connection Control Part.

Jens, Voigt, Jürgen Deissner, Johannes Hübner, et al., *Optimizing HSDPA Performance in the UMTS Network Planning Process,* IEEE Press, New York, 2005.

Johansson, Klas, and Anders Furusk, *Cost-Efficient Capacity Expansion Strategies Using Multiaccess Networks,* IEEE Press, New York, 2005.

Meng, Zhang, Tao Chen, and Jiancun Hu, *Throughput-Based and Power-Based Load Control in HSUPA,* IEEE Press, New York, 2005.

Tao Chen, Huibin Lin, Zhigang Yan, et al., "Code Utilization in HSDPA," Nokia Research Center, Nokia Networks, China Mobile (CMCC), University of Stuttgart, September 2005.

3GPP TS 22.060: General Packet Radio Service (GPRS); Service Description, Stage 1.

3GPP TS 23.002: Network Architecture (Release 1999).

3GPP TS 23.002: Network Architecture (Release 4).

3GPP TS 23.002: Network Architecture (Release 5).

3GPP TS 23.003: Numbering, Addressing, and Identification.

3GPP TS 23.009: Handover Procedures.

3GPP TS 23.018: Basic Call Handing; Technical Realization.

3GPP TS 23.060: General Packet Radio Service (GPRS), Service Description, Stage 2.

3GPP TS 23.101: General UMTS Architecture (Release 1999).

3GPP TS 23.101: General UMTS Architecture (Release 4).

3GPP TS 23.107: QoS Concept and Architecture.

3GPP TS 23.108: Mobile Radio Interface Layer 3 Specification, Core Network Protocols—Stage 2.

3GPP TS 23.110: UMTS Access Stratum; Services and Functions.
3GPP TS 23.205: Bearer Independent CS Core Network; Stage 2 (Release 4).
3GPP TR 23.930: Iu Principles.
3GPP TS 24.002: GSM—UMTS Public Land Mobile Network (PLMN) Access Reference Configuration.
3GPP TS 24.007: Mobile Radio Interface Signaling Layer 3; General Aspects.
3GPP TS 24.008: Mobile Radio Interface Layer 3 Specification, Core Network Protocols—Stage 3.
3GPP TS 25.101: UE Radio Transmission and Reception (FDD).
3GPP TS 25.104: UTRA (BS) FDD; Radio Transmission and Reception.
3GPP TS 25.211: Physical Channels and Mapping of Transport Channels onto Physical Channels (FDD).
3GPP TS 25.211: Physical Channels and Mapping of Transport Channels onto Physical Channels (FDD), version 5.5.0.
3GPP TS 25.212: Multiplexing and Channel Coding (FDD).
3GPP TS 25.213: Spreading and Modulation (FDD).
3GPP TS 25.214: Physical Layer Procedures (FDD).
3GPP TS 25.215: Physical Layer—Measurements (FDD).
3GPP TS 25.301: Radio Interface Protocol Architecture.
3GPP TS 25.302: Services Provided by the Physical Layer.
3GPP TS 25.304: UE Procedures in Idle Mode and Procedures for Cell Reselection in Connected Mode.
3GPP TS 25.306: UE Radio Access Capabilities.
3GPP TS 25.306: UE Radio Access Capabilities Definition, version 5.8.0.
3GPP TS 25.308: UTRA High-Speed Downlink Packet Access (HSDPA); Overall Description; Stage 2, version 5.5.0.
3GPP TS 25.321: MAC Protocol Specification.
3GPP TS 25.322: RLC Protocol Specification.
3GPP TS 25.323: PDCP Protocol Specification.
3GPP TS 25.331: RRC Protocol Specification.
3GPP TS 25.401: UTRAN Overall Description.
3GPP TS 25.410: UTRAN Iu Interface: General Aspects and Principles.
3GPP TS 25.411: UTRAN Iu Interface: Layer 1.
3GPP TS 25.412: UTRAN Iu Interface Signaling Transport.
3GPP TS 25.413: UTRAN Iu Interface: RANAP Signaling.
3GPP TS 25.414: Iu Interface Data Transport and Transport Signaling.
3GPP TS 25.415: UTRAN Iu Interface User Plane Protocols.
3GPP TS 25.855: High-Speed Downlink Packet Access; Overall UTRAN Description, version 5.0.0.
3GPP TR 25.931: UTRAN Functions, Examples of Signaling Procedures.
3GPP TR 25.944: Channel Coding and Multiplexing Examples.
3GPP TS 29.060: General Packet Radio Service (GPRS); GPRS Tunneling Protocol (GTP) across the Gn and Gp Interfaces.
3GPP TS 33.102: Security Architecture.
www.3gpp.org
"Release 7 HSPA+ For Mobile Broadband Evolution", Qualcomm, 2007
"Understand HSPA:High-Speed Packet Access For UMTS", Qualcomm, Nov 2006

CHAPTER 4
CDMA2000

CDMA2000 is a unique radio and network access system that is part of the *International Mobile Telecommunications 2000* (IMT-2000) specification suite of access platforms that make up what is known collectively as *third generation* (3G). The IMT-2000 specification from the *International Telecommunication Union* (ITU) defines one of its platform standards that make up the 3G suite of access platforms. CDMA2000 is unique in that while supporting 3G services and bandwidth requirements, it also enables a logical migration from the existing 2G platforms to 3G without forklifting the legacy system.

The CDMA2000 suite of systems includes CDMA2000 (IS-2000) as well as *Evolution Data Optimized* (EVDO, IS-856). The interrelationship between the various standards that are commonly referred to as *Code-Division Multiple Access* (CDMA) is best shown in Figure 4.1. For the rest of this chapter, CDMA2000 will be referred to as a 1.25-MHz *Frequency-Division Duplex* (FDD) system.

The IMT-2000 specification or vision for all the platforms supported has a common set of goals that all the standards are meant to achieve. The general specifications for the IMT-2000 are as follows:

- Support high-speed data services
- Global standard
- Worldwide common frequency band
- Flexibility for evolution
- Improved spectrum efficiency
- 2 Mbps for fixed environment
- 384 kbps for pedestrian use
- 144 kbps for vehicular uses

In reviewing this list, the underlying principle is that IMT-2000 is a high-speed packet-data network designed for mobility using the *Internet Protocol* (IP) as the enabling protocol.

Some of the 3G applications that are envisioned to be enabled by CDMA2000 are as follows:

- Wireless Internet
- Wireless e-mail
- Wireless telecommuting
- Telemetry

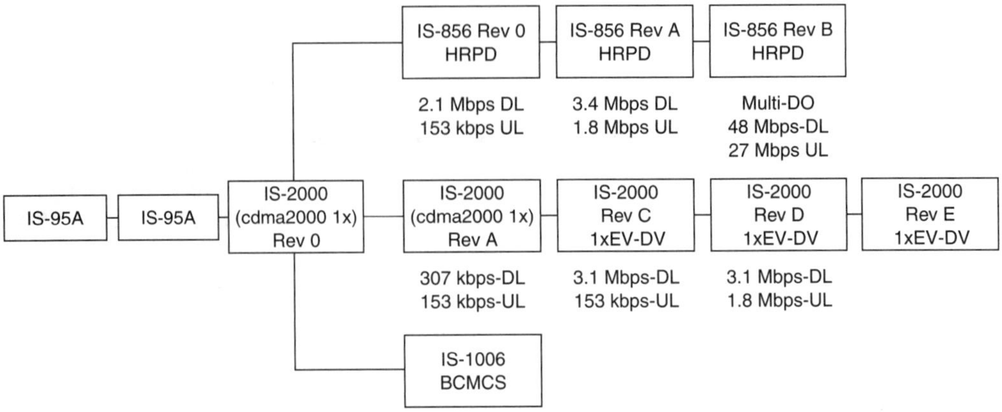

FIGURE 4.1 CDMA evolution.

- Wireless commerce
- Location-based services
- Longer standby battery life

CDMA2000 is standardized under the specification of IS-2000 and IS-856. IS-2000 (1xRTT) is backward-compatible with the IS-95A and IS-95B specifications, as well as with the J-STD-008 specifications, that collectively are called *cdmaOne.* The IS-95 and J-STD-008 specifications make up the CDMA mobility systems deployed prior to CDMA2000. CDMA2000, while being a 3G specification, is also backward-compatible with cdmaOne systems, allowing operators to make strategic deployment decisions in a graceful fashion.

Since CDMA2000 (1xRTT) is backward-compatible with existing cdmaOne networks, upgrades or, rather, changes to the network from a fixed network perspective can be done in stages. More specifically, the upgrades or changes to the network involve the *Base Transceiver Stations* (BTSs) with Multimode Channel Element cards, the *Base Station Controller* (BSC) with IP routing capabilities, and introduction of the *Packet Data Server Network* (PDSN). IS-856, also referred to as *EVDO,* is a packet-only wireless access platform that is meant to be overlaid onto an existing CDMA2000 (1xRTT) network. Alternatively, IS-856 can be deployed without a 1xRTT system being in place. The IS-856 packet service is backward-compatible with IS-2000 systems.

The radio-channel bandwidth is the same for CDMA2000-1x, IS-2000, and IS-856 as it is for existing cdmaOne channels, leading to a graceful upgrade. Of course, the subscriber units and mobile units need to be capable of supporting the CDMA2000 specification, but this can be done in a more gradual fashion because the existing cdmaOne subscriber units can use the new network.

The CDMA2000 specification stipulates the use of more than one carrier and was defined as IMT2000-MC and CDMA2000-MC. However, at this time, CDMA2000-MC is not being pursued as it was envisioned in the initial specification. As mentioned previously, several terms are used to describe CDMA2000 for the different radio-carrier platforms, some of which exist at present, whereas others are in the development phase.

However, the sequence of different CDMA2000 platforms or the migration path is as follows:

CDMA2000-1x (1xRTT)—IS-2000

1xEV-DO (IS-856)

CDMA2000-3x (3xRTT)—no longer in development

The 1xRTT and EVDO (Rev 0 and Rev A) systems use a single carrier requiring 1.25 MHz of radio spectrum, which is the same as the existing cdmaOne system's channel-bandwidth requirement. EVDO Rev B introduces the ability for EVDO to aggregate radio carriers to enhance the users data experience. EVDO Rev B is discussed in more detail later. However, the 1xRTT platforms can use a different vocoder and more Walsh codes, 256/128 versus 64, allowing for higher data rates and more voice conversions than are possible over existing cdmaOne systems. EVDO is a packet system that facilitates a host of IP applications and is a separate system from 1xRTT. Typically, an EVDO system is overlaid on a 1xRTT network, allowing 1xRTT and EVDO services to coexist.

Another important aspect of CDMA2000 is that it supports not only the IS-41 system connectivity, as does IS-95, but it also supports *Global System for Mobile* (GSM) communications and *Mobile Application Part* (GSM-MAP) connectivity requirements. This can lead to the eventual harmonization or dual-system deployment in the same market by a wireless operator wanting to deploy both *Wideband Code-Division Multiple Access* (WCDMA) and CDMA2000 concurrently.

Several key specifications are used to help define the particulars associated with a CDMA2000 system and are listed in Table 4.1.

4.1 Radio and Network Components

CDMA2000, whether 1xRTT or EVDO, requires upgrades to the radio and network architecture of the existing system. It is important to note that the migration path for a CDMA2000 operator will be from 1xRTT, with EVDO being overlaid. Currently, there is no clear harmonization path for 1xRTT and EVDO, as shown in Figure 4.1.

To understand which radio and network components are required for the successful implementation of a CDMA2000 system, whether it is 1xRTT, EVDO, or 1xRTT/EVDO, it is best to start with a simplified network layout for a cdmaOne system. Figure 4.2 shows a stand-alone cdmaOne system employing several BTSs that are homed to two BSCs. The BSCs are not collocated with the MSC in the figure but in reality could be collocated depending on the specific interconnection requirements and commercial agreements arrived at. The *Home Location Register* (HLR) is shown, but many of the supporting systems are left out of the figure for simplification purposes. The backhaul from the BTSs to the BSC and from the BSC to the MSC could be via microwave links or fixed facilities.

What follows is an example of a general CDMA2000 network, shown in Figure 4.3. The connectivity to other similar networks is not shown to keep the diagram less cluttered. In comparing Figures 4.2 and 4.3, it is apparent that new platforms are required to support the CDMA2000 network over a cdmaOne system.

What Figure 4.3 does not show are the platform additions needed for a 1xRTT and/or EVDO upgrade. However, Figure 4.4 indicates the various major platforms that either have upgrades performed or are essentially new to the CDMA2000 network as

TIA	3GPP2	Description
IS-2000-1	C.S.0001	CDMA2000 Introduction
IS-2000-2	C.S.0002	CDMA2000 Physical Layer
IS-2000-3	C.S.0003	CDMA2000 MAC Layer
IS-2000-4	C.S.0004	CDMA2000 Layer 2 LAC
IS-2000-5	C.S.0005	CDMA2000 Layer 3
IS-2000-6	C.S.0006	CDMA2000 analog
TIA/EIA-41-D		Cellular radiotelecommunications intersystem operations
IS-856	R0023	High-speed data enhancements for CDMA2000 1x—data only
TIA/EIA-97	C.S.0010	Base-station minimum standard
TIA/EIA-98	C.S.0011	Mobile station minimum performance
IS-127	C.S.0014	Enhanced Variable Rate Codec (EVRC)
TIA/EIA-637	C.S.0015	Short Message Service
TIA/EIA-683	C.S.0016	Over-the-air service provisioning
TIA/EIA-707	C.S.0017	Data services for spread-spectrum systems
TIA/EIA-733	C.S.0020	High Rate (13 kbps) speech SO
IS-801	C.S.0022	Location services (position-determination service)
IS-95A		Mobile-station–base-station compatibility standard for Dual-Mode Wideband Spread Spectrum Cellular System
IS-95B		Mobile-station–base-station compatibility standard for Dual-Mode Wideband Spread Spectrum Cellular System
	A.S.0001	Access network interfaces technical specification

Table 4.1 CDMA2000 Specifications

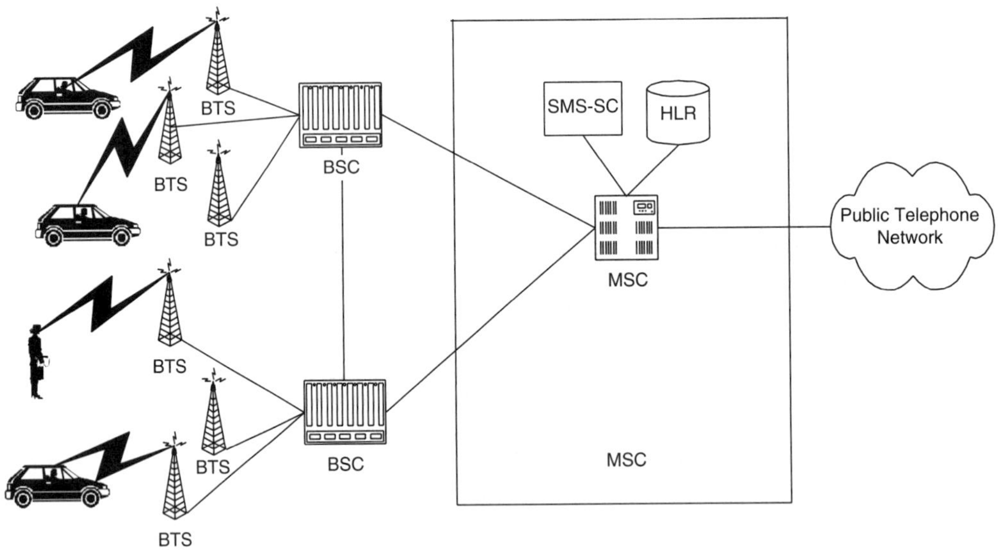

Figure 4.2 A cdmaOne simplified network.

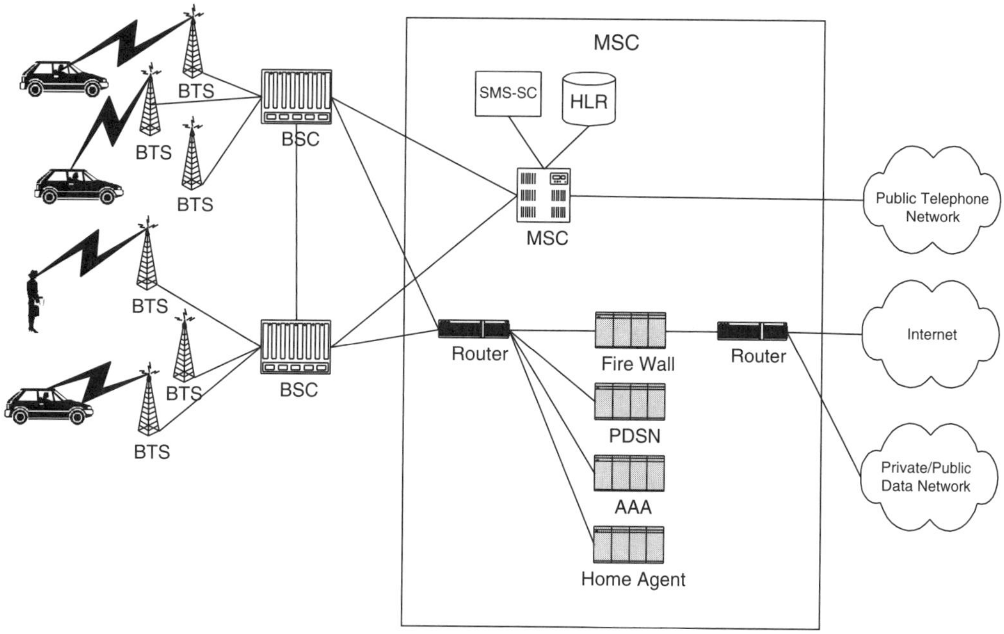

FIGURE 4.3 CDMA2000 system architecture.

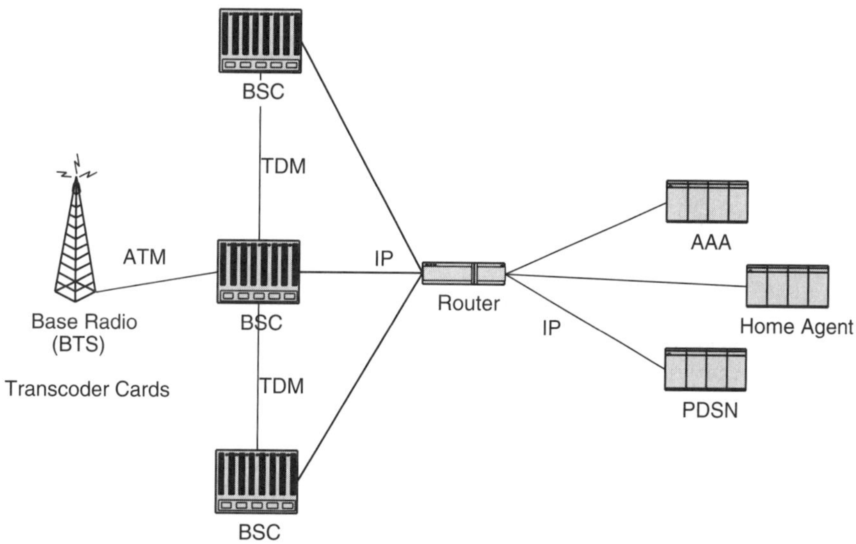

FIGURE 4.4 2.5G and 3G network element alterations.

compared with a cdmaOne system. It is important to note that an operator with a 1xRTT platform also will require upgrades for EVDO that are in the BSC, BTS, and PDSN.

The platform upgrades involve the BTS and BSC and can be facilitated by module additions or swaps depending on the infrastructure vendor that is being used.

The PDSN usually requires a firmware upgrade to support EVDO. Whether the system is new or upgrading from a cdmaOne system, the heart of the packet-data services for a CDMA2000 network is the PDSN.

4.1.1 Packet Data Serving Node (PDSN)

The PDSN is a new component associated with any CDMA2000 system as compared with cdmaOne networks. The PDSN is an essential element in the treatment of packet-data services that will be offered, and its location in the CDMA2000 network is shown in Figure 4.2. The purpose of the PDSN is to support packet-data services. Both 1xRTT and EVDO systems treat packet data in different ways, which is discussed later in this chapter. The PDSN performs the following major functions in the course of a packet-data session:

- Establishes, maintains, and terminates *Point-to-Point Protocol* (PPP) sessions with the subscriber.
- Supports both Simple and Mobile IP packet services.
- Establishes, maintains, and terminates the logical links to the *Radio Network* (RN) across the *Radio-Packet* (R-P) interface.
- Initiates *Authentication, Authorization, and Accounting* (AAA) for the mobile-station client to the AAA server.
- Receives service parameters for the mobile client from the AAA server.
- Routes packets to and from the external packet-data networks.
- Collects usage data that are relayed to the AAA server.

The overall capacity of the PDSN is determined by both the throughput and the number of PPP sessions that are being served. The specific capacity of the PDSN, of course, depends on the infrastructure vendor used as well as the particular card population that is implemented. It is important to note that capacity is only one aspect of the dimensioning process and that the overall network reliability factor must be addressed in the dimensioning process.

4.1.2 Authentication, Authorization, and Accounting (AAA)

The AAA server is another new component associated with CDMA2000 deployment. The AAA provides, as its name implies, authentication, authorization, and accounting functions for the packet-data network associated with CDMA2000 and uses the *Remote Access Dial-In User Service* (RADIUS) protocol.

The AAA server, as shown in Figure 4.3, communicates with the PDSN via IP and performs the following major functions in its role in a CDMA2000 network:

- Authentication associated with PPP and Mobile IP connections
- Authorization (service profile and security key distribution and management)
- Accounting

4.1.3 Home Agent

The *Home Agent* (HA) is the third major component to the CDMA2000 packet-data service network and should be compliant with IS-835, which is relevant to HA functionality within a wireless network. The HA performs many tasks, one of which is tracking the location of the Mobile IP subscriber as it moves from one packet zone to another. In tracking the mobile, the HA will ensure that the packets are forwarded to the mobile itself.

4.1.4 Router

The router shown in Figures 4.3 and 4.4 has the function of routing packets to and from the various network elements within a CDMA2000 system. The router is also responsible for sending and receiving packets to and from the internal network to the offnet platforms. A firewall, not shown in Figure 4.4, is needed to ensure that security is maintained when connecting to offnet data applications.

4.1.5 Home Location Register (HLR)

The HLR used in existing IS-95 networks needs to store additional subscriber information associated with the introduction of packet-data services. The HLR performs the same role for packet services as it currently does for voice services in that it stores the subscriber packet-data service options and terminal capabilities along with the traditional voice-platform needs. The service information from the HLR is downloaded in the *Visitor Location Register* (VLR) of the associated network switch during a successful registration process, the same process done in existing IS-95 systems and other 1G and 2G voice-oriented systems.

4.1.6 Base Transceiver Station (BTS)

The BTS is the official name of the cell site. It is responsible for allocating resources and both power and Walsh codes for consumption by subscribers with a 1xRTT system and time slots and modulation format for EVDO. The BTS also has the physical radio equipment that is used for transmitting and receiving the CDMA2000 signals.

The BTS controls the interface between the CDMA2000 network and the subscriber unit. The BTS also controls many aspects of the system that are directly related to performance of the network. Some of the items the BTS controls are the multiple carriers that operate from the site, the forward power (allocated for traffic overhead and soft handoffs), and of course, the assignment of Walsh codes.

With CDMA2000 systems, the use of several carriers per sector is possible, as with IS-95 systems. In addition, the use of 1xRTT and EVDO in the same site and sector is possible employing different carriers. Therefore, when a new voice or packet session is initiated, the BTS must decide how to best assign the subscriber unit to meet the services being delivered. The BTS in the decision process not only examines the service requested but also must consider the radio configuration, the subscriber type, and of course, whether the service requested is voice or packet. Thus, the resources the BTS has to draw on can be both physically and logically limited depending on the particular situation involved.

For example, with IS-2000, the BTS can perform a downgrade from a higher *Rate Code* (RC) or spreading rate to a lower RC or spreading rate if

- The resource request is not a handoff.
- The resource request is not available.
- Alternative resources are available.

Topic	Percentage	Resources
Total resources		64
Voice resources	70%	44
Data resources	30%	20
FCH resources	40%	8

TABLE 4.2 1xRTT Channel Resource-Allocation Example

In IS-2000 systems, the following are some of the physical and logical resources the BTS must allocate when assigning resources to a subscriber:

- The *Fundamental Channels* (FCHs) (the number of physical resources available)
- The FCH forward power (the power already allocated and that which is available)
- The Walsh codes required (and those available)

The physical resources the BTS draws on also involve management of the channel elements required for both voice and packet-data services. Although discussed in more detail later, handoffs are accepted or rejected on the basis of available power only with IS-2000 and on the basis of channel quality with EVDO.

Integral to the resource-assignment scheme is Walsh code management, covered in more detail later. For CDMA2000, a total of 128 Walsh codes can be drawn on, and these were expanded to a total of 256 in later releases.

For CDMA2000 1xRTT, the voice and data distribution is handled by parameters set by the operator that involve

- Data resources (percent of available resources, which includes FCH and *Supplemental Channels* [SCHs])
- FCH resources (percent of data resources)
- Voice resources (percent of total available resources)

This is best described by a brief example to help facilitate understanding of the issue of resource allocation in a 1xRTT system, as shown in Table 4.2.

Obviously, the allocation of data/FCH resources directly controls the amount of simultaneous data users on a particular sector or cell site.

4.1.7 Base-Station Controller (BSC)

The BSC is responsible for controlling all the BTSs under its domain. The BSC routes packets to and from the BTSs to the PDSN. In addition, the BSC routes *Time-Division Multiplexing* (TDM) traffic to the circuit-switched platforms, and it routes packet data to the PDSN.

4.2 Network Structure

The network structure for a CDMA2000 system that supports 2.5G and 3G has all the traditional voice elements associated with 2G wireless voice systems. However, introduction of a packet network requires the additional network equipment to provide the connectivity between the radio access network and the data network, whether it is public or private.

Obviously, or maybe it's not obvious, numerous IP network configurations can be, will be, and are being used for the support of 2.5G and 3G. The reason for many different configurations lies in the fact that the information is packet-based and therefore can be shipped between different company networks or kept localized. Of course, the issues of the required throughput and the physical interfaces also are location-dependent.

The packet network is often called the *IP network,* the *IP access network,* or the *carrier IP network* depending on the particular situation. However, the fundamental premise is that the packet network needs to support the transport and treatment of packets in the chosen configuration.

Because numerous implementation methods are available for configuring the packet network, only three main variants will be covered here. The main variants of a network configuration then can be modified to meet your particular requirements. For example, it might be best to send all Internet traffic to a local *Internet Service Provider* (ISP), and *Virtual Private Network* (VPN) applications, depending on the treatment required, can be brought to one centralized location for distribution on, say, an *Asynchronous Transfer Mode* (ATM) network when connection to a corporate *Local-Area Network* (LAN) is required.

The three main variants in configuring a CDMA2000 network are as follows:

- Distributed
- Regional
- Centralized

The regional and centralized variants are similar in concept, except that the centralized variant is an aggregation of several potential regional networks. Some of the determinations for deciding on which variant to implement are based on the following issues:

- Services supported
- Traffic volume
- Location of PDSN
- Commercial interconnection agreements
- Network reliability and availability

Regardless of which configuration is used, it will need a router for the backbone and, of course, a gateway for all the offnet service delivery and reception.

The centralized configuration is the most predominant network structure deployed presently. However, with the introduction of *IP Multimedia Service* (IMS) and *Voice over IP* (VoIP) as a fundamental service platform, employing an all-IP network enables an operator to realize distributed architecture.

Legacy systems rely on a hub-and-spoke configuration, whereas an IMS method using IP no longer requires the network platforms to be centrally located. The distribution of equipment and network resources allow for better service treatment, lower cost, and of course, better disaster recovery.

4.3 Packet-Data Transport Process Flow

CDMA2000 data services fall within two distinct categories: circuit-switched and packet. Circuit-switched data are handled the same as a voice call. For all packet-data calls with 1xRTT and EVDO, however, a PDSN is used as the interface between air-interface data transport and fixed-network transport. The PDSN interfaces to the *Base Station* (BS) through a *Packet Control Function* (PCF) that can be collocated with the BS.

CDMA2000 has three packet-data service states that need to be understood in the process:

- *Active/Connected.* Here, a physical traffic channel exists between the subscriber unit and the BS, with packet data being sent and received in a bidirectional fashion.
- *Dormant.* No physical traffic channel exists, but a PPP link between the subscriber unit and the PDSN is maintained.
- *Null/Inactive.* Neither a traffic channel nor a PPP link is maintained or established.

The relationship between the three packet-data states is best shown by the simplified state diagram in Figure 4.5.

CDMA2000 introduces to the mobility environment real packet-data transport and treatment at speeds that meet or exceed IMT-2000 system requirements. The voice call processing that is implemented by CDMA2000 is functionally the same as that of existing cdmaOne networks, with the exception that a vocoder change exists in the subscriber units. However, the key difference is that packet data now can be handled by the network.

The mobile unit initiates the decision as to whether the session will be a packet-data session, a voice session, or a concurrent session, meaning both voice and data. The network at this time cannot initiate a packet-data session with the subscriber unit, with the exception of the *Standard Management System* (SMS), which does require a packet-data session.

For call processing, the voice and data networks are segregated in general once the information, whether it is voice or data, leaves the radio environment at the BSC itself.

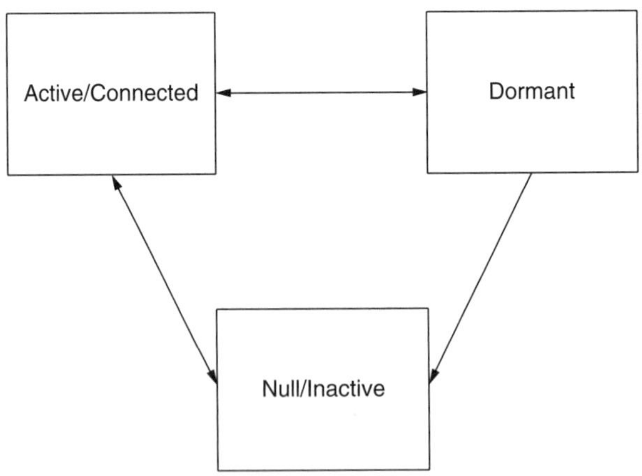

FIGURE 4.5 Packet-data states.

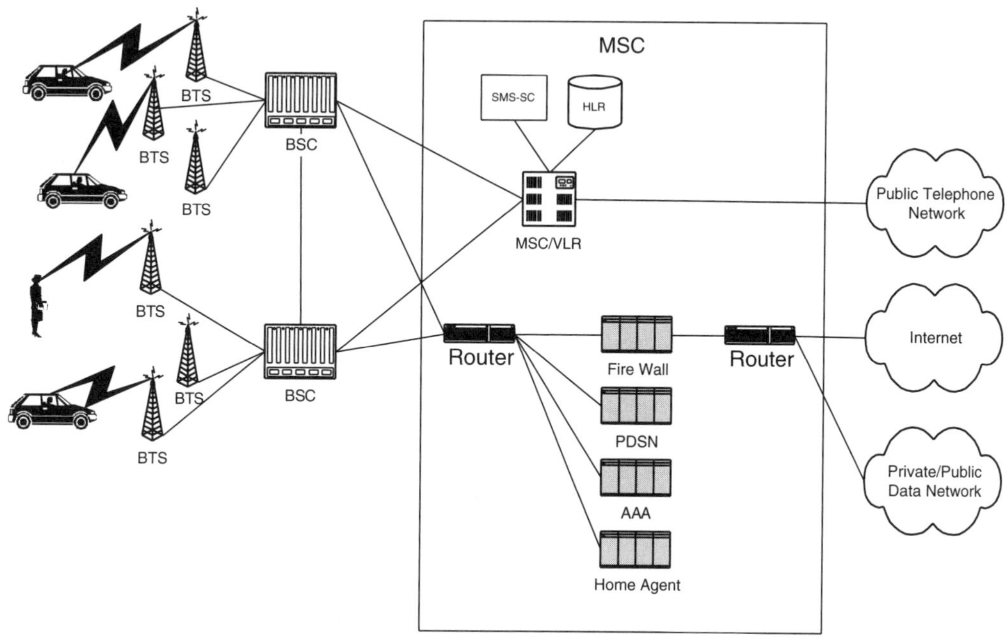

FIGURE 4.6 CDMA2000 network architecture.

Therefore, for packet data, the PDSN is central to all decisions. Figure 4.6 shows the general network architecture.

The PDSN does not communicate directly with voice-network nodes such as the HLR and VLR; instead, it is done via the AAA. As discussed previously, the voice and data networks normally are segregated once they leave the radio environment at the BSC. Additionally, in a CDMA2000 network, the system uses PPP between the mobile unit and the PDSN for every type of packet data that is transported and/or treated.

The PDSN is meant to provide several key packet-data services, including Simple IP and Mobile IP. Also, several variants, to be discussed shortly, are relative to each of these services. However, the concepts behind Simple IP and Mobile IP need to be explored first.

Simple IP is a packet-data service relative to CDMA2000 1xRTT and is where the subscriber is assigned a *Dynamic Host Configuration Protocol* (DHCP) address from the serving PDSN with its routing service provided by the local network. The specific IP address that the subscriber is assigned remains with the subscriber as long as it is served by the same radio network that maintains connectivity with the PDSN that issued the IP address. It is important to note that Simple IP does not provide for mobile terminations and therefore is an origination-based service only, that is, a PPP service using the DHCP.

In Mobile IP, the public IP network provides the mobile unit's IP routing service. In this functionality, the mobile unit is assigned a static IP address that resides with the HA. A key advantage of Mobile IP over Simple IP is that the mobile unit, owing to the static IP address, can hand off between different radio networks that are served via different PDSNs, which resolves the roaming issues that are part of Simple IP. Mobile IP, owing to the static IP address, also enables the possibility for mobile terminations.

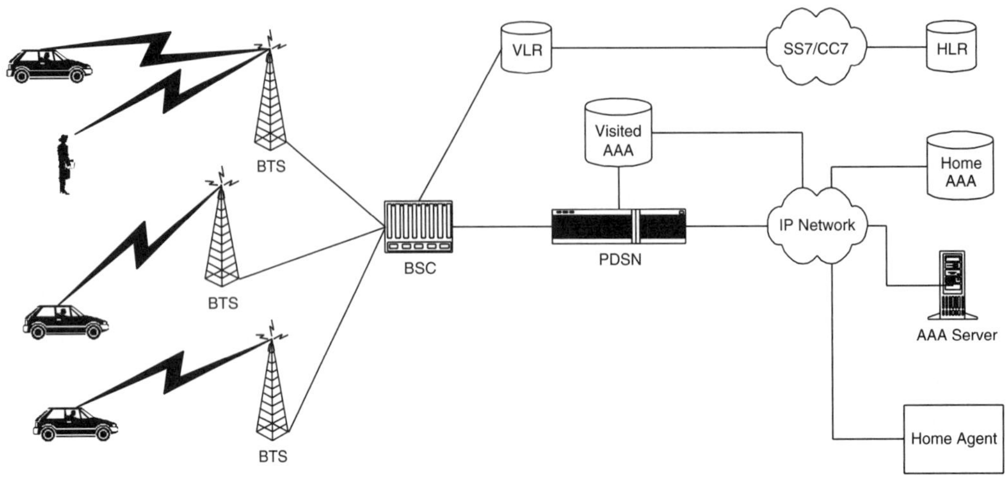

Figure 4.7 CDMA2000 packet-network nonhome.

Now, whether the packet service is Simple or Mobile IP, the notion of mobility is fundamental to the concept of CDMA2000. Figure 4.7 illustrates some of the internetwork communication that needs to take place to establish a packet-data session.

It is important to note that transport of the packets is not depicted in Figure 4.7. The figure shows just the elements in the network that need to communicate in order to establish which services the subscriber is allowed to have and how the network is going to meet the *Service-Level Agreement* (SLA) that is expected for the packet session.

The VLR is normally collocated with the MSC, as shown in Figure 4.7. When a subscriber initiates a packet-data session, the BSC via the MSC/VLR checks the subscriber subscription information prior to the system granting the service request to the mobile subscriber. This will take place prior to the PDSN being involved with the packet session.

The various packet sessions available for use within a CDMA2000 network are, of course, Simple IP and Mobile IP. However, along with each of these packet-session types are two variants, one of which uses a VPN and the other of which does not:

- Simple IP
- Simple IP with VPN
- Mobile IP
- Mobile IP with VPN

A more specific discussion of Simple and Mobile IP occurs in the following subsections.

4.3.1 Simple IP

Simple IP is similar to the dial-up Internet connections used by many people over standard landline facilities. In Simple IP, a PPP session is established between the mobile unit and the PDSN. The PDSN basically routes packets to and from the mobile unit in order to provide end-to-end connectivity between the mobile unit and the Internet. Figure 4.8 presents a diagram depicting Simple IP.

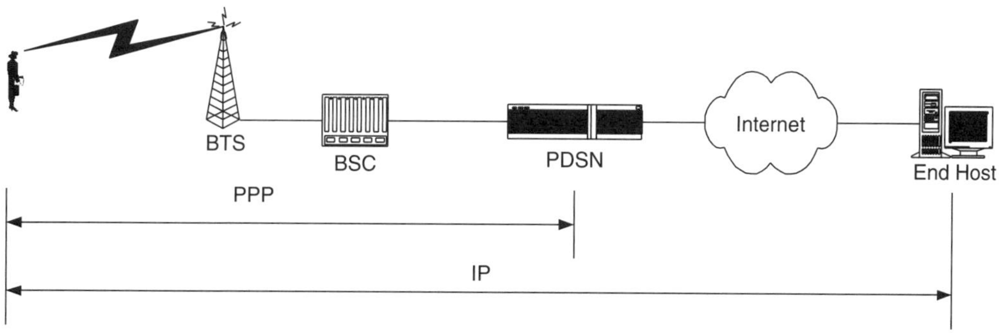

FIGURE 4.8 Simple IP.

When using Simple IP, the mobile unit must be connected to the same PDSN for the duration of the packet session. If the mobile unit, while in transit, moves to a coverage area whose BSC/BTSs are homed out of another PDSN, the Simple IP connection is lost and needs to be reestablished. The loss of the existing packet session effectively is the same as when the Internet connection on the landline is terminated and you need to reestablish the connection.

Let's look at Figure 4.7 again. Many of the details are left out, but the concept shows that the mobile unit is connected to the PDSN using a PPP connection in a best-effort data-delivery method at the agreed-on transfer rate. The transfer rate is determined by the subscriber profile, the radio resource availability, and the radio environment itself.

The IP address of a mobile unit is linked to the PDSN, which can be static or DHCP; for Simple IP, the choice is DHCP. A mobile unit with an active or dormant data call can transverse the network, going from cell to cell, provided that it stays within the PDSN's coverage area. Additionally, the PDSN should support both the *Challenge Handshake Authentication Protocol* (CHAP) and the *Password Authentication Protocol* (PAP).

Simple IP, as indicated, does not enable the subscriber full mobility with packet-data calls. When the subscriber exits the PDSN coverage area, it must negotiate for a new IP address from the new PDSN, which, of course, results in the termination of the existing packet session and requires a new session to begin.

Regarding the radio environment, the CDMA2000 radio network provides the mobile unit with a traffic channel that consists of a fundamental channel and possibly a supplemental channel for higher traffic speeds. To help explain the Simple IP process, a call flow or packet-session flowchart is given in Figure 4.9, which represents a subscriber operating in his or her home PDSN network. In Figure 4.10, the mobile unit is considered to be roaming.

4.3.2 Simple IP with VPN

An enhancement to Simple IP is the ability to introduce a VPN to the path for security and also to provide connectivity to a corporate LAN or other packet networks. With VPN, the mobile user should have the appearance of being connected directly to the corporate LAN.

The PDSN establishes a tunnel using the *Layer Two Tunneling Protocol* (L2TP) between the PDSN and the private data network. The mobile unit is effectively still

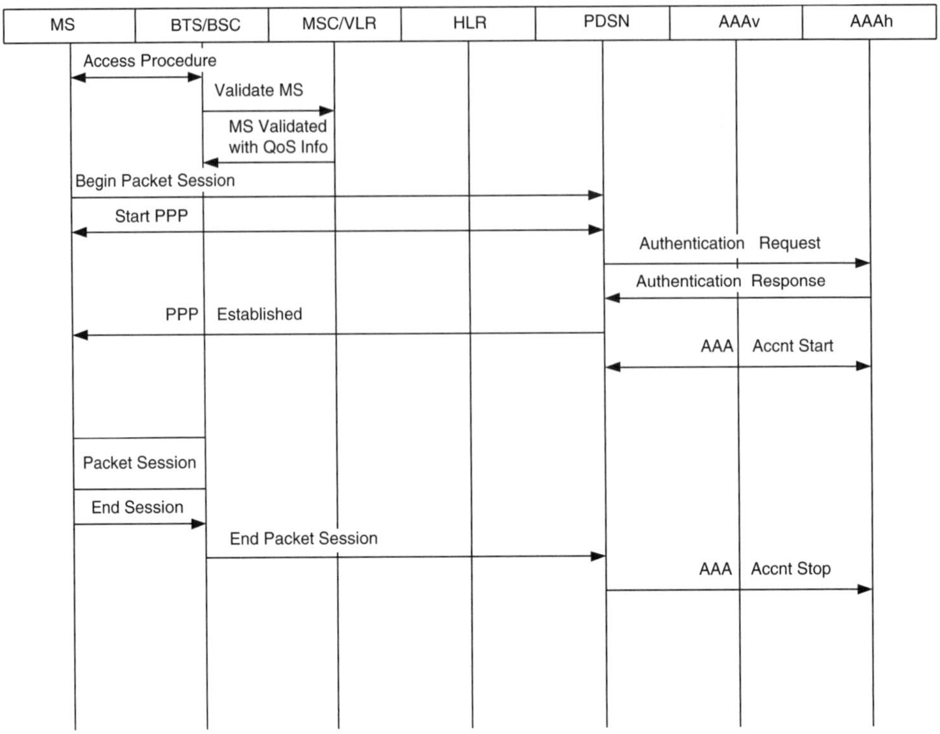

FIGURE 4.9 Simple IP flowchart.

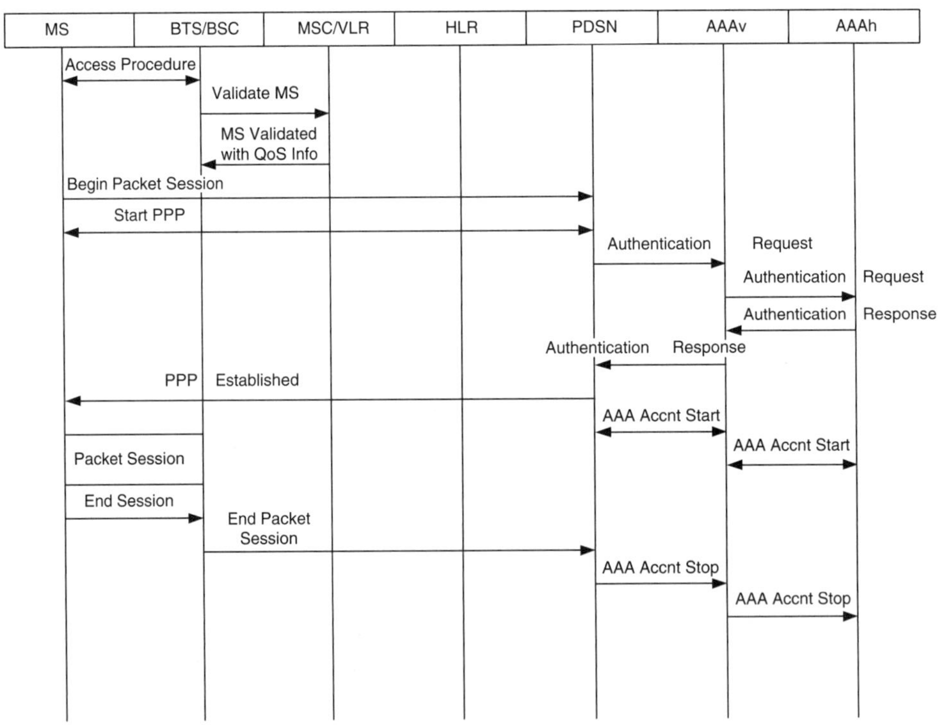

FIGURE 4.10 Simple IP roaming flowchart.

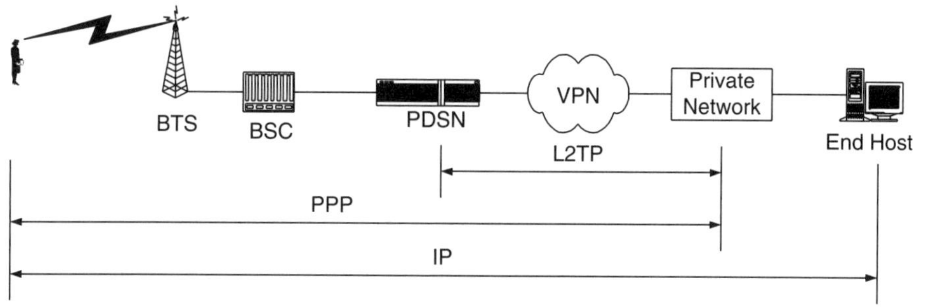

FIGURE 4.11 Simple IP with VPN.

using a PPP connection, but it is tunneled. The private network that the PDSN terminates to is responsible for assigning the IP address and, of course, authenticating the user beyond what the wireless system needs to perform for billing purposes.

Because of the specific termination and authentication that are performed by another network, the PDSN does not apply any IP services for the mobile, and all the system can provide is the predetermined speed of the connection.

Just as with Simple IP, the mobile unit still must be connected to the same PDSN for the packet session. If the mobile unit moves to another area of the network that is covered by a separate PDSN, the VPN is terminated, and the mobile unit must reestablish the session. A simplified diagram is shown in Figure 4.11.

The packet-session flowchart for Simple IP with VPN is shown in Figure 4.12 and assumes that the subscriber is not roaming.

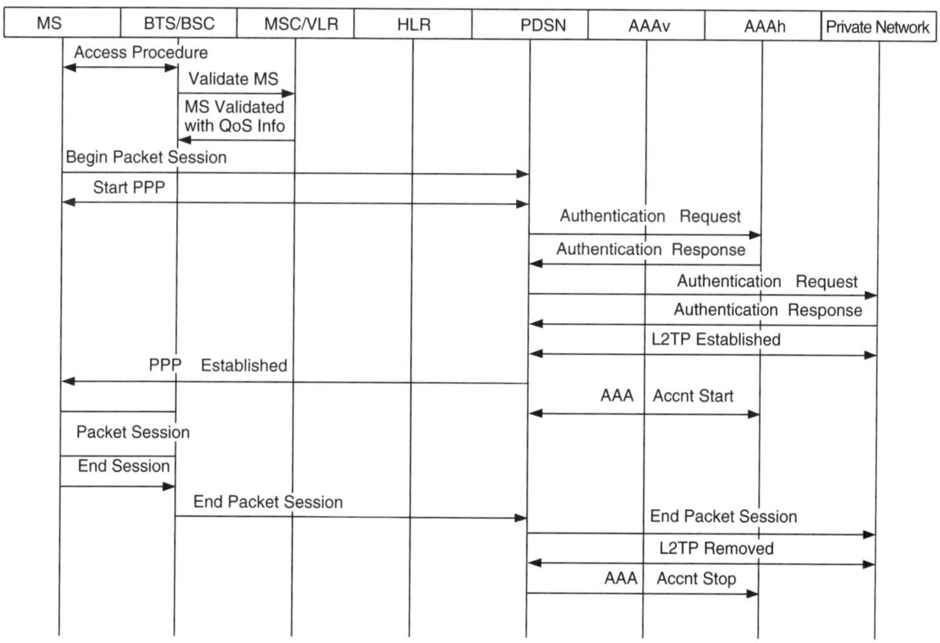

FIGURE 4.12 Simple IP with VPN flowchart.

4.3.3 Mobile IP (3G)

Mobile IP, while still a packet-transport method, is quite different from Simple IP in that it actually transports the data. Mobile IP uses a static IP address that can be assigned by the PDSN. The establishment of a static IP address facilitates roaming during the packet session, provided that the static IP address scheme is unique enough for the subscriber unit to be uniquely identified.

With Mobile IP, the PDSN is the *Foreign Agent* (FA), and the *Home Agent* (HA) is set up as a virtual HA. The mobile needs to register each time it begins a packet-data session, whether it is originating or terminating. Also, the PDSN on the visited network terminates the packet session using an IP-in-IP tunnel. The HA delivers the IP traffic to the FA through an IP tunnel.

The mobile unit is responsible for notifying the system that it has moved to another service area. Once the mobile unit has moved to another service area, it needs to register with another FA. The FA assigns the mobile unit a *Care-of Address* (COA).

The HA forwards the packets to the visited network for termination on the mobile unit. The HA encapsulates the original IP packet destined for the mobile unit using the COA. The FA, using IP-in-IP tunneling, extracts the original packet and routes it to the mobile unit.

The IP address assignment is done via DHCP and is mapped to the HA. However, PAP and CHAP are not used for Mobile IP as they are in Simple IP. In the reverse direction, the routing of IP packets occurs the same as on the home network and does not require an IP-in-IP tunnel unless the wireless operator decides to implement reverse IP tunneling.

In summary,

- The PDSN in the visited network always terminates the IP-in-IP tunnel.
- The HA delivers the IP traffic through the Mobile IP tunnel to the FA.
- The FA performs the routing to the mobile unit and assigns the IP address using DHCP.

Figure 4.13 is a simplified depiction of Mobile IP, and Figure 4.14 is an example of Mobile IP packet-session flow.

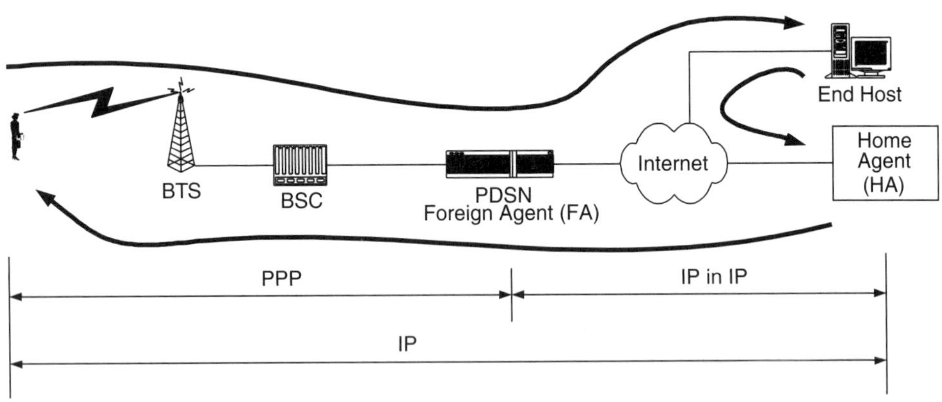

Figure 4.13 Mobile IP.

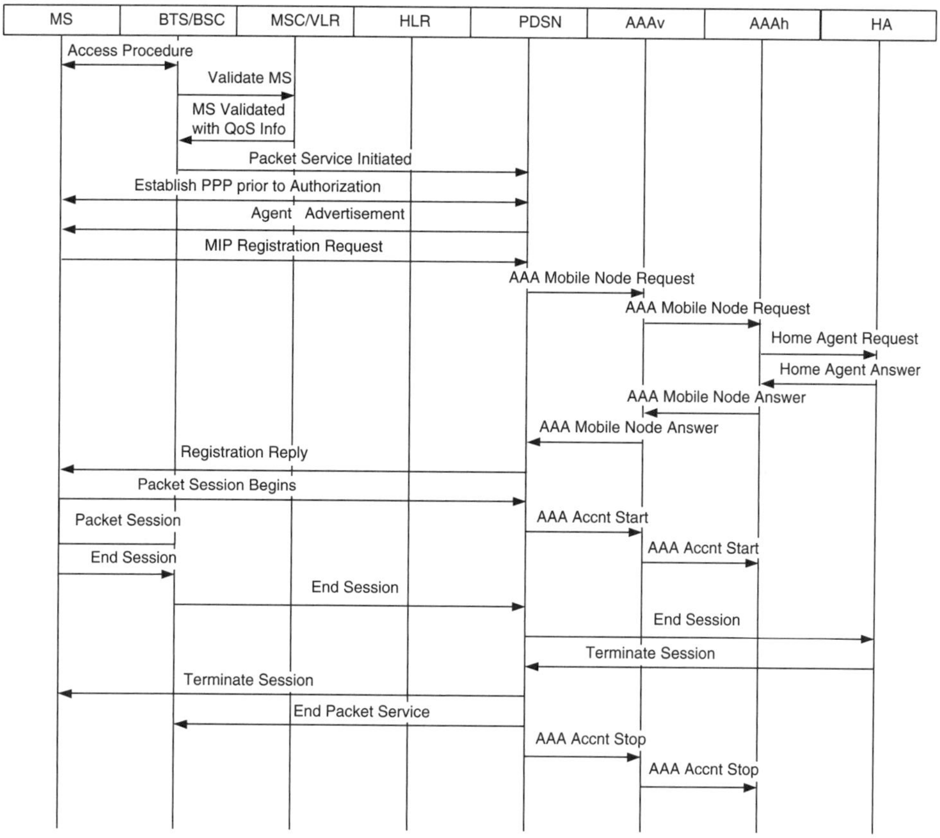

FIGURE 4.14 Mobile IP packet-session flow.

4.3.4 Mobile IP with VPN

The second variant to Mobile IP is Mobile IP with VPN. Mobile IP with VPN affords greater mobility for subscribers over Simple IP with VPN because it can maintain a session when the mobile unit moves from one PDSN area to another. As with Mobile IP, the IP address assigned to the subscriber is static; however, the private network to which the mobile unit is connected provides the IP address, which needs to be drawn from a predefined IP scheme that is coordinated. The PDSN provides a COA when operating in a nonhome PDSN for routing purposes. However, the IP packets in both directions flow between the HA and the FA using IP-in-IP encapsulation, and no treatment, with the exception of the throughput speed allowed, is performed by the wireless network. Figure 4.15 depicts the general packet flow for Mobile IP with VPN.

4.4 Radio Network (IS-2000 1xRTT)

The radio network for a CDMA2000 system has several enhancements over existing IS-95/J-STD-008 wireless systems. These enhancements involve better power control, diversity transmitting, modulation-scheme changes, new vocoders, uplink pilot channels,

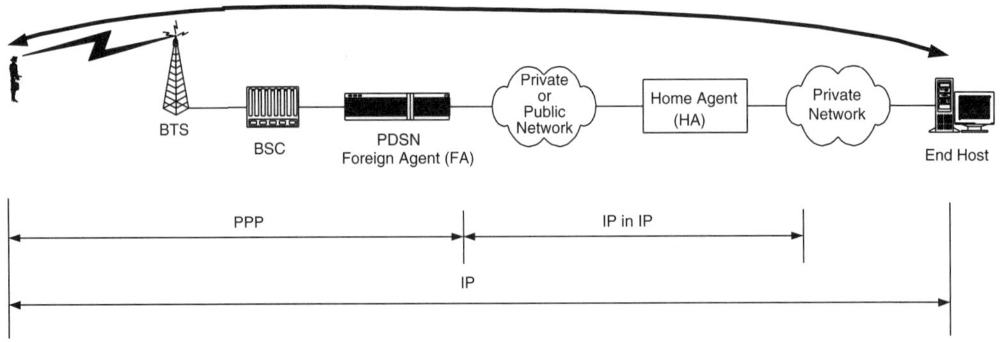

FIGURE 4.15 Mobile IP with VPN.

expansion of the existing Walsh codes, and channel-bandwidth changes and packet-data capability. The CDMA2000 radio system can include 1xRTT and EVDO radio access methods. The IS-2000 specification (1xRTT) is designed to provide an existing cdmaOne operator with a phased entrance into the 3G arena. EVDO (IS-856) is designed for packet data only and has several versions that build on each other, producing faster downlink and uplink data speeds.

The CDMA2000 (IS-2000) radio network, phase 1, implementation, also called *CDMA2000 1xRTT,* uses the same 1.25-MHz channel bandwidth as IS-95/J-STD-008 systems. In the specification, CDMA2000 phase 2, referred to as *CMDA2000 3xRTT,* is where multiple carriers now are used. A brief and simplified channel-bandwidth diagram is shown in Figure 4.16, which illustrates the radio-carrier differences between a CDMA IS-95, 1xRTT, and a 3xRTT system.

CDMA2000 (IS-2000) introduces several new channel types for the radio-access scheme. The new channel types are implemented in 1xRTT and are introduced to support high-speed data as well as enhanced paging functions. To accomplish the higher

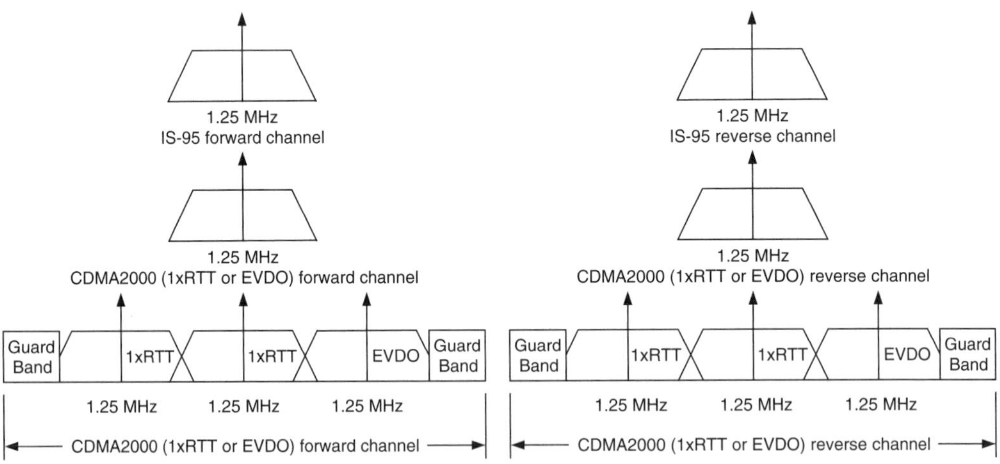

FIGURE 4.16 Radio-channel bandwidth.

data rates, CDMA2000 uses a combination of expanded Walsh codes and modulation and vocoder changes.

As depicted in Figure 4.16, a wireless operator can migrate to CDMA2000 from either IS-95A or IS-95B platforms using the same amount of existing spectrum when transitioning to a 1xRTT format. The common migration paths for implementing CDMA2000 are relative to operators using cdmaOne (IS-95A/B) platforms:

- cdmaOne (IS-95A)—CDMA2000 (1xRTT)—CDMA2000 (EVDO)
- cdmaOne (IS-95A)—cdmaOne (IS-95B)—CDMA2000 (1xRTT)—CDMA2000 (EVDO)

The CDMA2000 (1xRTT) radio-access scheme has several enhancements over the existing IS-95 systems, and they are as follows:

- Forward link:
  - Fast power control
  - *Quadrature Phase Shift* (QPS) keying modulation rather than dual *Binary Phase Shift* (BPS) keying
- Reverse link:
  - Pilot signal to enable coherent demodulation for the reverse link
  - *Hybrid Phase Shift* (HPS) keying spreading in the reverse link

Table 4.3 shows the various relationships between the IS-95 and CDMA2000 radio channels.

Platform	Description
IS-95A	Primarily voice with circuit switch speeds of 9600 bps or 14.4 kbps 64 Walsh codes SR1 (1.2288 Mbps)
IS-95B	Primarily voice, data on forward link, improved handoff 64 Walsh codes SR1 (1.2288 Mbps)
CDMA2000—phase 1 (1xRTT)	SR1 (1.2288 Mbps), voice and data (packet data via separate channel) 128 Walsh codes Closed-loop power control
CDMA2000—phase 2 (3xRTT)	SR3 (3.6864 Mbps) Data primarily Higher data rate 256 Walsh codes
EVDO (Rev 0/A)	(SR1) 1.2288 Mbps Packet data Closed power control

Table 4.3 CDMA2000

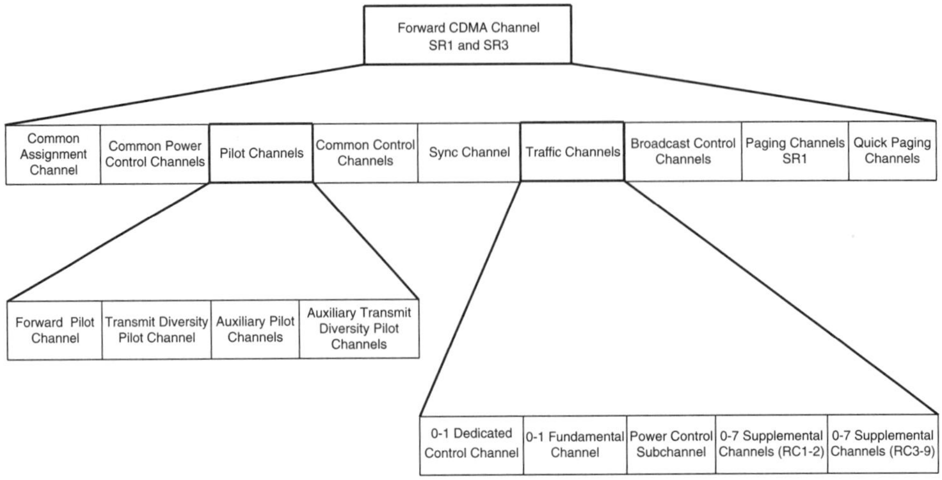

FIGURE 4.17 A forward CDMA channel transmitted by base station.

4.4.1 Forward Channel

The forward link for a CDMA2000 channel, whether for 1x or 3x implementation, uses the structure shown in Figure 4.17.

Reviewing the channel structure, the base station transmits multiple common channels as well as several dedicated channels to subscribers in their coverage areas. Each CDMA2000 user is assigned a forward traffic channel that consists of the following combinations (an important point to note is that F-FCHs are used for voice, whereas F-SCHs are for data):

1 *Forward Fundamental Channel* (F-FCH)
0–7 *Forward Supplemental Code Channels* (F-SCHs) for both RC1 and RC2
0–2 *Forward Supplemental Code Channels* (F-SCHs) for both RC3 and RC9

When the channel is associated with a 3xRTT implementation, the data for the subscriber are mapped to each of the three different carriers, enabling high throughput. However, the Walsh codes are the same for each carrier, meaning that they share the same throughput, distributing the traffic load evenly.

The CDMA2000 channel uses different modulation schemes depending on the radio configuration employed. The radio configurations are described later. However, the modulation scheme used for RC1 and RC2 is *Binary Phase Shift Keying* (BPSK), whereas *Quadrative Phase Shift Keying* (QPSK) is used for RC3 through RC9. For RC3 through RC9, the data are converted into a 2-bit-wide parallel data stream that initially would seem counterintuitive because it reduces the data rate for each stream by a factor of 2. Each data stream, however, is then spread by a 128 Walsh code to get the spreading rate up to 1.2288 Mbps, which effectively doubles the processing gain, allowing for greater throughput at the same effective power level.

The following are some forward-channel descriptions:

- *Forward Supplemental Channel* (F-SCH). Up to two F-SCHs can be assigned to a single mobile unit for high-speed data ranging from 9.6 to 153.6 kbps in Release 0 and 307.2 and 614.4 kbps in Release A. It is important to note that each F-SCH assigned can be assigned at different rates. The F5 SCH must be assigned with a *Reverse Supplemental Channel* (R-SCH) when only one F-SCH is assigned.

- *Forward Quick Paging Channel* (F-QPCH). The quick-paging channel enables mobile unit battery life extension by reducing the amount of time the mobile unit spends parsing pages that are not meant for it. The mobile unit monitors the F-QPCH, and when the flag is set, the unit looks for the paging message. There are a total of three F-QPCHs per sector.
- *Forward Dedicated Control Channel* (F-DCCH). This replaces the dim and burst and the blank and burst. It is used for messaging and control for data calls.
- *Forward Transmit Diversity Pilot Channel* (F-TDPICH). This is used to increase RF capacity.
- *Forward Common Control Channel* (F-CCCH). This is used to send paging, data messages, or signaling messages.

Table 4.4 helps to quantify the channel types and their number for CDMA2000, both 1x and 3x.

Channel Type (SR1)	**Maximum Number**
Forward Pilot Channel	1
Transmit Diversity Pilot Channel	1
Sync Channel	1
Paging Channel	7
Broadcast Control Channel	8
Quick Paging Channel	3
Common Power Control Channel	4
Common Assignment Channel	7
Forward Common Control Channel	7
Forward Dedicated Control Channel	1 per Forward Traffic Channel
Forward Fundamental Channel	1 per Forward Traffic Channel
Forward Supplemental Code Channel (RC1 and RC2 only)	7 per Forward Traffic Channel
Forward Supplemental Channel (RC3, RC4 and RC5 only)	2 per Forward Traffic Channel
Channel Type (SR3)	**Maximum Number**
Forward Pilot Channel	1
Sync Channel	1
Broadcast Control Channel	8
Quick Paging Channel	3
Common Power Control Channel	4
Common Assignment Channel	7
Forward Common Control Channel	7
Forward Dedicated Control Channel	1 per Forward Traffic Channel
Forward Fundamental Channel	1 per Forward Traffic Channel
Forward Supplemental Channel	2 per Forward Traffic Channel

TABLE 4.4 Forward and Reverse CDMA2000 Channel Descriptions

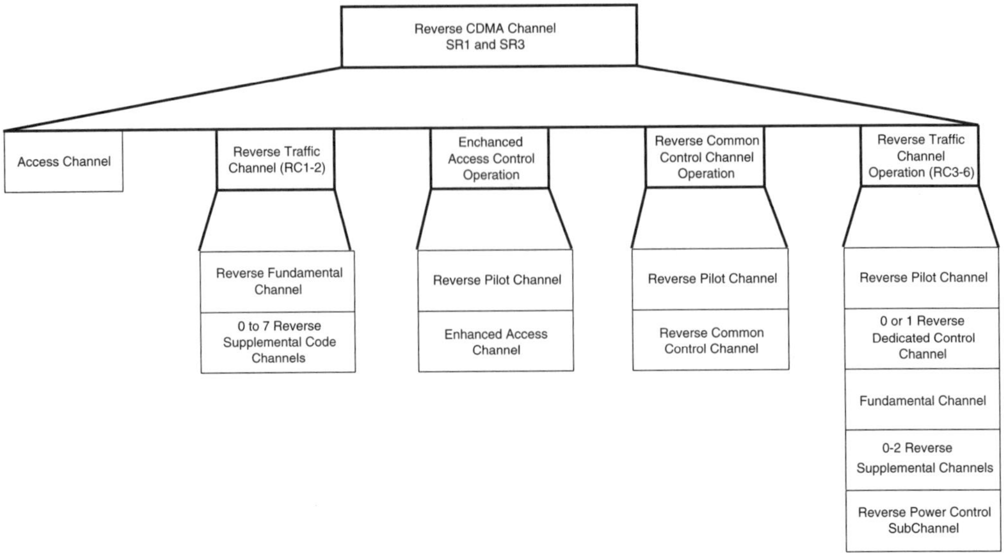

FIGURE 4.18 A reverse CDMA channel received at base station.

4.4.2 Reverse Channel

The reverse link or channel for CDMA2000 has many similar properties to the forward link and therefore differs significantly from that used in IS-95. One of the major differences, or rather, enhancements to CDMA2000 over IS-95 is the inclusion of a pilot on the reverse link. The structure of the reverse channel for CDMA2000 is shown in Figure 4.18.

Elaborating on the reverse channel, the subscriber, or mobile unit, is allowed to transmit more than one code channel to accommodate the high data rates. The minimum configuration consists of a *Reverse Pilot* (R-Pilot) channel to enable the base station to perform synchronous detection and a *Reverse Fundamental Channel* (R-FCH) for voice. The additional channels, such as the *Reverse Supplemental Channels* (R-SCHs) and the *Reverse Dedicated Control Channel* (R-DCCH), can be used to send data or signaling information. The association between the radio configuration and the spreading rates is shown in Table 4.7. It is important to note that the reverse channel for 3x is different from that in 1x in that it is a direct spread but can be overlaid over a 1x implementation. Depending on the subscribers operating in that sector, the appropriate SR and RC then are selected.

The following are some of the reverse link channel descriptions:

- *Reverse Supplemental Channel* (R-SCH). When data rates are greater than 9.6 kbps, an R-SCH is required, and an R-FCH is also assigned for power control. A total of one or two R-SCHs can be assigned per mobile unit.
- *Reverse Pilot Channel* (R-PICH). The R-PICH provides pilot and power-control information. The R-PICH enables the mobile unit to transmit at a lower power level and allows it to inform the base station of the forward power levels being received, enabling the base station to reduce power.
- *Reverse Dedicated Control Channel* (R-DCCH). This replaces the dim and burst and the blank and burst. It is used for messaging and control for data calls.

Channel Type (SR1)	Maximum Number
Reverse Pilot Channel	1
Access Channel	1
Enhanced Access Channel	1
Reverse Common Control Channel	1
Reverse Dedicated Control Channel	1
Reverse Fundamental Control Channel	1
Reverse Supplemental Code Channel (RC1 and RC2 only)	7
Reverse Supplemental Channel	2
Channel Type (SR3)	**Maximum Number**
Reverse Pilot Channel	1
Enhanced Access Channel	1
Reverse Common Control Channel	1
Reverse Dedicated Control Channel	1
Reverse Fundamental Control Channel	1
Reverse Supplemental Channel	2

TABLE 4.5 CDMA2000 Channel Types

- *Reverse Enhanced Access Channel* (R-EACH). This is meant to minimize collisions and therefore reduce the access channel's power.
- *Reverse Common Control Channel* (R-CCCH). Used by mobile units to send their data information after they have been granted access.

Table 4.5 helps to quantify the reverse channel types and their number for CDMA2000, both 1x and 3x.

4.4.3 SR and RC

CDMA2000 defines two spreading rates, referred to as *Spreading Rate 1* (SR1) and *Spreading Rate 3* (SR3). SR1 is used for IS-95A/B and CDMA2000 phase 1, 1xRTT implementations, whereas SR3 is destined for CDMA2000 phase 2, 3xRTT.

For CDMA2000, SR1 has a chip rate of 1.2288 Mbps and occupies the same bandwidth as cdmaOne signals. SR1 is a direct-spread method that follows the same concept as that used for IS-95 systems. However, for 3xRTT, an SR3 signal is introduced that has a rate of 3.6864 Mbps (3×1.2288 Mcps) and therefore occupies three times the bandwidth of a cdmaOne or 1xRTT channel. The SR3 system incorporates all the new coding implemented in an SR1 system while supporting even higher data rates. The 3xRTT channel scheme uses a multicarrier forward link and a direct-spread reverse link.

The IS-2000 specification also defines for both 1xRTT and 3xRTT radio access methods a total of nine forward and six reverse link radio configurations, as well as two different spreading rates. The radio configurations involve different modulations, coding, and vocoders, whereas the spreading rates address the usage amount of two different chip rates. The radio configurations are referred to as *Radio Configuration 1* (RC1).

RC1 is backward-compatible with cdmaOne for 9.6-kbps voice traffic and supports circuit-switched data rates of 1.2 to 9.6 kbps. RC3 is based on the 9.6-kbps rate and

Forward RC	SR	Data Rates	Characteristics
1	1	1200, 2400, 4800, 9600	$R = 1/2$
2	1	1800; 3600; 7200; 14,400	$R = 1/2$
3	1	1500; 2700; 4800; 9600; 38,400; 76,800; 153,600	$R = 1/4$
4	1	1500; 2700; 4800; 9600; 38,400; 76,800; 153,600; 307,200	$R = 1/2$
5	1	1800; 3600; 7200; 14,400; 28,800; 57,600; 115,200; 230,400	$R = 1/4$
6	3	1500; 2700; 4800; 9600; 38,400; 76,800; 153,600; 307,200	$R = 1/6$
7	3	1500; 2700; 4800; 9600; 38,400; 76,800; 153,600; 307,200; 614,300	$R = 1/3$
8	3	1800; 3600; 7200; 14,400; 28,800; 57,600; 115,200; 230,400; 460,800	$R = 1/4$ (20 ms) $R = 1/3$ (5 ms)
9	3	1800; 3600; 7200; 14,400; 28,800; 57,600; 115,200; 230,400; 460,800; 1,036,800	$R = 1/2$ (20 ms) $R = 1/3$ (5 ms)

Table 4.6 Forward Link RC and SR [16]

supports variable voice rates from 1.2 to 9.6 kbps while also supporting packet-data rates of 19.2, 38.4, 76.8, and 153.6 kbps, but it operates using an SR1.

Tables 4.6 and 4.7 help to illustrate the perturbations that exist with the different radio configurations and spreading rates. Table 4.6 is associated with the forward link, whereas Table 4.7 is associated with the reverse link.

Reverse Link RC	SR	Data Rates	Characteristics
1	1	1200, 2400, 4800, 9600	$R = 1/3$
2	1	1800; 3600; 7200; 14,400	$R = 1/2$
3*	1	1200; 1350; 1500; 2400; 2700; 4800; 9600; 19,200; 38,400; 76,800; 153,600; 307,200	$R = 1/4$ $R = 1/2$ for 307,200
4*	1	1800; 3600; 7200; 14,400; 28,800; 57,600; 115,200; 230,400	$R = 1/4$
5*	3	1200; 1350; 1500; 2400; 2700; 4800; 9600; 19,200; 38,400; 76,800; 153,600; 307,200; 614,400	$R = 1/4$ $R = 1/2$ for 307,200 and 614,400
6*	3	1800; 3600; 7200; 14,400; 28,800; 57,600; 115,200; 230,400; 460,800; 1,036,800	$R = 1/4$ $R = 1/2$ for 1,036,800
*Reverse pilot.			

Table 4.7 Reverse Link RC and SR [16]

4.4.4 Power Control

Power control is a major enhancement of CDMA2000 over IS-95 that enables higher data rates. The primary power-control enhancement is fast-forward link power control.

As discovered through practical implementation issues, CDMA systems are interference-limited, and reducing the interference results in an improvement in system capacity. Enabling better power control of both the forward and reverse links has several advantages:

- System capacity is enhanced or optimized.
- Mobile battery life is extended.
- Radio path impairments are compensated for properly or better.
- *Quality of Service* (QoS) at various bit rates can be maintained.

Obviously, with any wireless system that is interference-limited, it is important to ensure that all transmitters, whether mobile or located at a base station, transmit at the lowest power level while maintaining a good communication link. To achieve this, CDMA2000 uses fast-response closed-loop power control on the reverse link. In summary, the BTS measures the reverse link from the mobile unit and sends power-control commands to increase or decrease the mobile unit's power level, which is similar to IS-95. It is important to note that the mobile unit also can operate autonomously and make power corrections based on the *Frame Error (Erasure) Rate* (FER) of the forward link. From this, it infers what it needs to do for the reverse link in terms of power control.

Also, a refinement to the closed-loop power control is located on the reverse link, and this is where the base station performs an outer-loop power control, which is a refinement process for the inner-loop power-control process. Specifically, if the frame received from the mobile unit arrives without error, the base station instructs the mobile unit to power down, whereas if the frame arrives in error, the mobile unit is instructed to power up.

With CDMA2000, the use of power control on the forward channel is possible with the introduction of the reverse pilot channel. The reverse pilot channel for power control was introduced to help reduce the interference caused by forward energy. Effectively, the mobile unit measures the received power and compares it against a threshold that the mobile unit then feeds back to the base station. On receipt of the power information, the mobile unit then is instructed to power up or power down.

In addition, as with the reverse power link, an outer-loop power-control process dynamically adjusts the target *Energy per Bit per Noise Ratio* (E_b/N_o). This is done by comparing the FER with a target FER, and if the FER is greater than the target, the unit is instructed to power up. If it is below the target FER, the unit is instructed to power down.

4.4.5 Walsh Codes

CDMA2000 introduces an increase in the number of Walsh codes from 64 with IS-95 to a total of 256 with 3xRTT. As with IS-95, CDMA2000 uses PN long codes for both the forward and reverse directions. However, in CDMA2000, variable-length Walsh codes are introduced to accommodate fast packet-data rates.

The Walsh code chosen by the system is determined by the type of reverse channel. The R-SCH also uses a reserve Walsh code. If there is only one R-SCH, it uses a 2- or 4-chip Walsh code, but when there is a second R-SCH, it uses a 4- or 8-chip code. Therefore, in order to maintain or obtain the higher data rates on the F-SCH, the Walsh code must be shorter so as to maintain the same spreading rate.

	Walsh Codes							
	RC	256	128	64	32	16	8	4
SR1	1	Na	Na	9.6	Na	Na	Na	Na
	2	Na	Na	14.4				
	3	Na		9.6	19.2	38.4	76.8	153.6
	4	Na	9.6	19.2	38.4	76.8	153.6	307.2
	5	Na	Na	14.4	28.8	57.6	115.2	230.4
SR3	6		9.6	19.2	38.4	76.8	153.6	307.2
	7	9.6	19.2	38.4	76.8	153.6	307.2	614.4
	8		14.4	28.8	57.6	115.2	230.4	460.8
	9	14.4	28.8	57.6	115.2	230.4	460.8	1036.8

TABLE 4.8 Walsh Code Tree Table

Table 4.8 shows the relationship between Walsh codes, the SR, the RC, and of course, the data rates. One very important issue or, rather, effect with using variable-length Walsh codes is that if a shorter Walsh code is being used, then it precludes the use of longer Walsh codes that are derived from it. The table helps in establishing the relationship between which Walsh code length is associated with a particular data rate.

Table 4.9, a simplified table, shows the maximum number of simultaneous users for any data rate. For SR1 and RC1, a maximum number of users has individual Walsh codes equating to 64, a familiar number from IS-95A.

Looking at this table, if we had a total of 12 RC1 and RC2 mobiles under a sector, then that would allow for three data users at 153.6 kbps, 6 at 76.8 kbps, 13 at 38.4 kbps, 26 at 19.2 kbps, or 104 at 9.6 kbps. This relationship between the number of simultaneous users for a CDMA channel is depicted in Table 4.10. Obviously, the negotiated mobile data rate complicates determination of the total throughput traffic levels.

	Simultaneous Users							
	RC	256	128	64	32	16	8	4
SR1	1	Na	Na	9.6	Na	Na	Na	Na
	2	Na	Na	14.4				
	3	Na		9.6	19.2	38.4	76.8	153.6
	4	Na	9.6	19.2	38.4	76.8	153.6	307.2
	5	Na	Na	14.4	28.8	57.6	115.2	230.4
SR3	6		9.6	19.2	38.4	76.8	153.6	307.2
	7	9.6	19.2	38.4	76.8	153.6	307.2	614.4
	8		14.4	28.8	57.6	115.2	230.4	460.8
	9	14.4	28.8	57.6	115.2	230.4	460.8	1036.8

TABLE 4.9 Simultaneous Users with SR1 and SR3

	Data Rates							
	RC	256	128	64	32	16	8	4
SR1	1 & 2	Na	Na	9.6/14.4	Na	Na	Na	Na
	3	Na	Na	9.6	19.2	38.4	76.8	153.6
	Simultaneous Users							
SR1	1 & 2	Na	Na	12	Na	Na	Na	Na
	3	Na	Na	48	8	12	2	3

Table 4.10 Simultaneous Users for SR1 Only

The real issue behind this is the type of data that will be allowed to be transported over the network, which has a direct impact on the available users.

It is important to note that the shorter Walsh codes inhibit the use of longer Walsh codes because of the orthogonality required. Also, all channel requests are allocated from the same Walsh code pool on a per-sector basis. In addition, to achieve the higher data rate, not only is the Walsh code implementation modified, but the modulation scheme also has been changed.

Also, if there was a need for high-speed data for interactive video with phase 1 CDMA2000, the transport of 384 kbps of data would not be feasible with SR1, as indicated in Table 4.9.

4.5 EVDO

Evolution Data Optimized (EVDO) is also known as *High Rate Packet Data* (HRPD) and comes under the IS-856 specification. EVDO was implemented initially to optimize forward-link throughput, but several subsequent releases have focused on improvement not only to the forward link but also to reverse-link throughput. EVDO initially was a single carrier, 1x, in both Revision 0 and Revision A but will become a multicarrier with Revision B.

1xEVDO is a packet-data-only system and therefore has numerous advantages for handling data in both stationary and nonstationary environments. EVDO in Revisions 0 and A is a single carrier occupying 1.25 MHz spectrum in an FDD configuration. EVDO uses a chip rate of 1.288 Mcps, which is the same as 1xRTT, thereby facilitating its integration into an existing 1x system.

1xEVDO typically is deployed as an overlay system in conjunction with a 1xRTT network. 1xEVDO uses many of the same attributes as 1xRTT, for example, the active set where the terminal reports the strongest forward link pilots it can measure to the network, which enables the network to select an active set for the terminal. Interaction between 1xEVDO and 1xRTT systems is achieved via initial system acquisition, and when the data session is in progress, it is possible to hand off (hard handoff) from a 1xEVDO to a 1x-RTT system migrating down from the EVDO to a 1xRTT network. Presently, it is not possible to migrate a session from 1xRTT to EVDO.

However, there are fundamental differences between EVDO and 1xRTT in that in an EVDO network, transmission occurs only on the best link possible; that is, there are no soft and softer handoffs, and this process is referred to as the *Virtual Soft Handoff* (VSH). For a VSH, the mobile unit monitors the *Signal to Interference and Noise Ratio* (SINR)

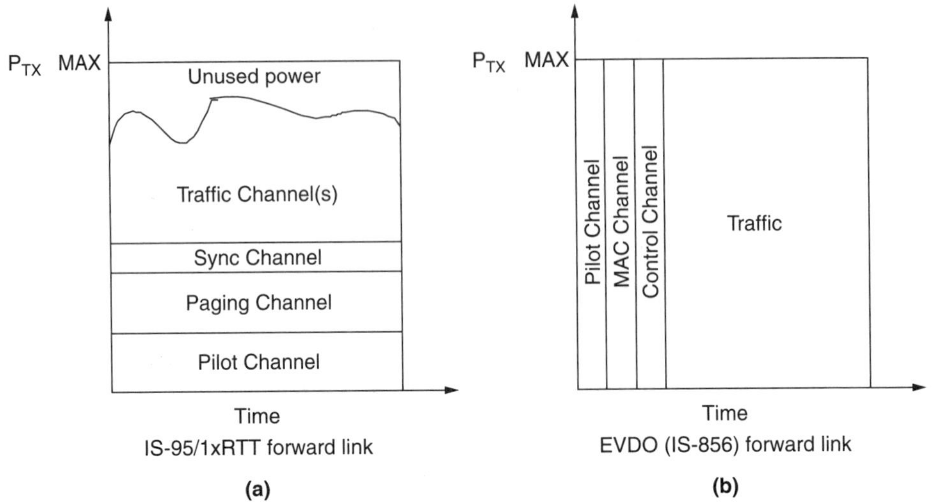

FIGURE 4.19 Forward link: (*a*) 1xRTT; (*b*) EVDO.

of the pilots in the active set and informs the network via the reverse channel the identity of the best serving sector or cell.

Rate adaption is used on the EVDO forward link as compared with power adaption. In other words, constant power is maintained in the forward link with EVDO, whereas in 1xRTT power adaption is used to allocate only the power needed to maintain communication. This fundamental difference is illustrated in Figure 4.19.

Figure 4.19*a* illustrates how power normally is distributed in a 1xRTT system. However, in Figure 4.19*b*, a constant-power envelope is displayed for an EVDO system, where power is constant and fully allocated for the communication. This is one of the fundamental differences from a power-budget perspective between an EVDO system and a 1xRTT network. Besides the power difference, there is also a fundamental change in resources for EVDO versus 1xRTT in that only one user is being communicated with at a time in an EVDO system as compared with multiple users in a 1xRTT system.

What we mean by this is that power normally is distributed between all users in an IS-95/1xRTT system, but in an EVDO system, only one user is in active communication with the system, meaning that all the power for the forward link is allocated to that single user. In an EVDO system, the network transmits only on the best forward link and allocates no power to any other user at that time, showing a full power envelope.

With EVDO, the mobile terminal plays a predominant role in transmission rate selection, involving closed-loop rate control. In closed-loop rate control, the terminal predicts the channel condition using a predicted SINR that is based on current packets received. In summary, the mobile predicts the channel condition based on SINR and then informs the serving cell or sector of its desired packet rate over the reverse-link feedback channel. When the network accesses a terminal, it selects the most recent rate fed back from the terminal itself.

Additionally, in EVDO Revision 0, traffic data are sent to a single user, and only one type of service is provided at a time. Since a sector transmits traffic data to a single user at any instant in time, a scheduling algorithm is used to allocate the time slots to the active users in the sector. This scheduling algorithm was changed for later releases starting with Revision A, where multiple services now can be sent to a user, such as http, e-mail, and VoIP services to be run at the same time.

The channel structure/coding for 1xEVDO is different from that of 1xRTT. In particular, 1xRTT has the ability to use 128/256 Walsh codes. Each Walsh code is used for a circuit-switched connection, usually voice. When applied for packet data, each Channel Element (CE) is concatenated at 9.6 kbps to achieve the requisite data rate. EVDO, however, does not use CEs and relies on data modems for transport. Also, regarding the data slots, there is no relationship with the code tree. Instead, the data rate is determined by the modulation scheme used.

EVDO's data channel structure uses a total of 16 slots for data that are encoded and spread with the appropriate Walsh codes. EVDO has available a total of 64 Walsh codes, of which the initial deployment of EVDO, Revision 0, facilitates the use of only 16 for data traffic. Later releases increase the Walsh codes available to 32 and then 64. EVDO to deliver packet traffic uses both the I and Q phases for data delivery in both the forward and reverse links. Since EVDO is designed for asymmetric traffic, the forward and reverse links for each data session can and do have different link rates.

An important issue is that the soft and softer handoffs are not used in the EVDO system, as they are in IS-2000 networks. Instead, VSH is introduced with EVDO. VSH uses many of the same attributes employed in soft and softer handoffs, except that only one connection is made at any given time. More specifically, once communication is established between the mobile unit and the network, the mobile unit continues to monitor the *Received Signal Strength Indication* (RSSI) of the pilot signals. The pilot is added to the mobile unit's active list when the pilot exceeds a predetermined threshold that is determined by the network.

As mentioned earlier, there is no soft handoff in IS-856, but the mobile unit can quickly reselect a better serving sector to adapt to the new channel conditions, which usually are determined by the strongest pilot RSSI. Hysteresis is used to control a ping-pong effect with a typical 3 to 5 dB being used.

The packet rate is adjusted continuously through a process called *rate adaption.* With rate adaption, the data rate is altered based on the RF environment in order to achieve the highest average data throughput with a constant average transmitted power.

Additionally, in EVDO systems, there is no forward power control because the channel transmits at full power all the time. Instead of power control, rate adaption is now used. With EVDO, the forward-loop control contains an inner- and outer-loop control used for the rate-adaption process. The inner loop is used to determine the highest data rate possible through the *Data Rate Control*'s (DRC's) channel-condition predictor. The outer loop is also used for the rate selection through measuring the SINR. This use of inner and outer loop is referred to as the *closed-loop control process* for EVDO systems.

Hybrid Automatic Repeat Request (HARQ) is used on the forward traffic channel to improve the data rate. HARQ uses a Turbo code of 1/3 or 1/5, with most of the data rates using a Turbo code of 1/5. The 1/5 Turbo code is used in IS-2000 as well.

The data scheduler is very important in EVDO systems in that it helps to maximize the overall throughput of the system, enhancing the user's experience. The scheduler helps to determine the data rate, the code allocated, and when the packets will be allocated in the queue. The maximum data rate is a function of the SINR and the code allocated.

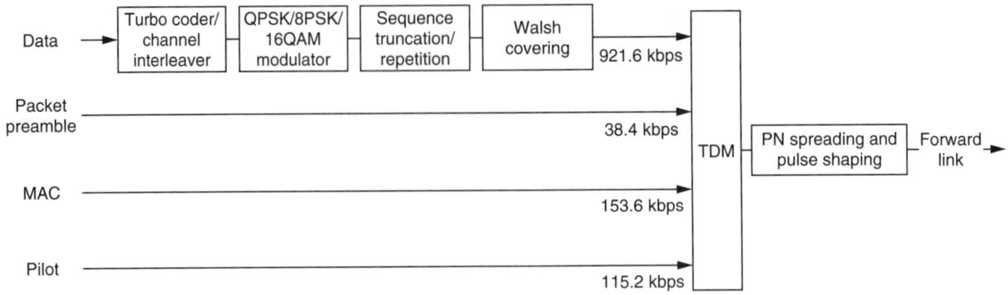

FIGURE 4.20 TDM forward link.

EVDO, IS-856, uses TDM along with codes to enhance the data experience by the user as compared with a IS-2000 network, which predominantly uses codes.

4.5.1 EVDO Forward Link

The forward link for EVDO consists of 16 slots. Each slot is 1.66 ms in duration, and a frame consists of 16 slots. There are a total of 2048 chips in each forward-link frame. The EVDO forward link has

- Pilot Channel
- Medium Access Control (MAC)
- Forward Traffic Channel
- Control Channel

Four effective data streams are used with EVDO for the forward link, and they are TDM-multiplexed together prior to spreading. The TDM method for the forward link is represented in Figure 4.20.

The Forward Traffic Channel is a shared medium that transmits to a single user at a time; a preamble sequence is transmitted to indicate the presence and stating point of the packet and also to indicate the intended target, the mobile terminal. Each frame contains 2048 chips. There are 1600 chips allocated to the Forward Traffic Channel/Control Channel, of which 192 are for the Pilot Channel, 256 for the MAC, and 64 for the Preamble, leaving 1536 chips for the data payload.

Figure 4.21 shows an EVDO channel. A slot where there is no traffic is considered an *idle slot*, and this is shown in Figure 4.22. If in *Idle*, no control or traffic data are sent.

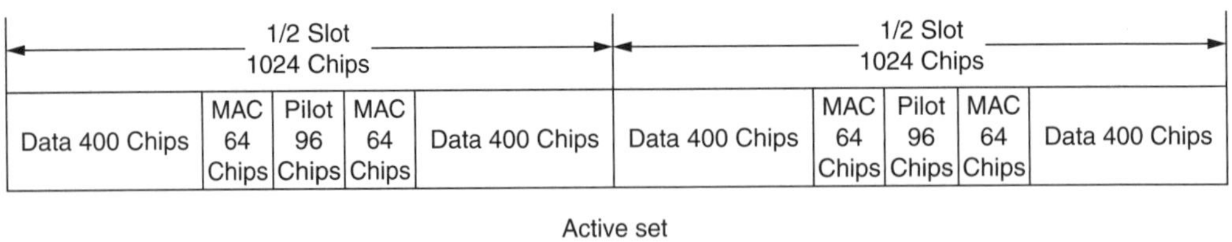

FIGURE 4.21 Active set.

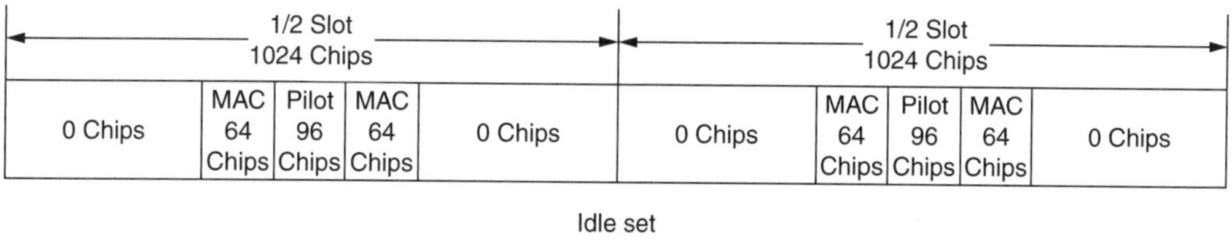

FIGURE 4.22 Idle set.

0, 1	Not used	Not used
2	Not used	76.8-kbps *Control Channel*
3	Not used	38.4-kbps *Control Channel*
4	RA Channel	Not used
5–63	RPC and DRC Lock Channel transmission	Forward and Traffic Channel transmissions

TABLE 4.11 MAC Index and Channel

In EVDO, the Pilot Channel of each sector is distinguished by the PN offset in increments of 64 chips, as is the case in IS-95.1xRTT systems. The only difference between 1xRTT and EVDO Pilot Channels is that 1xRTT systems transmit a continuous Pilot Channel, whereas EVDO uses a gated Pilot Channel because it is TDM-based.

The MAC Channel is transmitted in the 256 chips using a Walsh channel. The Walsh channel with even-numbered MAC Index values is assigned to in-phase modulation, whereas the odd-numbered MAC Index is assigned to the quadrature-modulation phase. The relationship between the MAC Index and the MAC Channel is shown in Table 4.11.

The MAC Channel tells the terminal how to access the network. It tells it the maximum data rate, or rather, rate limit on the reverse channel and also indicates reverse traffic channel interference received above a threshold requiring further rate adaption, *Reverse Activity* (RA). The RA Channel is used to control the total interference level received in a given sector.

4.5.2 EVDO Reverse Link

The reverse link for EVDO has access and traffic channels. The Access Channel consists of a Pilot Channel and a Data Channel that are used for sending signaling messaging when the access terminal is not in a connected state.

However, in the connected state, the reverse traffic channel contains Pilot, MAC, ACK, and Data channels. The MAC also consists of two channels: the *Reverse Rate Indicator* (RRI) and the *Data Rate Control* (DRC). The RRI is used to indicate whether or not the Data Channel is being transmitted on and its associated data rate. The DRC is used to indicate to the network the Forward Traffic Channel's data rate and best service sector for the forward link. The ACK channel is used to inform the network that the data packet sent on the Forward Traffic Channel was indeed received. The DRC provides accurate data-rate information to the forward-link adaptive scheduler but also is used to perform VSH.

The link rates for both the forward and reverse channels are revision-dependent. Therefore, the next two sections address the different releases in an attempt to help clarify the differences and/or similarities between the two. However, it is important to note that all the releases for EVDO are reverse-compatible, meaning that an EVDO-capable device that is Revision A–compliant can operate in a Revision 0 or 1xRTT environment.

4.5.3 EVDO Revision 0

Revision 0 is the first generation of EVDO and introduces many new capabilities for mobile data as compared with a 1xRTT system:

- HARQ
- Open-loop rate control
- Adaptive data schedule
- Closed-loop rate control
- Uplink rate control
- Uplink rate indication
- Closed-loop power control
- 2.4-Mbps DL
- 153-kbps UL

Figure 4.23 shows the overall structure in a block diagram of an EVDO Revision 0 forward channel, while Figure 4.24 represents the reverse channel.

Figure 4.25 is meant to depict the relationship between the frame used for EVDO and the slot assignment. The frame and slot relationship is the same for both the forward and reverse channels. The figure shows that there are a total of 16 slots and that each is 1.66 ms in duration. The reason I am mentioning the time slot duration will become apparent later in this chapter. In the figure, 4 slots are being used, and 12 of the 16 are not being used. The unused slots are a function of payload and the RF channel conditions coupled with rate adaption.

It is important to remember that the forward link channel is still 1.25MHz and is a direct-sequence spread rate of 1.2288 Mcps. In addition, an EVDO Revision 0 frame is 26.66 ms in duration, whereas an IS-2000 frame is 20 ms in duration.

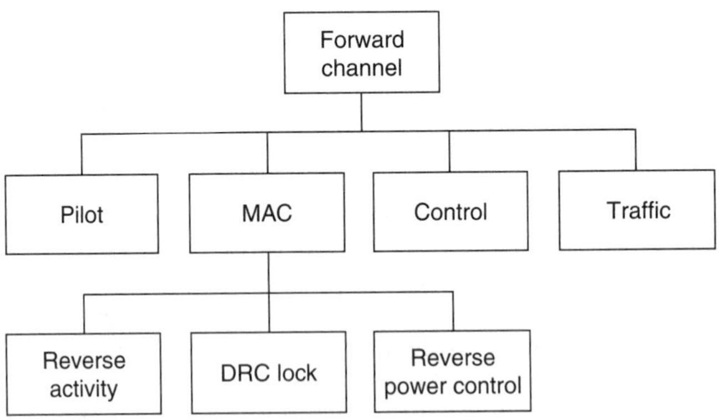

FIGURE 4.23 EVDO Revision 0 forward channel structure.

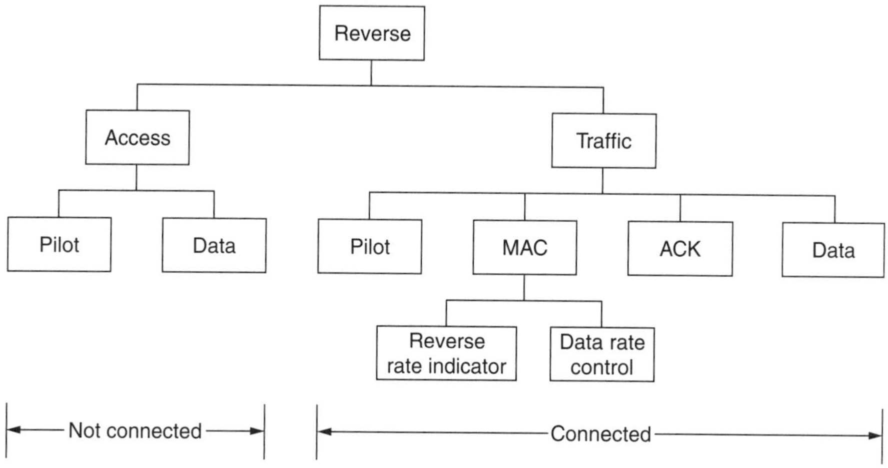

FIGURE 4.24 Reverse channel structure EVDO.

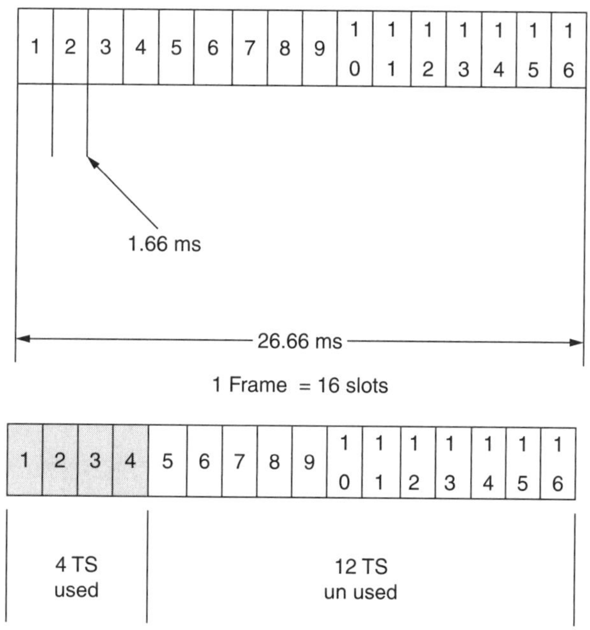

FIGURE 4.25 EVDO slots.

Table 4.12 indicates the various rates as a function of payload and slots used for EVDO in the forward link. Table 4.13 illustrates the various channel rates for EVDO Revision 0 in the reverse channel. Please keep in mind that the forward and reverse channel rates can support asymmetric traffic and that the forward rate is often different from the reverse rate.

Payload (Max)	Modulation Format	1 Slot	2 Slot	4 Slot	8 Slot	16 Slot
1024	QPSK	614.4	307.2	153.6	76.8	38.4
2048	QPSK	1228.8	614.4	307.2	NA	NA
3072	8 PSK	1843.2	921.6	NA	NA	NA
4096	16QAM	2457.6	1228.8	NA	NA	NA

TABLE 4.12 Revision 0 Forward Channel Rates

Payload (Max)	Modulation Format	4 Slot	8 Slot	12 Slot	16 Slot	Rev
256	QPSK	38.4	19.2	12.8	9.6	0
512	QPSK	76.8	38.4	25.6	19.2	0
1024	QPSK	153.6	76.8	51.2	38.4	0
4096	QPSK	614.4	307.2	204.8	153.6	0

TABLE 4.13 Revision 0 Reverse Channel Rates

4.5.4 EVDO Revision A

EVDO Revision A improves on the packet-handling capability of Revision 0 and is backward-compatible with 1xEVDO Revision 0, allowing for a seamless integration of the new service to the system. Revision A is an all-IP-based air interface that integrates voice (VoIP), high-speed packet data, and enhanced multimedia capabilities and adds higher data rates, higher system capacity, and improved QoS support for low-latency packet applications. Revision A also supports enhanced multicast capabilities for the delivery of high-quality bandwidth-intensive video and audio applications.

Revision A introduces many enhancements, including

- Higher-order modulation
- Improved QoS
- Radio-connection setup improvement (lower latency)
- New packet payload sizes
- Multiuser packers
- Efficient resource allocation
- Four-slot subpacket (reverse link)
- More forward and reverse data rates
- 3.1-Mbps DL
- 1.8-Mbps UL
- New channels
    - ARQ
    - Auxillary Pilot
    - DSC

EVDO Revision A invokes multiple service availability, enabling three unique and distinct services to be run at the same time. For instance, voice and data applications can run at the same time, whereas under 1xRTT or EVDO, only one type of session was possible. With EVDO Revision A, there are more subslots, enabling more granularity for data-rate throughput owing to efficiencies. In EVDO Revision 0, there were only four payload sizes for forward and reverse channels, whereas in Revision A, there are a total of 8 forward and 12 reverse channels, giving rise to the issue of better bandwidth efficiency for quicker throughput, as well as the ability to have the resource used by other users.

Revision A involves a new chip set that enables higher modulation schemes to be employed. The higher modulation schemes also require a more robust *Power Amplifier* (PA) at the BTS to support the higher power/bandwidth requirements. Although it is possible to use a PA that is designed to support EVDO Revision 0, the power limitation imposed necessitates the use of a new PA for EVDO Rev A.

Like Revision 0, EVDO Revision A is an overlay network. EVDO could be deployed in the core (*core* typically refers to *core network*) or high-data locations based on 1xRTT traffic information. Then it can be expanded to the rest of the most logical parts of the network where economics or other conditions prevail.

Most carriers have an imbedded cost for the Revision 0 deployment. Therefore, since Revision A is backward-compatible with Revision 0, it is possible to have an *N*:1 migration path strategy where high-traffic sites determined from Revision 0 data are migrated to Revision A. In addition, because EVDO Revision A is backward-compatible, it is possible to employ EVDO Revision 0 with Revision A hardware and then with software upgrades to Revision A.

Figure 4.26 shows the EVDO Revision A forward channel layout. Glancing at the figure shows that ARQ is now added to the forward link of Revision A as compared with Revision 0. *Hybrid ARQ* (HARQ) is used to improve power-control variations as well as multiple-access interference though early termination of packet transmissions, improving the overall throughput.

Figure 4.27 shows the fundamental structure of the EVDO Revision A reserve link. Both the Auxiliary Pilot Channel and the Data Source Control Channel have been added to the reverse link for EVDO Revision A and are used only while in the connected state. The Auxiliary Pilot Channel is used to improve channel estimation for high data rates, whereas the *Data Source Control* (DSC) Channel is used to minimize delays associated with queue transfer during cell switching.

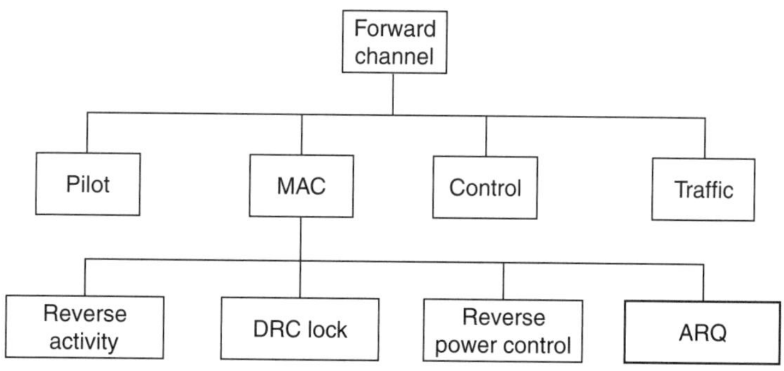

FIGURE 4.26 EVDO Revision A forward link.

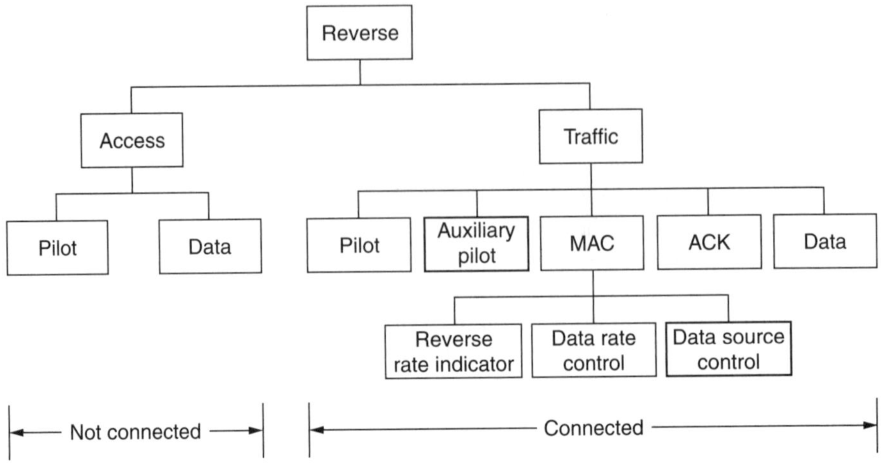

FIGURE 4.27 EVDO Revision A reverse link.

Table 4.14 shows the various data rates that are available on the forward link for EVDO Revision A. The right-most column indicates the revision that is used for backward compatibility with the network and/or device.

Table 4.15 shows the various data rates that are available on the reverse link for EVDO Revision A. The right-most column indicates the revision that it is used for backward compatibility with the network and/or device.

As indicated in Tables 4.14 and 4.15, Revision A uses higher-order modulation as compared with Revision 0 combined with the finer-rate quantization method used. For example, in EVDO Revision 0, there are 16 continuous slots (1.6 ms each) for a frame of 26.66 ms. However, in Revision A, 4 continuous slots (1.6 ms each) make up what is called a *subpacket frame,* and there are a total of 4 subpacket frames per user.

With Revision A, an early termination process is possible, where on receiving each subpacket the BTS sends an ACK or NACK to the terminals to indicate if it received the packet successfully. In addition, the ARQ channel on the forward link is involved with

Payload (Max) Bits	Modulation Format	1 Slot	2 Slot	4 Slot	8 Slot	16 Slot	EVDO Rev
128	QPSK	76.8	38.4	19.2	9.6	4.8	A
256	QPSK	153.6	76.8	38.4	19.2	9.6	A
512	QPSK	307.2	153.6	76.8	38.4	19.2	A
1024	QPSK	614.4	307.2	153.6	76.8	38.4	0
2048	QPSK	1228.8	614.4	307.2	NA	NA	0
3072	8 PSK	1843.2	921.6	NA	NA	NA	0
4096	16QAM	2457.6	1228.8	NA	NA	NA	0
5120	16QAM	3072	1536	NA	NA	NA	A

TABLE 4.14 EVDO Revision A Forward Rates

Payload (Max)	Modulation Format	4 Slot	8 Slot	12 Slot	16 Slot	Rev
128	QPSK	19.2	9.6	6.4	4.8	A
256	QPSK	38.4	19.2	12.8	9.6	0
512	QPSK	76.8	38.4	25.6	19.2	0
768	QPSK	115.2	57.6	38.4	28.8	A
1024	QPSK	153.6	76.8	51.2	38.4	0
1536	QPSK	230.4	115.2	76.8	57.6	A
2048	QPSK	307.2	153.6	102.4	76.8	A
3072	QPSK	460.8	230.4	153.6	115.2	A
4096	QPSK	614.4	307.2	204.8	153.6	0
6144	QPSK	921.6	460.4	307.2	230.4	A
8192	8 PSK	1228.8	614.4	409.6	307.2	A
12,288	8 PSK	1843.2	921.6	614.4	480.8	A

TABLE 4.15 EVDO Revision A Reverse Rates

the reverse-link closed-loop power update rate which is one-quarter of that used in Revision 0, going form 600 to 150 Hz once every four slots, allowing simultaneous users without increasing the MAC overhead.

Figure 4.28 shows the new slot structure for the reverse link with EVDO Revision A. What is important to note is the subframe rates enabling early termination.

4.5.5 Data Rate

The main question that comes to mind is, How can you equate the different data rates to the slots in both Revision 0 and Revision A? With Revisions 0 and A, each packet type is uniquely specified by its nominal data rate and nominal packet length. How this equates to actual data traffic is best illustrated with a quick example.

The maximum payload, data, is determined by the following:

Nominal data rate = physical layer packet size/nominal transport duration

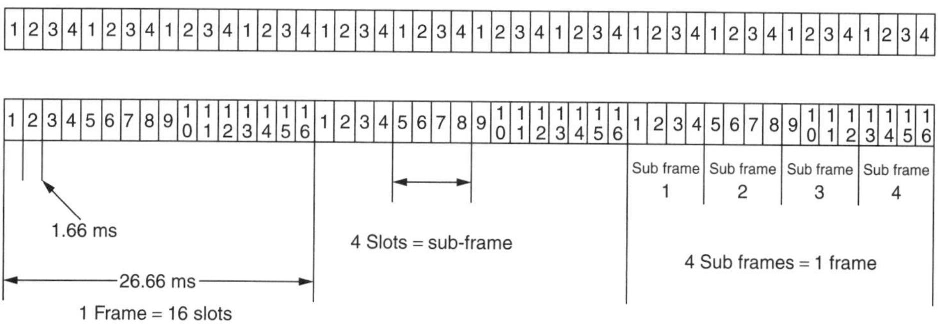

FIGURE 4.28 EVDO Revision A slot structure.

The physical layer packet size is the maximum payload in bits, and the nominal transport duration is the number of time slots used (which are 1.667 ms in length).

Therefore, looking at Table 4.14 and using a payload of 5120 with 2 time slots, the nominal data rate = 5120/(2 × 1.667 ms) = 1536 kbps, which is the maximum data rate for that payload. If we had used only one time slot, then the rate would be 3072 kbps.

Just how this plays into the actual data delivered goes like this: If you have a 300-bit packet to be delivered, the system will select a maximum payload of 512 using one time slot. This is so because 300/1.667 ms = 179.64 kbps, which can be supported by one time slot with a rate of 307.2 kbps. Obviously, if the channel conditions were different, that same amount of data could be handled with more time slots but at a lower rate.

The same concept can be applied to the reverse channel with one noticeable addition, the Revision A subpacket. Assuming the 300-bit payload example, the payload of 1024 would be sent at a rate of between 38.4 to 614.4 kbps for downlink in Revision 0. However, with Revision A, a payload of 512 could be used, improving the spectral efficiency. A key concept to note is that all 16 time slots would be used for the delivery in Revision 0 instead of the minimal amount in Revision A owing to the early-termination feature in Revision A.

For instance, looking at Figure 4.29, if this were Revision 0, then all 16 slots would be used for transmission of the payload. However, in Revision A, subframe 0 would be used to meet the goal, and right before sending the next packet, repeat and ACK would be sent, indicating successful reception of the data. If a NACK were received, then the information would be resent up to a total of four times.

4.5.6 EVDO Revision B

EVDO is also known as HRPD and comes under the IS-856 specification. EVDO was implemented initially to optimize forward-link throughput, but several subsequent releases have focused on improvement not only to the forward link but also to reverse-link throughput. EVDO initially was a single carrier, 1x, in both Revision 0 and Revision A, but will become a multicarrier with Revision B.

1xEVDO is a packet-data-only system and therefore has numerous advantages for handling data in both stationary and nonstationary environments. EVDO is designed to support multimedia services like streaming video, interactive emails and gaming, and telemetrics and support faster downloads of high-resolution images and large files. EVDO also supports simultaneous voice and data services; however, it was initially deployed for primarily data services and is backward-compatible with CDMA2000.

In Figure 4.30, various EVDO revisions are indicated. EVDO Release 0 was initially designed to optimize the forward link while EVDO Release A focused on the improvement

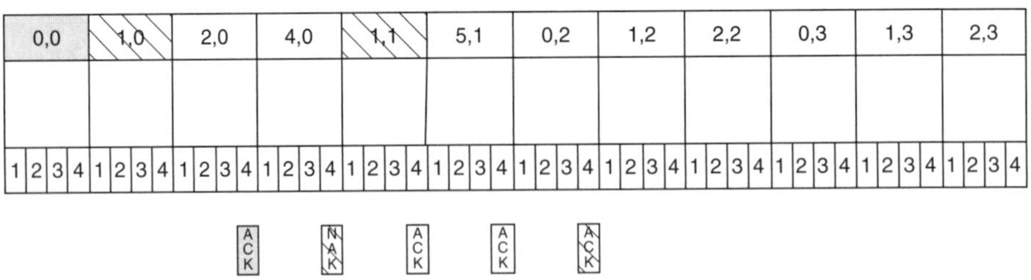

FIGURE 4.29 Subpacket.

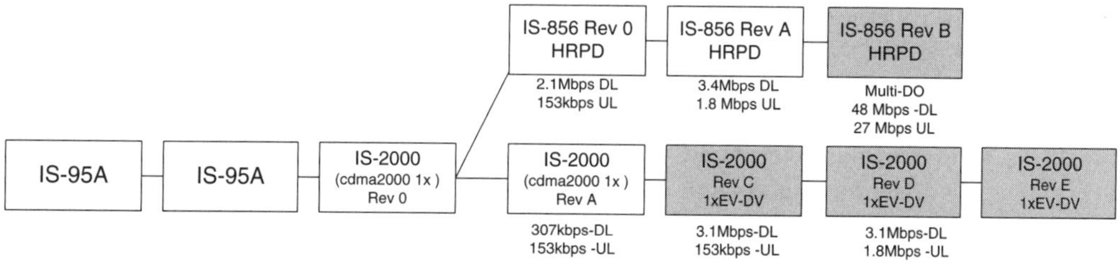

FIGURE 4.30 EVDO roadmap.

Peak Data Rates	Release 0	Release A	Release B
Forward link	2.4 Mbps	3.1 Mbps	4.8 Mbps
Reverse link	153.6 kbps	1.8 Mbps	2.4 Mbps

TABLE 4.16 EVDO Data Rates

to the forward link and reverse links throughputs with VoIP capabilities. EVDO is a single carrier in Revision 0 and Revision A; however with Revision B, which introduces the use of multicarriers, it enables even high data rates.

EVDO in Revisions 0 and A is a single carrier occupying 1.25 MHz spectrum in an FDD configuration. EVDO uses a chip rate of 1.288 Mcps, which is the same as 1xRTT, thereby facilitating its integration into an existing 1x system. Table 4.16 depicts the different data rates in both uplink and downlink for EVDO Revisions 0 and A.

Figure 4.31 is a high-level depiction of the new components for an EVDO network as compared to a CDMA 1xRTT network and is shown in the lower part of Figure 4.31.

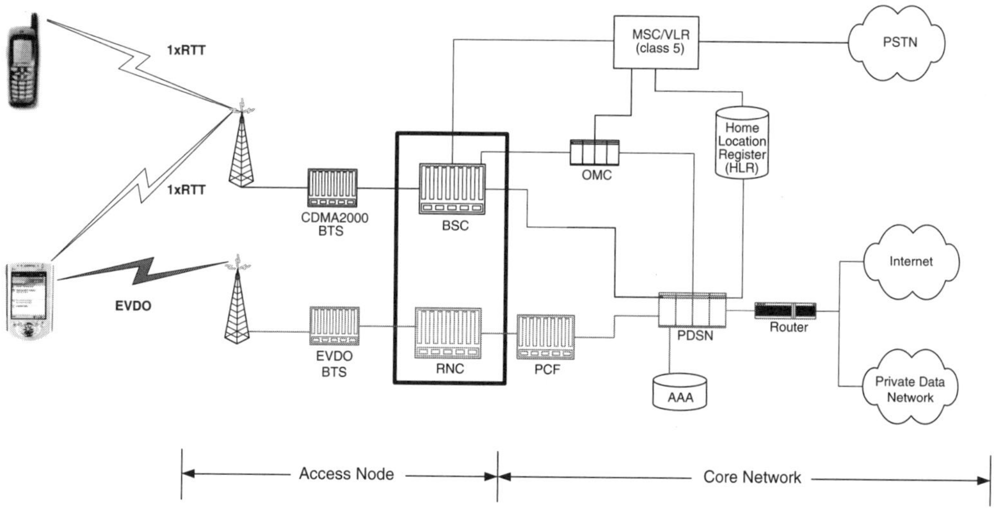

FIGURE 4.31 EVDO network.

The new components for the EVDO network into a CDMA system shown in Figure 4.31 involve the

- BTS (Base Transceiver Station)
- RNC (Radio Network Controller)
  - Session establishment and release
  - Frame selection
  - Radio Link Protocol (RLP) processing
- PCF (Packet Control Function)
  - Enables EVDO RNC to interface with PDSN
- PDSN (Packet Data Service Node)
  - Interfaces with Internet
  - Home/foreign agent for mobile IP
  - Terminates PPP connection with AT

1xEVDO typically is deployed as an overlay system in conjunction with a 1xRTT network. 1xEVDO uses many of the same attributes as 1xRTT, for example, the active set where the terminal reports the strongest forward link pilots it can measure to the network, which enables the network to select an active set for the terminal. Interaction between 1xEVDO and 1xRTT systems is achieved via initial system acquisition, and when the data session is in progress, it is possible to hand off (hard handoff) from a 1xEVDO to a 1xRTT system migrating down from the EVDO to a 1xRTT network. Presently, it is not possible to migrate a session from 1xRTT to EVDO.

However, there are fundamental differences between EVDO and 1xRTT in that in an EVDO network, transmission occurs only on the best link possible; that is, there are no soft and softer handoffs, and this process is referred to as the virtual soft handoff. For a virtual soft handoff, the mobile unit monitors the Signal to Interference and Noise Ratio (SINR) of the pilots in the active set and informs the network via the reverse channel the identity of the best serving sector or cell.

Rate adaption is used on the EVDO forward link as compared with power adaption. In other words, constant power is maintained in the forward link with EVDO, whereas in 1xRTT power adaption is used to allocate only the power needed to maintain communication. This fundamental difference is illustrated in Figure 4.19.

Figure 4.19*a* illustrates how power normally is distributed in a 1xRTT system. However, in Figure 4.19*b*, a constant-power envelope is displayed for an EVDO system, where power is constant and fully allocated for the communication. This is one of the fundamental differences from a power-budget perspective between an EVDO system and a 1xRTT network. Besides the power difference, there is also a fundamental change in resources for EVDO versus 1xRTT in that only one user is being communicated with at a time in an EVDO system as compared with multiple users in a 1xRTT system.

What we mean by this is that power normally is distributed between all users in an IS-95/1xRTT system, but in an EVDO system, only one user is in active communication with the system, meaning that all the power for the forward link is allocated to that single user. In an EVDO system, the network transmits only on the best forward link and allocates no power to any other user at that time, showing a full-power envelope.

With EVDO, the mobile terminal plays a predominant role in transmission rate selection, involving closed-loop rate control. In closed-loop rate control, the terminal predicts the channel condition using a predicted SINR that is based on current packets received. In summary, the mobile predicts the channel condition based on SINR and then informs the serving cell or sector of its desired packet rate over the reverse-link feedback channel. When the network accesses a terminal, it selects the most recent rate fed back from the terminal itself.

Additionally, in EVDO Revision 0, traffic data are sent to a single user, and only one type of service is provided at a time. Since a sector transmits traffic data to a single user at any instant in time, a scheduling algorithm is used to allocate the time slots to the active users in the sector. This scheduling algorithm was changed for later releases starting with Revision A, where multiple services now can be sent to a user, such as http, e-mail, and VoIP services to be run at the same time.

The channel structure/coding for 1xEVDO is different from that of 1xRTT. In particular, 1xRTT has the ability to use 128/256 Walsh codes. Each Walsh code is used for a circuit-switched connection, usually voice. When applied for packet data, each CE is concatenated at 9.6 kbps to achieve the requisite data rate. EVDO, however, does not use CEs and relies on data modems for transport. Also, regarding the data slots, there is no relationship with the code tree. Instead, the data rate is determined by the modulation scheme used.

EVDO's data channel structure uses a total of 16 slots for data that are encoded and spread with the appropriate Walsh codes. EVDO has available a total of 64 Walsh codes, of which the initial deployment of EVDO, Revision 0, facilitates the use of only 16 for data traffic. Later releases increase the Walsh codes available to 32 and then 64. EVDO to deliver packet traffic uses both I and Q phases for data delivery in both the forward and reverse links. Since EVDO is designed for asymmetric traffic, the forward and reverse links for each data session can and do have different link rates.

An important issue is that the soft and softer handoffs are not used in the EVDO system, as they are in IS-2000 networks. Instead, VSH is introduced with EVDO. VSH uses many of the same attributes employed in soft and softer handoffs, except that only one connection is made at any given time. More specifically, once communication is established between the mobile unit and the network, the mobile unit continues to monitor the RSSI of the pilot signals. The pilot is added to the mobile unit's active list when the pilot exceeds a predetermined threshold that is determined by the network.

As mentioned earlier, there is no soft handoff in IS-856, but the mobile unit can quickly reselect a better serving sector to adapt to the new channel conditions, which usually are determined by the strongest pilot RSSI. Hysteresis is used to control a ping-pong effect with a typical 3 to 5 dB being used.

The packet rate is adjusted continuously through a process called *rate adaption*. With rate adaption, the data rate is altered based on the RF environment in order to achieve the highest average data throughput with a constant average transmitted power.

Additionally, in EVDO systems, there is no forward power control because the channel transmits at full power all the time. Instead of power control, rate adaption is now used. With EVDO, the forward-loop control contains an inner- and outer-loop control used for the rate-adaption process. The inner loop is used to determine the highest data rate possible through the DRC's channel-condition predictor. The outer loop is also used for the rate selection through measuring the SINR. This use of inner and outer loop is referred to as the *closed-loop control process* for EVDO systems.

HARQ is used on the forward traffic channel to improve the data rate. HARQ uses a Turbo code of 1/3 or 1/5, with most of the data rates using a Turbo code of 1/5. The 1/5 Turbo code is used in IS-2000 as well.

The data scheduler is very important in EVDO systems in that it helps to maximize the overall throughput of the system, enhancing the user's experience. The scheduler helps to determine the data rate, the code allocated, and when the packets will be allocated in the queue. The maximum data rate is a function of the SINR and the code allocated.

EVDO IS-856 uses TDM along with codes to enhance the data experienced by the user as compared with an IS-2000 network, which predominantly uses codes.

4.5.7 Multicast

EVDO facilitates multimedia services through multicast. There are two versions of multicast for EVDO: gold and platinum. *Gold multicast* (Revision 0) allows for the distribution of multiple high-quality TV-like channels to many users at the same time. With gold multicast, channels can be transmitted at different rates, enabling multiple types of services to be offered to the subscriber. In addition, no physical-layer changes are needed to the channel, only a software upgrade.

Platinum multicast (Revision A) provides a method for operators to deliver a multitude of video and audio channels to a large subscriber population with three times the capacity of gold multicast. This higher capacity enables more channels or content with a higher resolution to be sent to the subscriber over a single 1xEVDO Revision A channel. Platinum multicast enables operators to deliver multiple content streams to many subscribers by dedicating any fraction of a 1.25-MHz carrier to multicast services.

4.5.8 MIMO

Multiple Input Multiple Output (MIMO) is one of the advances with EVDO. More specifically, EVDO can use a four-branch diversity to achieve higher data rates and higher voice capacity on the base stations that use four-branch receive diversity (two pairs of cross-polarized spatially separated antennas). This technique requires that receive diversity uses additional antennas and associated receive chains to enable the network to receive a stronger signal, increasing the potential footprint of the antenna space on the tower.

The four-branch system equates to a stronger signal, which enables higher data rates and increased system capacity. Receive diversity is backward-compatible with CDMA systems.

4.6 EVDO Session Flows

EVDO is a packet network. Figures 4.32 and 4.33 illustrate a PPP session establishment and a handover process with EVDO, respectively.

4.6.1 PPP Session Establishment

Figure 4.32 is a PPP session establishment ladder diagram for EVDO.

4.6.2 EVDO Mobility

Mobility is key to EVDO network. VSHs were introduced with EVDO and VSH uses many of the same attributes in soft and softer handoffs, except only one connection is made at any given time. Specifically in VSH only one forward sector sends data at one time;

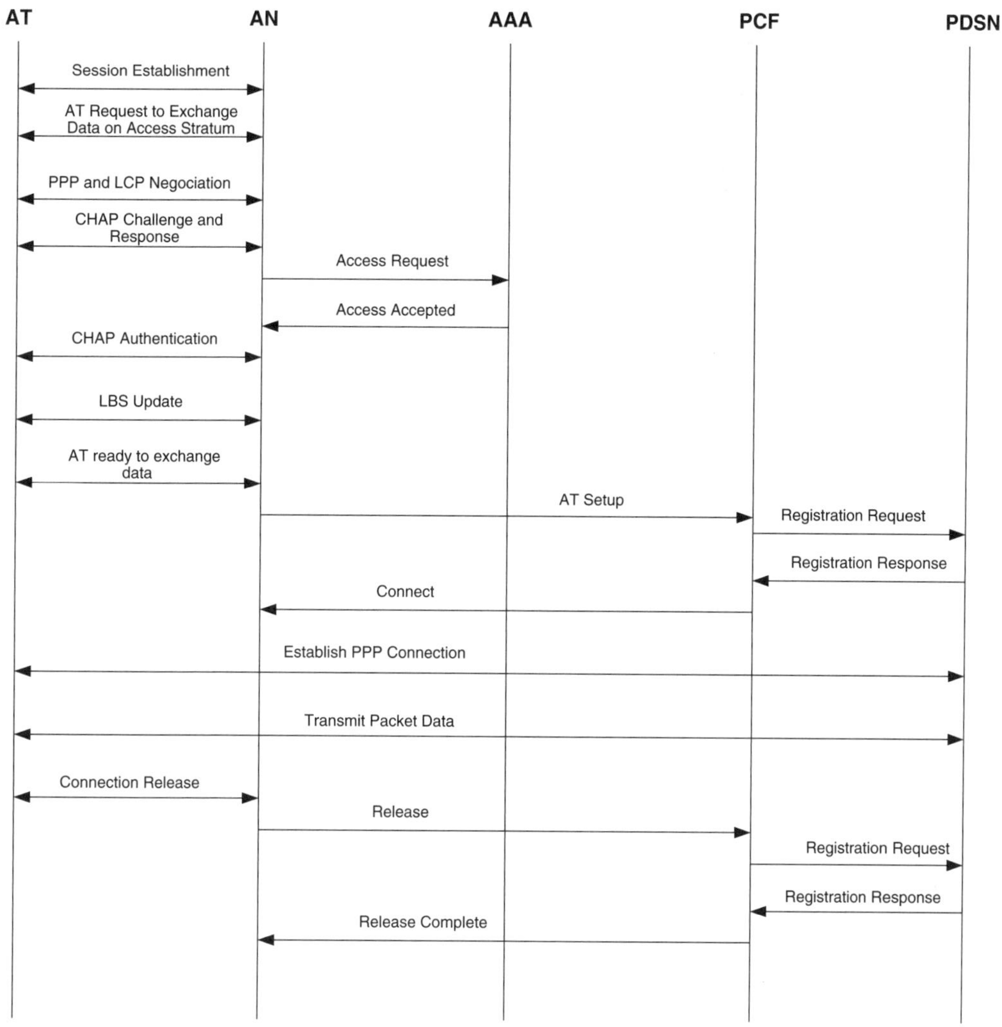

FIGURE 4.32 EVDO PPP session establishment.

this improves the overall system data throughput. Also with VSH the access terminal (AT) always dynamically selects to receive data from the one active sector based on the forward radio condition.

As part of the mobility process, once communication is established between the AT and the network, the AT continues to monitor the SINR of the pilot signals. The pilot for adjacent sites is added to the AT's active list when the pilot exceeds a predetermined threshold, which is determined by the network. The nominal hysteresis value of 3 to 5 dB is used to stop unnecessary handoffs from occurring, that is, the pingpong effect.

Also with EVDO, handoffs can also occur from EVDO to 1xRTT; however, they cannot go from 1xRTT to EVDO.

Figure 4.33 depicts the handover signaling in a ladder diagram format for EVDO.

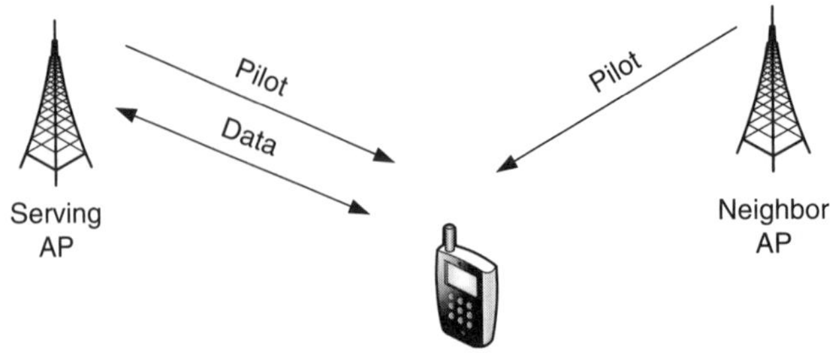

FIGURE 4.33(*a*) EVDO handover.

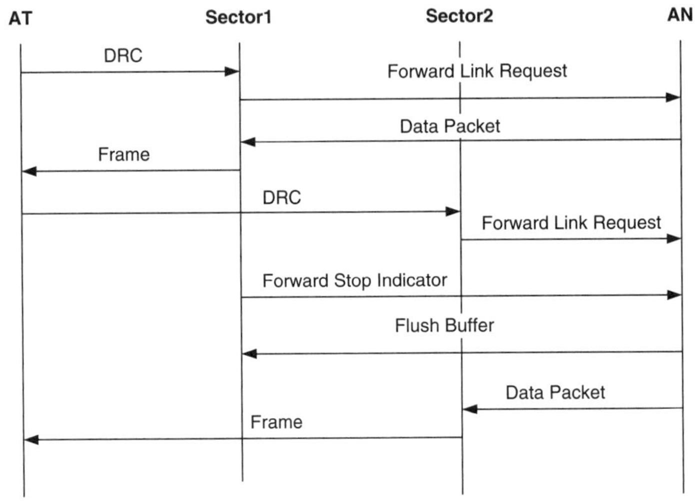

FIGURE 4.33(*b*) EVDO handover sequence.

Figure 4.38 is a simple depiction of a possible deployment of EVDO Revision B in an existing EVDO network.

4.7 EVDV

EVDV, also known as 1xEV-DV, means *Evolution for Data and Voice* and comes under the specification IS-2000 Revision C, as shown in Figure 4.1. EVDV is a major enhancement to CDMA2000's data capability, including QoS for the forward link and reverse-link enhancements.

EVDV will use channel-sensitive scheduling along with improved modulation schemes and coding, as well as *Hybrid Automatic Repeat Requests* (HARQs) to improve delays along with the *Radio Link Protocol* (RLP). Improvements to both the forward and reverse links are achieved with Revision D. The forward-link improvements include some new channels, and they are

- *Forward Grant Channel* (F-GCH). This is used to allow mobile access to the network.
- *Forward Indicator Control Channel* (F-ICCH). This is used for rate control.

- *Forward Acknowledgment Channel J* (F-ACKCHJ). This enables synchronous acknowledgments of reverse-link packets received and is able to support 192 users at a time in any sector.

The reverse link also includes the introduction of a few new channels, and they are

- *Reverse Packet Data Channel* (R-PDCH). This uses a fixed packet duration of 10 ms and allows for data ranges from 6.4 kbps to 1.8432 Mbps.
- *Reverse Request Channel* (R-REQCH). This is designed to improve reverse-link QoS, providing information about the mobile unit's transmit power and buffer status.
- *Reverse Secondary Pilot Channel* (R-SPICH). This is used to facilitate simple and fast detection, along with improving reverse-link power control.

Overall, EVDV is backward-compatible and also provides broadcast and multimedia capabilities for a 1x carrier.

4.8 CDMA Channel Allocation

The CDMA2000 channel allocations, just as with IS-95, have preferred locations and methods for deploying that are envisioned at this time to help facilitate the migration from 1x to 3x in the future. Tables 4.17 and 4.18 are for North America, but CS0002 has a channel plan for the conceivable band into which this technology can be implemented.

Please note that the channels defined in the tables are for 1.25-Mhz channel spacing. The asterisks denote where the first of three 1.25-MHz carriers are expected to be located for a 3x deployment. The first carrier is used in 3x for access, and this is used to help steer the subscriber to the correct carrier(s) to support the services being requested.

4.9 Deployment Scenario

The EVDO system uses the same 1.25-MHz radio channel bandwidth as IS-95 CDMA and is also a FDD system. When deploying EVDO with a CDMA2000 network where the EVDO system is A or 0, it is implemented as an overlay network.

Cellular System Carrier	Sequence	A	B
1	F1	283	384
2	F2	242*	425*
3	F3	201	466
4	F4	160	507
5	F5	119	548
6	F6	78	589
7	F7	37	630
8	F8 (Not advised)	691	777

Table 4.17 Cellular CDMA2000-1x and -3x Carrier Assignment Scheme

PCS System Carrier	A	B	C	D	E	F
1	25	425	925	325	725	825
2	50	450	950	350*	750*	850*
3	75*	475*	975*	375	775	875
4	100	500	1000	NA	NA	NA
5	125	525	1025	NA	NA	NA
6	150*	550*	1050*	NA	NA	NA
7	175	575	1075	NA	NA	NA
8	200	600	1100	NA	NA	NA
9	225*	625*	1125*	NA	NA	NA
10	250	650	1150	NA	NA	NA
11	275	675	1175	NA	NA	NA

TABLE 4.18 PCS CDMA2000-1x and -3x PCS Carrier Assignment Scheme

The choice of channels to use is dependent upon the spectrum portfolio that the wireless operator has in addition to its business plans for deployment. However, the overlay approach is done to account for the fact that EVDO will not be initially deployed everywhere that 1xRTT is deployed and with backward-compatibility EVDO devices can be served in a 1xRTT network, meaning that you do not need to build out the EVDO network 100 percent on day one.

Also regarding the spectrum selected for deployment of EVDO, should it be preferred to use the lower frequency bands that exist in the spectrum portfolio? However, this may not be practical due to existing services offered and the availability of *Customer Premise Equipment* (CPE) or simply that new spectrum was acquired and EVDO will be deployed in that part of the band first.

Figure 4.34 is a brief illustration of the introduction of an EVDO channel into a 5 × 5 MHz FDD frequency block. The single EVDO channel is chosen first and then as traffic increases, another 1xRTT channel can be retired in place of a new EVDO channel.

4.10 EVDO Revision B

EVDO Revision B brings to the table the ability to aggregate multiple EVDO Revision A carriers through *Multiradio Link Protocol* (MRLP) to improve overall throughput for an end user device. EVDO Revision B can be introduced in a phased approach in the network focusing on areas which have the highest data demand and then deciding later if additional areas or cells need to be enhanced.

EVDO Revision B can be realized in a wireless network currently using EVDO Revision A either through a software upgrade or hardware enhancements. However, a key component with EVDO Revision B is that it is backward-compatible with EVDO Revision A and Revision 0 as well as 1xRTT.

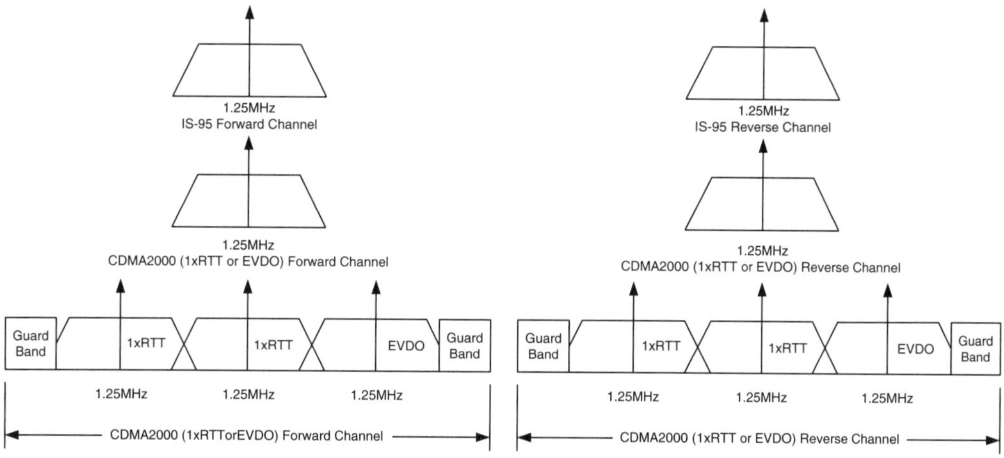

FIGURE 4.34 EVDO deployment in a 5 × 5 MHz FDD channel.

MRLP can be used effectively as a method to enhance throughput either as an augmentation to LTE deployments due to spectrum portfolio limitations or as a method to reduce capital expenditures in non-LTE areas, where improved data throughputs may be needed either to support new services or block competitive offerings.

Regardless, the software upgrade method is a logical first step in determining the true need for the multiple carrier support for the subscriber base. Because MRLP is a radio access layer improvement, the core network does not need to be upgraded except with the exception of backhaul. However, with the software upgrade approach new software supporting MRLP needs to be implemented in the channel cards as well as the RNC. The PDSN, however, does not need any modification again because MRLP is at the radio access portion of the network.

Depending on the demand for data services, EVDO Revision B also has a hardware upgrade path where a higher-modulation format, 64 QAM is supported with MRLP giving increased data rates with multiple carriers. The new data rate via hardware enables EVDO Revision B to have on a single carrier using a 1.25-MHz channel data rate of 4.9 Mbps in the downlink, and with MRLP a rate of 14.4 Mbps is possible using three EVDO Revision B carriers which are aggregated. The hardware upgrade for EVDO Revision B is only associated with new channel cards again because this is at the radio access portion of the network.

Obviously for EVDO Revision B to be realized, the edge devices, or handsets, need to be able to support MRLP as well.

Figure 4.35 is a high-level diagram of a CDMA2000/EVDO network where the network elements needing modification for EVDO Revision B are identified.

Figures 4.36 and 4.37 are brief examples of possible EVDO Revision B implementations for MRLP, where Figure 4.36 utilizes two EVDO carriers in a 5 x 5 FDD spectrum block, while Figure 4.37 uses three EVDO carriers in a 5 x 5 FDD spectrum block.

Figure 4.38 is a simple depiction of a possible deployment of EVDO Revision B in an existing EVDO network.

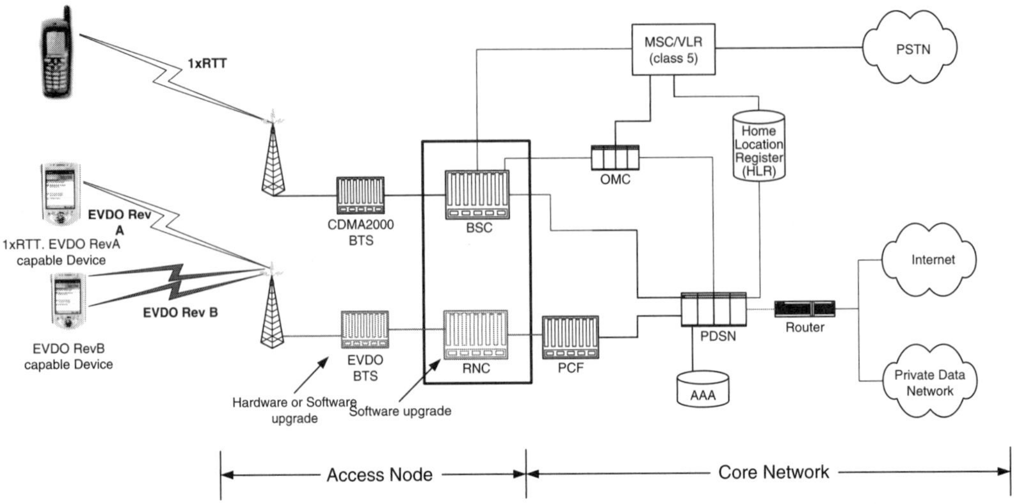

FIGURE 4.35 EVDO Revision B upgrade.

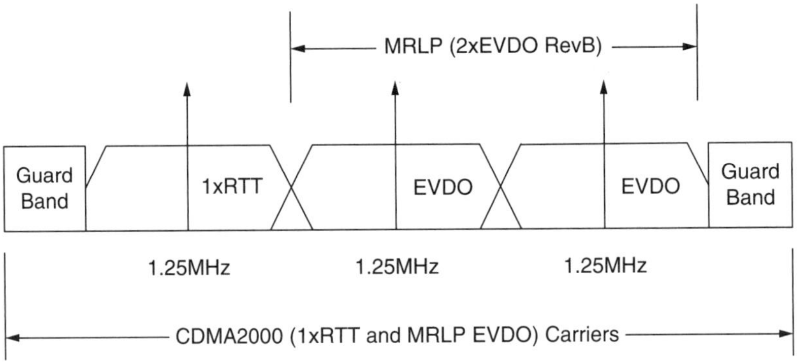

FIGURE 4.36 2 × EVDO.

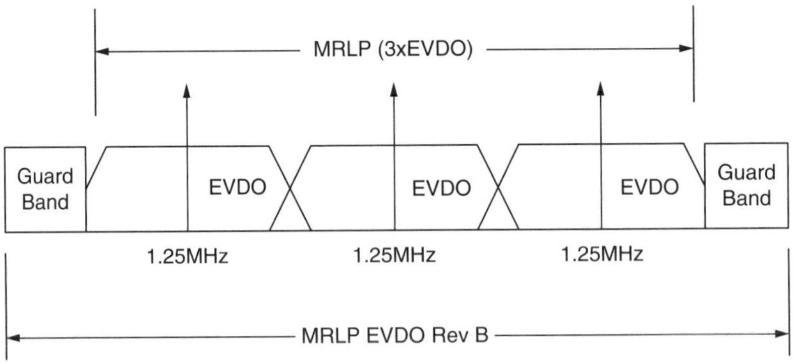

FIGURE 4.37 3 × EVDO.

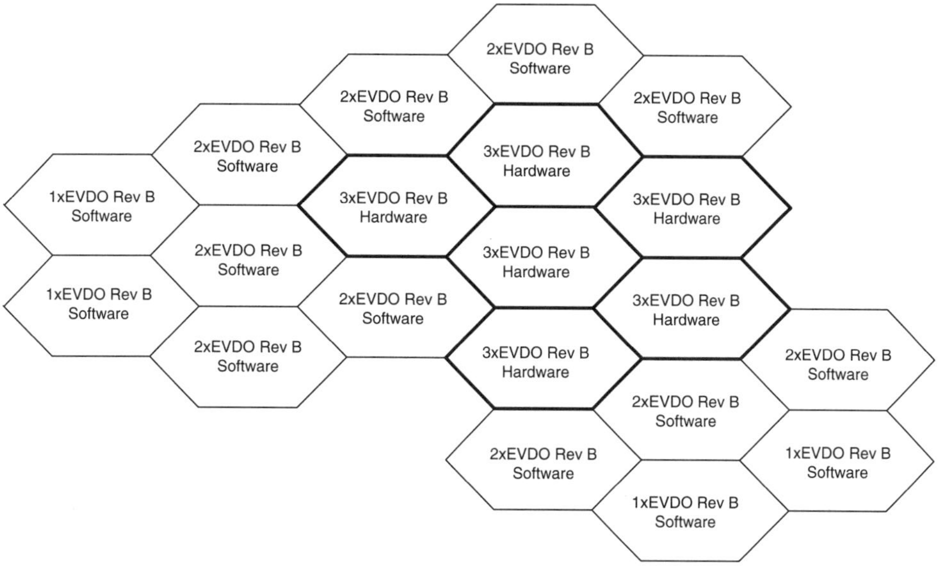

Figure 4.38 EVDO Revision B possible deployment scheme.

References

Agilent, "Designing and Testing cdma2000 Base Stations," *Application Note 1357.*

Agilent, "Designing and Testing cdma2000 Mobile Stations," *Application Note 1358.*

Andersen Consulting, Detecon, Telemate Mobile Consultants, "The GSM-CDMA Economic Study," February 16, 1998.

ANSI/EIA/TIA-553: Mobile Station–Land Station Compatibility Specification, September 1989.

ANSI/TIA-98-E-2003: Recommended Minimum Performance Standards for cdma2000 Spread Spectrum Mobile Stations, IEEE, February 26, 2003.

Barron, Tim, "Wireless Links for PCS and Cellular Networks," *Cellular Integration,* September 1995, pp. 20–23.

Bates, Gregory, *Voice and Data Communications Handbook,* McGraw-Hill, New York, 1998.

Black, Gregory, *TCP/IP and Related Protocols,* McGraw-Hill, New York, 1992.

Brewster, R.L., *Telecommunications Technology,* Wiley, New York, 1986.

Collins, Daniel, *Carrier Grade Voice over IP,* McGraw-Hill, New York, 2001.

DeRose, James E., *The Wireless Data Handbook,* Quantum Publishing, Mendocino, CA, 1994.

Dixon, Robert, *Spread Spectrum Systems,* 2d ed., Wiley, New York, 1984.

Held, Gil, *Voice & Data Interworking,* 2d ed., McGraw-Hill, New York, 2000.

IS-839

- TIA/EIA/IS-839: R-UIM Overview, Operation and File Structure Support in TIA/EIA-136-A, June 9, 2000.
- TIA/EIA/IS-841: Network Based Enhancements for User Identity Module (UIM), August 2000.

- TIA/EIA/IS-756-A, TIA/EIA-41-D: Enhancements for Wireless Number Portability Phase II, December 1998.
- TIA/EIA/IS-771: Wireless Intelligent Network, July 1999.
- TIA/EIA/IS-816: IDB Message Set Definitions for the Electrical Interface Between Portable Phone and Vehicle, 2000.
- TIA/EIA/IS-820: Removable User Identity Module (R-UIM) for TIA/EIA Spread Spectrum Standards, March 2000.
- TIA/EIA/IS-826, Wireless Intelligent Network Capabilities for Pre-Paid Charging, August 2000.

ITU M.3100: Generic Network Information Model, July 1995.

- TIA/EIA/J-STD-025: Lawfully Authorized Electronic Surveillance, 2000.
- TIA/EIA/J-STD-036: Wireless Enhanced Emergency Services, 2000.
- TIA/EIA/IS-835: cdma2000 Wireless IP Network Standard, December 2000.

McDysan, Spohn, *ATM Theory and Applications,* McGraw-Hill, New York, 1999.

Molisch, Andreas F., *Wideband Wireless Digital Communications,* Prentice-Hall, Englewood Cliffs, NJ, 2001.

Qualcomm, "An Overview of the Application of Code-Division Multiple Access (CDMA) to Digital Cellular Systems and Personal Cellular Networks," Qualcomm, San Diego, CA, May 21, 1992.

Smith, Clint, *Practical Cellular and PCS Design,* McGraw-Hill, New York, 1997.

Smith, Clint, *Wireless Telecom FAQ,* McGraw-Hill, New York, 2000.

3GPP2 A.R0003: Abis Interface Technical Report for cdma2000 Spread Spectrum Systems, December 17, 1999.

3GPP2 A.S0001-A: Access Network Interfaces Interoperability Specification, November 30, 2000.

3GPP2 A.S0004: Tandem Free Operation Specification, November 8, 2000.

3GPP2 C.S0007-0: Direct Spread Specification for Spread Spectrum Systems on ANSI-41 (DS-41) (Upper Layers Air Interface), June 9, 2000.

3GPP2 C.S0008-0: Multicarrier Specification for Spread Spectrum Systems on GSM MAP (MC-MAP) (Lower Layers Air Interface), June 9, 2000.

3GPP2 C.S0009-0: Speech Service Option Standard for Wideband Spread Spectrum Systems.

3GPP2 C.S0024: cdma2000 High Rate Packet Data Air Interface Specification, October 27, 2000.

3GPP2 C.S0025: Markov Service Option (MSO) for cdma2000 Spread Spectrum Systems, November 2000.

3GPP2 S.R0005-A: Network Reference Model for cdma2000 Spread Spectrum Systems, December 13, 1999.

3GPP2 S.R0021 Version 1.0: Video Streaming Services—Stage 1, July 10, 2000.

3GPP2 S.R0022: Video Conferencing Services—Stage 1, July 10, 2000.

3GPP2 S.R0023 Version 2.03: High-Speed Data Enhancements for cdma2000 1x—Data Only, December 5, 2000.

3GPP2 S.R0024 Version 1.0: Wireless Local Loop, September 22, 2000.

3GPP2 S.R0026: High-Speed Data Enhancements for cdma2000 1x—Integrated Data and Voice, October 17, 2000.

3GPP2 S.R0027: Personal Mobility, December 8, 2000.

3GPP2 S.R0032: Enhanced Subscriber Authentication (ESA) and Enhanced Subscriber Privacy (ESP), December 6, 2000.

3PP2 P.S0001-A-1 Version 1.03: Wireless IP Network Standard, December 15, 2000.

3GPP2 C.S0001-D Version 1.0: Introduction to cdma2000 Spread Spectrum Systems Revision D, February 2004.

3GPP2 R0023: High-Speed Data Enhancements for cdma2000 1x—Data Only.

3GPP2 S.R0048-A Version 3: 3G Mobile Equipment Identifier (MEID) Stage 1, September 2004.

3GPP2 P.S0001-B Version 2.0: cdma2000 Wireless IP Network Standard, September 2004.

3GPP2 S.R0023 Version 2.0: High-Speed Data Enhancements for cdma2000 1x—Data Only, December 5, 2000.

3GPP2 S.R0079-0 Version 1: Support for End-to-End, May 13, 2004.

3GPP2 S.S0083-A Version 1.0: Version Broadcast-Multicast Service Security Framework 126, August 2004.

3GPP2 S.R0021 Verson 1.0: Video Streaming Services—Stage 1, July 10, 2000.

3GPP2 S.S0078-A Version 2.0: Common Security.

TIA/EIA-41-D (Upgrade of TIA/EIA/IS-41-C Series): Cellular Radiotelecommunications Intersystem Operations.

TIA/EIA-98-C: Recommended Minimum Performance Standards for Dual-Mode Spread Spectrum Mobile Stations (Revision of TIA/EIA-98-B), November 1999.

TIA/EIA-637-B (Revision of TIA/EIA-637-A): Short Message Services for Wideband Spread Spectrum Systems, January 2002.

TIA/EIA-828: BTS-BSC Inter-Operability (Abis Interface), December 2001.

TIA/EIA IS-95-A: Mobile Station–Base Station Compatibility Standard for Dual-Mode Wideband Spread Spectrum Cellular System, May 1995.

TIA/EIA IS-97: Recommended Minimum Performance Standards for Base Stations Supporting Dual-Mode Wideband Spread Spectrum Cellular Mobile Stations, December 1994.

TIA/EIA IS-127: Enhanced Variable Rate Codec, Speech Service Option 3 for Wideband Spread Spectrum Digital Systems, September 1999.

TIA/EIA IS-637: Short Message Services for Wideband Spread Spectrum Cellular System, December 1995.

TIA/EIA IS-683-A: Over-the-Air Service Provisioning of Mobile Stations in Spread Spectrum Systems, May 1998.

TIA/EIA IS-707-A-3 (Addendum No. 3 to TIA/EIA/IS-707-A): Data Service Options for Spread Spectrum Systems Addendum 3—cdma2000 High Speed Packet Data Service Option 33, February 2003.

TIA/EIA IS-718: Minimum Performance Specification for the Enhanced Variable Rate Codec, Speech Service Option 3 for Spread Spectrum Digital Systems, July 1996.

TIA/EIA IS-733-1: High Rate Speech Service Option 17 for Wideband Spread Spectrum Communication System, September 1999.

TIA/EIA IS-736-A: Recommended Minimum Performance Standard for the High Rate Speech Service Option 17 for Spread Spectrum Communication Systems, September 6, 1999.

TIA/EIA IS-808: Incorporating UIM into 3G and IMT-2000 Systems, November 2000.

TIA/EIA IS-820: Removable User Identity Module (R-UIM) for cdma2000 Spread Spectrum Systems, June 9, 2000.

TIA/EIA IS-833: G3G CDMA-MC to GSM-MAP, March 2000.

TIA/EIA IS-834: G3G CDMA-DS to ANSI/TIA/EIA-41, March 2000.

TIA/EIA IS-2000-1: Introduction to cdma2000 Standards for Spread Spectrum Systems, June 9, 2000.

TIA/EIR IS-2000-2: Physical Layer Standard for cdma2000 Spread Spectrum Systems, September 12, 2000.

TIA/EIA IS-2000-3: Medium Access Control (MAC) Standard for cdma2000 Spread Spectrum Systems, September 12, 2000.

TIA/EIA IS-2000-4: Signaling Link Access Control (LAC) Specification for cdma 2000 Spread Spectrum Systems, August 12, 2000.

TIA/EIA IS-2000-6: Analog Signaling Standard for cdma2000 Spread Spectrum Systems, June 9, 2000.

TIA/EIA IS-2001: Interoperability Specification (IOS) for CDMA 2000 Access Network Interfaces, December 2000.

TIA-683-C (Revision of TIA/EIA-683-B): Over the Air Service Provisioning of Mobile Stations in Spread Spectrum Systems, March 2003.

TIA-801-A (Revision of TIA/EIA/IS-801): Position Determination Service for cdma2000 Spread Spectrum Systems.

TIA-820-B (Revision of TIA/EIA-820-A): Removable User Identity Module for Spread Spectrum Systems, May 2004.

TIA-864: Recommended Minimum Performance Standards for cdma2000 High Rate Packet Data Access Network Equipment, February 2002.

TIA-864-1[E]: Recommended Minimum Performance Standards for cdma2000 High Rate Packet Data Access Network Equipment, Addendum 1, January 2004.

TIA-866-1[E]: Recommended Minimum Performance Standards for cdma2000 High Rate Packet Data Access Terminal—Addendum, January 1, 2004.

TIA-870-1[E]: Test Data Service Option (TDSO) forcdma2000 Spread Spectrum Systems, Addendum 1, January 2004.

TIA-916: Recommended Minimum Performance Specification for TIA/EIA/IS-801-1 Spread Spectrum Mobile Stations, April 2002.

TIA-918: Signaling Conformance Standard for cdma2000 Wireless IP Networks, May 2002.

TIA-919: Signaling Conformance Standard for cdma2000: High Rate Packet Data Air Interface, May 2002.

TIA-926: Circuit Switched Video Conferencing Services, December 2002.

TIA-1006: CDMA2000 High Rate Broadcast-Multicast Packet Data Air Interface Specification GF, February 2004.

TIA-1015: File Formats for Multimedia Services for cdma2000 Spread Spectrum Systems, December 2003.

TIA-1016: cdma2000 Wideband Codec Algorithm Description, March 2004.

TIA-1030: Band Class Specification for cdma2000 Spread Spectrum Systems, March 2004.

TIA-1878: Interoperability Specification (IOS) for High Rate Packet Data (HRPD) Access Network Interfaces—Alternative Architecture, May 2003.

TIA-2000.1-D (Revision of TIA/IS-2000.1-C): Introduction to cdma2000 Spread Spectrum Systems, March 2004

TIA-2000.2-D (Revision of TIA/IS-2000.2-C): Physical Layer for cdma2000 Spread Spectrum Systems, March 2004.

TIA-2000.3-D (Revision of TIA/IS-2000.3-C): Medium Access Control (MAC) Standard for cdma2000 Spread Spectrum Systems, March 2004.

TIA-2000.4-D (Revision of TIA/IS-2000.4-C): Signaling Link Access Control (LAC) Standard for cdma2000 Spread Spectrum Systems, March 2004.

TIA-2000.6-D (Revision of TIA/IS-2000.6-C): Analog Signaling Standard for cdma2000 Spread Spectrum Systems, March 2004.

TIA-2001-C-1 (Supplement to TIA-2001-C): Interoperability Specification (IOS) for cdma2000 Access Network Interfaces Release C—Addendum 1, October 2003

Webb, William, "CDMA for WLL," *Mobile Communications International,* January 1999, p. 61.

Willenegger, Serge, "cdma2000 Physical Layer: An Overview," Qualcomm 5775, San Diego, CA. March 2000.

William, C. Y. Lee, *Lee's Essentials of Wireless Communications,* McGraw-Hill, New York, 2001.

William, C. Y. Lee, *Mobile Cellular Telecommunications Systems,* 2d ed., McGraw-Hill, New York, 1996.

www.fcc.gov

www.3gpp2.org

www.umtsworld.com

CHAPTER 5

TD-SCDMA

Time Division-Synchronous Code Division Multiple Access (TD-SCDMA) is a unique *third-generation* (3G) wireless mobile services standard that is being promoted by China. TD-SCDMA is also known as 3GPP TD-SCDMA (LCR) or low chip rate TDD. TD-SCDMA is a 3GPP standard which has a migration path to LTE-TDD. More specifically TD-SCDMA will evolve to HSxPA/EV and then 1xHSxPA/EV 1x will evolve to LTE-TDD. TD-SCDMA is perceived as a Chinese standard because the Chinese government owns the Intellectual Property Rights (IPR) associated with this technology. As of this writing, TD-SCDMA has seen only limited deployment and trial but no commercial deployment, even in China.

The TD-SCDMA standard is part of the *International Mobile Telephony 2000* (IMT-2000) group of specifications discussed previously in Chapter 1. TD-SCDMA differs from CDMA2000 and *Wideband Code Division Multiple Access* (WCDMA) in that it is a *Time-Division Duplex* (TDD) system as compared with *Frequency-Division Duplex* (FDD). The use of a TDD as the radio access method has several advantages over traditional FDD networks:

- TDD has no need for paired frequencies; it uses the same frequency for uplink and downlink transmission.
- TDD is suitable for asymmetric uplink and downlink transmission rates, especially for *Internet Protocol* (IP)–type data services.
- The major attribute of the TDD system is its spectral efficiency with asymmetric traffic such as IP. More specifically, TD-SCDMA adapts the uplink/downlink ratio according to the data load and provides data rates ranging from 1.2 kbps to 2 Mbps.

TD-SCDMA systems is one of the dominant mobile wireless access technologies in China.

For more extensive reading regarding TD-SCDMA (LCR), the following are excellent sources for information:

www.3GPP.org
www.tdscdma-forum.org
www.ccsa.org.cn

5.1 Generic TD-SCDMA Architecture

The fundamental architecture of a TD-SCDMA wireless system has a set of network components that are common to all 3G networks. TD-SCDMA uses a *Third Generation Partnership Project* (3GPP) core network while at the *radio access network* (RAN) and

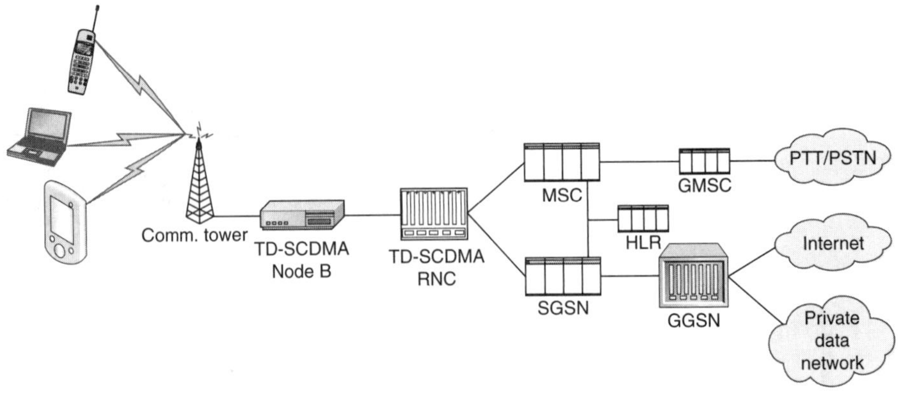

Figure 5.1 Generic TD-SCDMA network (single cell).

therefore can be connected to an existing *Global System for Mobile* (GSM) communications, WCDMA, or TD-CDMA network. However, the primary implication is to interface with an existing GSM network, providing a migration path to 3G. The RAN portion of TD-SCDMA, Node B, employs a combination of *Frequency-Division Multiple Access* (FDMA), *Time-Division Multiple Access* (TDMA), and *Code-Division Multiple Access* (CDMA) protocols.

Figure 5.1 provides a general depiction of a TD-SCDMA network that interfaces with a 3GPP core network. Only one TD-SCDMA cell is shown; in addition, other radio access systems are not shown either but could be included easily, provided that they interface at the *Mobile Switching Center* (MSC)/*Serving General Packet Radio Service (GPRS) Support Node* (SGSN).

TD-SCDMA has many unique characteristics. One such characteristic is the TD-SCDMA core network, which is an IP-based network following Releases 4 and 5 of the *Universal Mobile Telecommunications Service* (UMTS). The use of an IP-based core network facilitates many services and addresses the issue of interfacing with numerous appliances and services in the future.

Another unique aspect for TD-SCDMA is that it uses TDD as the access method, allowing for both synchronous and asynchronous operation. Because TD-SCDMA is a TDD network, it needs to reduce interference through a combination of interference-mitigation techniques that include the use of "smart" antennas and joint detection (mobile rake receiver) and the previously mentioned uplink synchronization. All these techniques help to reduce interference, thereby increasing pole capacity.

The RAN interface requires the use of asynchronous 1.6 MHz of spectrum, enabling it to have multiple carriers in the same bandwidth as a WCDMA, CDMA2000, or TD-CDMA carrier. There are two variations to TD-SCDMA: TD-SCDMA system for Mobile communication (TSM) and *low chip rate* (LCR). LCR requires that Release 4 be implemented in the core network.

TD-SCDMA is designed currently to operate in the 2010- to 2025-MHz frequency band, but since it is IMT-2000-compliant, it is possible, via the specification, to operate in all IMT-2000 bands.

5.2 Core Network

The core network associated with a TD-SCDMA wireless network should be either Release 4 or Release 5 depending on the legacy environment with which it must interface.

5.2.1 Release 4

Figure 5.2 shows the fundamental layout of a Release 4 core network. A Release 4 network introduces a significant enhancement to the core network architecture as it applies to the circuit-switched domain. The traditional MSC is broken into constituent parts and is allowed to be deployed in a distributed manner, as shown in the figure. Specifically, the MSC is divided into an MSC server and a *Media Gateway* (MGW). Release 4 removes the need for a class 5 circuit switch. Depending on the legacy system, the operator still may need to maintain a class 5 circuit switch so that the system can accommodate GSM voice and circuit-switched traffic.

The MSC server contains all the mobility management and call-control logic that would be contained in a standard MSC. It does not, however, reside in the media path. Rather, the media path is via one or more MGWs that establish, manipulate, and release media streams (e.g., voice streams) under control of the MSC server.

Control signaling for circuit-switched calls is between the *Radio Network Controller* (RNC) and the MSC server. The media path for circuit-switched calls is between the RNC and the MGW. Typically, the MGW takes calls from the RNC and routes those calls toward their destination over a packet backbone. The control protocol between the MSC server or *Gateway MSC* (GMSC) server and the MGW is the *International Telecommunications Union* (ITU) H.248 protocol, also known as *MEGACO* (standing for *Media Gateway Control*). The call-control protocol between the MSC server and the GMSC server can be any suitable call-control protocol.

Basically, the MSC is divided into an MSC server and an MGW. The MSC server contains all the mobility management and call-control logic that would be contained in

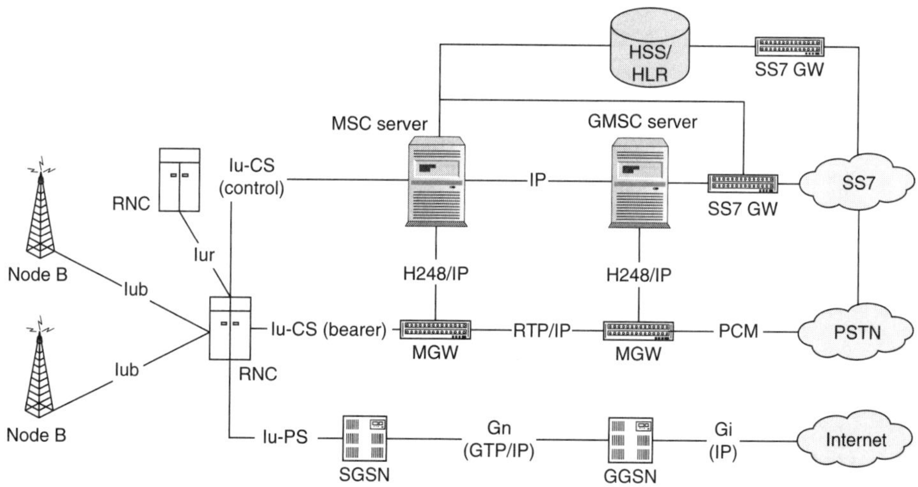

FIGURE 5.2 Release 4 core network.

a standard MSC. It does not, however, contain a switching matrix. The switching matrix is contained within the MGW, which is controlled by the MSC server and can be placed remotely from the MSC.

Control signaling for circuit-switched calls is between the RNC and the MSC server. The media path for circuit-switched calls is between the RNC and the MGW. Typically, an MGW will take calls from the RNC and route those calls toward their destination over a packet backbone. In many cases, that packet backbone will use the *Real Time Transport Protocol* (RTP) over the IP. Packet-data traffic from the RNC is passed to the SGSN and from the SGSN to the *Gateway GPRS Support Node* (GGSN) over an IP backbone. Given that data and voice both can use IP transport within the core network, a single backbone can be constructed to support both types of service. The control protocol between the MSC server or GMSC server and the MGW is the ITU H.248 protocol (MEGACO).

5.2.2 Release 5

Release 5 is the next step in the mobility evolution with the introduction of an all-IP multimedia network architecture. The architecture for a Release 5 core network is shown in Figure 5.3. The Release 5 core represents a change in the overall call model in which both voice and data are handled largely in the same manner all the way from the user terminal to the ultimate destination. This architecture can be considered the ultimate convergence of voice and data.

As we can see from Figure 5.3, there are no longer separate Iu-CS and Iu-PS interfaces as in Release 4. There is a single Iu interface that carries all media. Within the core network, that interface terminates at the SGSN—there is no separate media gateway.

There are a number of new network elements, notably the *Call State Control Function* (CSCF), the *Multimedia Resource Function* (MRF), the *Media Gateway Control Function* (MGCF), the *Transport Signaling Gateway* (T-SGW), and the *Roaming Signaling Gateway* (R-SGW).

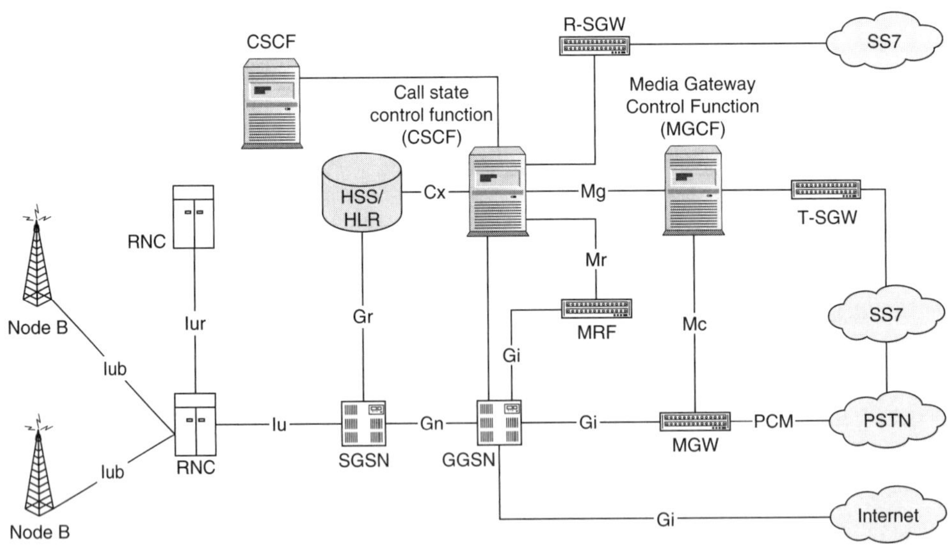

Figure 5.3 Release 5.

An important aspect of the all-IP architecture is the fact that the *user equipment* (UE) is greatly enhanced. Significant logic is placed within the UE. In fact, the UE supports the *Session Initiation Protocol* (SIP). The UE effectively becomes an SIP user agent. As such, the UE has far greater control of services than previously.

The SGSN and GGSN are enhanced versions of the same nodes used in Release 4. The difference is that these nodes, in addition to data services, now support services that traditionally have been circuit-switched, such as voice. Consequently, appropriate *quality-of-service* (QoS) capabilities need to be supported either within the SGSN and the GGSN or, at a minimum, in the routers immediately connected to them.

The MRF is a conference bridging function used for support of features such as multi- party calling and meet-me conference service. The T-SGW is a *Common Channel Signaling System 7* (CC7)/*Signaling System 7* (SS7) gateway that provides CC7 interworking with standard external networks such as the *Public Switched Telephone Network* (PSTN). The T-SGW will support SIGTRAN protocols. The R-SGW is a node that provides signaling interworking with legacy mobile networks that use standard SS7. In many cases, the T-SGW and R-SGW will exist within the same platform.

The MGW performs interworking with external networks at the media path level. The MGW in the 3GPP Release 5 network architecture is the same as the equivalent function within the 3GPP Release 4 architecture. The MGW is controlled by a *Media Gateway Control Function* (MGCF). The control protocol between these entities is ITU-T H.248. The MGCF also communicates with the CSCF. The protocol of choice for that interface is SIP.

5.3 Radio Network

The radio access network for TD-SCDMA has many unique characteristics that require some level of discussion. For more detail and specificity the Web site http://www.tdscdma-forum.org is recommended as an excellent starting point.

TD-SCDMA supports both circuit and packet services. Circuit-switched rates are defined as 12.2, 64, 144, 384, and 2048 kbps. Packet-data rates are defined as 9.6, 64, 144, 384, and 2048 kbps.

The TD-SCDMA radio access environment is shown in Figure 5.4. The most significant and interesting aspect with a TD-SCDMA system is the fact that the RAN uses a TDD method of access. The TDD RAN is unique in that it uses a 1.6-MHz-wide carrier, unlike the other 3G RAN networks, which use FDD paired 1.25 MHz (CDMA2000) or paired 5 MHz (WCDMA). Of course, there is another 3G RAN network that uses TDD, but it requires use of a 5-MHz channel.

5.3.1 Radio Spectrum

The radio spectrum required for a TD-SCDMA network can start with as little as 1.6 MHz of contiguous spectrum for an individual channel with possible use of a guard band of 100 to 200 kHz. The guard-band requirements will need to be specified when differing technologies are spectrally adjacent to each other. In many countries around the world, TDD spectrum has been allocated for potential use for mobility. In China, where TD-SCDMA is being heavily sponsored, a total of 155 MHz of frequency bandwidth has been allocated for TDD. Simple math indicates that with a 1.6-MHz carrier, a total of 96 unique TD-SCDMA carriers are possible.

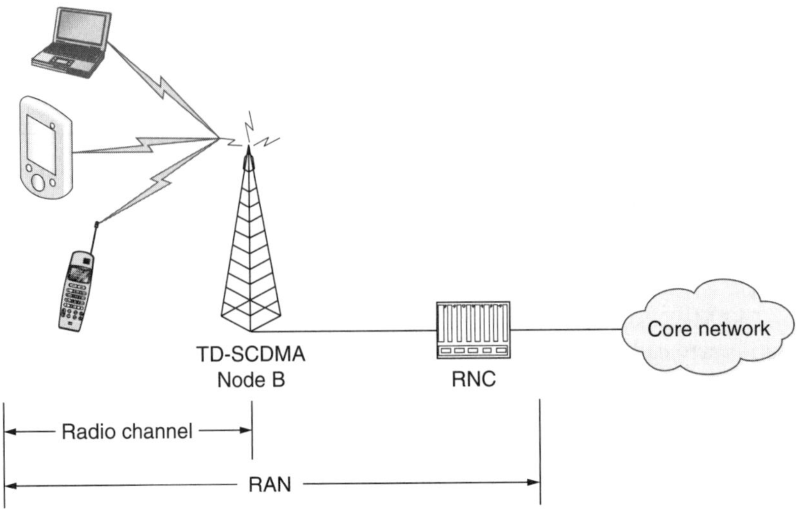

FIGURE 5.4 TD-SCDMA RAN.

Presently, the band being favored for TD-SCDMA is the 2010- to 2025-MHz band. However, TD-SCDMA is IMT-2000-compliant, so it is possible, via the specification, to operate in all IMT-2000 bands.

5.3.2 Channel Structure

The unique frame structure of the TD-SCDMA radio channel enables it to be more adaptive to network evolution. TD-SCDMA network construction follows the principle of a layered approach to design and implementation. The TD-SCDMA radio access scheme is different from those of other 3G systems in that it uses TDD, not FDD. Therefore, TD-SCDMA, because it uses a TDD mode, does not allocate resources in a symmetric fashion. Instead, it allocates resources for the uplink and downlink separately, even though the uplink and downlink are in the same carrier.

Figure 5.5 shows the channel structure for a TD-SCDMA carrier. There are a total of seven time slots for each TD-SCDMA carrier. Each of the carriers has a radio frame that is 10 ms in length, as shown in the figure. The radio carrier is made up of two separate subframes, and each subframe is made up of seven time slots.

Additionally, Figure 5.5 shows both TDD and codes in the diagram to help illustrate their relationship because TD-SCDMA uses time and codes. In each subframe shown in the figure, *Time slot 0* (TS0) acts as the *Broadcast Channel* (BCH). TS0 also sends downlink information and is always defined for downlink transmission. Conversely, TS1 is a dedicated uplink time slot. Both TS0 and TS1 use a forward and reverse pilot DwPTS for downlink and UpPTS for the uplink and are used for synchronization purposes. Gp in the figure is a guard period needed to prevent self-interference between uplink and downlink bursts.

TD-SCDMA uses a spreading factor of 16, and this is applied to each of the individual time slots. Therefore, each time slot has a total of 16 spreading codes.

The TDD principles permit traffic to be uplinked and downlinked using the same frame and different time slots. Additionally, for symmetric services used during circuit-switched

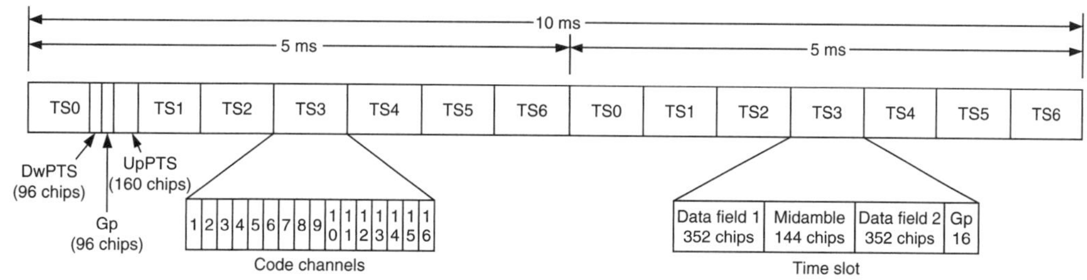

FIGURE 5.5 TD-SCDMA channel structure.

services such as telephone calls, the time slots are split equally between the downlink and uplink. This capability of adapting the uplink/downlink format according to the data loads within a single frequency band increases the overall capacity of the air interface. This makes TDMA/TDD operations ideal for 3G services.

5.3.3 Spreading Codes

A total of 16 spreading codes are used with TD-SCDMA. The codes are all orthogonal to each other and, like all CDMA codes, are part of a tree. The tree for the code sequence dictates the potential data throughput that can take place.

Figure 5.6 shows the current spreading-code tree for a TD-SCDMA system. The tree indicates the relationship between the various codes and the overall throughput and capacity for the radio carrier itself.

Remembering that there are two frames in a TDD sequence, the following is a quick illustration. If code 16,10 is in operation, the 8,5 through 8,8 and 4,4, 4,3, 2,2, and 1,1 are not available for use. Therefore, when a lower code on the tree is used, all codes above it are removed from possible use. Now, if we want 2048 kbps of data throughput, this will require the use of all the spreading codes. The 1.1 x2 will enable a data rate of 2048 kbps; if a lower-rate data user is operational on the radio channel at the same time, the effective data rate would drop to about 1536 kbps.

5.3.4 Logical Channel

There are numerous logical channels in a TD-SCDMA system. All the logical channels are basically one of two types. The two channel types are *control* and *traffic* channels, and they are shown in Table 5.1

The common control channel consists of the BCH and the PCH:

- *Broadcast Channel* (BCH). The BCH is allocated to at least one radio unit per cell or sector and is assigned to the first downlink time slot TS0 (see Figure 5.5). The BCH also operates at a higher power than the rest of the time slots, usually 9 to 11 dB higher.
- *Paging Channel* (PCH). The PCH is a special broadcast channel used for paging UEs and is assigned to the first downlink TS0, as is the BCH, and is transmitted with the same power level and antenna pattern as those of the BCH.
- *Forward Access Channel* (FACH). The FACH is used to respond to a UE's random access request and is paired with the *Radio Access Channel* (RACH). The FACH can be mapped to more than one radio unit and is assigned to the first (Td0) or

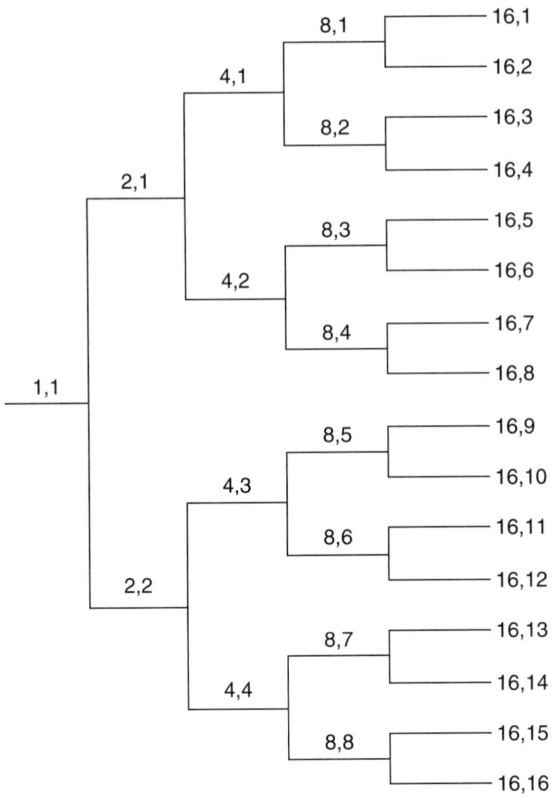

FIGURE 5.6 Spreading-code tree.

Type		Description
Control Channel (CCH)	SCCH	Synchronization Control Channel
	BCCH	Broadcast Control Channel
	PCCH	Paging Control Channel
	DCCH	Dedicated Control Channel
	CCCH	Common Control Channel
	ODCCH	ODMA Dedicated Control Channel
	OCCCH	ODMA Common Control Channel
	SHCCH	Shared Channel Control Channel
Traffic Channel (TCH)	DTCH	Dedicated Traffic Channel
	ODTCH	ODMA Dedicated Traffic Channel
	CTCH	Common Traffic Channel

TABLE 5.1 Logical Channels

second (Td1) downlink time slot per subframe. The location and number of the FACH are indicated on the BCH.

- *Random Access Channel* (RACH). The RACH is designed for uplink random access by the UE and is mapped to more than one radio unit; it is assigned to the first uplink time slot (Tu0) per subframe. The same time slot may be used for the RACH by more than one cell. Multiple transmissions using different codes could be received in parallel. The location allocated to the RACH in the time slots is indicated on the BCH. The RACH uses both power control and uplink synchronization control.

System information is transmitted on logical channel BCCH and then mapped to transmission channels BCH and FACH.

5.3.5 Codes

The TD-SCDMA system uses a variety of codes that are based on principles used in CDMA systems. The TD-SCDMA uses a chip rate of 1.28 Mcps as the heart of its coding scheme. There are three primary codes that are of immediate importance to the systems designer:

- Scramble
- Sync
- Sync1

The Scramble code is used to uniquely identify the Node B cells. A set of 32 codes is used as the Scramble codes. Because the Scramble code is cell-specific, the Sync code and the Scramble code are indexed together. The index is the same with the Sync code index (i.e., the Scramble codes and Sync codes are in one-to-one relationship).

The Sync code is used for downlink synchronization in the DwPTS, and there are a total of 32 unique codes. The Scramble/Sync codes are a gold code set and are designed to distinguish nearby cells for the purpose of easier cell measurement. The set of codes could be repeated in the cellular network. Typically, only 19 or 22 codes are used for traditional planning, with the remaining being held in reserve to resolve pilot pollution problems. The Sync1 codes, on the contrary, are used to achieve uplink synchronization and are associated with the UpPTS. There are a total of 128 Sync1 codes.

5.4 Interference-Mitigation Techniques

TD-SCDMA, because it uses a TDD method for access, has unique self-interference problems that need to be addressed in order to take advantage of system capabilities. To help mitigate interference, TD-SCDMA uses a combination of interference-mitigation techniques that include the use of "smart" antennas and joint detection, terminal (uplink) synchronization, and dynamic channel allocation. All these techniques help reduce interference, thereby increasing pole capacity.

- Smart antennas
- Joint detection
- Terminal synchronization
- Dynamic channel allocation

5.4.1 "Smart" Antennas

By design, all TD-SCDMA base stations are equipped with smart antennas. The use of smart antennas improves the signal-to-interference ratio of the mobile terminal by about 8 dB.

Effectively, a smart antenna can compress the interference among users that belong to the same time slot of a cell, reduce the transmit power, and improve system coverage and capacity. The smart antennas employed by TD-SCDMA technology are not conventional diversity beam-switching antennas but advanced beam-forming (and beam-steering) bidirectional adaptive antenna arrays.

The individual directivity between base stations and mobile terminals is achieved by a concentric array of eight antenna elements with programmable electronic phase and amplitude relations. Terminal tracking is performed by fast angle-of-arrival measurements in intervals of 5 ms 200 times per second.

The use of beam-steering antennas enables the systems to better track mobile usage through the cell and distribute the power only to cell areas where mobile subscribers are located. The focusing of energy reduces multiuser interference and therefore increases system capacity by minimizing intracell interference, increasing reception sensitivity, and lowering transmission power while increasing cell range.

The use of smart antennas along with *Time of Arrival* (TOA) mitigates the need for a *Global Positioning Satellite* (GPS) unit in the handset itself. Accuracy for location is within 20- to 30-m resolution with a 1-degree azimuth.

5.4.2 Joint Detection

Basically, *Joint Detector* (JD) allows the TD-SCDMA receiver to estimate the radio channel and works for all signals simultaneously. Through parallel processing of individual traffic streams, JD eliminates *Multiple-Access Interference* (MAI) and minimizes intracell interference, thus increasing transmission capacity. MAI occurs when the receiver cannot distinguish the desired signal from other signals and appears as increased background noise. MAI limits the traffic load per radio carrier.

To reduce the MAI, a matched-filter correlator called a *Joint Detector* (JD) is used. TD-SCDMA technology facilitates JD receivers in the Node B as well as in the terminal. The efficiency of the JD receiver in TD-SCDMA technology is based on the TDMA/TDD operation and on the limited number of codes employed per time slot.

5.4.3 Terminal Synchronization

Terminal synchronization is achieved through accurately tuning the transmission timing of each individual terminal uplink. In a TD-SCDMA system, this improves the knowledge of where the terminal is in relationship to the cell, thereby reducing time for position-location calculation and search time for handovers. Because of terminal synchronization, TD-SCDMA does not need soft handovers, which are used in CDMA2000 and WCDMA.

5.4.4 Dynamic Channel Allocation

Dynamic Channel Allocation (DCA) is also incorporated to reduce intercell interference. Because TD-SCDMA uses TDMA, FDMA, and CDMA, it is able to allocate radio resources in an adaptive fashion to minimize the interference. Four types of DCA are employed, correlated with each type of access method used:

- *Time Domain.* Traffic is allocated to the least-interfered time slots.
- *Frequency Domain.* Traffic is allocated to the least-interfered carrier when more than one is available.
- *Code Domain.* Traffic is allocated to the least-interfered codes.
- *Space Domain.* Smart antennas minimize interference by targeting the mobile unit.

5.5 RAN Traffic Planning

Proper RAN traffic planning for a TD-SCDMA system should be handled via a computer simulation coupled with real field data. Integral to the traffic-planning effort is a properly tuned propagation model that can spread the traffic under multiple interference and loading situations, including data and voice.

Besides the use of a good traffic-modeling program, the proper inputs are required. Table 5.2 lists some of the general system parameters pertaining to a TD-SCDMA system. Table 5.3 shows the various mobile and terminal power classes and their respective transmit powers.

Carrier bandwidth: 1.6 MHz
Duplex type: TDD
Multiple-access scheme: TDMA, CDMA, FDMA
Chip rate: 1.28 Mcps
Modulation: QPSK, 8-PSK
Maximum cell range: 40 km
Maximum voice capacity EFR: 55 Erlangs
Data throughput: 6 Mbps
Theoretical maximum data rate/user: 2 Mbps
System asymmetry (DL:UP): 1:6–6:1
Frequency reuse: 1
EFR = (Enhanced Full Rate): 12.2 kbps

Table 5.2 TD-SCDMA RAN System Parameters

Power Class	Max. Tx Power
1	+33 dBm
2	+30 dBm
3	+27 dBm
4	+24 dBm
5	+21 dBm
6	+10 dBm

Table 5.3 Mobile Class Power

	E_b/I_o			
	Static	Indoor Office, 3 km/h	Outdoor to Indoor and Pedestrian, 3 km/h	Vehicular, 120 km/h
Speech (BER × 10^{-3})	8	8	8	9
Data (BER × 10^{-6})				
64 kbps	8	8	8	10
144 kbps	10	10	10	12
384 kbps	10	10	10	NA
2048 kbps	12	NA	NA	NA

TABLE 5.4 E_b/I_o

The reported throughputs for data with TD-SCDMA are

- 8–384 kbps—high mobility
- Up to 2 Mbps—low mobility

However, to achieve these throughputs, the proper *radiofrequency* (RF) environment needs to be in place, assuming that the core network is designed to support the RF environment. The RF environment is defined through E_b/I_o. The E_b/I_o for various environments is listed in Table 5.4.

An important aspect with the RF planning and capacity management is the fact that with TD-SCDMA there is no soft or softer handoff, which typically is used in a CDMA network. This simple concept frees up resources at surrounding sites, enabling more capacity to be supported by the system itself.

Radio systems, regardless of technology, need to provide coverage to be successful. The coverage radius of TD-SCDMA is decided by two factors: (1) the transmit power and reception sensitivity and (2) the length of the guard period. With the adoption of smart antennas and joint detection, the requirements for transmit power and reception sensitivity in TD-SCDMA are reduced.

However, as with all CDMA-based systems, coverage and capacity are inversely related. With a TDD system, a guard period is needed between the uplink and downlink time slots that actually defines the radius of the cell site. With a cell offering maximum capacity, the maximum coverage radius of TD-SCDMA is about 11.25 km. When larger coverage areas are needed, such as for rural applications, capacity must be reduced to achieve greater coverage.

5.6 Handover

For the TD-SCDMA system, the following parameters are needed in the cell-selection monitoring set:

- SIR
- Path Loss

- Interference power
- Received power level on the BCH

The handover process is implemented in the mobile unit and the RNS. Measurements of serving radio connection downlink performance and candidate cell received signal strengths and quality are made in the UE. The RNS measures the uplink performance, as well as the position information for the UE being served, and uses these measurements in conjunction with defined thresholds and handover strategy to make a hand-over decision.

5.7 Implementation

Multiple scenarios exist for possibly implementing TD-SCDMA. However, all these implementation methods fall into one of two categories: new spectrum or overlay. Initially, TD-SCDMA deployment will not be as contiguous and extensive as that of existing wireless networks. It is also anticipated that the 3G RAN network will be an overlay on the 2G network and use the latter, in the minimum case, as a fallback to ensure continuity of service and maintain a good QoS as perceived by the user.

The new-spectrum approach involves having uncontested spectrum from which to implement a new system. Figure 5.7 presents an example of a new-system spectrum allocation where 5 MHz is allocated for possible use. The figure shows only a TDD spectrum allocation, but if an FDD spectrum allocation were shown, then this would equate to twice the spectrum for a TDD system. The system configuration for a new stand-alone system, however, resembles the one depicted in Figure 5.1.

The likely allocation for most operators will involve an overlay approach. Figure 5.8. illustrates the high-level configuration for an overlay approach for a TD-SCDMA system being placed into a GSM network. Figure 5.9 shows the spectrum allocation where the GSM system must integrate a TD-SCDMA network into its existing spectrum.

An important aspect of a TD-SCDMA overlay approach is that when the TD-SCDMA base stations are installed in an existing GSM system, the radio network layout can be retained, and existing transmission links can be reused. The obvious issue, however, with an overlay of a TDD system or any system into an existing network involves the displacement of spectrum and the need to preseed the subscriber base with dual-mode handsets, along with the need to expand the legacy system rapidly to accommodate the lack of capacity owing to the reduced spectrum availability.

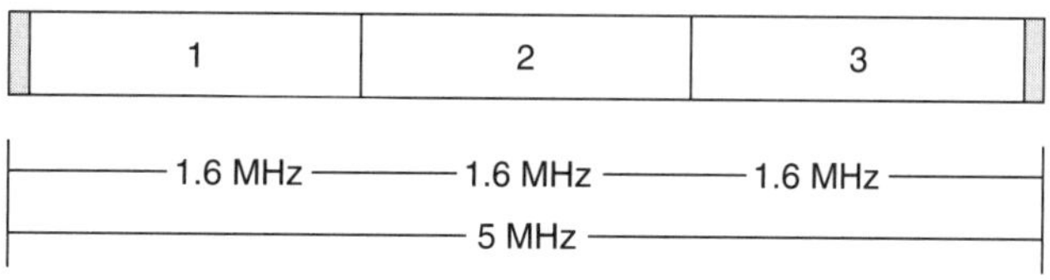

FIGURE 5.7 5-MHz spectrum allocation.

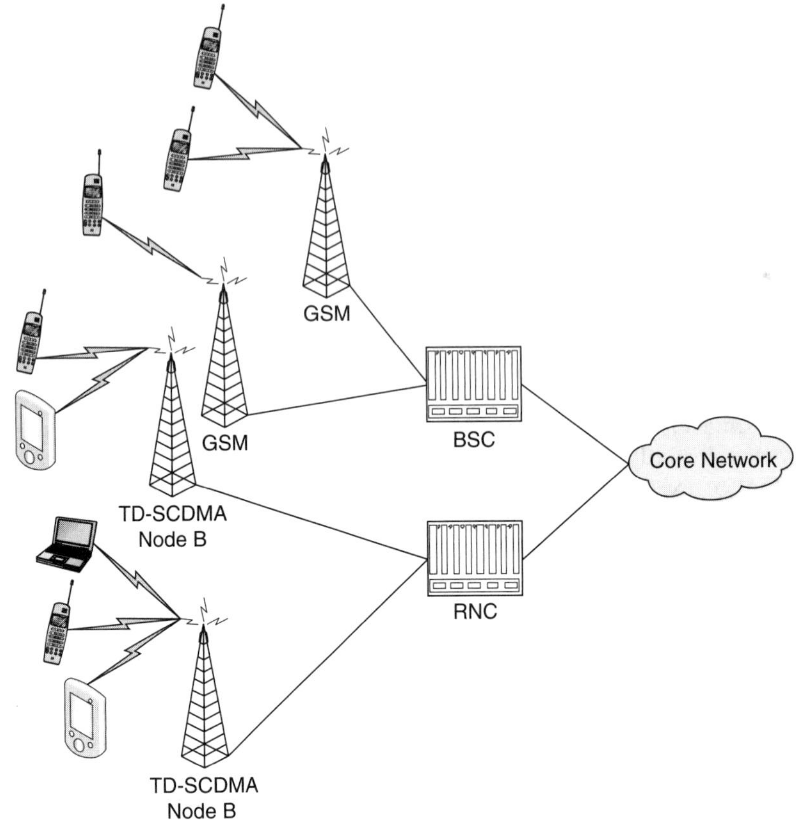

FIGURE 5.8 Overlay approach.

FIGURE 5.9 Spectrum segmentation.

References

Chen, Xinhua, and Liu Hu, *Technology Advantage of TD-SCDMA,* TDIA, February 2005.
CWTS TS C001 (Version 2.0.0): Radio Interface Protocol Architecture.
CWTS TS C002 (Version 2.0.0): Services Provided by the Physical Layer.

CWTS TS C003 (Version 2.0.0): UE Functions and Interlayer Procedures in Connected Mode.
CWTS TS C004 (Version 2.0.0): UE Procedures in Idle Mode.
CWTS TS C101 (Version 2.0.0): Physical Layer—General Description.
CWTS TS C102 (Version 2.0.0): Physical Channel and Mapping of Transport Channels onto Physical Channels.
ETSI GSM 05.08 (ETS 300 911): Digital Cellular Telecommunications System (Phase 2+); Radio Subsystem Link Control, Version 5.4.0.
Liu, Yuqin, and Xinhua Chen, *Market Advantage of TD-SCDMA,* TDIA, February 2005.
3GPP TS S25.221: Physical Channels and Mapping of Transport Channels onto Physical Channels.
3GPP TS 25.302 Version 2.3.0: Services Provided by the Physical Layer.
TS C001 Version 3.0.0 (1999-10): Radio Interface Protocol Architecture.
TS C002 Version 3.0.0 (1999-10): Services Provided by the Physical Layer.
TS C004 Version 2.1.0 (1999-10): UE Procedures in Idle Mode.
TS C101 Version 3.0.0 (1999-10): Physical Layer General Description (1999-10).
TS C102 Version 3.0.0 (1999-10): Physical Channels and Mapping of Transport Channels onto Physical Channels.
TS C103 Version 2.2.0 (1999-10): Multiplexing and Channel Coding.
TS C104 Version 3.0.1 (1999-10): Spreading and modulation.
TS C105 V3.0.0 (1999-10): Physical Layer Procedures.
TS C106 V2.0.0 (1999-10): Physical Layer—Measurements (TD-SCDMA).
TS C401 Version 3.0.0 (1999-10): TD-SCDMA (UE); Radio Transmission and Reception.
TS C402 Version 3.0.0 (1999-10): Node B, Radio Transmission and Reception.
TS C403 Version 1.2.0 (1999-10): RF Parameters in Support of Radio Resource Management.
www.ccsa.org.cn.
www.tdscdma-forum.org.
www.3gpp.org.

CHAPTER 6

TD-CDMA

Time Division-Code Division Multiple Access (TD-CDMA) is a *third-generation* (3G) *radio access network* (RAN) platform that uses the same core network as *Wideband Code Division Multiple Access* (WCDMA) and Time Division-Synchronous Code Division Multiple Access (TD-SCDMA). TD-CDMA and TD-SCDMA have an evolution path to LTE-TDD; however, LTE-TDD uses a different radio access modulation format covered later in this book than used by both TD-CDMA and TD-SCDMA. Time Division-Code Division Multiple Access (TD-CDMA) is also referred to as High Chip Rate (HCR) *Time-Division Duplex* (TDD) and *Universal Mobile Telecommunications Service* (UMTS) TDD and is an *International Mobile Telephony 2000* (IMT-2000)–compliant air-interface specification using packet data. Additionally, the TD-CDMA system interfaces with a *Global System for Mobile* (GSM) communications/UMTS core network, which allows it to be overlaid on top of the existing network or built as a new system.

The TD-CDMA RAN is well suited for high data traffic, which includes in-building applications. More specifically, it can be deployed in both licensed and unlicensed spectrum, facilitating the use of cordless phones and a host of wireless *Local Area Network* (LAN) applications. TD-CDMA is a robust access method that uses a combination of the three basic multiple-access schemes—*Frequency-Division Multiple Access* (FDMA), *Time-Division Multiple Access* (TDMA), and *Code-Division Multiple Access* (CDMA)—requiring a total of 5 MHz of frequency as compared to 10 MHz (5 × 5) for WCDMA.

The use of a TDD access method with TD-CDMA is designed to support asymmetric traffic such as *Internet Protocol* (IP) traffic. TD-CDMA systems use unpaired spectrum, enabling them to be more spectral efficient than a WCDMA network when supporting IP traffic, unlike circuit-switched voice traffic. TD-CDMA also supports circuit-switched traffic and *Quality of Service* (QoS) for *Voice over IP* (VoIP).

The TD-CDMA air interface consists of 15 time slots that can be assigned by the operator for uplink or downlink use. Of the 15 total slots, 12 typically are allocated for traffic and 3 for signaling with the subscriber units. Each of the time slots assigned to carry traffic also uses 16 unique spreading codes, one for each user of segment of bandwidth. Each code is effectively a separate channel, and each user or device can be allocated an individual code or group of codes, facilitating higher data throughput.

The TD-CDMA system also can assign a different modulation and coding scheme to each time slot in order to maximize the throughput for each of the subscribers in that sector in support of different services at the same time. With TD-CDMA, the uplink and downlink transmissions alternate on the same carrier, and the wireless operator actually allocates the ratio of uplink to downlink channels for each carrier based on actual or projected traffic patterns for the site/sector.

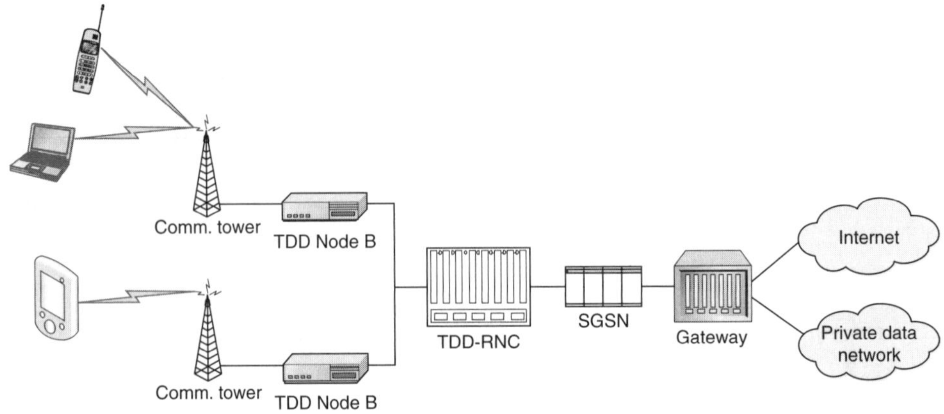

FIGURE 6.1 TD-CDMA system.

TD-CDMA has many similar characteristics to TD-SCDMA, except that it uses a chip rate of 3.84 Mcps, whereas TD-SCDMA uses 1.28 Mcps. TD-CDMA's use of 3.84 Mcps harmonizes it with WCDMA for a symbiotic deployment. When a 10-MHz channel is available, a chip rate of 7.68 Mcps is used, but the predominant implementation is the harmonized 3.84 Mcps.

TD-CDMA systems offer peak speeds of 7 Mbps per sector with a single carrier and up to 16 Mbps with two carriers. Higher rates are expected in future releases and enhancements of TD-CDMA, with rates expected to increase to 14 Mbps for a single carrier and 31.5 Mbps for a dual carrier. The dual carrier is important because 3G spectrum typically is allocated in a *Frequency-Division Duplex* (FDD) method. The throughput is achieved as a result of the on-demand method for bandwidth allocation and assignment of resources. Therefore, critical resources are allocated on an as-needed basis, and this is done on a shared basis in both the uplink and downlink directions, making it more efficient for bursty (IP) traffic.

Another variant of the TD-CDMA system is FDD-TDCDMA, which has increased bandwidth (10 MHz instead of 5 MHz). This is achieved by pairing channels for increased throughput by allowing more time slots to be concatenated.

6.1 Generic TD-CDMA Architecture

The TD-CDMA network architecture is very similar to that of other IMT-2000 networks. In fact, the RAN interface is the primary difference where the core network uses Release 99, Release 4, and Release 5. Figure 6.1 shows a sample TD-CDMA network that interfaces with a *Serving General Packet Radio Service (GPRS) Support Node* (SGSN), enabling it to be used as a data-solution offering for a GSM/GPRS/*Enhanced Data Rates for Global Evolution* (EDGE), WCDMA, and even an SD-CDMA system.

6.2 Core Network

The fundamental architecture of a TD-CDMA system follows the core-network architecture using Release 4 and Release 5 evolution, which was designed by the *Third Generation Partnership Project* (3GPP) community to accommodate flexibility for future

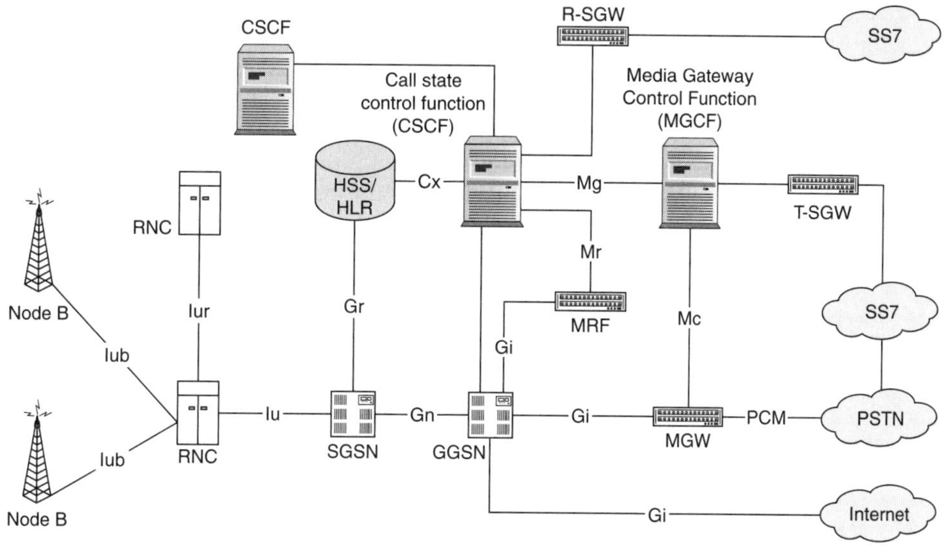

FIGURE 6.2 TD-CDMA IMS.

performance enhancements. Detailed descriptions of Release 99, Release 4, and Release 5 are included in Chapter 3.

The ability of TD-CDMA to use a host of core-network architectures allows it to be integrated into existing legacy systems without having to rip and replace the various core-network platforms. When reviewing Release 5, which is really an IMS core-network platform, the advantages of an all-IP network in conjunction with a TD-CDMA RAN can be exploited to deliver a host of IP-based services.

Figure 6.2 shows a TD-CDMA network that is an all-IP multimedia network, that is, IMS. The architecture of the core network shown in the figure represents a change in the overall call model in which both voice and data largely are handled in the same manner all the way from the user terminal to the ultimate destination. This architecture can be considered the ultimate convergence of voice and data.

As we can see in Figure 6.2, there are no longer separate Iu-CS and Iu-PS interfaces but a single Iu interface that carries all media, which is packetized. Within the core network, the interface terminates at the SGSN—there is no separate media gateway for voice and data.

There are a number of new network elements, notably the *Call State Control Function* (CSCF), the *Multimedia Resource Function* (MRF), the *Media Gateway Control Function* (MGCF), the *Transport Signaling Gateway* (T-SGW), and the *Roaming Signaling Gateway* (R-SGW).

An important aspect of the all-IP architecture is the fact that the *user equipment* (UE) is greatly enhanced. Significant logic is placed within the UE. In fact, the UE supports the *Session Initiation Protocol* (SIP). The UE effectively becomes an SIP user agent. As such, the UE has far greater control of services than previously.

The SGSN and *Gateway GPRS Support Node* (GGSN) are enhanced versions of the same nodes used in Release 4. The difference is that these nodes, in addition to data services, now support services that traditionally have been circuit-switched such as voice. Consequently, appropriate QoS capabilities need to be supported either within the SGSN and GGSN or, at a minimum, in the routers immediately connected to them.

The MRF is a conference bridging function used for support of features such as multiparty calling and meet-me conference service. The T-SGW is a *Common Channel Signaling System 7* (CC7)/*Signaling System 7* (SS7) gateway that provides CC7 interworking with standard external networks such as the *Public Switched Telephone Network* (PSTN). The T-SGW will support SIGTRAN protocols. The R-SGW is a node that provides signaling interworking with legacy mobile networks that use standard SS7. In many cases, the T-SGW and R-SGW will exist within the same platform.

The media gateway (MGW) performs interworking with external networks at the media path level. The MGW in the 3GPP Release 5 network architecture is the same as the equivalent function within the 3GPP Release 4 architecture. The MGW is controlled by a *Media Gateway Control Function* (MGCF). The control protocol between these entities is ITU-T H.248. The MGCF also communicates with the CSCF. The protocol of choice for this interface is SIP.

6.3 Radio Network

The RAN for TD-CDMA has many unique characteristics that require some level of discussion. It is recommended that for more detail and specificity, the Web site www.umtstdd.org is an excellent starting point.

TD-CDMA supports both circuit and packet services and is designed primarily to support the asymmetric characteristics of IP data, enabling broadband to the edge of the network. TD-CDMA uses TDD, and therefore, uplink and downlink traffic share the same physical radio channel. However, each radio channel has 15 time slots. Additionally, in each time slot, up to 16 separate codes, which are orthogonal to each other, can be transmitted, and each user can be allocated individual codes or groups of codes for higher data throughput.

Uplink and downlink channel allocation can be done by the system or the operator. A minimum of one uplink and one downlink always need to be allocated per TD-CDMA carrier.

In order to support both packet and circuit services, TD-CDMA uses both shared channels and dedicated channels. This is in comparison with an FDD system such as WCDMA, which supports only dedicated channels on the uplink. TD-CDMA, however, supports shared channels in both directions, enabling it to be more efficient with bursty traffic such as IP.

6.3.1 Radio Spectrum

TD-CDMA requires 5 MHz of radio bandwidth to operate a single channel. The 5-MHz carrier uses a chip rate of 3.84 Mcps, and this is meant to facilitate coexistence with WCDMA networks, enabling operators to deploy both technologies in the same market.

TD-CDMA can operate on all the IMT-2000 frequency bands. However, deployment of TD-CDMA has focused on the following frequency bands:

- 450–480 MHz
- 700 MHz (U.S. band)

- 1900–1920 MHz (IMT-2000 3G band)
- 2010–2025 MHz (IMT-2000 3G band)
- 2053–2082 MHz
- 2300–2400 MHz (China and Korea TDD band, WCS band in the United States, Australian MMDS band)
- 2500–2690 MHz (MMDS/ITFS band in the United States, IMT-2000 extension band internationally)
- 3400–3600 MHz (international FWA band)

6.3.2 Channel Structure

The channel structure for TD-CDMA radio access uses TDMA, FDMA, and CDMA. In addition, TD-CDMA uses the same chip rate, modulation, and bandwidth (nonduplex) that is used by WCDMA.

The heart of TD-CDMA radio access is the TDD access method. Each radio carrier is divided into 15 time slots, each containing 16 separate and unique codes; this is depicted in Figure 6.3.

The specific uplink and downlink transmissions alternate on the same *Radio-frequency* (RF) channel by allocating time slots to either the uplink or the downlink, with the ratio determining the relative uplink and downlink bandwidths.

An important aspect is that TD-CDMA supports both packet data (IP) and circuit-switched voice. Packet data is, of course, asymmetric in nature, having different uplink and downlink allocations. The resource-allocation scheme is different when supporting circuit-switched traffic in that uplink and downlink resources are allocated on a symmetric method, that is, the same allocations for uplink and downlink channels. The time slots for TD-CDMA, however, are not allocated on a dynamic basis; the changing traffic patterns need to be monitored, and uplink and downlink allocations need to be engineered and optimized on a regular basis.

Two modulation schemes are used by TD-CDMA systems: QPSK and 16QAM. Additionally, the downlink channel is synchronized through the use of GPS, whereas the uplink channel uses transmission control.

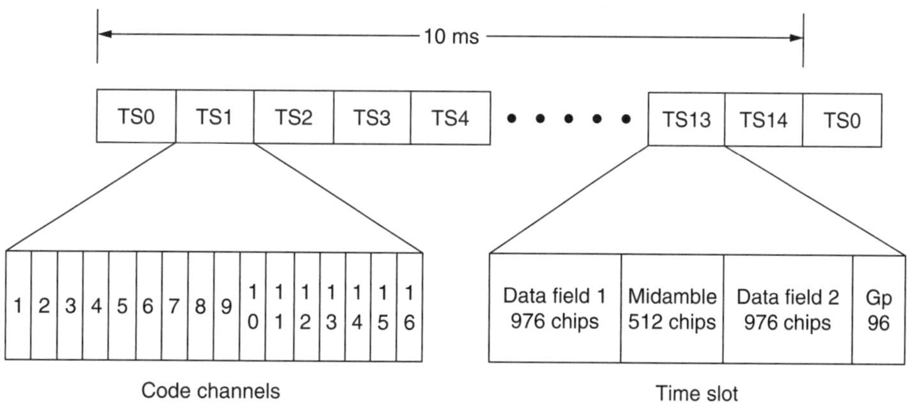

FIGURE 6.3 TD-CDMA radio channel structure.

6.3.3 Codes

Several coding schemes are used with TD-CDMA, including spreading, scrambling, and channelization codes. Spreading codes are an essential element in a TD-CDMA network. A total of 16 spreading codes are used, and all are orthogonal to each other and, like all CDMA codes, are part of a tree. The tree for the code sequence dictates the potential available data throughput that can take place and is shown in Figure 6.4.

TD-CDMA has both I and Q for data streams, and each of the two data flows uses the same scrambling code and two different channelization codes. A total of 128 scrambling codes having a short period of 16 are used to identify TDD cells in a system before reuse of those codes is required. The scrambling codes are grouped into four sets of 32 code groups.

Figure 6.4 shows the current *Orthogonal Variable Spreading Factor* (OVSF) code tree for a TD-CDMA system. The tree indicates the relationship between the various codes and overall throughput and capacity of the radio carrier itself. It is important to note that the code tree applies to each individual time slot, unlike other CDMA networks, where the code tree is part and partial to the entire carrier. Therefore, *time slot 1* (TS1) can use a different code on the code tree than TS3, allowing for better code management.

In TD-CDMA, a *Channel Unit* (CU) consists of one time slot and one code. Multiple CUs then are allocated when the traffic load exceeds one CU's capability, that is, requiring

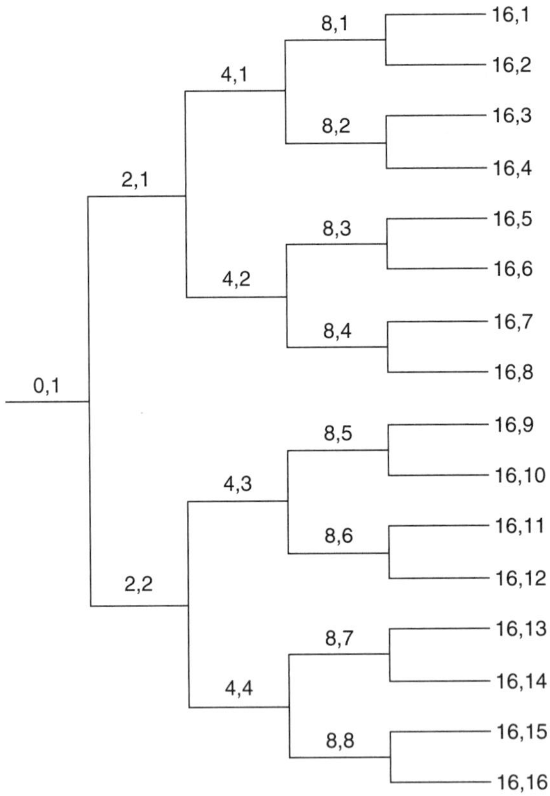

FIGURE 6.4 TD-CDMA code tree.

more throughput. For example, a call with a heavier load on the downlink than on the uplink requires a larger number of CUs for the downlink compared with the uplink.

Remembering that there are 15 frames in a TDD sequence, then the following is a quick illustration. If code 16,10 is in operation, then 8,5 through 8,8 and 4,4, 4,3, 2,2, and 1,1 are not available for use. Therefore, when a lower code on the tree is used, all codes above it are removed from possible use. Now, if we want 2048 kbps of data throughput, this will require the use of all the spreading codes to achieve this effort. The $1{,}1 \times 2$ will enable a data rate of 2048 kbps; if a lower data user is operational on the radio channel at the same time, the effective data rate would drop to about 1536 kbps.

6.3.4 Logical Channel

The TDD mode of UMTS, based on a TD-CDMA, has been designed to address specifically the needs of data traffic and has many unique features. A key concept is that the overall capacity of the radio interface is shared by common channels and data channels.

There are numerous logical channels in a TD-CDMA system. All the logical channels are basically one of two types—control or traffic—and are defined in Table 6.1.

Control Channel (CCH)	SCCH	Synchronization Control Channel	Used in downlink for broadcasting synchronization information
	BCCH	Broadcast Control Channel	Used in downlink for broadcasting system information
	PCCH	Paging Control Channel	Used in downlink for transferring paging information
	DCCH	Dedicated Control Channel	Point-to-point bidirectional channel that transmits dedicated control information between the UE and the network
	CCCH	Common Control Channel	Bidirectional channel used for sending control information between the BTS and the UE
	ODCCH	ODMA Dedicated Control Channel	Point-to-point bidirectional channel that transmits dedicated control information between UEs
	OCCCH	ODMA Common Control Channel	Bidirectional channel for transmitting control information between UEs
	SHCCH	Shared Channel Control Channel	Bidirectional channel that sends control information for uplink and downlink shared channels between the network and the UE
Traffic Channel (TCH)	DTCH	Dedicated Traffic Channel	Dedicated to a single UE, both uplink and downlink
	ODTCH	ODMA Dedicated Traffic Channel	Dedicated to one UE and exists in relay link
	CTCH	Common Traffic Channel	Point-to-multipoint unidirectional channel for transfer of user information from all or a specified group of UEs

Table 6.1 Logical Channels

In addition, there are some other signaling channels in TD-CDMA, namely, the *Forward Access Channel* (FACH) and the *Radio Access Channel* (RACH). The FACH is used to respond to the UE's random access request and is paired with the RACH. The FACH can be mapped to more than one radio unit, and the location and number of the FACH are indicated on the BCCH.

The RACH is designed for uplink random access by the UE and also can be mapped to more than one radio unit. The location allocated to RACH in the time slots is indicated on the BCCH. The RACH uses both power control and uplink synchronization control.

The first time slot of every frame is assigned to the RACH because it is always assigned for uplink traffic; the downlink common signaling channels (i.e., BCCH, PCCH, SCCH, and FACH) are in the last slot of the frame. The remaining slots are assigned on an as-needed basis for uplink or downlink usage. In TDD, the power control in TD-CDMA is done via a closed loop for the downlink and open loop for the uplink.

6.4 Interference-Mitigation Techniques

TD-CDMA overall throughput for both the uplink and downlink paths, as in any other radio access system, depends on the RF environment. TD-CDMA systems are designed primarily to support asymmetric traffic patterns such as IP.

In a *Non–Line of-Sight* (NLOS) environment, multipath reduces the orthogonality between the codes. To improve the orthogonality, TD-CDMA uses *Multiuser Detection* (MUD). MUD is employed in the receiver to restore orthogonality, resulting in coverage and data-throughput maximization for the RF environment. MUD for TD-CDMA constructively combines all the different radio paths received at the antenna to increase the usable signal level into the receiver and eliminate multipath interference.

6.5 RAN Traffic Planning

TD-CDMA RAN traffic planning can involve both circuit- and packet-switched services. However, TD-CDMA is a radio access system that is designed to handle asymmetric traffic such as IP. For planning purposes, a TD-CDMA cell or system can be stand-alone or can coexist within a WCDMA network or even a GSM network. The spectrum required for TD-CDMA is 5 MHz for one channel. Depending on the license issued, the radio spectrum could be allocated as a TDD or FDD system. If the spectrum allocation is TDD, then the number of radio channels available for possible use is determined strictly from the bandwidth of the license; that is, 10 MHz of TDD spectrum equates to two TD-CDMA channels without a guard band. However, if the spectrum is FDD, then a 5-MHz license also can support two TD-CDMA channels.

TD-CDMA systems are meant for high-data-usage areas such as urban or *Wireless Local Loop* (WLL) applications. Therefore, the coverage of a typical TD-CDMA cell is smaller than that of a typical macro cell. The coverage size allows for a lower power level, which reduces interference from an adjacent cell or mobile unit. The negative in terms of the smaller size is the potential for increased handoffs owing to the smaller coverage area.

The code schemes and variable power settings require that RAN traffic planning for a TD-CDMA system should be handled by a computer simulation coupled with real field data. Integral to the traffic-planning effort is a properly tuned propagation model that can spread the traffic under multiple interference and loading situations that include both data and voice.

The capacity of a TD-CDMA network is limited by the number of carriers, the time-slot allocations for uplink and downlink, and the radio environment itself. In a TD-CDMA system, signals coming from both user devices and base stations are all sources of interference because the system is TDD.

TD-CDMA systems are well suited for high traffic density and indoor coverage requiring high data rates. A direct example of a WLL application involves cordless phones and wireless *Wide-Area Network* (WAN) applications owing to the data capability.

The deployment of a TD-CDMA system follows many of the well-established guidelines for a CDMA-type of network deployment. A key element with code-based systems is the nonreliance on antenna orientation, enabling coverage steering or direction owing to antenna orientation and inclination adjustments.

Besides the use of a good traffic-modeling program, the proper inputs are required. The following are some of the parameters that should help to facilitate the planning process. TD-CDMA has the following throughputs:

- 1.5 Mpbs downlink and 1 Mbps uplink using an 8/4 split for channels
- Overall sector throughput of 2 Mbps downlink and 1 Mbps uplink

However, the specific throughput is defined by the uplink and downlink allocation scheme, as well as by the codes employed, besides the RF environment. Additionally, it has to be assumed that the transport layer from the Node B to the core network and beyond is dimensioned properly to support this service.

A TD-CDMA system is designed for use in the unpaired spectrum, unlike WCDMA, and its main advantage is improved efficiency in carrying asymmetric traffic. Another characteristic of TD-CDMA systems is their smaller cell sizes, which reduce the interference caused by other cells/users.

Additionally, for traffic planning, the expected maximum range for a TD-CDMA cell-site sector is 7.5 km and is not driven by propagation or link-budget issues but by the transmission control process involving a guard time for the UL path. The range can be extended but only through the removal of one of the uplink time slots available, enabling a larger guard time and extending the range to 29 km (at the obvious expense of overall thoughput).

Since the operator has the ability to determine the time-slot allocation for downlink and uplink to better match the traffic and service offering, a brief example will help to cement the relationship between time slot and uplink and downlink allocations.

There are a total of 15 time slots. Some time slots are needed, in that there are two dedicated for the downlink, namely, TS0 and TS1, and one for the uplink. The Pilot Channel is always associated with TS0 and the FACH with TS1. For the channel or time-slot allocation scheme, the downlink always leads the uplink, and this is done to facilitate synchronization and access control on the uplink.

The specific uplink time slot's position is determined based on downlink and uplink time-slot provisioning and, as mentioned earlier, is determined by the operator. To

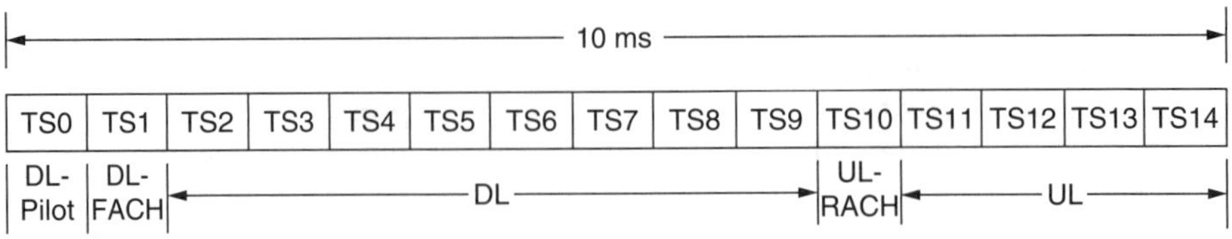

FIGURE 6.5 2:1 downlink-to-uplink allocation.

better illustrate this concept, Figure 6.5 shows a time-slot allocation that has a 2:1 relationship for downlink-to-uplink allocations. In this case, TS2 through TS9 (eight total) are allocated for the downlink, TS11 through TS14 (four total) are allocated for uplink traffic, and TS10 is set aside for the RACH.

Now, if we were to have a 1:1 downlink-to-uplink relationship, as shown in Figure 6.6, then TS2 through TS7 (six total) would carry downlink traffic, TS9 through TS14 (six total) would carry uplink traffic, and TS8 would have the uplink RACH.

An important issue is that downlink traffic time slots are always followed by the uplink. This is done for synchronization purposes because the downlink can use the *Global Positioning Satellite* (GPS) system for achieving its synchronization, whereas in the UL traffic, control is used and is not synchronized.

Addressing the code tree, a total of 16 orthogonal codes are used. Each time slot is independent of the next, meaning that each time-slot code is set by the traffic payload and RF conditions.

Because the spreading factor must remain the same, code 16 in the tree would be used for lower data throughputs, whereas code 1 would be associated with the maximum data throughput. *Spreading Factor 1* (SF1) is one-sixteenth that of SF16, meaning that the spread is achieved through the actual data payload for the time slot.

Throughput is important, and for traffic-planning purposes, a time slot can support 500 kbps in a 5-MHz channel. Using multiple TD-CDMA channels (two) for a 10-MHz allocation, the throughput for the time slot can get to 1 Mbps, with a 3-Mbps maximal download capability, provided, of course, that the RF environmental is able to support this level. Table 6.2 lists some of the general system parameters of a TD-CDMA system.

Table 6.3 shows the various mobile and terminal power classes and their respective transmit powers.

TD-CDMA data throughput is optimized through efficient on-demand assignment of resources, with combined downlink and uplink data rates. A single TD-CDMA

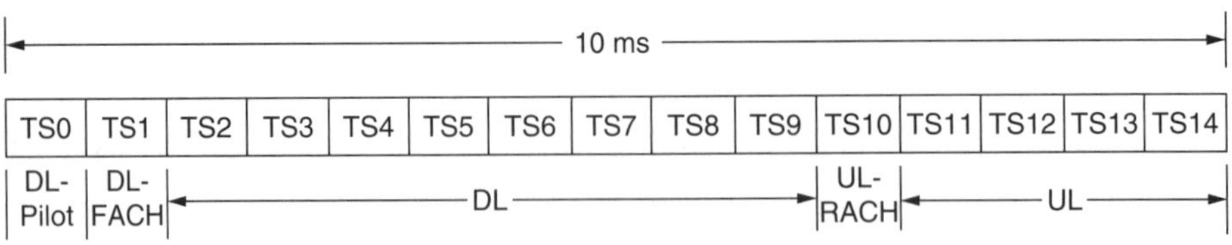

FIGURE 6.6 1:1 downlink-to-uplink allocation.

Carrier bandwidth: 5 MHz
Duplex type: TDD
Multiple access scheme: TDMA, CDMA, FDMA
Chip rate: 3.84 Mcps
Modulation: QPSK, 16QAM
Maximal cell range: 29 km (7.5 km nominal)
Theoretical maximal data rate/user: 2 Mbps
System asymmetry (DL:UP): 1:14–14:1
Frequency reuse: 1
EFR (Enhanced Full Rate): 12.2 kbps

Table 6.2 TD-CDMA RAN System Parameters

channel is able to support 5.5 Mbps in the downlink and 850 kbps in the uplink. When using two channels combined, the data rate is increased to 11 Mbps in the downlink and 1 Mbps in the uplink.

An important comment is that a single 5-MHz RF channel uses a 3.84-Mcps system, and one 10-MHz channel can support a 7.68-Mcps system. The technology also supports mobility, freeing the user terminal from being tied to a fixed location. Users can travel in excess of 120 km/h while maintaining network connectivity.

An important aspect of RF planning and capacity management is the fact that with TD-CDMA there is no soft or softer handoff, as typically used in a CDMA network. This simple concept frees up resources at surrounding sites, enabling more capacity to be supported by the system itself.

Radio systems, regardless of technology, need to provide coverage to be successful. The coverage radius of TD-CDMA is decided by two factors: the transmit power and reception sensitivity and the length of the guard period. However, with all CDMA-based systems, coverage and capacity are inversely related. With a TDD system, a guard period is needed between the uplink and downlink time slots that actually defines the

Power Class	Maximal Tx Power
1	+33 dBm
2	+30 dBm
3	+27 dBm
4	+24 dBm
5	+21 dBm
6	+10 dBm

Table 6.3 Mobile Class Power

radius of the cell site. With a cell offering maximal capacity, the maximal coverage radius of TD-CDMA is about 29 km but typically 7.5 km, leading it to a more urban and WLL application. When larger coverage areas are needed, such as for rural applications, capacity must be reduced to achieve greater coverage.

6.6 Handover

For a TD-CDMA system, the following parameters are needed in the cell-selection monitoring set:

- SIR
- Path loss
- Interference power
- Received power level on the BCH

The handover process is implemented in the mobile unit and *Radio Network Subsystem* (RNS). Measurement of serving radio connection downlink performance and candidate-cells received signal strengths and quality are made in the UE. The RNS measures uplink performance, as well as the position information for the UE being served, and uses these measurements in conjunction with defined thresholds and handover strategy to make a handover decision.

6.7 Implementation

Multiple scenarios are possible for implementing TD-CDMA. However, all these implementation methods fall into one of two categories: new spectrum or overlay. The specific methodology for implementing a TD-CDMA system largely depends on two elements: your business plan relating to the services and coverage area you want to service, and the spectrum that you have to use. There are, of course, other considerations, but without spectrum and a valid business plan defining what you want to provide, it is difficult to establish a proper implementation plan.

If the TD-CDMA system is a brand-new network with no legacy issues, then the configuration follows that shown in Figure 6.1 with the exception of the number of cells involved. Another, more likely scenario involves deployment of a TD-CDMA network into an existing GSM system where the operator has chosen to offer voice and legacy services with GSM and new data services via TD-CDMA. Figure 6.7 is a simple network configuration depicting a GSM and TD-CDMA system. Figure 6.8 illustrates a possible spectrum allocation in which TD-CDMA and GSM have split the spectrum.

Yet another scenario involves the overlay of TD-CDMA onto a WCDMA network, and this is shown in Figure 6.9. The possible spectrum allocation for a TD-CDMA network coexisting with WCDMA is shown in Figure 6.10*a*. Lastly, a possible spectrum allocation where TD-CDMA is deployed in an existing GSM and WCDMA network is shown in Figure 6.10*b*. Obviously, there are many more possible scenarios, but these simple examples are meant to provide a framework from which to begin planning.

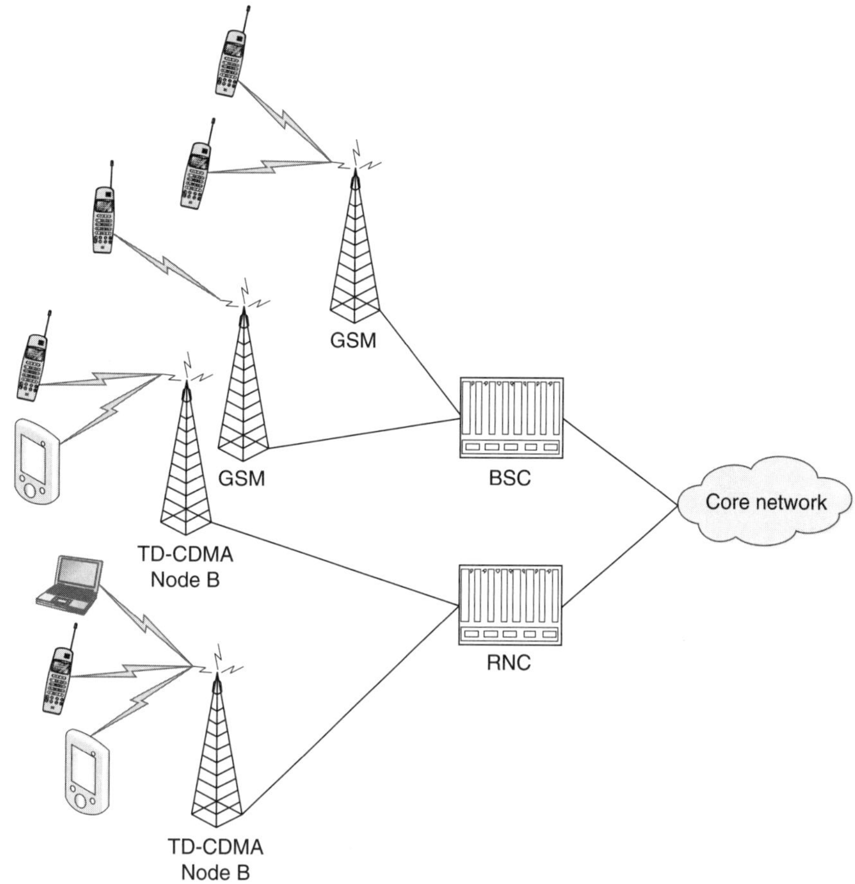

FIGURE 6.7 TD-CDMA overlay on a GSM network.

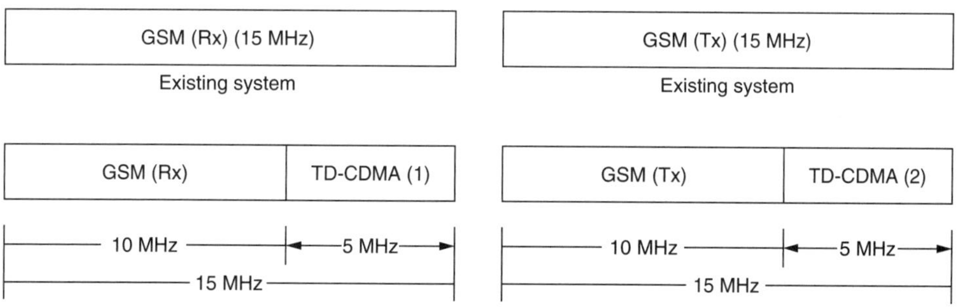

FIGURE 6.8 GSM and TD-CDMA spectrum allocation.

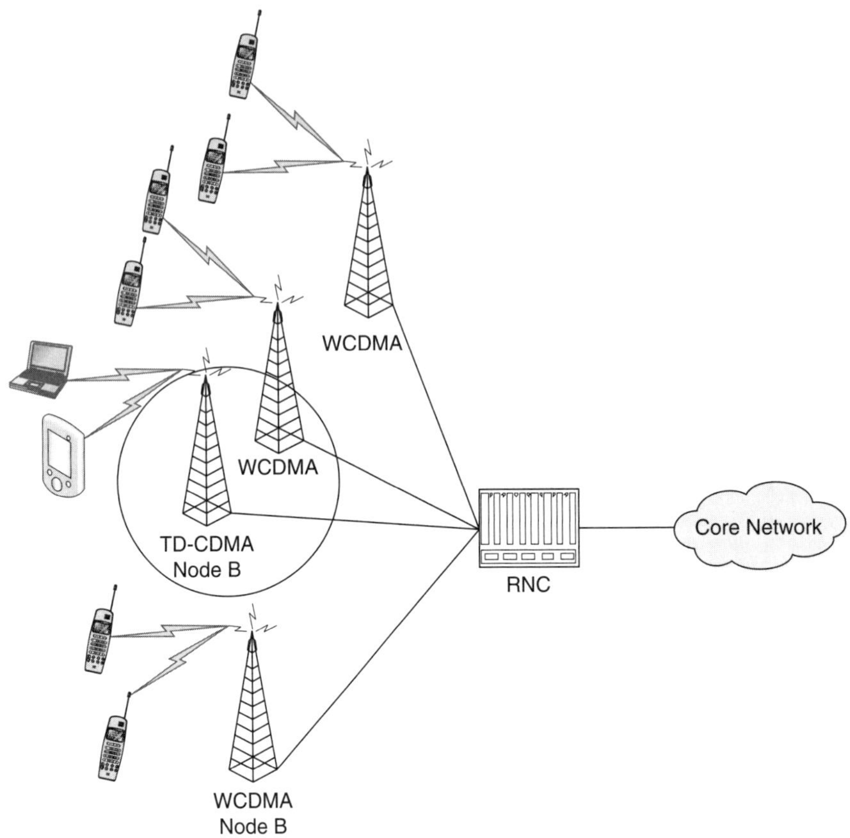

FIGURE 6.9 TD-CDMA overlay onto a WCDMA network.

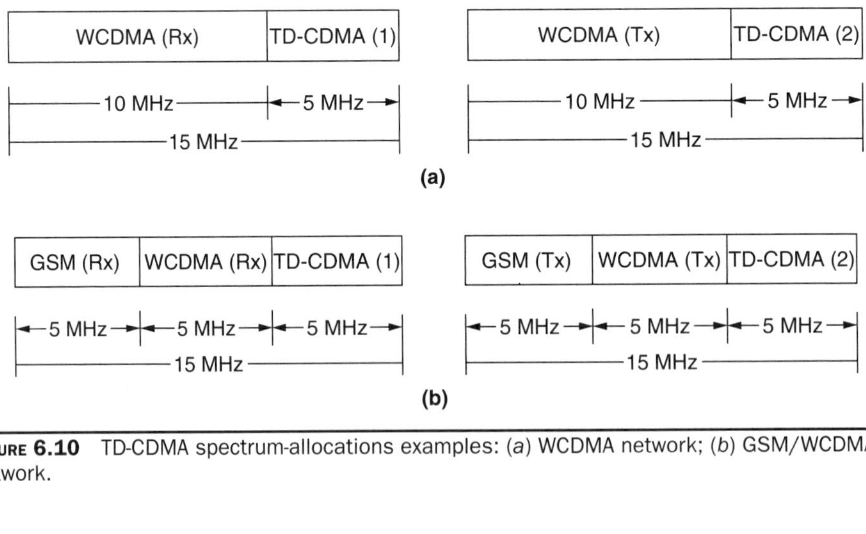

FIGURE 6.10 TD-CDMA spectrum-allocations examples: (*a*) WCDMA network; (*b*) GSM/WCDMA network.

	TD-CDMA	WCDMA	TD-SCDMA
Multiple access	HCR TDD	FDD	LCR TDD
Handoff	Hard	Soft	Hard
Modulation	QPSK	QPSK	QPSK/8PSK
Bandwidth	5 MHz	5 MHz	1.6 MHz
Chip rate	3.84 Mcps	3.84 Mcps	1.28 Mcps
Time slots per frame	15	7	7
Spreading factor	1, 2, 4, 8, 16	4–256	1, 2, 4, 8, 16
Receiver	Joint detection Rake (mobile)	Rake	Joint detection Rake (mobile)

Table 6.4 TD-CDMA, WCDMA, and TD-SCDMA

6.8 Comparison

Table 6.4 is a brief summary of the similarities and differences between TD-CDMA, WCDMA, and TD-SCDMA radio access systems.

References

Baier, P. W., M. Meurer, T. Weber, and H. Troger, "Joint Transmission (JT), an Alter-native Rationale for the Downlink of Time Division CDMA Using Multielement Transmit Antennas," IEEE 6th International Symposium on Spread-Spectrum Technology and Applications, NJIT, New Jersey, September 6–8, 2000.

Calin, Doru, and Marc Areny, "Impact of Radio Resource Allocation Policies on the TD-CDMA System Performance: Evaluation of Major Critical Parameters," *IEEE Journal on Selected Areas in Communications* 19(10), October 2001.

Dong, Geun Jeong and Wha Sook Jeon, "CDMA/TDD System for Wireless Multimedia Services with Traffic Unbalance Between Uplink and Downlink," *IEEE Journal on Selected Areas in Communications* 17(5), May 1999.

Forkel, Ingo, and Xin Jin, "Performance Comparison between UTRA-TDD High Chip Rate and Low Chip Rate Operation," Aachen University of Technology, Germany, June 2002.

Forkel, Ingo, Bernhard Wegmann, and Egon Schulz, *On the Capacity of a UTRA-TDD Network with Multiple Services,* Siemens AG, Munich, 2002.

Geher, Christina, Reinhard Kohn, Jorg Schniedenham, and Armin Sitte, *Layer 2 and Layer 3 of UTRA-TDD,* Siemens AG, Berlin, Germany, March 2000.

Ho-Shin, Cho, Jungchae Shin, and Yutae Lee, "Call Blocking Probability for Heterogeneous and Asymmetrical Traffics in a TD-CDMA System," *IEEE Communications Letters* 8(12), December 2004.

Ponnampalam, Vishakan, and Alan E Jones, "On Cell Parameter ID Assignment in UTRA-TDD," IPWireless, Inc., Chippenham, UK, March 2004.

Prätor, Oliver, Carsten Unger, Andre Zoch, and P. Gerhard, "Impact of Channel Estimation on the 3GPP-TD-CDMA," Dresden University of Technology, Dresden, Germany, September 2001.

TDD Ad-Hoc Group, "Relative Assessment of UMTS TDD and WLAN technologies," UMTS Forum Report 28, March 2003.

TP-040076, 3GPP TSG-T (Terminals) Meeting 24, Seoul, Korea, June 2–4, 2004.

Weber, Tobias, Johannes Schlee, Stefan Bahrenburg, et al., "A Hardware Demonstrator for TD-CDMA," *IEEE Transactions on Vehicular Technology* 51(5), September 2002.

Weckerle, Martin, Apostolos Papathanassiou, and Dieter Emmer, *The Benefits of Intelligent Antenna Arrays in TD-CDMA: A Study Based on Measured Channel Impulse Responses*, Siemens AG, Berlin, September 1998.

www.3gpp.org.

www.umtstdd.org.

Yin-Fu, Huang and Tsung-Yi Chiu, "Radio Resource Management for a Mobile Network with TD-CDMA," IEEE 6th CAS Symposium on Emerging Technologies: Mobile and Wireless Communications, Shanghai, China, May 31–June 2, 2004.

CHAPTER 7

Long-Term Evolution

Long-Term Evolution (LTE) is an exciting new wireless mobility technology that has thrust itself upon the industry. This chapter will cover many of the important areas of LTE itself. I have attempted to identify the key concepts that are important in order to understand LTE. However, justice cannot be done for a real deep dive into the technology since LTE encompasses a vast array of items. I have included many reference documents at the end of this chapter. In addition, for the continuous improvements and innovations that are taking place within the LTE industry, www.3gpp.org is an excellent source of very specific information related to LTE.

At this stage you might be thinking what is special about LTE as compared to wireless access technologies. While other wireless access technologies offer a multitude of advantages, LTE has several fundamental benefits. Unlike its predecessors for wireless access, such as WCDMA and CDMA which utilize circuit switching in some of the network architecture, LTE is an all IP network.

Some of the key attributes that can be associated with LTE are

- All IP network architecture
- Flat architecture
- Separation of the user control and user planes
- Improved Radio Access scheme
- Advanced Antenna solutions
- Spectrum bandwidth flexibility

The above attributes for LTE enable the wireless network operator to operate a distributed network, improve provisioning, reduce latency, provide high-speed broadband service as well as reduce the overall cost of ownership.

LTE is also capable of interworking with legacy wireless systems as well.

LTE is advancing quickly and the newer version of LTE, release 10/11, is called LTE Advanced and introduces new functionality, so LTE can offer even high data rates to the end user.

- Increased peak data rate, DL 3 Gbps, UL 1.5 Gbps
- Higher spectral efficiency, from a maximum of 16 bps/Hz in R8 to 30 bps/Hz in R10
- Increased number of simultaneously active subscribers
- Improved performance at cell edges, for example, for DL 2 × 2 MIMO at least 2.40 bps/Hz/cell

However, the main new functionalities introduced in LTE Advanced are *Carrier Aggregation* (CA), enhanced use of multiantenna techniques and support for *Relay Nodes* (RN), both of which will be discussed later in this chapter.

7.1 LTE Ecosystem

Knowing some of the key attributes of an LTE network leads to the question, just what makes up an LTE network? The LTE network, or rather architecture, is defined in a series of standards driven by the 3G Partnership Project (3GPP), www.3gpp.org.

LTE has many components that can make up the network; however, there are five main elements. The five main elements which make up any LTE network are listed next and depicted in Figure 7.1:

1. User Equipment (UE)
2. Evolved Universal Terrestrial Radio Access (E-UTRA)—the LTE air interface
3. Evolved Universal Terrestrial Radio Access Network (E-UTRAN)—the eNodeB
4. Evolved Packet Core (EPC)—the LTE core
5. Non-LTE—application servers, internet, legacy network (UMTS/CDMA)

At a high level, Table 7.1 tries to identify some of the major changes as compared to some of the other wireless access technologies. Because LTE is a flat architecture utilizing an all IP signaling scheme it is able to facilitate faster data rates as well as significantly reduced latency enabling it to be very suitable for video applications.

In the LTE ecosystem, once the UE has acquired the network it is basically an always-on connection. In the always-on mode even if the UE is in idle mode, it still has its IP address; and the connection between the *Serving Gateway* (SGW) and *Packet Data Network Gateway* (PGW) remains. However, the connection between the SGW and the eNodeB may be torn in the idle mode. The always-on connection even in idle mode enables fast session setup and is ideal for applications requiring a fast response time.

When looking at Figure 7.1 several new elements are defined. However, one key change is that the functionality that was associated with the RNC in CDMA and the

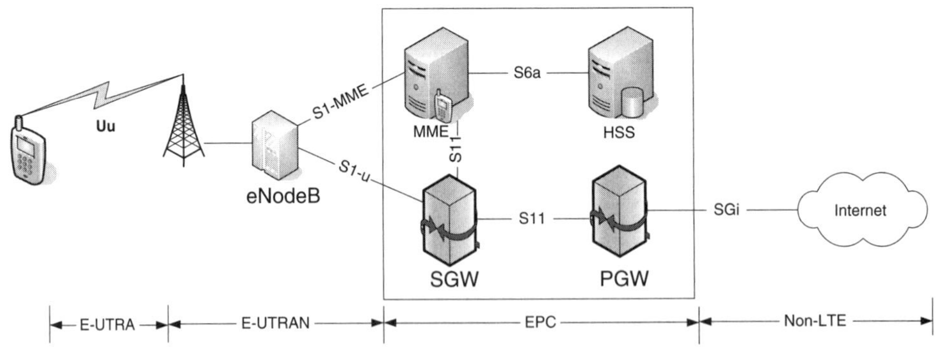

FIGURE 7.1 Basic LTE network.

	WCDMA	HSPA	HSPA+	LTE	LTE Advanced
Max downlink speed	384 k	14 M	28 M	100 M	1 G
Max uplink speed	128 k	5.7 M	11 M	50 M	500 M
Spectrum (FDD)	5 MHz	5 MHz	5 MHz	10 MHz	100 MHz
Latency round trip	150 ms	100 ms	50 ms (max)	~10 ms	~10 ms
3GPP release	Rel 99/4	Rel 5/6	Rel 7	Rel 8/9	Rel 10
Access method	CDMA	CDMA	CDMA	OFDMA/SC-FDMA	OFDMA/SC-FDMA

TABLE 7.1 Wireless Access Technology Comparison

NodeB in WCDMA/HSPA has now moved to the eNodeB in LTE. This puts a lot of the decisions for processing data sessions at the edge of the network in LTE.

LTE can operate either as a TDD or FDD system. LTE at this time is typically deployed as an FDD network. However the TDD configuration is also possible and enables LTE to be deployed in unpaired spectrum configurations like that associated with WiMAX.

The amount of spectrum, or rather bandwidth, the LTE can occupy is flexible. Specifically LTE is so structured that it can operate with a variety of common carriers (CC). The typical deployment for LTE involves a 10-MHz channel that is deployed in an FDD configuration as a 10×10 channel. However, LTE can have different common carriers. There are a total of 6 CC bandwidths associated with LTE and they are 1.4-, 3-, 5-, 10-, 15-, and 20-MHz-wide CCs. The designation of the bandwidth is for a single CC whether it is an FDD or TDD deployment. Therefore, a 5-MHz LTE CC for an FDD system is made up of two 5-MHz paired channels while a TDD system only has a single 5-MHz channel used jointly for transmit and receive.

Within LTE there is a more robust adaptation for different Quality of Service (QoS) is possible, where there are nine *Quality Class Identifiers* (QCI) defined as per standard.

7.2 Standards

LTE is a standards-based wireless access system. The standards for LTE are driven by the 3GPP organization. As with all standards, there are several major releases, with release 9 being the primary release at this time and release 10 in the near future. Additional releases will follow release 10; however, for the moment we will only briefly discuss release 9 and release 10, LTE Advanced.

Figure 7.2 is a brief diagram depicting the various LTE releases and how they correspond to releases with HSPA. Each LTE release has its unique attributes and benefits which are highlighted in Table 7.2.

7.3 Radio Spectrum

Like all other radio access systems, LTE utilizes radio spectrum. The radio spectrum that is designed for possible LTE use is broken down into specific bands, or rather band numbers. A wireless network operator based on its spectrum portfolio can have spectrum which includes one or several of the LTE-designated bands.

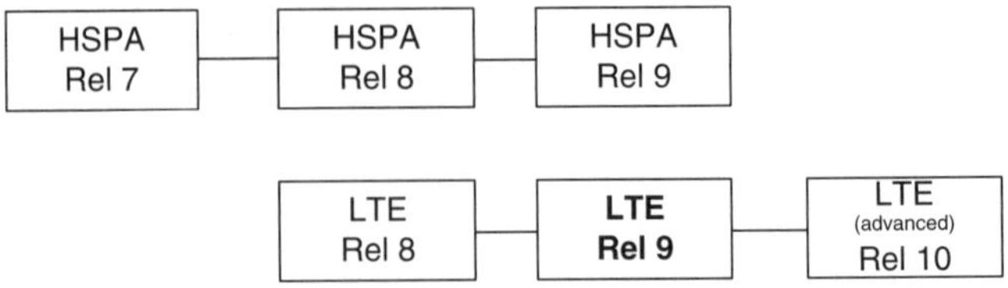

FIGURE 7.2 LTE releases.

Release 9	Multimedia Broadcast Multicast Service (MBMS) Home eNodeB (femtocells) Dual layer: UE-specific beam-forming Enhanced position location
Release 10	Carrier Aggregation (5 × 20MHz): FDD and TDD MIMO (8 × 4) Relaying Nodes (RN) to eNodeB Latency improvement • Reduced time to go from dormant to active state • Reduced time for RRC idle to connected state Self-Optimizing Network Enhancement • Max throughput: • DL: 1 Gbps • UL: 500 Mbps

TABLE 7.2 LTE Releases Major Features

There are over 20 different frequency bands which are associated with LTE. However, because a specific band is defined in the LTE specification does not imply that no other radio access technology can be used in that frequency band. For example, LTE band 4 is the *Advanced Mobile Phone Service* (AMPS) frequency band which has seen many wireless access technologies deployed in it.

Figures 7.3 and 7.4 are the current designations for LTE bands for FDD and TDD configurations, respectively. More bands can be and will be added, however, to the bands listed in Figures 7.3 and 7.4.

Figure 7.3 highlights the US AWS and 700-MHz bands which have seen the initial LTE deployment in an FDD configuration.

What is always important to consider when determining the band to operate in the wireless access is the selection of the *User Equipment* (UE). Most UEs are band-specific, even though the chip sets that reside in the devices can operate in any of the designed LTE bands. However, it is the *Radiofrequency* (RF) component of the UE which needs to match the frequency bands of operation. In reviewing the bands in Figures 7.3 and 7.4, the possible bandwidths are different. Additionally in Figure 7.3 the duplex spacing is unique for each band number; and this is important to note since some of the band numbers have transmit and receive reversed, bands 13 and 14, as compared to other bands.

FDD				
Band Number	Uplink (MHz)	Downlink (MHz)	Width of Band (MHz)	Duplex Spacing (MHz)
1	1920–1980	2110–2170	60	190
2	1850–1910	1930–1990	60	80
3	1710–1785	1805–1880	75	95
4	1710–1755	2110–2155	45	400
5	824–849	869–894	25	45
6	830–840	875–885	10	35
7	2500–2570	2620–2690	70	120
8	880–915	925–960	35	45
9	1749.9–1784.9	1844.9–1879.9	35	95
10	1710–1770	2110–2170	60	400
11	1427.9–1452.9	1475.9–1500.9	20	48
12	698–716	728–746	18	30
13	777–787	746–756	10	–31
14	788–798	758–768	10	–30
15	1900–1920	2600–2620	20	700
16	2010–2025	2585–2600	15	575
17	704–716	734–746	12	30
18	815–830	860–875	15	45
19	830–845	875–890	15	45
20	832–862	791–821	30	–41
21	1447.9–1462.9	1495.5–1510.9	15	48
22	3410–3500	3510–3600	90	100
23	2000–2020	2180–2200	20	180
24	1625.5–1660.5	1525–1559	34	–101.5
25	1850–1915	1930–1995	65	80

Figure 7.3 LTE FDD bands.

TDD		
Band Number	Allocation (MHz)	Width of Band (MHz)
33	1900–1920	20
34	2010–2025	15
35	1850–1910	60
36	1930–1990	60
37	1910–1930	20
38	2570–2620	50
39	1880–1920	40
40	2300–2400	100
41	2496–2690	194
42	3400–3600	200
43	3600–3800	200

Figure 7.4 LTE TDD bands.

7.4 LTE Architecture

The LTE architecture while being a flat IP network consists of many elements. Figure 7.5 is a more detailed illustration of an LTE network which includes several eNodeBs and a little more clarification regarding the core network components.

The various major elements comprising an LTE network will be defined. However, the key concept is that the eNodeB now has many of the functions that were associated with a Node B/BSC and the BTS/RNC. All non-LTE network components are referred to as non-3GPP.

In an LTE network an eNodeB can be connected to other eNodeBs via an X2 link which facilitates communication between the eNodeBs for handovers and other traffic management functions. Several eNodeBs are typically connected to an MME and SGW. The SGW handles the bearer traffic and communicates with a PGW, if required, for service delivery. However, the MME is only involved in the control plane portion and does not handle bearer traffic. As mentioned, several eNodeBs will be connected to an MME and SGW. There will be numerous MME and SGWs in an LTE network depending on the amount of eNodeBs and the network configuration. Additionally, there can be several PGWs depending on a particular service type or class associated with each PGW.

You can ultimately break down the LTE network into either the subscriber (UE), radio access network (eNodeBs) or the core network (EPC).

7.5 User Equipment (UE)

The UE is the device that the subscriber to the LTE wireless network interfaces with the most. Typically the UE can be thought of as either a handset, Smartphone, or aircard. However, many LTE UE devices are being integrated into appliances, vehicles, and clothing to mention a few alternative UE device locations and types.

Presently there are five different classes or categories of LTE UE devices, which are listed in Table 7.3.

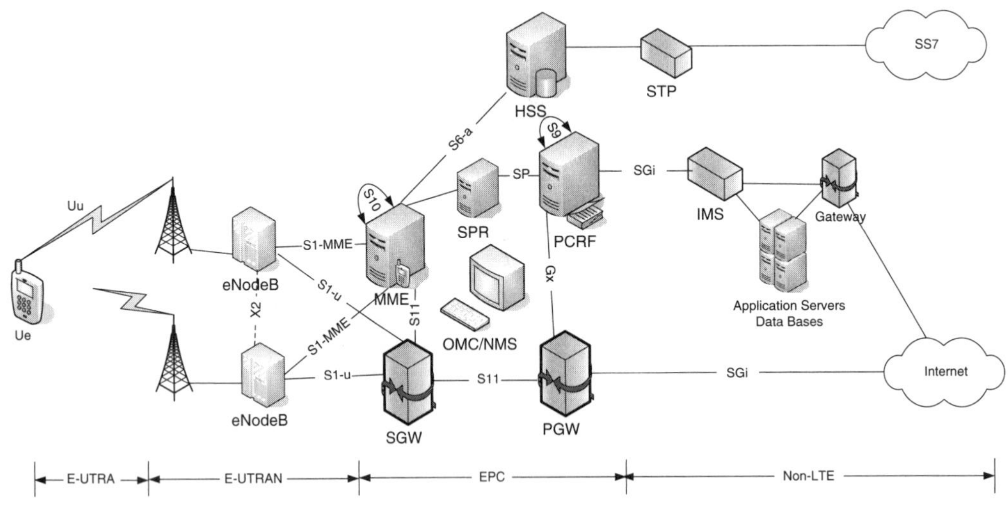

FIGURE 7.5 LTE architecture.

<table>
<tr><th rowspan="2">Category</th><th rowspan="2">Bandwidth (MHz)</th><th rowspan="2">MIMO</th><th rowspan="2">Duplexing</th><th colspan="2">Modulation</th><th colspan="2">Data Rate (Mbps)</th></tr>
<tr><th>UL</th><th>DL</th><th>UL</th><th>DL</th></tr>
<tr><td>1</td><td rowspan="5">1.4, 3,5,10
15 and 20</td><td rowspan="4">2 × 2</td><td rowspan="5">FDD
H-FDD
TDD</td><td rowspan="4">QPSK
16QAM</td><td rowspan="5">QPSK
16QAM
64QAM</td><td>5</td><td>10</td></tr>
<tr><td>2</td><td>25</td><td>51</td></tr>
<tr><td>3</td><td>51</td><td>100</td></tr>
<tr><td>4</td><td>51</td><td>150</td></tr>
<tr><td>5</td><td>2 × 2
and
4 × 4</td><td>QPSK
16QAM
64QAM</td><td>75</td><td>300</td></tr>
</table>

Peak downlink speed 64QAM (Mbps)	100 (SISO), 172 (2x2 MIMO), 326 (4x4 MIMO)
Peak uplink speeds (Mbps)	50 (QPSK), 57 (16QAM), 86 (64QAM)

Table 7.3 LTE UE Categories

All the LTE UEs can support different bandwidths in addition to TDD and FDD channel configurations. However, the UEs will not support every frequency band listed in the LTE bands. The limitation is not due to the LTE components for the UE, but due to the RF components that enable the UE to operate in a specific portion of the radio spectrum band.

7.6 Enhanced Node B (eNodeB)

The eNodeB can be defined as three major components involving the antennas system, radio elements, and digital processing components. The eNodeB therefore provides the physical radio connection between the LTE network and the UE. However, the eNodeB does much more than simply providing a physical connection.

The eNodeB performs all radio resource management functions, such as radio bearer control, radio admission control, radio mobility control, scheduling, and dynamic allocation of resources to UEs in both uplink and downlink. More specifically the eNodeB runs the scheduler, which is one of the most important components in resource allocations with an LTE Network.

Additionally the eNodeB is connected to both the MME and an SGW through S1 signaling links. The eNodeB routes all user plane data to the appropriate SGW and ensures that all data sent over the radio interface are encrypted.

Figure 7.6 depicts how a typical eNodeB is connected to an MME, SGW, and another eNodeB via an X2 interface.

The S1 interface is divided into two distinct parts. The S1-U and S1-MME. The S1-U is where the traffic data is routed and the S1-MME is where the control information, or, control plane protocol, transverses over. The X2 link, however, is shared directly between eNodeBs and is used to coordinate handovers and data transfers.

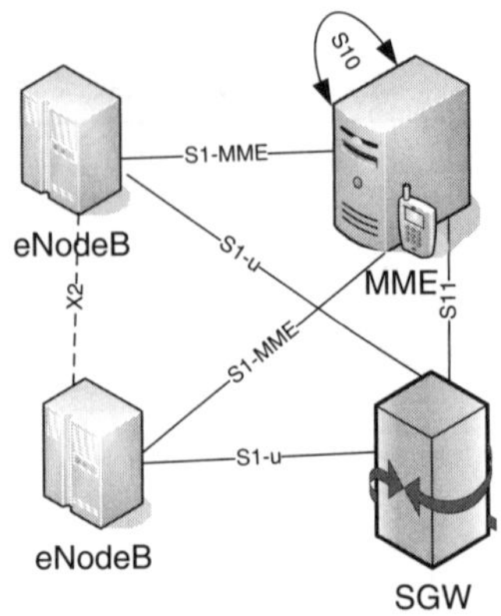

Figure 7.6 S1 interface.

The eNodeB utilizes two different modulation formats: *Orthogonal Frequency-Division Multiple Access* (OFDMA) for downlink and *Single-Carrier Frequency-Division Multiple Access* (SC-FDMA) for uplink.

7.7 Core Network (EPC)

Referring to Figure 7.6, the core network for LTE consists of several major elements, which are the *Mobility Management Entity* (MME), SGW, PGW and *Home Subscriber Server* (HSS).

7.7.1 Mobility Management Element (MME)

The MME is the control node that processes the signaling between the UE and the EPC. There are typically numerous MMEs within a larger LTE network. The MME plays a critical role in the ability of the UE to transverse an LTE network.

The MME keeps track of the UE that is associated with the eNodeBs that are connected to the MME. The UE can be in one of three states that the MME keeps track of.

1. Connected
   - UE has registered with the network.
   - UE has established a Radio Resource Connection (RRC).
   - Handovers are network controlled.
   - Measurement reports sent from UE to eNodeB.
2. Idle
   - UE has been inactive for an extended period of time.
   - Cell selection and reselection are under complete UE control.
   - UE is not connected to the network and is just monitoring the network for messages.
3. Dormant
   - UE is not connected or monitoring the network.
   - UE could be out of service area or powered off.

The MME also determines which SGW and PGW the UE will utilize based on information it obtains about the UE from the HSS. The MME also facilitates handovers either intra- or internetwork. The MME also monitors traffic and has the ability to avoid or resolve an overload situation through traffic shedding. The MME handles the ideal state mobility control for the UEs that are associated with the eNodeBs under the MME. The MME also handles charging and billing functions, as well as handling both the bearer management and authentication process for the UE.

7.7.2 Serving Gateway (SGW)

All IP packets sent to or from the UE are routed through the Serving Gateway, which serves as the local mobility anchor for the data bearers when the UE moves between eNBs. Typically, the UE remain connected to the same SGW as it moves from eNodeB to eNodeB in the network.

The SGW, therefore, is the local mobility anchor point for inter-eNodeB handover, downlink packet buffering, and initiation of network-triggered service requests. The SGW also is directly involved with lawful intercept for regulatory compliance of a wireless mobile network. The SGW also manages QCI granularity as part of the service delivery including UL/DL charging per UE.

The SGW also retains the information about the bearers when the UE is in the idle state (known as "EPS Connection Management—IDLE" [ECM-IDLE]) and temporarily buffers downlink data while the MME initiates paging of the UE to reestablish the bearers. In addition, the SGW performs some administrative functions in the visited network such as collecting information for charging like the volume of data sent to or received from the user.

The SGW also serves as the mobility anchor for interworking with other 3GPP technologies like UMTS.

7.7.3 Packet Data Network Gateway (PGW)

The PGW is responsible for IP address allocation for the UE as well as QoS enforcement and flow-based charging according to the rules it obtains from the *Policy Control and Charging Rules Function* (PCRF). The PGW is responsible for the filtering of downlink user IP packets into the different QoS-based bearers, which is accomplished through specific *Traffic Flow Templates* (TFTs). Therefore the PGW performs QoS enforcement for *Guaranteed Bit Rate* (GBR) bearers. It also serves as the mobility anchor for interworking with non-3GPP technologies such as CDMA2000 and WiMAX networks.

7.7.4 Home Subscriber Server (HSS)

The HSS is similar in function to a *Home Location Register* (HLR) except that it is IP-based. The HSS is a data base which contains all users' subscription data for the wireless network. The HSS that is accessed for obtaining the subscriber information can also be for a different wireless operator depending on the home network that the UE identifies itself with.

The HSS therefore provides all the relevant information necessary for the system to determine what services and functions the subscriber is allowed to use and receive. Included in the information from the HSS about the subscriber is their QoS profile amid possible roaming profile if they are visiting the network, and their current location/presence information that would include dynamic tracking of MME to which the UE is currently attached or registered so that services can be properly delivered to the subscriber.

The HSS may also integrate the *Authentication Center* (AUC), which generates the vectors for authentication and security keys.

7.7.5 Policy Control and Charging Rules Function (PCRF)

The PCRF while not technically part of the LTE core network plays an important role in delivering services for the subscriber.

The *Policy Control and Charging Rules Function* (PCRF) is responsible for the fundamental policy control decision making for the entire LTE network. It also is responsible for controlling the flow-based charging functionalities in the *Policy Control Enforcement Function* (PCEF), which resides in the PGW. The PCRF provides the QoS authorization, or rather QoS class identifier, QCI, and bit rate that decides how a certain data flow or session will be treated in the PCEF and ensures that this is in accordance with the user's subscription profile that was obtained from the HSS.

7.8 Radio Channel Components

With LTE as in all wireless access technologies the radio access medium is made up of many types of channels. Some of the radio channels are physical and others are logical. It is important to understand the fundamental channel types involved with the LTE network so that you can facilitate a good design or quickly troubleshoot a problem.

The radio access medium is the Uu, air interface layer, that is shown in Figure 7.1. The Uu is the air interface between UE and eNodeB and transports all signaling messages and data between the UE and the eNodeB.

The radio access medium involves both layer 1 and layer 2 components. The layer 1 is the actual radio communication or physical layer (PHY) that is responsible for functions like Modulation, Error Correction, interleaving, and HARQ.

The layer 2 components consist of three functional areas. The first is the *Packet Data Convergence Protocol* (PDCP), which provides header compression and encryption for the information sent between the UE and the eNodeB. Then there is the *Radio Link Control* (RLC), which is responsible for the successful segmentation and concatenation of the data that is sent between the UE and eNodeB. The third part of the layer 2 components is the *Medium Access Control* (MAC) and this is used for scheduling the uplink and downlink resources.

Figure 7.7 is a brief illustration of the association of the PHY, PDCP, RLC, and MAC.

Keeping in mind the layer 1 and layer 2 components, the next section will try to quickly define in a logical fashion the various channels used in LTE.

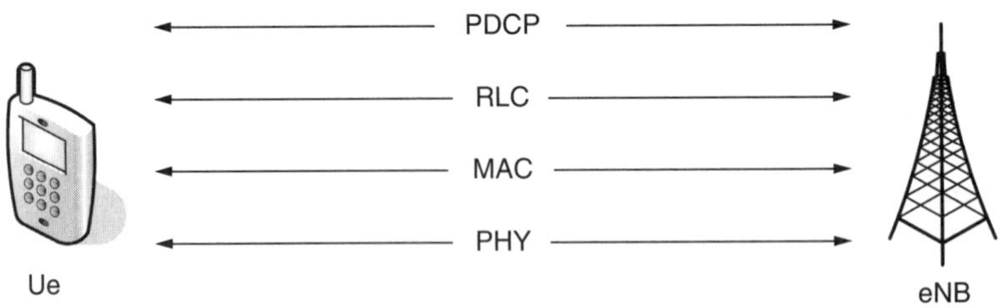

FIGURE 7.7 Layers 1 and 2.

7.8.1 LTE Channel Types

In LTE over the Uu interface, there are uplink and downlink channels which have many channels within them. These channels reside in the layer 2 portion discussed in the previous section. However, whether the channel is an uplink or downlink it can be categorized as either of the following three types of channels.

- Logical channels
    - Distinguish type of information to be carried
    - Can carry more than one data stream (multiple users)
- Transport channels
    - Indicates how information is formatted and coded
    - Carries one or more logical channels
- Physical channels
    - Defines the source and destination of the information
    - Can carry one or more transport channels

7.8.1.1 Downlink Channels

Figure 7.8 is a logical layout of the downlink channels for LTE. The interaction between each of the channels indicates that a particular channel can have several types of messages which it can convey regarding the transport or physical channel. It is important to note that the physical channel described is not at the layer 1 but at layer 2.

- Logical channels
    - *PCCH* (Physical Downlink Control Channel). This instructs mobile about the type of data and its scheduling including defining the resource blocks to use on the uplink.
    - *BCCH* (Broadcast Control Channel). This is local broadcast channel used to convey cell-specific information.
    - *DTCH* (Dedicated Traffic Channel). Each UE is assigned a dedicated traffic channel which is shared with other UEs.
    - *DCCH* (Dedicated Control Channel). This is used to send information for the management of the DTCH.

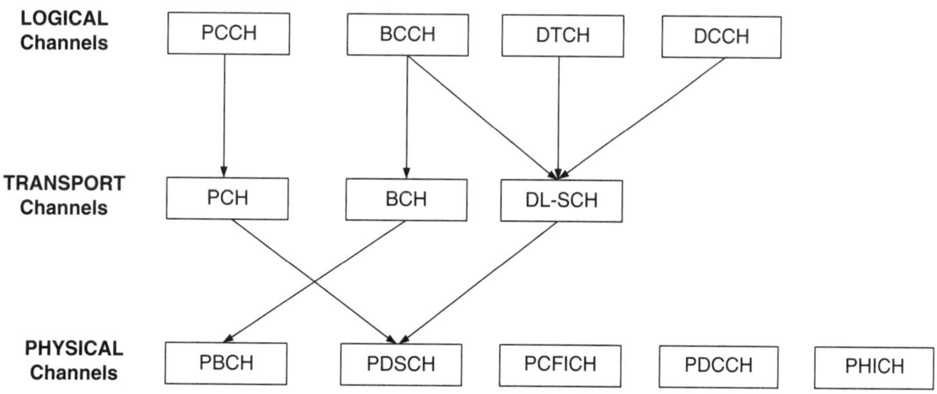

FIGURE 7.8 Downlink channel relationships.

- Transport channels
  - *PCH* (Paging Channel). This is used for sending pages.
  - *BCH* (Broadcast Channel). This carries cell-specific content and is used by all UEs on the site.
  - *DL-SCH* (Downlink Shared Channel). This multiplexes both the DTCH and DCCH as well as the BCCH on to the PDSCH.
- Physical channels
  - *PBCH* (Physical Broadcast Channel). This carries the Master Information Block (MIB) which contains the essential information for the UE to initially access the eNodeB.
  - *PDSCH* (Physical Downlink Shared Channel). This is the main data-bearing channel providing user data such as control signaling, broadcast information including System Information Blocks (SIB) and paging information.
  - *PCFICH* (Physical Control Format Indication Channel). This is used by eNodeB to inform UEs about the number of OFDMA symbols used for the PDCCH.
  - *PDCCH* (Physical Downlink Control Channel). This carries control information from the eNodeB to the UE, scheduling grants and UL and DL resource allocations, and UL power control.
  - *PHICH* (Physical Hybrid Automatic Repeat Request (ARQ) Indicator Channel). This is a physical channel that carries Hybrid ARQ Indicator (HI). The HI contains the ACK/NACK feedback to the UE pertaining to the uplink blocks.
  - *PMCH* (Physical Mulicast Channel). This carries the multicast/broadcast channel information for Multimedia Broadcast and Multicast Services (MBMS).

Within the downlink signaling there are some key signals which help in the UE in establishing and maintaining communication with the eNodeB. The key signals in the downlink involve both synchronization and reference signals.

- Synchronizaton signals
  - *Primary synchronization*. This is used by UE to obtain subframe synchronization. In the cell search process the UE determines the timing and center frequency of the downlink carrier.
  - *Secondary synchronization*. This is used by UE to obtain frame-level synchronization and allows the UE to obtain the cell-identity group of the eNodeB.
- Reference signals
  - Reference signals enable UE to mitigate amplitude, phase, and timing errors by providing a known reference signal so the UE can make corrections.
  - Three types of downlink reference signals are
    1. Cell-specific reference signals
    2. MBSFN reference signals
    3. UE-specific reference signals

7.8.1.2 Uplink Logical Channel Structure

Figure 7.9 is a logical layout of the uplink channels for LTE between the UE and the eNodeB. The interaction between each of the channels indicates that a particular

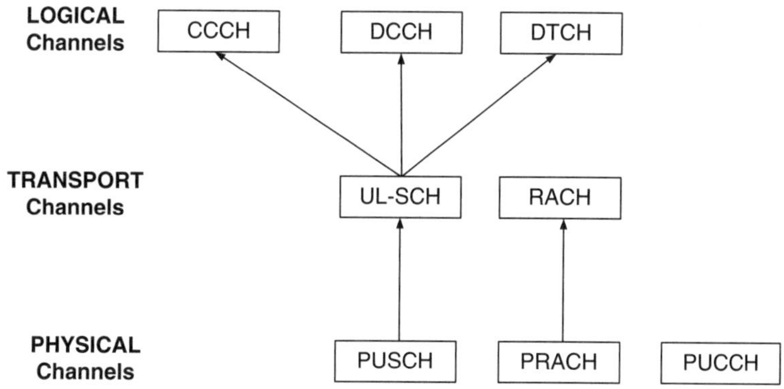

Figure 7.9 Uplink channel relationships.

channel can have several types of messages which it can convey regarding the transport or physical channel.

- Logical channels
  - *CCCH* (Common Control Channel). This controls information sent to UE when UE has no RRC connection with the network.
  - *DCCH* (Dedicated Control Channel). This is a point-to-point bidirectional channel that transmits dedicated control information to and from the UE.
  - *DTCH* (Dedicated Traffic Channel). Point-to-point dedicated channel to UE for the data traffic.
- Transport channels
  - *UL-SCH* (Uplink Shared Channel). Channel Quality Indicator (CQI), Precoding Matrix Indicator (PMI), Rank Indication (RI), and Redundancy Version (RV).
  - *RACH* (Random Access Channel). This is the transport channel for the PRACH.
- Physical channels
  - *PUSCH* (Physical Uplink Shared Channel). This carries data traffic, as well as the Uplink Shared Channel (UL-SCH) and the Uplink Control Information (UCI).
  - *PRACH* (Physical Random Access Channel). This is used for synchronization and initial uplink resource requests; also used for estimating time delay to the UE.
  - *PUCCH* (Physical Uplink Control Channel). This carries the uplink control information, which includes scheduling requests, HARQ ACK/NACK, and CQIs.

Just like the downlink, the uplink has some key signals which help in the UE in establishing and maintaining communication with the eNodeB.

- *DMRS* (Demodulation Reference Signal). This is used for synchronization and uplink channel estimation. There are two DMRSs: one for PUSCH and the other for the PUCCH.
- *SRS* (Sounding Reference Signal). This is used by the eNodeB to estimate uplink channel condition when PUCCH and PUSCH are not being used.

7.8.1.3 Hybrid Automatic Repeat Request (HARQ)

HARQ is used in LTE to facilitate good communication and ensure data is sent and received properly. This is essentially a Stop and Wait protocol which is sent in a subframe of a message and offset in time. HARQ works by sending a packet to the destination receiver and ACK/NACK is received before the next process is allowed to begin.

An interesting aspect with HARQ is that it can be configured in a TDD system; however, in an FDD system the process is not configurable by the network operator.

HARQ works slightly different in the uplink and downlink for LTE. In the downlink, it uses an asynchronous HARQ process requiring the receiver to be sent a HARQ identifier with the data. This allows for greater flexibility in the data transmission to each UE. However, in the uplink synchronous HARQ is used and is a fixed process where the HARQ process occurs in the 8th subframe. Synchronous HARQ does not require a HARQ identifier at the expense of flexibility.

Figures 7.10 and 7.11 illustrate the various signaling sequences associated with resource assignments.

7.8.1.4 QCI

QoS Class Identifier (QCI) and *Allocation and Retention Priority* (ARP) are two key parameters in LTE, which define the particular service treatment for a UE on a bearer level.

There are several QCI settings which are shown in Table 7.4. The QCI setting describes the errors and delay that are expected with a particular service treatment including whether the service is a *Guaranteed Bit Rate* (GBR) or is a *Nonguaranteed Bit Rate* (Non-GBR).

If a bearer is a GBR, then the GBR provisioned is an average bit rate that the user can expect. Coupled with the GBR is a maximum bit rate (MBR) and this is the maximum burst rate that can be provided for that service.

If the QCI assigned is a Non-GBR, then there are two primary parameters which help regulate or define its throughput. The first parameter is the per *APN Aggregate Maximum Bit Rate* (APN-AMBR) and is used to limit the total bit rate of non-GBR bearers for a UE on a particular eNodeB or access point. Coupled with the APN-AMBR is the

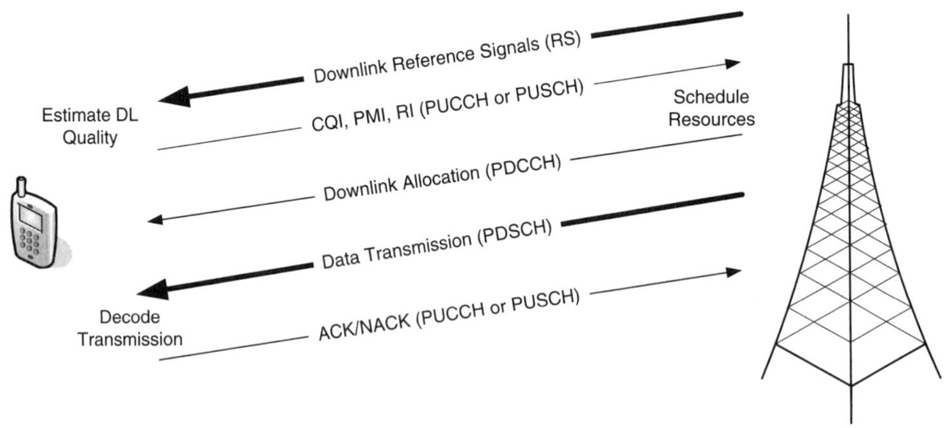

FIGURE 7.10 Downlink resource assignment.

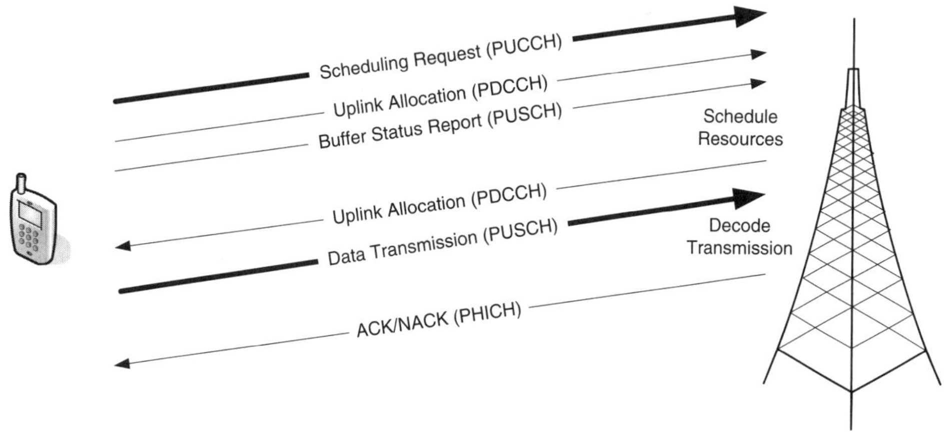

FIGURE 7.11 Uplink resource request and assignment.

QCI	Resource Type	Priority	Packet Delay (ms)	Packet Error Loss Rate	Service Description
1	GBR	2	100	10^{-2}	Voice
2		4	150	10^{-3}	Streaming Video
3		3	50	10^{-3}	Real-Time Gaming
4		5	300	10^{-6}	Buffered Video
5	Non-GBR	1	100	10^{-6}	IMS Signaling
6		6	300	10^{-6}	Buffered Video, e-mail, chat, ftp, HTTP
7		7	100	10^{-3}	Voice, Streaming Video, Interactive Gaming
8		8	300	10^{-6}	Buffered Video, email, chat, ftp, HTTP
9		9	300	10^{-6}	

TABLE 7.4 QCI

UE-AMBE or per UE aggregate bit rate limits the total bit rate of all non-GBR bearers for a particular UE.

The purpose of the QCI values is to provide uniformity of service delivered inside the network, on-net, or when using another network which adheres to the same QCI values. It is important to note that every bearer has a QCI; however, these QoS parameters are not mandatory and are in fact just guidelines for network operators.

The other partner to QCI is the *Allocation and Retention Priority* (ARP) parameter. These are used to either keep a bearer up or have it torn down based on congestion levels. There are 15 ARP values and each is associated with a particular UE or service profile. ARP 1 has the highest priority and ARP 15 has the lowest. However, a key point is that the ARP values are set by the network operator.

7.8.2 System Information Messages (SIB)

Within the downlink signaling from the eNodeB to the UE are numerous information messages that are meant to inform the UE about key attributes about the LTE system and the particular eNodeB. The attributes are located in several of the downlink channels and are referred to as System Information Messages (SIB).

The Master Information Block (MIB) is the basic system information and is continuously broadcast on the PBCH, and the UE needs the information provided by the MIB to acquire the correct LTE system. As suspected there are numerous other messages referred to as SIB messages, 12 currently, which made up the overall MIB; however, only some of them are broadcast depending on what attributes the LTE system has. Furthermore, depending on the configuration of the LTE network, various SIB messages are broadcast and each has a particular function.

The following is a brief summary of the various SIB messages.

- *SIB 1*. Cell access–related parameters and scheduling
- *SIB 2*. Common and shared channel configuration
- *SIB 3*. Parameters required for intra-frequency cell reselections
- *SIB 4*. Information on intra-frequency neighboring cells
- *SIB 5*. Information inter-frequency neighboring cells
- *SIB 6*. Information for reselection to UMTS (UTRAN) cells if no suitable LTE cell is available
- *SIB 7*. Information for reselection to GSM (GERAN) cells if no suitable LTE or UMTS cell is available
- *SIB 8*. Information for reselection to CDMA2000 systems (mostly for North America)
- *SIB 9*. Home eNodeB name—for future LTE femtocell applications
- *SIB 10 + 11*. ETWS (Earthquake and Tsunami Warning System) information
- *SIB 12*. Commercial Mobile Alerting System (CMAS) information

7.8.3 LTE DL RF Channel

The downlink RF channel used for LTE has many unique properties that enable it to provide high data rates in a very spectral efficient fashion.

LTE utilizes a modulation technique called *Orthogonal Frequency-Division Multiple Access* (OFDMA). OFDMA is similar to OFDM; however, there are some key differences that need to be highlighted. Figure 7.12 depicts both OFDM and OFDMA showing subscribers and symbols. What is unique is that in OFDM the subcarrier allocation is fixed per user; however, with OFDMA the subcarrier allocation is dynamic. The dynamic approach to assigning subcarriers leads to better trunking efficiency which equates to usable bandwidth that can be delivered to the subscriber.

OFDMA was chosen for the downlink modulation format for LTE because it has high spectral efficiency. Additionally OFDMA is also resilient to multipath propagation problems and it has the added benefit of reduced receiver complexity. The receiver complexity simplification is important for the UE aspect to keep the overall cost and battery consumption at a minimum.

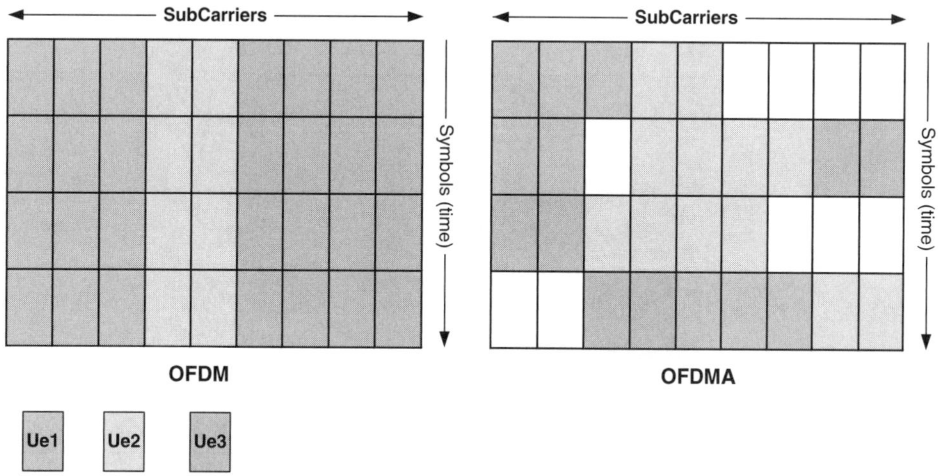

FIGURE 7.12 OFDM and OFDMA.

OFDMA also enables MIMO capability in the downlink. However, as indicated in Figure 7.12, with OFDMA resources can be both in the time domain (symbols) and the frequency domain (subcarriers) in addition to supporting parallel data streams to the UE.

Figure 7.13 shows the time and frequency relationship for the OFDMA signal.

With OFDMA in LTE each symbol is made up of multiple subcarriers and each subcarrier is 15 kHz in size. Because the allocation in the downlink channel with OFDMA is dynamic, a *Cyclic Prefix* (CP) is shown in Figure 7.13 where after each symbol a CP is inserted enabling the UE to be able to change from symbol to symbol.

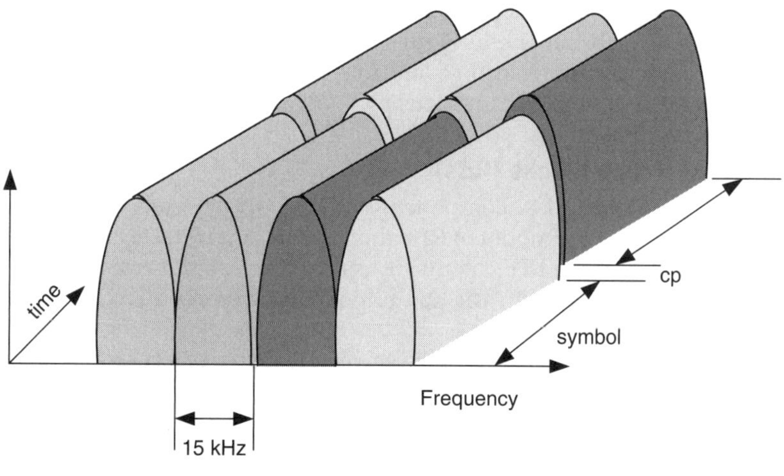

FIGURE 7.13 Downlink.

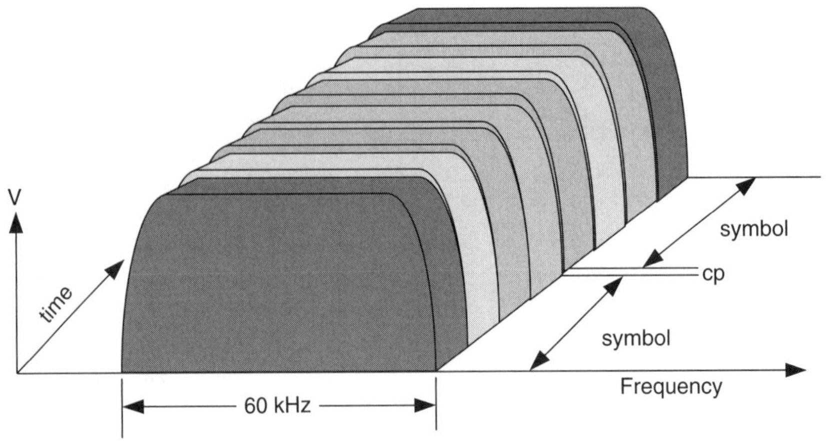

FIGURE 7.14 UL RF channel.

7.8.4 LTE UL RF Channel

In LTE the uplink RF channel utilizes a different modulation scheme than that used for the downlink RF channel. The uplink RF channel uses a Single-Carrier Frequency-Division Multiple Access (SC-FDMA) modulation scheme. In designing, LTE OFDMA was not used due to the high *Peak to Average Power Ratio* (PAPR). The high PAPR has a negative efficiency on the UE in terms of the RF power amplifier complexity that impacts both cost and power consumption. Therefore the SC-FDMA modulation scheme is used. SC-FDMA has similar multipath performance and flexible frequency allocation like OFDMA. However, SC-FDMA enables the UE to be manufactured at a lower cost and the battery life of the UE is extended as compared to using an OFDMA method.

Figure 7.14 depicts the uplink SC-FDMA format. In comparing the OFDMA scheme in Figure 7.13 with the one in Figure 7.14, similarities and differences stand out. By having a single carrier instead of multiple carriers, battery and UE cost savings come at some reduction in trunking efficiency.

7.8.5 Resource Blocks (RBs)

Resource blocks are a key component with LTE in the transport of data to and from the UE. However, the amount of RBs allocated has many factors involved. The amount of RBs available in the LTE channel is associated with the amount of subcarriers that are in the RF channel itself. The amount of subcarriers is directly related to the bandwidth of the LTE RF carrier.

Table 7.5 shows the relationship between resource blocks available and the LTE RF channel bandwidth.

The subcarriers for LTE channels are 15 kHz in size. However, in looking at Table 7.5 a brief explanation may be in order. Using a 10-MHz LTE channel the FFT size of which is 1024, which represents a possible 1024 subcarriers available; however, only 600 are available for use, meaning we do not use all of the available subcarriers for the delivery of payload data to the UE.

Bandwidth	1.4	3	5	10	15	20
Subframe Duration	1 ms					
Subcarrier Spacing	15 kHz					
FFT Size	128	256	512	1024	1536	2048
Resource Blocks	6	15	25	50	75	100
OFDM Symbols per Slot	6 or 7					
Occupied Sub Carriers	76	151	301	601	901	1201

Table 7.5 Resource Blocks

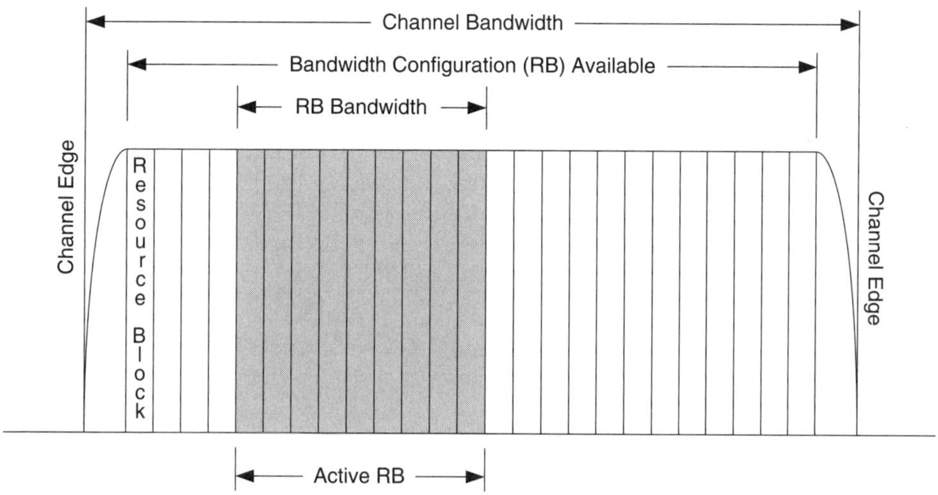

Figure 7.15 Active Resource Blocks.

Additionally, LTE can dynamically allocate resource blocks where not all the available resource blocks are allocated. For instance Figure 7.15 shows a 5-MHz LTE channel where out of the possible 25 RBs only 9 are active.

Figure 7.16 is a pictorial which relates the LTE frames to the subcarriers and then to a resource element. The resource element is the lowest common element in the delivery of data to and from the UE.

In Figure 7.16, the LTE frames used are defined in the SIB scheduling information. Resources used are allocated by Subframe (SF). The SFs are further broken down into Time Slots (TSs) and are used also for cell synchronization. The TSs make up an RB which has 12 subcarriers that are 15 kHz in size. The subcarriers have 7 symbols each and each subcarrier/symbol represents a *Resources Element* (RE).

Therefore, the various relationships are

- 1 Frame = 10 subframes
- 1 Subframe = 2 time slots
- 1 Timeslot = 6–100 RB (depending on bandwidth)

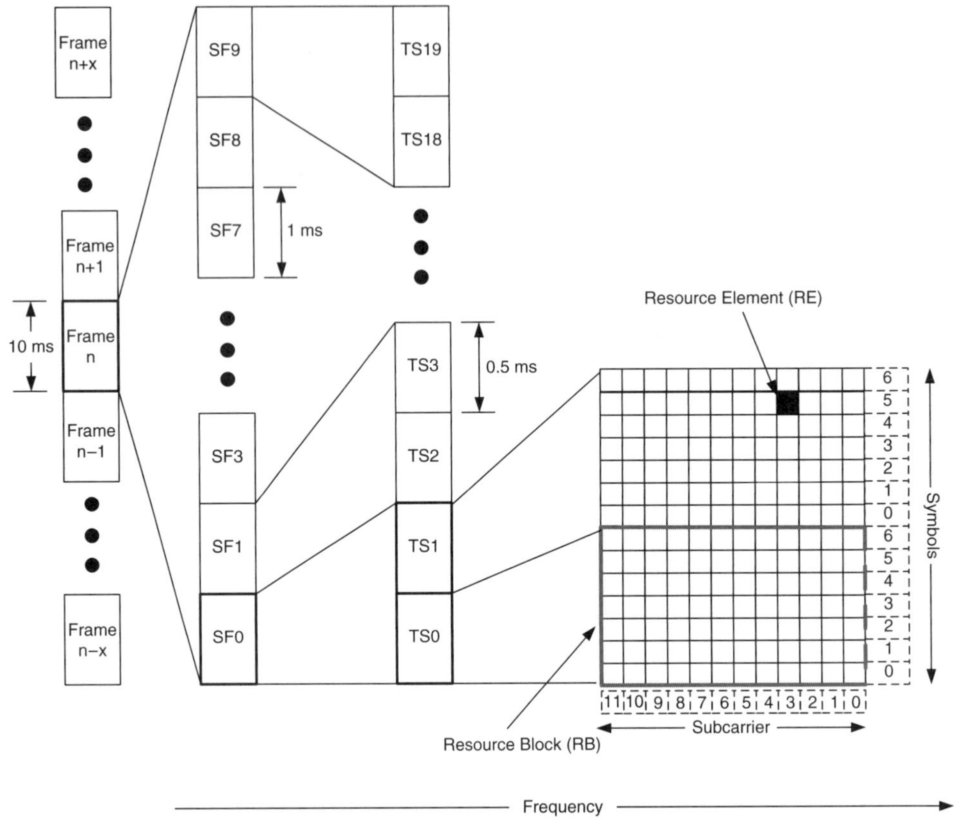

FIGURE 7.16 LTE FDD frame to resource element relationship.

- 1 RB = 12 subcarriers × 7 symbols
- 1 RE = 1 subcarrier × 1 symbol

7.9 TD-LTE

LTE can also be deployed as a TDD system. In LTE a TDD configuration is referred to as a Type 2 LTE network and with the TDD classification there are possibly seven different configurations possible.

Operating LTE in a TDD configuration enables several possible options. One key option is that LTE TDD can be deployed into existing TDD wireless access spectrum, which is currently using another technology as one possibility.

LTE TDD is in theory more spectral-efficient than an LTE FDD system due to the fact the same channel is used for uplink and downlink communication where an FDD system has dedicated channels, bandwidth, for both uplink and downlink. The spectral efficiency comes about due to the issue that data is not symmetric.

As suspected LTE TDD does not have the same air interface as LTE FDD due to the issue that the frame structure is different. However, unlike an FDD system the channel propagation characteristics are the same for both directions which enables transmit and receive to use one set of propagation parameters due to the duplex spacing for FDD channels as shown in Table 7.3.

In an LTE TDD system, the UL and DL capacity ratio can be changed dynamically to match demand allowing for better handling of data traffic.

However, in LTE TDD systems the base stations need to be synchronized with respect to the uplink and downlink transmission times. If neighboring base stations use different uplink and downlink assignments and share the same channel, then some level of mutual interference may occur between cells.

Figure 7.17 is a pictorial that relates the LTE TDD frames to the subcarriers and then to a resource element. The resource element again as in the LTE FDD configuration is the lowest common element in the delivery of data to and from the UE.

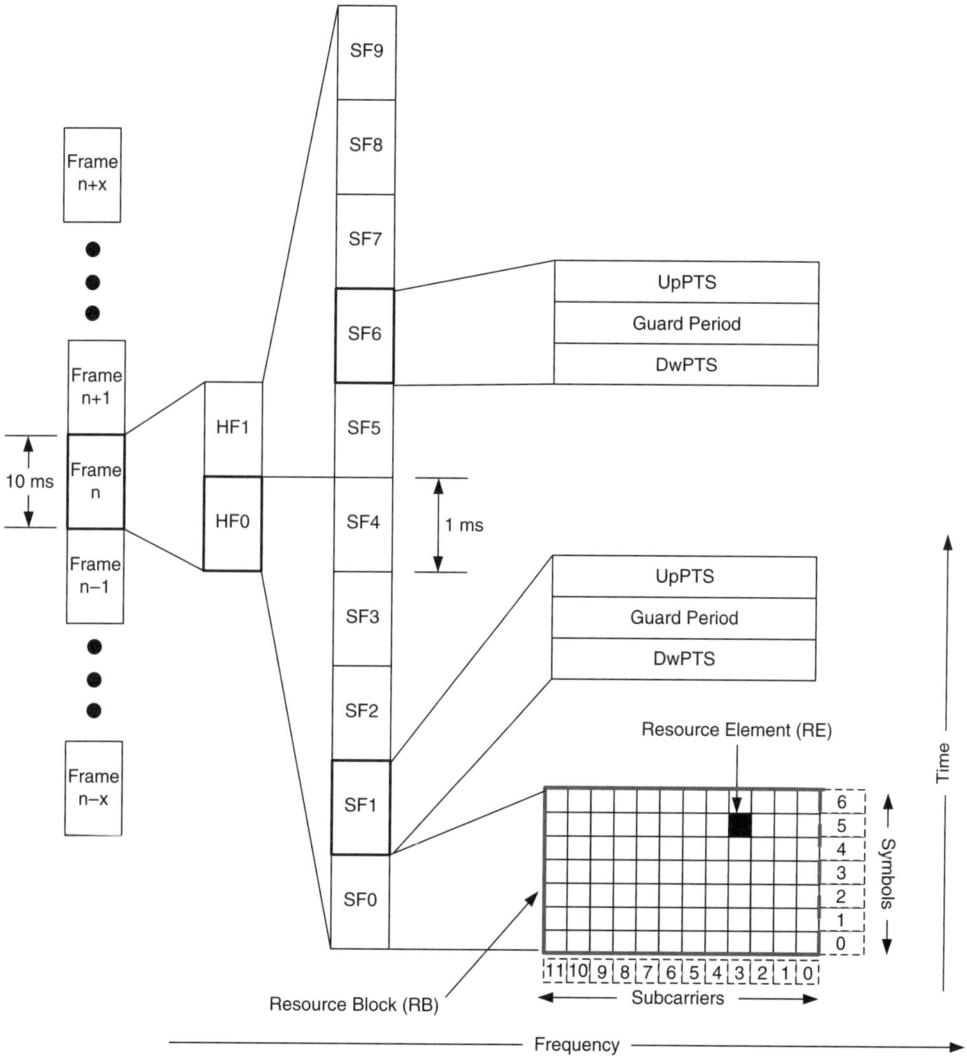

Figure 7.17 LTE TDD frame structure.

Uplink-downlink configuration	Peak Data Rates (DL/UL)	Subframe number									
		0	1	2	3	4	5	6	7	8	9
0	41 / 32 Mbps	D	S	U	U	U	D	S	U	U	U
1	62 / 22 Mbps	D	S	U	U	D	D	S	U	U	D
2	82 / 11 Mbps	D	S	U	D	D	D	S	U	D	D
3	64 / 15 Mbps	D	S	U	U	U	D	D	D	D	D
4	82 / 11 Mbps	D	S	U	U	D	D	D	D	D	D
5	94 / 5 Mbps	D	S	U	D	D	D	D	D	D	D
6	57 / 27 Mbps	D	S	U	U	U	D	S	U	U	D

D is a subframe for downlink transmission
S is a "special" subframe used for a guard time
U is a subframe for uplink transmission

Table 7.6 LTE TDD Configurations

When looking at Figure 7.17 and comparing it to Figure 7.16, many similarities are evident with a few noted differences. The frames are specially broken into hyperframes (HF) which then map to subframes (SF). The subframes then map to Resource Blocks.

Additionally, because this is a TDD system, some additional subframes, special subframes, were added to enable the same RF channel to be used for both transmit and receive. The special subframes consist of three fields and they are

- DwPTS—Downlink Pilot Time Slot
- GP—Guard Period
- UpPTS—Uplink Pilot Time Slot

7.9.1 LTE TDD Configuration

LTE TDD has several different configurations possible. Each of the seven configurations is listed in Table 7.6.

7.10 Multiple Input Multiple Output

Multiple Input Multiple Output (MIMO) is used extensively in LTE and refers to the ability of an eNodeB or UE to have multiple transmit and receive antennas. MIMO is also known as *Spacial Multiplexing* and can utilize one or several transmit antennas and one or more receive antennas, usually a minimum of two receive antennas for diversity. MIMO increases the overall bit rate through the transmission of two or more different data streams on two or more different antennas; however, the data streams use the same frequency and the same reference signal to differentiate them.

There are numerous possible configurations in LTE for MIMO. However, typical configurations utilized involve either

- 1Tx, 2Rx

or

- 2Tx, 2Rx

It is possible to have LTE operate with 8Tx and 4Rx per eNodeB leading to increased throughput at the expense of antenna real estate. However the more common configuration for an eNodeB is a 2Tx, 2Rx configuration, or 2x2.

UE devices are MIMO capable; however, their form factor limits the amount of available options and presently the mandatory configuration for UEs is a 1Tx and 2Rx configuration.

The rational for increased use of MIMO in LTE is that with MIMO RF signals emanating and received by different antennas experience different fading conditions. By increasing the amount of antennas for both uplink and downlink, this enables the LTE system to effectively increase the channel capacity without any change in available bandwidth.

MIMO also improves the RF channel characteristics by increasing the SNR through transmitting a unique bit stream using multiple antennas in the same channel (Spatial Multiplexing). When two or more antennas are used for transmit, this is referred to as *transmit diversity*.

Within MIMO there is single-user MIMO and multiple-user MIMO. In single-user MIMO (SU-MIMO), the data streams are sent to only one user. However, in multiple-user MIMO (MU-MIMO) the data streams are sent to multiple users at the same time, that is, broadcast.

Figure 7.18 shows one possible MIMO configuration for both the eNodeB as well as the UE. In Figure 7.18, the eNodeB is set up as 4×4 configuration; however, the UE can only support a 1×2 configuration. In this situation each of the UEs receives data streams from both receive antennas and transmit antennas of the eNodeB. However, the UE has one transmit antenna to send information on the uplink channel which is received by two of the eNodeB receive antennas; in reality all four eNodeB receive antennas can be utilized.

To help with MIMO communication several reference signals are sent from the eNodeB to help in the establishment of the communication link between the eNodeB and the UE. Specifically, the reference signals are related to the Resource Block (RB), which subcarriers are sent via different antennas.

In Figure 7.19, the reference signals are depicted with a 4Tx antenna system. As suspected, the specific reference signals are directly related to the MIMO configuration used.

7.11 LTE Scheduler

The scheduler for the LTE system is a critical component and directly influences the performance of the LTE network itself. Specifically, the scheduler located in the eNodeB is responsible for scheduling all LTE resources between the eNodeB and the UE. Scheduling is a network-controlled function at the eNodeB, and the UE can only make resource requests.

The eNodeB scheduler dynamically controls which resources in time and frequency are allocated to a certain user at a given time. The downlink control signaling informs the UE(s) what resources and respective transmission formats have been allocated. Additionally the eNodeB MAC sublayer is responsible for scheduling transmissions over the LTE air interface in both the downlink and uplink directions. This is because the MAC scheduler runs the scheduling algorithms that determine what gets sent, when, and to/by whom. The scheduler is also responsible for implementing the QoS characteristics assigned to radio bearers.

The output of the scheduler is a series of resource assignments for a downlink and uplink subframe. Resource assignments are defined in terms of resource blocks and the

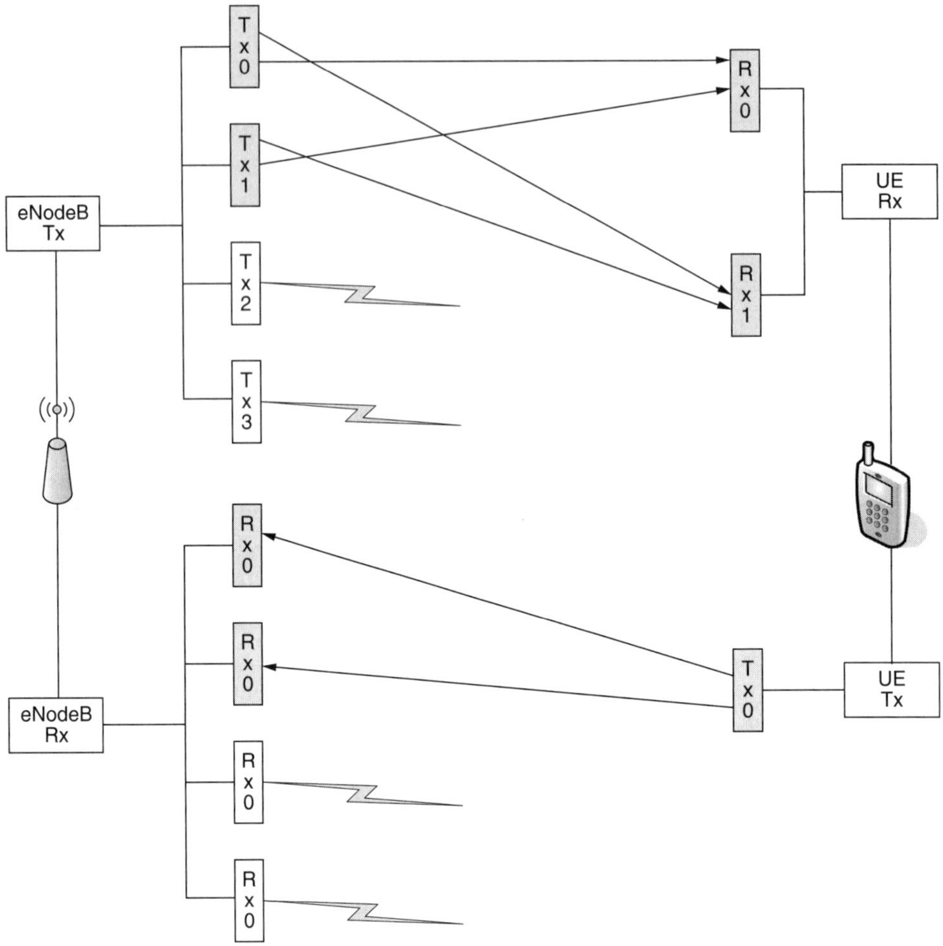

FIGURE 7.18 eNodeB 4 × 4 MIMO.

scheduler can dynamically choose the best multiplexing strategy from the available methods. Therefore the decision as to which users data and the order it is sent in within a given subframe is determined by the scheduler.

Data is transferred between the MAC sublayers in the UE and eNodeB using transport blocks which are sent via the downlink and uplink shared transport channels (DL-SCH and UL-SCH). In LTE the scheduling decisions utilize multiple inputs including link adaptation and HARQ.

When allocating resources and managing the QoS the scheduler makes numerous decisions based on its unique algorithm that it is coded with. Some of the variables that go into the scheduler decision involve

- System bandwidth
- Min and max data rate
- *Channel Quality Indicator* (CQI) reports from the UEs

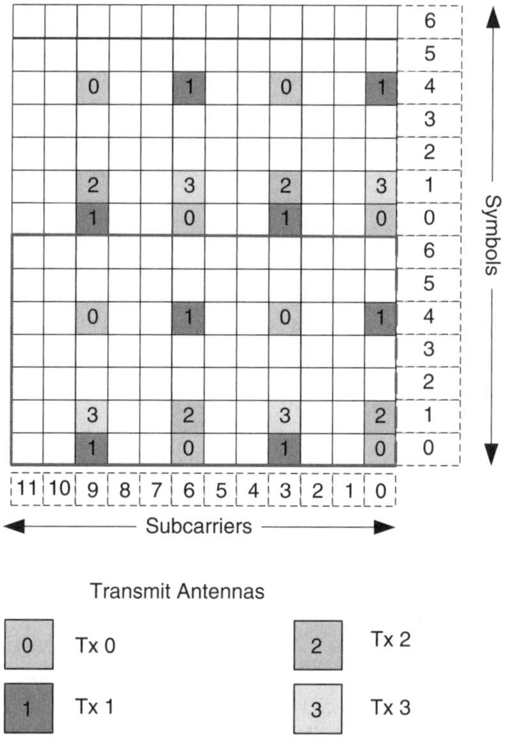

FIGURE 7.19 Reference signals.

- UE capabilities
- Pending retransmissions
- Quality-of-service parameters and measurements
- Available power allocation
- BER target requirements according to the service
- Latency requirement, depending on the service
- Payloads buffered in the eNodeB ready for scheduling
- UE sleep cycles and measurement gaps/periods

The scheduler optimizes the eNodeB throughput for the current loading condition using a scheduling policy. There are two scheduling policies that can be utilized.

1. Fair allocation scheme in which each UE (in DL or UL) is allocated the same amount of available *Proportional Resource Blocks* (PRBs). The number of allocated PRBs per UE changes only when the number of UE in the cell changes, that is, either increases or decreases.
2. Proportional allocation scheme in which the user bandwidth is adapted to the changing channel conditions.

The MME provides the inputs (variables) to the eNodeB scheduler.

7.12 Carrier Aggregation

Overall carrier aggregation will enable wireless operators to support extreme bandwidth in both downlink and uplink, thanks to a combination of carrier aggregation and enhanced MIMO. Specifically, carrier aggregation, as the name implies, allows for several distinct RF carriers to be used in collaboration for the delivery and reception of data services.

With Releases 8 and 9, carrier aggregation is not possible and the maximum channel bandwidth is 20 MHz. However, with carrier aggregation it is possible to obtain a 100-MHz-wide RF channel in both the uplink and downlink through aggregating five individual 20-MHz channels for a total of 100 MHz. Carrier aggregation is supported for both FDD and TDD systems. Therefore, as a brief example when carrier aggregation is utilized, 5 × 20 MHz for a 100-MHz-wide LTE FDD system the following is possible.

- Downlink—1 Gps (using 8 Tx antennas)
- Uplink—500 Mbps (using 4 Rx antennas)

Each aggregated carrier is referred to as a Component Carrier (CC). For carrier aggregation, a maximum of 5 CC are possible and each component carrier, typically referred to as an *LTE RF channel* will be considered individual channels until needed for intense bandwidth conditions. Therefore, component carriers can have bandwidths from 1.4 MHz to 20 MHz, depending on how each carrier is configured and the spectrum portfolio available. Furthermore, if you aggregated 5–20 MHz carriers then the maximum bandwidth possible is 100 MHz.

A key point is that the number of aggregated carriers can be different in DL and UL; however, the number of UL component carriers is never larger than the number of DL component carriers.

Additionally, due to the actual frequency of each carrier used in aggregation the coverage of the eNodeB may be affected. However, the coverage of the eNodeB does not only depend on the component carrier frequencies but also on power planning. Therefore, different component carriers can be planned to provide different coverage areas. In the case of interband carrier aggregation the component carriers will experience different path loss, which increases with increasing frequency.

The RRC connection in carrier aggregation is handled by one eNodeB and this is the primary serving cell providing the primary component carrier (DL and UL; PCC). The other component carriers are all referred to as *Secondary Component Carrier* (DL and UL; SCC), serving the secondary serving cells.

As expected, there are several configurations associated with carrier aggregation. Intraband contiguous is where the carriers used for aggregation operate in the same frequency band and are continuous. Intraband noncontiguous is where the carriers are in the same frequency band but are not continuous. Interband is where the carriers aggregated are operating in different frequency bands.

Figure 7.20 is a brief example of some carrier aggregation combinations. Figure 7.20(a) depicts a single LTE RF channel associated with Band 13. Intraband continuous carrier aggregation is shown in Figure 7.20(b) where two component carriers in the same frequency block are aggregated together. Intraband non-continuous carrier aggregation is shown in Figure 7.20(c), where the component carriers aggregated are not continuous. Lastly, Figure 7.20(d) is an example of interband channel aggregation where the CCs are in band 7 and band 13. Naturally, the bands can be any LTE-defined frequency band.

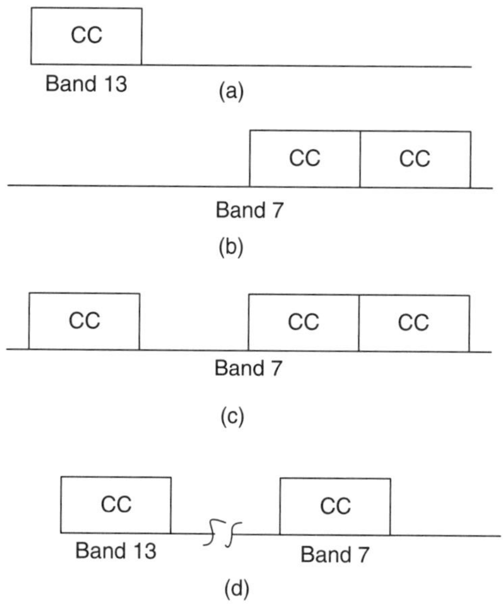

Figure 7.20 Carrier aggregation.

A very important attribute with carrier aggregation is that aggregating carriers can be mixed and matched so that you have different UL and DL carriers used for aggregation. Additionally, the carriers aggregated do not have to have the same bandwidth.

Also when using noncontinuous carriers, whether intraband or interband, the carriers will have different coverage areas and this needs to be factored into the design where the high-frequency carrier is the limiting case. Of course, this assumes that the power and other key attributes are the same.

Another point to factor in the design process is that the RRC messaging is handled by one of the CCs and this is typically referred to as the LTE, Primary CC (PCC). Other CCs are called Secondary CCs (SCC); however, each CC effectively forms a unique cell with different coverage areas.

7.13 Cell Search

Cell search happens when the UE initially powers up in an attempt to find a cell to camp on (attach). During cell search a potential cell, eNodeB, is considered to be suitable for the UE to use if

- The downlink signal strength is strong enough.
- At least one of the cells PLMN identities (broadcast in SIB1) must be in the UE's preferred PLMN list located in the USIM, which is in the UE.
- The cell's Tracking Area Identifier (TAI) must not be in the forbidden TAI roaming list.
- The cell status is not barred (blocked for any use) or restricted.

If the UE is unable to find a suitable cell, it may camp on a cell that does not meet the PLMN and TAI restrictions, called an acceptable cell. In this case the UE may not be able to access any services other than emergency calls in an acceptable cell.

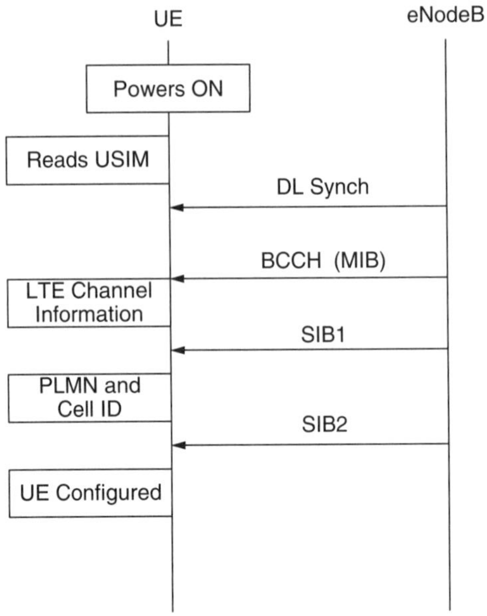

FIGURE 7.21 Cell search.

Figure 7.21 shows a simple cell search where the UE is able to find a suitable eNodeB. As part of the determination of the eNodeB's suitability, SIB information is read.

What is important to note about the cell search process is that upon the completion of the cell search the UE does not get attached to the network until it completes the RRC, Network Attach, and Authentication/Security. Once this is completed the UE can be assigned a default bearer and be allowed to send and receive data.

7.14 Cell Reselection

Cell reselection is a process where the UE is always scanning nearby cells in order to find the best cell to camp on. Once the UE camps on a cell (cell selection), it will continue to look for a better cell which is called cell reselection.

During the scanning for a possible cell reselection, the UE will consider a new cell if it is better than the current cell it is camped on provided that

- The current cell is not suitable.
- The downlink signal strength is insufficient.
- The new cell must also meet the normal PLMN, TAI, and status requirements.

7.15 Attach and Default Bearer Activation

Once the UE has selected a cell site to camp on, it will attempt to attach to the cell and if successful a default bearer will be assigned to the UE.

The overall process can be briefly summarized as

- Upon power up, UE scans channels for a suitable network.
- The UE synchronizes with the best cell to get system configuration information.

- UE then can request a *Radio Resource Connection* (RRC) to exchange signaling information with network.
- RRC is used to register the UE (attach).
- Network establishes the default bearer path.
- UE is assigned IP address.

Figure 7.22 shows attach and default bearer assignment process.

An important point to note is that when the UE no longer needs a data session, the default access bearer remains, that is, PGW to SGW. The SGW to eNodeB connection is torn down; however, the IP address associated with the UE is still maintained.

7.16 Handover (X2, S1, Inter-MME)

There are several types of handovers in LTE. The three types of handovers are

- S1
- X2
- Inter-RAT

As part of the handover process, the UE measures its neighboring cells to determine the best cell to be on. Which neighbor cells to monitor is contained in System Information Blocks (SIB) that is sent to the UE during the initial attach process.

The UE measurements about its neighbor cell is sent to the eNodeB that is currently serving the UE. The measurements that the UE sends to the eNodeB include

- Reference Signal Received Power (RSRR)
- Reference Signal Received Quality (RSRQ)

LTE uses a hard handoff process when handing off. However, the credentials for the UE are passed along, as well as packet buffering between the source and target eNodeBs to ensure that packets are not lost during the handover.

Handovers within an LTE network can be either intra- or inter-MME handovers. Essentially, an intra-MME handover is where the source and target eNodeB are under the same MME. Inter-MME handover involves handing off to another eNodeB which is under the control of a different MME.

Intra-MME handovers involve both X2- and S1-based handovers. However, the signaling paths take on different routes where the X2-based handover is a direct eNodeB-to-eNodeB communication while an S1-based handover includes the MME in the signaling coordination, eNodeB-MME-eNodeB. In both cases, the MME remains the same for both X2 and S1 handovers; however, it is possible that the SGW can change depending on the eNodeB used as the target.

Inter-MME handovers, however, involve changing the MME and it is likely the SGW will change as well again depending on which eNodeB is used as the target.

LTE also enables Inter Radio Access Technology (Inter-RAT) which involves handovers between LTE and 3GPP or non-3GPP systems.

7.16.1 X2 Handover

Figure 7.23 is a brief depiction of the call flows for an X2 handover. The X2 handovers are facilitated by eNodeBs, which are connected via the X2 link.

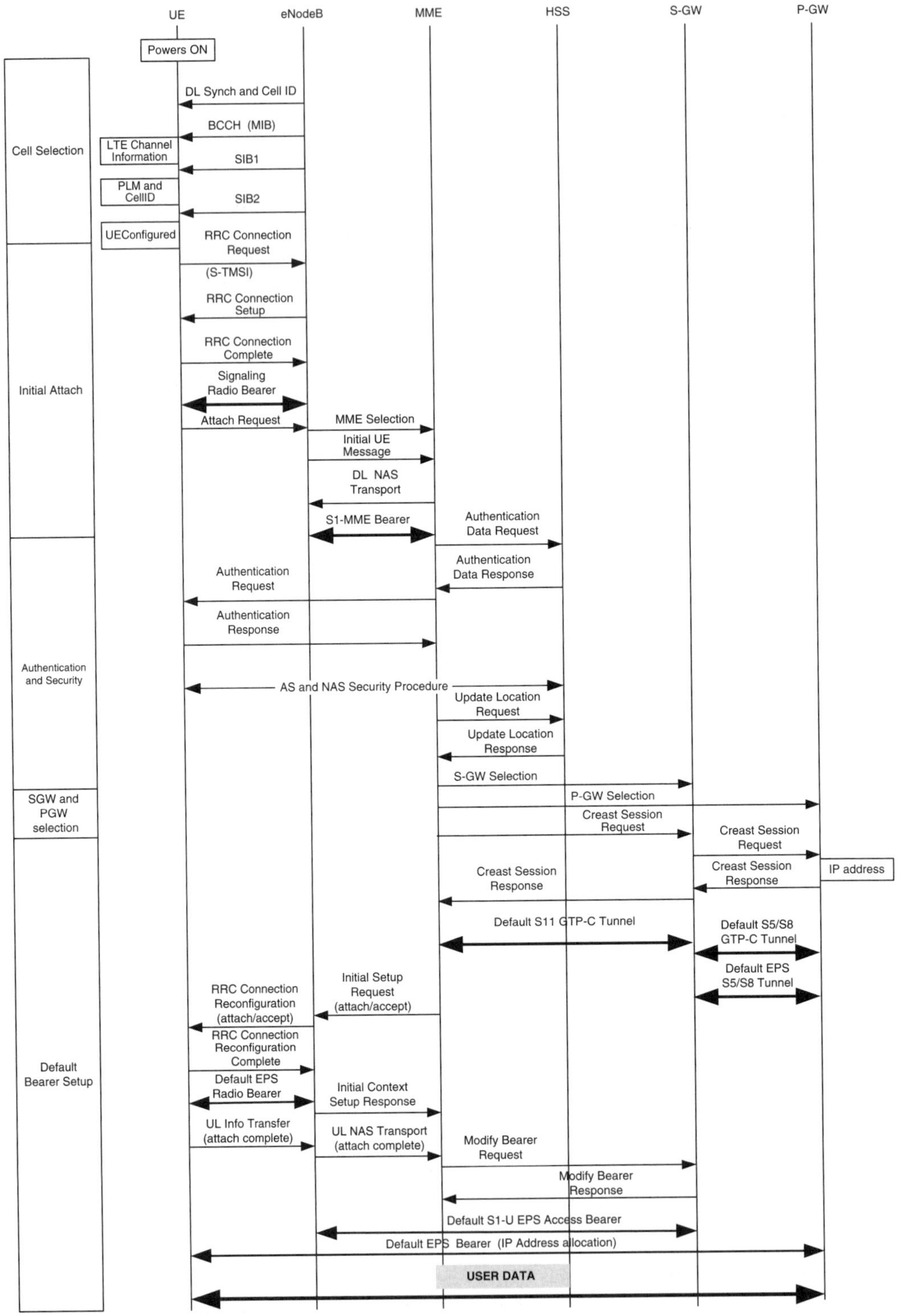

FIGURE 7.22 Attach and bearer assignment.

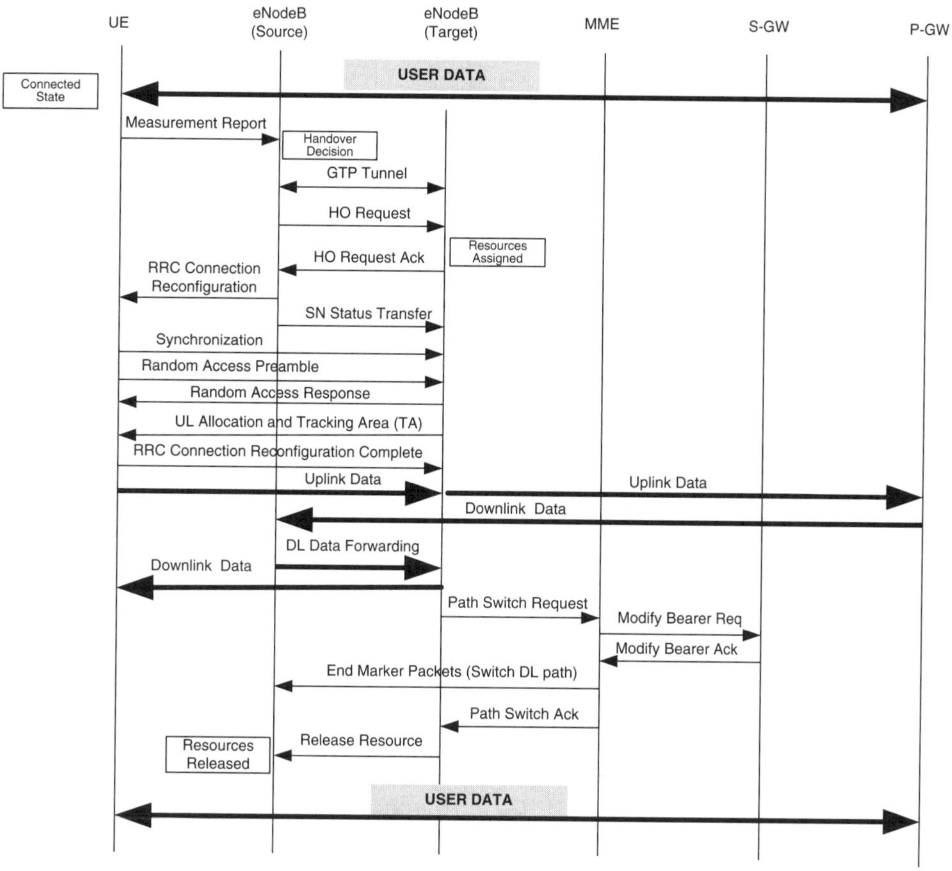

FIGURE 7.23 X2 handover.

7.16.2 S1 Handover

S1 handovers are the most prevalent handovers in LTE if the network topology is laid out correctly. Figure 7.24 shows a simple signaling flow for an S1 handover.

During an S1 handover, it is possible to change the MME depending on the target eNodeB. It is also possible for the MME to assign several different PGWs or a different PGW for the UE to use on the target eNodeB.

7.16.3 Inter-MME Handover

In LTE, inter-MME handovers is used where the target eNodeB resides on a different MME and this signaling flow is shown in Figure 7.25.

7.17 Self-Organizing Networks (SONs)

LTE supports the functions of SONs. It enables the SONs to automate network configuration and optimization processes for the introduction of a new node, as well as for the removal of the one on a permanent or temporary basis. A temporary removal of a node

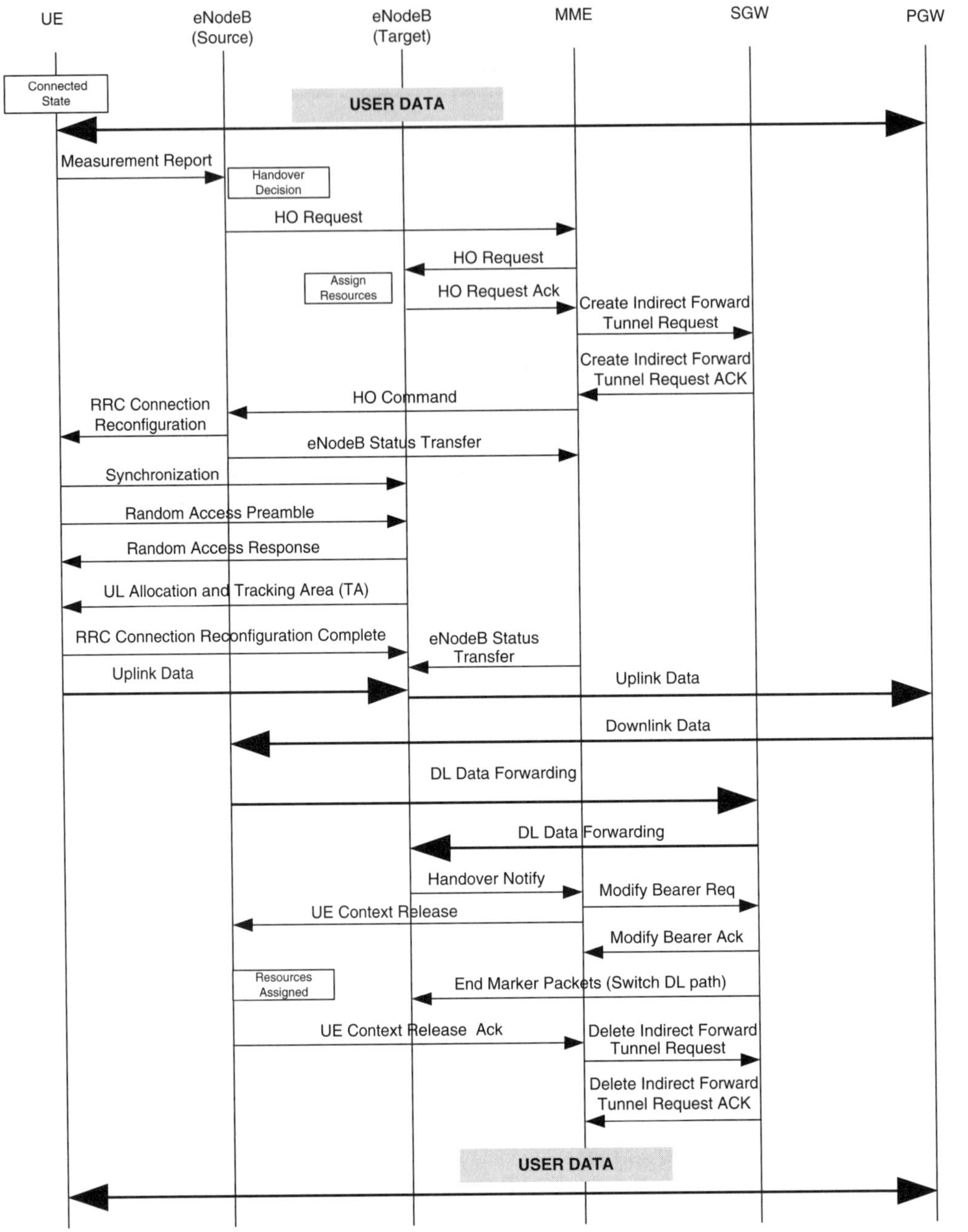

FIGURE 7.24 S1 handover.

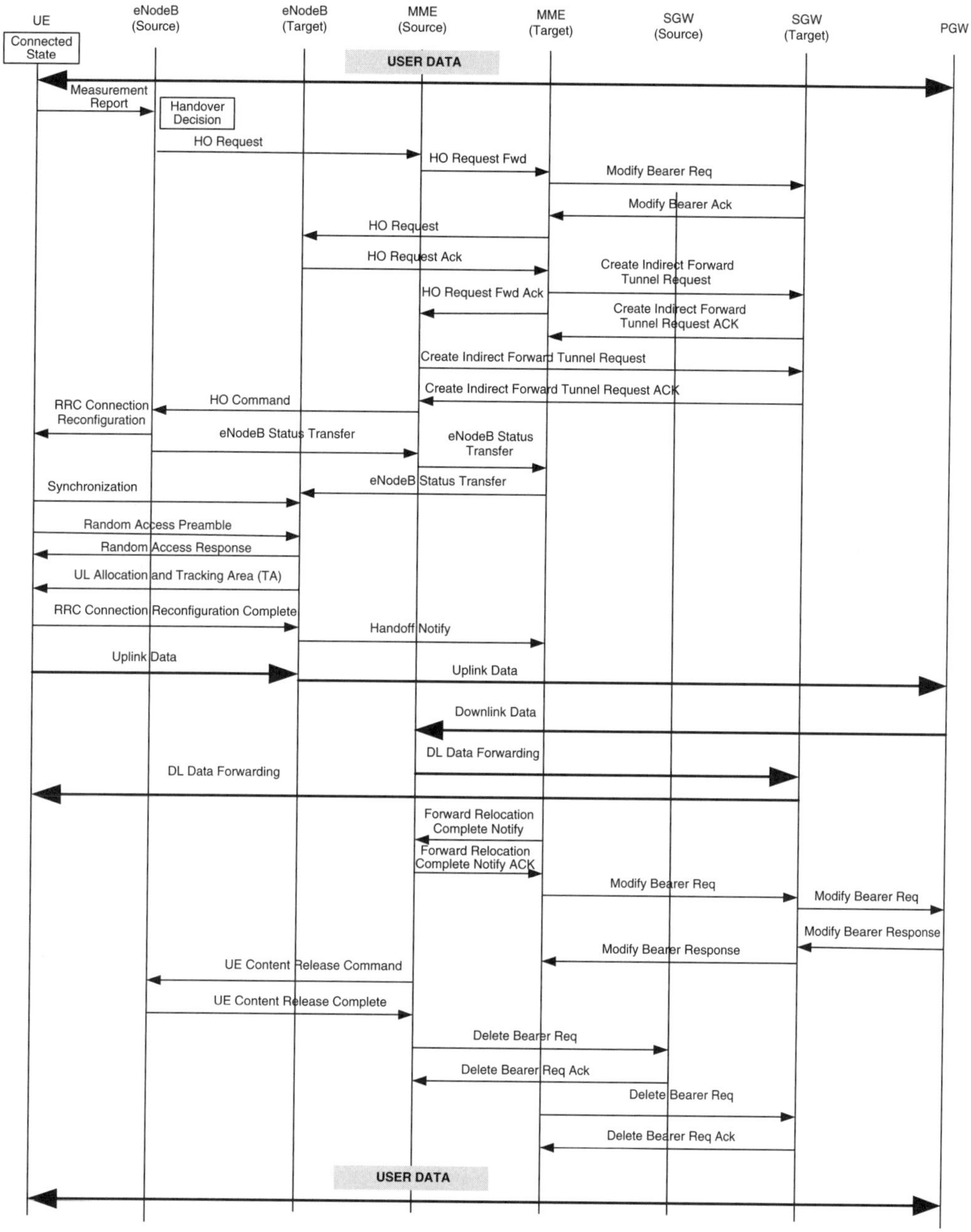

FIGURE 7.25 Inter-MME handover.

could be simply due to maintenance-related events. However, the key concept with SONs is that it can reduce the need for centralized planning and human intervention.

SONs functions in LTE are primarily focused on the exchange of information between neighbor eNodeBs; however, there are some other unique functions as well.

- Automatic Neighbor Relation function, which enables the eNodeB to establish its neighbor relations based on UE reports
- PCI selection, which enables the eNodeB to select its own PCI based on what the UE reports and information received from neighbor eNodeBs
- X2/S1 interfaces configuration, which allows the eNodeB to dynamically configure the S1-MME interface with the serving MMEs and the X2 interface with neighbor eNodeBs
- RACH parameters optimization, since the neighbor eNodeBs can exchange information about their used PRACH resources.
- Traffic shedding and redistribution between eNodeBs facilitating load balancing

However, SONs also enable the introduction of HeNBs.

7.18 Relay Cells

Relay nodes are introduced into the LTE portfolio for coverage improvements and throughput enhancements for localized areas. The LTE relays are different than small cells or femtocells in that they use an LTE channel for backhaul for a donner eNodeB cell site. The LTE relay sites also will have unique cell IDs as well.

Figure 7.26 shows one possible configuration for a relay cell. In the figure, the Relay Node (RN) is connected to a donner eNodeB (DeNB). The fact that the RN is connected to a DeNB differentiates it from a typical small cell, which is an independent cell site, that is, small/pico/femto.

Reviewing Figure 7.26, the RN is connected to the DeNB via an RF link. The two links shown for the GTP-U and S1-AP actually utilize the air interface for the connectivity since there is no physical connection between the RN and DeNB except through the radio link.

While Figure 7.26 shows only one RN connected to the DeNB, a DeNB can have multiple RNs associated with it.

There are several configurations available for an RN, and Type 1a is considered the more typical where the outbound channel Uu and Un are both different frequencies. However, it is possible to utilize the same radio carrier frequency for both Uu and the Un. However, there is a risk with self-interference. The self-interference can possibly be

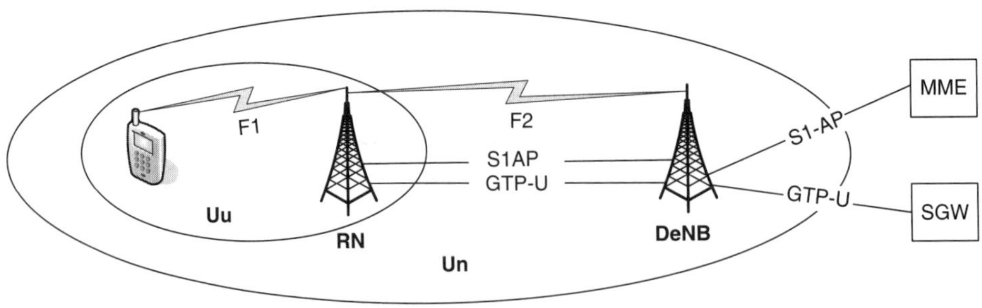

Figure 7.26 Relay Cell.

mitigated or at least reduced with the use of multiplexing, though at the cost of throughput and network complexity.

In Figure 7.26, one possible configuration could involve having Uu operating in AWS while the Un operates in the 700-MHz band.

With a Relay Node the S1 connection is terminated in the Relay Node itself which can lead to additional delays. The RN shares the same SGW and MME as the DeNB. However, terminating the S1 connection at the RN enables the RN to send out its own cell ID and Tracking Area (TA) for a unique service offering.

7.19 Heterogeneous Network (HetNET)

With LTE and other wireless mobility access technologies, the combination of macro, micro, pico, femto, and relay nodes creates heterogeneous network within the same network using the same access technology. This multitude of access points that can operate on different frequency bands enables a multilayer network to be deployed as compared to the traditional honeycombed approach. With HetNet, it is also possible and highly likely to have multiple wireless access methods at play within the same heterogeneous network. However, a key attribute or requirement for heterogeneous networks is the need to have solid radio resource management.

7.19.1 Small Cell

The term small cell has gained a lot of attention and will have a major impact on how LTE and other wireless access technology platforms are deployed. Specifically, a small cell is what some can classify as a pico or femto cell. However, small cells are meant to address the issues like spot capacity and coverage problems, at the same time requiring a small physical footprint for the infrastructure.

Small cells when included in a HetNet design can augment macro cell sites and in fact could possibly replace future macro cell deployments.

Presently, small cells are implemented in an underlay approach for a macro network. However, there is nothing that would preclude a large deployment of small cells from replacing the current macro cell deployment schemes in an urban environment.

Therefore, there are two views of what defines a small cell. One view is that physical footprint the small cell occupies from an infrastructure aspect. The other, and probably the best, definition is the coverage area that the small cell provides. By defining small cells based on their coverage area, the use of *Remote Radio Heads* (RRH) and femtocells plus other delivery platforms would easily be classified as a small cell.

7.19.2 HeNB

Key components to the advancement of a heterogeneous network architecture are femtocell or in LTE world the Home eNodeB (HeNB). The HeNB uses licensed spectrum; however, a HeNB is not a relay node primarily since it gets its backhaul from existing DSL/cable connections. A HeNB is mainly used for indoor coverage.

The HeNB includes the functions of an eNodeB as well as some additional HeNB-specific configuration/security functions. The HeNB accesses the core network and can support a large number of S1 interfaces with the use of a *HeNB-Gateway* (HeNB GW).

Three different access modes are defined for HeNBs. These are closed, hybrid, and open and are associated with the *Closed Subscriber Group* (CSG).

- *Closed access* mode provides services only to its associated CSG members.
- *Hybrid access* mode provides services to its associated CSG members and to nonmembers. In a hybrid mode, the CSG members are prioritized over nonmembers.
- *Open access* mode has the HeNB appear as a normal eNodeB.

The HeNB cells operate as either a closed or hybrid cell for access. The HeNB broadcasts its access configuration to help the UE in identifying the characteristics of the HeNB. The parameters that support the UE in the identification of closed/hybrid cell are

- CSG indicator
- CSG identity (CSG ID)
- HeNB name

7.20 Remote Radio Heads (RRH)

One important deployment method for LTE is the use of RRH. An RRH is typically meant for outdoor deployments; however, it can also be used for building systems in place of or augmenting DAS. The RRH is a radio head in an LTE transmitter which can be mounted close to the antenna and connected to an RRH controller through an Ethernet or fiber connection.

Figure 7.27 shows a high-level depiction of some possible locations where an RRH can be utilized. In essence, an RRH enables a very small physical footprint for the LTE base station or sector making it possible to be installed in locations that were previously prohibited due to space or other issues.

7.20.1 Common Packet Radio Interface (CPRI)

Common Packet Radio Interface (CPRI) is a fiber-optic communication standard that enables the use of RRH. CPRI defines the interface between an LTE eNodeB controller and a remote radio head for LTE. CPRI is essentially a layer 1 and layer 2 protocol that is IP-based utilizing an Ethernet connection. With CPRI the IP payload is referred to as an antenna carrier meant for a particular RRH. With multiple transmitters and receivers for LTE, each has its own CPRI antenna carrier payload.

7.21 VoLTE

VoLTE is a logical evolution or migration of voice services onto one platform. How LTE handles VoIP or VoLTE is important, since LTE is a packet-based system and voice services are effectively a circuit-based service. VoIP, as discussed in Chapter 9, is understood. However, the EPC for an LTE network does not have a VoIP gateway to interface to a TDM network.

Therefore, VoLTE uses an IP Multimedia Service (IMS) approach for VoIP treatment. The IMS platform is not part of the EPC, but as expected, referencing the EPC diagrams interfaces to the EPC via a PGW.

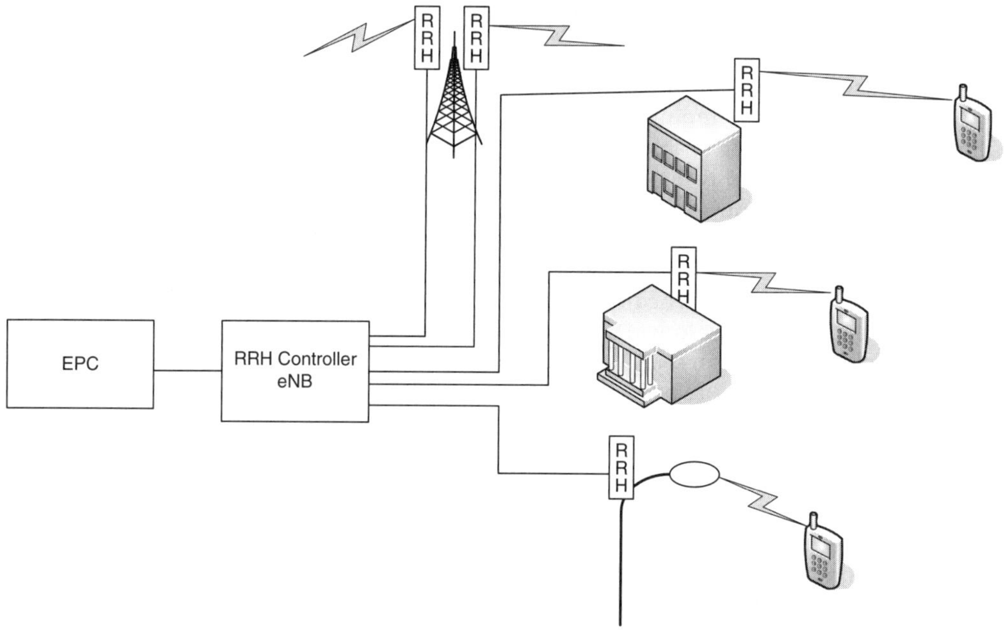

Figure 7.27 Remote Radio Head.

Further, there are some additional items which should be mentioned regarding voice services in an LTE network. Initially, when LTE was released it utilized a separate wireless network to deliver voice services. The two methods used involved *Circuit Switched Fall Back* (CSFB) and *Simultaneous Voice LTE* (SV-LTE). With CSFB when a voice call needed to be placed or received from a TDM network, the Ue would fall back to a 2G or 3G network. SV-LTE is slightly different than CSFB in that packet traffic is still handled by the LTE bearer and voice services use a CSFB. However, these techniques require two different radio access networks to serve the subscriber.

7.22 LTE Advanced

3GPP release 11 and beyond is referred to as LTE Advanced and is the next level or improvement for LTE. LTE Advanced has several key attributes associated with it, namely

- Both intra- and inter-frequency Carrier Aggregation, which increases throughputs in both the downlink and uplink
- Coordinated Multipoint (CoMP)
- Enhanced Physical Downlink Control Channel (E-PDCH), which is required to support CoMP including increased control channel capacity, frequency domain Intercel Coordinated Interference control (ICIC), and enhanced MIMO features like beam forming

The topics are discussed in several sections of this chapter.

7.22.1 Coordinated Multipoint (CoMP)

CoMP is meant to reduce inter-cell interference for cells that are operating on the same frequency, thereby maximizing the potential throughput from multiple distribution points which can be multiple eNodeBs, RRH, or other distribution points in the network. In short, CoMP enables multiple downlink and uplink transmissions for a mobile device, thereby not only improving throughput but also potentially improving the battery life of the end user device.

The multiple downlink and uplink transmission with CoMP are called Joint Transmission (JT) for the downlink and Joint Reception (JR) for the uplink. CoMP therefore enables a coordinated scheduling of traffic which involves the coordinated effort of multiple access points to work in harmony together by allowing the UE to select the best access points it can see.

CoMP is expected to be implemented first with Joint Transmission and then closely followed by Joint Reception.

References

"Dealing with Density: A move to Small Cells" Ruckus Wireless, Inc., February 2013.

Konstantinos Dimou, Min Wang, Yu Yang, Muhammmad Kazmi, Anna Larmo, Jonas Pettersson, Walter Muller, Ylva Timner, "Handover within 3GPP LTE: Design Principles and Performance," IEEE 2009.

Carl Eklund, Roger B. Marks, Kenneth L. Stanwood, and Stanley Wang, "IEEE Standard 802.16: A Technical Overview of the WirelessMAN™ Air Interface for Broadband Wireless Access," IEEE Communications Magazine, June 2002.

ETSI TS 101 220 V8.4 "Smart Cards; ETSI numbering system for telecommunication application providers (Release 8)", 2009.

ETSI TS 102 210 V1.1.1 (2003-11) Broadband Radio Access Networks (BRAN); HiperMAN; System profiles.

ETSI TS 102 177 V1.5.1 (2010-05) Broadband Radio Access Networks (BRAN); HiperMAN; Physical (PHY) layer.

ETSI TS 136 331 V8.10.0 (2010-07) LTE; Evolved Universal Terrestrial Radio Access (E-UTRA); Radio Resource Control (RRC); Protocol specification. ETSI TS 132 671 V10.0.0 (2011-04) Digital cellular telecommunications system (Phase 2+); Universal Mobile Telecommunications System (UMTS); LTE; Telecommunication management; Configuration Management (CM); State Management Integration Reference Point (IRP); Requirements.

ETSI TS 136 300 V10.3.0 (2011-04) LTE; Evolved Universal Terrestrial Radio Access (E-UTRA) and Evolved Universal Terrestrial Radio Access Network (E-UTRAN); Overall description; Stage 2.

ETSI TS 136 305 V9.5.0 (2011-01) LTE; Evolved Universal Terrestrial Radio Access Network (E-UTRAN); Stage 2 functional specification of User Equipment (UE) positioning in E-UTRAN.

ETSI TS 136 306 V10.1.0 (2011-04) LTE; Evolved Universal Terrestrial Radio Access (E-UTRA); User Equipment (UE) radio access capabilities.

ETSI TS 136 413 V10.1.0 (2011-04) LTE; Evolved Universal Terrestrial Radio Access Network (E-UTRAN); S1 Application Protocol (S1AP).

ETSI TS 136 355 V10.1.0 (2011-04) LTE; Evolved Universal Terrestrial Radio Access (E-UTRA); LTE Positioning Protocol (LPP) (3GPP TS 36.355 version 10.1.0 Release 10).

Seppo Hämäläinen, "Self-Organizing Networks in 3GPP LTE," 2009 IEEE Nokia Siemens Network, 2009. "Spectrum Analysis for Future LTE Deployments," Motorola 2007.

Mikio Iwamura, Hideaki Takahashi, and Satoshi Nagata, "Relay Technology in LTE-Advanced," August 2010. "MIMO in HSPA: The Real-World Impact," GSMA June 2011.

Naveen Kottapalli, "Diameter and LTE Evolved Packet System" Radisys, September 2011.

Anna Larmo, Magnus Lindström, Michael Meyer, Ghyslain Pelletier, Johan Torsner, and Henning Wiemann "The LTE Link-Layer Design," IEEE Communications Magazine, April 2009.

"Long Term Evolution (LTE): Overview of LTE Air-Interface Technical White Paper," Motorola 2007.

"Long Term Evolution Protocol Overview Document Number: LTEPTCLOVWWP Rev 0" October 2008, Freescale, October 2008.

Andreas Roessler, "Cell search and cell selection in UMTS LTE," Rhode & Schwarz, September 2009.

Roessler A. and Kottkamp M., LTE- Advanced (3GPP Rel.11) Technology Introduction White Paper 7.2013 – 1MA232_1E, Rhode& Schwarz. Rumney (editor), "LTE and the Evolution to 4G Wireless Design and Measurement Challenges," Agilent Press, 2009.

Martin Sauter, "From GSM to LTE," John Wiley & Sons, 2011.

Stefania Sesia, Issam Toufik, and Matthew Baker, "LTE – The UMTS Long-Term Evolution," John Wiley & Sons, 2011.

"The LTE Network Architecture | Strategic White Paper," Alcatel Lucent, December 2009.

3GPP TS 21.101 V4.13.0 (2006-03) Technical Specifications and Technical Reports for a UTRAN-based 3GPP system (Release 4).

3GPP TS 23.246: "Multimedia Broadcast/Multicast Service (MBMS); Architecture and functional Description."

3GPP TS 23.401 V9.2.0 (2009-09) General Packet Radio Service (GPRS) enhancements for Evolved Universal Terrestrial Radio Access Network (E-UTRAN) access (Release 9).

3GPP TS 23.401 V9.5.0 (2010-06) General Packet Radio Service (GPRS) enhancements for Evolved Universal Terrestrial Radio Access Network (E-UTRAN) access (Release 9).

3GPP TS 25.104, Technical Specification, "3rd Generation Partnership Project; Technical Specification Group Radio Access Network; Base Station (BS) radio transmission and reception (FDD) (Release 9)."

3GPP TS36 3212 V9.2 "Evolved Universal Terrestrial Radio Access (E-UTRA); Medium Access Control (MAC) protocol specification," 2010.

3GPP TS 36.213 v8.8.0. "Evolved Universal Terrestrial Radio Access (E-UTRA); Physical layer procedures."

3GPP TS36 323 V9.0 "Evolved Universal Terrestrial Radio Access (E-UTRA) Packet Data Convergence Protocol (PDCP) specification," 2009.

3GPP TS36 322 V9.1 "Evolved Universal Terrestrial Radio Access (E-UTRA); Radio Link Control (RLC) Protocol specification," 2010.

3GPP TS36 331 V9.2 "Evolved Universal Terrestrial Radio Access (E-UTRA); Radio Resource Control (RRC) Protocol specification," 2010.

3GPP TS36 806 V9.0 "Evolved Universal Terrestrial Radio Access (E-UTRA); Relay Architectures for E-UTRA (LTE-Advanced)," 2010.

3GPP TR 36.921 V9.0.0 (2010-03) Evolved Universal Terrestrial Radio Access (E-UTRA); FDD Home eNode B (HeNB) Radio Frequency (RF) requirements analysis (Release 9).

3GPP TR 36.922: "Evolved Universal Terrestrial Radio Access (E-UTRA); TDD Home eNode B (HeNB) Radio Frequency (RF) requirements analysis."

3GPP TS 37.104 V9.3.0 (2010-09) Technical Specification E-UTRA,, UTRA and GSM/EDGE; Multi-Standard Radio (MSR) Base Station (BS) radio transmission and reception.

3GPP TR 37.900 V9.0.1 (2010-07) Radio Frequency (RF) requirements for Multicarrier and Multiple Radio Access Technology (Multi-RAT) Base Station (BS) (Release 9).

Suyash Tripathi, Vinay Kulkarni, and Alok Kuma, "LTE E-UTRAN and its Access Side Protocols," Radisys, September 2011.

Mats Wernersson, "Scheduling of Multi-Media over 3GPP LTE," Lulea Tekniska Universitet, March 15 2007.

www.3gpp.org

www.ETSI.org

CHAPTER 8
WiMAX

8.1 Introduction

The *Worldwide Interoperability for Microwave Access* (WiMAX) is a broadband wireless access method that is based on the *Data over Cable Service Interface Specification* (DOCSIS) protocol and uses OFDMA as the modulation scheme. WiMAX is not a 3GPP standard, but an IEEE standard. WiMAX is based on the IEEE 802.16 set of standards for an IP packet-based system and can be deployed as an FDD or TDD configuration. WiMAX is also a complimentary broadband service to WiFi (IEEE 802.11).

The WiMAX IEEE 802.16 standards cover the following scenarios:

- WiMAX for backhaul solutions (point to point)
- Fixed wireless access WiMAX (point to multipoint)
- Mobile WiMAX (mobility)

Mobile WiMAX is also classified as a 3G (IMT-2000) and 4G (IMT-2000 advanced) wireless radio access technology. WiMAX using 802.16e comes under the 3G classification while WiMAX using 802.16m is classified as a 4G technology.

WiMAX, which originally began as a fixed wireless point-to-multipoint wireless network, was previously referred to as *Local Multipoint Distribution Systems* (LMDS). However LMDS systems were designed and deployed before the IEEE standard series of 802.16 (x) was established. Additionally LMDS was designed for fixed locations providing last mile broadband access using a variety of protocols. However with the advent of IEEE 802.16, the adaptation of WiMAX has greatly enhanced since it is no longer vendor-driven proprietary protocols but a standards-based solution.

The WiMAX IEEE 802.16, or 802.16, is an evolving IEEE standards that are applicable to a vast array of applications and defined spectrum bands currently ranging from 400 MHz to 66 GHz, which include both licensed and unlicensed (license exempt) bands.

WiMAX is used presently in many deployment configurations around the world ranging from last mile broadband access for both fixed and mobile as well as point-to-point links for backhaul.

Figure 8.1 briefly depicts the various areas within wireless access that WiMAX plays a role in.

In reviewing Figure 8.1, it is apparent that WiMAX meets a variety of wireless access requirements.

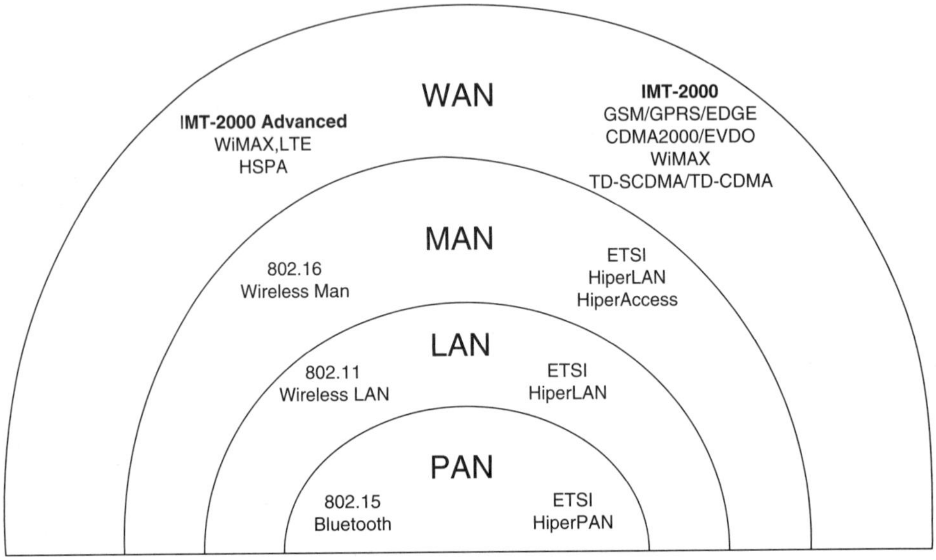

FIGURE 8.1 WiMAX wireless access relationship.

Some of the uses for fixed WiMAX include

- Last mile wireless broadband connections as an alternative for DSL and cable modems.
- Broadband access in developing countries as an alternative to installing fixed infrastructure like copper.
- Hotspots and high-speed connectivity for business customers.
- Backhaul connectivity as a point-to-point microwave link using licensed, lightly licensed, and unlicensed spectrum.

WiMAX also supports wireless mobility enabling

- Broadband mobility (3G and 4G)
- Nomadic connectivity, that is, hot spots and or hot zones

WiMAX enables the convergence of both mobility and fixed wireless access through a common radio access protocol.

An important attribute of WiMAX is the reliance on specific profiles for the *Access Node* (AN) and the *Subscriber Station* (SS). The profiles define the specific *Radiofrequency* (RF) characteristics, as well as the services that are supported. Profiles consist of both system and certification profiles. The system profiles define mandatory and options physical and MAC layer features, whereas the certification profiles define the operating frequency, RF channel bandwidth, and duplexing mode (TDD, FDD) and are tied to the system profile selected. When WiMAX equipment is certified it is certified based on interoperability using a given certification profile.

For a more detailed read on the specifications related to 802.16, there are several excellent sources for obtaining information and existing vendor interoperability, for example, www.wimaxforum.org and www.ieee802.org/16.

8.2 Standards

WiMAX is part of the IEEE 802 standards series. Figure 8.2 attempts to show some of the relationships between some of the more prevalent IEEE 802 standards and WiMAX.

Like all other standards within the 802 series like WiFi and Bluetooth/Zigbee, the 802.16 series, referred to as WiMAX has many sections which make up this standard. However, WiMAX can be grouped into two basic standard types, fixed and mobile. The WiMAX standards—like all standards—are continuously evolving and diligence is needed to keep up-to-date on the various advances that are being made with this unique series or protocols.

The unique aspect with WiMAX is that while the term is general its application can be widely different. For instance WiMAX can be used for point-to-point communication for backhaul or it can be used for mobility or fixed broadband. Because of the plethora of WiMAX configurations possible, a brief overview of the relevant 806.16 standards is shown in Table 8.1. There are other 806.16 specifications as part of the series that provide additional enhancements and functionality; however, they will not be covered here.

Further expanding upon Table 8.1 is Table 8.2 which briefly highlights some of the technical parameters associated with the mobility aspects for WiMAX.

The WiMAX Forum® like the IEEE plays an important role in the development and refinement of the WiMAX standards through promoting the standardization of WiMAX and obviously the compatibility and interoperability of devices that make up the WiMAX ecosystem. Details about interoperability and compatibility of WiMAX devices can be found at www.wimaxforum.org.

To help further discuss the differences between the various 802.16 standards, and there are many, the following is offered up.

802.16 is a point-to-multipoint protocol with a centralized base station and is a *Wireless Metropolitan Area Network* (WMAN) technology that provides both backhaul and an alternative for last mile broadband access as well as connecting 802.11 hot spots to the Internet. 802.16 standard specifies two convergent service layers that form the basis of protocol, and those two convergent service layers are ATM and Packet (IP).

802.16a focuses on spectrum that is below 10 GHz. In the United States the key spectrum for 802.16 is in the MMDS bands, mostly from 2.5 to 2.7 GHz. Worldwide, 3.5 GHz

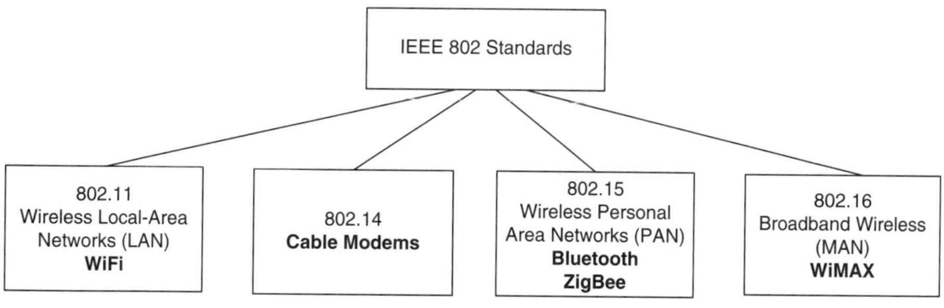

FIGURE 8.2 IEEE 802.

Standard	Frequency Band	Comments
802.16	10–66 GHz	• WirelessWAN, HiperAccess • LOS point-to-multipoint system • Original 802.16 standard (December 2001) • Wireless MAN-SC • Physical layer: Single-carrier architecture • MAC layer: TDM structure (FDD and TDD)
802.16a	2–11 GHz	• WiMAX, HiperMAN licensed bands • NLOS • OFDM physical layer and OFDMA MAC layer
802.16a (formerly b)	5–6 GHz	• Unlicensed band (mesh)
802.16e	450 MHz–3.6 GHz	• Mobility support • Nomadic support • 3G (IMT-2000)

802.16m	450 MHz–3.6 GHz	• Mobility
		• Enhancement to 802.16e
		• IMT-2000 Advanced (4G)
		• Backward-compatible with 802.16e
		• Multicarrier Support (100-MHz bandwidth)
		• Data Rates (per 20-MHz channel)
		    • DL: 300 Mbps
		    • UL: 135 Mbps

TABLE 8.1 806.16 variants

	802.16e	802.16m
Operating bandwidth	5–20-MHz single carrier	5–20 MHz per carrier Multiple carrier aggregation (100 MHz)
Duplex	TDD, FDD, H-FDD (MS)	TDD, FDD, H-FDD (MS)
MIMO	DL 1 × 1 (baseline), 2 × 2, 2 × 4, 4 × 4, 4 × 8, 8 × 8 UL 1 × 1 (baseline), 1 × 2, 1 × 4, 2 × 4, 4 × 4	DL 2 × 2 (baseline), 2 × 4, 4 × 4, 4 × 8, 8 × 8 UL 1 × 2(baseline), 1 × 4, 2 × 4, 4 × 4
Frame length	Variable; 2–20 ms	5 ms, (superframe = 20 ms)

TABLE 8.2 WiMAX Mobilty

and 10.5 GHz. Because *Non-Line-of-Sight* (NLOS) operation and lower component costs, these bands have seen good use as fixed wireless for both residential and small-business services. 802.16a has the same access method as HiperMAN.

The 802.16e standard added mobility and offers an alternative access method other than 3G mobility access methods.

The 802.16m standard is an enhancement to the 802.16e standard focusing on higher throughput capabilities and is defined as a 4G technology. 802.16m is also meant to support *Voice over IP* (VoIP).

- DL: 300 Mbps/20 MHz (4 × 4)
- UL: 135 Mbps/20 MHz (2 × 4)

Specifically, 802.16m has these additional attributes:

- Multicarrier operation (100-MHz bandwidth)
- Extended MIMO
- Fixed frame length to reduce round trip times (5 ms)
- Superframe structure

8.3 Generic WiMAX Architecture

The WiMAX is an IP packet-based network. The WiMAX *Network Reference Model* (NRM) as shown in Figure 8.3 contains the following major nodes.

- *Advance Mobile Station* (AMS) aka *Subscriber Station* (SS) or *Mobile Station* (MS)
- *Advanced Base Station* (ABS) aka *Access Serving Node* (ASN)
- *Core Serving Node* (CSN)

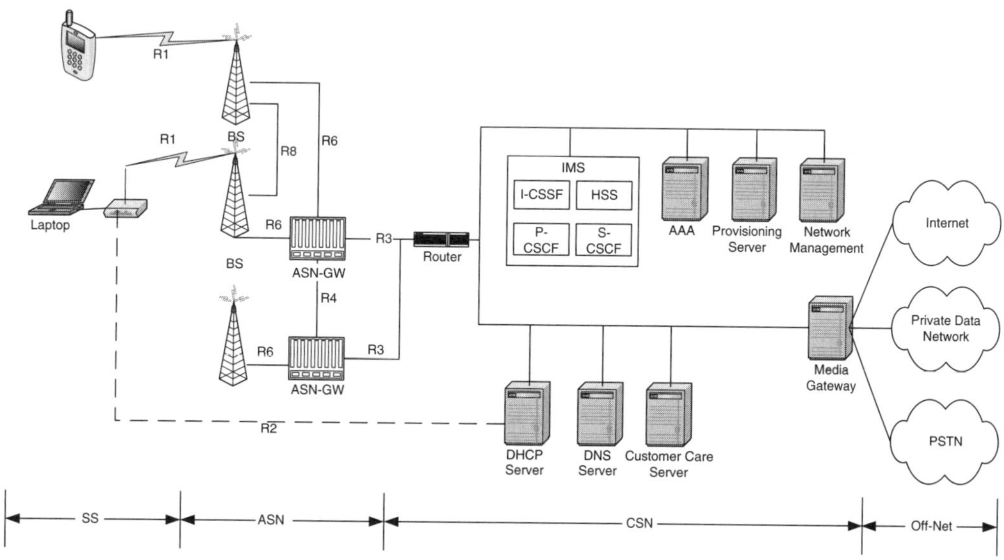

FIGURE 8.3 WiMAX Network Reference Model.

The WiMAX forum further defines several key components, plus some of the interconnections between the components called reference points, R1 to R5 and R8.

The components defined are

- SS/MS: the Subscriber Station/Mobile Station
- ASN: the Access Service Network
  - BS: Base station
  - ASN-GW: the ASN Gateway
- CSN: the Connectivity Service Network
  - HA: Home Agent
  - AAA: Authentication, Authorization, and Accounting Server
  - NAP: a Network Access Provider
  - NSP: a Network Service Provider

The SS can be a fixed device providing high-speed IP connectivity to a LAN or user device or it can be a mobile device like a Smartphone; however, most likely it is a USB dongle connected to an edge device. The SS is connected via RF to the *Base Station* (BS) and this is the R1 link. The SS can also communicate to the CSN via the R2 link as well which of course needs to transverse through the ASN first before reaching the CSN.

The ASN contains the BSs and the *Access Service Network Gateways* (ASN-GW). The ASN interfaces with the SS allowing the subscribers to connect to the WiMAX system. The ASN provides the radio access for the SS.

The BSs perform RF-related functions providing the layer 2 connectivity to the MS. Additionally the BSs can be connected to each other as well as using the R8 interface to facilitate the transfer of data context and handover information between BSs. The BSs consist of radio, controller, and antennas and provide air interface (RF link) to the MS (SS). The BS performs a variety of functions including handoff triggering and tunnel establishment, radio resource management, QoS (*Quality of Service*) policy enforcement, traffic classification, *Dynamic Host Control Protocol* (DHCP) proxy, key management, session management, and multicast group management.

The ASN-GW and BS are connected via IP connectivity, R6. The ASN-GW's aggregate data and control messaging from all the BSs related to issues like QoS, security, and, of course, mobility management. An ASN-GW can and does have multiple BSs associated with them. However as the MS moves from BS to BS the need to change ASN-GWs may be necessary and this is facilitated by the R4 interface, which connects several ASN-GWs. This also ensures that the IP session of the user remains the same. The ASN-GW performs intra-ASN location management and paging, radio resource management, admission control, the caching of subscriber profiles, encryption keys, and AAA client functionality. The ASN-GW also provides the establishment and management of mobility tunnel with base stations, QoS, and policy enforcement, foreign agent functionality for mobile IP besides routing to and from the selected CSN.

The ASN-GWs are also connected to the CSN via the R3 interface and provide layer 2 traffic aggregation for several base stations transporting bearer and control traffic.

The *Connectivity Service Network* (CSN) is principally concerned with interfaces between ASN and CSN to provide service to the SS. The CSN basically provides the means to connect SS with application services either on net or off net. The CSN provides the connectivity to the Internet, ASP, PLMN, and other networks. It provides *Authentication,*

Authorization, and Accounting (AAA) roaming between different *Network Service Providers* (NSPs). The CSN also performs many other functions including IP address management, location management, mobility management, connection for multiple services, policy and QoS management.

Defining a few of the nodes a little more:
The ASN provides the following high-level functions:

- L2 session/mobility management
- Network entry and handover
- Radio resource management and admission control
- QoS and policy enforcement
- Foreign agent
- Forwarding all data and messaging to and from the selected CSN
- Paging and location management

The ASN-GW, which is part of the ASN, performs the following high-level functions:

- Layer 2 traffic aggregation
- Management, network optimization
- Forwarding of all subscriber traffic to and from the MS and CSN
- Routing to the desired CSN, if more than one
- AAA client functionality
- Idle mode control

The CSN provides a host of functions of which some are:

- IP and traffic management
- QoS and policy
- Authentications for all SS devices
- Roaming support
- Billing
- O&M functions

The various interfaces for a WiMAX system are R1-R5 and R8 and they are described briefly.

- *R1* is the interface between the MS and the ASN and is the air interface specification for (PHY and MAC).
- *R2* is the interface between the MS and the CSN. The R2 interface enables authentication, service authorization, IP host configuration management, and mobility management. The R2 interface is a logical interface and is not a physical protocol interface between the SS and CSN.
- *R3* is the interface between the ASN and CSN and is meant to support AAA, policy enforcement, and mobility management. The R3 interface provides the bearer plane tunneling for IP data between the ASN and the CSN.

- *R4* interface is the control and bearer plane protocols within the ASN to coordinate MS mobility between ASNs.
- *R5* is a set of control plane and bearer plane protocols providing interworking between CSNs (not shown).
- *R6* is the communication interface between the BS and the ASN-GW for the control and bearer plane.
- *R8* is control plane messages for the bearer plane data flows between the BS for handovers.

8.3.1 Fixed

WiMAX can and is used as a point-to-point communication system for backhaul. The frequency band that is used for point-to-point communication can be any band. However there are some preferred bands that a point-to-point WiMAX link should utilize. Figure 8.4 is a high-level diagram of a simple point-to-point communication like using WiMAX.

Adherence to Fresnel zones is essential for good link quality and this is covered in Section 20.8.

8.3.2 Point to Multipoint (PMP)

Fundamentally 802.16 is the enabling technology or standard that is intended to provide WMAN access to locations, usually buildings, by use of exterior illumination typically from a centralized base station as shown in Figure 8.5. The simplified system shown in the figure could also have multiple subscribers that would be referred to as *Subscriber Stations* (SSs).

Local Multipoint Distribution Systems (LMDSs), *Fixed Wireless Point-to-Multipoint* (FWPMP), and *Multichannel Multiple-Point Distribution Systems* (MMDSs) have been in existence for quite some time now. However, each of the LMDS, FWPMP, and MMDS, while being superb in delivering a vast array of broadband services, has suffered from the proprietary systems which have not seen the reduction in total cost of deployment or ownership being reduced to a level that competes with the existing wired broadband services.

Wireless Internet Service Providers (WISPs) have also been active in delivering broadband services. WISPs have used both licensed and licensed exempt bands for service delivery.

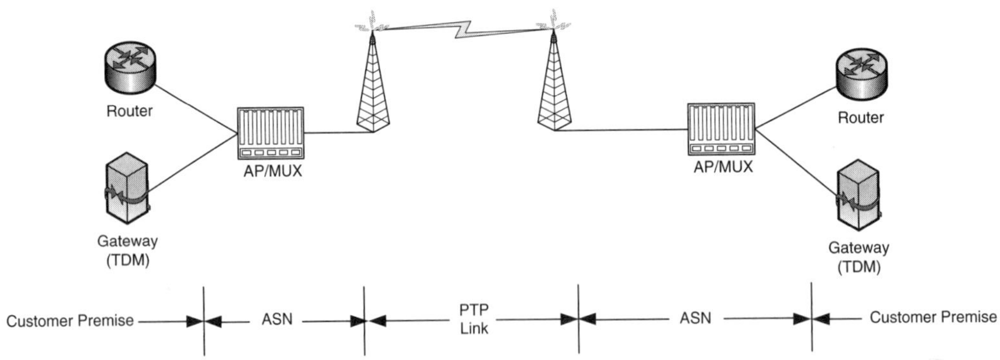

Figure 8.4 Point-to-point communication.

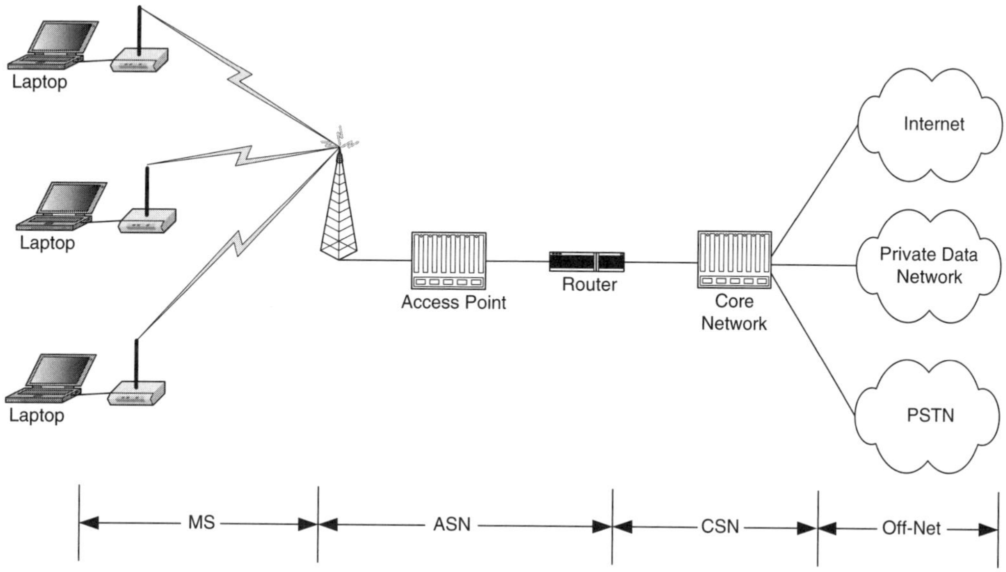

Figure 8.5 802.16 general configuration.

WISPs have been using point-to-multipoint or mesh systems along with 802.11 to serve their customers.

However, with the advent of the 802.16, standard equipment manufacturers and operators now had a standard for access profiles as well as known interoperability levels allowing for multivendor environments.

The PMP is a concept where multiple subscribers can access the same radio platform utilizing both a multiplexing method as well as queuing. 802.16 systems operate in the microwave frequency band and utilize similar radio technology as a point-to-point microwave system.

The point-to-multipoint protocol used is a connection-oriented system that can take on a star or mesh configuration using both FDD and TDD. In addition, 802.16 is in itself protocol-independent in that it can transport both ATM and IP depending on the content desired to be ported. In addition, 802.16 utilizes both contention and contentionless access supporting services that are *ATM Adaption Layer 1*(AAL1) to AAL5.

The system configuration used for an 802.16 system is designed to operate efficiently within the spectrum that is allocated. To achieve this, 802.16 is a wireless system that employs cellular-like design and reuse with the exception that there is no handoff.

However, 802.16 is different from 802.11 and wireless mobility systems like GSM, CDMA, and UMTS. 802.16 is a unique wireless access system whose purpose is to provide broadband access to multiple subscribers or locations within the same geographic area. It utilizes microwave radio as the fundamental transport media and is not fundamentally a new technology but rather an adaptation and standardization of existing technology for broadband service implementation.

For a wireless service provider 802.16 is one of the many tools that can be used to help improve the network and drive operating and capital costs downward.

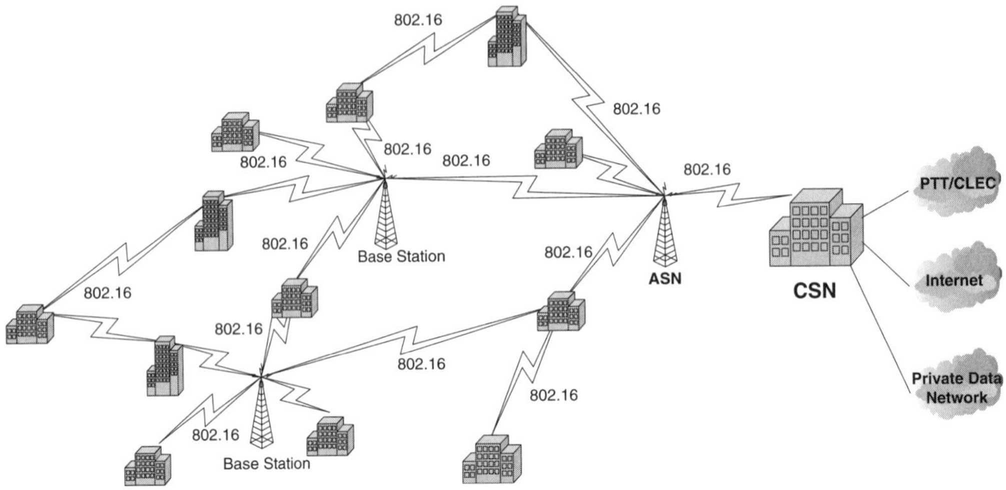

FIGURE 8.6 Mesh.

802.16 systems can be effectively deployed where

- Users are dissatisfied with the current packet and or network interface.
- Network operators need to reach customers cost-effectively without deploying fiber/copper.

8.3.3 Mesh

Figure 8.6 depicts a generalized mesh network for WiMAX. A mesh environment is very efficient for low network traffic levels since each node needs to transport its own traffic plus the associated node it is meant to relay, which can be multiple layers deep.

However the key advantage with a mesh environment is the flexibility it affords, as well as reducing some of the fixed infrastructure requirements when the customer base and true concentrations are not known from the beginning.

8.3.4 Mobility

A key criterion for WiMAX to be both a 3G and 4G wireless access media is the ability to provide mobility. Figure 8.7 is a very simplified depiction of a WiMAX mobility network.

8.4 Core Network

WiMAX, as stated previously, is an all-packet network. The core network is called the *Connectivity Service Network* (CSN) and includes many components. All ASNs in a WiMAX network need to connect to the CSN.

The CSN provides many functions including the ability to provide connectivity to the Internet and private data networks. The CSN also acts as a gateway providing interworking capability interfacing with circuit-switched services from the PSTN/PTT. The CSN gateway also enables connectivity with WiFi, 3GPP, and 3GPP2 wireless networks for roaming and interoperability.

The CSN also authenticates the WiMAX devices and users and provides the specific services that the subscriber is allowed to access as part of the WiMAX network.

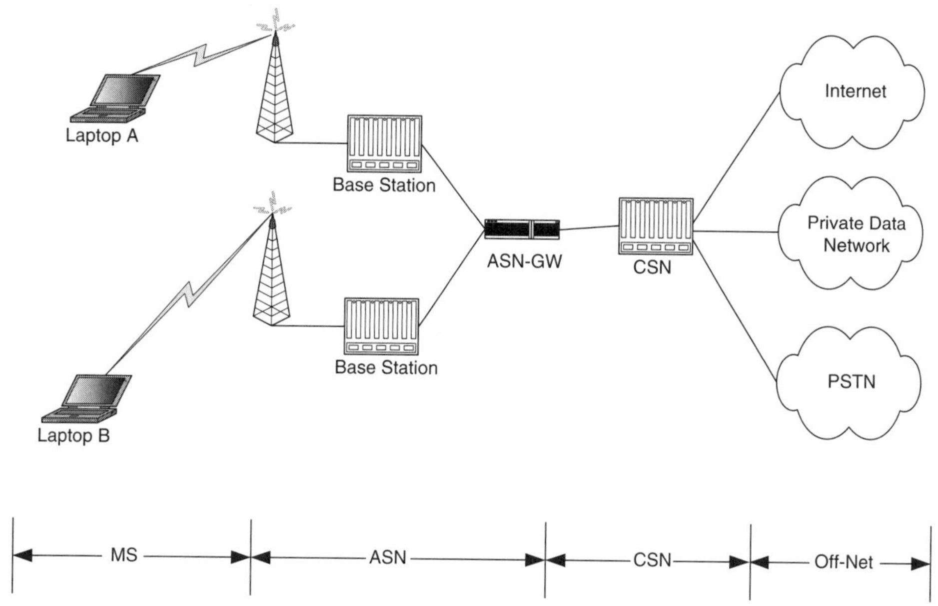

FIGURE 8.7 WiMAX mobile system.

Additionally the CSN also provides policy management for QoS and security for the network as well as per subscriber.

The CSN provides IP address management and for mobile WiMAX it also supports roaming between different NSP as well as facilitating location management of the edge device.

The main components for a CSN are

- Core packet switch (IMS)
- Authentication, Authorization, and Accounting (AAA) server
- Dynamic Host Control Protocol (DHCP) server
- Domain Name Service (DNS) server
- Network Time Protocol (NTP) server
- Home agent
- OSS/BSS
- Service Edge Router/Gateway

8.5 Radio Network

The WiMAX radio network is referred to as the *Access Network System* (ASN) shown in Figure 8.8 and includes the following main components:

- Spectrum
- Base station
- ASN radio equipment
- ASN-GW

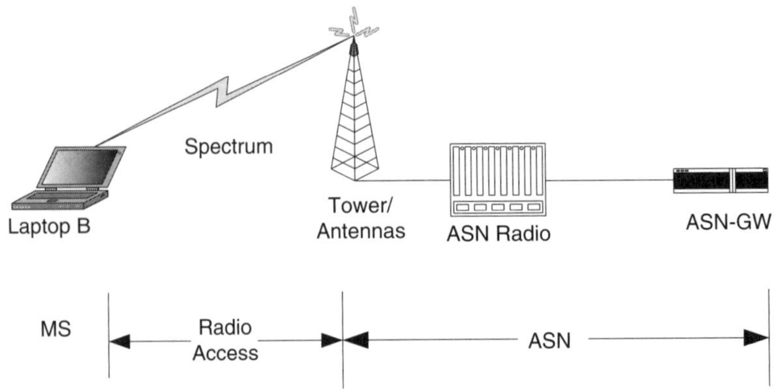

Figure 8.8 Radio access.

Within the functionality of the ASN is the scheduler. WiMAX draws upon the DOCSIS protocol for defining its service-scheduling techniques.

8.6 WiMAX Spectrum

WiMAX can and does operate in multiple frequency bands as well as offering a variety of related protocols for access. However, there is no uniform global licensed spectrum for WiMAX. Therefore since the most costly portion of any wireless network is the edge devices, the WiMAX forum has published three licensed spectrum profiles as a proactive method to drive standardization and decrease costs. The current profiles occupy the following frequency bands: 2.3 GHz, 2.5 GHz, and 3.5 GHz with the likelihood of additional standard profiles in the near term.

Presently, WiMAX has several frequency bands below 10 GHz that are associated with the mobility and edge devices. The frequency bands below 10 GHz currently meant for WiMAX are

- 450 MHz: 470 MHz (IMT-2000)
- 698 MHz: 960 MHz
- 1710 MHz: 2025 MHz (IMT-2000)
- 2110 MHz: 2200 MHz (IMT-2000)
- 2300 MHz: 2400 MHz
- 2500 MHz: 2690 MHz
- 3400 MHz: 3600 MHz

Additionally through the use of the IMT frequency bands WiMAX user terminals can have access all around the world. Specifically, 802.16-compliant equipment is meant to operate in multiple spectrum plans around the world. The two major spectrum channel plans are defined by the *Federal Communications Commission* (FCC) and *European Telecommunications Standards Institute* (ETSI). The channel plans, in addition to spectrum locations, are different for both the FCC and ETSI allocations. The typical FCC allocation for broadband systems is 20 or 25 MHz, while for Europe the spectrum

	Frequency	Channel Bandwidth (MHz)
MMDS	2150–2162 and 2500–2690	1.5, 3, 6, 12, 24
WCS	2305–2320 and 2345–2360	2.5, 5, 10, 15
ETSI	3410–4200, 10000–10680	1.75, 3.5, 7, 14, 28
UNII	5250–5360 and 5725–5825	Unspecified

Table 8.3 Lower 10 GHz WiMAX Spectrum

allotted is 28 MHz with both allocation methods favoring duplexed operation. The standard also supports *Frequency-Division Duplexing* (FDD) and *Time-Division Duplexing* (TDD). The inclusion of TDD is aimed at regulatory environments where structured channel pairs do not exist.

In the United States, the FCC channel allocation plan is done in increments of 5 MHz ranging from 5, 10, 15, 20, and 25 MHz channels that can be either FDD or TDD in nature. The FCC channel plan and spectrum also need to adhere to CFR 47 part 101 as well as parts 1, 15, and 17. While the ETSI channel plan for 802.16 has different channel bandwidths or channel plans than that used in the United States.

802.16 operating in the 10-GHz or higher frequency bands is meant for fixed applications. The channel plans are based on increments of 3.5 MHz ranging from 3.5, 7, 14, 28, 56, and 112 MHz depending on the country and spectrum that the system is to operate within.

However with respect to the bands below 10 GHz, WiMAX is deployed in both licensed and unlicensed bands. As previously mentioned, some of the typical bands for global WiMAX deployment are 2.3 GHz, 2.5 GHz, 3.5 GHz, and 5.7 GHz.

In the United States, the 2.5-GHz band is licensed spectrum and goes from 2.5 GHz to 2.7 GHz (2.690 GHz). This band has also been allocated in Canada, Mexico, Brazil, and parts of southeast Asia. In the United States, this band is referred to as the *Broadband Radio Service* (BRS) band, and since 1998 it is allowed to operate not only as a TDD but also as an FDD configuration for both fixed and mobile operation. In the United States the BRS is broken into eight licenses of 22.5 MHz each.

The 2.3-GHz band is a licensed band and has been allocated in the United States, Australia, New Zealand, and South Korea (WiBro). The 2.3-GHz band in the United States is called the *Wireless Communications Services* (WCSs) band and is allocated in 30-MHz blocks. However, a unique aspect with this band in the United States is its proximity to the DARSs (*Digital Audio Radio Services*) band, which is adjacent to the WCSs band that places significant restrictions on out-of-band emissions to protect DARSs.

The 3.5-GHz band is a licensed band also and has been allocated for fixed wireless broadband services in many countries (except the United States).

The UNII (*Unlicensed National Information Infrastructure*) or licensed license-exempt 5 GHz (sometimes referred to as 5.7- or 5.8-GHz band) is allocated for unlicensed operation in many countries. In the United States, the UNII band includes about 200 MHz of total spectrum.

However, Table 8.3 tabularizes the above information.

8.7 Modulation

WiMAX utilizes a variety of modulation schemes, QAM 4 (PSK), QAM-16, QAM-64, and QAM-256.

These modulation schemes are used with OFDM, OFDMA, and SOFDM. OFDMA is similar to OFDM; however, there are some key differences that need to be highlighted. Figure 7.12 depicts both an OFDM and OFDMA showing subscribers and symbols. What is unique is that in OFDM the subcarrier allocation is fixed per user; however, with OFDMA the subcarrier allocation is dynamic. The dynamic approach to assigning subcarriers leads to better trunking efficiency which equates to usable bandwidth that can be delivered to the subscriber.

OFDMA was chosen for use in WiMAX because it has high spectral efficiency. Additionally OFDMA is also resilient to multipath propagation problems and it has the added benefit of reduced receiver complexity. The receiver complexity simplification is important for the Ue aspect to keep the overall cost and battery consumption at a minimum.

OFMDA also enables MIMO capability in the downlink. However, as indicated in Figure 7.12 with OFDMA, resources can be both in the time domain (symbols) and the frequency domain (subcarriers) in addition to supporting parallel data streams to the Ue.

OFDM enables the RF channel bandwidth to be divided into multiple subcarrier frequencies that are orthogonal to each other. The amount of subcarriers that the RF channel bandwidth is divided into is directly related to the *Fast Fourier Transform* (FFT) for the channel.

WiMAX mobile and point-to-multipoint fixed utilize OFDMA for improved multipath performance in non-line-of-sight environments. OFDMA has multiple closely spaced subcarriers which are also grouped. OFDMA is able to assign a different subset of subcarriers to an individual user depending on the subscriber's capability, profile, and the RF channel bandwidth.

Additionally mobile WiMAX uses *Scalable Orthogonal Frequency-Division Multiple Access* (SOFDMA), which was introduced in the IEEE 802.16e amendment to support scalable channel bandwidths from 1.25 to 20 MHz. The scalability is supported by adjusting the FFT size while fixing the subcarrier frequency spacing at 10.94 kHz. In essence, SOFDMA adjusts the size of the number of FFTs while fixing the subcarrier frequency enabling better spectral efficiency through allocating smaller FFT sizes for lower bandwidth channels and larger FFT sizes to wider channels. However, with SOFDMA the capacity of each individual subchannel remains constant, which reduces system complexity for smaller channels and improves performance of wider channels through better spectral efficiency.

All versions of WiMAX utilize subchannels, and subchannelization techniques are a means to better manage network performance based on coverage and capacity requirements. The subcarriers based on the FFT size can be grouped together forming a subchannel. There are a few subchannel techniques that are used in WiMAX and those are adjacent and distributed.

Distributed subchannels are also called *Frequency Diverse Transmission* (FDT). The FDT can use the *Full Usage of Subchannels* (FUSC) and *Partial Usage of Subchannels* (PUSC) modes. With FDT, the subcarriers assigned to each logical subchannel are pseudorandomly distributed across the available subcarrier set and is meant to adjust for RF channel conditions for coverage and capacity optimization.

Adjacent subcarriers, also called *Frequency Selective Transmission* (FST), use both the *Band Adaptive Modulation* (BAM) and *Adaptive Modulation and Coding* (AMC) mode. BAM and AMC permit subchannel construction and allocation to be adjacent to each other. As part of the FST, the scheduler uses closed-loop channel feedback techniques to determine the optimal subchannels to be allocated to each SS based on the individual RF channel conditions.

However, with fixed WiMAX the OFDM has an FFT of 256 and the amount of FFTs is independent of the bandwidth that it occupies. OFDM bandwidths for WiMAX range from 1.25 MHz to 28 MHz. However, in both FDD and TDD modes there are really 200 usable carriers (FFT 256). Of the 256 FFTs, or rather subcarrier, since the number of FFTs determines the amount of subchannels, there are 192 subcarriers for data of which 8 are used for pilot and 56 for guard band. However, since the FFT size is fixed, the spacing between subcarriers is variable. Specifically for larger channel bandwidth the subcarrier spacing increases, which decreases the symbol time and is the fundamental tradeoff between spectral efficiency and delay spread that can occur.

Mobile WiMAX uses a Scalable OFDMA scheme (SOFDMA). The FFT size, however, is not fixed as in the case of fixed WiMAX but is variable ranging from 128 to 2048. The number of FFTs possible is directly related to the size of the channel bandwidth and, unlike fixed WiMAX, the subcarrier spacing in mobile WiMAX is constant. The amount of FFTs associated with several typical channel bandwidths in mobile WiMAX is

- 1.25 MHz (128 FFT)
- 5.0 MHz (512 FFT)
- 10 MHz (1024 FFT)
- 20 MHz (2048 FFT)

Subchannels can be used to group a set of carriers together so they can be assigned for different services or frequency reuse schemes. The subcarriers are differentiated by use of a subchannel index.

8.8 Channel Structure

WiMAX 802.16d, e, and m all have different frame structures and are not compatible with each other. There is legacy support for 802.16e within 802.16m; however, each has its own signaling protocols.

8.8.1 802.16e

In 802.16e the signaling protocol consists of slots and the frame structure. The slot is the minimum resource that can be allocated where one subchannel can have one, two, or three OFDM symbols. The user's data region is typically a contiguous series of slots that are assigned to the user. The frame, however, in a TDD mode is variable in size ranging from 2 ms to 20 ms. The frame is further divided into a *downlink* (DL) and an *uplink* (UL) subframes with several combinations of DL and UL supported. The DL and UL subframe in TDD are also separated by a guard time zone.

The overall structure for DL and UL is shown in Figure 8.9.

The frame structure for TDD consists of a DL and UL subframe in addition to a *Transmit Time Gap* (TTG) and *Receive Time Gap* (RTG). TTG and RTG is the guard time between DL and UL subframes respectively.

DL Subframe begins with the preamble, which is used for PHY layer procedures, such as time and frequency synchronization and initial channel estimation. Preamble occupies first symbol of the DL subframe.

The downlink preamble is followed by a *Frame Control Header* (FCH), which provides frame configuration information, such as the MAP message length, the modulation and coding scheme, and the usable subcarriers.

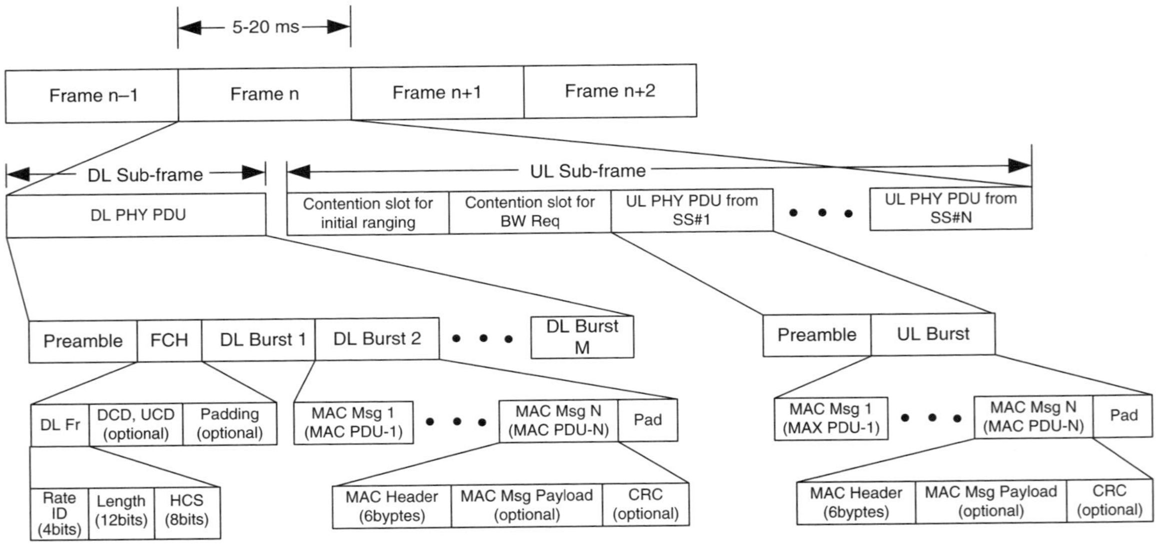

Figure 8.9 802.16e frame layout.

Users are allocated data regions within the frame and these allocations are specified in the MAP messages DL-MAP, downlink, and UL-MAP, uplink. The DL-MAP and UL-MAP messages are broadcast following the FCH in the downlink subframe. MAP messages include the burst profile for each user, which defines the modulation and coding scheme used in that burst.

The *UL subframe* consists of UL bursts, RNG channel, HARQ ACK channel, and CQI channels. The RNG channel is used for entry, periodic ranging, and bandwidth requests. The ACK channel is how the mobile stations provide HARQ acknowledgment messages. The CQI channel provides channel-quality indication which enables the scheduler to alter the modulation parameters to maximize system throughput.

8.8.2 802.16m

Both TDD and FDD schemes are supported with 802.16m. Additionally a method to increase the data rate of the use of multicarriers was introduced enabling more than one carrier to send and receive data. The multicarriers can either be adjacent to each other or they can be dispersed; however, in the same frequency band.

With 802.16m the channel bandwidths are 1.25, 2.5, 5, 10 and 20 MHz. However if five 20-MHz carriers are combined in a multicarrier approach, then a bandwidth of 100 MHz is achievable.

With 802.16m the frame length is 5 ms where in 802.16e the frame length is variable from 2 ms to 20 ms. Part of the rationale for the 5 ms frame length in 802.16m enable WiMAX to coexist with LTE.

Figure 8.10 shows the frame structure for an FDD 802.16m system. The example shown for Figure 8.10 is the same structure for UL and DL; however, only one is shown for illustrative purposes.

The fundamental layout of the frame is that there are eight subframes and each subframe can have six symbols associated with it.

Figure 8.11 represents the frame structure of a TDD system with 802.16m. With TDD each frame can contain a different number of subframes depending on the edge device as well as the bandwidth requirement needed. The subframes can be 5, 6, 7, or 8. Additionally the number of symbols can also vary with a subframe leading to many valid perturbations.

TDD, however, is the preferred WiMAX 802.16m configuration largely due to the spectrum allocation it occupies. MMDS, which is the former wireless cable band, is in 6-MHz channel blocks.

Additionally, in 802.16m the concept of a superframe structure was introduced and is shown in Figure 8.12. A Superframe consists of four frames for a total period of 20ms.

As expected, 802.16m uses a different method for resource assignments as compared to 802.16e. Specifically, 802.16m uses a Physical Resource Units (PRU) instead of slots and tiles. With 802.16m the PRU is assigned for both UL and DL. A PRU structure is shown in Figure 8.13, where a PRU has 18 subcarriers and 5 to 8 symbols.

Physical Resource Unit (PRU) is the basic physical unit for resource allocation. A PRU is made up of Psc consecutive subcarriers (18) and Nsym consecutive OFDMA symbols where Nsym can be 5 through 8 symbols. *Logical Resource Unit* (LRU or LPRU) is the basic logical unit for localized and distributed resource allocations. An LPRU can have the same number of symbols and carriers as the PRU. As such, an LRU is the fundamental resource allocation in the DL and UL.

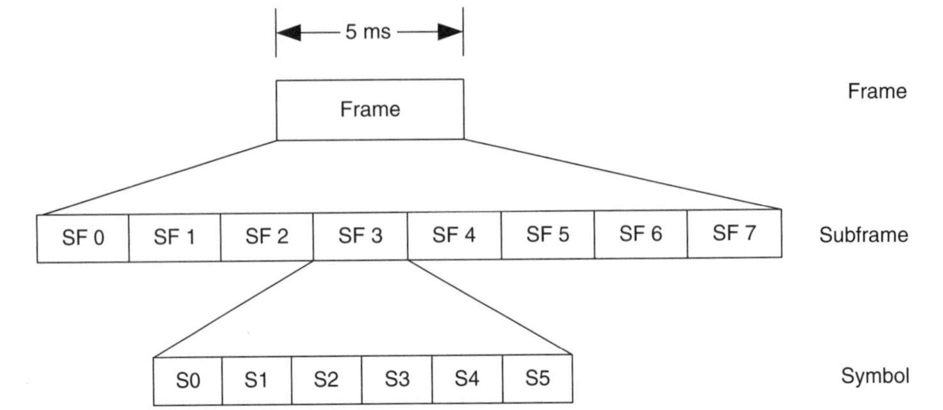

FIGURE 8.10 802.16m FDD structure either DL or UL.

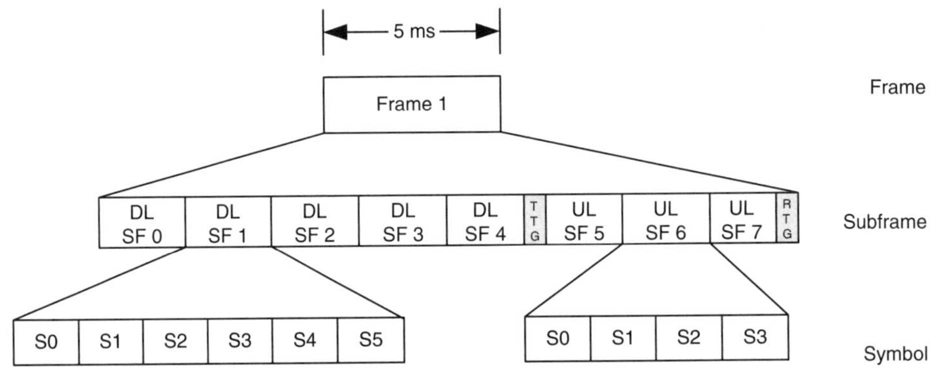

FIGURE 8.11 802.16m TDD.

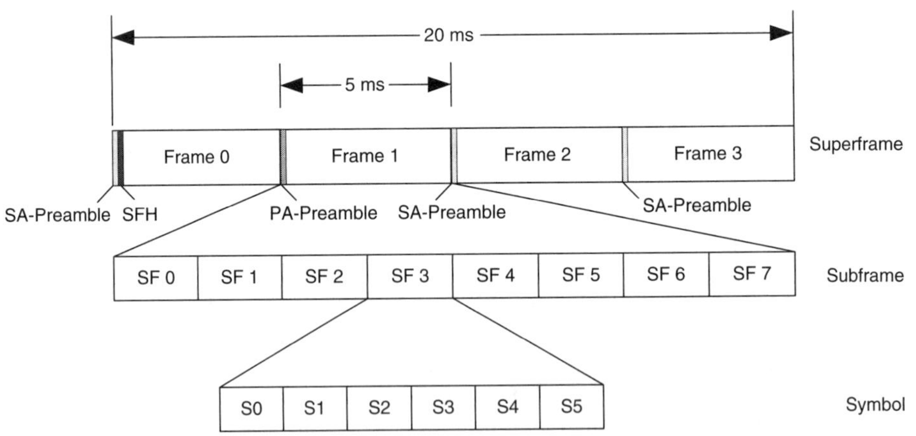

FIGURE 8.12 Superframe.

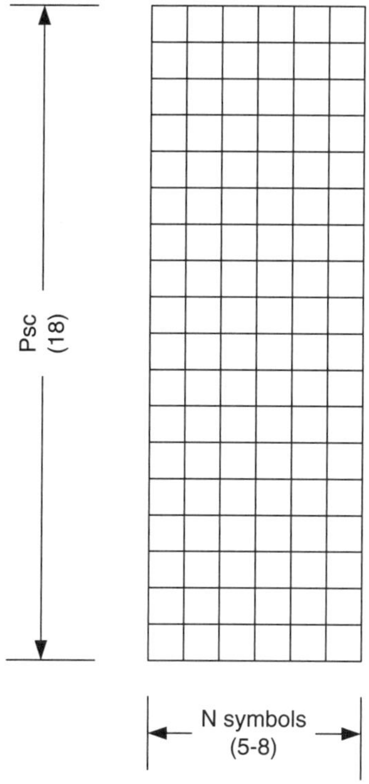

FIGURE 8.13 PRU.

In the UL, the LRU can be either localized (contiguous) or distributed (DRU), as shown in Figure 8.14. Because PRU and LRU are mapped together, if the distributed LRU is used then the PRU is also distributed in the same fashion.

802.16m has one fixed pilot per PRU and how the pilots are distributed in the PRU is dependent upon the amount of antennas deployed at the base stations and are common to all subscribers.

In the UL side, the pilots are dedicated for each user. For a contiguous LRU the pilots are the same as that for the PRU. However, in the distributed LRU the pilots are replicated in each LRU leading to a higher density of pilots for a distributed LRU scenario.

As expected 802.16m has channels to increase its functionality over 802.16e. These channels are described in Tables 8.4 and 8.5.

8.9 Mixed Mode

Mixed mode 802.16m and 802.16e is supported, enabling legacy and new systems to operate on the same RF carrier sharing radio resources dynamically. However, as described previously, both 802.16e and 802.16m have their own protocols and are not compatible with each other.

Figure 8.15 shows an example of how 802.16e and 802.16m can share the same RF carrier in a TDD configuration. In all cases for mixed mode operation the legacy protocol precedes the other.

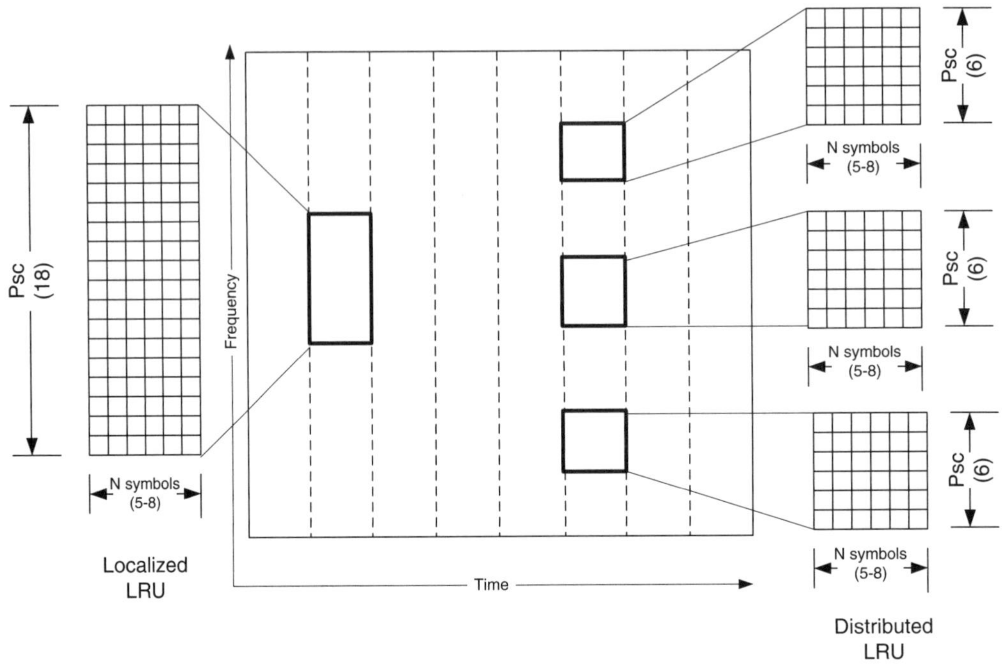

Figure 8.14 802.16m supports MIMO and as a minimum configuration it is a 2 × 2.

Downlink	Name	Description
A-Preamble	Advanced Preamble	Base Station ID, Time reference, frequency and frame synchronization, RSSI and channel sounding estimation
PA-Preamble	Primary Advanced Preamble	Initial acquisition, BS type, section ID, bandwidth, carrier config, and superframe synch.
SA-Preamble	Secondary Advanced Preamble	Cell/sector ID, channel estimation and synchronization refinement
SFH	Superframe Header	System parameters as well as P-SFH and S-P-SFH
P-SFH	Primary	Short-term control info and is sent with every superframe
S-SFH	Secondary	Contains DUIC
A-MAP	Advanced MAP	Resource assignment, scheduling info, power control, HARQ
E-MBS-MAP	Enhanced Multicast Broadcase Services MAP	Multicast service control info

Table 8.4 DL

8.10 Interference Mitigation Techniques

As with all wireless systems that reuse frequencies, interference mitigation is essential. In WiMAX, the primary method of interference mitigation is proper frequency planning and good cell site optimization. However, even great frequency planning and excellent optimization can still in some level of interference.

Uplink	Description
P-FBSCH	Primary fast feedback channel
S-FBSCH	Secondary fast feedback channel
UL HARQ FBCH	UL HARQ feedback channel
Ranging	UL-link synchronization
BW Req	Bandwidth request (3 steps and 5 steps)
CQI	Channel quality indicator regarding channel quality as seen by user
Power control	DL-MAP power control based on UL channel quality

Table 8.5 UL

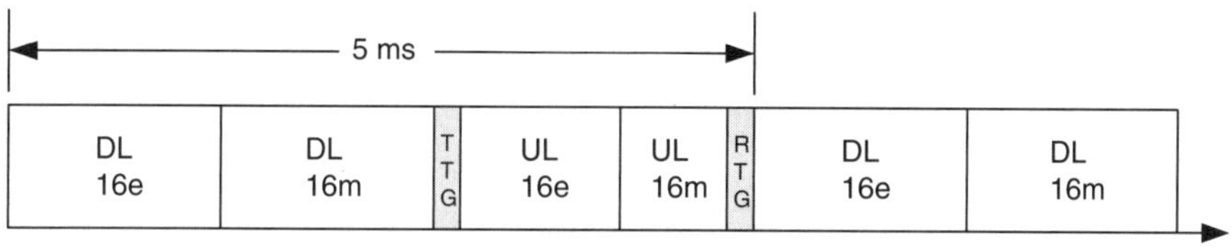

Figure 8.15 TDD multimode operation.

In a WiMAX system, various software and hardware techniques are used to minimize interference including

- Smart antenna systems (beam steering on a per-frame basis)
- Subgrouping of transmit and receive frames between adjacent sites (TDD)

Additionally, WiMAX uses adaptive modulation and coding that is based on the *Signal-to-Interference Plus Noise Ratio* (SINR). The modulation technique is used to match the link quality and dynamically adjust based on improving or degrading conditions.

Spatial multiplexing is also used where multiple independent streams are transmitted in parallel over several antennas. This enables the multipath channels to be somewhat independent by decorrelating them. This assists in high multipath environments that are NLOS. Spatial multiplexing also improves the data rates and overall capacity of the network based on the use of multiple antennas.

8.11 Frequency Planning

WiMAX frequency planning depends on the particular application that it is being used for. For instance, if WiMAX is being deployed as a point-to-point link for backhaul the channel assignments are usually fixed and can be licensed, unlicensed, or lightly licensed.

However, 802.16e and 802.16m address mobility which can also include fixed stations for providing DSL/cable alternatives. Regardless, WiMAX mobile utilizes frequency reuse to facilitate a greater amount of cell sites for coverage and follows a general *cellular* wireless architecture employing both cell site design and frequency reuse.

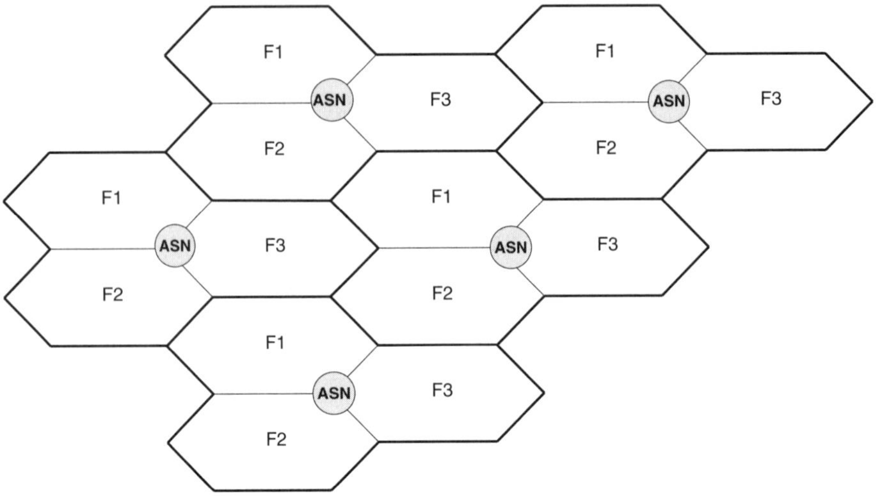

FIGURE 8.16 N = 3 frequency plan.

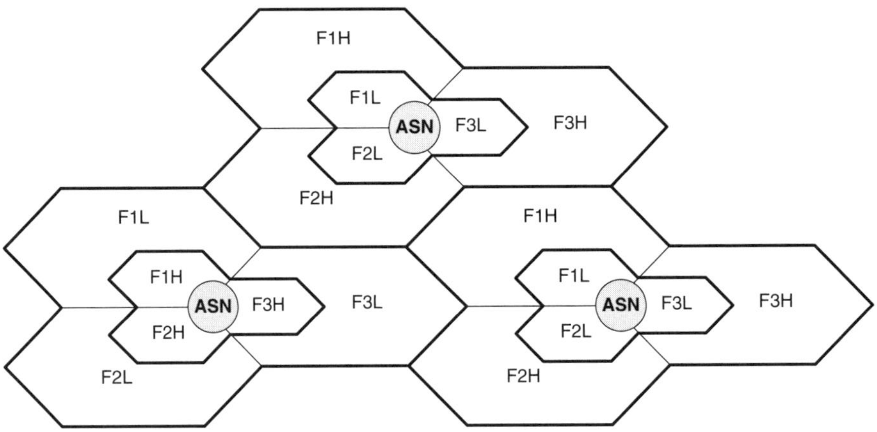

FIGURE 8.17 FFR reuse Example 1.

The frequency planning for mobile WiMAX follows an N = 3 reuse pattern, as shown in Figure 8.16. However, with fractional frequency reuse (FFR) it is possible to improve the performance for interference mitigation without resorting to obtaining more spectrum where the radio resources are broken into two groups, L and H, for each channel.

Two possible FFR reuse patterns are depicted in Figures 8.17 and 8.18.

8.12 Features and Applications

WiMAX for mobility has a variety of features that help enhance its ability to provide IP services to a mobile environment. Some of the features involve power management, multicast, and multihop relay.

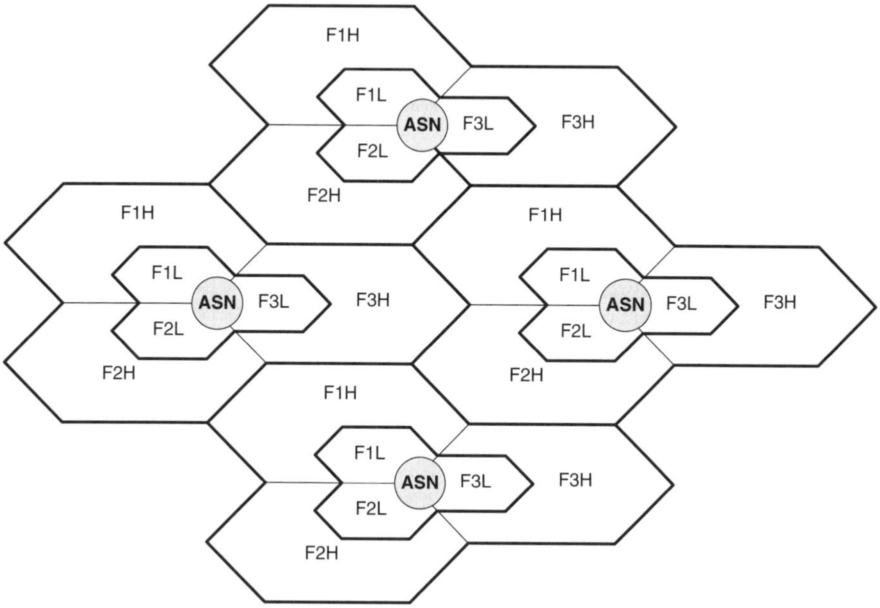

Figure 8.18 FFR reuse Example 2.

8.12.1 Power Management

As with all wireless mobility systems, power management is essential for both improving performance by reducing interference, as well as prolonging the battery life of the subscriber unit. With power management the subscriber unit is either turned off or turned on. When it is turned on, the device can either be in an active mode, sleep mode, or idle mode. In the active mode, the MS is sending and or receiving data, hence the term active.

When the MS is no longer sending and receiving data for a period of time it enters a sleep mode. After a period of time the MS comes back to active mode and is meant to conserve power this way. The ability to enter sleep mode is made by the base station.

The MS can also enter into an idle mode where it effectively enters a listen-only mode for the DL and does not communicate with the ASN anymore until it reestablishes for an active session.

8.12.2 Multicast and Broadcast Service (MBS)

This feature for WiMAX mobility enables the same data content to be sent to multiple users at the same time as a broadcast message. MBS has many of the same attributes as DVB-H, MediaFLO, and LTE. This is important for delivering the same streaming video to multiple users at the same time, therefore minimizing the resource consumption of the network.

8.12.3 Multihop Relay

Multihop relay is a method within WiMAX for increasing the coverage of the network by having relay sites acting as relays for the home base station to communicate with the mobile subscriber.

With multihop relay, the BS can communicate with one or more relay stations (RS). The traffic and signaling between the MS and the BS is relayed by the RS. The RS is connected to the BS via the WiMAX radio link. Additionally, an RS can also relay communication to the BS from other RS and therefore act as an intermediary relay. The relay functionality is defined under 802.16j.

8.13 Security

WiMAX in all its variants has a robust level of authentication and security. Security in WiMax is done not only at the time of system access but also during the data, session, ensuring all transmitted information is protected. The authentication and registration processes are part of the 802.16 MAC sublayer, called the *privacy sublayer*, and is based on the privacy key management protocol, which is part of the DOCSIS specification.

WiMAX authentication uses X.509 digital certificates, which is based on *Public Key Infrastructure* technology, and is provided by the manufacturer of the SS. The digital certificates allow the SS to uniquely authenticate itself back to the ASN base station. The ASN base station can then check to see if the SS is authorized to receive service. If the SS is authorized the ASN base station sends the SS an encrypted authorization key. The authorrization key along with the SS public key is used to encrypt and protect the sessions meeting DES security.

8.14 QoS

The 802.16 standard series was designed with a variety of traffic types such as VoIP and video or audio streaming as well as low-data-rate applications.

WiMAX QoS is implemented at the PHY layer and incorporates a number of mechanisms to provide QoS and some of them are

- Adaptive Modulation
- Fast Fourier Transform (FFT)
- Forward Error Correction (FEC)
- Frequency-Division Duplex (FDD)
- Time-Division Duplex (TDD)
- Orthogonal Frequency-Division Multiplexing (OFDM)

In addition to the above, WiMAX also utilizes another mechanism for QoS in the PHY level called *adaptive burst profiles*. Both TDD and FDD configurations are able to support adaptive burst profiling. With adaptive burst, the modulation and coding schemes are defined in the actual burst profile and the burst profile can be modified for each SS on a frame-by-frame basis based on the actual link conditions and are done for the UL and DL individually.

WiMAX has three main methods for QoS provisioning:

1. Service flow classification
2. Dynamic service establishment
3. Two-phase activation model

8.14.1 Service Flow Classification

With service flow classification, each packet within a service flow for WiMAX is connection oriented. Each connection is assigned a unique *Connection ID* (CID) as well as a *Service Flow ID* (SFID), which has an associated service class.

802.16 provides four scheduling services for 802.16e and an additional one for 802.16m, ertPS. These are referenced in Table 8.6.

8.14.2 Dynamic Service Establishment

Dynamic service establishment provides the ability for dynamically establishing service flows and requesting QoS parameters. There are three types of control messages for service flows:

- *Dynamic Service Activate* (DSA)—Activates a service flow
- *Dynamic Service Change* (DSC)—Changes an existing service flow
- *Dynamic Service Delete* (DSD)—Deletes a service flow

Dynamic service establishment enables an application to acquire more resources when required.

8.14.3 Two-Phase Activation

Activation of a service flow proceeds in two phases: Admit first, then activate. An Authorization Module in the ASN base station provides this function which can activate a service flow immediately or defer activation to a later time. Once the service flow has been admitted, both the ASN base station and SS can reserve resources for it. Therefore the two-phase activation allows service flows to be admitted before activation.

8.15 Profiles

At the heart of WiMAX and mobile WiMAX are the system profiles. As expected, there are numerous system profiles in WiMAX. However, WiMAX system profiles refer to a set of parameters on which a WiMAX network will operate. The system profile attributes include the frequency band, channel bandwidth, and duplexing scheme.

The various WiMAX system certification profiles meet a set of features in the appropriate 802.16 standard; however, those relevant features are selected by the WiMAX forum. A WiMAX system profile is essential, or rather, it is mandatory. Additionally

QoS Service Class	Description
Unsolicited Grant Service (UGS)	VoIP (without silence suppression)
Real-time Polling Service (rtPS)	MPEG video
Non-real-time Polling Service (nrtPS)	FTP
Best Effort (BE)	HTTP, e-mail
Extended Real-time Polling Service (ertPS)	VoIP with silence suppression

Table 8.6 QoS

the 802.16 standard indicates that a system certification or profile consists of five components:

- MAC profile
- PHY profile
- RF profile
- Duplexing method (TDD or FDD)
- Power class

The profiles for fixed WiMAX have channel sizes of 3.5 MHz, 5 MHz, 7 MHz, and 10 MHz, while those for mobile are 5 MHz, 8.75 MHz, and 10 MHz. The current fixed profiles are defined for both TDD and FDD profiles. Presently, all the mobile profiles are TDD only; however, this is possible to change in the near future.

Table 8.7 shows some of the WiMAX system profiles presently available; there are other profiles while new profiles can be and are added.

8.16 Origination

For a subscriber to gain access to the WiMAX network it needs to request a connection, or rather an attachment, to the network. As part of the process, the MS needs to communicate its capabilities to the base station and if compatible based on its profile keys are exchanged as part of the authentication process. Figure 8.19 is a high-level depiction of the basic origination/access request to a WiMAX network.

More details regarding the origination and access request process can be found in the specifications for WiMAX.

8.17 Handover

With WiMAX mobility there are three basic HO triggers defined; however, of the three—MS-initiated, serving base station initiated, and serving gateway initiated—only mobile initiated (MS-initiated) is utilized at this time.

As with all handoff/handover scenarios the process and procedures vary based on if the subscriber is handing off between base stations that are connected to the same ASN-GW or if an ASN-GW needs to be changed since the new base station has a different ASN-GW. Additionally, the handover process has some unique characteristics when

Frequency Band (GHz)	Duplexing	Channel Bandwidth (FFT Size)
2.3–2.4	TDD	5 MHz (512), 8.75 MHz (1024), 10 MHz (1024)
2.305–2.320	TDD	3.5 MHz (512), 5 MHz (512), 10 MHz (1024)
2.496–2.690	TDD	5 MHz (512), 10 MHz (1024)
3.3–3.4	TDD	5 MHz (512), 7 MHz (1024), 10 MHz (1024)
3.4–3.8	TDD	5 MHz (512), 7 MHz (1024), 10 MHz (1024)

Table 8.7 WiMAX System Profiles

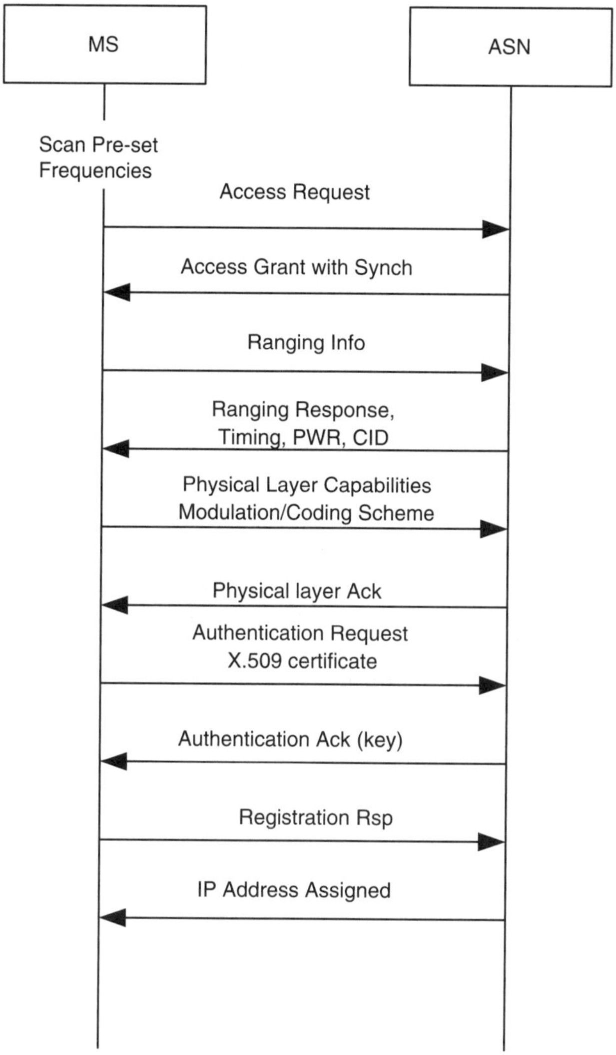

Figure 8.19 Access request.

handing over via InterRAT to another network. The various details of the process are included in great detail within the 802.16 specification.

However, for brevity, the handover process in WiMAX is a hard handover. Hard handovers are a break-before-make and this is typical for other wireless mobility data protocols like LTE, HSPA, and EVDO to mention a few.

With hard handover, the MS only communicates with one BS at a time (each time) and the connection with the old, or rather source, BS is broken before the new connection is established with the target BS. The handover is requested when the signal strength from the neighbor's cell exceeds the signal strength from the current cell site.

Figure 8.20 is a high-level depiction of the handover process with WiMAX.

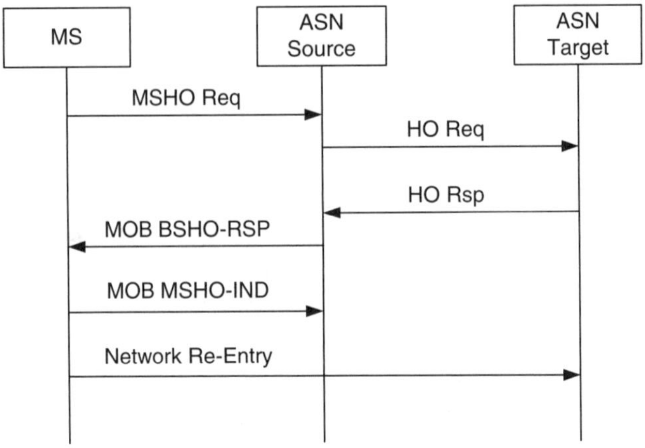

Figure 8.20 Handover process.

8.18 Femto and SON

Mobile WiMAX supports both femtocells and *Self-Organization Networks* (SONs). Femto cells are covered elsewhere in this book; however, they are low-powered cellular base stations typically deployed in residences or small offices using another broadband provider for the backhaul. SONs are also part of 802.16m where the various network nodes and cells are self-aware and can adapt to the network configuration as it changes, including the update of neighbor nodes and neighbor cells for fast reconfiguration either because of a new node or loss of a node.

References

Sassan Ahmadi, "Introduction to mobile WIMAX Radio access Technology:PHY and MAC Architecture," Intel, December 2006.

Sassan Ahmadi, "Overview of IEEE P802.16m Technology and Candidate RIT for IMT-Advanced," NIST, Janurary 2010.

Sassan Ahmadi, *Mobile WiMAX A Systems Approach to Understanding IEEE 802.16m Radio Access Technology*, Academic Press, New York, 2011.

Dean Chang and Subir Varma, "WiMax Rolls Interop Guidelines for 802.16a," Electronic Engineering Times, May 16, 2003.

Jianhong Chen, Wei Hong, Jianyi Zhou, Jianing Zhao, Jianjun Wang, and Wei Kang, "State Key Lab. of Millimeter Waves, Design of an Outdoor Unit for WiMAX," APMC2005 Proceedings, 2005.

Carl Eklund, Rodger Marks, and Kenneth Stanwood, "IEEE Standard 802.16: A Technical Overview of the Wireless MAN Air Interface for Broadband Wireless Access," *IEEE Communications Magazine*, June 2002, pp. 98–107.

Ohrtman Frank, "WiMAX Handbook Building 802.16 Wireless Networks," McGraw-Hill, New York, 2005.

IEEE 802.00: Comparison of Existing and Proposed Wireless Standards.

IEEE 802.16 hc-00 /05: USA Regulatory Constraints of a U-NII Broadband Wireless Access System Standard.

IEEE 802.16.4c-01/34: OFDMA Advantages for the 802.16b.

IEEE 802.16.4c-01/35: OFDMA for Mesh Topology.
IEEE 802.16-2001: "Part 16: Air Interface for Fixed Broadband Wireless Access Systems," IEEE Standard for Local and Metropolitan Area Networks, April 8, 2002.
IEEE 802.16-2004: IEEE Standard for Local and Metropolitan Area Networks, Air Interface for Fixed Broadband Wireless Access Systems, October 2004.
IEEE Std 802.16.2™-2004; Coexistence of Fixed Broadband Wireless Access Systems.
IEEE 802.16hc-00/01: Requirements for WirelessHUMAN Systems.
IEEE 802.16hc-00/02: Requirements for Broadband Wireless Access Systems in the UNII Bands.
IEEE 802.16hc-00/03: The Path Towards Efficient Coexistence in Unlicensed Spectrum.
IEEE 802.16hc-00/06: Comparison of QPSK/QAM, OFDM, and Spread Spectrum for 5-6 GHz PMP BWAS.
IEEE 802.16hc-00/07: Specifying TDD for the Proposed WirelessHUMAN Standard.
IEEE 802.16hc-00/08: Contribution for the WirelessHUMAN PAR.
IEEE 802.16hc-00/09: Applicability of IEEE 802.11 and HIPERLAN/2 for WirelessHUMAN Systems.
IEEE 802.16hc-00/12: A Proposed High Data Rate 5.2/5.8 GHz Point to Multipoint MAN System.
IEEE 802.16hp-00/04: FCC Rules in Regards to Unlicensed Spectrum Usage from 2 to 11 GHz.
IEEE 802.16t- 00/ 02: Coded Orthogonal Frequency Division Multiple Access.
IEEE S802.16e-03/23: 802.16e Requirements from an Operator's Perspective.
IEEE Std 802.16.1b™-2012: IEEE Standard for WirelessMAN-Advanced Air Interface for Broadband Wireless Access Systems Amendment 1: Enhancements to Support Machine-to-Machine Applications.
IEEE Std 802.16.1™-2012: IEEE Standard for WirelessMAN-Advanced Air Interface for Broadband Wireless Access Systems.
IEEE Std 802.16™-2012: IEEE Standard for Air Interface for Broadband Wireless Access Systems.
IEEE Std 802.16.1b™-2012; IEEE Standard for WirelessMAN-Advanced Air Interface for Broadband Wireless Access Systems Amendment 1: Enhancements to Support Machine-to-Machine Applications.
Sue O'Keefe, "TDD vs. FDD: The Next Hurdle," *Telecommunications*, December 1999, p. 40.
Roger B. Marks, Kenneth L. Stanwood, and Stanley Wang, "IEEE Standard 802.16: A Technical Overview of the WirelessMAN™ Air Interface for Broadband Wireless Access," *IEEE Communications Magazine*, June 2002, pp 98–107.
Clint Smith, and John Meyer, *WiMAX and WiFi*, McGraw-Hill, New York, 2005.
Clint Smith, *LMDS*, McGraw-Hill, New York, 2000.
wikipedia.org.
"WiMAX: IEEE 802.16e-2005 Introduction to OFDMA Measurements Application Note 1EF58," Rhode & Schwarz, May 2007.
"WiMAX: 802.16-2004, 802.16e,WiBRO Introduction to WiMAX Measurements Application Note 1EF57," Rhode & Schwarz, February 2006.
Loring Wirbel, "LMDS, MMDS Race for Low-Cost Implementation," Electronic Engineering Times, November 29, 1999, p. 87.
www.ieee802.org.
www.wimaxforum.org.
www.wirelessman.org.
Parviz Yegani, "WiMax Overview," Cisco, Nov 2005.

CHAPTER 9

Voice over IP (VoIP) Technology

As we have seen in previous chapters, the development of wireless technology involves a migration from the circuit-switched solutions of *First-Generation* (1G) and *Second-Generation* (2G) networks toward a completely packet-switched configuration with 4G networks for both voice and data. Although a number of packet-switching solutions could be leveraged, such as *Asynchronous Transfer Mode* (ATM) and Frame Relay, the ultimate goal is to use the *Internet Protocol* (IP). *Wideband Code-Division Multiple Access* (WCDMA), *Code-Division Multiple Access 2000* (CDMA2000), *Time Division Code Division Multiple Access* (TD-CDMA), and *Time Division-Synchronous Code-Division Multiple Access* (TD-SCDMA) and especially 4G with LTE and also WiMAX are all predicated on the concept of enabling IP at the edge of the network. For example, if we examine the migration from *Global System for Mobile* (GSM) communications to the *Universal Mobile Telecommunications Service* (UMTS) Release 5, we see the use of Frame Relay (the *General Packet Radio Service* [GPRS] Gb interface), followed by ATM, followed by IP. The same can be said for CDMA2000 because it was designed from the onset for mobile IP. However, with LTE being an all-IP network the use of VoIP with LTE (commonly referenced as VoLTE), the understanding of VoIP, and the network requirements to support this capability are important.

Although the IP transport of data is well-understood, the IP transport of voice is a relatively recent development, given that *Voice over IP* (VoIP) will be used in *Third-Generation* (3G) networks and is the fundamental enabler of *IP Multimedia Service* (IMS) besides being an integral item for LTE. In addition, the VoIP solution being pursued by all IMS platforms has the *Session Initiation Protocol* (SIP) at its foundation for voice traffic. However, *International Telecommunications Union* (ITU) H.323 is the predominant VoIP solution and has seen widespread deployment. It is essential for designing and managing 3G and 4G networks and their solutions to have a firm understanding of both SIP and H.323 VoIP methods.

Therefore, it is appropriate that we describe the solutions that make VoIP possible. This chapter is devoted to a brief overview of VoIP technology. It should be noted, however, that the explanations provided in this chapter are at a relatively high level and certainly are not detailed enough to provide a complete understanding of all aspects of VoIP. This is, after all, a book about 3G and 4G wireless. A more detailed discussion of VoIP is found in the reference section of this chapter.

9.1 Why VoIP?

IP clearly has a number of advantages over circuit switching. The most notable of these is the fact that it can leverage today's advanced voice coding techniques, such as the *Adaptive MultiRate* (AMR) coder used in *Enhanced Data Rates for Global Evolution* (EDGE) and UMTS networks. Thus voice can be transported with far less bandwidth than the 64 kbps used in traditional circuit-switched networks.

If we consider, for example, the network architecture of GSM, we find that speech from the mobile subscriber must be transcoded to 64 kbps before it enters the *Mobile Switching Center* (MSC). Thereafter, it is carried to the destination at 64 kbps. If the voice were to be carried most of the way with a packet transport such as IP, then the transcoding up to 64 kbps might not be needed at all or might be needed only very close to the destination. The bandwidth savings enabled by such technology can be significant. Of course, IP is not the only technology that can enable such bandwidth savings. ATM, for example, also can transport voice at rates of less than 64 kbps. IP, however, has other advantages.

Perhaps the biggest advantage of IP over technologies such as ATM is the fact that IP is practically everywhere. Not only is it supported by every PC on the market today, but it is also supported by handheld computers and personal organizers. ATM just does not have the same ubiquitous presence. Moreover, the availability of IP knowledge and experience is widespread, with numerous companies devoted to the development of IP-based applications. If IP resides in the handset and IP is used to carry both voice and data, then real voice-data convergence opportunities arise, offering the possibility of exciting new services.

9.2 The Basics of IP Transport

As shown in Figure 9.1, IP corresponds to Layer 3 of the *Open Systems Interconnection* (OSI) seven-layer protocol stack. At its most basic level, IP simply passes a packet of data from one router to another through the network to the appropriate destination,

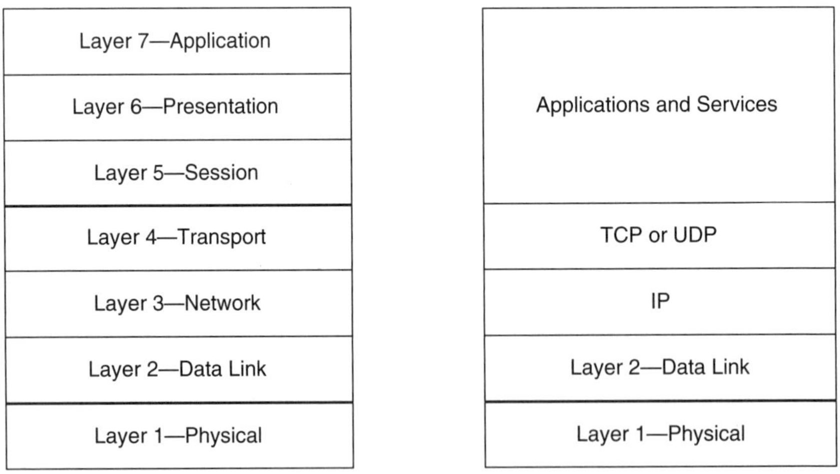

FIGURE 9.1 OSI and IP protocol stacks.

as identified by the destination IP address in the IP packet header. This simple operation means that IP is inherently unreliable. IP provides no protection against a loss of packets, which might happen if congestion occurs along the path from the source to the destination. Moreover, in a given stream of packets from the source to the destination, it is quite possible that packets will take different routes through the network, meaning that different packets can have different delays and also that packets may arrive at the destination out of sequence.

In data networks, in order to ensure an error-free in-sequence delivery of packets to the destination application, the *Transmission Control Protocol* (TCP) is used. This protocol resides on the layer above IP. When a session is to be set up between two applications, the application data first are passed to the TCP, where a TCP header is applied; the data then are passed to the IP, where an IP header is applied; and then the data are forwarded through the network. The information contained in the TCP header includes, among other things, source and destination port numbers, which identify the applications at each end; sequence numbers and acknowledgment numbers, which enable the detection of lost packets; and a checksum, which enables the detection of corrupted packets. TCP uses these information elements to request retransmission of lost or corrupted packets and to deliver packets to the destination application in the correct order. In order to do all this, TCP first establishes a connection between peer TCP instances at each end. This involves a sequence of messages between the TCP instances prior to the transfer of user data.

Instead of using TCP at Layer 4 in the stack, the *User Datagram Protocol* (UDP) is another option. This is a simple protocol that does little more than enable identification of the source and destination applications. It does not support recovery from loss or error and does not ensure an in-sequence delivery of packets. It is meant for simple request-response types of transactions rather than the sequential transfer of multiple packets. An application that would use UDP rather than TCP, for example, is the *Domain Name Service* (DNS), a classic one-shot request-response protocol.

9.3 VoIP Challenges

Good speech quality is a strong requirement of any commercial network, wireless or otherwise. Traditionally, this has been achieved through 64-kbps (G.711) voice coding and the use of circuit switching, which establishes a dedicated transmission path from the source to the destination. Nowadays, more advanced speech coding schemes can approach the quality of G.711 with a much lower bandwidth requirement. All the 3G platforms use codecs designed to improve spectral efficiency without sacrificing voice quality. The GSM Enhanced Full-Rate Coder is one such advanced coding scheme, and many others are available. Good speech coding is not the only requirement, however. Other requirements include low transmission delay, low jitter (delay variation), and the requirement that everything transmitted at one end is received at the other (low loss).

These requirements are somewhat contradictory when viewed from an IP perspective. For example, the requirement for low loss could be achieved through use of TCP at Layer 4. Such a solution, however, would cause excessive delay both at the start of the transfer when a TCP connection needs to be established and during the transfer when acknowledgments and retransmissions would cause a delay in the delivery of the voice packets. In order to minimize delay, one could use UDP at Layer 4. UDP, however, offers no protection against packet loss.

Given the choice between UDP or TCP, the issue is whether we consider minimizing delay to be more important than eliminating packet loss. The answer is that, for speech, excessive delay and excessive jitter are far more disturbing than occasional packet loss. Obviously, excessive packet loss is unacceptable, but a limited amount (<5 percent) can be tolerated without noticeable speech quality degradation. Consequently, when transporting voice, UDP is chosen at Layer 4 rather than TCP.

It is clear, however, that something more than UDP is required if VoIP is to offer reasonable voice quality. At a minimum, the destination application needs to know the coding scheme being used by the source application so that the voice packets can be decoded. The application also needs timing information so that packets can be played out to the user in a synchronized manner and help to mitigate against delay in the network. Moreover, the application needs to know when packets are lost so that a previous packet can be replayed to fill the gap, if appropriate.

In order to fulfill these needs, a protocol known as the *Real-Time Transport Protocol* (RTP) has been developed. This protocol resides above UDP in the protocol stack. Whenever a packet of coded voice is to be sent, it is sent as the payload of an RTP packet. This packet contains an RTP header, which provides information such as the voice coding scheme being used, a sequence number, a time stamp for the instant at which the voice packet was sampled, and an identification for the source of the voice packet.

RTP has a companion protocol, the *RTP Control Protocol* (RTCP). RTCP does not carry coded voice packets. Rather, RTCP is a signaling protocol that includes a number of messages that are exchanged between session users. These messages provide feedback regarding the quality of the session. The type of information includes such details as lost RTP packets, delay, and interarrival jitter.

Whenever an RTP session is opened, an RTCP session is also implicitly opened. This means that when a UDP port number is assigned to an RTP session for the transfer of voice packets (or any other media packets, such as video), a separate port number is assigned for RTCP messages. An RTP port number always will be even, and the corresponding RTCP port number will be the next highest number and hence odd. Thus, if we again consider the IP protocol stack for voice transport, it appears as shown in Figure 9.2.

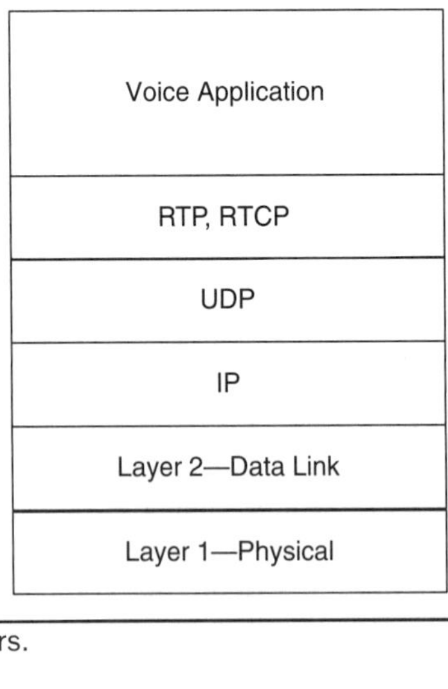

FIGURE 9.2 VoIP protocol layers.

It should be noted that RTP and RTCP do not guarantee minimal delays, low jitter, or low packet loss. In order to do this, other protocols are required. RTP and RTCP simply provide information to the applications at either end so that those applications can deal with loss, delay, or jitter with the least possible impact to the user.

9.4 H.323

In all telephony networks, specific signaling protocols are invoked before and during a call to communicate a desire to set up a call, to monitor call progress, and to gracefully bring a call to a conclusion. Perhaps the best example is the *ISDN User Part* (ISUP), a component of the *Signaling System 7* (SS7) signaling suite. In VoIP systems, signaling protocols also need to be used for exactly the same reasons. The first successful set of protocols for VoIP was developed by the *International Telecommunications Union* (ITU). This set is known as H.323 and has the title, "Packet-Based Multimedia Communications Systems." Although new protocols have emerged recently, notably the SIP, H.323 is still the most widely deployed VoIP signaling system.

9.4.1 H.323 Network Architecture

As is the case for most signaling systems, H.323 defines a specific network architecture, which is depicted in Figure 9.3. This architecture involves H.323 terminals, gateways, gatekeepers, and *Multipoint Controller Units* (MCUs). The overall objective of H.323 is to

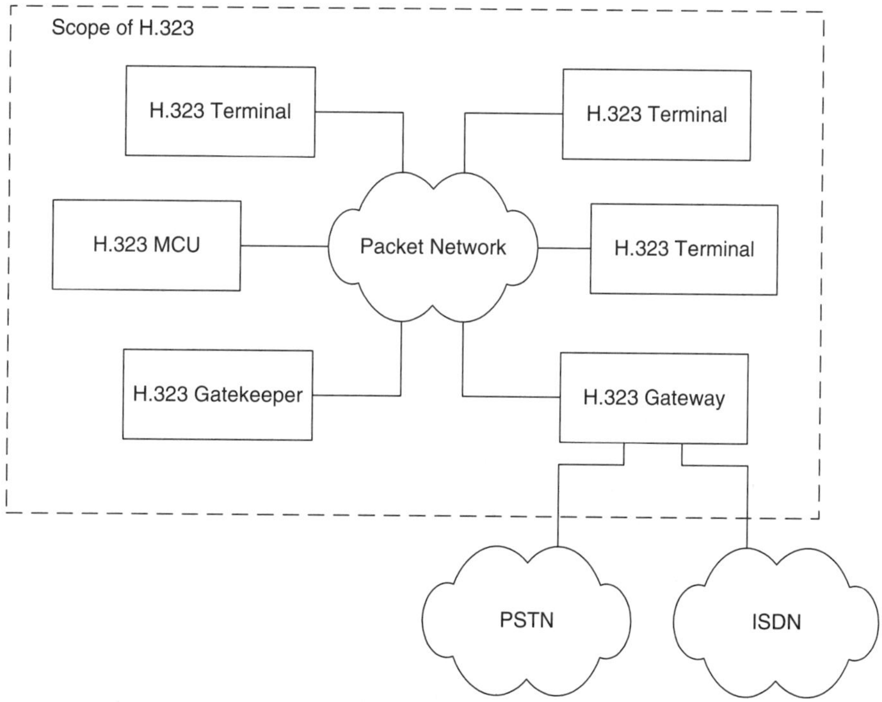

Figure 9.3 H.323 network architecture.

enable the exchange of media streams between H.323 endpoints, where an H.323 endpoint is an H.323 terminal, a gateway, or an MCU.

An H.323 terminal is an endpoint that offers real-time communications with other H.323 endpoints. It is typically an end-user communications device. It supports at least one audio codec and optionally may support other audio codecs and/or video codecs.

A gateway is an H.323 endpoint that provides translation services between the H.323 network and another type of network, such as an *Integrated Services Digital Network* (ISDN) or the *Public Switched Telephone Network* (PSTN). One side of the gateway supports H.323 signaling and terminates packet media according to the requirements of H.323. The other side of the gateway interfaces to a circuit-switched network and supports the transmission characteristics and signaling protocols of the circuit-switched network. On the H.323 side, the gateway has the characteristics of an H.323 terminal. On the circuit-switched side, it has the characteristics of a node in the circuit-switched network. A translation between the signaling protocols and media formats of one side and those of the other side is performed internally within the gateway. The translation is totally transparent to other nodes in the circuit-switched network and in the H.323 network. Gateways also may serve as a conduit for communications between H.323 terminals that are not on the same network, where the communication between the terminals needs to pass via an external network such as the PSTN.

A gatekeeper is an optional entity within an H.323 network. When present, it controls a number of H.323 terminals, gateways, and *Multipoint Controllers* (MCs). By *control,* we mean that it authorizes network access from one or more endpoints and may choose to permit or deny any given call from an endpoint within its control. It may offer bandwidth control services, which, if used in conjunction with bandwidth and/or resource management techniques, can help to ensure service quality. A gatekeeper also offers address translation services, enabling the use of aliases within the network. The set of terminals, gateways, and MCs controlled by a single gatekeeper is known as a *zone.* A zone can span multiple networks or subnetworks, and it is not necessary that all entities within a zone be contiguous.

An MC is an H.323 endpoint that manages multipoint conferences between three or more terminals and/or gateways. For such conferences, it establishes the media that may be shared between entities by transmitting a capability set to the various participants, and an MC may change the capability set in the event that other endpoints join or leave the conference. An MC may reside within a separate MCU or may be incorporated within the same platform as a gateway, a gatekeeper, or an H.323 terminal.

9.4.2 Overview of H.323 Protocols

Figure 9.4 shows the H.323 protocol stack. On examination, we find a number of protocols already discussed, such as RTP, TCP, and UDP. It is clear from the figure that the exchange of media is performed using RTP over UDP, and of course, wherever there is RTP, there is also RTCP.

In Figure 9.4 we also find two protocols that have not yet been discussed—namely, H.225.0 and H.245. These two protocols define the actual messages that are exchanged between H.323 endpoints. They are generic protocols in that they could be used in any number of network architectures. When it comes to the H.323 network architecture, the manner in which the H.225.0 and H.245 protocols are applied is specified by Recommendation H.323.

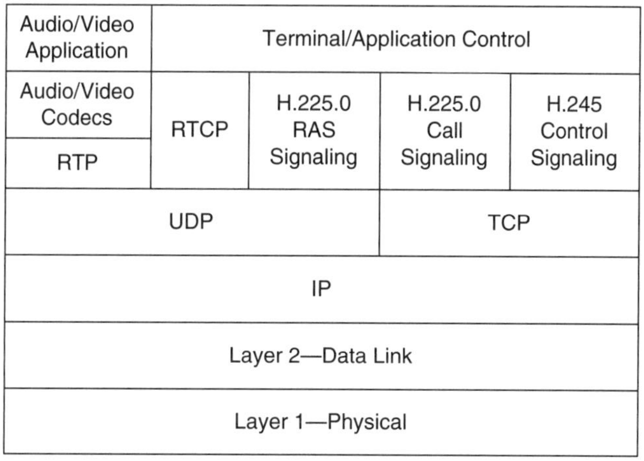

FIGURE 9.4 H.323 protocol layers.

H.225.0 is a two-part protocol. One part is effectively a variant of ITU-T Recommendation Q.931, the ISDN Layer 3 specification, and should be quite familiar to those with knowledge of ISDN. It is used for the establishment and teardown of connections between H.323 endpoints. This type of signaling is known as *call signaling* or *Q.931 signaling*. The other part of H.225.0 is known as *registration, admission, and status* (RAS) *signaling*. It is used between endpoints and gatekeepers and enables a gatekeeper to manage the endpoints within its zone. For example, RAS signaling is used by an endpoint to register with a gatekeeper, and it is used by a gatekeeper to allow or deny endpoint access to network resources.

H.245 is a control protocol used between two or more endpoints. The main purpose of H.245 is to manage the media streams between H.323 session participants. To this end, it includes functions such as ensuring that the media to be sent by one entity are limited to the set of media that can be received and understood by another. H.245 operates by the establishment of one or more logical channels between endpoints. These logical channels carry the media streams between the participants and have a number of properties, such as media type, bit rate, and so on.

All three signaling protocols—RAS, Q.931, and H.245—may be used in the establishment, maintenance, and teardown of a call. The various messages may be interleaved. For example, consider an endpoint that wants to establish a call to another endpoint. First, it may use RAS signaling to obtain permission from a gatekeeper. It then may use Q.931 signaling to establish communication with the other endpoint and set up the call. Finally, it may use H.245 control signaling to negotiate media parameters with the other endpoint and set up the media transfer. Figure 9.5 presents an example of the interaction between the different types of signaling.

9.4.3 H.323 Call Establishment

In the example of Figure 9.5, two terminals (H.323 endpoints) need to establish a VoIP call between them, and different gatekeepers control the two terminals. As a first step, the calling terminal requests permission from its gatekeeper to establish the call. This is

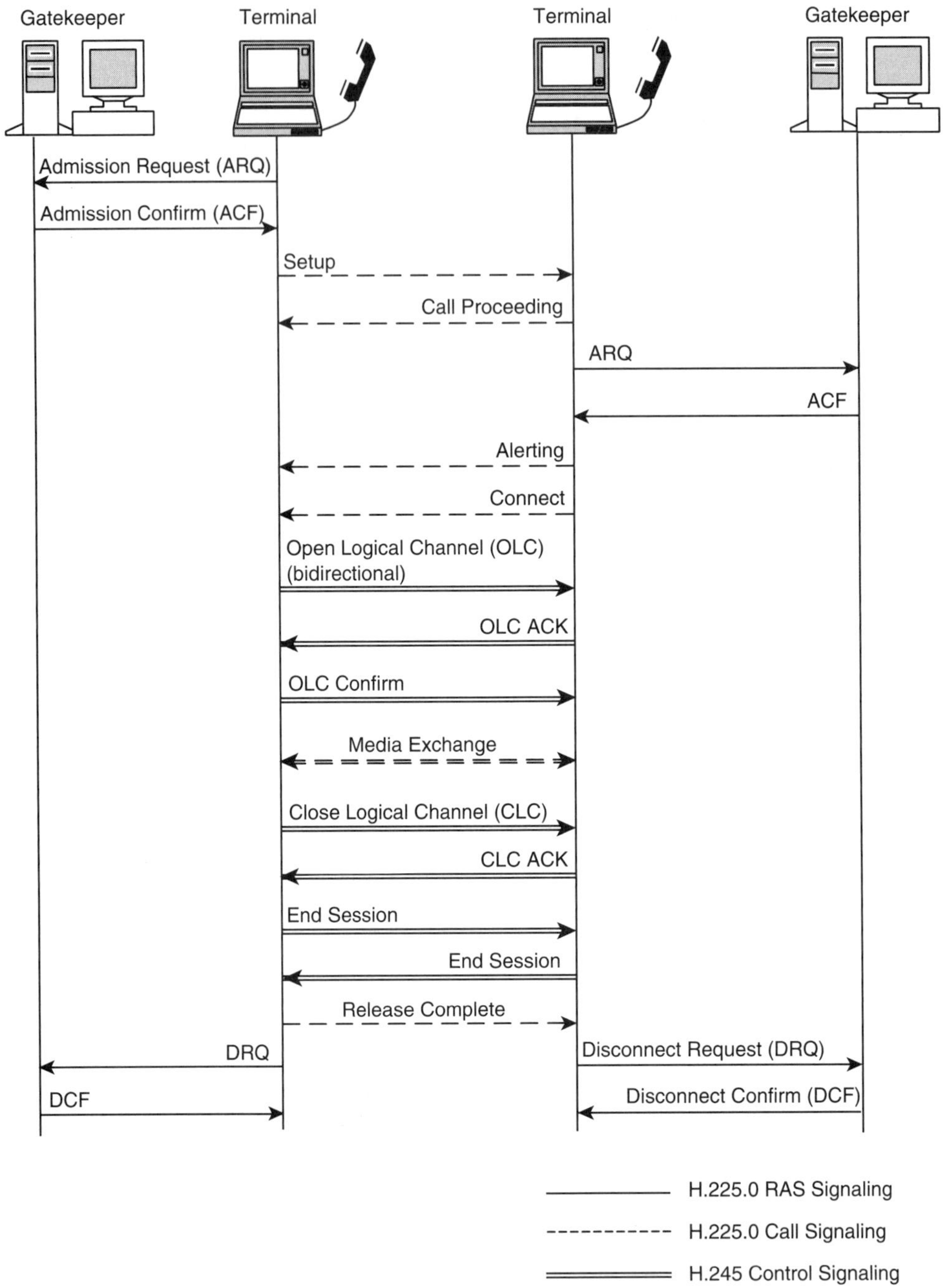

Figure 9.5 H.323 call establishment and release.

done with the *Admission Request* (ARQ) message. The terminal indicates the type of call in question (two-party or multiparty), the endpoint's own identifier, a call identifier (a unique string), a call reference value (an integer value also used in call signaling messages for the same call), and information regarding the other party or parties to participate in the call. The information regarding other parties to the call includes one or more aliases and/or signaling addresses. One of the most important mandatory parameters in the ARQ is the bandwidth parameter. This specifies the amount of bandwidth required in units of 100 bps.

Note that the endpoint should request the total media-stream bandwidth needed, excluding overhead. Thus, if a two-party call is needed, with each party sending voice at 64 kbps, then the bandwidth required is 128 kbps, and the value carried in the bandwidth parameter is 1280. The purpose of the bandwidth parameter is to enable the gatekeeper to reserve resources for the call.

The gatekeeper indicates a successful admission by responding to the endpoint with an *Admission Confirm* (ACF) message. This includes many of the same parameters that are included in the ARQ. The difference is that when a given parameter is used in the ARQ, it is simply a request from the endpoint, whereas a given parameter value in the ACF is a firm order from the gatekeeper. For example, the ACF includes the bandwidth parameter, which may be a lower value than that requested in the ARQ, in which case the endpoint must stay within the bandwidth limitations imposed by the gatekeeper.

Another parameter of particular interest in both the ARQ and the ACF is the *call Model* parameter, which is optional in the ARQ and mandatory in the ACF. In the ARQ, call Model indicates whether the endpoint wants to send call signaling directly to the other party or prefers that call signaling be passed via the gatekeeper. In the ACF, it represents the gatekeeper's decision as to whether call signaling is to pass via the gatekeeper or directly between the terminals. In the example in Figure 9.5, the calling gatekeeper has chosen not to be in the path of the call signaling.

The Setup message is the first call signaling message sent from one terminal to the other to establish the call. The message must contain the Q.931 Protocol Discriminator, a Call ReferenceSetup, a Bearer Capability, and the User-User information element. Although the Bearer Capability information element is mandatory, the concept of a bearer, as used in the circuit-switched world, does not map very well to an IP network. For example, no B-channel exists in IP, and the actual agreement between endpoints regarding the bandwidth requirements is done as part of H.245 signaling, where RTP information such as the payload type is exchanged. Consequently, many of the fields in the Bearer Capability information element, as defined in Q.931, are not used in H.225.0. Of those fields that are used in H.225.0, many are used only when the call has originated from outside the H.323 network and has been received at a gateway, where the gateway performs a mapping from the signaling received to the appropriate H.225.0 messages.

A number of parameters are included within the mandatory User-to-User information element. These include the call identifier, the call type, a conference identifier, and information about the originating endpoint. Among the optional parameters, we may find a source alias, a destination alias, an H.245 address for subsequent H.245 messages, and a destination call signaling address. The User-to-User information element is included in all H.225.0 call signaling messages. It is the inclusion of this information element that enables Q.931 messages, originally designed for ISDN, to be adapted for use with H.323.

The Call Proceeding message optionally may be sent by the recipient of a Setup message to indicate that the Setup message has been received and that call establishment procedures are underway. When sent, it usually precedes the Alerting message, which indicates that the called device is "ringing." Strictly speaking, the Alerting message is optional.

In addition to Call Proceeding and Alert, we also may find the optional Progress message (not shown). Ultimately, when the called party answers, the called terminal returns a Connect message. Although some of the messages from the called party to the calling party, such as Call Proceeding and Alerting, are optional, the Connect message must be sent if the call is to be completed. The User-User information element contains the same set of parameters as defined for the Call Proceeding, Progress, and Alert messages, with the addition of the Conference Identifier. These parameters are also used in a Setup message, and their use in the Connect message is to correlate this conference with that indicated in a Setup. Any H.245 address sent in a Connect message should match that sent in any earlier Call Proceeding, Alerting, or Progress messages. In fact, the called terminal must include at least an H.245 signaling address to which H.245 messages must be sent because H.245 messages are used to establish the media (i.e., voice) flow between the parties.

In the example in Figure 9.5, the H.245 message exchange begins after the Connect message is returned. This message exchange could, in fact, occur earlier than the Connect message. It is important to note that H.245 is not responsible for carrying the actual media. For example, there is no such thing as an H.245 packet containing a sample of coded voice. That is the job of RTP. Instead, H.245 is a control protocol that manages the establishment and release of media sessions. H.245 does this through messaging that enables the establishment of logical channels, where a logical channel is a unidirectional RTP stream from one party to the other.

A logical channel is opened by sending an *Open Logical Channel* (OLC) request message. This message contains a mandatory parameter called *forwardLogicalChannelParameters* that relates to the media to be sent in the forward direction, that is, from the endpoint issuing this command. It contains information such as the type of data to be sent (e.g., AMR-coded audio), an RTP session ID, an RTP payload type, and an indication as to whether silence suppression is to be used. If the recipient of the message wants to accept the media to be sent, then it will return an OpenLogicalChannelAck message containing the same logical channel number as received in the request and a transport address to which the media stream should be sent.

Strictly speaking, a logical channel is unidirectional. Therefore, in order to establish a two-way conversation, two logical channels must be opened—one in each direction. According to the description just presented, this requires four messages, which is rather cumbersome. Consequently, H.323 defines a bidirectional logical channel. This is a means of establishing two logical channels, one in each direction, in a slightly more efficient manner. Basically, a bidirectional logical channel really means two logical channels that are associated with each other. The establishment of these two channels can be achieved with just three H.245 messages rather than four. In order to do so, the initial OLC message not only contains information regarding the media that the calling endpoint wants to send, but it also contains reverse logical channel parameters. These indicate the types of media that the endpoint is willing to receive and where that media should be sent.

On receipt of the request, the far endpoint may send an Open Logical Channel ACK message containing the same logical channel number for the forward logical channel, a logical channel number for the reverse logical channel, and descriptions related to the media formats that it is willing to send. These media formats should be chosen from the options originally received in the request, thereby ensuring that the called end will send only media that the calling end supports.

On receipt of the Open Logical Channel ACK, the originating endpoint responds with an Open Logical Channel Confirm message to indicate that all is well. RTP streams and RTCP messages now can flow in each direction.

9.4.4 H.323 Call Release

Figure 9.5 also shows the disconnection of a call after media have been exchanged (a conversation has taken place). The first step in the process involves closing the logical channels that have been created by H.245 signaling—closing the RTP streams between the users.

Closing a logical channel involves sending a CloseLogicalChannel message. In the case of a successful closure, the far end should send the response message CloseLogical ChannelAck. In general, a logical channel can be closed only by the entity that created it in the first place. For example, in the case of a unidirectional channel, only the sending entity can close the channel. However, the receiving endpoint in a unidirectional channel can humbly request the sending endpoint to close the channel. It does so by sending the RequestChannelClose message, indicating the channel that the endpoint would like to have closed. If the sending entity is willing to grant the request, it responds with a positive acknowledgment and then proceeds to close the channel. When an entity closes the forward logical channel of a bidirectional logical channel, it also closes the reverse logical channel.

Once all logical channels in a session are closed, the session itself is terminated when an endpoint sends an EndSession command message. The receiving endpoint responds with an EndSession command message. Once an entity has sent this message, it must not send any more H.245 messages related to the session.

At this point, the call signaling comes to a close with the issuance of a Release Complete message. Unlike standard Q.931 ISDN signaling, no Release message is sent—just the Release Complete message, which is all that is needed to end call signaling.

Finally, each endpoint uses the *Disconnect Request* (DRQ) message to request permission from its gatekeeper to disconnect. The gatekeeper responds with the *Disconnect Confirm* (DCF) message.

9.4.5 The H.323 Fast Connect Procedure

It is clear from Figure 9.5 that H.323 call establishment and release can be quite cumbersome. Moreover, it is possible for a given gatekeeper to choose to be in the path of all call signaling and H.245 signaling in addition to RAS signaling. In such a scenario, the number of messages exchanged can be very large, which can extend the call setup time beyond acceptable limits. In order to speed things up, H.323 includes a procedure known as *Fast Connect,* a method that can reduce the amount of call establishment signaling significantly. The Fast Connect procedure is shown in Figure 9.6.

The Fast Connect procedure involves the setting up of media streams as quickly as possible. To achieve this, the Setup message can contain a *faststart* element within the User-User information element. The faststart element is actually one or more *Open Logical Channel* (OLC) request messages containing all the information that normally would be contained in such requests. It includes reverse logical channel parameters if the calling endpoint expects to receive media from the called endpoint.

If the called endpoint also supports the procedure, then it can return a faststart element in one of the Call Proceeding, Alerting, Progress, or Connect messages. This faststart element is basically another OLC message that appears like a request to open a bidirectional logical channel. The included choices of media formats to

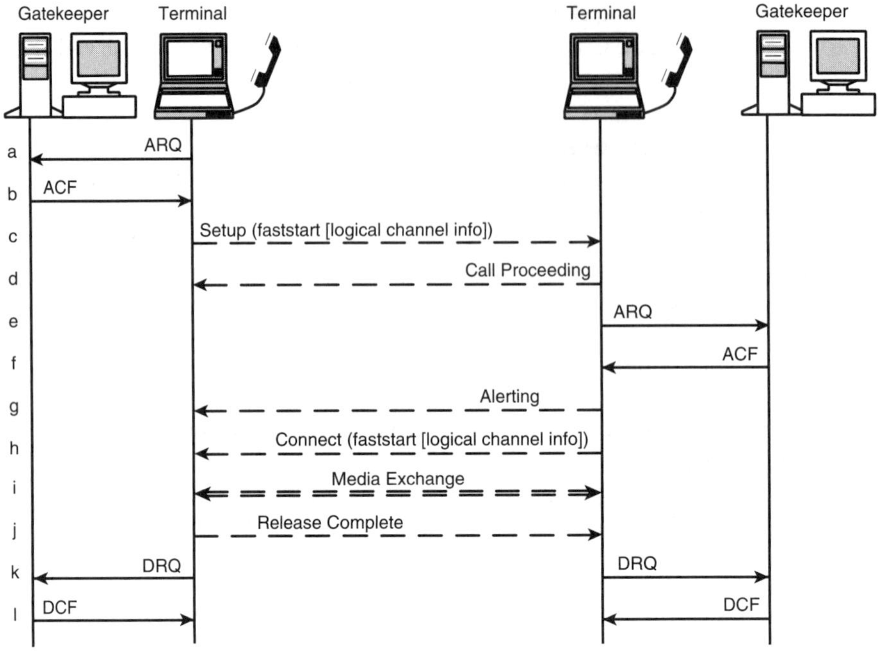

FIGURE 9.6 H.323 Fast Connect procedure.

send and receive are chosen from those offered in the faststart element of the incoming Setup message. The calling endpoint effectively has offered the called endpoint a number of choices for forward and reverse logical channels, and the called endpoint has indicated the choices that it prefers. The logical channels are now considered open as if they had been opened according to the procedures of H.245.

Note that a faststart element from the called party to the calling party may be sent in any message up to and including the Connect message. If it has not been included in any of the messages, then the calling endpoint assumes that the called endpoint either cannot or does not want to support faststart. In such a case, the standard H.245 methods must be used.

Use of the Fast Connect procedure means that H.245 information is carried within the call signaling messages and no separate H.245 control channel exists. Therefore, bringing a call to a conclusion is also faster. The call is released simply by the sending of the call signaling Release Complete message. When used with the Fast Connect procedure, this has the effect of closing all the logical channels associated with the call and is equivalent to using the procedures of H.245 to close the logical channels.

9.5 The Session Initiation Protocol (SIP)

The SIP is considered by many to be a powerful alternative to H.323 and is a fundamental component of all IMS platforms. SIP is considered to be a more flexible solution, simpler than H.323, easier to implement, better suited to the support of intelligent user devices, and better suited to the implementation of advanced features such as IMS. Although H.323 still may have a larger installed base than SIP initially, most people in the VoIP community believe that the future of VoIP revolves around SIP. In fact, the 3GPP has endorsed SIP as the session management protocol of choice for 3GPP Release 5, albeit with some enhancements.

Like H.323, SIP is simply a signaling protocol and does not carry the voice packets itself. Rather, it makes use of the services of RTP for the transport of the voice packets (the media stream).

9.5.1 The SIP Network Architecture

SIP defines two basic classes of network entities—clients and servers. Strictly speaking, a client, also known as a *user agent client*, is an application program that sends SIP requests. A server is an entity that responds to those requests. Thus SIP is a client-server protocol. VoIP calls using SIP are originated by a client and terminated at a server. A client may be found within a user's device, which could be, for example, an SIP phone. Clients also may be found within the same platform as a server. For example, SIP enables the use of proxies, which act as both clients and servers.

Four different types of servers are available—proxy servers, redirect servers, user agent servers, and registrars. A proxy server acts similarly to a proxy server used for Web access from a corporate *Local-Area Network* (LAN). Clients send requests to the proxy, which either handles those requests itself or forwards them on to other servers, perhaps after performing some translation. To those other servers, it appears as though the message is coming from the proxy rather than from some entity hidden behind it. Given that a proxy both receives and sends requests, it incorporates both server and client functionality. Figure 9.7 shows an example of the operation of a proxy server. It does not take much imagination to realize how this type of functionality can be used for call forwarding/follow-me services.

A redirect server is a server that accepts SIP requests, maps the destination address to zero or more new addresses, and returns the translated address to the originator of the request. Thereafter, the originator of the request may send requests to the address(es) returned by the redirect server. A redirect server does not initiate any SIP requests of its own.

Figure 9.8 presents an example of the operation of a redirect server. This can be another means of providing the call-forwarding/follow-me service that can be

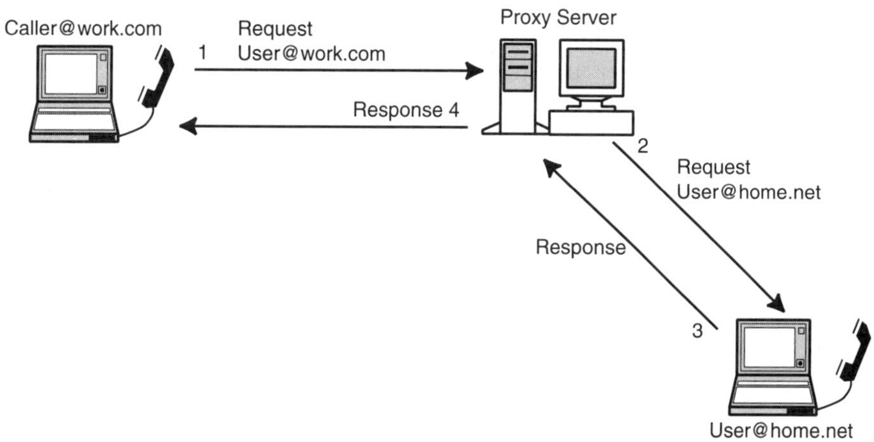

Figure 9.7 SIP proxy server.

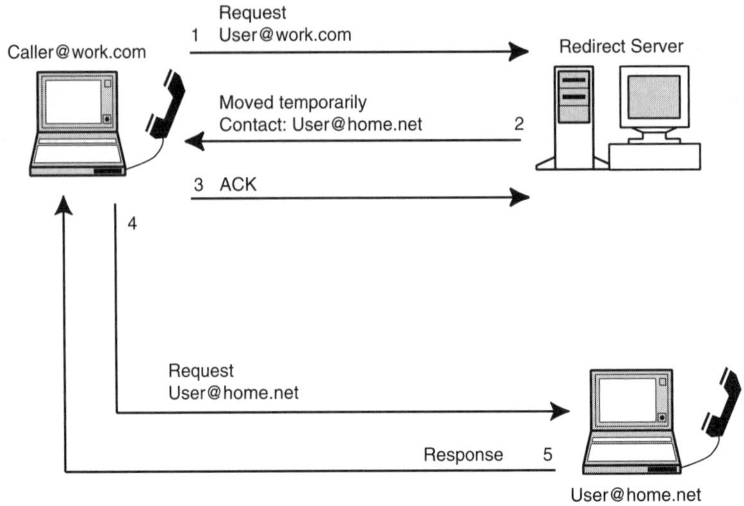

Figure 9.8 SIP redirect server.

provided by a proxy server. The difference is that in the case of a redirect server, the originating client does the actual forwarding of the call. The redirect server simply provides the information necessary to enable the originating client to do so, after which the redirect server is no longer involved.

A user agent server accepts SIP requests and contacts the user. A response from the user to the user agent server results in an SIP response on behalf of the user. In reality, an SIP device, such as an SIP-enabled phone, will function as both a user agent client and a user agent server. Acting as a user agent client, it is able to initiate SIP requests. Acting as a user agent server, it can receive and respond to SIP requests. In practical terms, this means that it is able to initiate calls and receive calls. This enables SIP, a client-server protocol, to be used for peer-to-peer communication.

A registrar is a server that accepts SIP Register requests. SIP includes the concept of user registration, whereby a user signals to the network that it is available at a particular address. Such registration is performed by the issuance of a Register request from the user to the registrar. Typically, a registrar will be combined with a proxy or redirect server. Registration in SIP serves a similar purpose to location updating in a GSM network; it is a means by which a user can signal to the network that he or she is available at a particular location.

Given that practical implementations involve the combination of a user agent client and a user agent server and the combining of registrars with either proxy servers or redirection servers, a real network may well involve only user agents and the redirection or proxy servers.

9.5.2 SIP Call Establishment

At a high level, SIP call establishment is very simple, as shown in Figure 9.9. The process starts with an SIP Invite message, which is used from the calling party to the called party. The message invites the called party to participate in a session—a call. Included with the Invite message is a session description—a description of the media that the

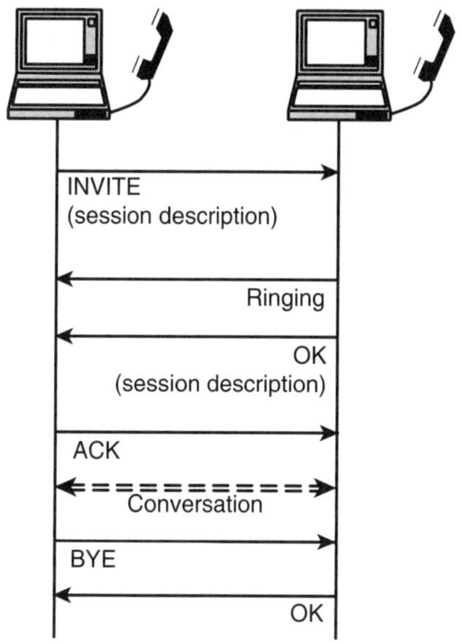

FIGURE 9.9 SIP basic call establishment and release.

calling party wants to use. This description includes the voice coding scheme that the caller wants to use, plus an IP address and a port number that the called party should use for sending media back to the caller.

A number of interim responses to the Invite may be sent prior to the called party accepting the call. For example, the caller might be informed that the call is queued and/or that the called party is being alerted (i.e., the phone is ringing). Subsequently, the called party answers the call, which generates an OK response back to the caller. The OK response is actually indicated by the status-code value of 200 in the response. In the example of Figure 9.9, the 200 (OK) response contains a session description indicating the media that the caller wants to use plus an IP address and port number to which the caller should send packets.

On receipt of the 200 (OK) response, the caller responds with ACK to confirm that the OK response has been received. At this point, media are exchanged. These media most often will be coded speech but also could be other media such as video. Finally, one of the parties hangs up, which causes a Bye message to be sent. The party receiving the Bye message sends 200 (OK) to confirm receipt of the message. At this point, the call is over.

All in all, SIP call establishment is quite a simple process. Of course, the signaling could well pass via one or more proxy servers, in which case the process becomes somewhat more complex. Nonetheless, it is clear that SIP call establishment is much simpler than the equivalent H.323 process.

9.5.3 Information in SIP Messages

Obviously, there is more to SIP signaling than the messages outlined in Figure 9.9. To start with, each SIP request or response contains addresses for the calling and called parties. Each such address is known as an SIP *Uniform Resource Locator* (URL) and has the format SIP:user@domain. This is somewhat similar to an e-mail URL, which has the format mailto:user@_domain. An SIP user might well want to have the same values for user and domain in his or her SIP and e-mail addresses, which would make it very easy to know how to contact an SIP user—much easier than having to remember a telephone number.

Several requests and many responses can be sent between SIP entities. For example, if, in the example of Figure 9.9 the called user were not available, then the response Temporarily Unavailable (status code 480) could have been returned rather than the 200 (OK).

Not only are there several requests and many responses, but many information elements also can be contained in those requests and responses. In SIP, these information elements are known as *header fields*. For example, when sending an Invite, the message contains not only a session description and the to and from addresses (contained in the To and From header fields), but it also can contain a Subject header field. This field indicates the reason for the call and can be presented to the called user, who may choose to accept or reject the call based on the subject in question. One can easily imagine this capability being used to filter out unwanted telemarketing calls.

Other header fields include, for example, Call ID, Date, Timestamp, In-reply-to, Retry-after, and Priority. The Retry-after header could be used, for example, with the 480 (Temporarily Unavailable) response to indicate when the caller should try the call again (if ever). One of the most important header fields is Content-type, which indicates the type of additional information included in the message. For example, when a user issues an Invite message, the message includes a session description. The Content-type field indicates how that session description is coded so that the receiver of the message can understand whether or not that type of session can be supported.

9.5.4 The Session Description Protocol

Clearly, SIP is used to establish sessions between users, which requires that the users agree on the type and coding of the information to be shared. For example, the two users must agree on the voice coding scheme to be used, which requires that they share session descriptions. These session descriptions are coded according to the *Session Description Protocol* (SDP).

SDP is simply a language for describing sessions. It contains information regarding the parties to be involved in the session, the date and time when the session is to take place, the types of media streams to be shared, and addresses and port numbers to be used. It is perfectly possible that a session description could refer to multiple media streams, such as in a videoconference where one media stream relates to coded voice and another media stream relates to coded video. Consequently, SDP is structured so that it can describe information related to the session as a whole (e.g., the name of the session) plus information associated with each individual stream (e.g., the media format and the applicable port number). Some of the information included in an SDP session description also will be included in the SIP message that carries the SDP description. This overlap is due to the fact that SDP is designed to be used by a range of other protocols, not just SIP.

Perhaps the best way to describe the combined usage of SIP and SDP is by example. Consider Figure 9.10, which is a more detailed version of the call establishment scenario presented in Figure 9.9. In this case, we see a call from User1@work.com, who is logged in at station1.work.com, to User2@work.com, who is logged in at station2.work.com.

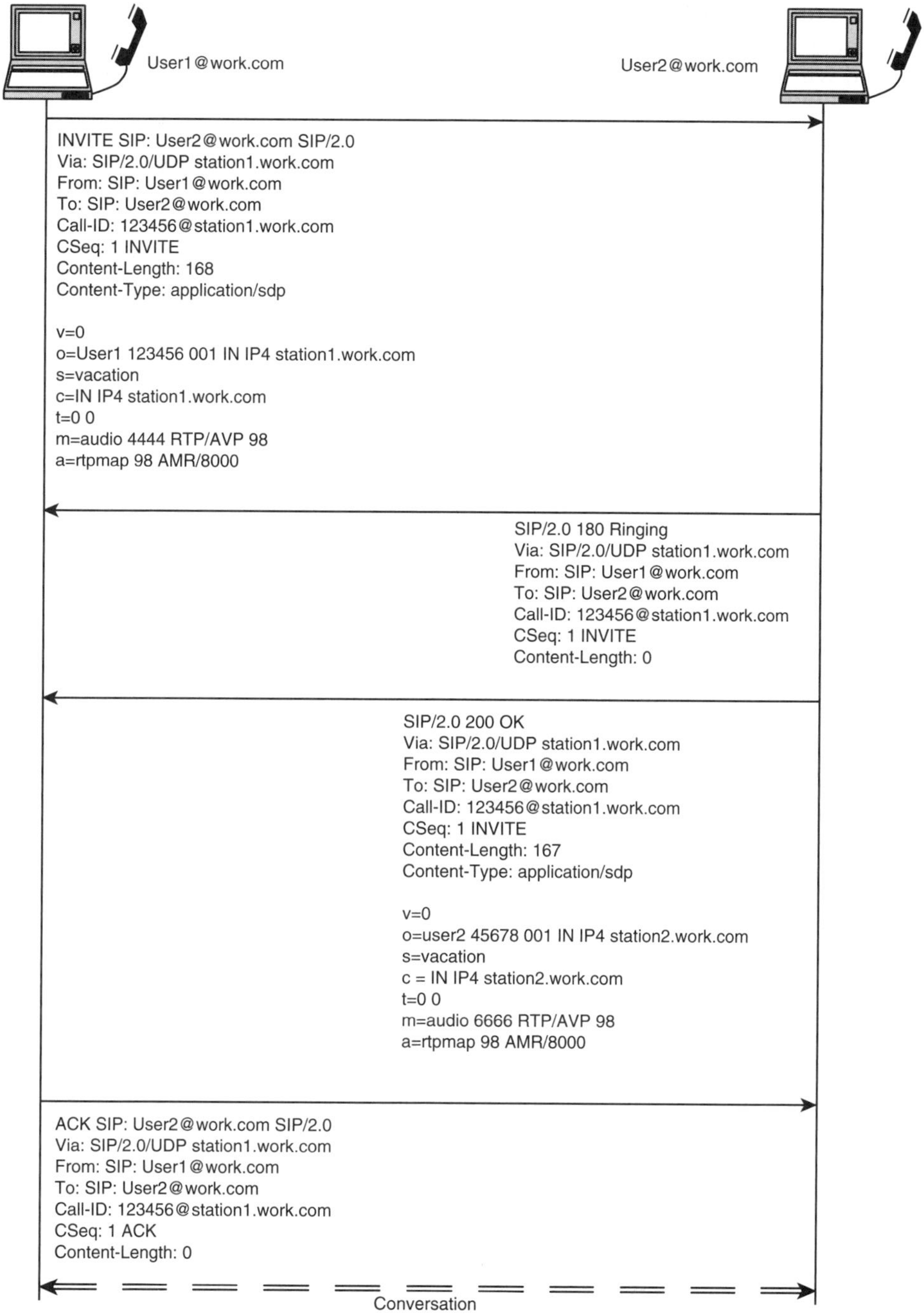

FIGURE 9.10 SIP call establishment showing message detail.

As with any SIP session establishment, the call begins with an Invite, which is indicated in the first line of the request. The first line also indicates the address of the entity to which the message is being sent, known as the request *Uniform Resource Indicator* (URI). In this case, the message is being sent directly to User2. If, however, there happens to be a proxy server between User1 and User2, then the request would go to the proxy first, in which case the request URI would indicate the proxy.

The Via header field is inserted by each entity in the chain from the source of a message to the destination. This is to ensure that the response can follow the same path back through the network as was taken by the original request. The From and To header fields indicate the initiator of the request and the recipient of the request. The Call ID is a globally unique identification. To ensure uniqueness, the Call ID should take the form indicated in the figure. The CSeq field refers to the command sequence. CSeq contains an integer and an indication of the type of request. The purpose of the CSeq header is to enable the initiator of a request to correlate a response with the request that generated the response.

Finally, we have two headers that provide information about the message body. The first, Content-Length, indicates the length of the message body. The second, Content-Type, indicates the type of message body. Strictly speaking, the message body could be any *Multipurpose Internet Mail Extension* (MIME)–coded type, such as text. In our example, the message body contains a session description code according to SDP.

The message body is separated from the SIP headers by a blank line. In SDP, it starts with a version identifier, which is version 0. Next, we find the Origin (o) field, which indicates the user name (User1), a session ID (123456 in our case) that does not have to match the SIP Call ID, a version for the session (001 in our example), the type of network (IN indicates Internet), the type of addressing used (IP4 indicates IP Version 4), and an address for the machine that initiated the session (station1.work.com in our example).

After the Origin field, we find the optional Subject (s) field and the Connection (c) field. The connection field provides information regarding where the user would like the media to be sent. In our case, it indicates that the type of network is Internet (IN), that the addressing uses IP Version 4 (IP4), and the address to which media should be sent. This address could be different from the address of the machine that created the session.

After the Connection field, we find the Time (t) field, which indicates the start and stop times for the session. In our example, these are both set to 0, which means that the session does not have any set start or stop time.

Next, we find the Media (m) field, which provides information about the media to be used and the port to which the media should be sent. In our example, the type of media is audio, and it should be received at port number 4444 (i.e., the far end should send the media to port number 4444). The Media field also indicates the type of RTP *Audio/Video Profile* (AVP) to be used, which is 98 in our example.

In the RTP, certain types of media stream codings are assigned specific values of payload types and are known as *static payload types.* For example, payload type 0 indicates G.711-coded voice. Thus, if the Media field of the SDP description indicated RTP/AVP value 0, then the far end would know that G.711-coded voice is required. The RTP also includes the concept of a dynamic payload type, where the payload type value is significant only within one session. Therefore, it is necessary to indicate additional attributes in order for the far end to understand the meaning of the payload type chosen. In our example, the attribute a = rtpmap AMR/8000 indicates that the payload type is adaptive multirate and sampled at 8000 Hz.

Figure 9.10 shows that the first response includes status code 180, indicating that the user is being alerted. It contains the same Via, To, From, CSeq, and Call ID header fields as the original request, and they enable the sender of the request to match the response with the request. This response does not contain any session description.

When the called user answers, a 200 (OK) response is generated. The SIP header fields are the same as the original Invite request, with the exception of the content length and the message body itself. This is so because the called device has included a session description of its own. This is quite similar to the session description in the Invite request but indicates a different address and port number. This makes sense because the address and port number indicate where User2 expects to receive the media stream.

Once User1 has received the 200 (OK) response, it sends an ACK message. The header fields in this message are identical to those of the original Invite, with the exception of the CSeq field, which now indicates the ACK request. At this point, media can flow between the two parties, and a conversation can take place.

9.6 Distributed Architecture and Media Gateway Control

The foregoing discussions regarding SIP and H.323 have focused primarily on the signaling needed to establish media streams between session participants. Although not clearly stated, it is implied that the entities generating the signaling are the same entities that will generate the actual media streams. In other words, we have not described a clear separation of media from call control.

If one looks carefully at the description of the SDP, however, one sees that it is possible to indicate different addresses for the entity that sends a session description and the entity that actually terminates the media stream. This indicates that the separation of media from call control is possible. Moreover, we have seen from the architectures of 3GPP Release 4 and Release 5 that the separation of media and call control is not only possible but often is desirable.

If we physically separate a call-control entity from an entity that handles media streams (such as a gateway that performs voice coding), then we need a protocol between those two types of entities so that the call-control entity can manage the media entity for the setup and teardown of calls. Provided that we have such a protocol, then there is no reason why one call-control device could not manage multiple media-handling devices. It simply would be a question of the processing power of the call-control device. In such a scenario, we can envision an architecture such as that shown in Figure 9.11, where a single controller manages multiple media-handling devices such as *Media Gateways* (MGs). In some quarters, this separation between call control and media is known as the *softswitch architecture.*

The advantages of such an approach are that MGs can be placed as close as possible to the source or sink of the media stream, which can be of great significance if voice is being carried on one side of the gateway at 64 kbps while it is carried at a much lower bandwidth over the IP network. Although we may place the MGs close to the edge of the network, we can centralize the call control and network intelligence. Depending on the processing power of the controller, the required size of the various gateways, and the cost of each type of node, it is possible to design a network that is very cost-efficient from both capital cost and operating cost perspectives. Of course, the critical requirement is that there be a fast, robust, and scalable control protocol between the controllers and the media devices.

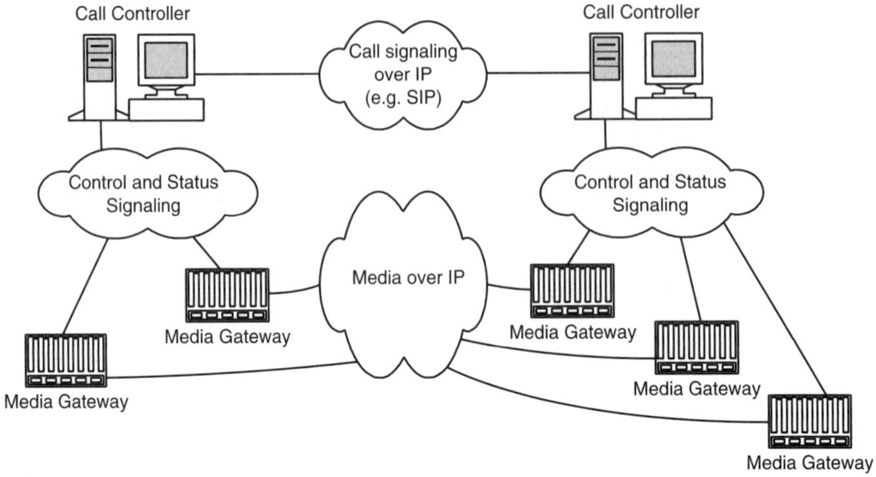

FIGURE 9.11 Separation of bearer and call control.

As it happens, several such protocols exist. The control protocol most widely deployed in VoIP networks today is the *Media Gateway Control Protocol* (MGCP), which was developed within the *Internet Engineering Task Force* (IETF). This protocol, however, has been superseded by a protocol known as *MEGACO/H.248*, which was jointly developed by the IETF and the ITU. In fact, it is known as MEGACO in the IETF community and as H.248 within the ITU; the terms *MEGACO* and *H.248* are interchangeable. MEGACO has been endorsed by the 3GPP as the protocol of choice for gateway control in 3GPP Release 4 and Release 5.

9.6.1 The MEGACO Protocol

The architecture associated with MEGACO defines MGs, which perform the conversion of media from the format required in one network to the format required in another. The architecture also defines *Media Gateway Controllers* (MGCs), which control call establishment and teardown within MGs.

The MEGACO protocol involves a series of transactions between MGCs and MGs. Each transaction involves the sending of a transaction request by the initiator of the transaction and the sending of a transaction reply by the responder. A transaction request consists of a number of commands, and the transaction reply consists of a corresponding number of responses. For the most part, transactions are requested by an MGC, and the corresponding actions are executed within an MG. However, a number of cases occur where an MG initiates the transaction request.

MEGACO defines *terminations*, which are logical entities on an MG that act as sources or sinks of media streams. Certain terminations are physical. They have a semi-permanent existence and may be associated with external physical facilities or resources. These would include a termination connected to an analog line or a termination connected to a DS0 channel or perhaps an ATM virtual circuit. Such terminations exist as long as they are provisioned within the MG. Other terminations have a more transient existence and exist only for the duration of a call or media flow. These are known as *ephemeral terminations* and represent media flows such as a stream of RTP packets.

These terminations are created as a result of a MEGACO Add command, and they are destroyed by means of the Subtract command.

Terminations have specific properties, and the properties of a given termination will vary according to the type of termination. It is clear, for example, that a termination connected to an analog line will have different characteristics than a termination connected to a TDM channel such as a DS0. The properties associated with a termination are grouped into a set of descriptors. These descriptors are included in MEGACO commands, thereby enabling termination properties to be changed according to instructions from MGC to MG.

A termination is referenced by a termination ID. This is an identifier chosen by the MG. MEGACO enables the use of the wildcards all (*) and any, and choose ($). A special termination ID is available called Root. This termination ID is used to refer to the gateway as a whole rather than to any specific terminations within the gateway.

MEGACO also defines *contexts,* where a context is an association between a number of terminations for the purposes of sharing media between those terminations. Terminations may be added to contexts, removed from contexts, or moved from one context to another. A termination may exist in only one context at any time, and terminations in a given gateway may exchange media only if they are in the same context.

A termination is added to a context through the use of the Add command. If the Add command does not specify a context to which the termination should be added, then a new context is created as a result of the execution of the Add command. This is the only mechanism for creating a new context. A termination is moved from one context to another through use of the Move command and is removed from a context through use of the Subtract command. If the execution of a Subtract command results in the removal of the last termination from a given context, then that context is deleted.

The relationship between terminations and contexts is illustrated in Figure 9.12, where a gateway is depicted with four active contexts. In context C1, we see a simple two-way call across the MG. In context C2, we see a three-way call across the MG. In contexts C3 and C4, we see a possible implementation of call waiting. In context C3, terminations T6 and T7 are involved in a call. Another call arrives from termination T8 for termination T7. If the user wants to accept this waiting call and place the existing call on hold, then this could be achieved by moving termination T7 from context C3 to context C4.

The existence of several terminations within the same context means that they have the potential to exchange media. However, the existence of terminations in the same context does not necessarily mean that they can all send data to each other and receive data from each other at any given time. The context itself has certain attributes. These include the topology, which indicates the flow of media between terminations (which terminations may send media to others or receive media from others). Also, the priority attribute indicates the precedence applied to a context when an MGC must handle many contexts simultaneously. An emergency attribute is used to give preferential handling to emergency calls.

A context is identified by a context ID, which is assigned by the MG and is unique within a single MG. As is the case for terminations, MEGACO enables wildcarding when referring to contexts such that the all (*) and any and choose ($) wildcards may be used. The all (*) wildcard may be used by an MGC to refer to every context on a gateway. The choose ($) wildcard is used when an MGC requires the MG to create a new context.

A special context known as the *null context* also exists. This contains all terminations that are not associated with any other termination—that is, all terminations that do not

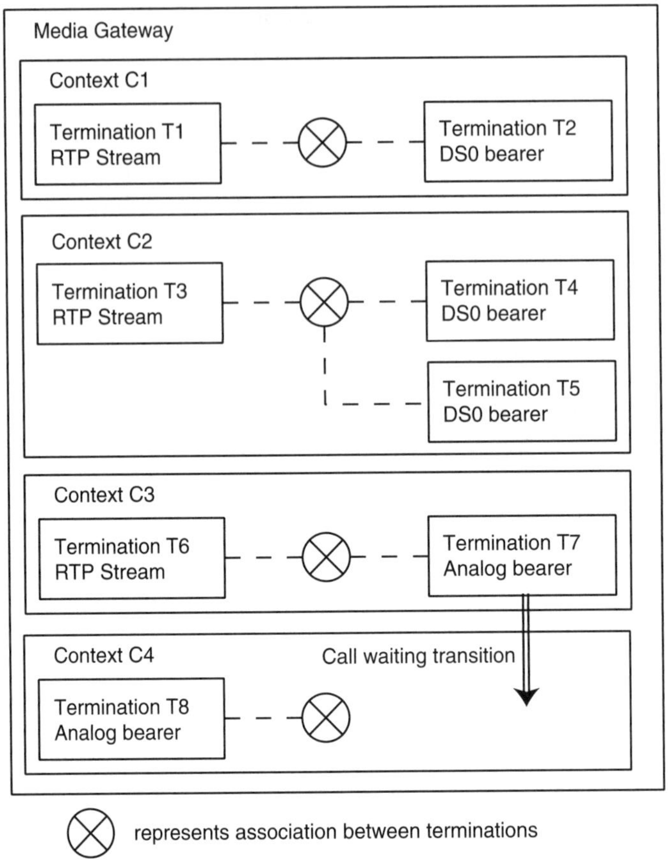

FIGURE 9.12 Contexts and terminations.

exist in any other context. Idle terminations normally exist in the null context. The context ID for the null context is simply "-".

9.6.1.1 MEGACO Transactions

MEGACO transactions involve the passing of commands and the responses to those commands. Commands are directed toward terminations within contexts. In other words, every command specifies a context ID and one or more termination IDs to which the command applies. This is the case even for a command that requires some action by an idle termination that does not exist in any specific context. In such a case, the null context is applicable.

Multiple commands may be grouped together in a transaction structure whereby a set of commands related to one context may be followed by a set of commands related to another context. The grouped commands are sent together in a single transaction request. This can be represented as

Transaction Request (Transaction ID)
ContextID1 {Command, Command,_._._._Command},

ContextID2 {Command, Command,_._._._Command},
ContextID3 {Command, Command,_._._._Command} })

No requirement specifies that a transaction request contain commands for more than one context or even contain more than one command. It is perfectly valid for a transaction request to contain just a single command for a single context.

On receipt of a transaction request, the recipient executes the enclosed commands. The commands are executed sequentially in the order specified in the transaction request. On completed execution of the commands, a transaction reply is issued. This has a similar structure to the transaction request in that it contains a number of responses for a number of contexts. A transaction reply may be represented as

TransactionReply (TransactionID)
ContextID1 {Response, Response,_._._._Response},
ContextID2 {Response, Response,_._._._Response},
ContextID3 {Response, Response,_._._._Response} })

9.6.1.2 MEGACO Commands

MEGACO defines the following eight commands. Most of the commands are sent from an MGC to an MG. The exceptions are the Notify command, which is always sent from an MG to an MGC, and the ServiceChange command, which can be sent from either an MG or an MGC.

- *Add.* The Add command adds a termination to a context. If the command does not specify a particular context to which to add the termination, then a new context is created. If the command does not indicate a specific TerminationID but instead uses the choose ($) wildcard, the MG will create a new ephemeral termination and add it to the context.
- *Modify.* The Modify command is used to change the property values of a termination, to instruct the termination to issue one or more signals, or to instruct the termination to detect and report specific events.
- *Subtract.* The Subtract command is used to remove a termination from a context. The response to the command is used to provide statistics related to the termination's participation in the context. These statistics depend on the type of termination in question. For an RTP termination, the statistics may include items such as packets sent, packets received, and jitter. If the result of a Subtract command is the removal of the last termination from a context, then the context itself is deleted.
- *Move.* The Move command is used to move a termination from one context to another. It should not be used to move a termination from or to the null context because these operations must be performed with the Add and Subtract commands, respectively. The ability to move a termination from one context to another provides a useful tool for accomplishing the call-waiting service.
- *Audit Value.* The Audit Value command is used by the MGC to retrieve current values for properties, events, and signals associated with one or more terminations.

- *Audit Capabilities.* The Audit Capabilities command is used by an MGC to retrieve the possible values of properties, signals, and events associated with one or more terminations. At first glance, this command may appear very similar to the Audit Value command. The difference between them is that the Audit Value command is used to determine the current status of a termination, whereas the Audit Capabilities command is used to determine the possible statuses that a termination might assume. For example, Audit Value would indicate any signals that are currently being applied by a termination, whereas Audit Capabilities would indicate all the possible signals that the termination could apply if required.
- *Notify.* The Notify command is issued by an MG to inform the MGC of events that have occurred within the MG. The events to be reported will have been requested previously as part of a command from the MGC to the MG, such as a Modify command. The events reported will be accompanied by a RequestID parameter to enable the MGC to correlate reported events with previous requests.
- *Service Change.* The Service Change command is used by an MG to inform an MGC that a group of terminations is about to be taken out of service or is being returned to service. The command is also used in a situation where an MGC is handing over control of an MG to another MGC. In this case, the command is first issued from the controlling MGC to the MG to instigate the transfer of control. Subsequently, the MG issues the Service Change command to the new MGC as a means of establishing the new relationship.

9.6.1.3 MEGACO Descriptors

Associated with each command and response are a number of descriptors. These descriptors are effectively the parameters or information elements associated with each command or response. The content of a given descriptor will depend on the termination in question.

Many such descriptors exist, but one in particular is worth noting. This is the *media descriptor,* which describes media streams. It contains two components—the termination state descriptor and the stream descriptor. The stream descriptor consists of three components—the local control descriptor, the local descriptor, and the remote descriptor. This structure can be represented as follows:

- Media descriptor
  - Termination state descriptor
  - Stream descriptor
    - Local control descriptor
    - Local descriptor
    - Remote descriptor

The termination state descriptor indicates whether the termination is currently in service, out of service, or in test. It also provides information about how events detected by the termination are to be handled.

The stream descriptor is identified by a stream ID. Stream ID values are used between an MG and an MGC to indicate which media streams are interconnected. Within a given context, streams with the same stream ID are connected. A stream is created by specifying a new stream ID on a particular termination in a context.

The local control descriptor is used to indicate the current mode of the termination, such as *send-only, receive-only,* or *send-receive,* where these terms refer to the direction from the context to the outside world. Thus the term *receive-only* means that a termination can receive media from outside the context and pass it to other terminations in the context, but it cannot send media to anywhere outside the context.

The local descriptor and remote descriptor are basically SDP session descriptions related to the local end and the far end of a media stream, respectively. Imagine, for example, a VoIP gateway (gateway A) that is communicating with another VoIP gateway (gateway B) across an IP network. The local descriptor for gateway A specifies the media formats that gateway A wants to receive and the address and port number to which that media (i.e., the RTP stream) should be sent. The remote descriptor for gateway A indicates the media formats that gateway B wants to receive and the address and port to which that media should be sent.

9.6.1.4 Call Establishment

Call establishment with MEGACO is based on the foregoing high-level descriptions, and we are in a position to describe how a basic call can be established using MEGACO. Figure 9.13 shows a scenario where a call is to be established between two terminations, T1 and T4, that reside on two different MGs. In this example, the two MGs are controlled by the same MGC.

The MGC has determined, through call-control signaling (not shown), that a call needs to be established between termination T1 on MG A and termination T4 on MG B. It first requests MG A to add T1 to a new context. The fact that it is a new context is indicated by the $ wildcard. It also requests that the MG create a new ephemeral termination (indicated by the wildcard associated with the second Add command) and add that termination to the same context. The MGC specifies that the new termination should be able to receive media from the far end but not send media to the far end. This is reasonable because the MG has not yet received any information as to where the media should be sent.

The MGC also makes a suggestion as to the media coding that the new termination should use. This can be seen in the local descriptor, which contains an SDP description. In this example, the suggestion is that the new termination use audio coded according to AMR and use the dynamic RTP payload type value of 98. Note that the MGC has not specified any IP address or port number because these are associated with a termination that MG A has yet to create. Note also that the session description provided by the MGC is merely a suggestion. The MGC is not required to suggest any format. If it does suggest a format, then the MG should comply with that suggestion, if possible, but it does not have to.

MG A responds to the MGC using the same transaction ID. It indicates that it has created a new context with ContextID 5 1001. It has added termination T1 to the context as requested. It also has created termination T2 and added it to the same context. Associated with termination T2 is an SDP session description. Unlike the suggested session description received from the MGC, this session description (included in the local descriptor) includes an IP address (311.311.1.1) and port number (1199) at which termination T2 expects to receive the RTP stream.

The MGC now requests MG B to set up a new context and add two terminations to that context—termination T4 and a new termination that MG B must create. For the new termination, the MGC makes a suggestion as to the content of the local descriptor. It also specifies the exact content of the remote descriptor. Although the information for

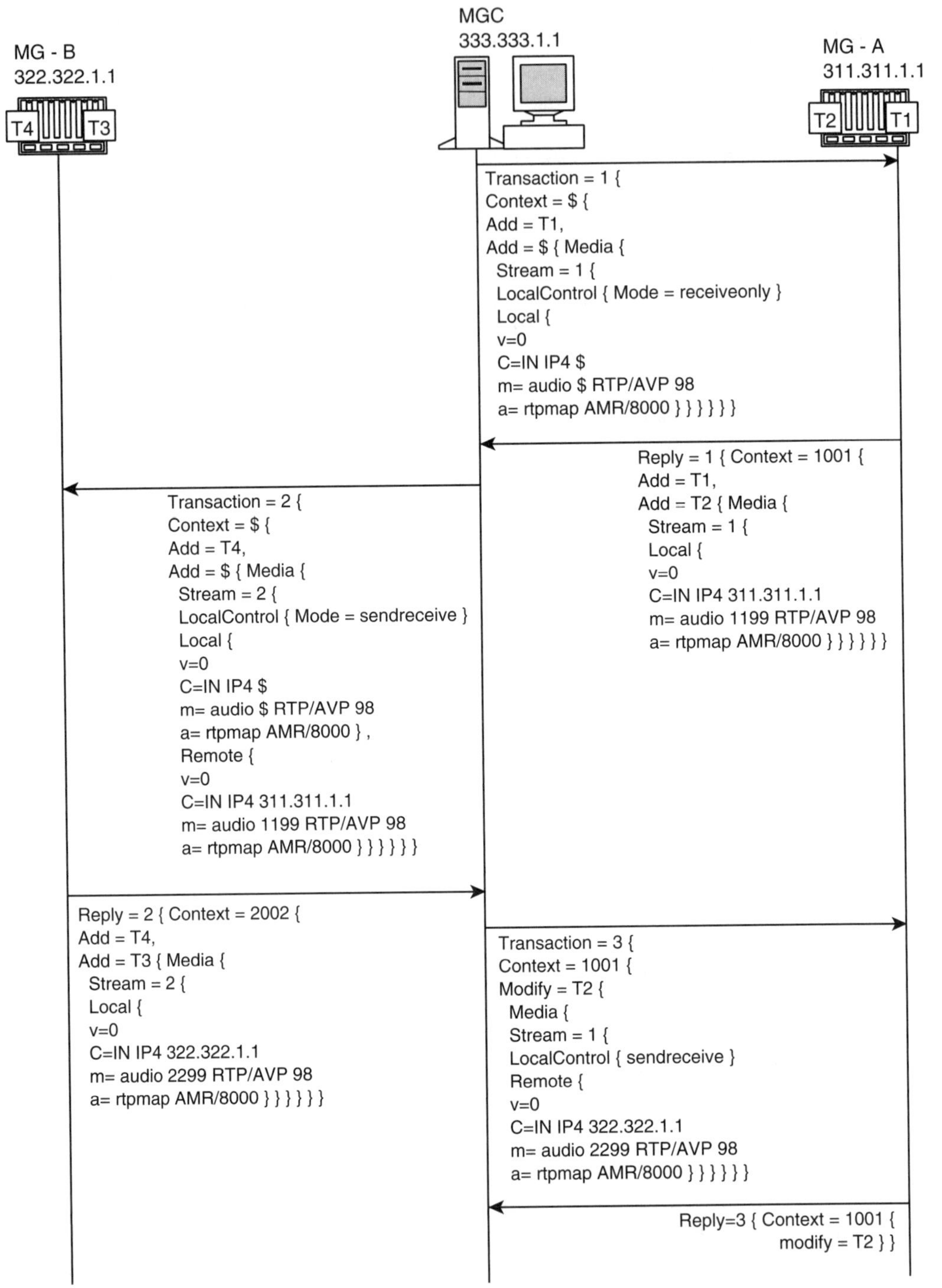

FIGURE 9.13 Call establishment using MEGACO/H.248.

the local descriptor is simply a suggestion, the information in the remote descriptor is what the new termination must use. The content of the remote descriptor is, after all, the content of the local descriptor for termination T2 on MG A. In other words, the local descriptor specifies which media format the new termination should send and where it should send it.

Note the use of the local control descriptor. In this case, the MGC specifies that the mode should be send-receive. This is so because the far end is ready to receive RTP packets and will soon know where to send them, even though it does not know this quite yet.

MG B creates the new context (Context ID 5 2002) and adds termination T4 to that context. It also creates termination T3 and adds it to the context. For the new termination, it specifies a local descriptor, which includes the media format it wants to receive, and the address and port number to which the packets should be sent.

The MGC takes the local descriptor information related to termination T3 and, using the Modify command, sends it to MG A as a remote descriptor for termination T2. It also specifies the mode for termination T2 to be send-receive. Termination T2 now knows where to send RTP packets and has permission to send them.

The chain is now complete. Terminations T1 and T2 are in the same context, so a path exists between them across MG A. Equally, terminations T3 and T4 are in the same context, so a path is created between them across MG B. Finally, T2 and T3 have established a bidirectional RTP stream between them. Thus a path is available from T1 to T4, as originally intended.

9.6.1.5 MEGACO and SIP Interworking

Imagine the case where the two MGs in Figure 9.13 happen to be controlled by separate MGCs. In this case, a protocol needs to be used between the two MGCs. The obvious choice for this protocol is SIP. Once a local descriptor is available at the gateway where the call is being originated, this can be passed in an SIP Invite as an SIP message body. On acceptance of the call at the far end, the corresponding session description is carried back as an SIP message body within the SIP 200 (OK) response and is passed to the originating-side gateway. Once the gateway on the originating side has acknowledged receipt of the remote session description, then the MGC can send an SIP ACK to complete the SIP call setup.

9.7 VoIP and SS7

Although new signaling solutions, such as H.323 and SIP, exist for VoIP networks, the standard in traditional telephony and in mobile networks is SS7. Therefore, if a VoIP-based network is to communicate with any traditional network, not only must it interwork at the media level through media gateways, but it also must interwork with SS7. To support this, the IETF has developed a set of protocols known as *Signaling Transport* (SIGTRAN).

In order to understand SIGTRAN, it is worth considering the type of interworking that needs to occur. Imagine, for example, an MGC that controls one or more MGWs. The MGC is a call-control entity in the network and, as such, uses call-control signaling to and from other call-control entities. If other call-control entities use SS7, then the MGC must use SS7, at least to the extent that the other call-control entities can communicate freely with it. This means that the MGC does not necessarily need to support the whole SS7 stack—just the necessary application protocols.

Consider Figure 9.14, which shows the SS7 stack. The bottom three layers are called the *Message Transfer Part* (MTP). This is a set of protocols responsible for getting

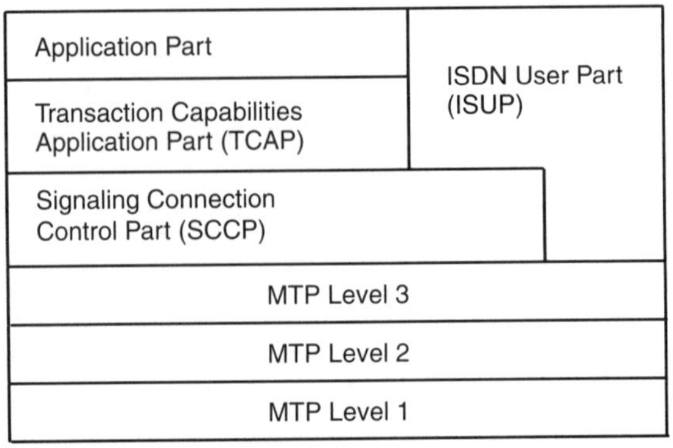

Figure 9.14 SS7 protocol stack.

a particular SS7 message from the source signaling point to the destination signaling point. Above the MTP, we find either the *Signaling Connection Control Part* (SCCP) or the *ISDN User Part* (ISUP). ISUP generally is used for the establishment of regular phone calls. SCCP also can be used in the establishment of regular phone calls, but it is used more often for the transport of higher-layer applications, such as the GSM *Mobile Application Part* (MAP) or the *Intelligent Network Application Part* (INAP). In fact, most such applications use the services of the *Transaction Capabilities Application Part* (TCAP), which, in turn, uses the services of SCCP.

SCCP provides an enhanced addressing mechanism to enable signaling between entities even when those entities do not know each other's signaling addresses (known as *point codes*). This addressing is known as *global title addressing*. Basically, it is a means whereby some other address, such as a telephone number, can be mapped to a point code either at the node that initiated the message or at some other node between the originator and destination of the message.

Figure 9.15 provides some examples of communication between different SS7 entities. Consider scenario A. In this case, the two entities, represented by point codes 1 and 4, communicate at Layer 1. At each layer, a peer-to-peer relationship exists between the two entities. Scenario B has a peer-to-peer relationship at Layer 1, Layer 2, and Layer 3 between point codes 1 and 2, 2 and 3, and 3 and 4. At the SCCP layer, a peer-to-peer relationship exists between point codes 1 and 2 and between point codes 2 and 4.

At the TCAP and Application layers, a peer-to-peer relationship can take place only between point codes 1 and 4. In other words, the application at point code 1 is only aware of the TCAP layer at point code 1 and the application layer at point code 4. Similarly, the TCAP layer at point code 1 is aware only of the application layer above it, the SCCP layer below it, and the corresponding TCAP layer at point code 4. It is not aware of any of the MTP layers. Equally, if we consider communication between point codes 2 and 4, the SCCP layer at each point code knows only about the layer above (TCAP), the layer below (MTP3), and the corresponding SCCP peer. As far as the SCCP layers are concerned, nothing else exists. Therefore, SCCP neither knows nor cares that point code 3 exists.

Consider scenario C, where point code 3 is replaced by a gateway that supports standard SS7 on one side and an IP-based MTP emulation on the other side. Point code 4 does not support the lower SS7 layers at all—just an MTP emulation over IP.

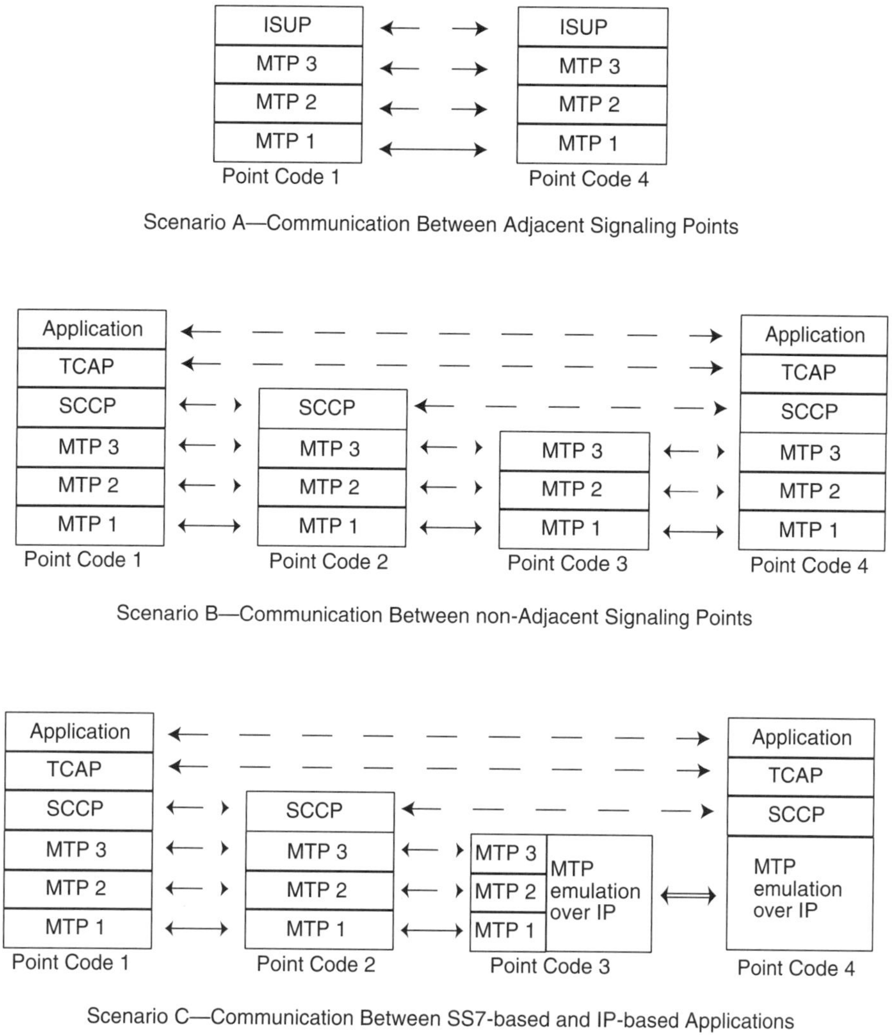

FIGURE 9.15 Example SS7 communication scenarios.

Provided that the MTP emulation at point code 4 appears to the SCCP layer as standard MTP, then the SCCP layer does not care, nor do any of the layers above the SCCP. Equally, the SCCP layers at point code 1 and 2 do not care. Consequently, it is possible to implement SS7-based applications at point code 4 without implementing the whole SS7 stack. This is the concept behind the SIGTRAN protocol suite.

9.7.1 The SIGTRAN Protocol Suite

Figure 9.16 shows the SIGTRAN protocol suite and the relationship between the SIGTRAN protocols and standard SS7 protocols. Above IP, we find a protocol known as the *Stream Control Transmission Protocol* (SCTP). The primary motivation behind development of the SCTP is the fact that neither UDP nor TCP offers both the speed and

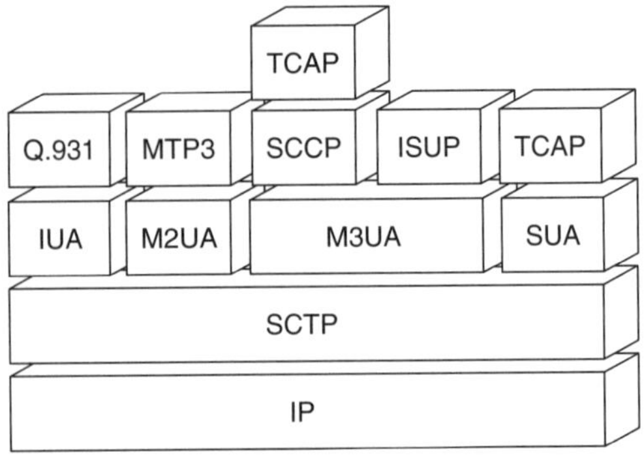

FIGURE 9.16 SIGTRAN protocol suite.

reliability required of a transport protocol used to carry signaling. The design of SCTP is an attempt to make such reliability and speed available to the users of SCTP.

In the SCTP specification, such a user is known as an *Upper Layer Protocol* (ULP). A ULP can be any of the protocols directly above the SCTP layer, as illustrated in Figure 9.16. Each of the protocols above the SCTP is an adaptation layer. For example, M3UA is the MTP3 User Adaptation Layer. Thus we could have ISUP over M3UA over SCTP. Each of the adaptation layers uses the same primitives to and from the layer above as are used by the equivalent SS7 layer. Thus, the layer above does not see any difference between the adaptation layer and its SS7 equivalent. Furthermore, if we have ISUP over M3UA, the ISUP layer believes the M3UA to be standard MTP3 and does not know that the transport is IP-based.

The following adaptations layers are defined:

- *SS7 MTP2-User Adaptation Layer* (M2UA) provides adaptation between MTP3 and SCTP. It provides an interface between MTP3 and SCTP such that standard MTP3 may be used in the IP network without the MTP3 application software realizing that messages are being transported over SCTP and IP instead of MTP2. For example, a standard MTP3 application implemented at an MGC could exchange MTP3 signaling network management messages with the external SS7 network. In the same manner that MTP2 provides services to MTP3 in the SS7 network, M2UA provides services to MTP3 in the IP network.
- *SS7 MTP3-User Adaptation Layer* (M3UA) provides an interface between the SCTP and those applications that typically use the services of MTP3, such as ISUP and SCCP. M3UA and SCTP enable seamless peer-to-peer communication between MTP3 user applications in the IP network and identical applications in the SS7 network. The application in the IP network does not realize that SCTP over IP transport is used instead of typical SS7. In the same manner that MTP3 provides services to applications such as ISUP in the SS7 network, M3UA offers equivalent services to applications in the IP network.
- *SS7 SCCP-User Adaptation Layer* (SUA) provides an interface between the SCCP user applications and the SCTP. Applications such as TCAP use the services of SUA in the same way that they use the services of SCCP in the SS7 network. In fact, those applications do not know that the underlying transport is different in

any way. Hence, transparent peer-to-peer communication can take place between applications in the SS7 network and applications in the IP network.

- *ISDN Q.921-User Adaptation Layer* (IUA) is the SIGTRAN equivalent of the Q.921 Data-Link Layer that is used to carry Q.931 ISDN signaling. Thus, Q.931 messages may be passed from the ISDN to the IP network, with identical Q.931 implementations in each network, and neither of them will recognize any difference in the underlying transport.

9.7.1.1 Stream Control Transmission Protocol (SCTP)

The SCTP provides for the reliable and fast delivery of signaling messages. It is reliable because it includes mechanisms for the detection and recovery of lost or corrupted messages. It is faster than TCP, however, because it avoids head-of-line blocking, which can occur with TCP, and it also has more efficient retransmission mechanisms than TCP.

Head-of-line blocking is avoided in the SCTP through the use of streams. A *stream* is a logical channel between SCTP endpoints. It also may be thought of as a sequence of user messages between two SCTP users. When an association is established between endpoints, part of the establishment of the association involves each endpoint specifying how many inbound streams and how many outbound streams are to be supported. If we think of a given association as a one-way highway between endpoints, then the individual streams are analogous to the individual traffic lanes on that highway. The advantage of the stream concept is that resources (or queues) are allocated individually to each stream rather than to the complete set of packets that might pass between two endpoints. Consequently, a message from one stream does not have to wait in a queue behind a message from another stream.

Retransmission in SCTP is based on the fact that SCTP packets carrying user data (known as *chunks*) include a *Transmission Sequence Number* (TSN). The receiver of the chunks checks to make sure that all chunks have been received by ensuring that no gap exists in the TSNs. If a gap is found, then the SCTP enables the receiver to specify which TSNs are missing, and it is only those TSNs that need to be retransmitted, which is more efficient than TCP.

Consider, for example, the situation depicted in Figure 9.17. Chunks with TSNs 1 to 4 have been received correctly, the chunk with TSN 5 is missing, the chunk with TSN 9 is missing, and the chunks with TSNs 8 and 11 have been received twice. If TCP were to deal with this situation, then all chunks from TSN 5 onwards would be retransmitted. The SCTP, however, has the means for the receiver to clearly specify to the sender what is missing and what is duplicated so that the minimum retransmission takes place.

Not only does SCTP support fast transmission and efficient retransmission, but it also supports congestion avoidance and network-level redundancy. Congestion avoidance is achieved through the use of a parameter in SCTP messages called the *Advertised Receiver Credit Window.* This parameter indicates to the far end how much buffer space the receiver has for the receipt of new messages. This helps to avoid flooding a receiver with more messages than it can handle. Redundancy is achieved through the fact that a given endpoint can be logically distributed across multiple platforms with multiple IP addresses. If a given platform fails, then another platform can take over. SCTP includes messages for monitoring the reachability of a given endpoint and failover messages for one endpoint to indicate to another that a different IP address should be used for future messages.

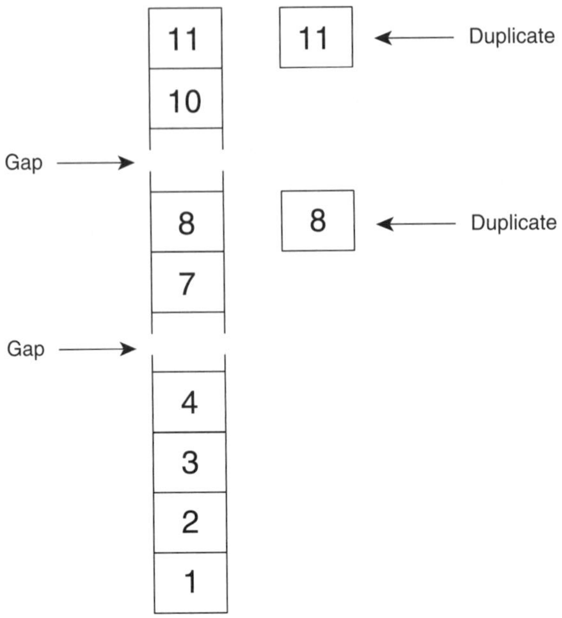

Figure 9.17 Example of lost and duplicated SCTP chunks.

9.7.2 Example of SIGTRAN Usage

Figure 9.18 provides an example of how IP and SS7 networks can interwork using SIGTRAN. The IP-to-SS7 connectivity diagram shows how an SIP device could be connected to an MG and an MGC such that it can communicate with a standard telephone in the PSTN. The IP-to-SS7 protocol interworking diagram shows how protocol interworking can take place via a *signaling gateway* (SG). The net effect is that the nodes, such as a PSTN switch, in the SS7 network can communicate with the SIP terminal via the SG and MGC and MG without realizing that the SIP terminal is not a standard telephone connected to a standard SS7-enabled switch.

Of course, the MGC must be able to translate SIP messages to ISUP messages and vice versa. Although these two protocols are different, the messages of SIP and those of ISUP do serve similar functions, and it is possible to map from one protocol to the other. For example, the ISUP *Initial Address Message* (IAM) maps quite well to the SIP Invite. The SIP 183 (Session Progress) response, an extension to the original SIP specification, maps to the ISUP *Address Complete Message* (ACM). The SIP 200 (OK) response maps to the ISUP *Answer* (ANS) message.

9.8 VoIP Quality of Service

Perhaps the biggest issue with VoIP is ensuring that the *quality of service* (QoS) is comparable with the QoS achieved in traditional circuit-switched telephony. As we have seen, IP and UDP provide no quality guarantees whatsoever. Although RTP and RTCP provide QoS-related information (such as jitter, number of lost packets, and so

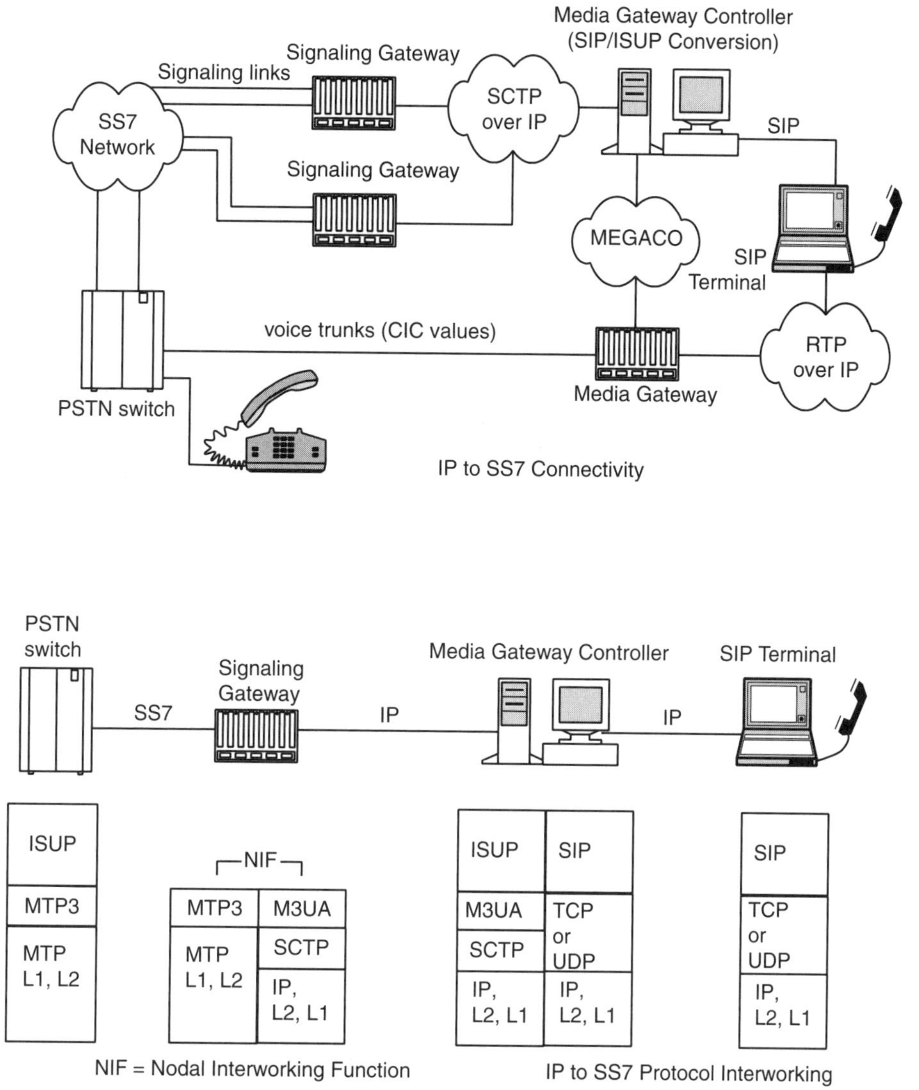

FIGURE 9.18 IP/SS7 interworking example.

on), they do not provide any assurance of quality. In order to ensure that VoIP is not a low-quality service, specific solutions must be implemented in the network.

One way to help ensure that VoIP offers high quality is to ensure that more than enough bandwidth is available—in terms of both throughput on transmission facilities and processing power within routers. By overprovisioning the network, one can reduce the likelihood of congestion and thereby improve quality. This, however, is an expensive option that leaves much of the network capacity unused much of the time. Moreover, it does not guarantee quality. Thus one needs technical solutions within the network. Additionally, LTE has unique QoS capabilities that can be leveraged to support VoIP. The LTE QoS parameters can be used in place of the traditional method of overprovisioning.

The following subsections provide a brief overview of some QoS techniques. For more detailed explanations, the reader is referred to the applicable IETF specifications.

9.8.1 The Resource Reservation Protocol

Resource reservation techniques for IP networks are specified in RFC 2205, the *Resource Reservation Protocol* (RSVP), which is part of the IETF integrated services suite. It is a protocol that enables resources to be reserved for a given session or sessions prior to any attempt to exchange media between the participants. Of the solutions available, it is the most complex, but it is also the solution that comes closest to circuit emulation within the IP network. It provides strong QoS guarantees, a significant granularity of resource allocation, and significant feedback to applications and users.

RSVP currently offers two levels of service. The first is *guaranteed*, which comes as close as possible to circuit emulation. The second is *controlled load*, which is equivalent to the service that would be provided in a best-effort network under no-load conditions.

Basically, RSVP works as depicted in Figure 9.19. A sender first issues a Path message to the far end via a number of routers. The Path message contains a *traffic specification* (TSpec) that provides details of the data that the sender expects to send, in terms of the bandwidth requirement and packet size. Each RSVP-enabled router along the way establishes a "path state" that includes the previous source address of the Path message (i.e., the next hop back toward the sender). The receiver of the Path message responds with a *Reservation Request* (RESV) that includes a flowspec. The flowspec includes a Tspec and information about the type of reservation service requested, such as controlled-load service or guaranteed service.

The RESV message travels back to the sender along the same route that the Path message took (in reverse). At each router, the requested resources are allocated, assuming that they are available and that the receiver has the authority to make the request. Finally, the RESV message reaches the sender with a confirmation that resources have been reserved.

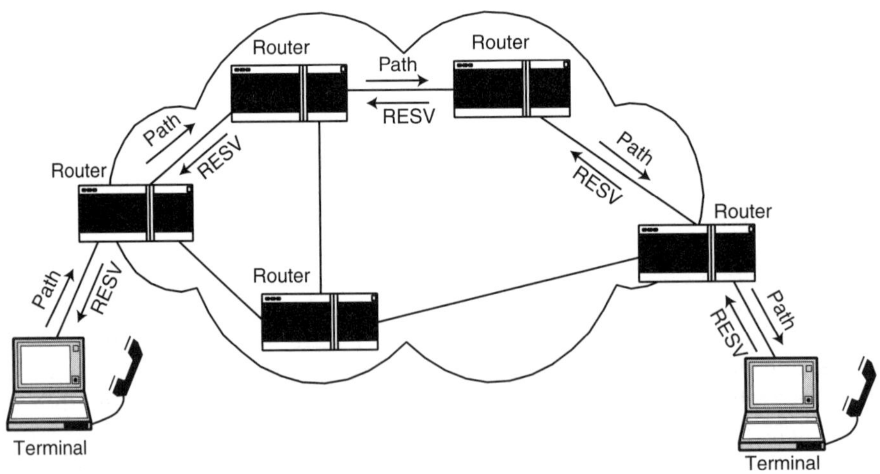

Figure 9.19 Resource reservation.

One interesting point about RSVP is that reservations are made by the receiver, not by the sender of data. This is done in order to accommodate multicast transports where there may be large numbers of receivers and only one sender.

Note that RSVP is a control protocol that does not carry user data. The user data (e.g., voice) is transported later using RTP. This occurs only after the reservation procedures have been performed. The reservations that RSVP makes are *soft,* which means that they need to be refreshed on a regular basis by the receivers(s).

9.8.2 Differentiated Service

Differentiated Service (DiffServ) is a relatively simple means for prioritizing different types of traffic. The DiffServ protocol is described in RFC 2475. Basically, DiffServ makes use of the IPv4 *Type of Service* (TOS) field contained in the IPv4 header and the equivalent IPv6 Traffic Class field. The portion of the TOS/Traffic Class field used by DiffServ is known as the *DS field.* This field is used in specific ways to mark a given stream as requiring a particular type of forwarding. The type of forwarding to be applied is known as *per-Hop Behavior* (PHB), of which DiffServ defines two types. These are *Expedited Forwarding* (EF) and *Assured Forwarding* (AF).

EF is specified in RFC 2598. It is a service whereby a given traffic stream is assigned a minimum departure rate from a given node, one that is greater than the arrival rate at the same node, provided that the arrival rate does not exceed a pre-agreed maximum. This ensures that queuing delays are removed. Since queuing delays are a major cause of end-to-end delay and are the main cause of jitter, this ensures that delay and jitter are minimized. In fact, EF can provide a service that is equivalent to a virtual leased line.

AF is defined in RFC 2597. This is a service whereby packets from a given source are forwarded with a high probability, provided that the traffic from that source does not exceed some pre-agreed maximum. AF defines four classes, with each class allocated a certain amount of resources (buffer space and bandwidth) within a router. Within each class, a given packet may have one of three drop rates. At a given router, if congestion occurs within the resources allocated to a given AF class, then the packets with the highest drop rate values will be discarded first so that packets with a lower drop rate value receive some protection. In order to work well, it is necessary that the incoming traffic does not have packets with a high percentage of low drop rates. After all, the purpose is to ensure that the highest-priority packets get through in the case of congestion, and this cannot happen if all the packets have the highest priority.

9.8.3 Multi-Protocol Label Switching

Label switching is something that has seen significant interest from the Internet community, and significant effort has been made to define a protocol called *Multi-Protocol Label Switching* (MPLS). In some ways, it is similar to DiffServ in that it marks traffic at the entrance to the network. However, the primary function of the marking is not to allocate a priority within a router but to determine the next router in the path from the source to the destination.

MPLS involves the attachment of a short label to a packet in front of the IP header. This is like inserting a new layer between the IP layer and the underlying link layer of the OSI model. The label contains all the information that a router needs to forward a packet. The value of a label may be used to look up the next hop in the path and forward to the next router. The difference between this and standard IP routing is that the match is an exact one and is not a case of looking for the longest match (i.e., the match with the longest subnet mask). This enables faster routing decisions within routers.

The label identifies something called a *Forwarding Equivalence Class* (FEC). This term is chosen because it means that all packets of a given FEC are treated equally for the purposes of forwarding. All packets in a given stream of data, such as a voice call, will have the same FEC and will receive the same forwarding treatment. It is therefore possible to ensure that the forwarding treatment applied to a given stream can be set up such that all packets from *A* to *B* follow the same path. If this stream has a particular bandwidth requirement, then that bandwidth can be allocated at the start of the session. This can ensure that a given stream has the bandwidth that it needs and that the packets that make up the stream arrive in the same order as transmitted. Hence, a higher QoS is provided.

In many ways, MPLS is as much of a traffic engineering protocol as it is a QoS protocol. It is somewhat analogous to the establishment of virtual circuits in ATM and can lead to similar QoS benefits. It helps to provide QoS by helping to better manage traffic. Whether it should be called a traffic engineering protocol or a QoS protocol hardly matters if the end result is better QoS.

References

Camarillo, Gonzalo, *SIP Demystified*, McGraw-Hill, New York, 2002.

Collins, Daniel, *Carrier Grade Voice over IP*, 2d ed., McGraw-Hill, New York, 2003.

IETF Draft: SS7 MTP2-User Adaptation Layer, Work in Progress.

IETF Draft: SS7 MTP3-User Adaptation Layer (M3UA), Work in Progress.

IETF Draft: SS7 SCCP-User Adaptation Layer (SUA), Work in Progress.

IETF RFC 768: User Datagram Protocol (STD 6).

IETF RFC 791: Internet Protocol (STD 5).

IETF RFC 793: Transmission Control Protocol (STD 7).

IETF RFC 1889: RTP: A Transport Protocol for Real-Time Applications.

IETF RFC 1890: RTP Profile for Audio and Video Conferences with Minimal Control.

IETF RFC 2205: Resource ReSerVation Protocol (RSVP)—Version 1: Functional Specification.

IETF RFC 2327: SDP: Session Description Protocol.

IETF RFC 2475: An Architecture for Differentiated Services.

IETF RFC 2543: Session Initiation Protocol (SIP).

IETF RFC 2597: Assured Forwarding PHB.

IETF RFC 2598: An Expedited Forwarding PHB.

IETF RFC 2701: Media Gateway Control Protocol (MGCP), Version 1.0.

IETF RFC 2719: Architectural Framework for Signaling Transport.

IETF RFC 2805: Media Gateway Control Protocol Architecture and Requirements.

IETF RFC 2960: Stream Control Transmission Protocol.

IETF RFC 3031: Multiprotocol Label Switching Architecture.

IETF RFC 3051: MEGACO Protocol.

ITU-T H.225.0: Call-Signalling Protocols and Media Stream Packetization for Packet-Based Multimedia Communication Systems.

ITU-T H.245: Control Protocol for Multimedia Communication.

ITU-T H.248: Media Gateway Control Protocol.

ITU-T H.323: Packet-Based Multimedia Communications Systems.

ITU-T Q.931: ISDN User-Network Interface Layer 3 Specification for Basic Call Control.

Mueller, Stephen, *API's and Protocols for Convergent Network Service*, McGraw-Hill, New York, 2002.

Shepard, Steven, *IMS Crash Course*, McGraw-Hill, New York, 2006.

Sinnreich, Johnston, *Internet Communications Using SIP*, Wiley, New York, 2001.

CHAPTER 10
WiFi

The mobile wireless community is upgrading and deploying a host of wireless platforms that are focused on lowering the overall cost of ownership for the operator as well as provide new and improved packet services. WiFi is also being used as an effective method to off-load wireless networks and will continue to play a major role in the end users broadband experience.

Mobile phones are now becoming so commonplace that it is the primary communication device for a customer or family with traditional landline services either not being used or simply providing backup. While concurrently the proliferation of WiFi, IEEE 802.11, is beginning to influence the customers overall data experience by untethering them. WiFi has seen widespread acceptance in both commercial and residential markets.

The wireless data revolution is not relegated to mobility by itself. There are numerous types of wireless data systems that can and do complement a mobile wireless network. However, unlicensed broadband is a critical component in the rollout of both *Third-Generation* (3G) and *Fourth-Generation* (4G) networks. Unlicensed broadband includes a host of radio access methods that use unlicensed spectrum.

Some of the more prevalent systems for wireless data involve

- WiFi
- Bluetooth
- *Ultrawide Band* (UWB)

This chapter will focus on WiFi as the leading unlicensed broadband or rather license-exempt broadband access technology.

10.1 WiFi (802.11)

There are currently a plethora of radio access methods supporting broadband using license-exempt spectrum. However, WiFi (802.11) has seen widescale commercial success for mobile data and is the primary common access method that can be leveraged by wireless mobile operators to augment their service offerings. In most cases to provide mobile data services, the *Radio Access Network* (RAN) becomes secondary to the data entry method and the physical display, both of which fall under the form factor of the device.

WiFi is a wireless LAN based on the 802.11 standards and is a wireless access method used to connect devices and appliances to either a LAN or the Internet via a wireless connection. WiFi is integrated with 2.5G/3G and 4G networks. Specifically

WiFi will continue to provide a robust off-loading technique for data services freeing up licensed spectrum capacity.

WiFi, however, is not classified as a *Wireless Mobile Broadband Network*, although recent advances in 802.11 have introduced mobility features to improve the user's experience. Unlike 3G and 4G mobility networks, WiFi is designed for smaller coverage areas.

Radio access in WiFi is achieved by using an *Access Point* (AP) to connect wireless devices to a LAN/WAN. WiFi is also referred to as a *Wireless Local Area Network* (WLAN) and can act as a stand-alone network, can be integrated into an existing corporate LAN, or act as a WiFi hot spot to off-load 3G/4G wireless data traffic; depending on the services and requirements, a WLAN can require several levels of authentication or require no authentication.

There are two basic components to an 802.11 WLAN or hot spot as shown in Figure 10.1.

- Access Point (AP)
- Wireless Adaptors (CPE)

Access point has the following characteristics:

- Provides Wireless Access for Wireless Adaptors (CPE)
- Provides Authentication for wireless adaptors
- Provides in some cases access to a wired LAN
- Is a hot spot
- Contains all the functions of wireless adaptors
- Can operate in a trusted or untrusted mode
- Can be combined with other APs

Wireless Adaptor (CPE) has the following characteristics:

- Wireless adaptor provides radio access between AP and other WiFi CPE.
- Wireless adaptors can be integrated into an appliance.
- Wireless adaptors can be external to an appliance (USB, ExpressCard, PCMCIA, etc.).

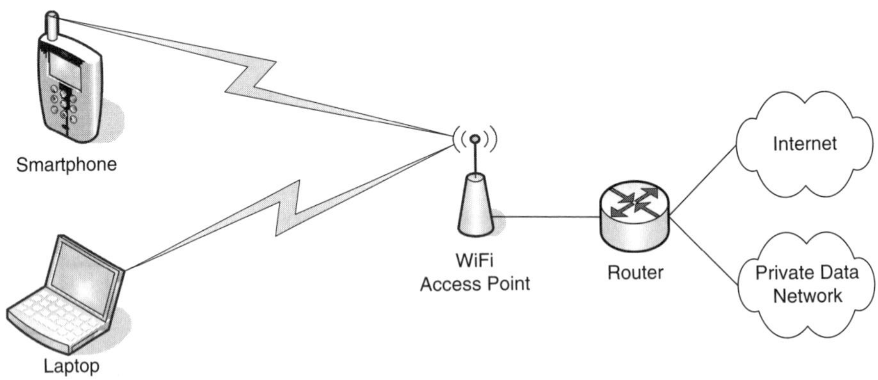

Figure 10.1 WiFi components.

As expected, WiFi like any wireless access technology, has many advantages and disadvantages:

- Advantages
  - Wireless LAN
  - Established standards
  - Universally accepted (laptops, PDA)
  - Excellent broadband traffic off-load platform for 3G/4G
- Disadvantages
  - Limited coverage from AP
  - Security
  - Unlicensed frequency

WiFi equipment costs for the end device as well as the infrastructure, APs, is relatively inexpensive, unlike its 2.5G/3G and 4G counterparts. Its service is easy to deploy and affordable. Additionally, the *Wireless Ethernet Compatibility Alliance* (WECA) provides a certification process for all 802.11 components in order to ensure that the AP and CPE from multiple vendors can interoperate with each other.

However, as always the real issue regarding ease of deployment and affordability is that it depends on what the objective or service offering really is. For instance, if WiFi is deployed for a home network for linking a computer or computers to the Internet then the low cost and ease of installation is real. But if the WiFi network is to support more than the home usage and provide ubiquitous access to public and private packet data networks, that is, the Internet and corporate LANs, then the cost and installation concerns take on another level of complexity. Add in the component of wireless mobile services and the decisions that need to be made, and then it becomes daunting. They are daunting because not all services are the same and you need to have prior knowledge of what you want to do before selecting the access method.

Some of the key issues that need to be addressed or factored into the decision process are

- Coverage
- Roaming
- Backhaul
- Security
- Device

The coverage question comes down to a simple question: where are you going to use the service and how is it going to be used? If it is desired to have medical data available to visiting RNs for a large metropolitan area, then the use of WiFi by itself will need to be augmented by a 2.5G/3G, 4G system. However, if the objective is to have coverage in the visiting conference area, then a stand-alone WiFi system could be used.

The roaming issue that needs to be answered is akin to the issue of coverage. It is desired to have the workforce use different WiFi networks or hot spots that may or may not be integrated into a 2.5G/3G, 4G network. 802.11, however, elevates some of the roaming issue with multiple APs in different locations due to the ability to broadcast a common *Service Set Identifier* (SSID) establishing an *Extended Service Set* (ESS).

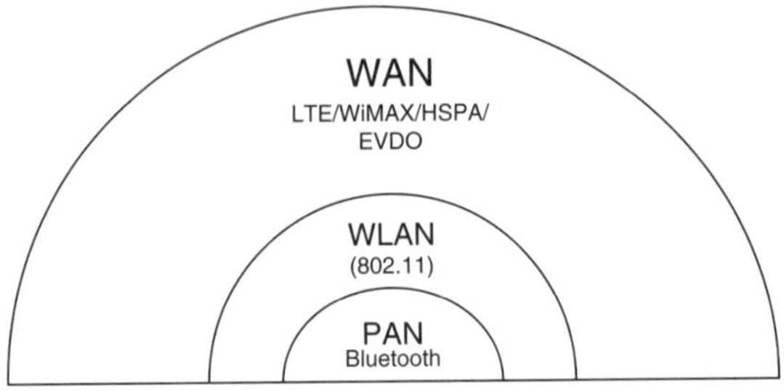

Figure 10.2 WAN, WLAN, PAN.

The issue of backhaul is important from a wireless service provider's aspect as well as the end user. The service provider needs to determine and support the requisite backhaul transport level to ensure the service offered, at the defined data rate, is achievable and can be sustained via wireless or wired connections. However, if the WiFi AP is a corporate LAN or a hot spot then best effort service may be the desired service level. As mentioned previously, wireless operators are basically compelled to try and off-load heavy data traffic onto another RAN to preserve or better utilize the RAN for less bandwidth intensive services or make them available for true mobility needs.

Figure 10.2 illustrates where WiFi plays in the wireless access scheme.

There are numerous standards that make up 802.11 with the most popular being referred to as WiFi. This chapter's objective is to provide an overview of the various flavors of 802.10. In addition, this chapter will introduce and better define the interaction of 802.11 and wireless mobility.

The convergence of 802.11 with wireless mobility has been described as the real "killer application." The killer application for mobility is that it will truly allow the subscriber to take advantage of all the applications available on the *World Wide Web* (WWW) while at the office or home office or on the road at some unknown location, provided, of course, that there is coverage. The issue of security and provisioning to make this a reality is not a trivial matter if true transparency is desired with the intranet of a company by its sales and support staff.

It is not intended to present a detailed definition of the 802.11 specifications, since they can be found in the appropriate IEEE documents, www.ieee.org. Instead this chapter will attempt to make sense out of the plethora of information and options available for a mobile wireless operator to integrate 802.11 into the wireless data and overall service offerings.

10.2 802.11 Standards

The 802.11 standard is part of a family of IEEE standards devoted to defining local and wide area networks. This family of standards deals with the physical and data link layers as defined by the *International Organization for Standardization* (ISO), *Open Systems Interconnection* (OSI), and basic *Reference Model* (RM) (ISO/IEC 7498-1:1994).

The OSI/RM is used for the following reasons:

- The OSI/RM provides a common vehicle for understanding and communicating the various components and interrelationships of the standards.
- The OSI/RM defines terms.

Standard	Topic Addressed
802	Overview and architecture of the family of standards
802.1B	LAN/MAN remote management
802.1D	MAC bridging (interconnecting of 802 LANS)
802.1E	System loading and QoS
802.1F	Procedures for management information of 802 LANs
802.1G	Remote MAC bridging (non-LAN technologies)
802.2	Logical link control
802.3	CSMA/CD Ethernet physical layer
802.4	Token bus access method and physical layer
802.5	Token ring access method and physical layer
802.6	Dist. queue dual bus access method
802.9	Integrated services LAN interface MAC and physical layers
802.11	Wireless LAN MAC and physical layers (WiFi)
802.12	Demand priority access method
802.7	Broadband LANs
802.14	Cable TV protocols
802.15	Wireless PAN
802.16	Wireless LAN/MAN
802.20	3G wireless data

Table 10.1 802 Standards

- The OSI/RM provides a convenient framework to aid in the development and enhancement of the standards.
- The use of the OSI/RM facilitates a higher degree of interoperability than might otherwise be available.

For a detailed view of this family of standards, see IEEE STD 802-2001. However, Table 10.1 is a short overview of pertinent 802 standards.

802.11 is part of the 802 standards that define the physical and link layers for various media access to be transparent to upper levels of the OSI RM. The 802.11 standard describes the functions and services required by an IEEE 802.11 compliant device including the requirements for privacy, security, and authentication. This basically allows for the interconnection of various LANs without regard to their physical characteristics such as being wireless.

Other standards of interest associated with WiFi include 802.3, which describes the Ethernet that is employed in many wired LANs. However, the complete definition of the 802 specifications family can be found in the appropriate IEEE documents (www.ieee.org).

10.3 WiFi Protocols

Several protocols fall into the WiFi/WLAN arena, not all of which are compatible with each other, leaving the possibility of having local islands being established. The most prevalent WLAN protocol is IEEE 802.11, but Bluetooth is also being referred to as a WLAN protocol. 802.11 is an IEEE specification that is made up of several standards; some of the more prevalent are 802.11a, 802.11b (WiFi), 802.11g, and 802.11n, to mention a few.

What is interesting is that 802.11a operates in the 5-GHz *Unlicensed National Information Infrastructure* (UNII) band, whereas 802.11b and 802.11g operate in the 2.4-GHz *Industrial, Scientific, and Medical* (ISM) band along with Bluetooth, which also operates in the 2.4-GHz ISM band. However, 802.11n operates in both the UNII and ISM bands. 802.11g specifically is meant to increase the data rate to 54 Mbps while providing backward-compatibility for 802.11b (WiFi) equipment. What this means is that 802.11g equipment operating in the 2.4-GHz band can operate at speeds previously enjoyed by 802.11a equipment in the 5-GHz band. 802.11n can operate in both the 2.4-G and 5-GHz bands and can achieve speeds of 540 Mbps due to improved modulation techniques and increased channel bandwidth.

To complicate matters, there are a host of other 802 specifications, all of which either exist or are in the process of being standardized. The 802.11 specifications were designed initially as a wireless extension for a corporate LAN for enterprise applications, and numerous devices have been manufactured to this specification. For example, the 802.11b protocol is a shared medium and uses *Collision Sense Multiple Access/Collision Avoidance* (CSMS/CA), which is a listen-before-talk protocol with standards for collision sense multiple access and collision avoidance.

Table 10.2 shows a simple comparison between key 802.11 protocols along with Bluetooth. Both 802.11b and Bluetooth use the ISM band, although their format and purpose are different. However, 802.11a operates in the UNII band and can operate at a much greater *Effective Radiated Power* (ERP). Basically, 802.11 devices are meant to cover a wider area than Bluetooth devices, and 802.11 devices have the potential of higher throughput. The data rate in the chart for 802.11a and 802.11b shows a range in speeds that are, of course, dependent on the modulation format used, power, and the interference experienced.

Why 802.11 is important for wireless mobility is because it provides direct mobile data interoperability between the LAN of a corporation and the wireless operator's system. Inclusion of the ability to extend the corporate *Internet Protocol–Public Branch Exchange* (IP-PBX) has great potential. Presently, there have been many demonstrations and some operational systems involving this integration of wireless mobility and wireless LANs that require the need for *Application-Specific Protocols* (ASPs) to enable the interoperability.

There is also another specification, HiperLAN/2, that is a WLAN specification that has been developed under *European Telecommunications Standards Institute* (ETSI). HiperLAN/2 has similar physical layer properties as 802.11a in that it uses *Orthogonal Frequency-Division Multiplexing* (OFDM) and is deployed in the 5-GHz band. However, the *Media Access Control* (MAC) layers are different, hence the different technology specification, in that HiperLAN uses a *Time-Division Multiple Access* (TDMA) format as compared with 802.11a, which uses OFDM. 802.11 has four main components: 802.11b, 802.11g, 802.11a, and 802.11n.

WLAN	801.11a	802.11b	802.11g	802.11n	Bluetooth
Frequency band	5-GHz UNII	2.4-GHz ISM	2.4-GHz ISM	5-GHz UNII 2.4-GHz ISM	2.4-GHz ISM
Max data rate	54 Mbit/s	11 Mbit/s	54 Mbit/s	540 Mbit/s	1 Mbit/s (v1.2) 3 Mbit/s (v2.0) 24Mbps (v3.0)
Number of none overlapping channels	11	3	3	3/11	79
Range	*	50 m	50 m	50 m*	1–10 m
Modulation/ coding	OFDM	CCK,HR/DSS HR/DSS/short HR/DSSS/ PBCC	DSSS/CCK or OFDM	OFDM	FHSS
Power	0.05/0.25/1W	+20 dBm		0.05/0. 25/1W UNII +20 dBm ISM	0 dBm

Table 10.2 WiFi Comparison Table

* If used with an external antenna, the WLAN can be extended beyond the immediate office environment.

10.3.1 802.11b

This is the most widespread variant of the standards. It is the one typically found in "hot spots," with 802.11g quickly becoming more dominant. This 802.11b standard was published in 1999 and has been adapted widely by manufacturers of infrastructure, such as APs, routers, and bridges. It is also adapted widely by vendors of interface devices for laptops, desktops, and PDAs. 802.11b operates in the ISM band at 2.4 GHz and specifies data rates of up to 11 Mbps. The standard specifies *Direct Sequence Spread Spectrum* (DSSS) and several modulation schemes, including *Complimentary Code Keying* (CCK) and *Packet Binary Convolutional Coding* (PBCC).

10.3.2 802.11g

WiFi specification 802.11g provides higher data rates (up to 54 Mbps) than 802.11b. The 802.11g standard employs DSSS/*Frequency Hopping Spread Spectrum* (FSSS) and OFDM. 802.11g is backward-compatible with 802.11b. This means that any 802.11g device must be able to coexist with 802.11b devices. An 802.11g-equipped laptop must work in an 802.11b Access Point (AP) coverage area. In addition, if an 802.11b laptop comes into an 802.11g AP coverage area, then 802.11g AP must be able to serve the device. A drawback to this is the fact that when an 802.11b device is detected in the 802.11g serving area, additional overhead is introduced. This overhead diminishes actual throughput by as much as 25 percent.

10.3.3 802.11a

WiFi systems using the 802.11a specification operate in UNII band, which enables systems using this particular network to operate not only at higher speeds but also at high power, enabling more reach. 802.11a operates in the UNII band at 5 GHz and uses OFDM as its modulation scheme. 802.11a is designed to provide data rates of up to 54 Mbps.

It is important to note that while 802.11a and 802.11b/802.11g are not compatible, it is not unusual to use them both in an enterprise network. Most users may be employing 802.11b/802.11g, whereas power users may be assigned 802.11a. These networks would be overlapping and not interoperating. Although not required by any standard, APs are available that offer both 802.11b/802.11g and 802.11a.

10.3.4 802.11n

The next generation for 802.11 is 802.11n. The 802.11n protocol is designed to effectively replace 802.11a, b, and g for local area networking. 802.11n enables speeds of 540 Mbps through improved modulation schemes and increased channel bandwidth. The increased channel bandwidth is achieved by combining two channels therefore increasing the bandwidth from 20 MHz to 40 MHz.

In addition, 802.11n uses multiple antennas to both send and receive information. The multiple antenna system is typically referred to as *Multiple Input Multiple Output* (MIMO). The use of MIMO not only increases the range of the 802.11n network, but also the throughput as well.

802.11n systems are also meant to be backward-compatible, therefore incorporating legacy devices like 802.11a, b, and g networks.

The vision for 802.11n systems is to render the wired networks in the home unnecessary, allowing for high-definition streaming video and other broadband-intensive applications to be leveraged in the home and office environment.

10.3.5 802.11ac

802.11ac is a logical extension, or rather evolution, of the 802.11n standard that is widely deployed. 802.11ac is compatible with 802.11n with the exception that it only operates in the 5-GHz band, which is typically less congested than the 2.4-GHz band. What 802.11ac brings to the table are several key attributes that enable it to deliver higher data rates than those available with 802.11n.

The higher data rates are possible with 802.11ac through the use of higher modulation schemes where 256 QAM is achieved. Additionally, the radio channel size was increased to 80 MHz and 160 MHz. Also, 802.11ac is able to provide multiple spatial streams at the same time, up to 8 as compared to 802.11n which can only do 1. The spatial stream for a single user or device is referenced as SU-MIMO while the multiple user spatial stream is referenced as MU-MIMO.

10.4 Frequency Allocation

The 802.11 standards address two frequency ranges. 802.11b and 802.11g operate in the 2.4-GHz band (ISM), while 802.11a and 802.11ac operate in the 5-GHz band (UNII). Each has a set of channel numbers associated with it in the appropriate band. The availability of specific channels varies among regions. For example, in the United States there are 11 channels available to 802.11b and 802.11g. The reality (discussed later in the chapter) is that only three of the channels can be used together without overlap.

Channel	Frequency (GHz)	US	Europe	Spain	France
1	2412	X	X		
2	2417	X	X		
3	2422	X	X		
4	2427	X	X		
5	2432	X	X		
6	2437	X	X		
7	2442	X	X		
8	2447	X	X		
9	2452	X	X	X	X
10	2447	X	X	X	X
11	2462	X	X		X
12	2467		X		X
13	2472		X		
14	2484				

Table 10.3 802.11b,g Channel Chart

10.4.1 802.11b and 802.11g

Table 10.3 shows channel allocation within the ISM band for 802.11b,g services for a number of regions around the world. It is important to note that the spectrum or channel allocations are not uniform throughout the world.

10.4.2 802.11a

The 802.11a frequency and corresponding channel number are shown below in Table 10.4. These channels correspond to the 5-GHz UNII band.

Band	Channel Number	Center Frequency
Lower	36	5180
	40	5200
	44	5220
	48	5240
Center	52	5260
	56	5280
	60	5300
	64	5320
Upper	149	5745
	153	5765
	157	5785
	161	5805

Table 10.4 802.11a Channels

10.5 Modulation and Coding Schemes

There are several coding schemes used by the WiFi variants of 802.10. While WiFi is a CDMA technology, the modulation and coding schemes are significantly different than those of 2.5/3G and 4G technologies used by mobile wireless systems.

10.5.1 802.11b

In order to develop higher data rates for 802.11b, several modulation schemes were added to the standard.

10.5.1.1 Complementary Code Keying (CCK)

The addition of this modulation scheme enables the higher data rates of 5.5 Mbps and 11 Mbps. The higher data rate capability is known as *High-Rate Direct Sequence Spread Spectrum* (HR/DSSS). The same preamble and header are used as the earlier basic rate DSSS method. This allows the HR/DSSS and basic DSSS to coexist in the same *Basic Services Set* (BSS).

The standard also offers optional features that can be used to increase data throughput. These features allow WiFi vendors to differentiate their product offerings. They include HR/DSSS/short—essentially HR/DSSS with a short preamble—and HR/DSSS/PBCC. This optional implementation can coexist under limited conditions. For example, this can be implemented on different channels.

10.5.2 802.11g, 802.11a, and 802.11n

These standards both incorporate the modulation scheme of OFDM. This technique breaks down a wide carrier into several smaller subcarriers. Data is transmitted on each of the subcarriers and then combined into one code-division channel.

10.6 Network Architecture

WiFi network architecture can either be a single AP or one that has multiple APs. In addition, the network architecture variations and capabilities depend on whether the WiFi AP is connected to a LAN or Internet or not connected to any backhaul or external network (Figure 10.3).

However, the fundamental difference between wireless LANs and their wired LAN counterpart is that the users do not have to be in a fixed location. Therefore, the user device can appear in any BSS and then needs to establish an association with that BSS. Because the wireless device is mobile or capable of mobility, a more complex form of authentication is therefore required.

As part of the architecture, there are three basic distinctions for 802.11 networks.

1. *Basic Services Set* (BSS). This is a simple WiFi network where it is simply the association of two stations in an ad hoc network.
2. The *Distribution System* (DS) is another network architecture that involves *Communication between a Station* (CPE) and an Access Point (AP). A DS can and does handle the interworking with multiple BSSs.
3. The third general type of architecture is the *Extended Services Set* (ESS) that allows for communication between two or more DSs.

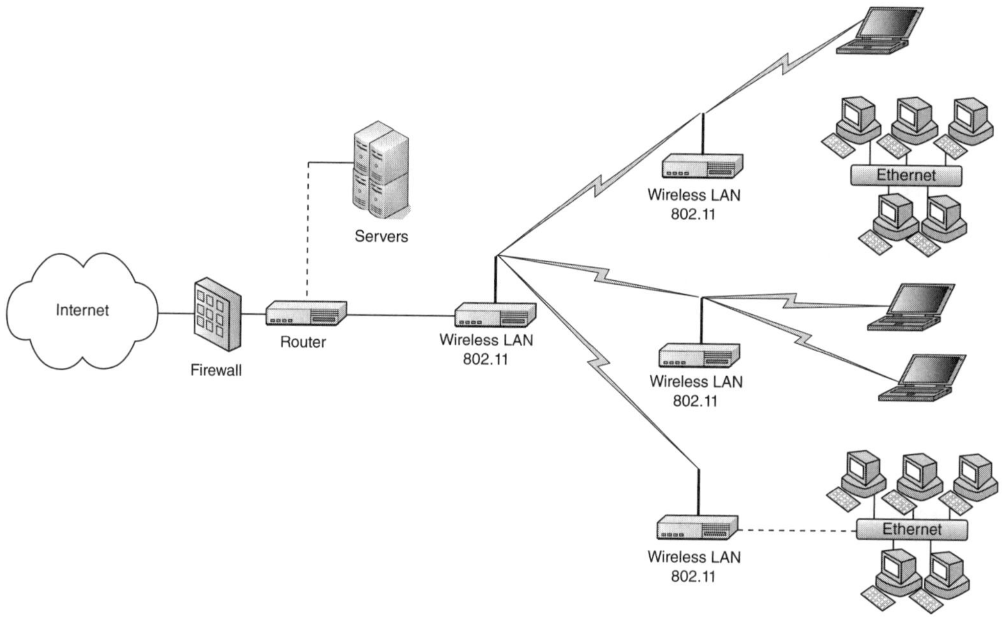

Figure 10.3 Wireless local area network.

Many combinations of the above three items are possible in a WiFi architecture. For example, BSSs can have overlapping coverage areas or may be configured not to have any overlap.

The two common types of network are the ad hoc and the infrastructure topology.

10.6.1 Network Components

There are two basic components to an 802.11 WLAN: AP and the station. The AP allows for communications among stations as well as access (in some cases) to a wired LAN. The access point contains all the functions of a station (station services) as well as DSS.

Functions available in an AP:

- Authentication
- Association
- Deauthentication
- Disassociation
- Distribution
- Integration
- Privacy
- *MAC Services Data Units* (MSDU) delivery
- Reassociation

Figure 10.4 Ad hoc network.

Functions available in a station device (CPE):

- Authentication
- Deauthentication
- Privacy
- MSDU delivery

10.6.2 Ad Hoc Network

A WLAN formed by two 802.11 stations communicating directly with one another is called ad hoc (Figure 10.4). No AP is required in this configuration. The two stations establish point-to-point connectivity.

This makes it possible for two users with compatible 802.11 stations to instantly set up a network between them. This could prove useful, for instance, in a meeting where two users wish to share data.

10.6.3 Infrastructure Network

The infrastructure configuration involves an AP or wireless router in addition to the station's elements (Figure 10.5). This is a typical SOHO implementation. A user wishes to create a wireless network available to members of his small community. This could be family members of employees in a small business.

In addition, a number of overlapping APs can form a WLAN. This configuration not only handles portable stations but also mobile stations. That is to say, a station can move from the coverage area of one AP to another (Figure 10.6). This method allows the expansion of coverage within or around a building. It also provides additional capacity.

Lastly, an 802.11 LAN can interface with a wired LAN through the use of a router as a portal (Figure 10.7). This allows the combination of the users wired LAN and his WiFi LAN.

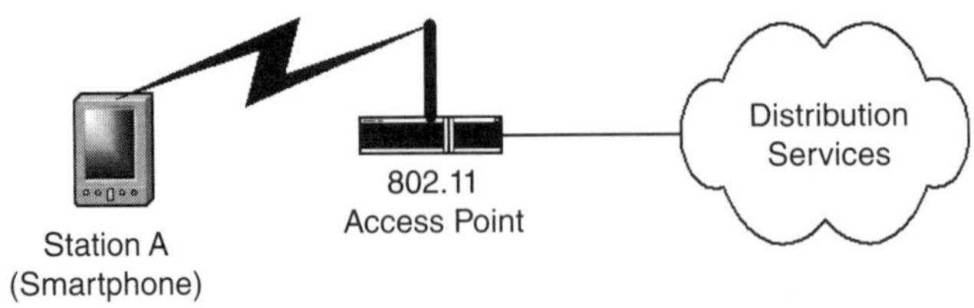

Figure 10.5 Infrastructure mode.

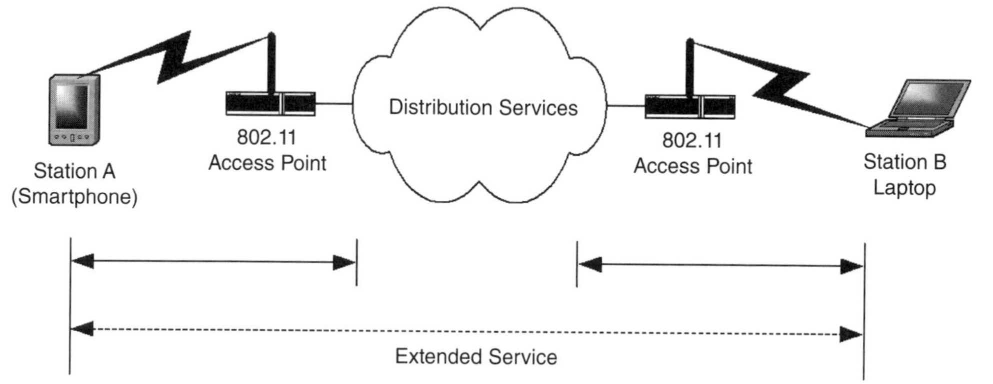

Figure 10.6 ESS.

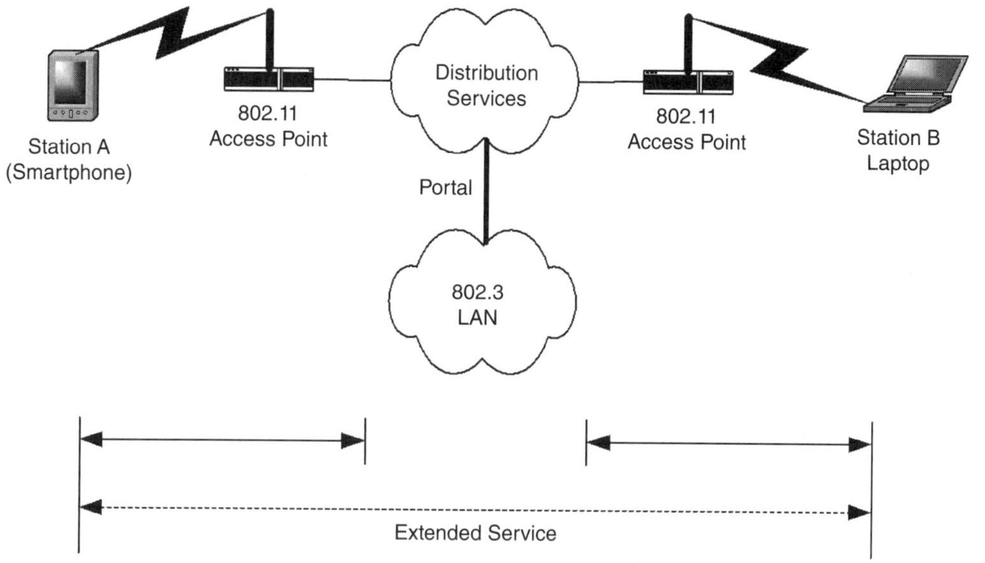

Figure 10.7 Portal.

10.7 Typical WiFi Configurations

The following section goes over some of the issues of different WiFi configurations, which are not connected directly to wireless mobility networks. However, as a design engineer or manager in telecom and when desiring to integrate these WiFi network types with the overall wireless mobility system, it is essential that a basic understanding of these typical configurations is known.

Three basic WiFi configurations will be covered and they are

- SOHO WiFi Network
- Enterprise application
- Campus deployment

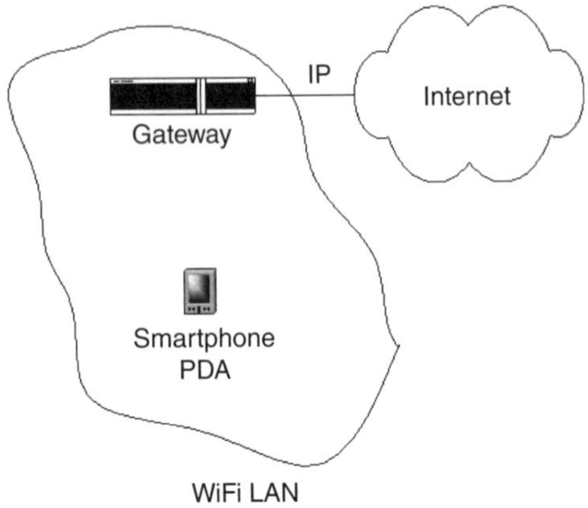

Figure 10.8 SOHO configuration.

10.7.1 SOHO WiFi Network

A sample SOHO application is depicted in Figure 10.8. This network typically serves as a home or small office; possibly only one room. There are no frequency reuse issues and probably no capacity issues. In the case of a SOHO network, there are very few design concerns.

10.7.1.1 Design Issues

1. *Technology.* The first obvious issue is which flavor (a, b, g, n) to choose. Depending on any legacy devices that may be present the desired protocol is 802.11n, and you can then differentiate services between 802.11b/g and 802.11a if desired.

   Due to the backwards compatibility of 802.11g, it shares the previous advantage of hot spot access. Lastly, 802.11a offers higher data rates and "cleaner" spectrum. The spectrum for 802.11a is in the UNII band and while it is unlicensed it has more rules for control and has far fewer technologies using it at present. Due to lack of compatibility this is probably not a good choice if you intend to be mobile or portable with your WiFi devices, since many mobile devices do not support 802.11a, or rather the 5.8-GHz portion of 802.11n.
2. *Coverage and capacity.* Coverage issues should be minimal unless you are trying to cover several rooms. In that case, coverage could be poor and an AP used as a coverage extender may be required.
3. *Security.* Due to the small number of users a few simple security measures should be employed. Since the community is small the distribution of *Wired Equivalency Privacy* (WEP) keys is not an issue. While these can be broken by hackers, they provide privacy and security from casual outside users.

The use of MAC address is also practical in this situation and adds another level of difficulty to hackers. Lastly, you should customize your SSID. This value is typically set to a default well known to hackers. While these actions are fairly simple, a vast number of WiFi users don't even implement these security features.

10.7.2 Enterprise Application

The task of designing a WiFi network for a business or commercial application is far more complex than the SOHO example in Section 10.7.1. In addition to technical issue, the enterprise will be more demanding in several areas:

- Return on investments
- Network management
- Security/monitoring
- Adaptability to capacity requirements

A number of design considerations come into play as well. An enterprise network layout is depicted in Figure 10.9.

10.7.2.1 Design issues

1. *Technology.* Should 802.11a, b, or g be implemented? As before, this will be driven by a number of issues including the needs of the user community. Users should be surveyed for their requirements:
   - Applications used
   - Capacity requirements
   - Current capacity utilized
   - Mobility within the location and traveling
   - Telecommuting

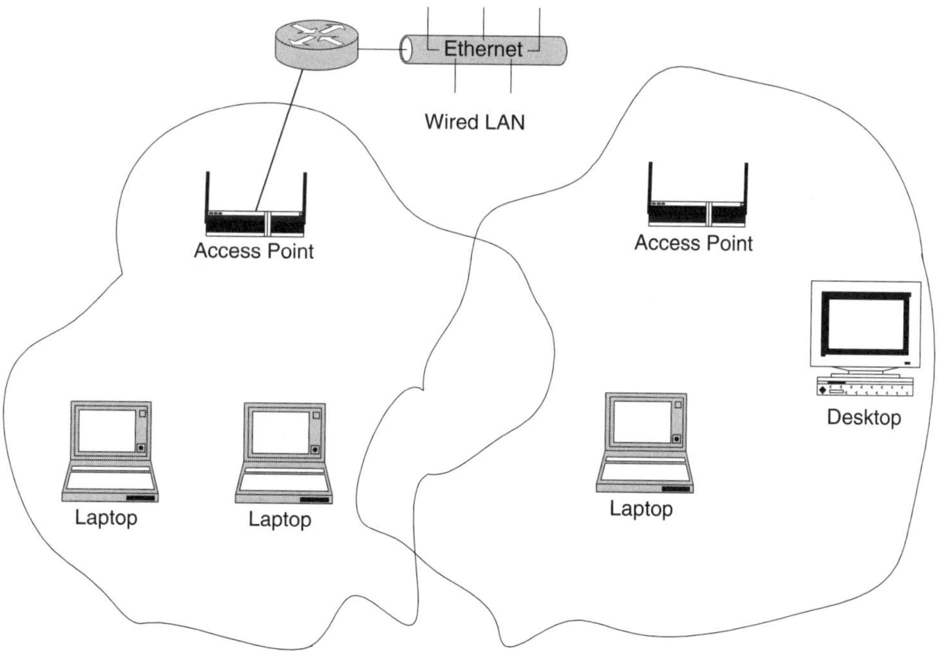

FIGURE 10.9 Enterprise deployment.

It may turn out that a combination of flavors will be required. You could designate that some users utilize 802.11g or b while other "Power Users" could be served by 802.11a. Power users often required higher bandwidth. By providing them essentially an overlay of 802.11a, they get the bandwidth they require and do not burden the other WLAN. If a user is a heavy traveler or telecommuter, 802.11b or g is a better choice in most cases.

2. *Coverage and capacity.* The number and types of AP will be driven by both coverage and capacity requirements. The simple addition of APs as coverage extenders may suffice. In some areas the need for more capacity may dictate the use of multiple channels requiring rudimentary frequency planning.

   Frequency planning becomes necessary when channel reuse is required. For example, in the United States there are 11 channels available to 802.11b/g. These channels overlap so that in reality only three of them can be used without interference caused by overlap. These are channels 1, 6, and 10. In the case where capacity demands multiple APs, these APs need to be set to one of these three unique channels. If more than three APs are required for capacity reuse, these channels need to maximize *Radiofrequency* (RF) separation to minimize interference. For a more in-depth look at radio engineering issues, see Chapter 11.

   In addition to coverage, survey may be required. In the simplest case, the installation of an access point in conjunction with software and a laptop may be sufficient. Once the AP is up and running, signal strength measurements can be recorded throughout the service area.

   Site survey issues that factor into the design process include

   - Likely point for installing APs
   - Available power
   - Floor plans, electrical diagrams, and blueprints if available
   - Obstructions to coverage such as metal walls
   - Test transmitter locations and AP testing plan (RF coverage)

3. *Security.* Some or all of the methods discussed in the SOHO example may be implemented. In the case of a large enterprise with numerous users, WEP and MAC address techniques are too cumbersome and difficult to administer. In the case of an enterprise, WiFi network additional security methods need to be considered.

In addition to privacy and security, the enterprise network should be monitored to detect and locate all wireless APs. It should then be determined that the APs have proper security implemented and are not simply rouge devices. In other words, make certain they were installed by and sanctioned by the IT department.

10.7.3 Campus Deployment

In a campus deployment, WiFi access is required in two or more discrete locations. Coverage may or may not also be desirable outdoors on the campus. Much of this will be driven by the type of campus. In the case of a corporate campus outdoor coverage may be undesirable. On the other hand, a university campus may require it.

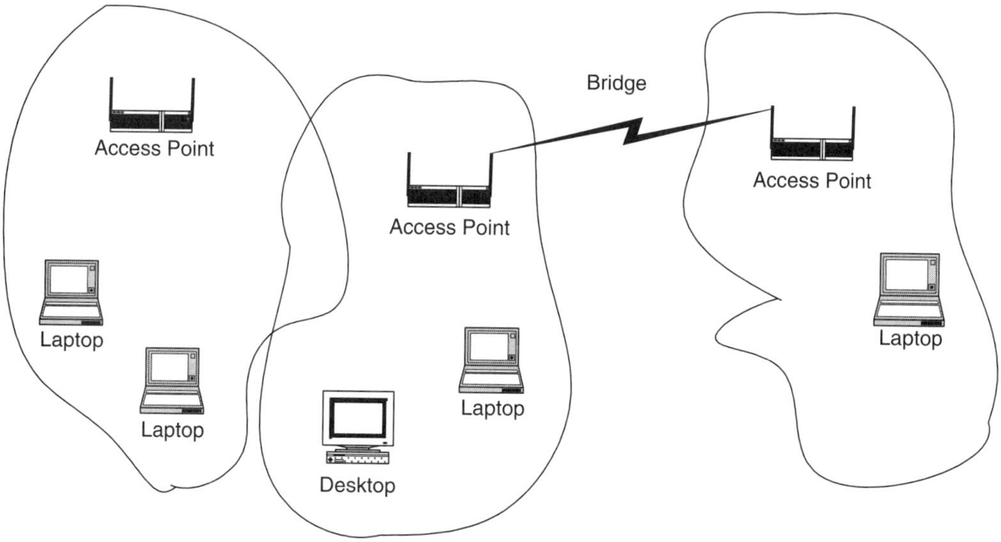

Figure 10.10 Campus deployment.

10.7.3.1 Design Issues

For the most part design issues for the campus environment are very similar to those of the enterprise. There is, of course, a potentially much larger serving area in the campus environment.

The chief differentiator is the requirement for interconnecting the various location or buildings throughout the campus. This is often accomplished through the use of additional 802.11 compliant solutions. Other methods could include: wired facilities such as DSL, cable modems, T1 circuits, unlicensed microwave solutions, or even optical solutions. A network diagram of a campus implementation is shown in Figure 10.10.

10.8 Security

The intention of the 802.11 standard is to require and specify authentication and privacy sufficient to bring wireless LANs into parity with their wired counterparts. 802.11 security includes a variety of security protocol choices, for example, WPA-PSK, WPA2, IEEE 802.1x/RADIUS, WDS, WEP, TKIP, and CCMP (AES) encryption. However, for the most part this is accomplished through the use of WEP.

WEP provides a level of authentication that (if implemented) will prevent communications with just any station that comes along. For example, if WEP is enabled, your neighbor will not be able to communicate with your LAN unless he has the key. Of course, if he or she is a hacker it is a different story. When additional security is required other forms of security and authentication should be implemented.

In the case of mobile users VPN can be a very effective method for providing security between the mobile user and the corporate network. *Virtual Private Network* (VPN) in its many flavors is discussed later in this chapter.

10.8.1 WEP

There are two types of authentication in 802.11: Open authentication and shared key authentication.

1. *Open systems authentication.* In the case of open authentication, the only requirement for authentications is that both stations have the dot11 authentication field set to open systems authentication. This is commonly used in hot spot implementation to avoid the need for distributing or sharing keys.
2. *Shared key authentication.* Shared key authentication requires that WEP be enabled in the station. It operates using a secret shared key that has been delivered to the stations by an alternative mechanism to 802.10. In other words, user of the shared WEP key must obtain it prior to enabling WEP. Each user requiring access to the LAN must enter the WEP key into his station. Shared key authentication is invoked only if the dot11 *privacy option implemented* is "true." Keys are shared in a write-only *Management Information Base* (MIB) attribute.

The authentication sequence:

- A management message is sent from the requestor to the responder indicating that shared key authentication is enabled.
- A second management message is sent (from the responder to the requestor) with the 128 octet generated by the WEP pseudorandom number generator (PRNG). This is the authentication challenge text.
- A third message is sent from requestor back to the responder. It contains the challenge text sent by the requestor encrypted using the WEP algorithm.
- The final message denotes success or failure of authentication. The result will be "success" if (after deencryption) the challenge text returned matches that sent in the second message. Otherwise the response will be "unsuccessful."

10.8.2 Encryption with WEP

WEP was designed to prevent casual eavesdropping. The objective was to provide equivalent privacy to that of a wired LAN. In other words, if you were not on the LAN you would not see messages. The WEP algorithm is far from unbreakable. Depending on the length of the key used; the frequency of changing the key; and the ability to key the secret key, secret WEP provides a protection from brute force attacks.

WEP is self-synchronizing which makes it less susceptible to message delivery problems inherent in the wireless environment. It can be implemented in hardware as well as software allowing it to be very efficient. It is illegal to export many encryption algorithms from the United States. Every effort was made by the standards body to make WEP exportable.

Use of WEP is optional within the standard.

10.8.3 Alternatives to WEP

As mentioned earlier, WEP is not intended to provide ultimate security. In fact, it has been proven to be easily breached. It is only prudent when transmitting sensitive data to implement one or more additional forms of security like AES.

10.8.4 SSID

Two other screening elements in the 802.11 standard can be used to increase privacy and security. One is the SSID. Use of SSID is intended to direct stations to the correct AP in cases where multiple APs exist. It can be used to keep casual users from accessing your network. Unfortunately, even if the SSID is set to a secret value it can be easily learned by hackers.

In many cases users leave their SSID set to the default value. Since these defaults are widely known they leave little protection for the WLAN.

10.8.5 MAC Address

This technique involves the use of the client's physical MAC address. While this is effective it too can easily be breached. The hacker simply observes network traffic and assumes the MAC address and IP address of a genuine user. Once that user drops off the network the hacker gains entry with the false ID.

In addition to the weakness of this method it is also cumbersome for an administrator to maintain.

10.8.6 Extensible Authentication Protocol (EAP)

With this method a client is unable to gain access to the LAN without first going through a log-in procedure the process follows:

- The client first supplies a user name and password to a RADIUS server.
- The server sends a challenge message to the client.
- The client then uses his password to supply a response to the RADIUS server's challenge.
- The server then authenticates the client.
- The process is then repeated in reverse.
- The client and server then create a WEP key.
- This key is delivered to the AP over the wired LAN.
- The AP then sends the WEP key to the client and this key is used for the session.

Using this authentication scheme has several advantages. First of all, the use of the WEP key is reduced to only this one session. This, of course, makes WEP key discovery much less likely. Secondly, it adds an additional degree of control to LAN management.

10.9 802.11 Services

The term services has many meanings. However, for 802.11 there are agreed-upon definitions. The services defined by the 802.11 standard address the delivery of MSDU as well as access and privacy.

1. *Distribution services*. This is the primary service used by stations to send data messages. The 802.11 standard does not address how messages are distributed. It does, however, provide the distribution systems with enough information to deliver the message.

2. *Integration*. The integration function delivers messages from the Distribution System Media to an integrated LAN. This service is used when the destination or intended recipient is a member of an integrated LAN.

3. *Association services*. Association is a distribution system service that associates a station with an AP. The association service is always initiated by the mobile station. A station can only be associated with one AP (at a time), while an AP can be associated with numerous stations.

   Stations associated themselves with the appropriate AP based on the scanning and synchronization algorithms specified in the specification. Essentially, a station associates with an AP with the proper SSID and signal.

4. *Reassociation*. This service is used to move a station's association from one AP to another. It is used when a station is mobile. Reassociation can also be used to change association attributes while on the same AP. This service is always invoked by the station.

5. *Disassociation*. Disassociation tells the distribution service to void a current association. This service may be invoked by either the station or the AP.

6. *Control services*. Two services are introduced by 802.11 to provide the security and privacy that is inherent in wired LANs.

7. *Authentication*. Two methods are implemented in the standard: open authentication and shared key authentication. The standard does not provide for end-end authentication; this is left to a higher level of the network protocol. The intention is only to bring wireless to the level of wired LANs with respect to privacy and security.

   Open authentication is essentially no authentication. The only requirement is that the two stations involved be set to "open authentication."

   Shared key authentication involves the sharing of a secret key by both stations. In 802.11 this is implemented through WEP. In this case both systems must know the WEP key in order to authenticate 802.11, which requires authentication prior to association taking place.

8. *Preauthentication*. Preauthentication can take place prior to reassociation. This speeds up the process of transitioning between APs.

9. *Deauthentication*. Deauthentication is a notification that a (non-AP) station wishes to be deauthenticated. This also results in disassociation since authentication is a prerequisite to association.

10. *Privacy*. As with authentication, the purpose of the privacy service is to bring 802.11 up to equivalency with a wired LAN. It is not intended to be a robust form of security. The privacy services are achieved using WEP, as mentioned earlier. If stations are set for encryption, the WEP key is used to encrypt all messages. This makes it possible for stations sharing the key to read these messages.

11. *Relationship between stations/services*. Each station keeps track of what authentication and association level it has with all other stations. There are three levels:

    - State 1 Neither authenticated nor associated
    - State 2 Authenticated only
    - State3 Both associated and authenticated

The state of a station controls what messages are acceptable. Stations will accept any message at or below the present state. In state 3 all messages are accepted. In state 2 both level 1 and 2 messages, and in state 1 only level 1 message.

Thus only those messages allowed due to the level of authentication and association will be seen by a station. If a station receives unauthorized frames it will send the appropriate disauthentication or disassociation message to the sending station.

10.10 Hot Spots

Numerous entities are developing WiFi (802.11) networks around the globe. These networks, known as "hot spots," allow instant access to anyone with a WiFi device and of course, access permission. Most of these hot spots are 802.11b systems.

While many hot spots simply link the user to the Internet (provided that he has access), it is only a matter of time before more sophisticated services become available, for example, services such as the ability to establish VPN with their inherent security between the mobile user and his corporate Network.

Additionally WiFi hot spots have become increasingly important to wireless carriers to off-load their existing licensed spectrum traffic to a hot spot which is typically off-net. However roaming agreements have been extended to the wireless mobile carriers enabling combined "cellular" and WiFi-capable devices to hook up to the hot spot while in its coverage area. Under a roaming agreement it is possible for a "handing off" to the cellular network when leaving the hot spot.

10.10.1 Security at Hot Spots

For the most part security is nonexistent at hot spots. This is primarily due to the transient nature of the users and the lack of support at the locations. For example, it would be difficult to manage the use of WEP keys since everyone would have the key. However, other authentication methods can be used involving simple login procedures can be used to minimize access.

10.11 Virtual Private Networks (VPNs)

VPNs make using the Internet more like a private network. This is a similar concept to that of using WEP to make a wireless network similar in privacy to a wired network. Privacy and security are added to each IP packet, which is encapsulated inside another IP packet. This is known as tunneling. In the case of a VPN, the actual data is contained inside the IP packet visible to the Internet.

A very common implementation of VPN is *IP Security* (IPSec). While IPSec does not provide authentication it does provide confidentiality. With IPSec each user must have an IPSec client on his PC.

Placing a VPN gateway or firewall between wireless networks and your network provides additional levels of control and security to the network.

10.11.1 Proprietary Vendor Solutions

There are a number of manufacturers who provide solutions to the inherent weakness in WEP. These solutions add security but usually require the use of one manufacturer's equipment throughout the network. This, of course, removes the commercial advantages of adopting an industry standard such as WiFi (802.11).

10.12 Mobile VPN

The concept of virtual private networks predates data networking. The term is used by traditional phone companies to refer to services offered to various entities. In some cases, it is as simple as a leased line facility. In others it is a complete outsourced network that is carved out of the "public network" for entities' exclusive use.

A virtual network is essentially a connection that appeases to the users as a dedicated end-to-end connection. It is not important for the user to know the actual route taken by the data. In fact the route will often change during the course of the session. In order to make the network private (in an IP environment) other services must be implemented.

To ensure privacy several things must happen. The user must be verified to be eligible to use the VPN. This is accomplished through the use of RADIUS servers and shared certificate methods. The data must be protected or made private. This is typically accomplished through various tunneling techniques.

With a private network, users expect a certain level of service. This requires the ability to implement *Quality of Service* (QoS) schemes. While this can be accomplished it becomes more complex in an IP environment with varied paths and providers. In a Mobile IP environment it is further complicated by mobility and roaming issues.

10.12.1 Advantage to Using VPNs

There are a number of compelling reasons to use VPN. In particular, for telecommuters. The rapid increase in telecommuters is forcing IT departments to provide remote access to more and more users. Traditionally this access was provided through dial-up procedures; however, with ever-increasing costs and the relatively low speeds users are moving to high-speed access.

10.12.2 Increased Security and Control of Users

Not only does VPN make corporate data private and secure, it also controls who has access to the private network. In addition, levels of access can be assigned to different users depending on their need to access sensitive information. VPNs also enable the network to determine who is dialing in.

10.12.3 Cost and Maintenance

Dial-up, of course, requires a modem pool. This is not only an added cost but also a source of hardware failures. Moving from dial-up to high-speed access reduces the requirements for modem pools.

10.12.4 PSTN Costs

In many cases an employer supplies a form of 800 access to avoid phone charges to the employee. Obviously this cost can become quite high. For this and the reasons listed above, employers are beginning to demand telecommuters acquire high-speed access or move back to the office. The choice is clear.

10.13 VPN Types

There are three basic types of VPN: voluntary, compulsory, and chained. Each has its advantages and disadvantages. A brief description of these configurations follows.

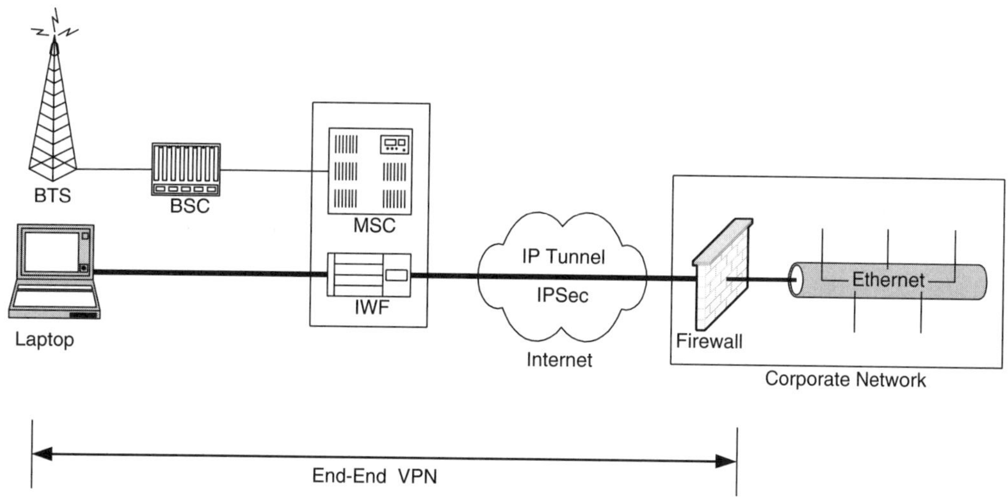

Figure 10.11 Voluntary VPN.

10.13.1 Voluntary VPNs

Voluntary VPNs are set up by the end user and require little intervention by the network administrators. It typically has a public IP address and the data is hidden from end to end. However QoS implementation is typically not feasible.

A voluntary VPN has the advantage of simplicity to the network operators; however, the need for end-end VPN puts special burdens on the cellular carrier and the end user device. The additional overhead required to take the VPN through the RAN uses scarce spectrum resources. In addition the mobile device must contain the processing power and software to implement the VPN. See Figure 10.11.

10.13.2 Compulsory VPN

In the case of compulsory VPNs any user requesting access to a network or service will utilize a VPN that is controlled by the operator (Figure 10.12). In the case of a cellular system this VPN may only extend from the private network being accessed to the mobile switching center. This method removes the disadvantages of spectrum utilization and end user device requirements of voluntary VPNs. It also removes the requirement for Public IP addresses. Public IP addresses are becoming a very scarce resource. On the downside, a portion of the data path is unprotected as well as unpredictable. This makes the assurance of security, privacy, and level of service impossible.

10.13.3 Chained VPN

In the case of chained VPNs, essentially two unique VPN implementations are joined together (Figure 10.13). This allows, in the case of cellular, the use of a VPN method geared to the mobile environment combined with a more prevalent VPN implementation in the nonmobile environment.

10.14 WiFi Integration with 3G/4G

There are multiple ways to link a 2.5G/3G or 4G network to one or more WiFi networks in order to provide seamless broadband service. The method chosen depends on the

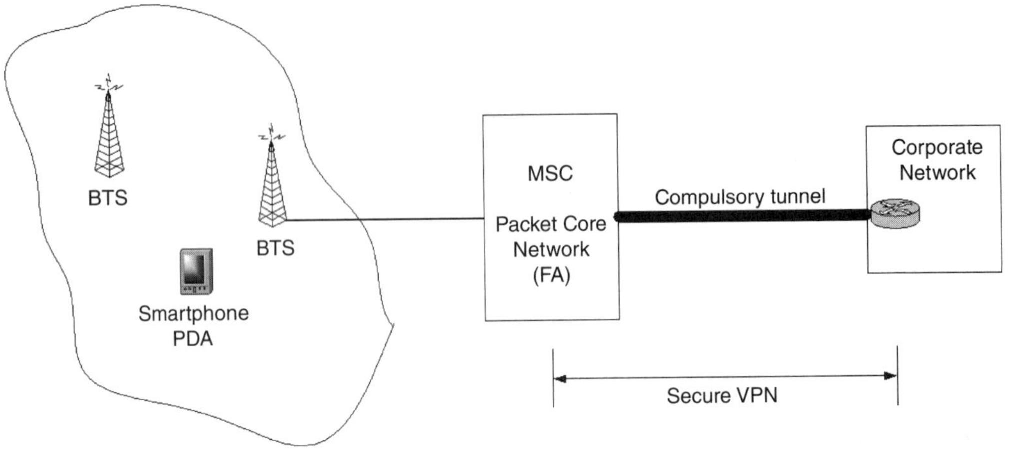

FIGURE 10.12 Compulsory VPN.

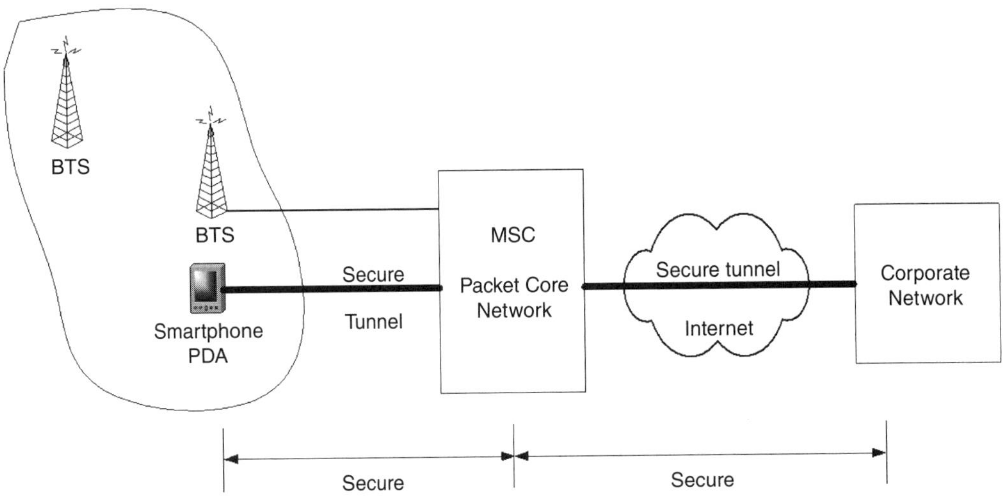

FIGURE 10.13 Chained VPN.

2.5G/3G and or 4G RAN and the types of customers being pursued. WiFi is a very effective method of off-loading packet traffic from the wireless network.

3G/4G/WiFi interoperability, therefore, has been a focus of the wireless industry and is evident through the use of Mobile IP. Mobile IP, as discussed earlier in other chapters, facilitates the continuity of the data session when the user is mobile by facilitating handoffs between disparate networks so that they're not noticeable to users or programs, that is, e-mail, when their IP connection changes.

There are two general classifications of mobile data users, casual and business. It is important to note that data requirements are very specific to the individual and cannot be a one-service-fits-all. Data requirements are unique and are vertical. For instance, your specific data requirements are different than my data requirements. Some of this is evident in the type of device that is selected to be used, the applications selected, and the bandwidth required.

The ability to link a 3G/4G network with WiFi networks is driven at this time by the network and the user device(s). Linking 3G/4G networks with WiFi networks involves both business and technical considerations. The business considerations focus on the service management plan which includes at its heart a service level agreement (SLA) and possible charging rates that can or will be used. This is in addition to the determination of what services specifically will be used or available.

The technical considerations involve design, deployment, provisioning, and ongoing support. Technical issues generally involve billing and authentication and increase in complexity when the networks are owned by different companies. Therefore the demarcation and how the networks will interact with each other in the delivery of services is another area of technical interaction. But the ongoing support is of course a critical decision that involves problem identification, resolution, and enhancements.

The wireless data provider must therefore be able to authenticate the user, track the usage, and then deliver that information in a format compatible with the operator's billing systems.

Mobile wireless providers are trying to reduce churn by creating sticky content to keep their customers. One such method is the use of an integrated bill besides data services, and the possibility of having a single monthly bill for 3G/4G, and WiFi is a convenience that could be one more differentiator to reduce churn.

The convergence of 802.11 and wireless mobility has several possible directions that it can take, some of which will be discussed next.

10.14.1 Loose and Tight Coupling

The following section details two potential scenarios for combining WiFi and mobile service. The first example is defined as loosely coupled. In this implementation the user experiences what appears to be ubiquitous coverage from his mobile carrier; however, the actual data session (on the WiFi network) is independent of the mobile operator's core network. There is, of course, the need for authentication of the user. The basic configuration for loosely coupled is shown in Figure 10.14.

The tightly coupled case has the WiFi LAN directly connected to the mobile operator's core packet network. It functions as just another 3G/4G network in the eyes of the mobile network. Some of the drawbacks include the difficulty for independent hot spots and corporate networks to be part of this implementation due to the direct access into the 3G/4G core network. As shown in Figure 10.15, AAA services are combined in this implementation.

10.14.2 GSM and WiFi

Figure 10.16 is an example of how 802.11 can and is converged with an existing GSM network that utilizes GPRS and/or EDGE. The 802.11 systems can be integrated directly into the operator's network or through a roaming agreement providing access to the GSM subscriber.

10.14.3 CDMA/EVDO and WiFi

Figure 10.17 depicts a high-level configuration where an EVDO network is connected to both loose- and tight-coupled WiFi networks. This configuration is similar in concept to that of Figure 10.15 with the exception that the underlying RAN is CDMA2000/EVDO.

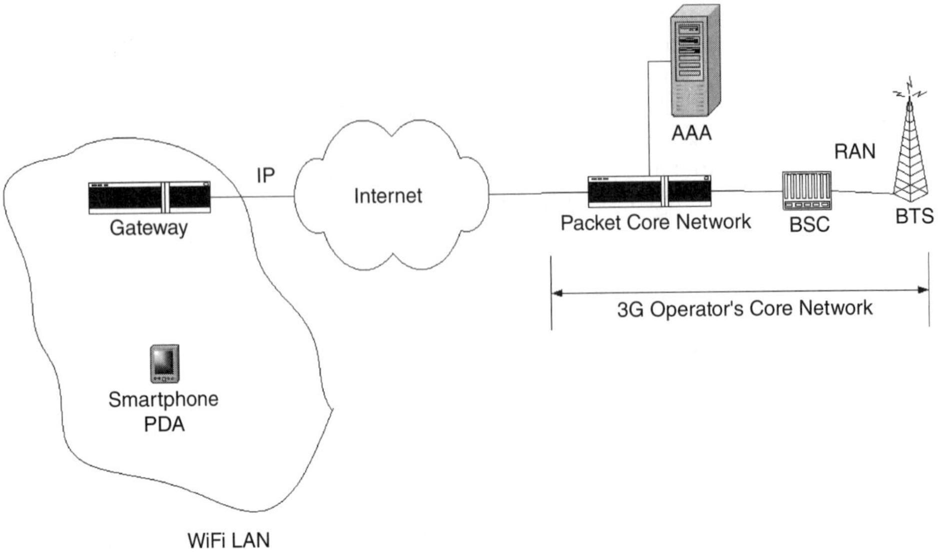

Figure 10.14 Loose integration.

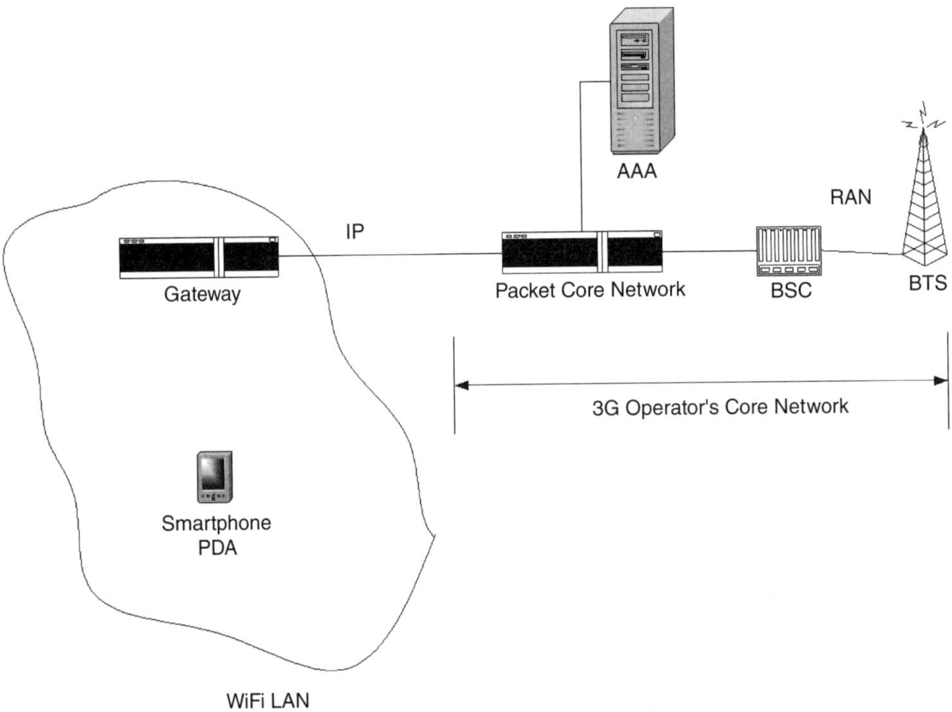

Figure 10.15 Tight integration.

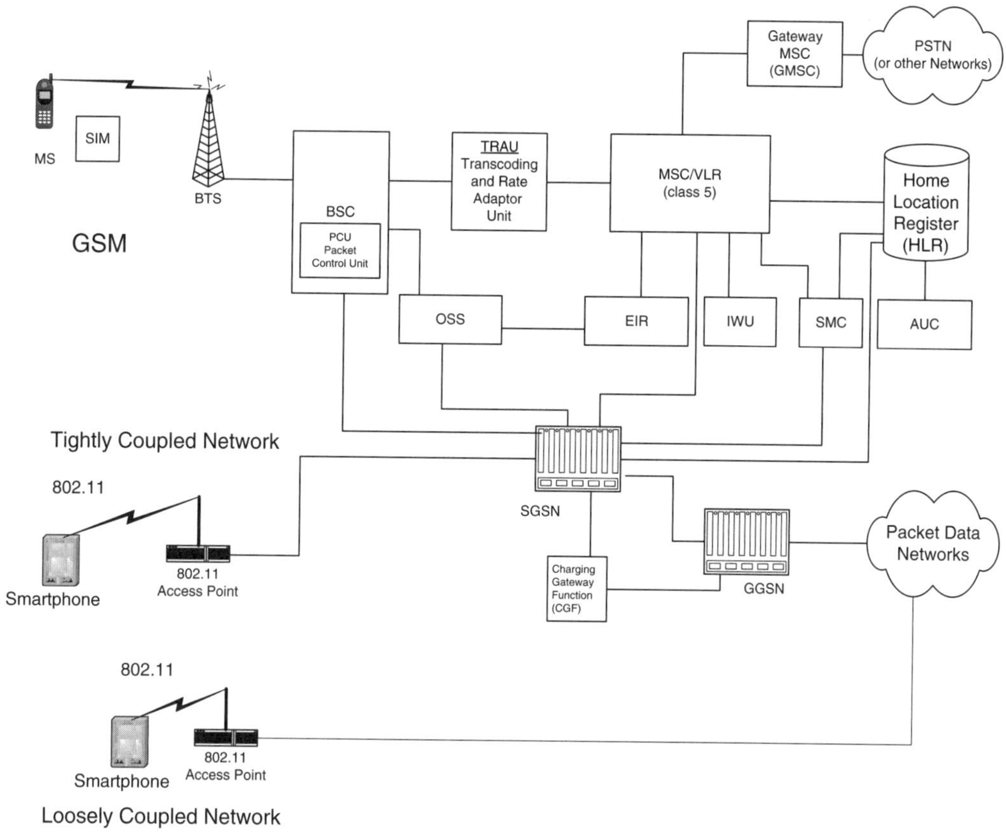

FIGURE 10.16 GSM and 802.11 convergence.

10.14.4 GSM/HSPA and WiFi

Figure 10.18 is an example of how a wireless operator that migrates from GSM to UMTS/HSPA integrates with an 802.11 system. UMTS/HSPA connectivity and integration with 802.11 systems is relatively the same as that for GSM due to the commonality of the core packet network.

10.14.5 LTE and WiFi

Figure 10.19 is an example of how a wireless operator having LTE as the RAN integrates with an 802.11 system.

10.15 Benefits of Convergence of WiFi and Wireless Mobile

To some extent WiFi can be thought of as a competitor to mobile services. In particular, the 3G/4G high-speed data access capabilities. On the other hand, the services are very complementary. While WiFi provides higher data rates, mobility provides much greater roaming and voice services on a global basis. There are several benefits to the customer as well as the mobile operator when the two technologies are combined.

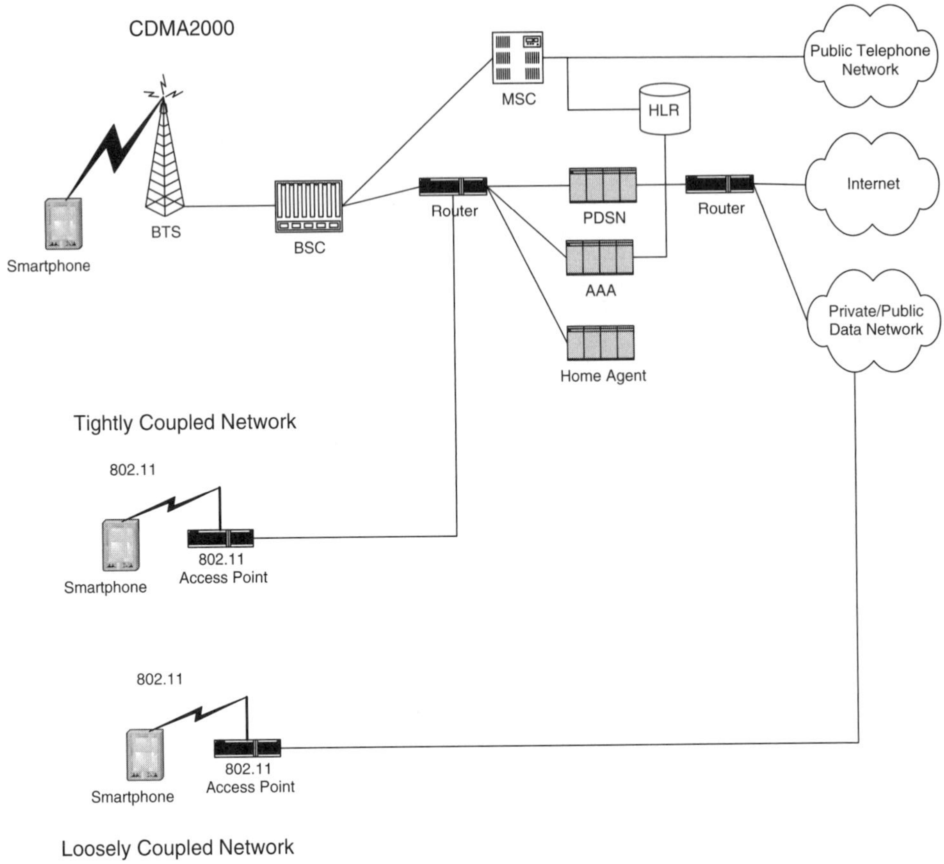

Figure 10.17 CDMA2000/EVDO and 802.11 convergence.

The customer, for example, can be a corporation if they desire to replace much, if not all, of their premise telephone system with IP-based WiFi-enabled handsets. If these handsets are also capable of mobile service, they have a single solution for both land and mobile phone service for the workforce. This provides high-speed access on less expensive WiFi equipment while in the building and mobility and 3G data services on the road.

The wireless operator will benefit form churn reduction. If a customer bundles his mobile and local phone solution, he is far less likely to change service providers. Another benefit is the reduction in use of the operator's spectrum. Spectrum is of course a very scarce resource.

In conclusion, as you can see and imagine, there are numerous possibilities related to the methods and possibilities for convergence to take place with 802.11 and wireless mobility. The key concepts that need to be kept in mind with convergence are

- What are you really trying to accomplish?
- What is the optimal technology?

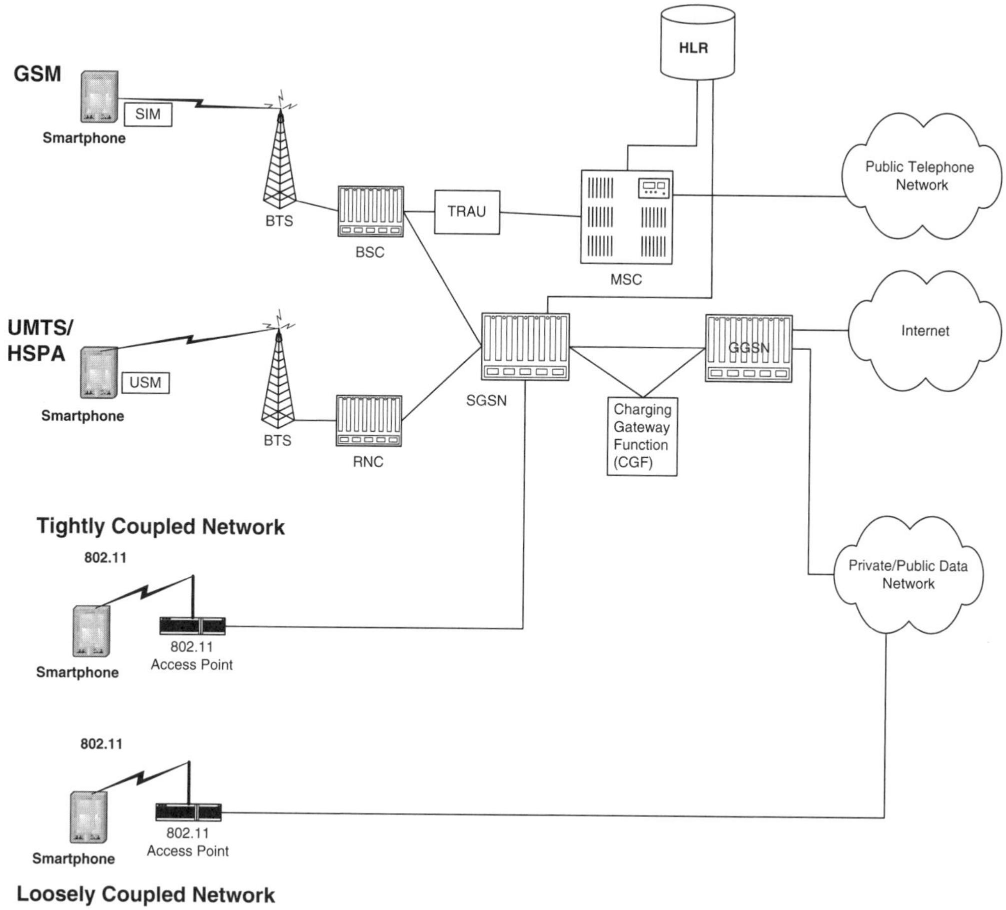

FIGURE 10.18 UMTS/HSPA and WiFi.

- When is the solution required?
- How will you achieve it, that is, the deployment plan?

References

ANSI/IEEE Std 802.11, 1999 Edition [ISO/IEC 8802-11: 1999].

ANSI/IEEE Std 802.1B and 802.1k [ISO/IEC 15802-2] LAN/MAN Management.

ANSI/IEEE Std 802.1D [ISO/IEC 15802-3] Media Access Control (MAC) Bridges.

ANSI/IEEE Std 802.1G [ISO/IEC 15802-5] Remote Media Access Control (MAC) Bridging.

ANSI/IEEE Std 802.2 [ISO/IEC 8802-2] Logical Link Control.

Davis, S. Kent III, "Crafting software-defined radio for 3G," *Electronic Engineering Times*, Nov 17, 2003, p. 77.

Dhir, Amit, "The ABC's of 2.4 and 5GHz Wireless LANS," Aug 1, 2001, Xilnx.

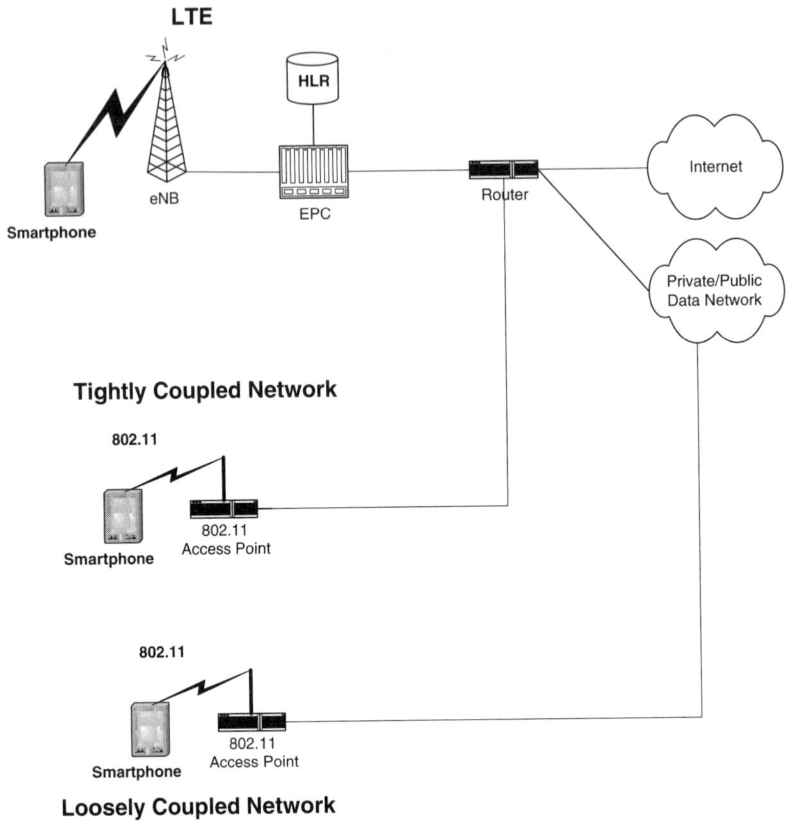

Figure 10.19 LTE and WiFi.

"802.11ac: The Fifth Generation of Wi-Fi," Cisco, June 2014.

Gary, James, "Welcome to the Jungle," *EDN*, Oct 30, 2003, p. 39.

IEEE Std 802.1F Common Definitions and Procedures for IEEE 802 Management Information.

IEEE Std 802.11™-2012 Part 11: Wireless LAN Medium Access Control (MAC) and Physical Layer (PHY).

IEEE Std 802®-2012 (Revision of IEEE Std 802-1990).

IEEE Std 802.11™-2012 Part 11: Wireless LAN Medium Access Control (MAC) and Physical Layer (PHY) Specifications.

Mannion, Patrick, "Uwb Fate Still Up in the Air," *Electronic Engineering Times*, Dec 8, 2003, pp. 18–26.

Mannion, Patrick, "Wireless Mesh Networks Gain Traction," *Electronic Engineering Times*, Nov 17, 2003, p. 56.

Marek, Sue, "Fragmented Wi-Fi Industry Begins to Coalesce," *Wireless Week*, Oct 15, 2003, pp. 7–8.

Mayor, Tracy, "Wi-Fi Hot Spots Head Up" *Electron Business*, May 1, 2003, pp. 14–16.

Miller, Stewart, *WiFi Security*, McGraw-Hill, New York, 2003.

Minoli, Daniel, *Hot Spot Networks*, McGraw-Hill, New York, 2002.

Muller, Nathan, *WiFi for the Enterprise*, McGraw-Hill, New York, 2002.
O'Shea, Dan, "A marriage of convenience: Where Wi-Fi & Mobile merge." Ohrtman, Frank and Konrad Roeder, *Wi-Fi Handbook*, McGraw-Hill, New York, 2003.
P24-36, Telephony, March 17, 2003.
Qualcomm Inc, "3G/Wi-Fi Seamless Offload," Qualcomm Inc, March 2010.
Shneyderman, Alex and Allesio Casatir, *Mobile VPN*, Whiley, New York, 2003.
Smith, Clint, and Dan Collins, *3G Wireless Netwo*rk, McGraw-Hill, New York, 2002.
Ward, Lisa, "802.11ac Technology Introduction White Paper 1MA192," Rhode& Schwarz, March 2012.
www.ieee.org/.
www.wballiance.com
www.wi-fi.org

CHAPTER 11

Broadband System RF Design Considerations

The *Radiofrequency* (RF) design process is very important with any radio access system. *Third-Generation* (3G) and *Fourth-Generation* (4G) wireless access systems have introduced several new radio access methods that need to employ good RF design practices. Good RF design practices are predicated on the RF design criteria that are established and followed. The RF design criteria are a set of rules or parameters used by the RF engineering department not only to design the network and the new components that are added, such as cell sites, but also to improve the performance of the network. The values included for each of the design criteria topics are driven by the desire to offer the best service within available monetary and technological constraints.

Therefore, the design criteria for the radio access part of a 3G/4G system are extremely important to establish at the onset of the design, whether for a new system, for migrating to a new platform, or expanding an existing system. Many aspects are associated with an RF design, and surprisingly, they are common in concept in any radio access platform that is being used by a wireless operator.

This chapter will attempt to consolidate many of the most important issues concerning the generation and execution of design criteria associated with the radio access portion of a system. The topics that will be discussed in this chapter include

- RF system design procedures
- Methodology
- Propagation models
- Link budget
- Tower-top amplifiers
- Cell-site design
- RF design report

The chapter concludes with a recommended format for presenting the design criteria in a formalized report that will list the design criteria, assumptions, and other key issues.

In summary, RF design for a wireless network is an ongoing process of refinement and adjustment based on a multitude of variables, most of which are not under the control of the engineering department. The RF design process involves both RF and network engineering efforts and implementation, operations, customer care, marketing, and of course, operations. However, it is important to note that although many issues are outside the control of the technical services group of a wireless company, the need to stipulate a design and its associated linkages is essential if there is any desire to obtain an operating system that meets the system objective of fulfilling customer requirements.

Therefore, the RF system design process that should be followed is listed here in summary form. The process can be used for an existing system or a new system because the material needs to be revisited for each of the topics when any system design takes place:

- Marketing requirements
- Methodology
- Technology decision
- Defining the types of cell sites
- Establishing a link budget
- Defining coverage requirements
- Defining capacity requirements
- Completing RF system design
- Issuing a search area*
- Site qualification test (SQT)*
- Site acceptance/rejection*
- Land-use entitlement process
- Integration
- Handover to operations

It is important to note that the design guidelines involve not only establishment of the criteria but also the realization of the design itself.

The information needed for a system design varies from market to market, and of course, nuances can be noticed between the different technology platforms. However, commonality exists between markets and technology platforms. The following is a brief listing of the most important pieces of information needed for a system design:

- Time frames for the report to be based on
- Subscriber growth projections (current and future by quarter)
- Subscriber voice usage projection (current and forecasted by quarter)
- Subscriber packet usage projection (current and forecasted by quarter)
- Subscriber types (mobile, portable, packet capable, blend)
- New features and services offered
- Design criteria (technology-specific issues)
- Baseline system numbers for building on the growth study
- Cell-site construction expectations (ideal and with land-use entitlement issues factored in)

---

*Applicable to new sites.

- Fixed network equipment (FNE) ordering intervals
- New technology deployment and time frames
- Budget constraints
- Due date for design
- Maximum and minimum off-loading for cell sites when new cells are added to a design

Of course, many sources and types of information are required for an RF design. The basic inputs usually obtained from the marketing and sales organization within a wireless network are listed in this chapter. The output from the RF design process will determine the requirements and fundamental structure of the radio access aspects of a wireless system. A simplified radio access structure is shown in Figure 11.1 but can apply to any situation involving the expansion of individual components relative to the different technology platforms used.

In order either to design a new system or to establish the migration path for an existing system, the RF design is relegated to determining the specific access method that the subscriber will have with the wireless system. The subscriber and base stations (*Base Transceiver Stations* [BTSs]) both have a transmitter and receiver incorporated in their fundamental architecture. Figure 11.2 shows the various components that need to be factored into the design of a system. This figure can be used for any technology platform, and the specifics of the various network elements make up part of the propagation analysis, where *Carrier-to-Interferer* (C/I) or *Energy per Bit per Noise* (E_b/N_o) values are used to determine the performance criteria necessary for the successful transmission and reception of the information being delivered.

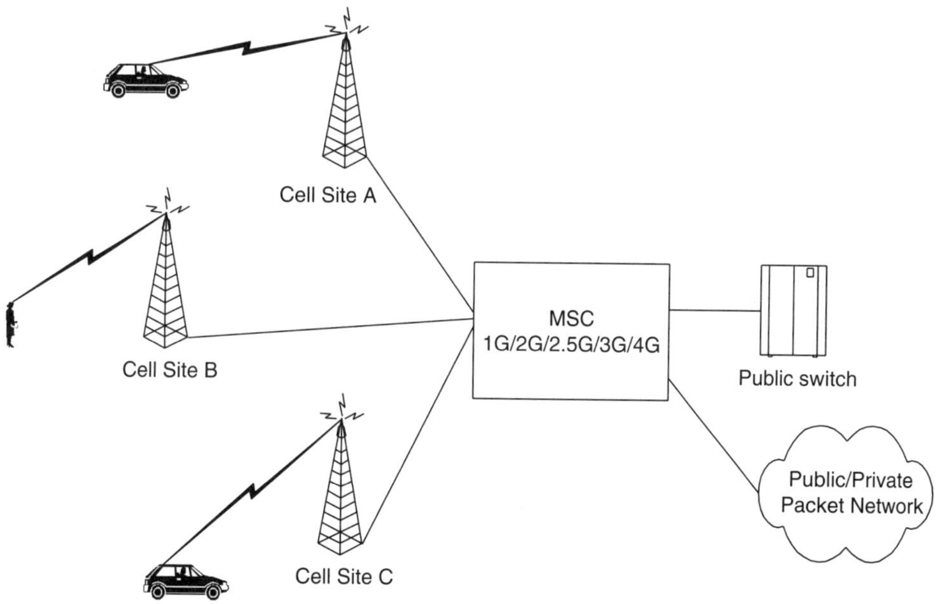

FIGURE 11.1 Generic radio access system.

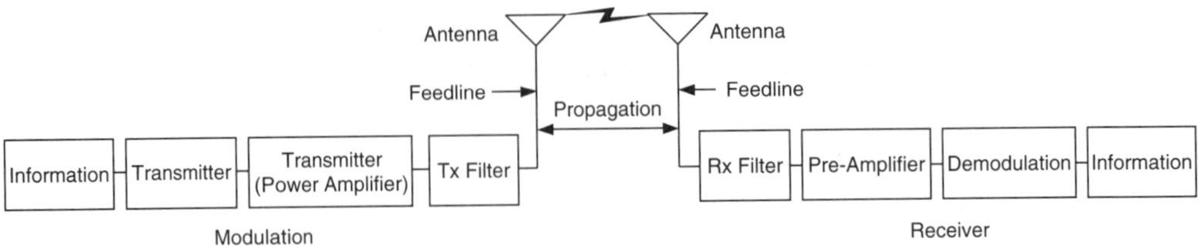

FIGURE 11.2 Generic radio system.

11.1 RF System Design Procedures

The RF system design procedures associated with a 3G/4G system design are similar to those followed for a *Second-Generation* (2G) or even a *First-Generation* (1G) wireless system. There are amazing similarities for implementing 2.5/3G into an existing system, as was the case when 2G was introduced into cellular systems. 4G, however, is an all packet system and has some unique requirements and capabilities over 3G. For instance the one major difference in many of the design decisions is the inclusion of the PDSN/SGSN to support wireless data mobility. However with 4G only a packet network exists and there is no longer any circuit switching involved.

Fundamentally, a wireless communication system has three possible system designs:

- Existing system expansion (no new access platforms)
- New system design (700 MHz, *Advanced Wireless Services* (AWS), etc.)
- Introduction of a new technology platform to an existing system

The system design needs to factor into the process all the components that make up the path the radio signal takes, as well as how the individual base stations are integrated into a larger system. Additionally, the method of voice and data service treatment needs to be included. The specific procedures that need to be followed vary depending on the market, the individual technology platform being installed, and the type of legacy system that is in place, if any. It is important to restate that you need to know what your objective is from the onset of the design process and that the objective needs to be linked to the business and marketing plans of the company. Following the direction of design discovery ("We will build it and they will come") has resulted in some very negative consequences in the wireless industry to date.

With this said, the following is a brief list of the general design procedures that need to be performed whether the system is a new or existing system. If, as in most cases, you first migrate from a 2.5G to a 3G platform and then from a 3G to a 4G platform, the design procedure to follow is that of introducing new technology for both scenarios.

11.1.1 Planning Process Flow

With any good design, a process is required. The recommended planning process flow is shown in Figure 11.3. Several concepts need to be stressed. The first is that for the design to have any relevance, there is a requirement for external inputs from various departments within the company. Specifically, what services to offer and their volume and

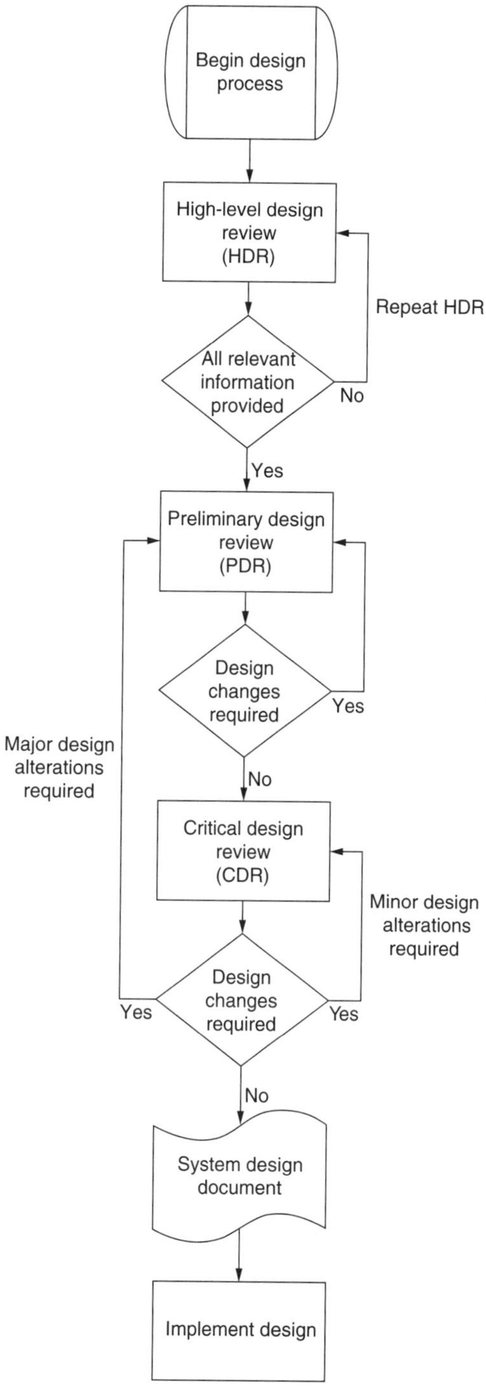

FIGURE 11.3 Design process.

treatment requirements need to be defined at the onset of the design process. Without critical marketing, sales, customer-care, and legal/regulatory guidance, the technical design in all likelihood will not support the desired services the company wants to sell.

Therefore, each of the functional disciplines within the company needs to be involved with the overall high-level design review, but only after the elements can be broken down by functional group. Of course, there would be several layers within each of the efforts involving a *Preliminary Design Review* (PDR) and a *Critical Design Review* (CDR). Regardless of the number of sublayers within the PDR, there also would be a final design review meeting where the output of all the different functional groups is combined into one unified package that has both network and RF components.

Additionally, the network components also should include the transmission planning aspects, not just the infrastructure requirements.

It is important to note that while operations and implementation are not included in the design process by name, they are involved in the system planning. Although the engineering departments are responsible for generation of the technical design, they will need to coordinate their information with operations and implementation to ensure that the design is possible and that specific network requirements from operations are not excluded by mistake. If your organization has the performance group incorporated into the operations division or department, then that group will need to be involved directly in the RF planning aspects.

11.1.2 New Wireless System Procedure

The RF design process for a new 3G or 4G system has many of the same basic processes and is used for legacy wireless system designs. However, subscriber usage and treatment methods for packet data (QoS) need to be factored into the design as well as the inclusion of voice services and whether they will be handled by circuit switching (3G) or packet (4G). The steps to do this are as follows:

1. Obtain a marketing plan and objectives.
2. Establish a system coverage area.
3. Establish system on-air projections.
4. Establish technology platform decisions.
5. Determine the maximum radius per cell (link budget).
6. Establish environmental corrections.
7. Determine the desired signal level meeting the QoS design criteria.
8. Establish the maximum number of cells to cover the area.
9. Generate the coverage-propagation plot for the system.
10. Determine subscriber usage.
11. Determine usage per square kilometer (voice and packet).
12. Determine the maximum number of cells for capacity.
13. Determine if the system is capacity- or coverage-driven.
14. Establish the total number of cells required for coverage and capacity.
15. Generate the coverage plot, incorporating coverage and capacity cell sites (if different).

16. Reevaluate the results and make assumption corrections.
17. Determine the revised (if applicable) number of cells required for coverage and capacity.
18. Check the number of sites against the budget objective; if it exceeds the number of sites, reevaluate the design.
19. Using a known database of sites, overlay this onto the system design and check for matches or close matches.
20. Adjust the system design using site-specific parameters from known database matches.
21. Generate propagation and usage plots for system design.
22. Evaluate the design objectives with time-frame and budgetary constraints and readjust if necessary.
23. Issue search rings.

11.1.2.1 Migration RF Design Procedure

The process for introducing a new radio access platform into an existing wireless system needs to account for the impact reallocation of the spectrum will have on the legacy system. Also, the design needs to address the new platforms and modifications to existing platforms that are needed to facilitate introduction of the new system. Therefore, the following is a brief summary of the main issues that need to be addressed when integrating a new platform into an existing system:

1. Obtain a marketing plan.
2. Establish a technology platform introduction time table.
3. Determine new technology implementation tradeoffs.
4. Determine a new technology implementation methodology (footprint and 1:1 or 1:N).
5. Identify coverage problem areas.
6. Determine the maximum radius per cell (link budget for each technology platform).
7. Establish environmental corrections.
8. Determine the desired signal level (for each technology platform).
9. Establish the maximum number of cells to cover an area(s).
10. Generate the coverage-propagation plot for the system and the areas, showing before-and-after coverage.
11. Determine the subscriber usage (existing and new, packet and voice).
12. Determine the subscriber usage by platform type.
13. Allocate the percentage of system usage to each cell.
14. Adjust the cells' maximum capacity by the spectrum reallocation method (if applicable).
15. Determine the maximum number of cells for capacity (technology-dependent).

16. Establish which cells need capacity relief.
17. Determine the new cells needed for capacity relief.
18. Establish the total number of cells required for coverage and capacity.
19. Generate the coverage plot, incorporating the coverage and capacity cell sites (if different).
20. Reevaluate the results and make assumption corrections.
21. Determine the revised (if applicable) number of cells required for coverage and capacity.
22. Check the number of sites against the budget objective; if it exceeds the number of sites, reevaluate the design.
23. Using the known database of sites, overlay on the system design and check the matches or close matches.
24. Adjust the system design using site-specific parameters from known database matches.
25. Generate the propagation and usage plots for the system design.
26. Evaluate the design objectives with time-frame and budgetary constraints and readjust if necessary.
27. Issue search rings (if needed).

The preceding two procedures can be crafted easily into a checklist for the design team to follow. Obviously, the lists are generic and need to be tailored to the specific situation. However, when using the lists, whether for a new system or for migration from 2G to 2.5G or 3G or from a 2.5G/3G to 4G, they should provide sufficient guidance to organize the process for a successful design that will meet customer and business objectives for the wireless company.

11.2 Methodology

Although more subjective at times, the methodology that is followed for the RF design process is essential in establishing an RF design that correlates with the business plan of the wireless company. The methodology that is followed for the RF design process involves a look at which services need to be supported, where they will be supported, and how they will be supported. Answering the four fundamental questions of what, where, when, and how will determine the methodology for the design:

- *What* defines what you are trying to accomplish with the design.
- *Where* clarifies the issue of where the service will be introduced.
- *When* defines a time frame to follow.
- *How* clarifies the concept of how this will be realized.

More specifically, if the plan is to offer high-speed packet-data services for fixed applications only and medium-speed packet services for mobility, then the methodology for implementing the packet-data services will have a direct impact on RF system design. Another issue is a decision to bifurcate the system by introducing two new platforms. Yet another issue is whether the choice is to plan for a 4G implementation or just a 3G platform with the proviso that migration to a full 4G platform will not be considered.

Regarding implementation of the technology, a decision needs to be made as to how it will be introduced. Two possible alternatives involve a direct 1:1 overlay on the existing system or introduction of the new technology with a limited footprint only, say, in the core of the network.

Therefore, the methodology chosen determines the fundamental direction of the RF design itself. The methodology obviously should not be left to the purview of the technical services group but needs to have direct involvement from senior management from marketing, sales, customer service, new technology, operations, implementation, network engineering, and, of course, RF engineering.

11.2.1 RAN Migration Methodology

The methodology used for migrating, or rather integrating, a new radio access platform into a legacy network is complicated and always takes longer to complete than desired or planned. The complication is set forth in terms of market pressure, as well as physical logistics, such as lease amendments or obtaining a building permit, let alone having a decision made within the company itself.

Several factors need to be accounted for when introducing a new radio access system into a legacy system, and they are

- Objective (services and timeline)
- Spectrum
- Access method or methods
- Voice and data requirements
- Legacy system
- Subscriber migration process

Obviously, you may be thinking of a few other items that need to be factored into this mix, but the preceding items are an excellent starting point.

First in the process, as with all design issues, is determination of the fundamental objective with services—sometime general concepts—along with a timeline. The timeline needs to be compared against reality for completion owing to supplier issues, preseeding customer base, and or land-use acquisition issues, to mention three items.

11.2.2 Spectrum

Spectrum is interesting in that in most cases introduction of a new system into a legacy environment requires the freeing up of spectrum. The freeing up of spectrum requires that the existing infrastructure now handle the same offered traffic with fewer spectrum resources. Some operators have sufficient spectrum to accomplish this. However, more times than not, the operator needs to clear spectrum in order to introduce the new radio access platform unless unused or underutilized spectrum is acquired.

The clearing of spectrum usually coincides with an expansion of the existing legacy system to handle the traffic load. To make a simple example, if you had 10 MHz of *Frequency Division Duplex* (FDD) spectrum operating *Global System for Mobile* (GSM) communications and wanted to introduce *Universal Mobile Telecommunications Service* (UMTS, i.e., WCDMA), this would require the GSM system to operate on only half the spectrum it had before. The obvious impact of this decision is the increase in GSM sites needed to handle the traffic load. If the operator is also operating a 2G network such as IS-136, the complication increases owing to additional system support and migration planning. Additionally, if you deploy say 4G LTE as a 10 × 10 FDD system and you have only a 10 × 10 block of spectrum used for UMTS (5 × 5) and GSM, then alternative spectrum allocations need to be sought.

The recommended method for obtaining spectrum involves only a few viable options:

- Expand the legacy system to accommodate the traffic shift.
- Acquire more spectrum (auctions or acquiring a competitor).
- Establish a *Mobile Virtual Network Operator* (MVNO) arrangement and migrate customers to that system, allowing the new network to be built.

Obviously, when parsing up the spectrum for multiple technologies, it is essential that the spectrum remains contiguous so as to prevent further complications in the future as the 3G or 4G network expands and the legacy system is methodically decommissioned.

Another important issue is preseeding the customer base with multiband/mode handsets. The problem with relying on this approach, however, is that the customer base may or may not be uniformly distributed in the network, leading to a nonuniform traffic shift.

11.2.3 Radio Access Method

The radio access method chosen has a large impact on the migration methodology used. However, before leaping to the technical issues alone, there are marketing and business issues related to services offered that are at the heart, or should be, of the decision regarding which *Radio Access Network* (RAN) to choose.

As indicated, the RAN used has a direct impact on the migration process. Referring to Figure 11.4, which shows the various migration paths for the radio access method, you will notice that IS-136 has no migration path to 3G. Additionally, a GSM network has a different RAN than a *Wideband Code Division Multiple Access* (WCDMA) system requiring an overlay. The same argument can be made with a *Time Division–Synchronous Code-Division Multiple Access* (TD-SCDMA) or *Time Division–Code-Division Multiple Access* (TD-CDMA) network with GSM or a CDMA2000 1xRTT network with *Evolution Data Optimized* (EVDO)/*Evolution Data and Voice* (EVDV).

For instance, using TD-SCDMA, if it is introduced into the GSM network, then eight GSM channels (8 × 200 kHz) need to be cleared for a total of 1.6 MHz. This method facilitates the ability to introduce a second TD-SCDMA channel as well because it is *Time-Division Duplex* (TDD) and the GSM network is *Frequency-Division Duplex* (FDD). Therefore, if you had 5-MHz FDD of radio spectrum, this would require migration of 36 percent of the GSM band for the new service.

TD-CDMA requires 5 MHz of spectrum, and this requires a total of 25 GSM channels (25 × 200 kHz). Again, since TD-CDMA is a TDD network, both portions of the FDD system need to be cleared. Obviously, if you had a 5-MHz network, then to introduce TD-CDMA, you would require elimination of the GSM service. Alternatively, if you had 10 MHz of FDD spectrum, then a single TD-CDMA channel would require that 50 percent of your usable spectrum be cleared for this new RAN.

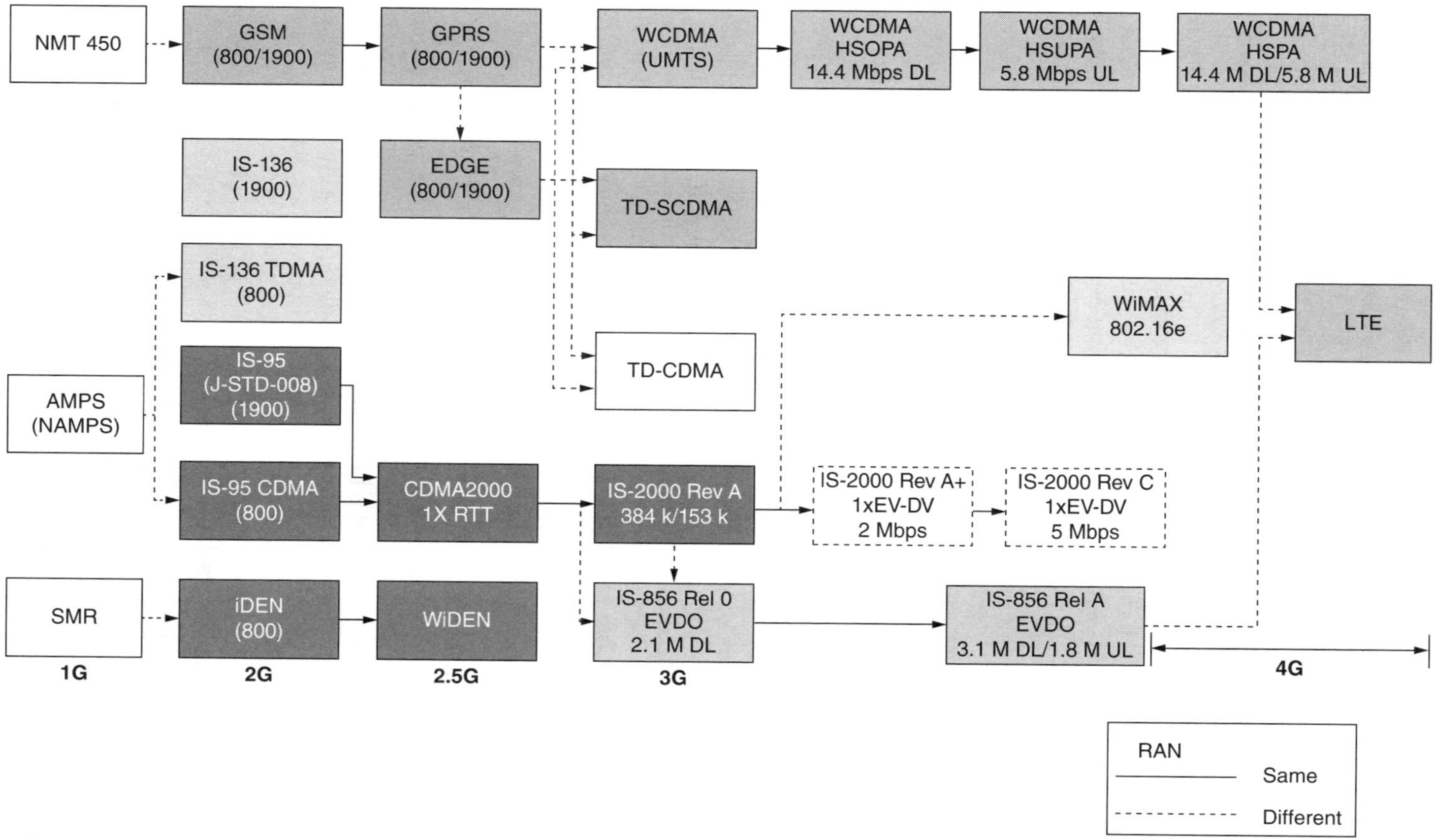

Figure 11.4 RAN migration paths.

Continuing with one other example, if the desire is to introduce EVDO into a CDMA2000 (1xRTT) network, then a CDMA2000 (1xRTT) carrier would need to be displaced. If, for example, you had a 5-MHz license allowing for three 1xRTT carriers, then 33 percent of the band would need to be cleared to support a single EVDO carrier. The one key advantage with EVDO, though, is that it is backward-compatible with 1xRTT, allowing for a gradual introduction and not the harsh clear-first-then-populate with subscribers.

If introducing LTE into a WiMAX wireless system, it is important to note that WiMAX is deployed as a TDD system using 10-MHz channels.

The introduction of LTE into a system is typically deployed as a 10 × 10 FDD channel; however, it also can be deployed as a TDD system. LTE again is a different radio access method than existing 2G/3G. However, LTE can be deployed in different channel bandwidth configurations facilitating legacy migration strategies. For instance, LTE can be deployed as a 5 × 5 MHz system displacing maybe only one UMTS channel or possibly four EVDO channels. This would facilitate the existing subscriber base as well as provide the platform to migrate the customers onto the new system.

11.2.4 Traffic Forecast

Voice and data requirements are a key element in a migration path strategy. For instance, if the desire is to have 3G systems for packet-only services, then service offerings and appliances need to be put in the customers' hands to ensure that this can take place. If you are deploying a 4G network, then it is packet-based and has no circuit switching. It is recommended that to facilitate migration from legacy to 3G and 4G, voice services should be available only on the legacy network, new data offerings, improved ring tones, e-mail, *Short Message Service* (SMS), and the like should be available only on the 3G network. An important point to bring up is that if the desire is to just offer voice services on 3G, then the objective of deploying a 3G network needs to be reviewed. With the above said, the migration to packet voice for 4G systems is being implemented. The amount of type of voice services you desire to have on the new 4G network, as well as the migration strategy for transitioning circuit switch voice to a pure packet-based method, needs to be included in the forecast process.

11.2.5 Legacy System

The support of the legacy system is as important as or even more important than deployment of a new system. The rationale for the support of the legacy system is that the customer base is using the legacy system and is funding the expansion of the 3G and 4G network.

The legacy system needs to have a decommissioning plan, as well as a method and dedication to operating the legacy system, so that it does not degrade. More often than not the new system gets the attention, and the legacy system is left to its own devices.

11.2.6 Subscriber Migration Process

One of the often-overlooked issues for migrating to a 3G and/or 4G network is establishment of a viable plan by sales and marketing that is in concert with the technical community for migrating customers to the new system rationally. Some of the methods that can and should be used involve

- Preseeding the existing subscriber base with multimode/band devices
- Incenting customers to acquire the new multimode/band devices

- Incenting the sales/retail/wholesale distribution channels to no longer distribute legacy devices

Obviously, the key to any process, migration and new service or platform introduction, is to be able to offer new services that customers can use at the right price. Without having thought out in advance the service delivery and applications that will be offered, you run the risk of deploying technology for technology's sake.

11.3 Link Budget

The establishment of a link budget is one of the first tasks that the RF engineer needs to perform when beginning the design process. The link budget can be established only after a decision has been made as to which technology platform(s) to use. When introducing, say, a 2.5G platform into a 3G or 4G system, it will be necessary to have a link budget for each of the individual technology platforms involved. In addition, with the introduction of packet data, the higher data rates have a direct influence on the range of the site and/or its capacity.

What exactly is a link budget? A *link budget* is a power budget that is one of the fundamental elements of a radio system design. The link budget is the part of the RF system design where all the issues associated with propagation are included. Simply put, the link budget can either be forward- or reverse-oriented; it must account for all the gains and losses that the radio wave will experience as it goes from transmitter to receiver.

The link budget is the primary method that an RF engineer must first determine to ascertain if a valid communication link can and does exist between the sender and the recipient of the information content. The link budget, however, incorporates many elements of the communication path. Unless the actual path loss is measured empirically, the RF engineer has to estimate, or rather predict just how well the RF path itself will perform. The many elements involved in the communication path incorporate assumptions made regarding various path impairments.

Figure 11.5 shows which part of the radio communication path the link budget tries to account for. The link budget has two paths: uplink and downlink. The uplink path is the path from the subscriber unit to the base station. The downlink path is the path from the base station to the subscriber unit. Both the uplink path and the downlink path are

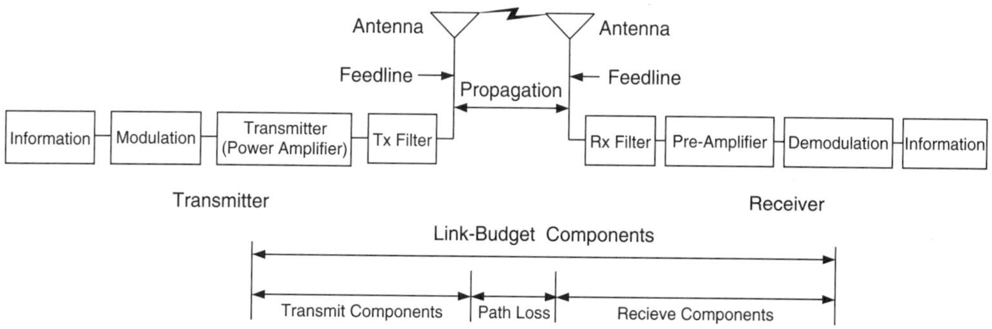

FIGURE 11.5 Radio path components.

	Downlink	Uplink
Transmit (ERP)	50 dBm	36 dBm
Rx antenna gain	3 dBd	12 dBd
Cable loss	2 dB	3 dB
Rx sensitivity	116 dBm	116 dBm
C/N ratio	17 dB	17 dB
Max path loss	150 dB	144 dB

Table 11.1 1G Link Budget

reciprocal, provided that they are close enough in frequency. However, the actual paths should be the same, with the exception of a few key elements that are hardware-related. The actual loss in the path the radio wave transverses from antenna to antenna is the same whether it is an uplink or downlink path.

The maximum path loss, or *limiting path,* for any communication system determines the effective range of the system. Table 11.1 illustrates a simplistic calculation of a link budget associated with a 1G system and is used for determining which path is the limiting case from which to design. In this example, the receiver sensitivity value has thermal noise, bandwidth, and noise figures factored into the final value presented.

The uplink path, defined as mobile to base, is the limiting path case. As shown in the table, the talk-back path is 6 dB less than the talk-out path. The limiting path loss then is used to determine the range for the site using the propagation model for the network.

However, with introduction of a 2.5G and/or 3G platform into a wireless system, the issue of the components of the link budget becomes more complicated. The complications arise because of differing modulation techniques, bandwidth as well as process gain, and finally, the E_b/N_o or *Carrier-to-Noise* (*C/N*) ratio values required for a proper *Bit Error Rate* (BER) or *Frame Error Rate* (FER). The link budgets for 3G technologies are included in their respective chapters and will not be referenced here; instead, because many issues are associated with 3G systems and their associated legacy systems, the fundamental issues involved in a link budget will be discussed.

When putting together a link budget, it will be common to have more than one link budget based on the morphology and, of course, the technology platform used. However, the morphology variation in the actual link budget is included in the propagation analysis for the particular site and varies depending on local particulars. The link budget itself involves establishment of the maximum path loss, either in the uplink or the downlink, that the signal can attenuate while still meeting the system design requirements for a quality signal.

When calculating the actual link budget, the items in Tables 11.2 and 11.3 are recommended to be included in the calculation. These items should have more items included in them than what might be used in the physical system being installed. However, all the items that can affect the link budget are included for reference. It is also highly possible that other devices can be added in the path to either enhance or potentially degrade the performance of the network.

These tables help to define the forward and reverse radio path components that make up the forward and reverse link budgets. One final note on the path is that certain

Downlink Path		Units
Base-station parameters	Tx PA output power	dBm
	Tx combiner loss	dB
	Tx duplexer loss/filter	dB
	Jumper and connector loss	dB
	Lightning arrestor loss	dB
	Feedline loss	dB
	Jumper and connector loss	dB
	Tower-top amp Tx gain or loss	dB
	Antenna Gain	dBd or dBi
	Total power transmitted (ERP/EIRP)	**W or dBm**
Environmental margins	Tx diversity gain	dB
	Fading margin	dB
	Environmental attenuation (building, car, pedestrian)	dB
	Cell overlap	dB
	Total environmental margin	**dB**
Subscriber unit parameters	Antenna gain	dBd or dBi
	Rx diversity gain	dB
	Processing gain	dB
	Antenna cable loss	dB
	C/I or E_b/N_o	dB
	Rx sensitivity	dB
	Effective subscriber sensitivity	**dBm**

Table 11.2 Generic Downlink Link Budget

wireless access technologies use different modulation formats on both the uplink and the downlink paths. If this is the case, then some of the reciprocity may not be applicable.

Because the link budget is such an integral part of the RF design process, the link budget used for the system design needs to be documented and made available for the design community to use.

11.4 Propagation Models

The use of propagation modeling is a requirement in the RF design process. The propagation modeling techniques used are meant to determine the attenuation of the radio wave as it transverses from the transmitter antenna to the receiver's antenna. The propagation model therefore is meant to characterize the radio path shown in Figure 11.6.

As with all aspects of radio design, numerous methods are used in the course of arriving at the desired result (i.e., how much attenuation did the signal experience, and does it exhaust the values defined in the link budget?).

Some of the most popular propagation models used are Hata, Carey, Elgi, Longley-Rice, Bullington, Lee, and cost231, to mention a few. Each of these models has advantages

Uplink		Units
Subscriber unit parameters	Tx PA ouput	dBm
	Cable and jumper loss	dB
	Antenna gain	dBd or dBi
	Subscriber unit total Tx power (ERP, EIRP)	**W or dBm**
Environmental margins	Tx diversity gain	dB
	Fading margin	dB
	Environmental attenuation (building, car, pedestrian)	dB
	Total environmental margin	**dB**
Base-station parameters	Rx antenna gain	dBd or dBi
	Tower-top amp net gain	dB
	Jumper and connector loss	dB
	Feedline loss	dB
	Lightning arrestor loss	dB
	Jumper and connector loss	dB
	Duplexer/Rx filter loss	dB
	Rx diversity gain	dB
	C/I E_b/N_o	dB
	Processing gain	dB
	Rx sensitivity	dBm
	Base-station effective sensitivity	**dBm**

TABLE 11.3 Generic Uplink Link Budget

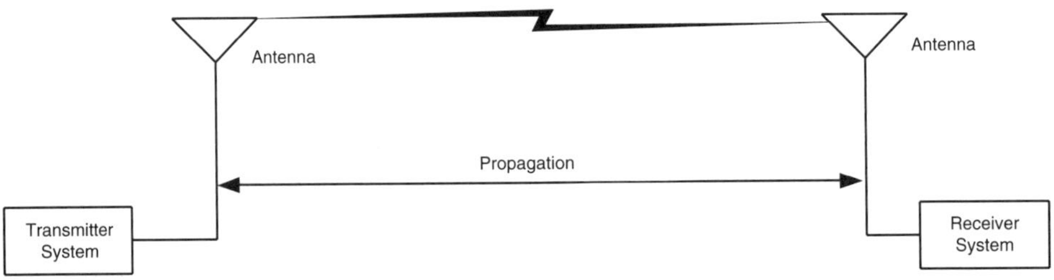

FIGURE 11.6 Propagation path.

and disadvantages. Specifically, some baseline assumptions are used with any propagation model and need to be understood prior to employing the model. Most cellular operators use a version of the Hata model for conducting propagation characterization. The Carey model, however, is used for submitting cell-site filing information to the *Federal Communications Commission* (FCC). Cellular and *Personal Communication Services* (PCS) operators use either Hata or cost231 as their primary methods for determining path loss. With the introduction of 3G, cost231 is the model of choice for application to any of the spectrum allocations defined by the *International Telecommunications Union* (ITU).

Regardless of the frequency band of operation, the model used for predicting coverage needs to factor in a large number of variables that have a direct impact on actual RF coverage prediction. The positive attributes affecting coverage are the receiver sensitivity, transmit power, antenna gain, and antenna height above average terrain. The negative factors affecting coverage involve line loss, terrain loss, tree loss, building loss, electrical noise, natural noise, antenna pattern distortion, and antenna inefficiency, to mention a few.

With the proliferation of cell sites, the need to predict the actual path loss experienced in the communications link is becoming more and more critical. To date, no overall theoretical model has been established that explains all the variations encountered in the real world. However, as the cellular PCS, UMTS, and AWS communication systems continue to grow, a growing reliance is being placed on propagation-prediction tools. Reliance on a propagation tool is intertwined in the daily operation of the wireless communication system. The propagation model employed by the wireless operator has a direct impact on the capital build program of the company for determining the budgetary requirements for the next few fiscal years. Therefore, it is essential that the model used for the propagation prediction tool be understood. The model should be understood in terms of what it actually can predict and what it cannot predict.

Over the years, numerous articles have been written about propagation modeling in the cellular communications environment. With the introduction of PCS and UMTS, there was and continues to be an intense focus on refining the propagation models to assist in planning out the networks. However, no one model can predict every variation that will take place in the environment. To overcome this obstacle, some operators have resorted to using a combination of models depending on the environmental conditions relevant to the situation.

In addition to which model would be the best to use, other perturbations to the model need to be considered. One of the most basic considerations is determining the morphology to which the model will be applied. Morphologies normally are defined in four categories: dense urban, urban, suburban, and rural. The selection of which morphology to use at times is more of an art than a direct science, and this often leads to gross assumptions being made for a geographic area. The morphologies generally are defined using a rough set of criteria:

- *Dense urban.* This is normally the dense business district for a metropolitan area. The buildings for the area generally are 10 to 20 stories or above, consisting of skyscrapers and high-rise apartments.
- *Urban.* This type of morphology usually consists of building structures that are from 5 to 10 stories in height.
- *Suburban.* This morphology is a mix of residential and business premises, with the buildings ranging from one to five stories but mainly consisting of one- to two-story structures.
- *Rural.* This morphology, as the name applies, generally consists of open areas with structures not exceeding two stories and that are sparsely populated.

From these morphologies, it may seem obvious that classifying an area is rather ambiguous because the geographic size of the area is left to the engineer to define. (The use of population and employment databases helps to standardize the definitions of each morphology category. With the small cell sizes required to provide data-service mobility, we are finding that for greenfield designs, the morphology definitions and service-area-footprint requirements are key to the business plan.)

As mentioned earlier, several propagation models are used currently throughout the industry, and each of the models has pros and cons. It is through understanding the advantages and disadvantages of each of the models that a better engineering design actually can be created for a network.

11.4.1 Free Space

Free-space path loss is usually the reference point for all the path loss models employed. Each propagation model points out that it more accurately predicts the attenuation experienced by the signal over that of free space. The equation used for determining free-space path loss is based on a 20 dB/decade path loss. The free space equation is as follows:

$$L_f = 32.4 + 20\log(R) + 20\log(f)$$

where R is in kilometers, f is in megahertz, and L_f is in decibels.

The free-space path loss equation has a constant value that is used for the air-interface loss and a distance and frequency adjustment. Using some basic values, the different path-loss values can be determined for comparison with other models that will be discussed.

11.4.2 Hata

The most prolific path-loss model employed in cellular presently is the empirical model developed by Hata or some variant of it. The Hata model is an empirical model derived from the technical report made by Okumura, so the results can be used in a computational model. The Okumura report is a series of charts that are instrumental in radio communication modeling. The Hata model is as follows:

$$L_H = 69.55 + 26.26\log(f) - 13.87\log(h_b) - a(h_m) + (44.9 - 6.55\log h_b)\log R$$

where $f = 150$–1500 MHz frequency

$R = 1$–20 km distance from the site

$h_m = 1$–10 m height of receive antenna above ground

$h_b = 20$–200 m height of base station above ground

$L_H = \text{dB}$

It should be noted that some additional conditions are applied when using the Hata model as compared with the free-space equation. The values used depend upon the range over which the equation is valid. If parameters outside the values are used, the equation is defined for the results and will be suspected to error.

Therefore, the Hata model should not be employed when trying to predict a path loss less than 1 km from the cell site or if the site is less than 30 m high. This is an interesting point to note because cellular sites are being placed less than 1 km apart and often below the 30-m height.

In the Hata model, the value h_m is used to correct the mobile antenna height. The interesting point is that if you assume a height of 1.5 m for the mobile, that value nulls out or becomes 0 in the equation.

A critical point to mention here is that the Hata model employs three correction factors based on the environmental conditions under which path-loss prediction is evaluated. The three environmental conditions are urban, suburban, and open.

The environmental correction values are calculated easily but vary for different values of mobile height. For the following values, a mobile height of 1.5 m has been assumed:

Urban	0 dB
Suburban	29.88 dB
Open	28.41 dB

11.4.3 Cost231—Walfish/Ikegami

The cost231 Walfish/Ikegami propagation model is used for estimating the path loss in an urban environment for wireless communication systems. The cost231 model is a combination of empirical and deterministic modeling for estimating the path loss in an urban environment over the frequency range of 800 to 2000 MHz. The cost231 model is used primarily in Europe for GSM modeling and for some propagation models used for cellular in the United States.

The cost231 model consists of three basic components:

1. Free-space loss
2. Roof-to-street diffraction loss and scatter loss
3. Multiscreen loss

The equations that make up the cost231 model are listed next:

$$L_c = \begin{cases} L_f + L_{RTS} + L_{ms} \\ L_f \text{where } L_{RTS} + L_{ms} \leq 0 \end{cases}$$

$$L_f = \text{Free space loss}$$

$$L_{RTS} = \text{Rooftop-to-street diffraction and scatter loss}$$

$$L_{ms} = \text{Multi-screen loss}$$

$$L_f = 32.4 + 20 \log R + 20 \log_{10} f_c$$

where R = range or distance (km)

f_c = carrier frequency (MHz)

$$L_{RTS} = -16.9 - 10 \log_{10} W + 10 \log f_c + 20 \log \Delta h_m + L_o$$

where ω = street width, m

$\Delta hm = h_r - h_m$

$$L_o = \begin{cases} -10 + 0.354\phi & 0 \leq \phi \geq 35 \\ 2.75 + .075(\phi - 35) & 35 \geq \phi \geq 55 \\ 4.0 - 0.114(\phi - 55) & 55 \leq \phi \geq 90 \end{cases}$$

where ϕ = the incident angle relative to the street.

$$L_{ms} = L_{bsh} + k_a + k_d \log R + k_f \log f - 9 \log b$$

where b = the distance between buildings along the radio path.

$$L_{bsh} = -18 \log (1 - \Delta h_b) \text{ when } h_b > h_r$$
$$= 0 \text{ when } h_b < h_r$$

and

$$k_a = 54 \qquad \text{when } h_b > h_r$$
$$= 54 - 0.8\, h_b \qquad \text{when } d >= 500 \text{ m and } h_b <= h_r$$
$$= 54 - 1.6\, h_b \times R \qquad \text{when } d < 500 \text{ m and } h_b <= h_r$$

Both L_{bsh} and k_a increase the path loss with a lower base station antenna.

And for the final k factor,

$$k_f = 4 + 0.7(f/925 - 1) \text{ for a midsized city and suburban area with moderate tree density}$$
$$= 4 + 1.5(f/925 - 1) \text{ for a metropolitan center}$$

As with the Hata equation, the equation is designed to operate within a useful range, which is shown here:

$$f = 800\text{–}2000 \text{ MHz}$$
$$h_b = 4\text{–}50 \text{ m}$$
$$h_m = 1\text{–}3 \text{ m}$$
$$R = 0.02\text{–}5 \text{ km}$$

Some additional default values apply to the cost231 model when specific values are not known. The default values can and will alter the path-loss values resulting from the model significantly. The recommended default values are listed in the following section.

$$B = 20\text{–}50 \text{ m}$$
$$W = b/2$$
$$h_r = 3 \times (\#\text{ floors}) + \text{roof}$$
$$\text{roof} = 3 \text{ m for pitched and 0 for a flat roof}$$
$$\phi = 90 \text{ degrees}$$

Figure 11.7 helps to bring the cost231 equation variables into perspective. In the preceding equations that make up the cost231 model, it is important, as always, to know what the valid ranges are for the model.

11.4.4 Cost231–Hata

The cost231 Hata model has been tailored for the PCS 1900-MHz environment and is being used by many PCS operators in establishing their system design. The equation

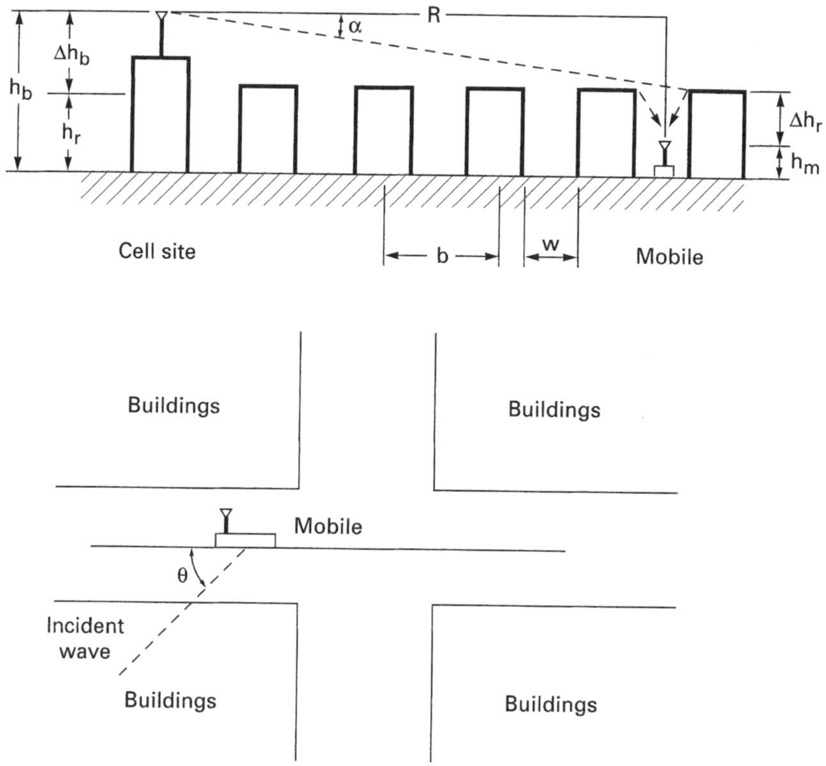

Figure 11.7 Cost231 parameter diagram.

used for cost231 Hata is shown here. The cost231 Hata model is similar to the Hata model with the exception that frequency and correction factors are added based on the morphology to which the model is applied.

$$L_{CH} = 46.3 + 33.9 \log f - 12.82 \log h_b + (44.9 - 6.55 \log h_b)\log d + c$$

where c is

13 dB	Dense urban
0 dB	Urban
–12 dB	Suburban
–27 dB	Rural

11.4.5 Quick

The quick model is a down-and-dirty estimate that can be used to determine the general propagation expectations for the area. The model is rather simplistic and straightforward. The advantage of this model is its quickness for use in roughly estimating the situation at hand. The disadvantage is that it lacks the refinement of the other models. The quick method should be used when conducting a generalized approach to a cell design that requires a rough answer.

The quick method uses two equations, one for cellular (880 MHz) and the other for PCS (1800–2000 MHz).

$$880\text{ MHz } P_L = 121 + 36\log(\text{km})$$

$$1800 - 2000\text{ MHz } P_L = 130 + 40\log(\text{km})$$

The quick method gives a reasonable approximation for a propagation prediction over a variety of morphologies and can be used when details regarding the particular environment may not be readily available. Additionally, the quick method can be used for both 700 MHz as well as AWS where the 880 MHz is used for 700 MHz and the 1800–2000 MHz is used for AWS.

Regardless of which model you use for your analysis, it should undergo continuous examination to ensure it is truly being a benefit to the company as a whole. The propagation model employed by the engineering department not only determines the capital build program but also plays a direct role in the performance of the network. The RF design is affected directly by the propagation model chosen and particularly by the underlying assumptions that accompany the particular model.

The propagation model is used to determine how many sites are needed to provide a particular coverage requirement for the network. In addition, the coverage requirement is coupled with traffic-loading requirements. These traffic-loading requirements rely on the propagation model chosen to determine the traffic distribution or off-loading from an existing site to new sites as part of the capacity-relief program. The propagation model helps to determine where the sites should be placed to achieve an optimal position in the network. If the propagation model used is not effective in helping to place sites correctly, the probability of incorrectly justifying and deploying a site into the network is high.

Reiterating, although no model can account for all the perturbations experienced in the real world, it is essential that you use one or several propagation models for determining the path loss of your network.

11.5 Tower-Top Amplifiers

Tower-top amplifiers have been deployed in numerous communication sites and are anticipated to be used in the introduction of 2.5G and 3G as well. Tower-top amplifiers occasionally have been misapplied in that the gain exhibited by them is added directly to the link budget. However, the purpose of a tower-top amplifier is to improve the noise figure for the receive system.

The noise figure is improved by having the first amplification stage placed as close as possible to the antenna itself, thereby eliminating the loss owing the feedline that connects the antenna to the rest of the receive system. The location of a tower-top amplifier is shown in Figure 11.8.

The tower-top amplifier has to have a minimal gain of 10 dB, and because the feedline usually has 2 to 4 dB of loss, the additional gain needs to be attenuated by the insertion of a resistive pad, as shown in Figure 11.8.

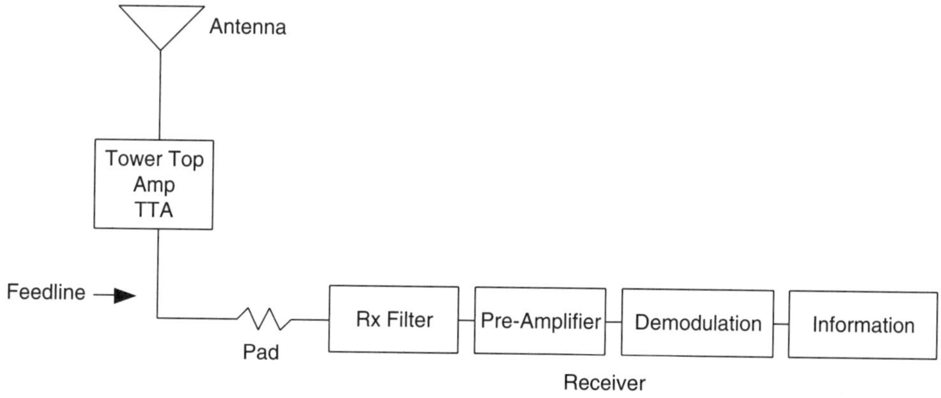

FIGURE 11.8 Tower-top amplifier.

The typical improvement in the receive path due to the introduction of a tower-top amplifier is equal to the line loss that would have been attributed to the feedline, nothing else. The negative issues with tower-top amplifiers include the requirement for power, in dc, to be supplied to the unit and increased maintenance issues in the event of a failure. Another problem is the increased system noise owing to the amplifier having a less-than-optimal receive filter because of size and weight restrictions imposed on installing the unit on a tower.

11.6 Dual Band Coupler (Diplexer)

The proliferation of radio access platforms as well as the plethora of spectrum allocations places extreme pressure on the RF design process. Specifically, the one element that all RF design engineers are faced with is, should you utilize existing feedlines and antennas for the new platform? Obviously the cost advantage with using existing assets is desired; however, there is no "free lunch."

The use of dual-band couplers, also called *diplexers,* is possible to enable multiple-frequency bands to utilize an existing feedline of a cell site. Now if the new radio access platform uses fiber for connecting the radio head to the controller at the site where the radio heat is next to the antenna, then sharing the feedline is not viable.

The use of a dual-band coupler is shown in Figure 11.9, where the antenna feedline is shared between different radio technologies. However, when using a dual-band coupler you may lose some diagnostic capabilities, but more important you have introduced an additional component into the RF path, which contributes to noise and possibly will be a source of *Passive Intermodulation* (PIM). When using dual-band couplers several key criteria need to be addressed in its selection including the power levels the dual-band coupler can support and its linearity over the frequency bands used. Additionally, knowledge of the group delay performance is needed in order to support packet data.

11.7 RF Design Guidelines

No true RF engineering can take place without some RF design guidelines, whether formal or informal. However, with the level of complexity introduced when integrating a 3G or 4G platform into an existing system, the need for a clear, definitive set of design

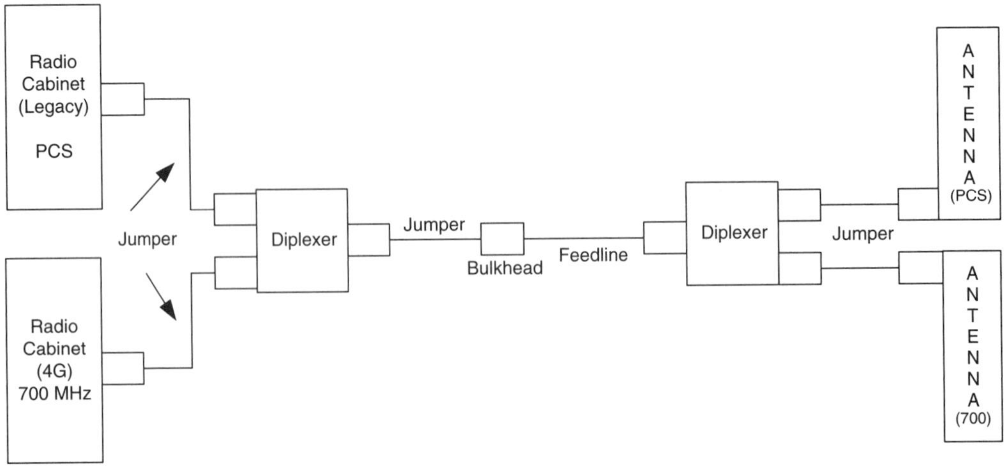

FIGURE 11.9 Diplexer/dual-band coupler.

guidelines is paramount for success. Although this concept seems straightforward and simple, many wireless engineering departments, when pushed, have a difficult time defining what exactly their design guidelines are.

The actual format, or method, of how it is conducted, should be structured in such a fashion as to facilitate ease, documentation, and minimization for formal meetings. For most of the design reviews, a formal overhead presentation is not required; instead, a meeting with the manager of the department is the level of review necessary. It is also important that another qualified member of the engineering staff reviews the material in order to prevent the common or simple mistakes from taking place. Ensuring that a design review process is in place does not eliminate the chance of mistakes occurring. Design reviews ensure that when mistakes do occur, the how, why, and when issues needed to expedite the restoration process are already in place.

It is highly recommended that the department's RF design guidelines be reasonably documented and updated on a predetermined basis, yearly at a minimum. The use of design guidelines will facilitate the design review process and establish a clear set of directions for the engineering department to follow. The RF design guidelines also will ensure that a consistent approach is maintained for designing and operating the capital infrastructure that has been or will be put into place within the network.

The actual design guidelines that should be used by the RF engineers need to be well-documented and distributed. The design guidelines, however, do not need to consist of voluminous amounts of data. The design guidelines should consist of a few pages of information that can be used as a quick reference by the engineering staff. The design guidelines sheet has to be based on the system design goals and objectives set forth in the RF system design.

The actual content of the design guidelines can and will vary from operator to operator. However, it is essential that a list of design guidelines be put together and distributed. The publication and distribution of RF design guidelines will ensure that a minimum level of RF design specifications exists in the network.

The proposed RF design guidelines are shown in Table 11.4 for a generic wireless system. The guidelines can be crafted easily to reflect the particular design guidelines

System Name:				
Date:				
RSSI	**ERP**	**Cell Area**	**Antenna Type**	
Urban 80 dBm	16 W	3.14 km^2	12 dBd	90H/14E
Suburban 85 dBm	40 W	19.5 km^2	12 dBd	90H/14E
Rural 90 dBm	100 W	78.5 km^2	10 dBd	110H/18E
E_b/N_o	7 dB (90th percentile)			
Frequency reuse or $N = 1$				
Maximum no. carriers per sector				
Maximum no. traffic channels per carrier				
Maximum kbps per sector				
Sector cell orientation	0, 120, 240			
Antenna height	100 ft or 30 m			
Antenna pass band	XXX–XXX MHz			
Antenna feedline loss	2 dB			
Antenna system return loss	20–25 dB			
Diversity spacing	$d = h/11$ (d = receive antenna spacing, h = antenna AGL)			
Receive antennas per sector	2			
Transmit antennas per sector	1			
Roof height offset	$h = x/5$ (h = height of antenna from roof, x = distance from roof edge)			
Performance criteria				
Lost call rate	<2%			
Attempt failure	<2%			
RF blocking (voice)	1–<2%			
FER	<1%			

Table 11.4 RF Design Guidelines (Per Radio Access Platform)

used for the market where it will be applied. In addition, a need exists to have RF design criteria for each technology platform with links to each other to ensure that one platform is not factored over another by mistake.

11.8 Traffic Projections

There is a process and methodology for conducting RAN traffic engineering, that is, determining the amount of physical and logical resources that need to be in place at different points and nodes within the network to support the current and future traffic. The determination of existing traffic loads is rather more straightforward in that you have existing information from which to make decisions. For future forecasts, the level of uncertainty grows exponentially the further the forecast or planning takes you into the future. However, many elements in the network require long lead times, ranging from 3 weeks to over 1 year to implement. Obviously, the goal of traffic engineering is to design the network and its subcomponents not only to meet the design criteria, which should be driven by both technical and marketing and sales information, but also to achieve a good design in a cost-effective manner. It is not uncommon to have conflicting objectives within a design, that is, ensure that the customers have the highest *quality of service* (QoS)/*grade of service* (GoS) for both voice and packet data but yet have a limited amount of capital from which to achieve this goal. Therefore, it is important to define the details at the onset of the design process and have some interim decision points where the design process can be reviewed and altered, if required, either by increasing the capital budget, revisiting the forecast input, or altering the QoS/GoS expectations.

Since different circuit- and packet-switched services will be offered, the variations are vast. However, there are some commonalities that can be drawn on. Several methods can be used to calculate the required or estimated traffic for the network. We cannot provide an infamous F1 button approach for traffic engineering or forecasting with the different wireless technology platforms. This is so because of the number of perturbations that exist between technology platforms and service offerings by each wireless operator. However, what we can provide is a guide from which you can craft the traffic projections to meet your individual market requirements.

It is essential to note that there are several key points within the network where the traffic engineering calculations need to be applied:

- BTS to subscriber terminal (RF)
- BTS to BSC/RNC (RF)
- BSC to PDSN/SGSN
- BSC to MSC/Media Gateway
- MSC/Media Gateway to PSTN/SGSN
- PDSN/SGSN to private and public data network

I have simplified the above 3G. However, with 4G lets say for LTE instead of the BSC/RNC you have the MME/SGW etc. As a good design practice you should have available for quick reference the entire ecosystem of the wireless network so you can easily determine the demarcation points for off-net/on-net traffic engineering.

There are a number of scenarios and unknown perturbations that can occur in the estimation of traffic for a system. In an ideal world, the traffic forecast would be projected by integrating the marketing plan with the business plan coupled with the products that should be integral to both the marketing and business plans. However, reality is much harsher, and usually, very little information is available to the technical team.

Initially, packet-data traffic is low, but at this time, all the pundits indicate that it will become more predominant. One method to increase usage, of course, is to reduce the usage fee, encouraging more loading.

For this chapter, we will focus on the RAN aspects, but it is obvious that traffic engineering crosses multiple disciplines and organizational elements, and failure at any one point affects all the other nodes, with the most important node being the customer itself.

11.8.1 Traffic Tables (Appendix)

There are several tables that need to be used for putting together the RF growth plan. For voice traffic, the most important traffic table is the Erlang B table (see Appendix). Use of an Erlang B table is extremely valuable in putting together any projected equipment requirements by defining the number of individual channels, or time slots, depending on the technology used. It is important to note, however, that the Erlang B model is to be used for circuit-switched, that is, voice, and not packet-based (IP) services.

To facilitate inclusion of an Erlang B table in a typical growth plan for any wireless network, the equation is shown below. The Erlang B equation lends itself to easier insertion into a spreadsheet through use of a look-up table, which is a standard function in all spreadsheet programs available commercially.

$$E = [(A^N/N!)/(\Sigma^N_{n=0}(A^n/n!)]$$

In this equation, E is the Erlang traffic, and N is the number of trunks (voice channels) in the group.

If the traffic is data traffic, how it is projected and accounted for in the estimation of network elements depends on whether it is contention-based or not. If the data traffic is contention-based, that is, subject to radio blocking, one could make the argument that use of an Erlang C table would be appropriate. This is so because the *Internet Protocol* (IP) traffic or other similar traffic can be placed in a queue. However, the inclusion of an Erlang C table in spreadsheets has been problematic owing to its complexity. We have reverted typically to use of a direct look-up and large table entry into a spreadsheet for the look-up table.

Additionally, for contention-based traffic, the issue of overbooking, depending on the service offering, needs to be accounted for in the process. We will assume that the traffic is symmetric, that is, the same downlink and uplink. Therefore, a simple example of overbooking would be to offer 56 kbps as the Mobile IP for, say, 100 subscribers that could be active simultaneously for that sector. Using the linear approach, this would equate to 5.6 Mbps being required for the facility. However, if one were to use a 20:1 service offering and everyone is demanding bandwidth at the same instant, this then becomes a radio network requirement of 280 kbps. Obviously, the issue of overbooking and active subscribers needs to be defined, besides the service offering requirement.

Now, if the traffic is not contention-based, the issue lies in the format in which the data will be encapsulated. You have to determine if the traffic will be transported as IP or encapsulated in another format, such as Frame Relay or ATM, or have dedicated *Time-Division Multiplexing* (TDM) bandwidth, that is, a T1/E1.

A simple example of TDM usage for IP traffic would be a situation in which you have the 280 kbps of traffic and consider it the peak load for any time. TDM, however, is dedicated, and once it is assigned for this effort, that is all it is allowed to do, and this is rather inefficient. The issue of how to calculate the amount of bandwidth for a TDM circuit is always a question.

Using the E1/T1 DS0 as the fundamental transport bandwidth size of 64 kbps, the 280 kbps requires a total of 4.375 DS0s, but this is really 5 DS0s out of a possible 24/30 usable time slots. The reason we mention this is that this could lead to the sharing of facilities, reducing backhaul requirements through the use of drop and insert equipment.

With packet data in 3G and primarily in 4G circuit switch services are not included. Therefore the issue is determining the amount of data traffic the radio access network needs to support as well as where it needs to support it. If you have a 3G/4G network already deployed, then the location and traffic patterns of the packet services is somewhat known. If there is no network, then the concern will be how large of a backhaul do you need to deploy initially that can be grown quickly. Regardless you need to determine the packet data demand.

A simple example is where you have an existing packet network and you have 100 Gbs of data usage during a busy hour with 100k 4G subscribers. If your plan is to keep the same data services, let's say best effort services, and you want to grow the subscriber base by 25 percent in one year then the new traffic load will be 1.25 × 100 Gbs = 125 Gbps, and if you have 1000 RF sites then that is 1.25 Mbps per site assuming uniform distribution.

Packet voice also should be included in the packet data calculations. However, the RF conditions needed for voice packet communication is UDP requiring the RF design process to use either the packet voice requirements as the design criteria, or possibly could include the packet voice services as additional data offerings. However, one key point is that if you have an existing network delivering packet voice then the packet data values are already at your disposal.

11.8.2 Radio Traffic Projections

There are several methods for estimating the amount of voice and packet traffic an operator can pursue with regard to planning a network. The air interface, which represents the scarcest resource in the network, is dimensioned with the highest blocking probability. Typically, network designers dimension the air interface according to a 2 percent blocking probability (Erlang B). Voice metrics are well-understood in terms of usage and forecasting using mErlang's/sub which, when coupled with an Erlang B table, nets the amount of radios, TCHs, or CEs required to support the traffic load for that particular sector or cell. However, when dealing with packet-data forecasting for a system, without historical usage patterns for the demographic segment associated with the market, forecasting takes on a more artistic and unscientific approach.

The projection of voice traffic for a wireless mobility system is a reasonably well-documented process. One of the biggest factors in determining voice usage for any growth plan is obtaining both good baseline data, that is, what the current capacity is, and fairly accurate projections about voice traffic growth that includes usage patterns. Owing to forecasting timing differences, reliance on historical data with linear trending is the most accurate method for the initial forecast and dimensioning. The linear trending method works well with *Advanced Mobile Phone Service* (AMPS), IS-136, CDMA, GSM, and *Integrated Digital Enhanced Network* (iDEN) regardless of some of the specific technical issues. However with 4G since it is a packet network, voice traffic calculations are not based on circuit-switched methods.

For packet-data forecasting, the key point is that without specific applications that the system is trying to address, the estimation of traffic is rather dubious because it bases several assumptions on each other. For example, streaming video is the desired application, but the real service is more e-mail and outlook synchronization. Therefore, for packet-data usage forecasting, there are two approaches: the forecast and discovery approaches.

The forecast approach involves a detailed analysis of existing data traffic, and working with the marketing and subscriber sales forces a take rate and estimated bandwidth for each subscriber. This then is distributed across the regions or appropriate BTSs to arrive at the appropriate forecasted packet-data volume, which then can be equated into the requisite radio and network elements necessary to support the estimated load.

The other approach is the discovery approach, where the whole system or just the core of the network—one or more packet channels—is deployed, which can be individual TCHs or 1xRTT- or EVDO-capable channels. There are two ways to implement the discovery approach. The first involves a method similar to the forecasting, except that there are no actual usage data to draw on. Therefore, you will need to determine the number of subscriber units that are or will be packet-data-capable and then multiply that number by a kilobits-per-second value and assume that each will be operational during the busy hour initially and weight the traffic volume over the RF access points involved.

The other discovery method, and it is actually a more widely used method, is to deploy packet-data-capable network elements in the system, making the packet service available. Normally, this approach is implemented when traffic data forecasting is not available or was never attempted. For *General Packet Radio Service* (GPRS) systems, what is typically done involves allocating a few TCHs for packet data on the initial channel that has the *Broadcast Control Channel* (BCCH). For a CDMA2000 1xRTT system and/or EVDO-capable channels, channel elements are deployed, enabling packet services to be available for those cells or sectors. The objective with the GPRS or CDMA discovery approach is to enable the system to be packet-data-capable and then measure the actual usage once services are being employed by subscribers and not system testers.

Obviously, the forecasting method for voice and especially packet is the desired approach to traffic projections. However, reality often deals a different hand, and the discovery approach is more or less the method that is used initially until real usage patterns have been established. An important concept to note is that the usage part of the forecasting method is probably the easiest to solve by offering the service free; it is equating a usable revenue level that will pay for the infrastructure and recurring expenses that is the true challenge.

In our experience, it has been a rare occasion that anyone ever has all the material needed, either in content, quality, or both. The reason that this seems to be an elusive target is the simple fact that most of the projection information needs to be supplied by another division within the organization. The issue is not the organizational competence per se but is driven by the fact that the technical community needs marketing forecasts about three or six months prior to that division being required to forecast that expectation.

11.8.3 Radio Voice Traffic Projections

The projection of radio voice traffic for a wireless mobility system was previously covered in this chapter. However, I cannot over stress the need in determining the voice usage for any growth plan is obtaining both good baseline data, that is, what the current capacity is, and fairly accurate projections about voice traffic growth that includes usage patterns. And this is still important in a packet-based network since it still needs to interface with a circuit-switched network, whether landline or other legacy wireless networks.

The radio growth forecast method will in almost every instance rely on historical data with linear trending. The linear trending is primarily used because of the inability to obtain traffic projections that are either timely or accurate enough. However, the primary problem with using the trending method is the inherent inaccuracy of the loading that is projected because it assumes a linear and even subscriber growth pattern.

The linear trending is insensitive to local marketing changes and subscriber hot spots. Therefore, linear trending can be used in the absence of other inputs. However, it is important to recognize the potential errors that can occur, and with all good plans there is a strong need to periodically review and modify the plan to better reflect reality.

11.8.4 Radio Data Traffic Projection

The inclusion of mobile data traffic in a wireless system can be a very complex issue. The mobile data traffic volume depends directly on the type and quality of services that will be offered and how they will be transported. The traffic estimation process involves not only the radio link but also the other fixed facilities that make up the network. Additionally, the inclusion of 802.11, WiFi, services that use the unlicensed bands, commonly referred to as *license exempt*, complicates the situation for estimating wireless data usage in a LAN environment.

Mobile data traffic is a contention-based service, and therefore, the issue of how much blocking, that is, denied access owing to radio facilities, to allow will continue to be difficult. The difficulty lies in the nature of mobile data, in that it is packet-based and not circuit-switched, meaning that you can and should overbook the service.

Therefore, the forecasting of the Mobile IP traffic would be much more simplified if the usage patterns were understood better. In light of this issue, however, the forecast, or growth, could be extrapolated from the business plan or simplified marketing plans, which would specify a certain desired growth level.

The equation to follow for an existing system would be

$$\text{Total traffic} = \text{existing traffic} + \text{new traffic expected}$$

The new traffic expected could be a simple multiplication of the existing traffic load. For instance, if the traffic is 25 Mbps and the plan is to increase the traffic by 25 percent over the next year, then the one-year traffic forecast would be altered by increasing the current load at each node by that amount and determining the requisite amount of logical and physical elements needed in addition to any load sharing that might be achievable.

The ultimate question that the designer must answer is how to plan on supporting the traffic, both data and voice, with their prescribed services. Since numerous types of services are available for both circuit-switched and packet data, some generalizations need to be made in order to have a chance at arriving at the conclusions necessary for input into the design phase. For a new or existing Mobile IP system, the issue of where to begin is always the hardest part. However, one of the key parameters you need to obtain from marketing and/or sales is the penetration rate, or take rate, for each of the service types offered. This can be determined by several methods. You could use a general approach where a standard percentage is used for, say, packet services. Or you could base the number of packet-data subscribers on the number of handsets expected to be purchased for resale in the market.

When *Voice over IP* (VoIP) services need to be included, the determination is driven by the fact that the wireless system, if offering packet services, most likely also will offer voice services, and the spectrum and radio resources need to account for this.

Regardless, the first step in any traffic study is to determine the population density for a given market. In the case of an existing system, the population density and primary penetration rates are already built into the system owing to known loading issues. However, especially for new services such as packet data, the process of determining the population density for a given area, followed by the multiplication of this by the penetration rate, will greatly help in the determination of expected traffic load on which to design the system.

- *Population density.* This is a measure of the number of people who potentially could use the service for a given geographic area. When determining population density, it is important to note that for the same geographic area there could be different population densities. For example, an area could have 100,000 pedestrians per square kilometer but only a vehicle density of $3000/km^2$.
- *Penetration rate.* This is a measure or an estimate of the number or percentage of the people in the population density that will use a particular service. For instance, of the 100,000 possible users, only 5 percent may want a particular service; therefore, the possible usage may be experienced by only 5000 people for that service offering. An important issue is that each service offering likely will have a different penetration, and thus it is very possible that based on the amount of services offered, the total penetration rate could exceed 100 percent because of the various service offerings.
- *Cell-site area.* The geographic area that a cell site or its sector will cover is determined either by computer simulation or from a rough estimate by a two-dimensional approach. The equations for determining the area of a cell are shown in Table 11.5. The radius for the cell site is determined from the link budget and depends on numerous issues. However, the use of a standard cell radius for a given morphology is recommended in the initial design phase. The cell coverage area for later phases in the design process can be determined through computer simulations that should factor in the cell breathing issues that are evident in CDMA and WCDMA systems or dynamic allocation of TCHs for GSM/GPRS systems.
- *QoS. Quality of service* is a term used by many and has a great many meanings. For this discussion, QoS is a description of the bearer channel's ability to deliver a particular *Grade of Service* (GoS). The GoS typically is defined as a blocking criterion, and for circuit-switched data, it is defined by Erlang B, Erlang C, or Poisson equations, whereas for packet data the relationship for QoS/GoS is blocking, Erlang C, and delay, to mention some of the key attributes.

With the introduction of packet data, traffic modeling for packet-switched data involves an interaction of the following items:

- Number of packet bursts per packet session
- Size of packets
- Arrival time of packet burst within a packet session
- Arrival times for different packet sessions

Cell Type	Cell Area	Sector Area (Three-Sector Cell)
Circular	πR^2	$\pi R^2/3$
Hexagonal	$2.598R^2$	$2.598R^2/3$

Table 11.5 Cell Area Equations

Packet usage understanding is still evolving, and as of this writing, the "killer application" appears to be wireless mobility, that is, transport ubiquity.

The issue of where, when, and how much you dimension a system for packet data always will be a debate between marketing and technical teams. However, in light of the fact that packet-data usage is at its infancy, there is little guidance on how to go forth and design the network. *International Telecommunication Union Radiocommunication* (ITU-R) M.1390 has some guidelines for dimensioning, and Tables 11.6 and 11.7 are extracted from that specification. The values in these tables should be used as a guide to establish packet loading for dimensioning when market-specific data are not available for numerous reasons.

It is important to note that all the services defined in these tables are either symmetric or asymmetric. Of the services listed, only MMM and HMM are asymmetric; the rest are symmetric service offerings.

Please note that the penetration rates for the services are the same and come to more than 100 percent for any location. Looking at Table 11.8, it is also important to note that the numbers above indicate that 73 percent of system usage is expected to be voice-oriented and that 10.8 percent is for SM; 3.51 percent for both SD, MMM, and HMM; and 6.75 percent for HIMM. Tables 11.9 and 11.10 help to provide additional insight into possible traffic dimensioning criteria.

The next step is to determine the traffic forecast by user service type. The method for achieving this value involves an equation for each of the service types and locations defined, building/pedestrian/vehicular,

$$\text{Traffic/user} = \text{BHCA} \times \text{call duration} \times \text{activity factor} = 0.9 \times 120 \times 0.5 = 54 \text{ calls/s}$$

during the system busy hour for downlink or uplink voice service for a building environment.

Service Type	Downlink (kbps)	Uplink (kbps)
S	16	16
SM	14	14
SD	64	64
MMM	384	64
HMM	2000	128
HIMM	128	128

Table 11.6 Net User Bit Rate

Service Type	Building	Pedestrian	Vehicular
S	73	73	73
SM	40	40	40
SD	13	13	13
MMM	15	15	15
HMM	15	15	15
HIMM	25	25	25

Table 11.7 Penetration Rates

Service Type	Building	Pedestrian	Vehicular
S	0.9	0.8	0.4
SM	0.06	0.03	0.02
SD	0.2	0.2	0.02
MMM	0.5	0.4	0.008
HMM	0.15	0.06	0.008
HIMM	0.1	0.05	0.008

TABLE 11.8 Busy Hour Call Attempts (BHCA)

Service Type	Building	Pedestrian	Vehicular
S	180	120	120
SM	3	3	3
SD	156	156	156
MMM	3000	3000	3000
HMM	3000	3000	3000
HIMM	120	120	120

TABLE 11.9 Call and Session Duration (s)

Service Type	Downlink	Uplink
S	0.5	0.5
SM	1	1
SD	1	1
MMM	0.015	0.00285
HMM	0.015	0.00285
HIMM	1	1

TABLE 11.10 Activity Factor

The number of circuits required for circuit-switched voice, switched data, and HIMM services is determined via Erlang B, whereas the remaining packet-data services are determined by Erlang C.

Now, the next step is to define the next set of variables that need to be established to help dimension the rest of the packet network. For symmetric services, dimensioning is straightforward, as well as straight as it can get. However, for asymmetric services, a few more details are required that are used for selection and performance of the PDSN. Table 11.11 provides some dimensioning information for the uplink and downlink directions involving MMM and HMM services.

Type	Description	Downlink	Uplink
NPCPS	No. of packet calls per session	5	5
NPPPC	No. of packets per packet call	25	25
NBPP	No. of bytes per packet	480	90
PCIT	Packet call interarrival time	120	120
PIT	Packet interarrive time	0.01	0.01

TABLE 11.11 Packet Data

How all the items interrelate in Table 11.11 is shown in the next set of equations, which can be used for traffic dimensioning.

$$\text{Transmission time (s)} = \text{NPCPS} \times \text{NPPPC} \times \text{NBPP} \times (8\ \text{b/B})/1024\ \text{kbps}$$

$$\text{Total session time} = \text{packet transmission} + [\text{PCIT} \times (\text{NPCPS} - 1)] + [\text{PIT} \times (\text{NPPPC} - 1)]$$

$$\text{Activity factor} = \text{packet transmission time}/\text{total session time}$$

11.9 Cell-Site Design

Although this is not necessarily the first step in any design process, it is one of the most important for the RF engineering department. The cell-site design is critical because that is where the bulk of the capital is spent. The cell-site design guidelines can be used directly or modified to meet your own particular requirements.

The use of a defined set of criteria will help to facilitate the cell-site build program by improving interdepartmental coordination and providing the proper documentation for any new engineer to review and understand the entire process with ease. Often, when a new engineer comes onto a project, all the previous work done by the last engineer is reinvented, primarily owing to a lack of documentation and/or design guidelines from which to operate.

The cell-site design process has many facets, and each company's internal processes are different. However, no matter what internal processes you have, the following items are needed as a minimum:

- Search area*
- *Site Qualification Test* (SQT)*
- Site acceptance*
- Site rejection*
- FCC guidelines
- E911 guidelines
- FAA guidelines
- *Electromagnetic Force* (EMF) compliance

*Applicable to new sites.

The RF design process still needs to be done when introducing a new radio access platform since space, power, and possibly new antennas and or spectrum will be involved.

11.9.1 Search Area

The definition of a search area is a critical first step in the cell-site design process. The search-area request is a key source document that is used by the real estate acquisition department of the company. The selection and form of the material presented should not be taken lightly because more times than not RF engineers rely heavily on the real estate group to find a suitable location for the communications facility. If the search-area definition is not done properly in the initial phase, it should not be a surprise when the selection of candidate properties is poor.

The search areas issued need to follow the design objectives for the area based on the RF system design objectives. The search area should be put together by the RF engineer responsible for the site's design. The final paper needs to be reviewed and signed by the appropriate reviewer, usually the department manager, to ensure that checks and balances are used in the process. The specifications for the search-area document need to meet not only the RF engineering department's requirements but also the real estate and construction groups' needs. Therefore, the proposed form must be approved by the various groups but be issued by the RF engineering department. It is imperative that the search-area request undergoes a design review prior to issuance. An example of a search-area request form is shown in Figure 11.10.

Search Area Request Form

Search Area Code:______		Capital Funding Code:________
Issuance Date:		
Target On-Air Date:		
Search Area Map		
# Sectors:		
AGL: ASML:		
Size of Equipment Room: (ft^2 or m^2)		
# Antenna		
Maximum Cable Length (ft,m)		
Comments:		
		Search Area Request Document #______ Date:_________ Design Engineer:________________ Reviewed by:____________________ Revision:______________________

FIGURE 11.10 Search-area request (SAR) form.

11.9.2 Site Qualification Test (SQT)

The *Site Qualification Test* (SQT) used to be an integral part of any RF system design. Even in the age of massive computer modeling, it is still essential that every system has some form of transmitter or site qualification test conducted. The fundamental reason behind requiring a test is to ensure that the site is a viable candidate before a large amount of company capital is spent on building the site. This test is also required to make sure that the site will operate well within the network. The financial implications associated with accepting or rejecting a transmitter necessitates a few thousand dollars expended in the front end of the build process. If a site is accepted that will not perform its intended mission, additional capital will need to be spent to accomplish it.

Based on the number of sites required within a specified time frame, it may not be possible to physically test every cell-site candidate. Therefore, it is essential that the number of sites to be physically tested is defined. This goal for physically testing or using a propagation model evaluation will help to establish the risk factors associated with building of the network.

Regardless of whether a site is to be physically tested or evaluated through a computer simulation, there are several stages to the process. It is very important that a proper investigation be performed because this will determine the cost of the facility, which could range from \$500,000 to \$1 million.

It is strongly recommended that the RF engineer responsible for the final site design visit the location prior to the site being accepted. This site visit will facilitate several factors. First, the engineer will have a better idea of the potential usefulness of the site and its ability to be built.

It is strongly recommended that the RF engineer does not design the test on the fly by telling the testing team where to place the transmitter and which routes to drive. The desired approach is to have the engineer determine where to place the transmitter, either as part of the tower or rooftop, and the location for the crane. The RF engineer then puts together his or her test plan, identifying the location of the transmitter antenna, the ERP, the drive routes, and any particular variations. The test plan then is submitted to the manager of the department for approval and is passed to the SQT team.

11.9.3 Site Acceptance (SA)

Once a site has been tested for its potential use in the network, it is determined to be either acceptable or not acceptable. For this section, the assumption will be that the site is acceptable for use by the RF engineering department as a communications facility. It is imperative that the desires of the RF engineering department be communicated properly to all the departments within the company in a timely fashion. The communication can be verbal at first, based on time constraints, but a level of documentation must follow that will ensure that the design objectives are communicated properly.

The forms outlined later in this chapter are meant to be general guidelines and may need to be modified based on your specific requirements. Before the *Site Acceptance* (SA) is released, it is imperative that it goes through the design review process to ensure that nothing is overlooked. The SA will be used to communicate the RF engineering department's intention for the site and will be a key source document used by real estate, construction, operations, and the various subgroups within the engineering department itself.

The SA also will need to be given a document control number to ensure that changes in personnel during the project are as transparent as possible. Figure 11.11 is an example of a *Site Acceptance Form* (SAF). The form again is meant as a guideline, and specific

Site Acceptance Form (SAF)

Search Area Code:		Capital Authorization Code:
SAF Document #:		
Site Address:		
Lat:	AGL	
Long	ASML	
Ptp Backhaul (Y/N)		
Regulatory Issues		
PtP License Needed (Y/N)		
PtP Frequency Secured (Y/N)		
FAA Approval Attached (Y/N)		
FAA Marking and Lighting Required (Y/N)		
Site Specific Information		
a) Antenna Configuration Attached		
b) Radio Equipment Location Defined		
c) Network Equipment Location Defined		
d) Equipment Room and Location sketch attached		
Radio Equipment		
Network Equipment (if required)		
e) Antenna Structure (roof, tower, monopole, water tank)		
f) Equipment Room: (Prefab, Interior Fitup (TI), Exterior)		
g) Approx Cable Length (ft,m)		
Type and Quantity of Antenna (include PtP)		

Sector	Antenna Type	# Antennas	Orientation	ERP

Existing Transmitters On Structure (Y/N)
If Yes state Freq, EIRP, Call Sign and physical location for each antenna and service
Qualification Information
Coverage Objective Obtained (Y/N)
% of Area site will cover verse design objective
IMD Study Complete
EMF Study Complete
Site Particular Comments:

Figure 11.11 Site acceptance form (SAF).

modifications relative to the individual company and local conditions, besides design criteria, most likely will result in some modification.

11.9.4 Site Rejection (SR)

In the unfortunate event that a potential site has been tested and is determined not to be suitable for use in the network, a *Site Rejection* (SR) form needs to be filled out. The issuance of an SR form may seem trivial until a change of personnel occurs, and the site is tested again at a later date. The SR form serves several purposes. The first is that it formally lets the real estate acquisition team know that the site is not acceptable for engineering to use, and the team needs to pursue an alternative location. The second purpose is that this process identifies why the site does not qualify as a potential communications site. The third purpose ties into future use when the SQT data are stored and when the site might be more favorable for the network.

It is recommended that the SR process includes a design review with a sign-off by the manager. This is to ensure that the reasons for rejecting the site are truly valid and that the issues are communicated properly. The SR form shown in Figure 11.12 needs to be distributed to the same parties that the SA would be sent to. The reason is that if a site does not meet the design criteria specified at this time, this does not mean that it always will be unsuitable. Therefore, it is imperative that the *Site Investigation* (SQT) information collected for this site be stored in the search area's master file. The storing of SQT information in the central file will assist later design efforts that could involve the capacity or a relocation of existing sites to reduce lease costs.

11.9.5 Site Activation

The activation of a cell site into the network is exciting. It is at this point that a determination is made about how effective the design of the cell site is. Numerous steps must be taken after the site acceptance process. The degree of involvement with each of these steps largely depends on the company resources available and the interaction required between the engineering and construction departments.

At a minimum, these two groups should perform site visits together. These site visits involve the group responsible for the cell site's architectural drawings and the overall design of the site's structure. Regardless of the interaction between groups, when it comes to "show time," it is imperative to have a plan of action.

Site Rejection Form
RF Engineering

Search Area Code:

The (name of test location) was visited on (date of test) and did not meet the design criteria for the search area defined.

The location did not meet the design criteria for the following reasons (state reasons).

RF Engineer: ______________________

Engineering Manager: ________________

FIGURE 11.12 Site rejection (SR) form.

11.9.6 FAA Guidelines

Federal Aeronautics Administration (FAA) compliance is mandatory for all the sites within a system. Verification of whether the site is within FAA compliance should be covered during the design review process. If a site does not conform to FAA guidelines, then a potential redesign might be in order to ensure FAA compliance.

The key elements that need to be followed for compliance are as follows:

- Height
- Glide slope
- Alarming
- Marking and lighting

Verification of the height and glide-slope calculations is needed for every site. It is recommended that every site have its FAA compliance checked and included in the master site reference document. If no documented record has been made for a site regarding FAA compliance, it is strongly recommended that this be done immediately. Checking FAA compliance does not take long and can be done within a week for a several-hundred-cell system.

11.9.7 EMF Compliance

EMF compliance needs to be factored into the design process and continued operation of a communications facility. The use of an EMF budget is strongly recommended and can ensure personnel safety and government compliance. A simple source for the EMF compliance should be the company's EMF policy.

Establishment of an EMF power budget should be incorporated into the master source documents for the site and be stored on the site itself, identifying the transmitters used, the power, who calculated the numbers, and when it was last done. As a regular part of the preventive maintenance process, the site should be checked for compliance and changes to the fundamental budget calculation.

The method for calculating the compliance issue is included in the IEEE C95.1-1991 specification, and measurement techniques are included in C95.3. Both cellular and PCS use the same C95.1-1991 standard. Currently, different guidelines are used for different wireless services. It is recommended that the C95.1-1991 specification be used for all wireless services. However, be sure that the license under which you are operating complies with the applicable EMF standard.

11.9.8 E911

E911 in the United States refers to the ability to process an emergency call placed by a subscriber and comes under several different names depending on the specific country in which the wireless system is or will be operating. E911 by itself is a U.S.-based regulatory requirement that enables a subscriber who dials 911 to be connected to a *Public Safety Answering Point* (PSAP). Depending on the network configuration of both the operator and the PSAP, different levels of information can be displayed, facilitating the PSAP with allocating resources. The two primary items of information involve the phone number and position location information.

The ability to deliver 911 service is a legal requirement for mobile wireless providers. However, there is little regulation regarding the specific technology that needs to be used to deliver E911 services; only the content is defined. Also, each PSAP has different interface requirements, meaning that a PSAP in New York could have different interface requirements from a PSAP in Louisiana.

The levels or types of wireless 911 service can be classified into three areas:

- Phase 0
- Phase 1
- Phase 2

Phase 0 is also called *basic 911 service.* For Phase 0, a wireless carrier is required to transmit all 911 calls to a PSAP without regard for validation procedures intended to identify and intercept calls from nonsubscribers or a subscriber who has a billing dispute. Under the rules, both subscribers and nonsubscribers can dial 911 and reach emergency assistance providers without having to prove their subscription status. No information is passed to the PSAP; the PSAP operator must talk with the subscriber and ascertain who he or she is and where he or she thinks he or she is located.

Phase 1 provides much more information than Phase 0. For Phase 1, the wireless operator must not perform validation checks on the subscriber. In addition, the location of the cell site or sector must be provided in addition to the subscriber's phone number *Automatic Number Identification* (ANI). This basic location of the cell site is defined in a database or series of maps provided to the PSAP by the operator. This information provides some general location information for the call being received and permits the PSAP to reestablish a connection with the caller if the call is disconnected.

All wireless services provided in the United States meet Phase 1 at this time. Phase 2 is currently in various stages of deployment, with many urban areas already in compliance.

Phase 2 builds on the enhancements in Phase 1. More specifically, Phase 2 requires that the wireless operator provide *Automatic Location Identification* (ALI). ALI uses location-based services that enable the wireless carrier to locate the subscriber's handset using a variety of technologies. The two primary methods for ALI involve *Uplink Time Difference of Arrival* (U-TDOA) and the *Global Positioning Satellite* (GPS). Initially, there were more than seven competing technologies for ALI.

Using ALI enables the PSAP to determine more accurately where the subscriber is located, thereby eliminating errors usually relayed by the subscriber during an emergency situation. ALI can be implemented either as a network-based solution or as a handset-based solution. The requirements for network and handset solutions are different in that handset-based solutions require more accuracy.

A network-based solution requires the position reported to be accurate within 100 m for 67 percent of calls and 300 m for 95 percent of calls. However, for handset-based solutions, the accuracy is 50 m for 67 percent of calls and 150 m for 95 percent of calls.

U-TDOA is a network-based solution, whereas GPS is a handset-based solution. CDMA2000 uses a GPS handset solution, whereas GSM/TDMA uses a network-based solution. WCDMA in the United States also will use a network-based solution. ALI needs to be very accurate if it is to be used in conjunction with incident scene communications.

11.10 RF Design Report

As with any good effort, the results need to be documented and communicated to the respective parts of the wireless organization. The following is a proposed guideline that can be used to help construct such a report. The report is not inclusive of the circuit-switched and packet-fixed networks; it just applies to the RF portion of the system.

11.10.1 Cover Sheet

The cover sheet of the report should include the following items:

- The system it is meant for (such as "New York Metro")
- Date of issuance (June 21, 2014)
- Revision number
- Who or which group issued the report
- Confidentiality statement (this should be on every page of the document, usually in the footer)

11.10.2 Executive Summary

This is a one- or two-page summary that includes the findings of the report, and it is meant to serve as a base from which most management decisions can be made.

11.10.3 Revision

This is meant to document which version of the report this particular version is. The sign-off section that is included is meant to ensure that the version that is under scrutiny is the current one and has undergone a design review.

The format of the revision page should be as follows:

Date	Originator	Reviewed by	Comments	Revision Number

11.10.4 Table of Contents

The contents section is meant to serve as a simple reference point so that anyone picking the package up can find a particular section quickly without having to read the whole document.

The suggested format for the contents is shown here:

- Page
- Introduction
- Revision
- Coverage objectives

11.10.5 Introduction

This is where a description of the objective is presented. Specifically, the topics here should cover the market, the general types of equipment, and for whom this document is intended. Also included with the introduction is the time frame this report is meant to cover. Specifically, if this is a new system, then the time frame for the validity of this report could be one year. However, if this is an existing system or one that is particularly built out, then the time frame also may be one year but really should be two years as a minimum.

11.10.6 Design Criteria

The design criteria used for establishing the design should be listed here. Inclusion of the link budget and propagation-modeling assumptions need to be listed here as well.

11.10.7 Existing System Overview

This section is meant to describe which areas the system incorporates. A map showing the physical boundaries also will be necessary for this section. The key elements that need to be included here are the technology used and the changes envisioned in the future.

11.10.8 Coverage Objectives

This section is meant to describe the coverage objectives for the system. The following are suggested points that need to be covered in this section:

- What is the current coverage of the system?
- What are the coverage requirements?
- Which areas need coverage?

This information should be derived from the marketing, operations, RF, and system performance engineering departments.

This section should include a map of the geographic area that encompasses the system. The map should include which type of coverage objective is desired and its approximate differentiation on a map. The map could be either of the existing system or of an area that is currently being contemplated for building a system.

If multiple phases are associated with the build program, then this also should be reflected in the coverage objective section. In the event of multiple phases, a map showing the overall plan differentiating the various phases should be included as well. Such a map will be used as the foundation for the deployment of resources that will be tied into the overall system design document.

If the system currently exists, then a coverage map also should be included with this section. The coverage map should convey where the current system coverage problems are.

11.10.9 Coverage Quality

This section is meant to describe the coverage-quality requirements for the system. The coverage quality is a series of parameters that will be used to clearly define the link budget requirements for the system and the geographic areas within the network.

The coverage quality is meant to define the different morphology requirements that will be used in determining how much of an area will need to be satisfied by the coverage requirements. The coverage quality also could include not only the cell-edge coverage requirements but also the overall coverage requirements for the cell itself depending on the morphology to which the cell is referenced.

11.10.10 RF System Growth Requirements

Cell-site growth projections in a network are an ongoing process of refinement and adjustment based on a multitude of variables, most of which are not under the control of the engineering department. However, cell-site growth analysis can be used to help direct the limited resources of the company.

The final output of the cell-site growth section of the plan is to identify the number of cell sites required for the network and their required on-air dates. In addition to the number of new cells, any needed radio equipment expansion should be noted.

11.10.11 Intersystem Coverage

This section of the report includes the requirements to provide contiguous coverage into another market. Specifically, this would be applicable to an area of the system that, say, interfaces with another *Basic Trading Area* (BTA), *Metropolitan Trading Area* (MTA), *Cellular Geographic Service Area* (CGSA), or *Rural Service Area* (RSA), and the ability to handle roaming traffic is desired.

The coverage objective here defines what the overlap should be in terms of decibels and where the coverage objective should be. A map indicating the desired geographic areas would be directly applicable here. Also, any comments with regard to the other system's build program and coverage objectives should be listed here also.

11.10.12 Link Budget

The calculations and assumptions that make up the link budget need to be included in this section. The link-budget format shown in Tables 11.2 and 11.3 should be used for this section.

11.10.13 Analysis

This is where analysis is put into the report. Issues that are included pertain to spectrum utilization, channels selected, and migration strategy.

11.10.14 Summary of Requirements

This section is the end result of the design work and should include a summary table that indicates the amount of capital, either in product and dollars or just in product.

References

American Radio Relay League, *The ARRL 1986 Handbook,* 63d ed., American Radio Relay League, Newington, CT, 1986.

American Radio Relay League, *The ARRL Antenna Handbook,* 14th ed., American Radio Relay League, Newington, CT, 1984.

AT&T, *Engineering and Operations in the Bell System,* 2d ed., AT&T Bell Laboratories, Murray Hill, NJ, 1983.

Carr, J. J., *Practical Antenna Handbook,* Tab Books, McGraw-Hill, Blue Ridge Summit, PA, 1989.

Fink, Beaty, *Standard Handbook for Electrical Engineers,* 13th ed., McGraw-Hill, New York, 1995.

Fink, Donald, and Donald Christiansen, *Electronics Engineers Handbook,* 3d ed., McGraw-Hill, New York, 1989.

Jakes, W. C., *Microwave Mobile Communications,* IEEE Press, New York, 1974.

Johnson, R. C., and H. Jasik, *Antenna Engineering Handbook,* 2d ed., McGraw-Hill, New York, 1984.

Kaufman, M., and A. H. Seidman, *Handbook of Electronics Calculations,* 2d ed., McGraw-Hill, New York, 1988.

Lathi, *Modern Digital and Analog Communication Systems,* CBS College Printing, New York, 1983.

Lynch, Dick, "Developing a Cellular/PCS National Seamless Network," *Cellular Integration,* September 1995, pp. 24–26.

MacDonald, V. H., "The Cellular Concept," *Bell Systems Technical Journal,* 58(1), 1979, p. 15-41.

Miller, Nathan, *Desktop Encyclopedia of Telecommunications,* McGraw-Hill, New York, 1998.

Pautet, Mouly, *The GSM System for Mobile Communications,* Pautet and Mouly, France, 1992.

Sams, Howard, *Reference Data for Radio Engineers,* 6th ed., Howard W. SAMS & Co, Inc., New York, 1983.

Schwartz, Mischa, William Bennett, and Seymour Stein, *Communication Systems and Technologies,* IEEE Press, New York, 1996.

Smith, Clint, *Practical Cellular and PCS Design,* McGraw-Hill, New York, 1997.

Smith, Clint, *Wireless Telecom FAQ,* McGraw-Hill, New York, 2000.

Smith, Gervelis, *Cellular System Design and Optimization,* McGraw-Hill, New York, 1996.

Steele, Raymond, *Mobile Radio Communications,* IEEE Press, New York, 1992.

Stimson, G.W., *Introduction to Airborne Radar,* Hughes Aircraft Company, El Segundo, CA, 1983.

Webb, Hanzo, *Modern Amplitude Modulations,* IEEE Press, New York, 1994.

Webb, William, *Introduction to Wireless Local Loop,* Vol. 2: *Broadband and Narrowband Systems,* Artech House, Boston, 2000.

White, Duff, "Electromagnetic Interference and Compatibility," Interference Control Technologies, Inc., Gainesville, GA, 1972.

William, C. Y. Lee, *Mobile Cellular Telecommunications Systems,* 2d ed., McGraw-Hill, New York, 1996.

Williams, Taylor, *Electronic Filter Design Handbook,* 3d ed., McGraw-Hill, New York, 1995.

Winch, Robert, *Telecommunication Transmission Systems,* 2d ed., McGraw-Hill, New York, 1998.

www.fcc.gov.

Yarborough, R.B., *Electrical Engineering Reference Manual,* 5th ed., Professional Publications, Belmont, CA, 1990.

CHAPTER 12

Network Design Considerations

Chapter 11 addressed the *Radiofrequency* (RF) design issues related to implementation of a *Third-Generation* (3G) network. While Chapters 7 and 8 focused on LTE and WiMAX (both 4G networks) design considerations, respectively. This chapter, however, focuses on design of the non-RF aspects of the network, which is where the service treatment and delivery take place. The network designs are becoming more and more complex when factoring in legacy networks often leading in having three types of wireless access networks that the core network has to support. There unfortunately is not one single and standard design solution for your network design. Obviously, having only one type of access technology to support would greatly reduce the complexity of a network design needed to support all the access technologies. However, the vision of one radio access protocol needing to be supported is the designed approach where everything is packet-based and is effectively a LAN.

For the moment, as the various *Second-Generation* (2G)/2.5G networks migrate to 3G, and 3G networks migrate to 4G and beyond, the vision of an all–Internet Protocol (all-IP) network comes closer to reality. Thus, we consider such issues as placement and dimensioning of *Mobile Switching Centers* (MSCs), *Base-Station Controller* (BSC), *Serving GPRS Support Node* (SGSN), *Packet Data Serving Node* (PDSN), and so on. We also address the connectivity and transport requirements among the various network elements.

In general, design of the core network involves striking a balance among three requirements—meeting or exceeding the capacity needed to handle the projected demand, minimizing the capital and operational cost of the network, and ensuring high network reliability/availability. In short, we can refer to these three issues as cost, capacity, and quality. Of course, meeting one or more of these requirements often means making sacrifices elsewhere such that it is impossible to divorce one network design consideration from any of the others. For example, a lower cost might well mean lower network capacity or lower network quality. Thus we will never get a network that is remarkably cheap to implement and operate while still offering high capacity, high quality, and a robust set of service features. Instead, we must aim to establish some "happy medium" where we satisfy at least the most important criteria.

The network system design, in particular, the mobile data, is unique for wireless systems in that it may, depending on the services offered and the technology platform chosen, involve multiple protocols. The use of multiple protocols within a telephony and wireless system is not unique. However, the fundamental choice as to which platform will be the predominant force in the network design can and does lead to many perplexing situations.

The complication is further magnified based on the amount of on-net and off-net traffic the system will need to handle and the types of protocols such traffic entails. The backhaul from the access point and/or base station to switching/packet concentration node adds another wrinkle in this effort based on traffic volume and interconnect facilities that are available in the time frame desired or at the cost anticipated or both.

Figure 12.1 illustrates a generic wireless mobility network that represents any mobility technology. The figure attempts to illustrate some of the complexities involved with establishing a network design since most if not all network designs involve both on-net and off-net components.

Additionally, the complexity of the wireless mobile network increases either as a result of service offerings with their various treatments or the integration or deintegration of legacy networks the need to know what components and signal flows are on-net and those that are off-net. A suggestion is that you maintain a brief network map that is updated on a quarterly basis. The network map with on-net and off-net boundaries indicated will enhance your network design capabilities.

The telecommunications industry is moving toward convergence of the plethora of service protocols. The choice of which protocol will be used, that is, migrated, has yet to be determined. Legacy services also will need to be supported until convergence really takes place. Convergence is occurring on the fixed access network as well as the *Radio Access Network* (RAN). The convergence in the fixed access network involves *Time-Division Multiplexing* (TDM) and packet data, where packet data are *Asynchronous Transfer Mode* (ATM) and IP. Both ATM and IP have their advantages and are used for different applications.

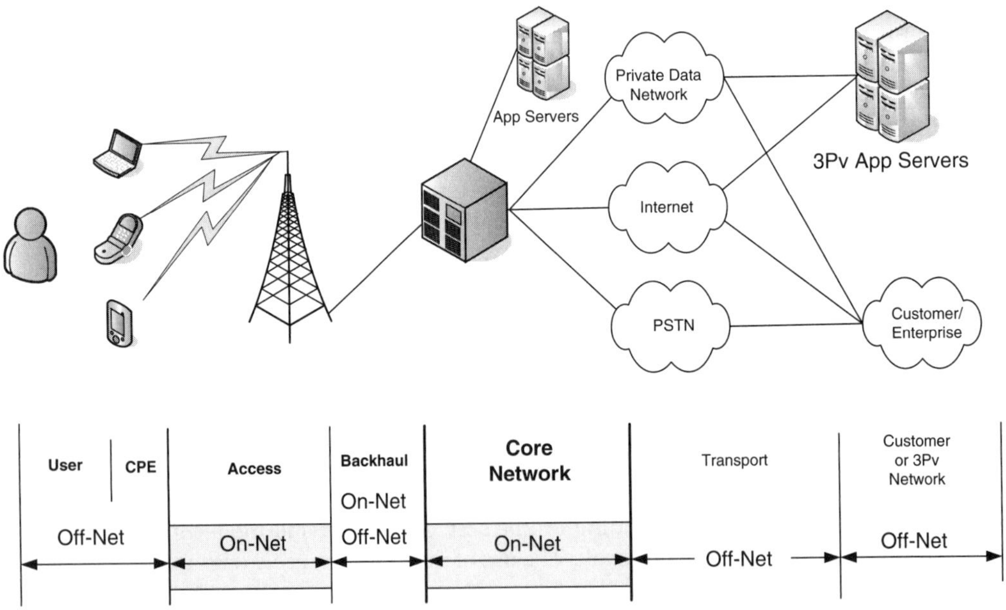

FIGURE 12.1 Generic wireless network.

Thus, the perplexing question is: Which platform should we use that will be "future proof" and will not require additional capital investments owing to technology obsolescence? The answer to this question depends on which part of the network you are referring to. If the part of the network is the edge, then the convergence is toward IP. However, for the core of the network, the convergence is toward IP also, however, there are a host of legacy protocols making the transition more complex. As always, the solution is not based on a single "killer protocol" but on the proper application of each toward obtaining the desired solution.

A decision will need to be made regarding the packet- and circuit-switching network. In particular, the decision will need to be made as to whether or not to lease capacity from another provider, thereby expediting the time to market and reducing operating expenses, but all at the cost of control.

However, there is no standard network topology type used by all wireless broadband networks and many wireless networks employ various network topologies based on the access or transport technology utilizes. Therefore, the wireless network topology is often defined by the legacy network's prior configuration, operational considerations, as well a budgetary (cost) reasons.

There are four basic network topology types that are used in a wireless mobile network and they are

1. Star
2. Ring
3. Mesh
4. Flat

Wireless networks typically have circuit and packet switch networks whose topology is different resulting in the use of different topologies for the core network depending on the RANs used in addition to the various types of service treatments as well as the backhaul and other interconnectivity requirements.

However, the most important decision that drives all the others relates to which services you intend to offer either at system launch or in 1 or 2 years. The decisions made will dictate the network configuration, which, if not chosen well, will result in excessive capital expenditures in the future to compensate for the incorrect decision.

Since there are multiple platform decisions to make, this chapter will attempt to cover the issues the network designer will need to address or at least consider in the design process.

12.1 Traffic Forecasts

Obviously, we need to design a network that will support the projected traffic demand for both circuit-switched and packet-data services. Consequently, projecting subscriber usage is a critical first step in the network design process. This projection often involves a certain amount of upfront guesswork, particularly if this is the first network of a given type in a given market that also includes an overlie network or any new service offering. If one is building a network to compete with someone else's established network, then one can forecast subscriber growth based on the competitor's subscriber numbers, which often are available publicly, but the specific usage data are not. If, however, one is building the first network in a given market or one is building a network very soon after a competing network has been launched, then less data are available. In such a situation, one needs to make educated estimates based on factors such as average household income, existing penetration of mobile voice service (such as 2G service), average Internet usage in the market, and similar data.

Traffic forecasts need to address several considerations, including total subscriber numbers, per-subscriber voice usage, per-subscriber data usage, and signaling demand. All forecasts can be grouped into

- Transactions (per subscriber or system node)
- Erlangs (per subscriber or cell/sector)
- Packet data (kbps per subscriber, cell, or system)

Although high level, a design engineer and his or her management need to understand the transactions and usage (Erlang/packet) of the system and the various nodes to ensure a good design and establish or adhere to dimensioning guidelines that ensure that capacity requirements are met, capital and operating expenses are minimized, and the customer experience is maximized.

12.1.1 Subscriber Forecast

For a given market, an estimate of total subscriber population is needed. Ideally, this should be broken down on a monthly basis so that we have an understanding of how subscriber numbers will grow over time. This is necessary because the design of the network will involve a certain amount of build-ahead. For some nodes, three-month intervals are sufficient granularity, whereas others require smaller intervals.

If there are to be a number of different commercial offerings, then there may well be a number of different subscriber categories, in which case forecasts are needed for each type of category. For example, a network operator may choose to offer some combination of services involving voice-only, voice and data, and data-only. Moreover, the data services may be further broken down depending on commercial offerings and subscriber devices. For example, one data offering might be limited to Web browsing, and another might include Web browsing, e-mail service provided by the network operator, and some other service such as Web space. Yet another data service might be aimed at telematic devices. Forecasts are needed for each type of user category.

12.1.2 Voice Usage Forecast

A voice usage forecast involves an estimation of the amount of voice traffic generated by the average voice user. Ideally, this also should be provided on a monthly basis. The voice profile should include the distribution of traffic in terms of mobile-to-land, land-to-mobile, mobile-to-mobile, and mobile-to-voicemail services. For the mobile-to-land aspect, there also should be a breakdown of what percentage is local and what percentage is long distance. Ideally, the voice usage profile information should include the average number of calls per subscriber in the busy hour and the *Mean Holding Time* (MHT) per call. Quite often, however, marketing organizations are likely to provide information in terms of *Minutes of Use* (MoUs) per subscriber per month. In this case, it is up to the engineering organization to derive the busy-hour usage. The following is an example of how this can be done.

Example: An average user has 400 MoUs per month. Assume, for example, that 90 percent of the traffic occurs during work days (i.e., only 10 percent on weekends).

Assume 21 work days per month. Assume that in a given day, 10 percent of voice traffic occurs during the busy hour. Then the average busy-hour usage (in MoUs) per subscriber is given by

$$\text{MoUs per month} \times \text{fraction during working days} \times \text{percentage in busy hour/work days per month}$$

Thus, in our example, we get $400 \times 0.9 \times 0.10/21 = 1.71$ MoU/subscriber/busy hour. Dividing by 60 gives the number of Erlangs, which is 0.0286 in our example, or 28.6 milliErlangs. If we multiply this number by the total number of subscribers, then we can determine the total busy-hour Erlang demand, which is a critical network dimensioning factor.

What we also need, however, is the total number of call attempts because some network elements are limited more by the processing effort involved in call establishment than by the total throughput. If we assume that most calls are completed (which is often the case in today's world of voice mail), then determining the number of *Busy-Hour Call Attempts* (BHCAs) is done by the following formula:

$$\text{BHCA} = \text{traffic in Erlangs} \times 3600/\text{MHT in seconds}$$

If in our example we assume an MHT of 120 seconds, then we get a per-subscriber BHCA of

$$0.0286 \times 3600/120 = 0.86$$

Thus the average subscriber makes 0.86 call attempts in the busy hour.

The increased deployment of *Voice over IP* (VoIP) for voice service delivery will necessitate the allocation of voice traffic to traditional circuit-switching facilities or to an IP network that can support VoIP services properly.

12.1.3 Data Usage Forecast

As mentioned earlier, we need to address the various categories of data users and forecast for each user type and the amount of data throughput. We also need to forecast where the throughput begins and ends. If, for example, a given user has Web browsing service plus operator provided e-mail, then a certain amount of traffic will terminate on an e-mail server within the operator's network, whereas a certain amount of traffic will be sent to and from the Internet. The dimensioning of the interfaces to the e-mail system and the Internet will depend on the amount of traffic related to those services. Moreover, the e-mail system will need to be dimensioned to meet requirements for total number of users, total storage, and total traffic in and out.

For each type of user and data service, we perform a similar analysis to determine busy-hour usage. We assume, for example, a certain amount of usage during work days and a certain percentage in the busy hour. From this we calculate the average throughput per user and per type of service in the busy hour. This throughput should be calculated in bits per second, and the uplink/downlink split should be specified. For most services, we will find that the downlink traffic is far greater than the uplink traffic, with an 80 percent/20 percent split being common. Once we have determined the busy-hour usage, we need to add some buffer to allow for burstiness or peaks within the busy hour.

The amount of buffer to be added will depend on the amount of build-ahead factored into the network design. If, for example, there is already a build-ahead of 12 months, then the system will be purposely overdimensioned at the beginning, in which case a further buffer would be wasteful. On the other hand, if little build-ahead has been factored in, then a 25 percent buffer for data traffic peaks could well be appropriate.

It should be noted that the busy hour for voice traffic and the busy hour for data traffic might not coincide. Given that for many network technologies different core network nodes are used for voice and data, whether the two busy hours happen to be the same often will not be an issue for network node dimensioning. For example, in the *Third Generation Partnership Project* (3GPP) Release 1999, voice traffic is handled by an MSC and data traffic by an SGSN. Similarly, in CDMA2000, voice traffic is handled by an MSC, and data traffic is handled by a PDSN. Therefore, for dimensioning of an SGSN or PDSN, whether the voice busy hour is coincident with the data busy hour is of no consequence.

As the networks converge toward an IP-only network with IMS, data (IP) dimensioning will be the predominant method of service delivery for all the services offered. The access network has unique dimensioning requirements and is RAN-specific. In addition, the backbone transport network and the systems resiliency need to be established from the onset. On the access network, for example, the capacity of a BSC or *Radio Network Controller* (RNC) will be determined both by the voice usage and by the data usage. In the core network backbone, we may wish to use VoIP (such as with 3GPP Release 4), in which case the backbone network will carry both voice and packet data, in which the issue of coincident busy hours is important. Using historical data, it is wise to assume the worst case—that is, that the voice busy hour and data busy hour are coincident. However, in most systems, voice and data (non-VoIP) traffic have different peak times, but this depends entirely on the services offered and the tariff charged to the customer.

12.2 Build-Ahead

It makes no sense to design a network to support the traffic demand that we expect today. This means that we must return tomorrow to enhance the network capacity. Instead, we need to design the network to support the demand that we expect at some point in the future so that we are not enhancing network capacity on a daily basis. Moreover, a reasonable build-ahead provides extra capacity so that the network is prepared to handle extra traffic in case subscriber growth is greater than projected. Build-ahead also provides a buffer in case of a sudden change in marketing tactics. For business reasons, it may be necessary to introduce new pricing plans or incentives that can change subscriber numbers or usage patterns significantly. It is wise to have the network prepared in advance for such eventualities.

So how much build-ahead is reasonable? Typically, it is wise to design the network to support the traffic demand expected 6 to 12 months in the future. If, for example, we launched a network in December of 2012, then a 12-month build-ahead would mean that we used the subscriber forecasts and usage projections applicable to December 2013 as input to the network design process. In general, the build-ahead can be larger at the beginning and can be reduced over time. If, for example, we include a 12-month build-ahead at the beginning, we might want to reduce this to a 6-month build-ahead after 2 years because we will have a better understanding of traffic growth patterns, and usage forecasts (assuming that they are updated regularly) will be more dependable.

A typical method is to have a 1-, 2-, and 5-year forecast in which years 1 and 2 are in monthly intervals and years 3 through 5 are relaxed to 6-month intervals. Additionally, the forecast needs to be updated on a continuous basis, with a maximum span between updates of 6 months.

12.3 Network Node Dimensioning

In order to determine the number of nodes of each type in the network, we first must understand the dimensioning rules associated with each type of node. If we understand the capacity limits of a given node type, then we can determine the minimum number of nodes of that type required. For a number of reasons, it is likely that we will deploy greater than the minimum number, but at least we must know the starting point.

Of course, for a given node type, the dimensioning rules and capacity limits will vary from vendor to vendor. Any examples provided in the following sections should be considered examples only and do not necessarily reflect the characteristics of any given vendor's implementation.

12.3.1 BSC Dimensioning

Typically, a BSC will have a number of capacity limitations. The following types of limitations are typical:

- Maximum number of transceivers (TRXs) (such as 256 or 512)
- Maximum number of base stations (such as 128, 256, or 512)
- Maximum number of cells (i.e., sectors) (such as 256 or 512)
- Maximum number of packet-data channels (such as 2000)
- Maximum number of physical interfaces (such as 128)

It is therefore essential to know the design limits and plan for expansion at a predetermined threshold, typically 70 percent, that allows for network reconfigurations as well as localized capacity issues.

For example, in many cases, a BSC from a given vendor has a fixed capacity based on a combination of the previous limitations. In determining the number of BSCs required, one analyzes the RF design in the market and calculates the number of BSCs based on which limitation imposes the greatest restraint. Imagine, for example, a GSM network that for a given market has 200 sites, each with three sectors and one TRX per sector. Consider a BSC model that can support up to 256 sites, 256 sectors, and 512 TRXs. Then, based on the site counts, we need two BSCs; based on the sector count, we need three BSCs; and based on the TRX count, we need two BSCs. Therefore, it is necessary to deploy at least three BSCs. This example is simplistic, but it illustrates that one or several variables often determine the BSC dimensioning requirements. The same multidimensional approach is required for all nodes within a wireless network.

12.3.2 UMTS RNC Dimensioning

Although an RNC generally is limited by the number of RF network elements (e.g., sites, sectors, and TRXs) that can be supported, the capacity of an RNC tends to be limited by traffic or throughput. This is so because of the fact that an RNC can be involved in traffic handling for base stations that it does not control directly. For example, an

RNC can act as a serving RNC or drift RNC during soft handovers. In such cases, the RNC may be handling traffic to or from a base station that it does not control. Thus the number of controlled base stations becomes less important and the amount of traffic handled is of greater significance to the capacity of the RNC.

The capacity of an RNC typically is limited by a combination of the following factors:

- Total Erlangs
- Total Data Traffic
- Total BHCAs
- Total voice subscribers
- Total data subscribers
- Total Iub-interface capacity (Mbps)
- Total Iur-interface capacity (Mbps)
- Total Iu-interface capacity (Mbps)
- Total switching capacity (Mbps)
- Total number of controlled base stations
- Total number of RF carriers

The determination of the number of RNCs required in a given market will be based on which of these limitations is the most restrictive.

Unlike the situation for BSCs, it is more common for RNCs from a given vendor to be offered in a variety of configurations. For example, the Iu interface might be offered using different transmission interface capacities (such as E1, T1, or STM-1). Moreover, a given vendor's RNC might come in several multicabinet configurations such that one can start with a small configuration and expand capacity by adding additional cabinets.

Determination of the number of required RNCs is more complex than the equivalent determination of the number of required BSCs. In particular, the effect of soft handover needs to be considered. Imagine, for example, that there are two RNCs supporting a number of base stations. One RNC limitation will be the total switching capacity. If there is a great deal of inter-RNC soft handover, then switching capacity is consumed on both RNCs. In fact, switching capacity can be consumed on both RNCs even after the soft handover is finished if *Serving Radio Network Subsystem* (SRNS) relocation has not yet taken place. Thus, determination of the number of RNCs needs to consider not just the RF elements and not just the overall traffic load, but it also must consider the effects of soft handover. For this reason, calculation of the number of required RNCs should be done in close cooperation with the RF design effort.

In most cases, we find, however, that the most limiting factor is the Iub-interface capacity.

12.3.3 CDMA2000 BSC

The CDMA2000 BSC has many of the same dimensioning requirements as a UMTS RNC. The specific dimensioning requirements are vendor platform–specific and, of course, require a valid forecast and design methodology. The capacity of a CDMA2000 BSC typically is bounded by the following:

- Total Circuit-switched traffic
- Total Packet traffic

- BHCAs
- Number of subscribers (voice and data)
- BTS-to-BSC links
- BSC-to-circuit-switched links and capacity
- BSC-to-packet-network links and capacity
- Number of BTSs
- Number of RF carriers (1xRTT, EVDO)
- Registration (transactions and borders)

12.3.4 MSC Dimensioning

In the case where the network technology involves BSCs (or RNCs) that are separated from the MSC, then the MSC capacity generally has two limitations—maximum BHCAs and maximum traffic throughput (Erlangs and Mbps). In networks where BSC/RNC functionality is included within the same machine as the MSC, then there also will be limitations in terms of RF elements supported (such as sites, sectors, and TRXs).

The separation of BSCs or RNCs from the MSC is the most common configuration in 3G networks. Consequently, the MSC capacity generally is not limited by RF-specific factors. Thus the capacity is limited by Erlangs/Mbps or BHCAs. (There may be a total cell limit, but this generally is sufficiently large that it is not a limiting factor.)

Although we say that the capacity of an MSC is typically limited by Erlangs or BHCAs, the reality is that the BHCA limit is the real bottleneck. Although BHCAs and total throughput are closely related, Erlangs reflect the switching capacity and port capacity of the MSC, whereas BHCAs reflect the processing power of the MSC. In general, the number of supported Erlangs can be increased by adding extra MSC hardware, whereas the maximum BHCAs for a given release of MSC typically is fixed. Thus it is usually possible to add hardware and increase the supported Erlangs until such time as the BHCA limit is reached. Adding extra hardware after this point provides no extra capacity. Thus, when determining the number of MSCs required to support a given market, the calculation is BHCA-based.

When we come to distributed architectures (IMS), such as the MSC server–*Media Gateway* (MGW) architecture of 3GPP Release 4, many of the same dimensioning rules still apply. In this case, the MSC server is most likely to be BHCA-limited, whereas the MGW is likely to be Erlang-limited.

Today's MSCs typically have BHCA limitations on the order of 300,000 to 500,000 BHCAs. As technology advances, these numbers will increase, and capacities of up to 1 million BHCAs will be common in the next few years.

For most vendors, the configuration of a given MSC in a given market is custom-designed. In other words, the size of the switching matrix and the numbers and types of ports are designed specifically to meet the market requirements. If one is building a limited number of markets at a given time, this is the optimal approach. On the other hand, if one is attempting to build a large network (e.g., a nationwide deployment) with many MSCs, custom design of the hardware configuration for each MSC may be overly time-consuming and may jeopardize a timely launch. In such a situation, it is often wise to work with the MSC vendor to define a number of network-specific standard configurations, such as small, medium, and large configurations, depending on the types of markets to be supported. Thus, in a large metropolitan city one might need to deploy two large MSCs, although in a smaller city one might need only a single medium-sized MSC or small-sized MSC. Although this approach is not optimal from a hardware perspective, judicious determination of the different configurations is likely to ensure that there is not a great deal of overdimensioning. The resulting ease of cookie-cutter design and easier ordering and delivery may well result in savings in the design effort and more rapid deployment.

12.3.5 SGSN and GGSN Dimensioning

In UMTS, we continue to use SGSNs and *Gateway GPRS Support Nodes* (GGSNs) largely because they are used in standard *General Packet Radio Service* (GPRS), and the dimensioning rules that apply in UMTS are similar to those that apply in GPRS. The dimensioning limits applicable to an SGSN generally are as follows:

- Total number of simultaneously attached subscribers
- Total number of active Packet Data Protocol (PDP) contexts
- Total number of Gb or Iu-PS interfaces
- Total number of routing areas
- Total throughput

It is common to find that the real bottlenecks will involve the total number of attached subscribers or the total throughput. Of course, the limitations will vary from vendor to vendor, but typical values will range from 25,000 to 150,000 attached subscribers. As with any technology, these limits tend to increase over time so that much higher capacities will be available in one or two year's time.

For a GGSN, the typical limitations are the total throughput and the number of simultaneous PDP contexts. Typical systems of today have limitations on the order of 100,000 simultaneous PDP contexts, but we can expect significant capacity enhancements over the coming years.

12.3.6 PDSN and Home Agent Dimensioning

Typically, the capacity of a PDSN is limited by the total throughput it can support and the total number of simultaneous *Point to Point Protocol* (PPP) sessions. One is likely to find that the PPP session limit is reached before the throughput limitation. With today's technology, limits on the order of 50,000 PPP sessions are common.

For a home agent, the most limiting factor often is the number of supported Mobile IP binding records. Values of 100,000 to 200,000 binding records are common. Note that in some implementations the PDSN and home agent may be combined within one physical machine.

The PDSN typically has these items associated with its dimensioning criteria:

- Number of simultaneous sessions (attached)
- Number of interfaces
- Number of routing areas (RAs)
- Total throughput (Gbps)
- Number of transactions per second

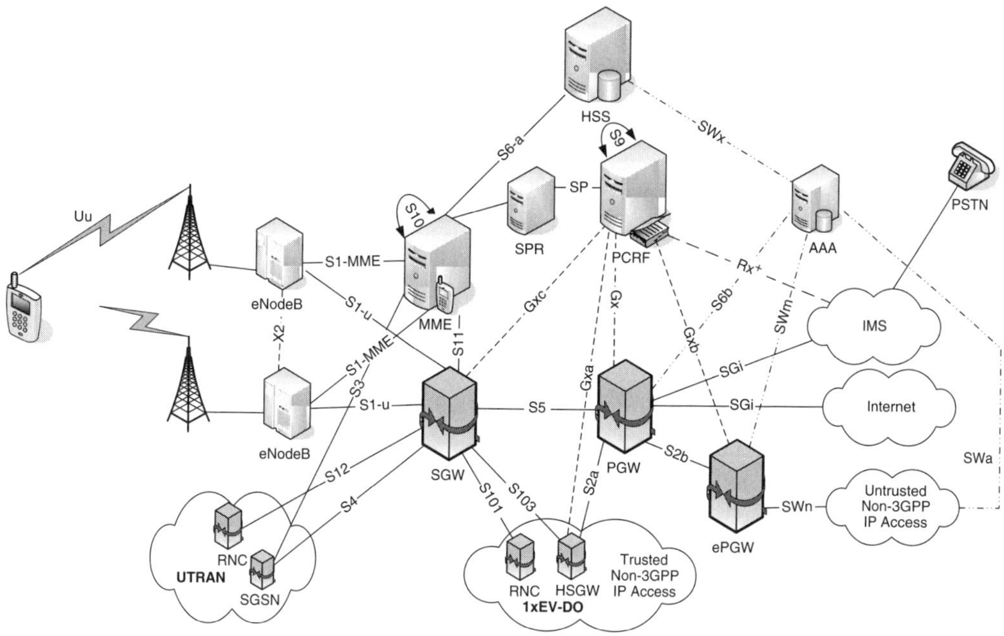

Figure 12.2 LTE network.

12.3.7 LTE Core Elements

The network dimensioning for LTE core elements is included in Chapter 7. However, for completeness the Figure 12.2 is a diagram depicting a high-level diagram of an LTE network with signaling links included for reference. If more detail is desired, 3GPP TS 23.401 and 3GPP TS 23.402 should provide sufficient information or direct you to the appropriate specification for the information desired.

In the LTE design phase, the core network components in the *Enhanced Packet Core* (EPC) are the

- *Mobility Management Entity* (MME)
- *Serving Gateway* (SGW)
- *Packet Gateway* (PGW)
- *Policy Control and Rating Function* (PCRF)

In purest terms, however, the PCRF is not part of the EPC. Regardless, Figure 12.2 illustrates that there are numerous aspects, or rather networks that the LTE system can interface with. The concepts of on-net and off-net are very important in this aspect. Specifically, a *UMTS Terrestrial Radio Access Network* (UTRAN) environment that interfaces with an LTE network may be controlled by the same wireless network provider; however, this is not an on-net LTE system but is on-net for a wireless service provider. Additionally, the IMS ring that the LTE network can and usually does interface with again falls into the same quasi definition for on-net and off-net as the UTRAN example. From the LTE system aspect they are off-net, therefore it is recommended that you consider the non-LTE network components as interim-net and the rest as true off-net components.

12.3.7.1 MME

The MME design parameters are not all inclusive. The values are all in subscriber/hour interval. It is important to remember that you should not exceed the recommended utilization levels for any of these design parameters defined by the *Original Equipment Manufacturer* (OEM) vendor you are using.

- MME design parameters (subscribers/hour):
- Initial UE attach/detach
- Idle-active transactions
- Active-idle transactions
- Bearer channel activation/activation per session
- # Bearers per session
- Session setup/tear down
- # Sessions
- Ingress paging
- Egress paging
- Intra-MME S1 handover SGW relocation
- Intra-MME S2 handovers without SGW relocation
- X2 handovers
- Inter-MME handovers
- # Intra-MME TAUs
- # Inter-MME TAUs

12.3.7.2 SGW and PGW

The design parameters for the SGW and PGW again are not all inclusive; however, they do cover most of the important dimensioning aspects needed to be considered. As with all network components, it is recommended that you do not exceed the recommended utilization levels for any of these design parameters defined by the OEM vendor you are using.

- # Simultaneous active subscribers
- Percent of subscribers sending and receiving data
- DL average packet size (Mbps)
- UL average packet size (Mbps)
- # Active sessions (subscribers/hour)
- # Active bearers (subscribers/hour)
- # Bearers setup and torn down (subscribers/hour)
- # Idle-active and active-idle transactions (subscribers/hour)
- # Intra-SGW handovers (subscribers/hour)

- # Inter-SGW handovers (subscribers/hour)
- # Inter-system handovers (subscribers/hour)
- # Subscribers using IP4 (# and %)/hour
- # Subscribers using IP6 (# and %)/hour
- # Gx transactions (connections/hour)
- # Gy transactions (connections/hour)

12.3.8 WiMAX Core Elements

The core elements that make up a WiMAX network are shown in Figure 12.3. The core elements are referred to as a CSN and the dimensioning of those core elements is driven by numerous inputs. Some of those elements pertaining to the CSN and its dimensioning are

- Core platforms chosen
- Regulatory requirements
- Software release schedule
- IMS design
- VoIP/SIP design
- HSS design
- Transmission and interconnection plans
- DNS/DHCP design
- CC7/SS7 design
- Synchronization plan
- Power plant (AC/DC and generator) design
- Disaster recovery plan/requirements
- Service treatment methods (subscriber features)

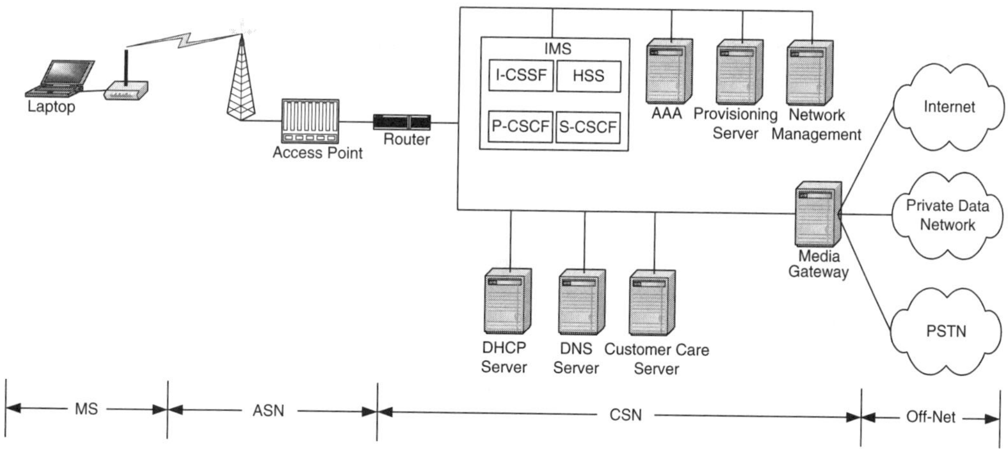

FIGURE 12.3 WiMAX network diagram.

All of the above plus some additional elements make up the design criteria for the WiMAX CSN.

However, as part of the design criteria, there are service treatment methods for packet data. The inputs for service treatment can be summed up into three general categories, all centered on service types. The services offered ultimately determine the dimensioning requirements and with WiMAX the three general service types are

1. Packet usage (TCP/UDP)
2. Voice usage (VoIP)
3. Advanced services

12.3.8.1 Packet Usage (TCP/UDP)

Forecasting the packet usage involves estimating the amount of packet traffic generated by the average data user and special treatment for *Quality of Service* (QoS) that needs to be applied. The forecast should include the distribution of traffic by user type, the amount of estimated data throughput, and the QoS associated with the various types of packet usage. Essential to the estimation in the forecast is where the packet throughput begins and ends so that all the nodes that the packet touches can be factored into the design and dimensioning process.

12.3.8.2 VoIP Usage

When voice service is offered as part of WiMAX, an estimation of the amount of voice-related traffic generated by the average subscriber over a defined interval is needed. The VoIP estimate should include the distribution of traffic in terms of each user type, the amount of MOU, and the distribution of VoIP calls, either MS to PSTN, PSTN to MS, or MS to MS.

12.3.8.3 Advanced Services

Advanced services being defined here for WiMAX are those services that you want to offer as an additional offering over the basic packet data and VoIP services. For example, some advanced services could involve video conferencing, multimedia streaming, web hosting, and others. The primary function of identifying advanced services is to help quantify their potential bandwidth as well as any specific QoS, platform, and software requirements that may be placed on the network to support this service.

However, the dimensioning requirements for CSN elements in a WiMAX network not only need to address current and future service treatment requirements but need to factor in dimensioning criteria pertaining to the various network elements and their respective interconnection links used for service delivery including those used for provisioning and/or network management.

As expected, the specific dimensioning requirements are unique for each platform or network element that comprises the CSN system. However, the following should help bound the dimensioning requirements. The suggested topics listed next are not all-inclusive but are meant as a good starting point or even a reference against what you are using presently.

Therefore when dimensioning any element in a CSN the following topics need to be answered.

- Total number of simultaneously attached subscribers
- Total number and types of *Access Points* (APs)

- Total throughput (on-net and off-net).
- Services offered by QoS
- System and service totals with relevant QoS by ASN and type
  - Average kbps
  - Peak kbps
  - Average PDP size
  - Peak PDP size
  - Session time
- Mobility requirements (if required)
  - # Simple and mobile IP connections
  - # Handovers/handoffs per session

VoIP dimensioning in a WiMAX network will also need to account for latency with the respective interconnection links used for service delivery, provisioning, or management. The specific dimensioning requirements used for VoIP are unique for each platform or network element that comprises the CSN. However, the following is a more specific list that will aid in the design of the VoIP portion of the CSN.

- Service features/design criteria
- # Subscribers (net)
- BHCA
- BH MOU or mE/subscriber
- Monthly MOU or mE/subscriber
- MS to PSTN (outbound) %
- PSTN to MS (inbound) %
- MS to MS (on-net) %
- Number of blocks assigned
- # Subscribers per IMS switch

I used mE/sub for VoIP calls simply because when interfacing with TDM circuits the need to have Erlangs, usually Erlang C, is needed. However when on the WiMAX network based on the vocoder that is used a kbps/subscriber value will need to be established and that is then provided to the RF engineering department as part of their radio dimensioning effort.

12.3.9 CDMA2000/EVDO Core Elements

Figure 12.4 depicts a high-level representation of a CDMA2000/EVDO network. For EVDO and CDMA2000 the radio access portion involves the BTSs as well as the BSC/RNC. However the BSC/RNCs are typically collocated in the MSC even though they could be distrusted across the network. However the issue of establishing and maintaining multiple concentration nodes where the BSC/RNCs can reside leads to operational constraints for servicing and maintenance leading them to be collocated at an MSC.

The core elements for an EVDO system involve the PCF and PDSN and all the traffic is IP based.

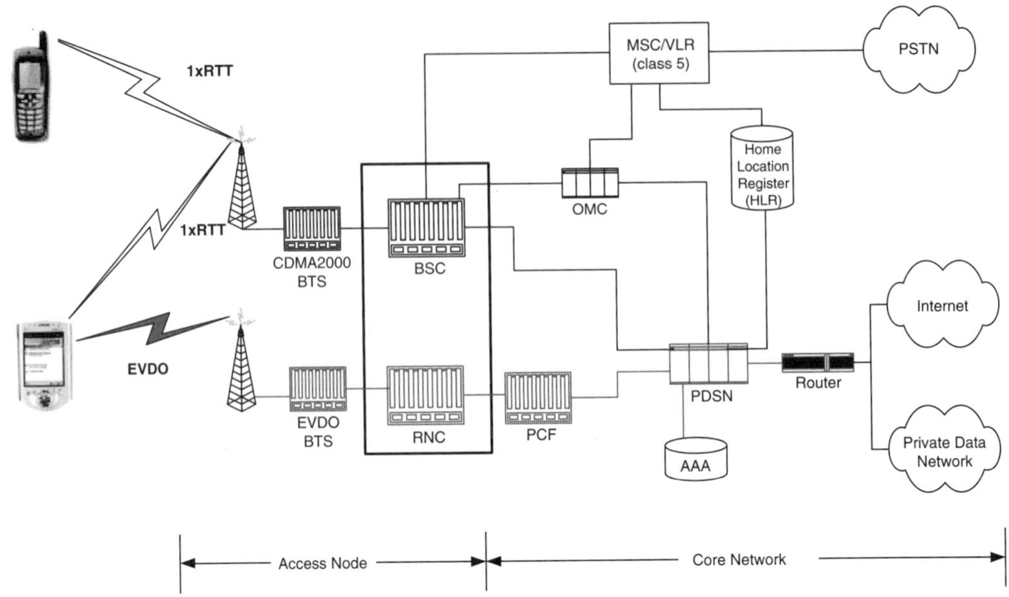

Figure 12.4 CDMA/EVDO network diagram.

12.3.10 UMTS/HSPA Network

Figure 12.5 depicts a high-level diagram of a UMTS/HSPA network.

12.3.11 Dimensioning of Other Network Elements

In previous descriptions we have provided the basic dimensioning information for a number of central nodes in a 3G and 4G network. There are, however, many other

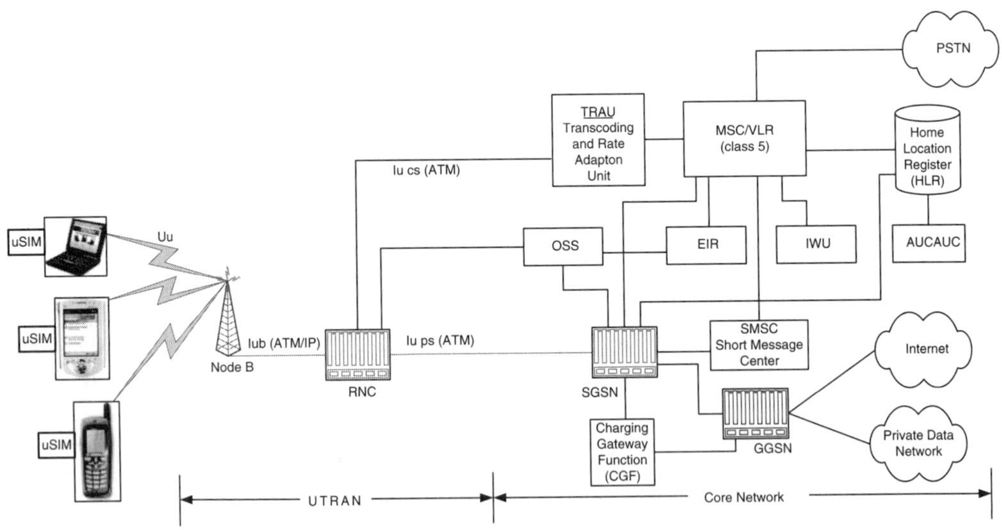

Figure 12.5 UMTS/HSPA network diagram.

network elements that need to be sized correctly. These are nodes, such as *Home Location Registers* (HLRs), voice-mail systems, *Short Message Service Centers* (SMSCs), and others. Each such node type has its own dimensioning limitations. For example, an HLR typically is limited by the number of subscriber records and transactions it can support. A voice-mail system often is limited both by the number of subscriber mailboxes (of a given size) that it can support, plus the number of message deposits or retrievals in the busy hour. An SMSC typically is limited by the number of messages per second that can be supported. In the case of *Global System for Mobile* (GSM) communications, UMTS, and CDMA2000 networks, it should be noted that *Short Message Service* (SMS) is used as the delivery mechanism for voice-mail notifications, which means that the number of short messages supported by an SMSC needs to be factored into the dimensioning.

For all the other network elements that need to be deployed—such as an *Equipment Identity Register* (EIR), an *Intelligent Network* (IN), a *Service Control Point* (SCP), an e-mail system, a *Hypertext Transfer Protocol* (HTTP) gateway, a Wireless Application Protocol (WAP) gateway, an *Authentication, Authorization and Accounting* (AAA) server, and so on, one needs to acquire from the particular vendor the specific dimensioning rules and capacity limitations. Keep in mind that regardless of the node and the specific vendor dimensioning requirements, the following four items need to be considered:

- Number of transactions
- Throughput (Gbps)
- Memory and CPU utilization
- Number of interfaces

12.4 Interface Design and Transmission Network Considerations

In general, determination of the amount of bandwidth required for a given interface is a relatively straightforward process. For example, if we expect a given RNC to support a given number of Erlangs, then we can easily determine the required bandwidth on the Iu-CS interface. Similarly, if we size some RNCs to support a given amount of data traffic in the busy hour, then we can determine the bandwidth required for the Iu-PS interface, and so on. Thus, once we have used traffic forecasts and dimensioning rules for the access network elements, we can determine the bandwidth requirements from the access network to the core network.

For example, if a given RNC is expected to carry 2000 Erlangs of voice traffic in the busy hour, and if we dimension the Iu-CS interface at a 0.1 percent grade of service, then Erlang B tables tell us that we need approximately 2100 circuits. Assuming that traffic between the RNC and the MSC is carried at 16 kbps, we need 2100/4 DS0s, which equates to 525 DS0s, or approximately 22 T1s. This bandwidth is carried over ATM, so we must add approximately 201 additional overhead for ATM.

Similarly, if a given RNC is expected to carry 50 Mbps of user data, then we can determine the Iu-PS bandwidth requirements directly. It typically will be about 120 to 130 percent of the user-data bandwidth to enable for *GPRS Tunneling Protocol* (GTP) overhead. Thus, for 50 Mbps of user data, we would need a bandwidth of 65 Mbps on the Iu-PS interface. Again, ATM overhead must be added.

Dimensioning of interfaces between RNCs (or between BSCs) will depend on the specifics of the radio network. RF design input regarding the number of soft handover traffic is critical. For example, if RF designers estimate that 20 percent of all voice calls will involve inter-RNC soft handover, then we can use this information to determine the bandwidth requirement. For example, we can assume that 20 percent of the voice traffic will be carried across a given Iur interface.

Overall, the dimensioning of bandwidth requirements for individual interfaces is not overly complicated, provided that we have determined the number of network elements and have established detailed traffic-demand estimates. The next step, however, is the design of a transport network that supports those bandwidth requirements in an efficient but reliable manner. Such a design effort can involve more complex issues and a greater degree of design expertise. Consider, for example, a scenario such as the one shown in Figure 12.6. In this example, there is a large local market and a remote medium-sized market. It has been determined that one MSC and three SGSNs should be placed in the large market. These are connected to four collocated RNCs that serve the local market. In addition, two RNCs are placed in the remote market. Thus we need Iu-CS and Iu-PS connections from the remote market to the local market. We also may need one or more Iur interfaces between the local market and the remote market, particularly if the RF coverage of an RNC in one market borders the RF coverage of an RNC in another market, as might be the case along a highway between the two cities.

In addition, in North America at least, there will need to be connections from the MSC in the local market to the *Public Switched Telephone Network* (PSTN) in the remote market for support of PSTN calls to or from subscribers whose dialable numbers belong in the remote market. Imagine, for example, a subscriber in the remote market who makes a local call. That call is first carried to the MSC and then must be carried back to the PSTN in the remote market. This can be done through direct trunks to that PSTN carrier, as shown in Figure 12.6, or the calls can be handed over to a long-distance carrier.

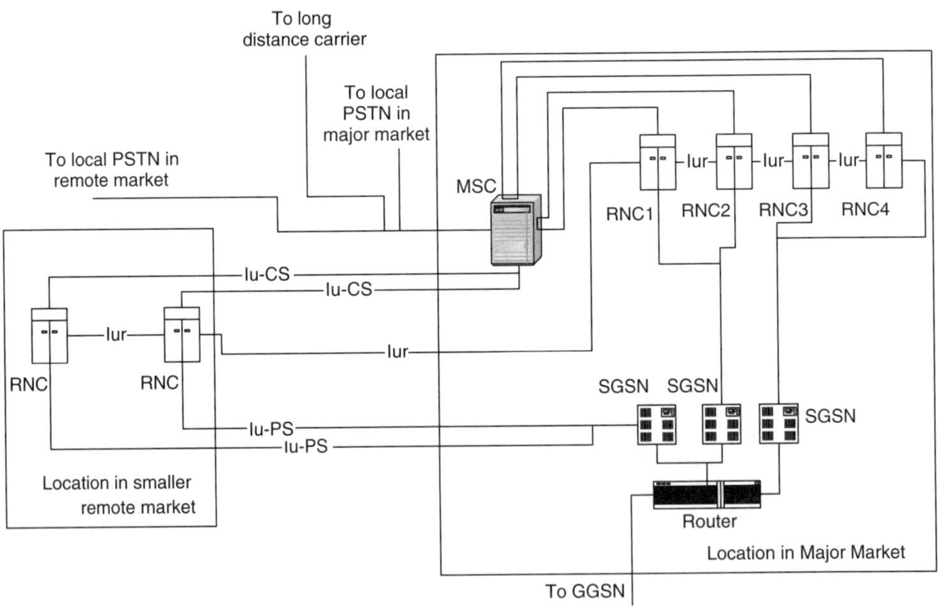

FIGURE 12.6 3GPP Release 1999 example network.

Finally, of course, there needs to be hundreds of Iub interfaces from the RNCs to the base stations in each city. All these interfaces and their associated bandwidth requirements need to be supported by an integrated transmission design that provides the necessary bandwidth and reliable transport.

In most cases, the MSC will be placed on a fiber ring. The total capacity of the ring will depend on the total transmission in and out of the MSC site, but it will be divided into a number of discrete capacities such as a number of DS3s or OC-3s. Typically, the ring will have a number of nodes, including hubbing nodes that belong to the ring provider. The individual links from the base stations in the local market will be connected to such points on the ring where they will be multiplexed onto DS3s or OC-3s for transport to the RNCs at the MSC site.

At the remote market, there also will be a significant number of Iub interfaces from base stations. Depending on the number of such interfaces, the availability of transport facilities, and the cost of those facilities, the remote RNC site also might be placed on a ring. In fact, if the distance is not too great, the remote RNC location might be a node on the same ring as the MSC site. In many cases, however, the distance between the cities may be too great to justify the cost of extending the ring to the remote city. In any event, the traffic from the remote RNC location to the MSC location will need to be protected. This generally will mean that there are diverse transmission facilities between the remote city and the MSC location. These diverse facilities must be sized to support the Iu-CS, Iu-PS, Iur, and PSTN connectivity requirements. This diversity will involve extra cost. That extra cost, however, generally will be justifiable. After all, the traffic demand in the remote city will be significant, or it would not have made sense to dedicate RNCs in that remote city.

Often the availability of transport facilities is a major factor in the timely deployment of a network. In the United States, most transport uses transmission facilities leased from local and long-distance carriers. Depending on the carriers, it might take 6 months before a ring can be installed at an MSC location. Moreover, it may be necessary to wait until the ring is installed before ordering individual circuits on that ring. Consequently, the earlier the transmission network requirements can be established, the better.

12.5 Placement of Network Nodes and Overall Network Topology

The example in Figure 12.6 assumes that the number of nodes in each city was established already and that the transmission network design was based on that established network element distribution. Often, however, one must consider a multitude of factors before making the decision to place equipment in a given location.

12.5.1 Cost Optimization

Among other considerations, one should not determine the placement of network elements without considering the transmission requirements and the likely transmission cost. For example, in Figure 12.6, one equally could have determined that it would be better to place an MSC (and perhaps some SGSNs) at the remote market in addition to RNCs. This would greatly reduce the transport requirements between the two cities. On the other hand, however, there would be greater capital cost involved in placing an MSC in the remote city. Alternatively, one could have decided that it would be better to completely serve the remote city from equipment housed in the larger local city. This likely would reduce the total RNC cost and would avoid the need for a suitably conditioned building in the remote city to house RNC equipment. The capital cost reduction in such a situation could be considerable. On the other hand, the additional transport required between the remote city and the MSC site could be very great and could mean a large cost. (After all, there likely will be at least a T1 from every site to the serving RNC regardless of how heavily used that site happens to be. On the other hand, the Iu-CS and Iu-PS interfaces are sized based on use only.)

This discussion is applicable for UMTS but easily could be applied to a CDMA2000, LTE, WiMAX, TD-SCDMA, or TD-CDMA networks. Having said this, there will need to be a certain amount of transport from the MSC to the PSTN in the remote market in any case. It may well be that the size of that transmission facility is such that extra capacity is available "for free" or that additional capacity can be added at a reasonably low cost. Thus the cost structure for transmission bandwidth also must be considered. For example, although a DS3 supports 28 DS1s, the cost of a DS3 is approximately 8 to 10 times that of a DS1. Thus, if one needs 12 DS1s, one is better off to lease a DS3 and get up to 20 DS1s "for free." Similarly, an OC-3 costs less than 2 DS3s, even though it supports up to 3 DS3s.

Finally, one must consider future technology evolution and the expected costs and capacities of future network elements. If one were not anticipating an upgrade to 3GPP Release 4, then the capital cost of an MSC in the remote city might be justified if it could be depreciated over a 7- or 10-year period and the effective cost compared with the transmission cost of placing just BSCs or RNCs in the market. Imagine, however, that one is deploying 3GPP Release 1999 and expecting to upgrade the network to 3GPP Release 4 within a 2-year time frame. In this case, one could delay the deployment of switching equipment in the remote city until such time as MGWs are available, provided, of course, that those MGWs are sufficiently scalable and sufficiently inexpensive compared with a traditional MSC. It might make more financial sense to absorb the cost of transmission between the two cities until the more efficient architecture is available.

Similar issues need to be considered in the placement of other network nodes such as SGSNs or PDSNs, GGSNs, and so on. Let us take a UMTS example. An SGSN is at the same level as an MSC in the network hierarchy. Consequently, it generally makes sense for SGSNs and MSCs to be collocated, as is the case with the PDSN as well. What about the placement of GGSNs? Well, this question comes down to the types of data services that the network operator wishes to offer and the relative use of those services. If, for example, a great deal of user traffic goes to and from the Internet, then it would make sense to place GGSNs at or close to the SGSNs and connect to the Internet relatively close to the user. This can save bandwidth. On the other hand, if one expects that subscribers will make a lot of use of operator-provided services, such as e-mail, where those services are housed in a limited number of centralized locations, then it can make sense to place the GGSNs nearer to those centralized locations. Although this approach can mean greater transmission overhead (because of the tunneling overhead between SGSN and GGSN), it also may mean a net fewer number of GGSNs in the network. Given that a GGSN, or cluster of GGSNs, needs to have other associated equipment, such as Dynamic Host Configuration Protocol (DHCP) servers and firewalls, a reduction in the number of GGSNs or the number of GGSN locations may mean a considerable reduction in capital cost. Again, we are faced with the issue of striking a balance between capital expense and operating expense.

In the case of placement of data nodes, there also may be special cases that need to be considered. Imagine, for example, that a given network operator establishes a relationship with a large corporate customer in a given city. The individual subscribers from that customer may have a totally different usage profile from other subscribers. They might, for example, use the wireless data service exclusively for access to the corporate network. In such a case, it could be appropriate to dedicate one or more GGSNs in a specific location for the use of those subscribers.

12.5.2 Considerations for All-IP Networks

As we move toward all-IP network architectures (IMS), then we will find the situation where we can establish just a single IP-based backbone network for the support of voice, data, and signaling. This amalgamation can mean a more efficient and cost-effective network. It is important to remember, however, that different *Quality-of-Service* (QoS) requirements will apply to such categories of service. In fact, there will be different QoS requirements for different data services. Consequently, it is not sufficient simply to size the core network backbone to meet the expected bandwidth requirements. We also must ensure that QoS mechanisms are built into the network so that quality is not jeopardized. Specifically, one wishes to ensure that each service is provided with the required quality without adversely affecting any other service. The first step in doing this is to ensure that the backbone network has sufficient bandwidth. Not only does this mean ensuring sufficient bandwidth in transmission facilities, but it also means that core network routers must have the switching capacity to handle the routing and switching of millions of packets.

Once we have established that we have the necessary bandwidth and packet-switching capacity in place, we then must make sure that each service and perhaps each service user can have reasonable access to that capacity in accordance with desired QoS objectives. This means that no one service can hog capacity at the expense of others; it may mean that one service can preempt another; and it may mean that traffic shaping is necessary at the edge of the network to ensure that the traffic entering the core network is in accordance with an agreed-on profile. A number of QoS solutions are available. To begin with, ATM has the capability to provide QoS guarantees. A number of other techniques are also available, such as the *Resource Reservation Protocol* (RSVP) and *Multi-Protocol Label Switching* (MPLS). The various techniques each have different advantages and disadvantages. For example, ATM is a Layer 2 protocol. If one decides to use ATM at Layer 2 in the network, then one can take advantage of the QoS mechanisms it can provide. On the other hand, one may not wish to be forced to choose ATM at Layer 2, particularly if other options exist (such as packet over SONET), and QoS guarantees can be achieved in other ways. RSVP can provide strong QoS guarantees and comes very close to circuit emulation. It has the disadvantage, however, of being processing-intensive and does not scale well to support very large networks. For many, MPLS holds the greatest promise. It offers strong QoS capabilities, can scale better than RSVP, and can be used with any Layer 2 protocol, including ATM. MPLS is likely to be the most flexible solution for most large networks. Further discussion of IP QoS techniques is provided in Chapter 9.

12.5.3 Network Reliability Considerations

Clearly, one would like to build a network that supports the expected demand and do so at the lowest overall cost—including both capital and operating costs. Reducing capital cost often involves a centralized design in which equipment is deployed in fewer locations, thereby taking advantage of efficiencies of scale. This also helps to reduce some aspects of operating costs because it reduces the number of locations and the number of network elements that need to be managed. On the other hand, a centralized design can lead to greater operational costs caused by increased transmission requirements, particularly if transmission facilities are leased and paid for on a monthly basis.

Cost is not the only important factor, however. Network reliability also plays a big role. The fewer the number of locations used to support a given subscriber base, the greater is the impact if one of those locations becomes inoperative owing to some major catastrophe. If, for example, a single city is served by a single location and that location suffers some catastrophe, such as an earthquake or tornado, it is likely that service to the city will be degraded, if not halted completely. If, however, that location also serves a number of other cities, then service in those cities will suffer equally. Thus, in parts of California, one might want to limit the number of markets served by a particular location so that damage is somewhat contained in the event of an earthquake. Similar considerations might apply in parts of the eastern United States that are subject to hurricanes.

12.6 Service Treatments

Wireless data services offered need to be realized. One of the areas where realization takes place is with the network design. The network engineering design is interested in the types of services offered, where they originate, where they terminate, if any treatment is needed, the provisioning and monitoring methods, and of course, the *Committed Information Rate* (CIR) with an overbooking factor that will be promoted with each service.

Just what services can be offered depends largely on the radio infrastructure and not the network engineering design. What we mean is that the network engineering design, being fixed, can support and deliver any protocol and service offering requested given the time and resources to accomplish the task. However, if the RAN interface to the customer is only able to provide 15 kbps of *Unspecified Bit Rate* (UBR) traffic, then offering *Real Time Variable Bit Rate* (rt-VBR) for video conferencing is not viable because of the RF environment, not the fixed network.

Supporting the services and what this means again constitute a vast area that has multiple meanings. For instance, the service offering simply could be a wireless operator offering connectivity and no content, that is, a "pipe provider." Another example could be a service provider offering a *Local-Area Network* (LAN) and *IP–Private Branch Exchange* (IP-PBX) extension. But what exactly is service? Is service the delivery of the bandwidth only, or do you provide adjunct services to support the primary bandwidth provision, that is, sell *Consumer Premise Equipment* (CPE), cable the customer, configure their routers, provide *Application Specific Programs* (ASPs), and so on?

Often the services offered will change with time owing to varying market conditions brought on by competition. Therefore, the platforms used by network engineering design need to be flexible enough to account for the vast array of unknown changes. This is a seemingly daunting task, but in reality, the issue is really a scaling issue.

The three major platforms that need to be supported are

- ATM
- IP
- TDM

With each of these platforms, the types of services that are to be offered will be proposed by the marketing department in conjunction with assistance from the technical community. Associated with each service offered, the following five issues need to be addressed:

- Committed Information Rate (CIR)
- Peak Information Rate (PIR)
- QoS
- Overbooking factor
- Fixed or variable billing

These five issues are really derivatives of the services offered, with the noted exception of the overbooking factor.

However, the direction that the industry is migrating to is IP at the end terminal, that is, mobile phone, with ATM as the backbone. The use of TDM circuits has to be considered in the interconnection method when connecting to another service provider for further delivery options. However, IP is the prevalent platform that will govern mobile data deployments.

12.7 TDM/IP/ATM Considerations

Well, how do you decide which platform to use, TDM, IP or ATM, and associated with this, how do you decide the dimension or proportion that each have to be within the network? The answers, of course, are not simple because experience dictates that no one platform is a solution for all situations and requirements. Most of the time, the interface is defined by the infrastructure vendor. However, as always when you have multiple interfaces or protocols and services to transport and treat, the issue of what is the transport protocol to use needs to be addressed.

There are several things to consider when selecting the platform used for the fixed network layout:

- What services do you need to support, keeping in mind that the key to mobile data success is targeting a niche market and not providing a general service to customers with no sticky applications?
- What is the RAN protocol used and its bandwidth-delivery capability?
- What is the bandwidth required for connecting the base station or access point back to the concentration node, for example an MSC?
- Can wireless be used instead of a *Postal Telegraph and Telephone* (PTT)/*Competitive Local Exchange* (CLEC) for connectivity between the BTS and the MSC?
- What is the reliability required between the various RANs and the MSC?

Table 12.1 indicates the types of platforms, that is, the protocols, that are best used for different types of services offered. In the table, NA refers to the fact that the protocol does not have a standard method of supporting that particular service.

A fundamental difference between ATM and IP and Frame Relay is that ATM uses fixed-length cells, whereas IP and Frame Relay use variable-length packets. Of course,

Protocol	TDM Circuit	Voice	Packet Data	Video
ATM	1st	1st	3d	1st
IP	NA	3d	2d	2d
Frame Relay	NA	2d	1st	NA

Table 12.1 Protocol Efficiency Comparison

TDM can handle ATM, IP, and Frame Relay services but not efficiently, that is, cost-effectively; in addition, you must realize that unless you want to become the next PTT, which is your biggest competitor, building a TDM system may not be a desired solution.

However, with this said, voice services are still a large percentage of the wireless telecommunications market. Therefore, the use of class 5 switches is still prevalent throughout the industry. However, the reliance on class 5 switches is being supplemented through the use of soft switches and VoIP using the *Session Initiation Protocol* (SIP) with IMS.

12.8 TDM Switching

The role of the TDM switch in the network has grown along with the importance and complexity of the network itself. The exact role TDM switches will have in the future is unclear, but they play an important role in wireless mobility systems today. The industry migration from TDM to IP/ATM platforms should be considered heavily when moving forward with the TDM switching decision.

However, no matter what is said, TDM switching will continue to play a major role in wireless mobility. The role of TDM switches began with first the manual exchange (also referred to as just a switch), followed by the rotary exchange, the crossbar exchange, and eventually, the modern electronic *Stored Program Control* (SPC) exchange. Even among the newer switches, there are various designs and functional applications depending on who the switch manufacturer is and what specific role the switch serves in the network, that is, central office, tandem, *Private Branch Exchange* (PBX), etc.

Regardless of type, all these switches have the same basic function: to route call traffic and to concentrate subscriber-line traffic. Some of the more common switch concepts and designs are described briefly below, along with a few example network applications.

12.9 Switching Functions

There are numerous functions of a switch within the network. These functions can be categorized into three basic groups: elementary functions, advanced functions, and intermediate functions (Table 12.2).

Elementary switch functions include the process of connecting individual input and output line circuits (trunks) within the switch itself and the ability to control the distribution of communication traffic across clustered groups of line circuits (trunk groups). The ability to interconnect individual line circuits allows the transfer of voice or data signals between end subscriber units or between network nodes in a controlled and selective manner. The switch's ability to direct or route traffic between individual

Switching Types	Functions
Elementary switching	Interconnection of input and output line circuits
	Control of communication traffic across line-circuit groups
Advanced switching	Digit analysis
	Call record generation
	Route selection
	Fault detection
Intermediate switching	Monitor subscriber-line circuits

TABLE 12.2 Switching Functions

line circuits based on larger defined groups of line circuits allows for more efficient and reliable control of large volumes of system traffic.

The more advanced functions are digit analysis, generation of call records, route selection, and fault detection. Digit analysis is the process of receiving the digits dialed by the customer, analyzing them, and determining what action the switch should perform based on this information, that is, attempt to place a call to another party, connect the caller to an operator for calling assistance, provide a recorded announcement stating that the digits dialed were in error, etc. The process of generating a record or multiple records for any calling activity taking place within the switch is crucial to creating the corresponding billing records for these calls, which, in the end process, results in the final bill being completed for the customer. Therefore, it is important that the switch produce an accurate account (record) of all call-processing activity it engages in.

The route-selection function directs all system traffic within the switch to a specific set of facilities (e.g., transmission circuit, service circuit, etc.) based on routing tables developed and maintained by the equipment vendor and the system operator. Finally, the detection of errors or problems occurring within the switch's own hardware and software, plus the identification of failures with any of the interconnected facilities, is a required function of the switch to ensure the operating quality of the network.

An example of an intermediate switch function would be the monitoring of subscriber lines. This function involves the completion of regularly scheduled checks of all line circuits interconnected to the switch for proper operation; that is, is the circuit still functional and able to carry system traffic? If a circuit is not performing properly, then it will be taken out of service, and the fault-detection function will be notified for alerting the operations staff.

12.10 Circuit Switches

Telephone networks use circuit switches for the processing and routing of subscriber calls. Circuit switching can be either the space-division type or the time-division type or a combination of these two designs. These designs are used in the matrix of the circuit switch. The matrix is where the actual switching of line circuits or trunks takes place.

12.10.1 Space-Division Switching

In space-division switches, the message paths are separated by space within the matrix. Thus, the name is derived. Figure 12.7 shows a simple space-division matrix. At each

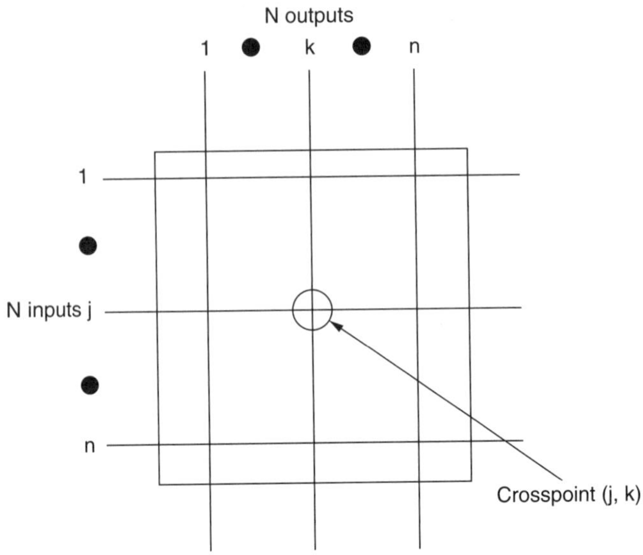

FIGURE 12.7 An $N \times N$ space-division matrix.

end of the matrix are the wires (i.e., actual subscriber lines, line circuits, trunks, etc.) available for switching. This is represented as $1 - N$ input lines and $1 - N$ output lines. In this example, the input j is connected to output k by closing the crosspoint jk (i.e., relay, contact, semiconductor gate, etc.). The concept here is that only one row of input can be connected to a column of output. In wireless networks using space switches, each call has its own physical path through the network.

12.10.2 Time-Division Switching

In time-division switches, the message paths are separated in time, hence the name. A simple diagram of a time-division switch is shown in Figure 12.8 and will serve as a simple means to explain this switching concept. In Figure 12.7, subscriber units J_1 through J_n are in conversation with subscriber units K_1 through K_n by means of a time-division switch. The actual input (originating subscriber) and output (terminating subscriber) line circuits are opened and closed by individual switching devices such as A_1 through A_n and B_1 through B_n. (The use of the input and output line circuits does not matter in this explanation. It is used to show some similarity with the space-division switch example.) In the time-division matrix, the connection of subscribers takes place by controlling the operation of selected switching devices A_1 through A_n and B_1 through B_n.

For digital time-division switches to operate, the incoming voice signals for every phone call must be in a digitized and encoded format. A T-carrier transmission system can provide the proper format to allow direct interconnection to a digital time-division switch without any additional conversion equipment.

In summary, space-division switching involves the switching of actual circuit interconnections, whereas time-division switching involves the switching of actual digitized voice samples within the switch's matrix.

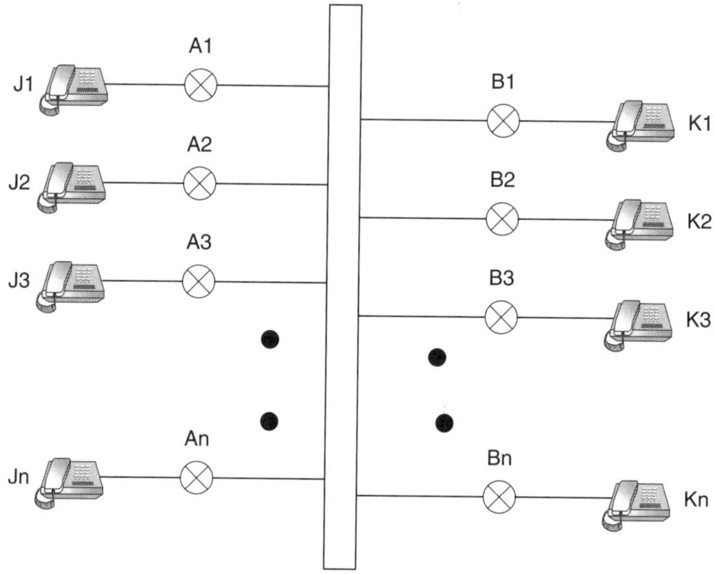

Figure 12.8 Time-division switch example.

12.11 Circuit-Switching Hierarchy

Within the North American PSTN, sometimes referred to as the *land-line telephone system,* there are five classes of switches. These classes can be divided up into the central office class 5 level switch and the remaining tandem-type switches of classes 1 through 4. The basic difference between these two categories of switches is that a class 5 local switch (also referred to as an *end office*) provides the ability to interconnect or interface directly with a subscriber's terminal equipment, whereas a tandem-type switch will only provide interconnection to other switching equipment or systems. Thus the local class 5 switch has the ability to terminate a call to one of its subscriber units, whereas a tandem switch can only route calls to other destined nodes and never act as a final call delivery point (Figure 12.9).

12.12 Packet Switching

The main function of the packet core network in a wireless mobile system is to provide the overall IP connectivity. The packet network for fixed and wireless mobility is post-RAN. The RAN is integral to the overall success and functionality of the wireless system. However, once information is in IP format, then only the source and destination are required, enabling the packet network to be RAN-technology-agnostic. For wireless mobility systems, the packet core network consists of the SGSN/GGSN or the PDSN packet network complex.

Packet networks are fundamentally different from traditional voice, or circuit-switched, communications systems. Packet switching is not connection-oriented, whereas circuit switching is, by definition, connection-oriented.

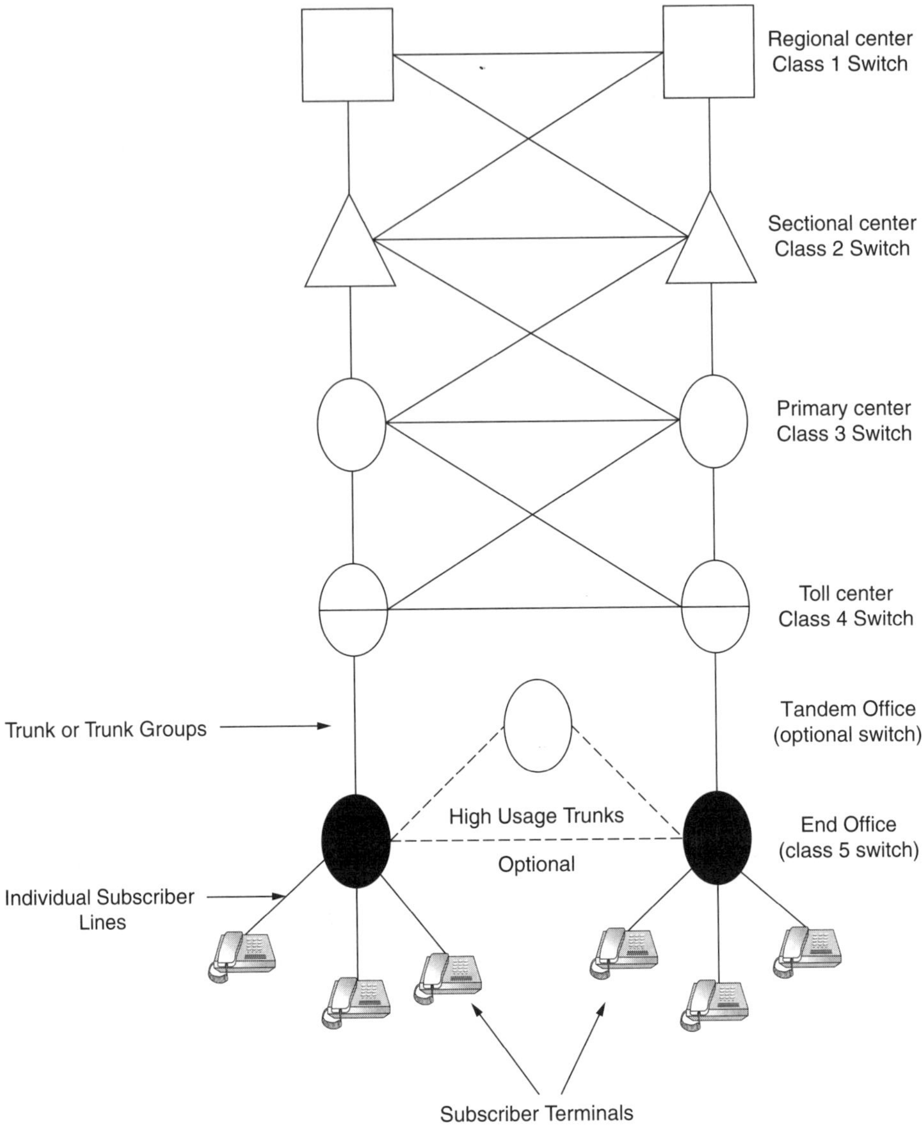

FIGURE 12.9 North American telephone switching hierarchy.

Traditional voice communications systems use circuit switches to provide the switching function of voice line circuits. In data communications systems, packet switches are used to perform the switching of data packets between the various nodes and computers in the network. Unlike the longer-duration calls that require circuit switching in a telephone network, packet switching is better suited for the short-burst-like transmissions of the data network. Packet switching involves sorting data packets from a single line circuit and switching them to other circuits within the network.

The packet network contains the functional network elements that work together for Internetworking gateways to external networks such as the public data network, corporate enterprise intranets, and *Radio Access Networks* (RANs). The core network also provides the interface into network management and connects RANs with mobility management, security, etc. The core network is designed to provide the operator with the ability to enhance the system offerings, commission features from third parties, and reduce overall operating costs.

With IP functionality at the mobile handset, applications such as *Voice over IP* (VoIP) and Push to Talk functionality are becoming more widespread. The RAN and the proper handset are essential for the implementation of IP services; however, it is the packet network that resides after the RAN that treats and delivers the various IP-based services to the end user.

12.13 IP Networks

The IP network has gained wide-scale usage with the introduction of the graphic interface for browsing the Internet, a true "killer application" or "killer enabler." The IP network can be a LAN, a *Wide-Area Network* (WAN), an intranet, or the Internet, to mention a few possibilities. The vastness of what an IP network can be has fostered some of the misunderstanding. Basically, an IP network is another protocol that is another enabler allowing for more information to be transported; how and where the data are transported can and does take on many forms.

Associated with the LAN or WAN is the use of an intranet, representing an internal network where members of the same campus, corporation, or whatever share resources, at least some, but these resources are not shared with anyone outside the users. Typically, the LAN/WAN involves connecting a series of computers to a hub, which, in turn, might or might not be connected to an internal server, whether used for file sharing, database, Web, or all the above.

Numerous design books have been written about IP networks, and each has its own specific slant, service-provider or vendor-driven. Web sites such as www.cisco.com are excellent sources of information on IP design and questions about routing and IP address designs.

However, the heart of an IP network is the fostering of an Application Specific Program (ASP) because providing a large pipe by itself will not result in additional revenue over the customer life cycle. The biggest issue with wireless data for the network engineer is to ensure that there is sufficient bandwidth to support the various applications and services offered and the correct *Central Processing Unit* (CPU) memory to support the required additional services.

The delivery of IP data and the associated benefits of this exciting transport method are available with all 2.5G and 3G RANs, as well as with 802.11, 802.16, and 802.20 and of course 4G systems. The key difference is how they deliver the IP data, which is defined by their latency and throughput.

In addition, the choice of using static or dynamic IP addressing schemes needs to be established. The use of static or dynamic IP addresses can be achieved with both public and private IP addresses. A static address is one in which the user is assigned the address and effectively owns the address forever, or until he or she stops paying his or her bills. Dynamic DHCP involves assigning an address to the user for a limited time, referred to as a *lease,* and once the time is expired, the IP address is reinserted into the pool for use by someone else. With DHCP, the user does have a different IP address at different times when he or she uses a system that is either public or private.

The use of static public IP addresses, while preferred by end users, is not an efficient use of this scarce resource. Therefore, dynamic addressing is the preferred method in the IP. In addition, private IP addresses should be used instead of public IP addresses.

12.14 IP Addressing

The issue of IP addressing is important to understand in any wireless data system design. No matter what the RAN is or who the infrastructure vendor is, the use of IP addresses is essential to system operation. It is imperative that the IP addresses used for the network be approached from the initial design phase to ensure a uniform growth that is logical and easy to maintain over the life cycle of the system.

The use of *Internet Protocol Version 4* (IPv4) formatting is shown below; IPv6 or IPng is the next generation and allows for QoS functionality to be incorporated into the IP offering. However, this discussion will focus on IPv4 because it is the protocol used today and has legacy transparency for IPv6.

Every device that wants to communicate using IP needs to have an IP address associated with it. The addresses used for IP communications have the general format shown in Table 12.3.

There are, of course, public and private IP addresses. Public IP addresses enable devices to communicate using the Internet, whereas private IP addresses are used for communications in a LAN/WAN intranet environment. A wireless data system can use both public and private IP addresses in order to facilitate the implementation of all its nodes.

The following tables represent the valid range of public and private IP addresses that can be used. The private addresses will not be recognized on the public Internet system, and this is why they are used. Also, it is possible for a wireless data system to reuse private addresses within sections of its network, profound as this may sound. The concept is that the system can be segregated, and the segregation allows for the reuse of private IP addresses, ensuring a large supply of a seemingly limited resource. This is applicable with connecting various mobile data networks together. The public addresses are broken down into class A, B, and C addresses with their ranges as shown in Table 12.4. The private addresses that should be used are shown in Table 12.5.

Network number	Host number
Network prefix	Host number

Table 12.3 IP Address

Network Address Class	Range
A (/8 prefix)	1.xxx.xxx.xxx through 126.xxx.xxx.xxx
B (/16 prefix)	128.0.xxx.xxx through 191.255.xxx.xxx
Class C (/24 prefix)	192.0.0.xxx through 223.255.255.xxx

Table 12.4 Public IP Addresses

Private Network Address	Range
10/8 prefix	10.0.0.0 through 10.255.255.255
172.16/16 prefix	172.16.0.0 through 172.31.255.255
192.168/16 prefix	192.168.0.0 through 192.168.255.255

TABLE 12.5 Private IP Addresses

Mask	Effective Subnets	Effective Hosts
255.255.255.192	2	62
255.255.255.224	6	30
255.255.255.240	14	14
255.255.255.248	30	6
255.255.255.252	62	2

TABLE 12.6 Subnets

To facilitate the use of IP addressing, the use of subnetting further helps to refine the addressing by extending the effective range of the IP address itself. The IP address and its master subnet directly affect the number of subordinate subnets that can exist and, from those subnets, the number of hosts, or users, that also can be assigned to that subnet. Table 12.6 shows the various numbers of users that a subnet can support.

It is important to note that the IP addresses assigned to a particular subnet include not only the host IP addresses but also the network and broadcast addresses. For example, the 255.255.255.252 subnet, which has two hosts, requires a total of four IP addresses to be allocated to the subnet, two for the hosts, one for the network, and the other for the broadcast address. Obviously, as the number of hosts with a valid subnet range increases, the more efficient the use of IP addresses becomes. For instance, the 255.255.255.192 subnet allows for 62 hosts and uses a total of 64 IP addresses.

Therefore, you might say, "Why not use the 255.255.255.255.192 subnet for everything?" However, this would not be efficient either, so an IP address plan needs to be worked out in advance because it is extremely difficult to change once the system is being or has been implemented.

Just what is the procedure for defining IP addresses and subnetting. The following rules apply when developing the IP plan for the system, and the same rules are used for any LAN or *Internet service provider* (ISP) that is designed. There are four basic questions that help to define the requirements:

- How many subnets are needed presently?
- How many are needed in the future?
- What is the number of users or APs on the largest subnet presently?
- What is the number of users or APs on the largest subnet in the future?

Therefore, using the preceding method, an IP plan can be formulated. It is important to note that the IP plan should factor into the design not only the end customers' needs but also the wireless data providers' needs as well.

Specifically, the wireless operator needs will involve IP addresses for the following platforms at a minimum. The platforms requiring IP addresses are growing in number constantly as more and more functionality for the devices is done through *Simple Network Management Protocol* (SNMP).

- Base stations
- Radio elements
- Microwave point to point
- Access Points (APs)
- Customer terminals (i.e., mobile phones)
- Routers
- ATM switches
- Workstations
- Servers

This list can and will grow when you tally up all the devices within the network from both hardware and network management perspectives. Many of these devices require multiple IP addresses in order to ensure their functionality of providing connectivity from point *A* to point *B*. It is extremely important that the plan follows a logical path.

A suggested methodology is to

- List all the major components that are, will be, or could be used in the network over a 5- to 10-year period.
- Determine the maximum number of these devices that could be added to the system over 5 to 10 years. A suggestion is to do this calculation by BSC or BTS.
- Determine the maximum amount of packet-data devices per sector that can be added.
- Determine the maximum number of customers that can be connected to each of the packet-data devices (could be more than one).
- Determine the maximum number of physical sectors that could be used per base station.
- Determine the maximum number of AP radios and their type for each sector and base station.

Keep in mind that if there are multiple customers per data terminal, the number of IP addresses required in all likelihood will increase depending on the type of service offered. For instance, the idea of using public addresses for all the devices is not practical, but some public addresses will be required. Therefore, the use of private IP addresses will be required to ensure that all the required devices have an associated IP address that has the correct network and subnet associated with the particular service.

Depending on how the network is laid out, the reuse of IP addresses can take place if each city or region had its own domain name and, of course, *Network Address*

Translation (NAT). The objective is to keep as many of the devices on the same network so that the number of retranslations and hops is kept to an extreme minimum.

For discussion purposes, the IP assignment scheme used for APs should be defined prior to the initial deployment. The IP assignment process could be designed so that the IP addresses are associated with a particular grid number instead of a sector of a particular cell. However, assigning IP addresses by sector probably will prove more user-friendly for operations than having a grid and then another look-up cross-reference table to the cell and sector. Naturally, your particular requirements will be different, as will be the IP addressing method you implement. However, the concept is make the plan and work the plan.

12.14.1 IPv6

IPv6 is an improvement over IPv4 and offers many advantages. However, the IPv4 will remain an integral part of any wireless data network for the foreseeable future.

However IPv6 and IPv4 are not compatible and this needs to be factored into any IP address scheme that is used in a wireless network. IPv4 is based on a 32-bit address scheme, which has a possibility of 232 IP addresses, whereas IPv6 is based on a 128-bit address scheme with 2128 IP addresses possible.

The obvious scale and flexibility offered with IPv6 as compared to IPv4 are evident if you have ever had the opportunity to establish an IP scheme for your wireless network.

A simple comparison of IPv4 versus IPv6 formats may help illustrate the differences.

- IPv4 IP Address: 192.168.0.1
- IPv6 IP Address*:* xxxx:xxxx:xxxx:xxxx:xxxx:xxxx:xxxx:xxxx

IPv6 also implements additional features not present in IPv4. However the main advantage of IPv6 over IPv4 is its larger address space. A key attribute of IPv6 besides its number of IP addresses is that it is designed to minimize packet header processing by router.

Some very good news is that most protocols used for IPv4 can ride on IPv6 with no changes and this is done via tunneling. Additionally, IPv4 to IPv6 compatibility addresses have been defined.

When thinking of IPv6 and also transitioning from IPv4 to IPv6, it is best to think about where they may need to be implemented first and look at those nodes. However, here are high-level IPv4 to IPv6 address schemes:

- *IPv4-compatible address* is used by IPv6/IPv4 nodes that are communicating using IPv6.
    - IPv6/IPv4 nodes are nodes with both IPv4 and IPv6 protocols.
- *IPv4-mapped address where IPv4 node addresses* are mapped to an IPv6 node.
- *IPv6 to IPv4 address scheme* is used between two nodes running using IPv4 and IPv6 over the Internet.
- *Intra-Site Automatic Tunnel Addressing Protocol* (ISATAP) defines addresses used between nodes running both IPv4 and IPv6 over a private intranet.
- *Teredo address* schemes are used for communicating between two nodes running both IPv4 and IPv6 over the Internet and the IPv4 node is using NATed.

12.15 Soft Switches

Soft switches are making their way into wireless mobile networks. Soft switches have several significant advantages over traditional legacy circuit switches. But just what is a soft switch?

There is no single answer to this question because there is no clear definition of what a soft switch is, unlike descriptions for traditional switches such as class 5 or class 4 switches. A soft switch is a distributed architecture that is packet-based. The soft-switch distributed switching architecture provides convergence between traditional voice and packet (IP) services.

The distributed architecture is possible largely because it is a packet-based network, and therefore, the various elements, BSCs, and servers do not need to be centrally located, as is done traditionally. Traditionally, at the heart of a wireless mobile system is a class 5 switch that performs all the traditional voice-centric service treatments and is a large concentration node.

Soft switches also interface with traditional circuit-switched services through standard interfaces providing TDM functionality. Therefore, a soft switch can support both class 4 and class 5 service applications and deliver traditional PSTN, ISDN, VoIP/*Fax over IP* (FoIP), and 3G services over both wireline and wireless networks.

Soft switching enables the wireless operator to distribute the call-processing functionality, thereby enhancing disaster recovery and removing single points of failure such as the traditional class 5 switch in a mobile network. The soft switch also affords the operator the opportunity to better load balance the system, as well as introduce new features and functionality in a more efficient method. This ability is due to the inherent structure of a soft-switch architecture, in which the capacity is very scalable and can be grown in a modular and incremental fashion.

With any new or emerging technology, numerous variations or types of soft switches are available. Pressure continues for wireless operators to exploit packet-data offerings. Therefore, interoperability becomes paramount as different communications networks begin to converge. With the increased processing power and memory for computing, soft switches rely on ATM and IP as their fundamental platform.

Soft switching is based on open interfaces that allow for standard off-the-shelf equipment, such as commercial computing platforms and industry-standard network components to be used, enabling a mix-and-match solution that can, if done properly, lower the total cost of operation significantly compared with traditional proprietary closed solutions. A soft switch requires less room and less power to operate and can be deployed much more quickly than traditional legacy circuit switches, further improving the flexibility and usefulness of this architecture.

In addition, soft-switch architectures allow rapid feature development and high core networking flexibility. Both circuit and packet networks are converging, requiring both voice and data support with minimal impact on the network or its components. This will create new vertical and geographic markets that take advantage of new trends for economic viability. Convergence to a single network will merge today's separate voice and data networks into multipurpose packet networks supporting voice, data, and multimedia. Operating, administering, and maintaining one network significantly reduces maintenance costs for telephone operators. This unique solution eliminates or reduces the need to backhaul voice/data traffic for local calls and reduces operational costs. Open-standard technologies allow for greater flexibility.

Packet technology greatly reduces the bandwidth needed to support a voice call. This means that additional call capacity is increased with existing bandwidth, thereby decreasing the need to build out these telecommunications facilities. Point-to-point routing eliminates routing through an intermediate centralized switch. With soft switches, wireless operators can continue to offer and deploy traditional wireless voice services.

Most important, though, the deployment of soft switches is a clear way to "future-proof" the core network. This is so because soft-switch technology brings all the benefits of 3G networks to current 2G and 2.5G using remote intelligent MGWs to adapt standard base stations to the 3G and 2.5G core network.

12.15.1 IP Multimedia Subsystem (IMS)

An *IP Multimedia Subsystem* (IMS) network architecture utilizes an open standard and uses the SIP protocol for all session control. SIP-enabled IMS platform can support services from many different access technologies because it is access-neutral or independent. Additionally, IMS utilizes a common core that can deliver various services to any end user or system. A key feature of the IMS architecture is that it is based on a horizontal architecture (flat) for service delivery instead of the traditional stove. Figure 12.10 is a high-level diagram showing an IMS network.

IMS is referred to as the *Next Generations Network* (NGN) technology and involves three distinct functions.

1. Applications
2. Session control
3. Access

12.15.1.1 Applications

The applications are the various applications you want to deliver on the network. Applications are delivered utilizing SIP.

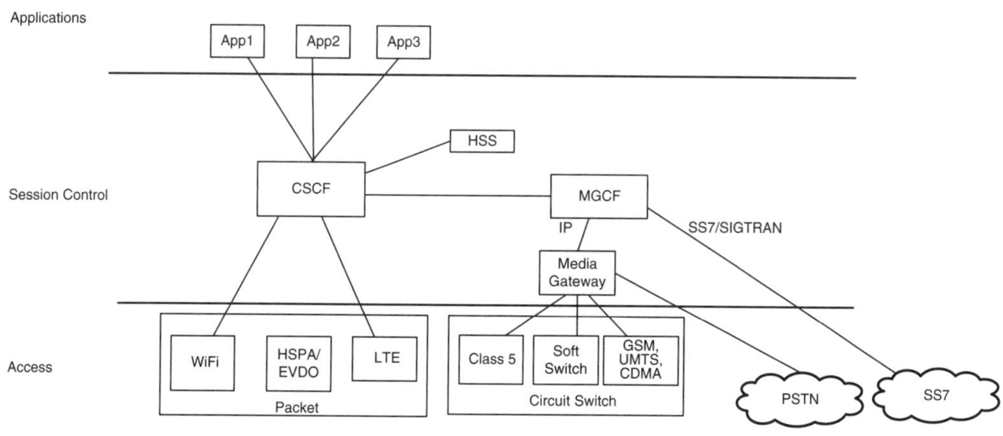

FIGURE 12.10 IMS network.

12.15.1.2 Session Control

Session control is the heart of the IMS platform since it directs, routes, and connects the access portion with the applications. The session control involves numerous subsystems or functions within the IMS network. The session control functions include the following:

- *Call Session Control Function* (CSCF). The CSCF handles session establishment, modification, and release of call sessions using the SIP/SDP protocol. The CSCF is the central feature of IMS architecture and it enables the registration of devices or endpoints and the routing of SIP signaling messages to the appropriate application server. Additionally, the CSCF also inter-works with network access and transport layers to guarantee quality of service across all services.
- *Media Gateway Control Function* (MGCF). The MGCF manages the call control protocol conversion between SIP and ISUP. It is interfaced with signaling gateway and media gateway. The MGCF is the central node of the PSTN gateway and is responsible for controlling the media resources used when traffic needs to flow between networks using different media, typically between a *Time-Division Multiplexing* (TDM) network and an IP-based network.
- *Media Gateway* (MG). The MG typically interfaces to 5E or wireless mobility switches, a softswitch or the PSTN for the media flow from PSTN to IP or vice versa. The MG provides functions such as media conversion (circuit to packet, packet to circuit) and echo control as some of its functions. The MG also provides for a centralized or decentralized connectivity scheme and is an IP-based interface that uses VPNs to connect to remote locations that have an MG card to interface to traditional TDM circuits.
- *Home Subscriber Server* (HSS). The HSS may be part of the IMS platform or the IMS platform can interface as needed to an HSS using SS7/SIGTRAN signaling. The HSS database contains the subscriber-related information to support call/session handling. The home network may contain one or more number of HSSs. In case of multiple HSSs, the subscriber's location function (SLF) selects the proper HSS.

12.15.1.3 Access

Access to an IMS platform can occur via a variety of methods using IP. Therefore, there are multiple broadband access methods that can and do use an IMS platform including WiFi, EVDO, HSPA, WiMAX, and LTE, to mention a few. However, IMS also can support GSM, UMTS, and CDMA2000 along with other circuit-switched services like plain old telephone service (POTS), Centrex/PBX , H323 and non-SIP-based VoIP protocols by using a media gateway for the protocol conversion. Therefore, an IMS platform enables non-paired wireless and non-wireless access methods to be used in conjunction with each other.

12.15.2 IMS and Softswitch Comparison

The following is a brief high-level comparison or rather listing of key attributes for an IMS and softswitch which can be used to determine which type of platform may meet your specific network design requirements. Figure 12.11 is a side-by-side high-level diagram of both a softswitch and an IMS platform.

- Softswitch
  - Transitional architecture change
  - Yields IP cost advantages

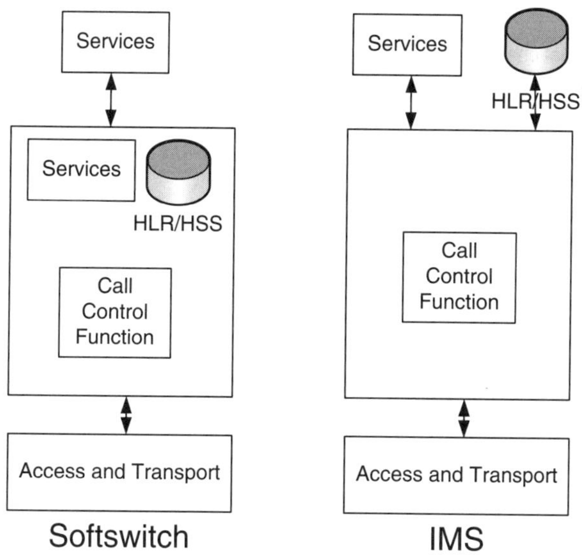

FIGURE 12.11 IMS and softswitch.

  - Supports current PSTN services
  - Makes use of *Intelligent Network* (IN) services
- IMS
  - Fundamental architecture change
  - Yields IP cost advantages
  - Current PSTN services may not be transparent (media gateway dependent)
  - Does not use IN
  - Platform for non-PSTN services
  - Achieves fixed and mobile core network convergence

12.16 ATM

ATM transport has been used by the PSTN for some time now, but with the introduction of mobile data and the need to transport data, as well as possibly voice, a transport medium that can handle both continuous traffic and bursty traffic is essential. With the continuous push for more IP-related services, the network designer for a mobile data system is faced with the dilemma of ensuring different *Quality of Service* (QoS) while at the same time not overdimensioning the various pipes within the network so as to reduce recurring facility costs.

Entire books are written on the design and functionality of ATM switching. It is not our intent to rewrite those excellent books, which can be found in the reference section of this chapter. Instead, the objective is to show how ATM platforms fit into wireless mobile service platforms, depending, of course, on the services and functions offered by the system.

ATM is by default a connection-oriented transport method that is excellent for transporting all types of information content, especially data traffic, and also can transport

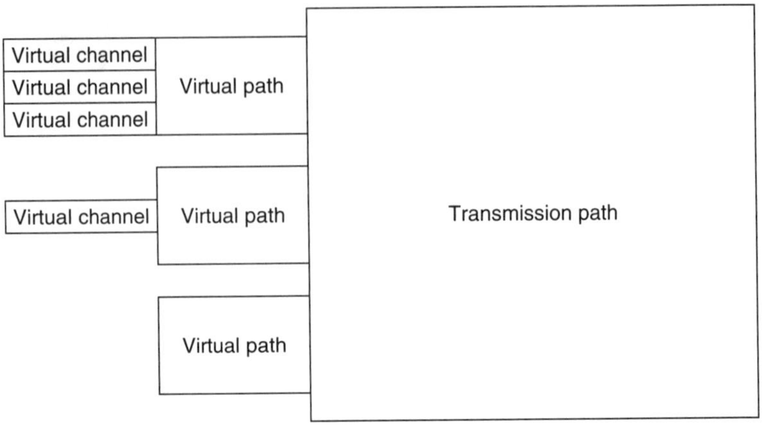

Figure 12.12 Virtual path.

voice services effectively. The ATM switch performs three major functions—routing, self, and label—in addition to header translation.

The ATM routing types are *Sequential Routing* (SR), *Random Alternative Routing* (RAR), *Least Loaded Routing* (LLR), and *Minimum Cost Routing* (MCR). In an ATM switch, the connections are referred to as *Virtual Circuits* (VCs), but the VC is just a container for tributary paths called *Virtual Paths* (VPs). There can be multiple VPs within a VC, as illustrated in Figure 12.12.

ATM is able to provide greater throughput in the network because it uses fixed-length packets called *cells.* A cell consists of 53 bytes, of which 48 bytes are for data and 5 bytes are for the header. Figure 12.13 illustrates the makeup of an ATM cell.

The various fields in the figure have specific functions. More specifically,

- *Generic Flow Control* (GFC). The purpose of this field, as implied by its name, is to control the traffic flow on ATM connections. This field is also used for CBR services for the purpose of controlling jitter.
- *Virtual Path Identifier* (VPI). The basic function of this field is to identify the VPI used for connecting the Virtual Channel Identifier (VCI).

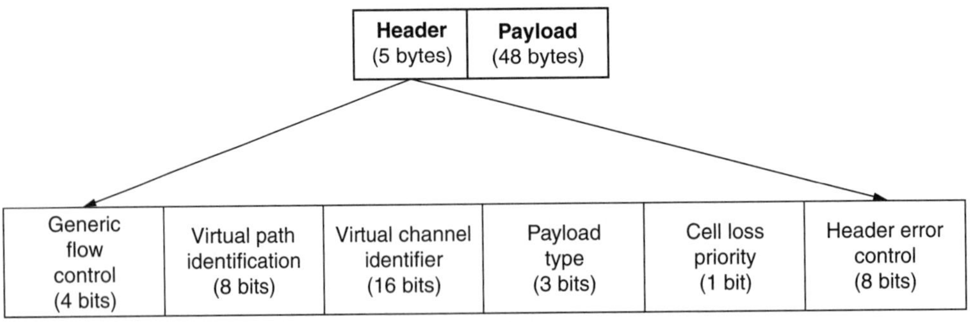

Figure 12.13 ATM cell format.

- *Payload Type* (PT). This provides the type of data contained within the cell. Types of information the PT will identify include user data, signaling data, congestion information, and maintenance data.
- *Cell Loss Priority* (CLP). This field identifies if the cell should be disregarded during congestion.
- *Header Error Control* (HEC). This field provides the error correction.

The ATM transport method allows for different service classes, and those classes are directly related to the *ATM Adaptation Layer* (AAL) layer. There are, of course, five AALs, and each is meant to transport a particular type of service. The type of service for each of the AALs and the connection type are listed in Table 12.7 for easy reference.

Table 12.8 is interesting, but without relating the specific AAL levels to possible service offerings, it has little value. Therefore, the following is meant to help further refine that table:

- *Constant Bit Rate* (CBR). This supports applications that require a fixed data rate. Examples of where CBR is applied include TDM circuits and voice and leased lines such as T1/E1.
- *Real-Time Variable Bit Rate* (rt-VBR). This service supports applications requiring real-time data flow control. Examples of where rt-VBR is used include videoconferencing. The main difference between CBR and rt-VBR is that rt-VBR has tighter timing controls than CBR.
- *Non-Real-Time Variable Bit Rate* (nrt-VBR). This service supports applications that do not require exact timing between source and destination. Some applications for nrt-VBR include e-mail and multimedia applications.
- *Available Bit Rate* (ABR). This service is really an enhancement of UBR in that a minimum and peak cell rate can be defined. ABR has priority over UBR traffic. Effectively, ABR is a managed best-effort service.

Service Class	AAL	Bit Rate	Timing Relationship between Source and Destination	Connection Type	Applications
A (1)	1	Constant	Required	Connection-oriented	CES CBR video
B (2)	2	Variable	Required	Connection-oriented	rt-VBR Audio/video (multimedia)
C (3)	5 (3)	Variable	Not required	Connection-oriented	Frame Relay FTP
D (4)	5 (4)	Variable	Not required	Not connection-oriented (i.e., connectionless)	IP, SMDS

TABLE 12.7 AAL

Class/Type	Bandwidth Guarantee	Real-Time Traffic	Bursty Traffic	Congestion Feedback
CBR (Constant Bit Rate)	Yes	Yes	No	No
rt-VBR (Variable Bit Rate, Real Time)	Yes	Yes	No	No
nrt-VBR (Variable Bit Rate, Non Real Time)	Yes	No	Yes	No
ABR (Available Bit Rate)	Yes/no	No	Yes	Yes
UBR (Unspecified Bit Rate)	No	No	Yes	No

Table 12.8 ATM Class/Type

- *Unspecified Bit Rate* (UBR). This is also referred to as a best-effort service. This service supports applications that can tolerate variable delays between source and destination plus the possibility of cell loss. UBR traffic uses the bandwidth that is left over after CBR and VBR services have taken their share of the pipe. The UBR service allows more use of the ATM network by passing UBR traffic at different rates between CBR and VBR allocations.

The next logical question that should arise is, what are the performance parameters that should be used for evaluating the health and well-being of an ATM network and its components? Section 12.16.2 discusses the key ATM parameters that measure system performance.

12.16.1 ATM Networks

Just what is an ATM network? Well, to start with, an ATM network can consist of a single ATM switch or multiple ATM switches. A key concept to keep in mind is that when you want to leave the ATM media world, it will require converting to an appropriate protocol such as TDM or IP, which is usually done with an ATM Edge switch.

ATM switches are being deployed in base stations as a method of concentrating TDM and IP traffic onto a single pipe for backhaul transport. The issue that needs to be considered by the network engineer is the introduction of ATM switching platforms that are effectively protocol converters that increase operating expenses.

With this said, ATM systems support two general types of interfaces:

- *User-Network Interfaces* (UNIs)
- *Network-Network Interfaces* (NNIs)

Figure 12.14 shows the locations of the different interfaces. Basically, a private NNI is used when connecting to ATM switches within your own network, and a public UNI is used when you connect to the public ATM switch or cloud. The reason this is mentioned is that *Private Network-to-Network Interface* (PNNI) signaling is not available with a UNI interface.

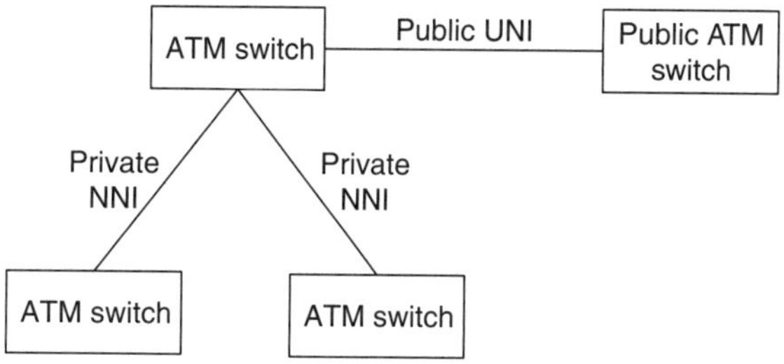

FIGURE 12.14 ATM network interfaces.

There are also several types of NNIs for ATM networks:

- *Private Network-to-Network Interfaces* (PNNIs)
- *Broadband Inter-Carrier Interfaces* (B-ICIs)
- *Broadband ISDN Services User Part* (B-ISUP)
- *Interim Interswitch Signaling Protocol* (IISP)

ATM routing is centered around circuits of two types: *Virtual Paths* (VPII) and *Virtual Connections* (VCII). These are made up of SVCs and PVCs:

- *Switched Virtual Connection* (SVC). This is a connection that is set up dynamically based on the need for the connection. Specifically, every time service is requested, the path taken from the source to the destination can change based on resources available. SVCs make best use of the network's facilities by increasing utilization.
- *Permanent Virtual Connection* (PVC). This is a predefined route that is programmed into the ATM switch via a craftperson. Figure 12.15 shows all the paths that could be considered PVCs as long as the VPI/VCI and ports remain the same.

Table 12.9 is an illustration of the routing functions that are associated with an ATM switch. Naturally, there is more than the simple port mapping illustrated in Table 12.9, but the diagram clarifies the issue of VP and VC mapping.

12.16.2 ATM Design Aspects

The design aspects for ATM switches focus on several key attributes:

1. Traffic characteristics
   - *Burstiness*—commonly used to measure how infrequently the traffic volume and rate are between the source and destination. Burstiness = peak rate/average rate.
   - *Traffic delay tolerance*
   - *Response time*
   - *Capacity throughput*

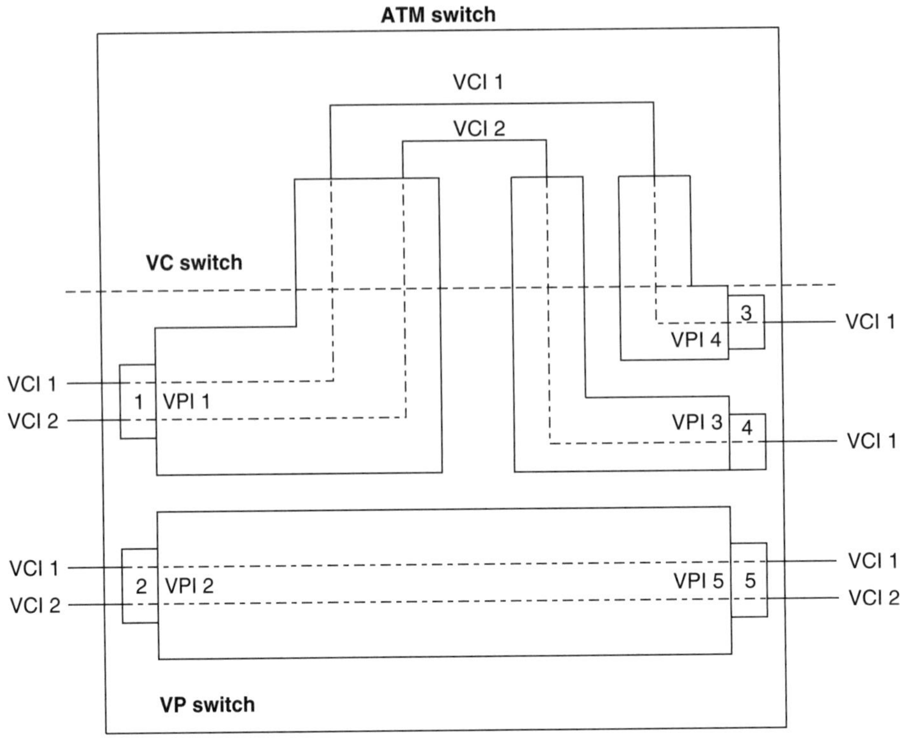

FIGURE 12.15 ATM switch.

Input			Output		
Port	VPI	VCI	Port	VPI	VCI
1	1	1	3	4	1
1	1	2	4	3	1
2	2	1	5	5	1
2	2	2	5	5	2
5	5	1	2	2	2
5	5	2	2	2	2
3	4	1	1	1	1
4	3	1	1	1	2

TABLE 12.9 ATM Switch Mapping

2. Cell delay
3. Cell loss
4. Congestion
5. Ports available

The first traffic characteristic is important to factor in because an ATM switch is designed to handle CBR, VBR, and UBR traffic. The type of traffic, the QoS associated with it, and of course, the burstiness of the traffic all factor into the overall throughput for the switch. Typical design limits put the desired throughput for the design at 70 percent of the allowable limit of the ATM switching platform. Throughput is not the same as the number of ports.

The other topics, such as cell delay, factor into the ability to deliver particular time-sensitive traffic. Cell delay is important for the individual switch but is really related to overall network design and in particular the path the cell has to transverse from start to finish.

Cell loss is important for many protocols that do not have error correction. Additionally, if there is a lot of cell loss, the link loads increase owing to retransmissions, and this has a direct effect on the throughput and cell delay factors.

Congestion by itself is an important issue to avoid in an ATM switch design. During high congestion periods, heavy cell loss can occur by design, resulting in more congestion owing to the number of retransmissions that take place.

The port issue is important when connecting to different ATM networks and providing circuit emulation to a TDM platform. Frequently the number of ports is the driving issue for ATM edge switches and not the other design parameters. The ports for an ATM switch can run at 100 percent utilization. However, in the initial design phase with unknown traffic projections, the desired level is 70 percent of the estimated load for whichever ATM switch platform is chosen at the design end point. This is based on the premise that the ATM platform chosen has the available card slots and pods to accommodate this potential growth.

12.17 Facility Sizes

Table 12.10 indicates some of the different facilities that a wireless data operator may have to interface with and the bandwidth with which each is associated. Table 12.11 is a quick reference regarding the similarities between a *Synchronized Optical Network* (SONET) and *Synchronous Digital Hierarchy* (SDH) fiber system.

Signal Level	Carrier System	Number of DS1 Systems	Mbps
DS0	DS0	1/24	0.064
DS1	T1	1	1.544
DS1C	T1C	2	3.152
DS2	T2	4	6.312
DS3	T3	28	44.736
DS4	T4	168	274.76
OC1	OC1	28	51.84
OC3	OC3	84	155.52
OC12	OC12	336	622.08
OC48	OC48	1344	2488.32

TABLE 12.10 Facility Sizes

Signal Type	Mbps	SDH
STS-1/OC1	51.84	—
STS-3/OC-3	155.52	STM-1
STS-12/OC-12	622.08	STM-4
STS-24/OC-24	1244.16	STM-8
STS-48/OC-48	2488.32	STM-16

TABLE 12.11 Fiber Sizes

12.18 Demand Estimation

Determining the demand that a system will need to transport and the necessary bandwidth can be calculated given historical data intermixed with marketing forecast data. The marketing forecast data are important for both new and existing systems because they are the best way to estimate the amount of traffic that a system will need to transport and, of course, the media types associated with the traffic load.

The key difference between a new and an existing system design lies in the issue of having a baseline from which to begin the forecast. If the system or service is new, then there is no baseline. However, if the system is in operation, some traffic is carried by the system, and therefore the marketing forecast is an addition to the current traffic being carried or designed:

$$\text{Design traffic} = \text{current carried traffic} + \text{forecasted traffic}$$

The key point is to always try to determine how far in advance the study needs to be done and the frequency of the study. The recommendation is that the study should not be done in detail for more than 2 years. The first year should be broken down by quarter, whereas the second year should be done in 6-month intervals. The traffic estimation will need to be revisited on a 3- or 6-month basis for the life of the system to ensure proper dimensioning and to account for the "hot spots" or lack of take rates estimated.

For forecasting, there are several key elements that the network engineer must factor into the forecast and design:

- Types of services
- Volume of traffic for each service
- Platform requirements and growth
- Connectivity between base station or access point and central office or concentration node
- Connectivity between concentration node and the various transport providers (i.e., PTT, CLEC, IP)
- Time frame of study
- Current traffic utilization

This list can and should be tailored to your specific requirements and needs.

12.19 VoIP

VoIP has been gaining in acceptance in recent years as the benefits of this transport method for voice are being developed. Overall, the transport of data via IP is used widely, but transporting voice using IP is not as well-known or understood. VoIP has been used for many years and has seen the most usage as a backbone, core transport mechanism and not focused on the end or edge user or application.

VoIP is an exciting application that is starting to be exploited by mobile and fixed wireless service providers. However, the proper treatment, or discussion, of the details with VoIP is best found in the references included at the end of this chapter.

The original standards for VoIP were defined in H.323, which has the title, "Packet-Based Multimedia Communication Systems." This standard was a direct result of Microsoft's offering VoIP as freeware. However, there is an alternative standard that is currently in competition with H.323, and this is the *Media Gateway Control Protocol* (MGCP), also called the *Single Gateway Control Protocol* (SGCP) and the *Session Initiation Protocol* (SIP).

IP has a number of advantages over traditional circuit switching. The most notable of these is the fact that it can leverage today's advanced voice coding techniques, such as the *Adaptive Multirate* (AMR) coder use in EDGE and UMTS networks, as well as other 2.5G/3G systems. Thus, voice can be transported with far less bandwidth than the 64 kbps used in traditional circuit-switched networks.

IP is a Layer 3 protocol in the *Open Systems Interconnection* (OSI) seven-layer protocol stack. IP by itself is inherently unreliable and provides no protection against loss of packets or delays. To combat this, the *Transmission Control Protocol* (TCP) is used to ensure error-free in-sequence delivery of packets to the destination application. This protocol resides on the layer above IP.

When a VoIP session is to be set up, the session or application data are first passed to TCP, where a TCP header is applied; then are passed to IP, where an IP header is applied; and then are forwarded through the network. The information contained in the TCP header includes, among other things, source and destination port numbers, which allow for identification of the applications at each end; sequence numbers and acknowledgment numbers, which allow for detection of lost packets; and a checksum, which allows for detection of corrupted packets.

At the heart of VoIP is the desire to provide good speech quality. Speech quality is essential, and traditional voice coding uses 64 kbps of bandwidth based on G.711, which is a circuit-switched approach. Alternatives to the G.711 coding scheme are seen in mobile wireless systems through the use of vocoders that employ advanced coding schemes to emulate or achieve the quality of G.711. Other requirements include low transmission delay and low jitter (delay variation) and the requirement that everything transmitted at one end is received at the other (i.e., low loss).

In order to minimize delay, *User Datagram Protocol* (UDP), Layer 4, is used. UDP, however, offers no protection against packet loss. Given the choice between UDP and TCP, the issue is whether we consider minimizing delay to be more important than eliminating packet loss. The answer is that for speech, excessive delay and excessive jitter are far more disturbing than occasional packet loss. Consequently, when transporting voice, UDP is chosen at Layer 4 rather than TCP.

It is clear, however, that something more than UDP is required if VoIP is to offer reasonable voice quality. In order to fulfill this need, a protocol known as the *Real-Time Transport Protocol* (RTP) was developed. This protocol resides above UDP in the protocol stack. Whenever a packet of coded voice is sent, it is sent as the payload of an RTP packet.

This packet contains an RTP header, which provides information such as the voice coding scheme being used; a sequence number; a timestamp for the instant at which the voice packet was sampled; and an identification for the source of the voice packet.

RTP has a companion protocol, the *RTP Control Protocol* (RTCP). RTCP does not carry coded voice packets. Rather, RTCP is a signaling protocol that includes a number of messages that are exchanged between session users. These messages provide feedback regarding the quality of the session.

However, RTP and RTCP do not guarantee minimal delay, low jitter, or low packet loss. In order to do that, other protocols are required. RTP and RTCP simply provide information to the applications at either end so that those applications can deal with loss, delay, or jitter with the least possible impact on the user.

If the VoIP session stays within a data network, then the use of traditional landline class 5 switches is not required. In many instances, though, the call could originate as a VoIP call but terminate as a circuit-switched call at a residential or business phone that does not use VoIP. In order to achieve communication between the IP client and the traditional circuit-switched user, an MGW is required, which is essentially a protocol converter that ensures that speech quality is achieved at the interface.

12.20 OSI Levels

The OSI levels are important to understand when converging technologies and commingling access platforms. The OSI consists of seven layers, 1 through 7, as shown in Table 12.12. Layer 7 is the highest OSI layer, whereas Layer 1 is the lowest. OSI Layer 1 is the physical layer and is the starting point for the OSI layer hierarchy.

12.21 Signaling System 7 (SS7)/Common Channel Signaling (CC7)

SS7/CC7 is a global standard data communications transfer protocol used to provide out-of-band signaling for processing calls in both the PSTN/PTT and wireless mobility networks. SS7/CC7 is used to support

- Traditional telephony land call delivery
- Mobile/wireless communications (roaming, authentication)

OSI Layer	Layer Name	Comments
7	Application	Used for connecting the application program or file to a communications protocol.
6	Presentation	Performs the encoding and decoding functions.
5	Session	Establishes and maintains the connection for the communication processes in the lower OSI layers.
4	Transport	Error correction and transport; both Tx and Rx are performed here.
3	Network	Switching and routing functions for the MSC are done here.
2	Data Link	Receives and sends data over the physical layer.
1	Physical	Actual media used for sending and receiving the communications; radio waves and fiberoptic wires are two examples.

Table 12.12 OSI Layers

- Advanced intelligent networking
- VoIP

The SS7/CC7 protocol has three major advantages over in-band signaling. The first advantage is the improved post-dial delay for faster call setup times. This network improvement is very noticeable to the end user. The second advantage is the ability to signal in the reverse direction improving communication between various nodes in the network. And, a third advantage over conventional in-band signaling is the improvement in the network fraud prevention due to the data (voice) messages and signaling messages being sent over separate routes.

However, SS7/CC7 is a separate network from internet and circuit switch (voice) and should not be confused with SIP. SS7/CC7 networks use a separate network to communicate and establish voice calls.

As implied there are two basic variants to this out-of-band communication protocol.

- SS7: North America—follows the ANSI T1.11x series
- CC7—follows the ITU Q.700 standards; ETSI also has a similar standard

Both SS7 and CC7 are similar and utilize packet-based (x.25) signaling. However, they have different protocols and therefore cannot communicate directly with each other. In order to facilitate communication between SS7 and CC7 networks the use of a protocol translator is required.

For the remainder of this section, I will use SS7 as the key term for simplicity. However, CC7 has the same structure and functionality.

An SS7 system consists of three basic components:

1. *STP* (Signaling Transfer Point). Responsible for transferring SS7 messages between interconnected SS7 nodes based on the information in the SS7 address fields.
2. *SCP* (Signaling Control Point). Processes data base and performs data processing functions only. Typical SCPs include HLR/HSS and billing systems.
3. *SSP* (Service Switching Points). Provides the connection between the terminal equipment and the network.

This can seem daunting when initially looking at an SS7 network. However I have found it easier to think of the STP as a router, the SCP as a database, and the SSP as a computer.

Figure 12.16 represents a typical SS7 network. Referring to the figure, the following is a brief description of the various elements and links depicted.

- *Signaling Service Points (SSPs) and Signaling Points (SPs).* SSPs and SPs serve as the connecting points for end subscriber units such as land line and mobile phones. These nodes perform call processing on calls that originate, terminate, or tandem at the location.
- *Signal Transfer Points (STPs).* STPs transfer SS7 messages between interconnected nodes based upon information contained in the SS7 address fields.
- *Signaling Control Points (SCPs).* SCPs operate in a similar manner as the SSPs and SPs but only process database information functions. A *Home Location Register* (HLR) is an example of an SCP.

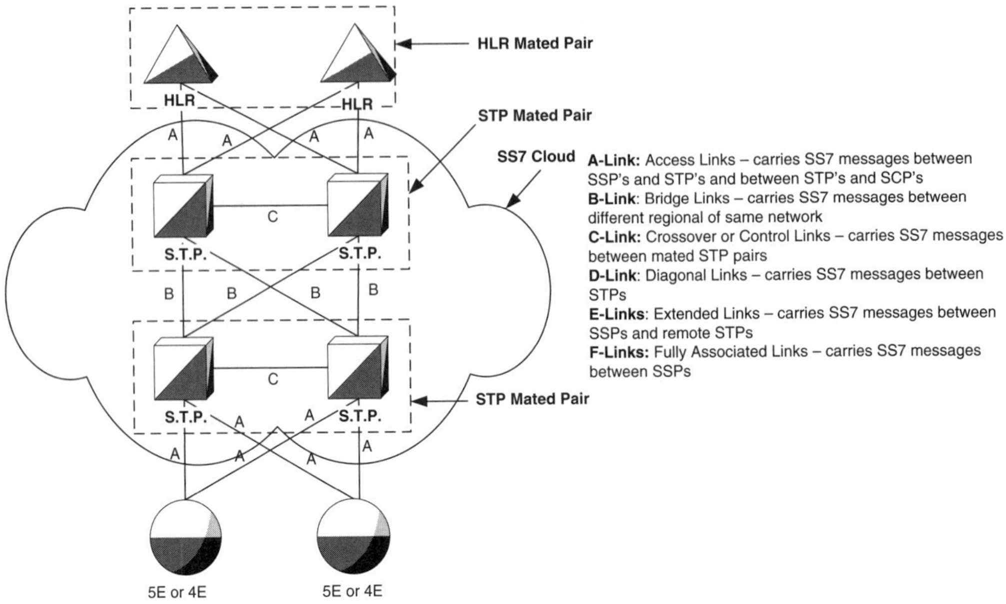

Figure 12.16 SS7 network.

- *Access Links ("A" links).* Access links carry SS7 messages between SSPs and STPs and between STPs and SCPs.
- *Bridge Links ("B" links).* Bridge links carry SS7 messages between STPs in different regions of the same network.
- *Control Links ("C" links).* Control links carry SS7 messages between mated STPs.
- *Diagonal or Quad Links ("D" links).* Diagonal or quad links carry SS7 messages between STPs in networks.
- *Extended Links ("E" links).* Extended links carry SS7 messages between SSPs and remote STPs.
- *"F" Links ("F" links).* "F" links carry SS7 messages between SSPs.
- *Linkset.* A linkset consists of two or more links connecting adjacent nodes and sharing the same routing information.

12.21.1 Intelligent Network (IN)

The network elements or entities in Figure 12.16 make up the *Intelligent Network* (IN). This network concept extends beyond the prior telecommunications systems in that it provided considerably more flexibility in services and features by allowing them to be altered and tailored to meet the needs of the customers. Further, the customer can make these changes themselves without the intervention of the carriers. Once the customer has a facility (i.e., trunk circuit) they can begin to utilize these new services.

The IN-based network relies on the SS7 protocol and network to form its interconnecting backbone to all the system nodes. Later developments in this networking concept led to the design of the *Advanced Intelligent Network* (AIN). AIN contains an additional component not found with an IN-based system, the *Service Creation Environment* (SCE). The SCE function is to define the interface between a user and the end office switches for designing new services and features.

As mentioned above, the SCP interfaces to network databases in the support of system services. The most commonly utilized mobility databases are listed:

1. *Home Location Register* (HLR) stores and provides customer data for mobile services. This data includes the mobile phone number, the mobile unit's electronic serial number, and activated features.
2. *Visitor Location Register* (VLR) is a temporary database resident on the mobile switch that is used to store subscriber data downloaded from the HLR. This database provides the mobile switch with the necessary information to offer wireless services to the subscriber.
3. *Local Number Portability* (LNP) function provides the ability for a customer to retain their same telephone number when changing to a different service provider within a local service area. This database provides the necessary phone number assignments to support this type of service.

Additional telephony databases include

1. *Call Management Services Database* (CMSDB). It provides data regarding call processing, network management, and call sampling.
2. *Line Information Database* (LIDB). It provides subscriber data such as calling card service, third-party billing instructions, etc.
3. *Calling Name* (CNAM). It provides the name of the calling party based on the information from the LIDB databases.

An example of an intelligent wireless network diagram is shown in Figure 12.17.

12.21.2 SS7 Data Network Protocols

There are a number of protocols that can be supported by the SS7 network. While the SS7 network protocol provides the reliable transport of data messages between network nodes, there exist additional protocols used to build and decode these various messages based upon the particular application in the network. The use and function of these additional protocols of an SS7 network follows the open system interface model, which consists of seven layers as described below.

1. *Physical layer*. This layer provides the electrical and mechanical interface to the network transmission facilities. This layer defines the protocol for actually transmitting a stream of serial data bits between two network nodes.
2. *Data link layer*. This layer provides control for the transmission, framing, and error correction functions over a single transmission data link facility.
3. *Network layer*. This layer provides the functions to establish clear, logical, and physical connections (if required) across the network. Included in these functions are network routing (addressing) and flow control functions across the computer to network type interfaces.

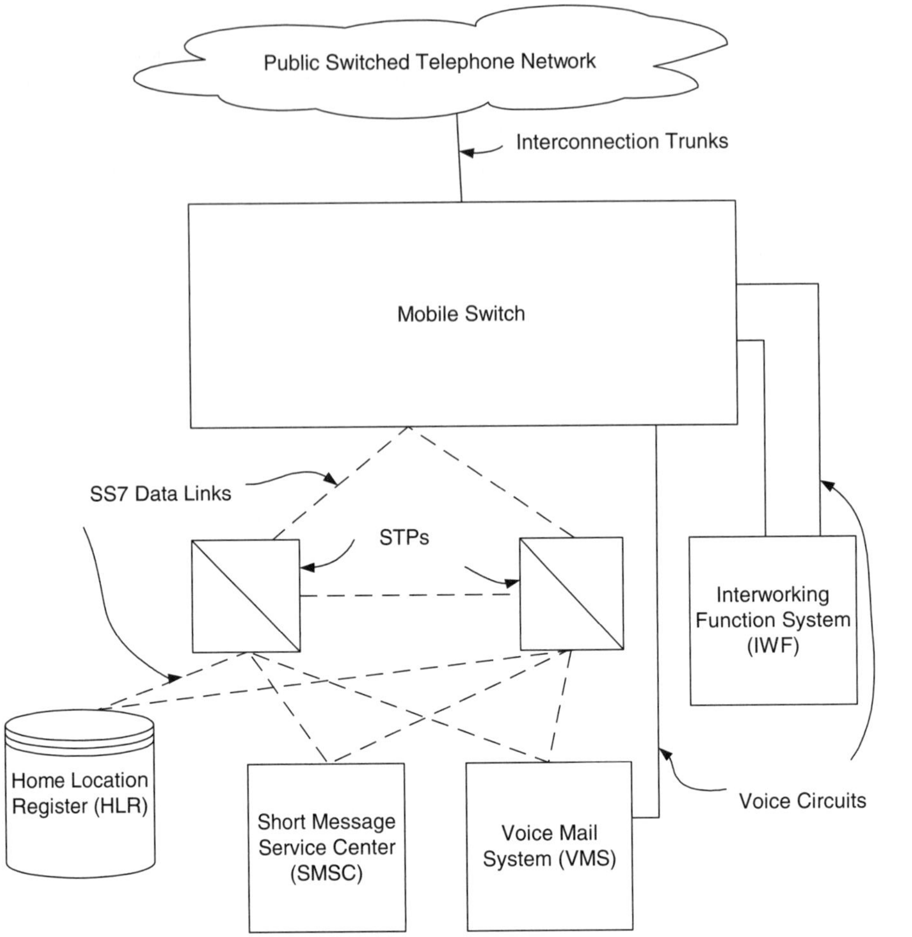

FIGURE 12.17 Example of an intelligent wireless network.

4. *Transport layer*. This layer provides independent and reliable message interexchange functions to the upper three application-orientated layers. This layer basically acts as an interface between the bottom three layers and the upper three layers.
5. *Session layer*. This layer provides the functions that control the data exchange process. It basically sets up and clears communication channels between two communicating network entities.
6. *Presentation layer*. This layer provides the syntax translation between two applications' communication over the network.
7. *Application layer*. This layer provides the user interface to a variety of network services and also manages the communication between applications on two communicating systems. Example applications include *File Transfer Access and Management* (FTAM) and electronic mail.

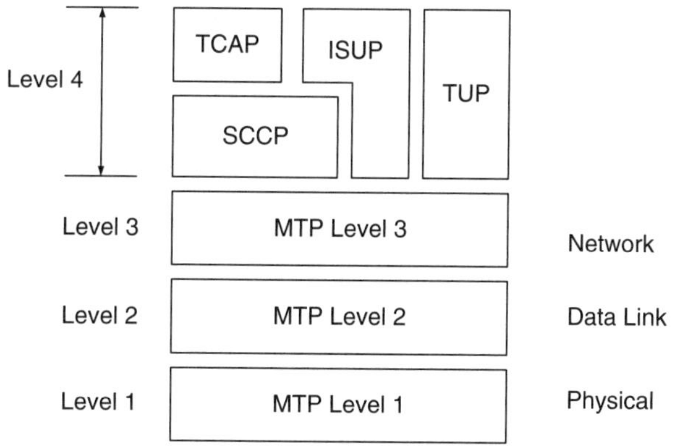

FIGURE 12.18 SS7 OSI model.

The OSI layers pertaining to the SS7 network are shown in Figure 12.18.

The different levels shown in Figure 12.18 represent different functionalities that are defined next.

- *TCAP (Transaction Capability Part).* It uses the SCCP for the message transport. TCAP supports the transfer of messages not associated with circuit control, that is, billing.
- *SCCP (Signaling Connection and Control Part).* It provides support for both connectionless and connection-oriented network services.
- *ISUP (ISDN User Part).* It provides the interoffice ISDN services and uses either the SCCP or MTP for message transport.
- *TUP (Telephone User Part).* It provides for call setup and teardown under the SS7 network control.
- *MTP (Message Transfer Part).* Lowest level and most basic layer. All other parts of SS7 network use MTP for basic transport between SPs. MTSP is comprised of the physical (data link) layer and the network layers of the OSI model.

The inter-switch standard for providing automatic call delivery and handoffs is the IS41 protocol. This protocol resides in the TCAP portion of SS7 network structure and uses SCCP as a message transport.

Additionally the *Operations Maintenance and Administration Part* (OMAP) provides the application protocols to monitor and coordinate all the network resources.

For reference, X.25 is used in traditional SS7 and typically the connectivity in an SS7 network uses a 64-kbps link and large SS7 connections may be twice this size, 128 kbps.

12.21.3 SIGTRAN

SIGTRAN is another variant of SS7 and serves the same purpose. However SIGTRAN is IP-based and is effectively the convergence of SS7 with IP. The IP protocol that SIGTRAN uses is *Stream Control Transmission Protocol* (SCTP).

The fundamental difference between SS7 and SIGTRAN is that, while offering the same functionality, SIGTRAN requires different hardware and offers more flexibility since it uses IP instead of a separate protocol.

12.22 Timing (Clocking)

Timing for any wireless network is important. Timing is important so that all the nodes are in synch with each other and that time stamps used for messaging is the same. Obviously IP communication is not synchronous. However when interfacing with external devices like a PBX, billing or providing a clocking source for the eNodeB", BTS, or EVDO is essential. Therefore timing is used in many places in the network and it is essential that you establish a good timing tree, or rather distribution scheme from the onset.

The timing system tree that is used is referred to as a *primary reference source* (PRS) and this means when two networks interface each other one has to be primary and the other secondary. Usually the network or master network that interfaces with multiple nodes is the master.

The PRS is the clocking source and provides the overall timing or clock reference point. A good timing source is needed for all broadband wireless operators, and having the same timing reference for the network facilitates handovers and billing, to mention a few easy reasons why a good clock source is needed.

The PRS has four levels referred to as different Stratum levels. The stratum clock levels are Levels 1 through 4 with the associated name of Stratum 1, Stratum 2, Stratum 3, and Stratum 4. Stratum 1 is the highest and Stratum 4 is the lowest level of the stratum clock systems. The hierarchy is based on Stratum 1 being the master reference for the network. A sample timing tree is shown in Figure 12.19.

As you would expect the farther down the timing tree you are from the master or the primary PRS the less accurate the timing reference becomes. It is essential that you limit the amount of branches for a timing hierarchy; however, not all nodes in a wireless network require the same level of timing accuracy.

The following is a brief breakdown of the different stratum levels and their most likely location in a wireless network.

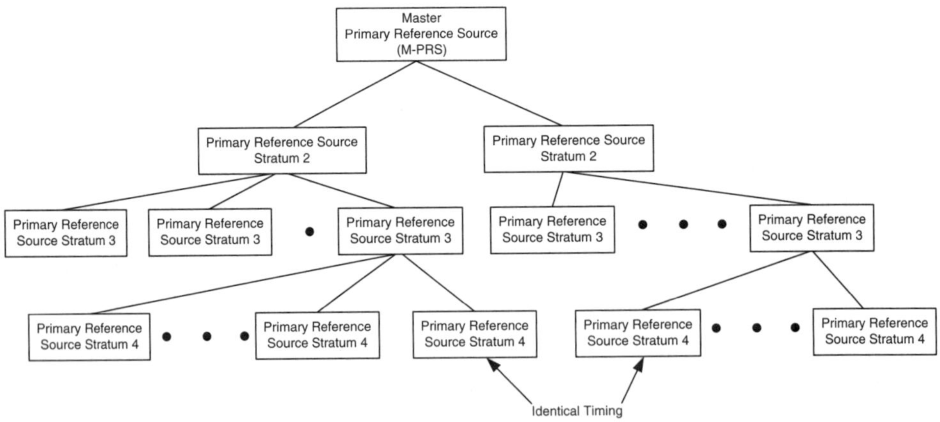

Figure 12.19 Timing tree.

12.22.1 Stratum 1

This is the top or the highest level in the stratum hierarchy and is the primary reference source, (PRS). It serves as a master clock for the entire network and utilizes a traceable timing reference and has the highest holdover period in that when it loses its source it will keep its accuracy for the longest period of time of all the stratum levels and is accurate to within 1X10-11 or better.

In a wireless system, a GPS and/or Loran-C clock source is used as its stratum 1 source. GPS and Loran-C, however, are not true stratum 1 sources since they are steerable, that is, they achieve their timing from an outside source. However, even though a GPS or Loran-C is not classified as a stratum 1 source they still are classified as a PRS within the network.

12.22.2 Stratum 2

Stratum 2 is the second highest level in the stratum hierarchy and is normally found in tandems and STPs. Stratum 2 is typically found in a wireless operator MSC and is referred to as a transit node clock.

12.22.3 Stratum 3

This is the third highest level in the stratum hierarchy and is normally associated with *Digital Access Carrier System* (DACS) and digital switches utilized for wireless and landline applications.

12.22.4 Stratum 4

Stratum 4 is the lowest level in the stratum hierarchy. Stratum 4 clocks are normally associated with a PBX, channel bank, echo cancellers, or similar network elements.

12.23 Precision Time Protocol (PTP) and Synchronous Ethernet (SyncE)

The migration from circuit switch to packet-based transport network creates some challenges of providing the proper synchronization requirements as defined for 3GPP, 3GPP2 (LTE), and IEEE 802.16e (WiMAX) networks. Specifically, a TDM network is best for handling symmetric traffic unlike packet networks, which are bursty and asymmetric traffic. However packet networks are becoming more prevalent than TDM connections.

There are two primary methods for delivering synchronization for packet traffic and that is either IEEE 1588 or SyncE. Both 1588 and SyncE are methods used to achieve synchronization in Ethernet networks as compared to the traditional method found in TDM networks. Both 1588 and SyncE are able to provide synchronization over an Ethernet network; however, there are some differences. 1588 defines a best master clock algorithm, while SyncE uses a quality level (QL) that enables every device in the network to select the best clock source.

IEEE-1588 (2008) is called *1588v2* or *Precision Timing Protocol* (PTP). 1588v2 has many similarities to SONET's *Network Time Protocol* (NTP), which uses the master and slave endpoints to transport the timing information and not the intermediate nodes that exist in a network. 1588v2 is also able to provide time and day synchronization that is needed for TDD mobile networks.

SyncE, on the other hand, has many advantages in that it is a scalable layer 1 timing solution for a packet network. With SyncE, the QL used is independent of the traffic payload because it is at the layer 1, meaning that usual packet quality factors like utilization, latency, and packet loss are involved with the quality level. SyncE provides a frequency reference; however, it is not ideal for TDD networks since it does not provide any phase or time of day references.

Regardless of which synchronization scheme is used, there are several reasons why it is necessary for synchronization and clocking, to be delivered over a packet network instead of a TDM network. First, it eliminates the need to have GPS receivers at all receiving nodes. Second, it provides a reduction in capital and operating expenses.

So, which synchronization method should you use for your wireless packet network? The answer depends on what you are trying to do. Both 1588v2 and SyncE provide synchronization. However, SyncE uses the physical layer for delivering a frequency reference, while 1588v2 delivers phase and frequency references independent of the physical layer.

When implementing either 1588v2 or SyncE, there are some differences. 1588v2 can be put into legacy Ethernet networks and equipment; however, *Packet Delay Variations* (PDVs) may negatively impact its performance. SyncE does not have PDV concerns; however, it cannot be deployed into existing Ethernet networks unless all the nodes support SyncE.

There are some things to consider when choosing the synchronization method since both LTE Advanced and TD-LTE require phase and time of day synchronization. 1588v2 can have impairments on the link causing synchronization issues; however, it is possible to have the network as a SyncE that transports 1588v2. However the drawback to this is the SyncE infrastructure requirements.

12.24 Roaming IS41 and MAP

Roaming is an essential feature and function for wireless mobility. In short, roaming is the ability of a mobile subscriber to automatically send and receive data, make and receive telephone calls, or access other services while using another operator's network.

Therefore roaming can be either in-market, national, or international. In-market could involve using another network being shared for particular features, coverage, or capacity. National roaming typically is geographical areas where the home network does not have coverage. International roaming is when the subscriber uses another network in a foreign country as compared to its home network.

Mobile Application Part (MAP) is used to facilitate roaming and is SS7-based signaling. MAP protocol is used to provide interoperability between differing networks. Additionally, MAP uses TCAP over SCCP and MTP as a transport mechanism.

MAP functions include

- Authenticate a subscriber for service including features allowed
- Location management information to HLR from other network
- Detect the presence of a mobile subscriber in a visited system
- Track and manage the subscribers state (idle, active, dormant)
- SMS call delivery to other networks
- Handover measurement and process between adjacent networks
- Call release management

However there are two primary types of MAP used:

1. GSM-MAP (GSM/UMTS/LTE)
2. EIA/TIA IS-41 (CDMA/EVDO)

12.24.1 EIA/TIA IS-41

EIA/TIA IS-41 is an inter-switch standard protocol utilized for providing automatic call delivery and handoffs for North American wireless carriers. This protocol, or better defined as an application, utilizes the TCAP and SCCP SS7 components for message transmissions and processing. When two wireless carriers form an agreement to allow each other's customers to use their mobile service, an interface must be established between the two systems for the exchange of subscriber data and network information. This interface or data link uses the SS7 protocol in conjunction with the IS-41 application to allow customers to "roam" onto the other network.

As an example of this interaction, if a mobile from carrier "A" roams to the carrier "B" network, the first step toward providing service is a query from the carrier "B" mobile switch to the HLR of carrier "A." The HLR is a network element that stores all the necessary customer data needed to support the carrier's mobile services. This data will include the customer's phone number, the electronic serial number of the mobile unit, the list of features the customer has been given, etc. When a roaming mobile turns on their phone the unit will request service from the system. The system will take the mobile phone number, electronic serial number, and authentication key and send it to the home system. The home system will check to make sure the subscriber is valid (has paid their bill and is active in the home system). Once this step is completed the home system will send the customer's data (user's profile information) to the roaming system. The roaming system will download this data into the *Visitor Location Register* (VLR) located on the mobile switch. Next, the roaming system will allow the roaming mobile service on their network.

As shown in Figure 12.20, the IS-41 protocol allows for the exchange of information between network databases and switches. It provides a method where mobile switches can share customer data amongst different carriers. Where two carrier networks share border cell sites the IS-41 protocol provides the ability for calls to hand-off between systems. In this application the IS-41 protocol provides for the exchange of information between the mobile switches for allocating RF channels and interconnecting trunks or voice circuits.

12.24.2 GSM MAP Protocol

The GSM, UMTS, and LTE technology uses the GSM MAP protocol, standardized by the *European Telecommunications Standards Institute* (ETSI), to provide a large selection of wireless services. The specifications can be divided into two main components, the switching portion and the radio portion. The radio portion includes the mobile unit and the antenna system, or cell sites, used to receive and concentrate all RF signaling within the defined service area. Cell sites contain the antenna, transceiver, and interface equipment necessary for voice path assignments and call control.

The core network portion includes the MSC/EPC, the HLR/HSS, and *Visitor Location Register*, the *Equipment Identity Register*, and the *Authentication Center* (AC).

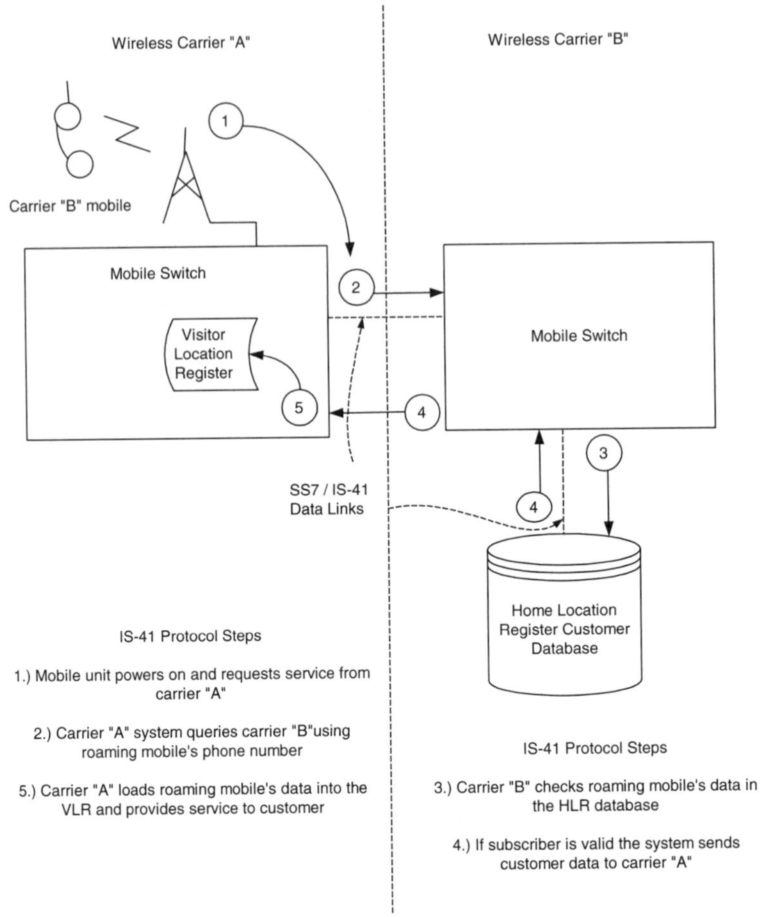

FIGURE 12.20 IS-41 protocol function.

The interface between the core network and the radio access network for GSM is called the A-bis interface. This is a 64-kbps link that utilizes four protocols for signaling between these two nodes. These protocols include

1. Link Access Procedure on the D Channel (LAPD)
2. Base Transceiver Station Management (BTSM)
3. A-bis Operations and Maintenance (ABOM)
4. Direct Transfer Application Part (DTAP)

For the A-bis interface the LAPD protocol functionality provides for layer 2 transport of data packets between these network elements. The BTSM protocol provides for control of the base station radio equipment and the interconnecting interface to the MSC. The DTAP protocol supports the transfer of data and additional signaling from the base station to the MSC.

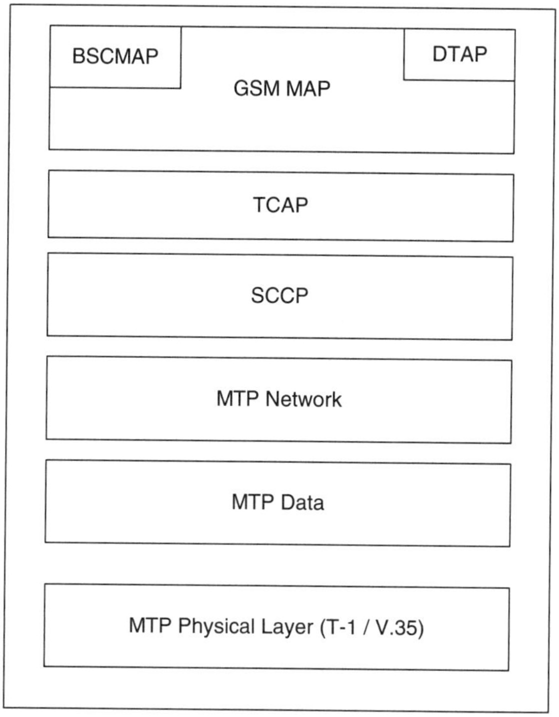

Figure 12.21 MAP protocol stack.

In the network, SS7 protocols are used for the transport of data between network elements that support circuit allocation and control as well as to access database information. These SS7 protocols include the following (Figure 12.21):

1. Message Transfer Part (MTP)
2. Signaling Connection Control Part (SCCP)
3. Mobile Application Part (MAP)
4. Base Station Subsystem Mobile Application Part (BSSMAP)
5. Transaction Capabilities Application Part (TCAP)

12.25 Diameter

Diameter protocol is used extensively in 3GPP starting with release 5. Diameter uses IP networks as its medium, and runs on top of TCP (*Transport Control Protocol*) or SCTP (*Signaling Control Transport Protocol*). The Diameter protocol provides an AAA framework for Mobile-IP, NASREQ, and ROAMOPS.

Both Diameter and SS7/CC7 are signaling protocols used in telecommunication systems. However, Diameter is used in 3GPP wireless networks for AAA services (*authentication, authorization, and accounting*), while SS7/CC7 is used in call management and other services management.

Diameter is essentially a peer–to-peer communication protocol with high security and has since replaced many other protocols in wireless mobility. Diameter is derived from RADIUS (*Remote Authentication Dial-In User Service*) protocol with a number of improvements and is defined in RFC 6733. Diameter is also backward-compatible with RADIUS to ease migration from and with legacy platforms.

Essentially, Diameter is a policy control protocol, which is used to perform policy functions, resource control, as well as AAA, enabling a single server to handle multiple policies associated with multiple services. More specifically, Diameter in implementation looks like a server client and is meant to utilize existing RADIUS dictionaries and user profiles.

With Diameter every host or node can act as either a client or a server depending on what its function is in the wireless network.

In Diameter, a node can be defined as several things including

- Client
- Server
- Agent

A *Diameter Client* is a node that is at the edge of the network that performs access control. For example, the node that receives the user connection request will act as the client. Two examples of a Diameter Client are the *Non-Access Stratum* (NAS) and *Foreign Agent* (FA).

A *Diameter Server* authenticates the user based on the information provided. If the authentication process succeeds, then the user's access privileges based on predefined policies are included in the response message that is sent to the Diameter Client.

Additionally, in Diameter there is a node called an *agent* that can act as any one of the following:

1. *Relay agents.* They forward Diameter requests and responses based on routing-related *Attribute-Value Pairs* (AVPs) and the associated routing table entries. However, a relay agent does not make policy decisions and does not examine or alter non-routing AVPs. Additionally, relay agents do not originate messages.
2. *Proxy agents.* These are used to forward Diameter requests and response messages and also make policy decisions relating to resource usage and provisioning.
3. *Redirect agents.* These enable clients and servers to communicate directly. Redirect agents do not alter any AVPs and also do not originate any messages. However, a redirect agent is capable of handling any message type it gets.
4. *Translator agent.* This is a node that acts as a translator between a Diameter node and a RADIUS node.

12.26 Regulatory Considerations

When designing or growing a core network or any wireless network, the inclusion of regulatory considerations needs to be factored into the design. The regulatory considerations are of course specific to each country where their wireless network has a physical presence. Since regulatory requirements are country-specific, I will utilize the United States as an example for some of the regulatory considerations that you may need to consider. However, I would recommend that if you are designing the network from scratch or growing it that you consult with your regulatory counterpart and determine what requirements you need to meet or those requirements that you will most likely need to meet in the near future.

In the United States, regulatory requirements involve both federal- and state-driven requirements. Specifically, in the United States many states have specific telecom requirements that are governed by the PUC/PSC (*Public Utility Commission/Public Service Commission*). However not all states have a PUC or PSC.

The following is a brief list of areas you need to address for a wireless network pertaining to regulatory compliance. The rules for some of these can and do change requiring vigilance.

- FCC—licensing and regulatory oversight for commercial, state, and local government radio licenses
- PUC/PSC (not in every State)
- NTIA—licensing and regulatory oversight for federal government radio licenses
- FAA—air navigation protection (tower height, marking, and lighting)
- E911
- CALEA (*Communications Assistance for Law Enforcement Act*)
- EMF (RF Emissions)
  - Compliance and monitoring required
  - CFR47 part 1
  - Listed as an *Environmental Assessment* (EA)
- Net neutrality
- Local Number Portability (LNP) Part of Telecom Act 1996
- Note: Wireless industry initially argued it did not apply to wireless mobility systems

12.26.1 E911

Enhanced 911 (E911) is a requirement that wireless operators need to implement. E911 involves the ability of a subscriber to dial 911 and be connected to a local *Public Service Answering Point* (PSAP) when they dial 911. In other countries the dialing sequence is different. With E911 the subscriber is required to be connected to the local PSAP regardless of their current billing or subscriber record.

However, as expected there are several phases to E911 for wireless mobility networks. In all cases, however, a wireless operator is required to connect to 911 tandems and this is typically a direct connection to each of the 911 tandems in the market itself.

- Phase 1
  - Basic identifications
  - *Public Service Answering Point* (PSAP) dependent
  - Caller ID
  - Cell/sector has unique translated phone number
  - Cross reference, database, is used to identify general location only
- Phase 2
  - Referred to as E911
  - Meant to improve effectiveness and reliability of 911 service

- Location-specific info accurate within 50–100 m
- Multiple methods used (no FCC or federal guidance)

What is additionally interesting with E911 is that there is no defined standard for E911 from a RAN perspective. In addition, there is no federal standard for how the E911 is to be delivered to the PSAP.

Each state, county, and possible city has its own unique E911 connectivity and interface requirements that need to be understood.

Phase 1 involves simple location identification. A cell site or sector has a unique number assigned to it, that is, telephone number, and this is how the geographic area the distressed caller is defined to be within. The obvious issue is location accuracy and the overall area a cell can cover, since it depends on many issues. It is better than no identification and often when a distressed caller places a E911 call they cannot identify their location like mile marker, intersection, address, etc., unlike a landline 911 call where a physical station location can be identified to an address.

Phase 2 involves more specific location identification. However, there is no standard for the RAN. There are, or were, over 7 competing E911 RAN techniques being pursued in the United States. 50 m is for A-GPS and 100 m is for GSM/UMTS, which were deployed using *Angle of Arrival* methods. LTE utilizes A-GPS.

However, pure data connections do not need to support E911 for the moment in the United States. However this requirement is likely to change in the near future.

12.26.2 Lawful Intercept

Lawful intercept is commonly called wire tapping and is formally known as *Communications Assistance for Law Enforcement Act* (CALEA) in the United States. However the typical conception of wire tapping is called Title III and is not the common method used for lawful intercept.

Lawful intercept encompasses many areas of a wireless network. However with lawful intercept there are three basic elements with any intercept.

1. Agency—refers to any federal, state, or county law enforcement entity that would normally be expected to require assistance from the service provider in their efforts to lawfully intercept electronic communications provided by the carrier to one or more of its customers
2. Wireless operator—the network operator
3. Target—refers to the communications provider's customer designated by the agency as the subject of a lawful intercept effort (the bad guys)

With lawful intercept the methods used for obtaining information are one of the four methods listed next, with method 1 being the most prevalent and method 4 being the least.

1. Billing information—origination/destination
2. Pen register—origination
3. Trap and trace—termination/destination
4. Title III—content

12.27 Design Considerations

There are numerous design considerations to factor into any core network design. I have a brief list of the major salient points to consider for a core network design that I use and with luck they will also provide some initial guidance in putting together a list of all the items that you need to include in your core network planning.

- **Core Network Planning Considerations**
  - Marketing plan
  - Financial plan
  - Sales plan
  - Customer care
  - Legal/regulatory
  - Technical platforms current performance
- **Core Network Design Considerations**
  - Design and dimensioning criteria
  - Service treatment methods (end user features)
  - Platforms chosen (new and legacy)
  - Regulatory requirements
  - Software release schedules
  - Radio access method used
  - Authentication design
  - SS7/CC7 design
  - HLR/HSS design
  - Transmission plans (backhaul)
  - Interconnection plan
  - Synchronization plan (clocking)
  - Facility diversity plan
  - Power plant (AC/DC and generator) design
  - Disaster recovery plan/requirements
  - Network security design
  - Billing and remediation
  - Auxiliary platform design
  - Off-network system interface and traffic requirements
  - Equipment-ordering intervals
  - Network management
  - Budget

12.28 Final Report

The final report for the network growth plan is the part of the project where all the efforts put forth to date are combined into a uniform document. Many methods and formats can be used for putting together a growth plan report. When crafting the final network growth report, it is exceptionally important to remember who your target audience is. The report itself will be used by both upper management and the engineering department to conduct the actual planned network growth.

It is essential that the network plan be coordinated and possibly combined with the RF plan, making it a cohesive and harmonized approach. Regardless, the output of the design effort will be a report that will need to follow a uniform structure that captures the salient issues as well as facilitates the design review process. The outline that follows is a recommended format to follow. However, your particular system requirements in all likelihood will demand some variations to the proposed structure.

1.0 Executive Summary
2.0 Introduction
3.0 Subscriber and Usage Forecast (Circuit and Packet)
4.0 Design Criteria
5.0 Existing Configuration with Exhaustion Points Defined
6.0 Expansion or Migration Plan
7.0 Circuit Switch
8.0 SGSN/PDSN
9.0 Interconnection
10.0 IP Scheme
11.0 Implementation Plan
12.0 Headcount Requirements
13.0 Budget

Regardless of the actual format used, the process is important to follow and, as repeated numerous times, essential for success.

12.29 Summary

We understand that the foregoing discussion does not provide any real rules for determining the placement of network elements and for establishing the overall network topology. In reality, there are no hard and fast rules that can be applied to any network. Therefore, we have attempted to provide a description of the issues that need to be considered in the network design exercise. Each network or operator will vary in terms of geographic service area, quality and reliability objectives, capital and operating expenditure limitations, offered services, service packaging, equipment capacities, technology roadmap, and so on. All these aspects must be considered in determining the initial network design and how that design should evolve over time. It is possible to include some of these factors in software-based network design models, to which network design experience should be added in developing the optimal design.

References

3GPP TS 29.171 V9.2.0 (2010-09) Technical Specification Group Core Network and Terminals; Location Services (LCS); LCS Application Protocol (LCS-AP) between the Mobile Management Entity (MME) and Evolved Serving Mobile Location Centre (E-SMLC); SLs interface (Release 9). Azzam, Albert, *High-Speed Cable Modems*, McGraw-Hill, New York, 1997.

Bass, Stryland, "Fiber Optic Handbook," McGraw-Hill, New York 2002. Bates, Gregory, *Voice and Data Communications Handbook*, McGraw-Hill, New York, 1998.

Bates, Gregory, "Voice and Data Communications Handbook," McGraw-Hill, New York, 1998.

Black, Uyless, *TCP/IP and Related Protocols*, McGraw-Hill, New York, 1992.

Collins, Daniel, *Carrier Grade Voice Over IP*, McGraw-Hill, New York, 2001.

Dudendorf, Vern, *Wireless Data Technologies*, Wiley, New York, 2003.

Frank Hiatt, "SONET Synchronous Optical Networking *Technical Overview*," Lucent Technologies, January 1999.

Frank Ohrtman, "WiMax Handbook Building 802.16 Wireless Networks," McGraw-Hill, New York, 2005. Goralski, Walter, *ADSL and DSL Technologies*, McGraw-Hill, New York, 1998.

Goralski, Walter, "ADSL and DSL Technologies," McGraw-Hill, New York, 1998.

Guizani, Rayes, *Designing ATM Switching Networks*, McGraw-Hill, New York, 1999.

"Introduction to MPLS," Cisco April, 2000. McDysan, Spohn, *ATM Theory and Applications*, McGraw-Hill, New York, 1999.

McDysan, Spohn, "ATM Theory and Applications," McGraw-Hill, New York, 1999.

Naveen Kottapalli, "Diameter and LTE Evolved Packet System," Radisys, September 2011. Ohrtman, Roeder, *Wi-Fi Handbook*, McGraw-Hill, New York, 2003.

Ohrtman, Roeder, "Wi-Fi Handbook," McGraw-Hill, New York, 2003.

Pahlavan, Krishnamurthy, *Principles of Wireless Networks*, Prentice-Hall, Englewood Cliffs, NJ, 2002.

Russell, Travis, *Signaling System 7*, 2d ed., McGraw-Hill, New York, 1998.

Sassan Ahmadi, "Mobile WiMAX A Systems Approach to Understanding IEEE 802.16m Radio Access Technology," Academic Press, NY 2011 Smith, Clint, *LMDS*, McGraw-Hill, New York, 2000.

Smith, Clint, *Wireless Telecom FAQ*, McGraw-Hill, New York, 2000.

Smith, Gervelis, *Cellular System Design and Optimization*, McGraw-Hill, New York, 1996.

"SONET Telecommunications Standard Primer," Tektronix, August 2001.

Stefania Sesia, Issam Toufik, Matthew Baker, "LTE – The UMTS Long Term Evolution," John Wiley & Sons, 2011.

"The LTE Network Architecture | Strategic White Paper," Alcatel Lucent, December. 2009.

Torsten Musiol, "LTE Transport," Nokia Siemens, 2009. Winch, Robert, *Telecommunication Transmission Systems*, 2d ed., McGraw-Hill, New York, 1998.

CHAPTER 13

Backhaul

Backhaul Network for wireless networks is part of the *nonaccess stratum* and continues to be a critical component of wireless mobile network. Backhaul, however, has become even more important in the deployment of broadband wireless services with 3G, 4G, and what lies ahead.

Backhaul is different than interconnect even though interconnect and backhaul can and do use similar type circuits. However, for this discussion backhaul will include the portion of the wireless network that concerns itself with connecting the radio access portion of the network to the core network. Interconnect is slightly different in that it focuses on the ability to deliver and receive services from off-net resources like the Internet and *Public Switched Telephone Network* (PSTN) to mention just two.

Figure 13.1 is a high-level depiction of a wireless network which can be 3G, 4G, or the next G that becomes viable. Figure 13.1 has several terms or definitions that will help in bounding the discussion for backhaul. Off-net, as shown in Figure 13.1, involves anything that the information the wireless network operator needs to transport that goes over a third-party vendor's (3PV) network. While on-net involves those components of the network over which the wireless operator has direct control of, that is, he/she directly owns and/or maintains it.

Backhaul, which is the ability to connect the radio access portion of the wireless network to the core network, has been getting more attention as compared to backhaul used for 2G, 2.5G, and some 3G. The reason for the increased focus on 3G and primarily 4G is the increased throughput on the radio access portion of the wireless network necessitates that the bandwidth for backhaul needs to be large enough to support the packet data traffic.

Typically, circuits for 2G and 2.5G would be 2 or 3 T1/E1s, while 3G was seeing backhaul circuits of a DS3 level. However, with 4G the typical backhaul levels are 30 to 100 Mbps and with the introduction of LTE advanced the need for a 1 Gbps has arrived.

As expected there are many backhaul types and services that a wireless operator can and does use depending on the radio site's location, existing media to use for backhaul purposes, and of course the cost to build or lease the backhaul method chosen. It is also common to have a combination of owned and leased backhaul circuits used in any wireless operation.

The *Broadband Wireless Network* (BWN) typically has numerous transmission connections methods which comprise the network itself consisting of owned and leased facilities. The BWN can consist of both wired and radio broadband connections used for backhaul purposes. The typical backhaul types for use are

- Copper (DS3)
- Fiber optic
- Microwave (licensed and unlicensed)

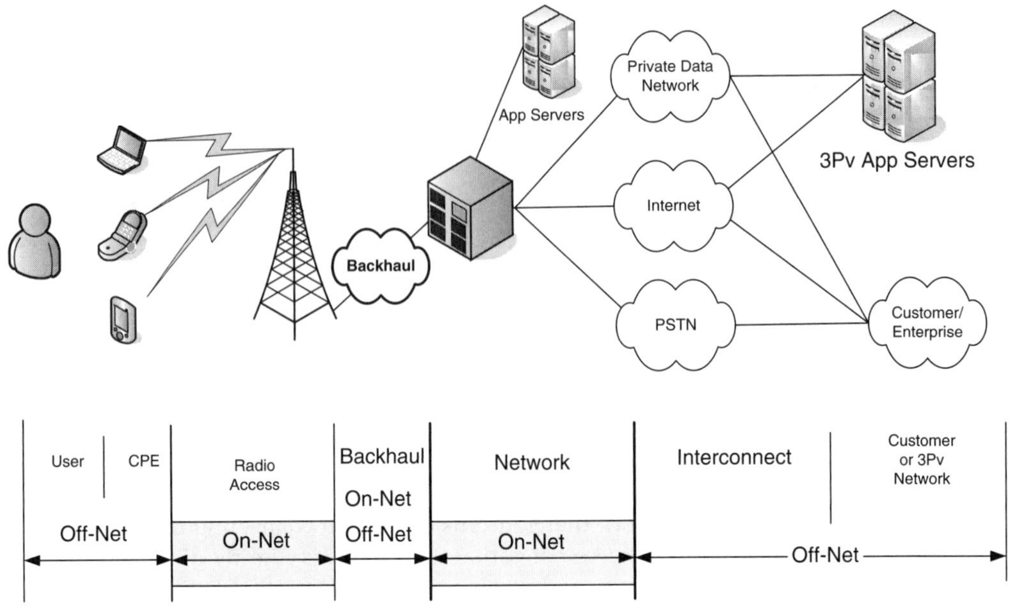

FIGURE 13.1 On-net vs. off-net.

Fiber optics is the preferred backhaul method; however, fiber is not available everywhere that cell sites are deployed.

Because every wireless network has radio access sites, the configuration is unique for every operator and location. However, most backhaul configurations, architectures, or topologies can be defined as either

- Flat
- Star
- Mesh
- Ring

Figure 13.2 illustrates a flat network architecture. One of the key attributes associated with flat architectures is that it utilizes *Internet Protocol* (IP) to achieve connectivity with all the devices. The flat architecture is used in a 4G network and improves the provisioning and management of end devices since there are less intermediate devices required. The flat architecture also is more resilient than other architectures for adding, removing, and element failures which do occur. Effectively a flat IP architecture can be thought of as a big LAN that is private, though technically there is no reason why it has to be private except for security reasons.

The star configuration shown in Figure 13.3 is a typical configuration associated with 2G/2.5G and 3G wireless mobility networks. As shown in the figure, there are several concentration points where all nodes or devices are directly connected to a common element. With a star configuration, a link failure to a node will isolate that node from the rest of the network computer; however, all other nodes will continue to operate as normal.

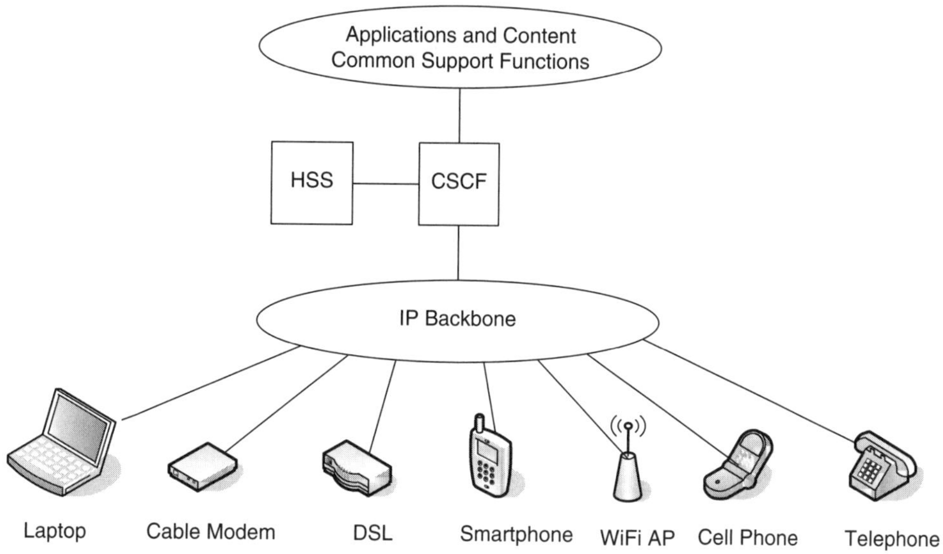

Figure 13.2 Flat architecture.

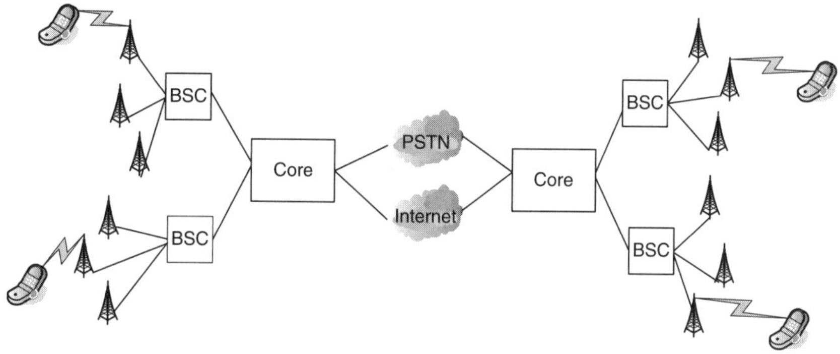

Figure 13.3 Star configuration.

However, if a failure occurs at a central concentration point or common node like the BSC/RNC or even the core, this will disrupt the network.

A wireless mesh network is shown in Figure 13.4 and is essentially an array of wireless devices that can communicate to each other via multiple paths. The mesh architecture has the advantage of providing multiple routes for an edge device to achieve communication. However this resiliency comes at a price involving increased signaling overhead and the requirement to over provision links used for upstream traffic in order to handle the different traffic routes that are possible.

Figure 13.5 shows a typical ring topology which provides two paths for the data to travel. A ring topology improves resiliency for the network by providing an alternative path in the event that a failure in one link occurs. The disadvantage is network

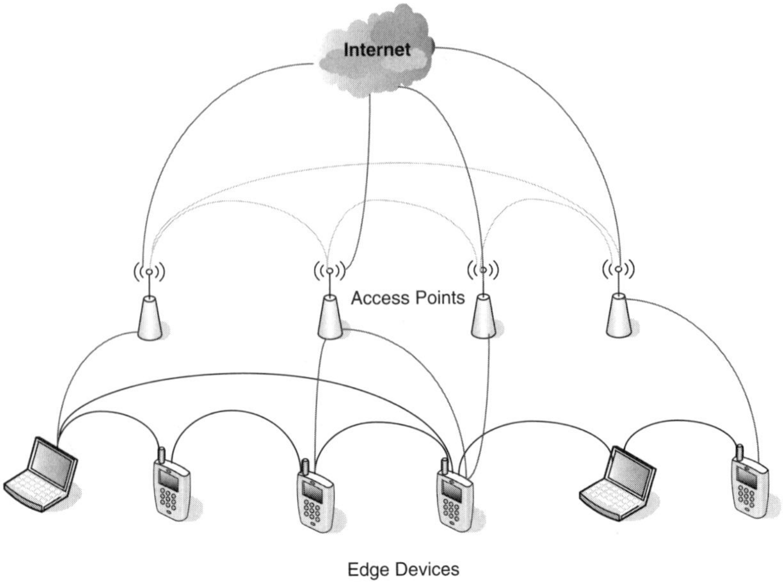

Figure 13.4 Mesh.

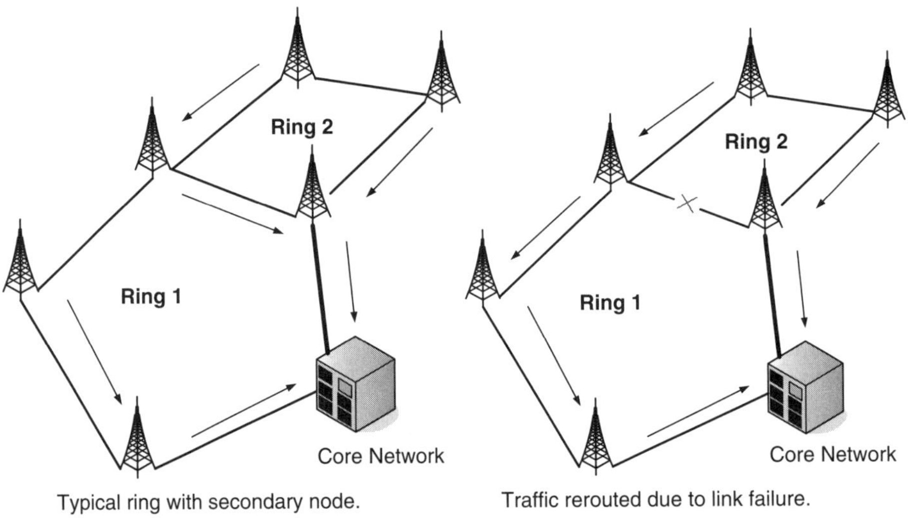

Figure 13.5 Ring architecture.

complexity and increased bandwidth requirements for each node to transport all of the data traffic. Ring topologies are associated with *Synchronous Optical Network* (SONET) networks, though not exclusively. In a ring topology, traffic is redirected when a link failure occurs as shown in Figure 13.5 (right illustration).

Which architecture(s) will be used for a backhaul configuration(s) depends on a multitude of conditions and criteria that include

- Backhaul design requirements
- Existing infrastructure
- Growth requirements
- New technology requirements
- Legacy system requirements
- Resiliency/availability (99.99x%)
- Regulatory requirements
- Budgetary considerations
- Time/schedule

Typically a BWN backhaul configuration includes both terrestrial and microwave circuits and is predominantly off-net using leased lines (i.e., not owned by the network operator).

A key concept for backhaul is that backhaul is the transport of data and voice traffic between *Radio Access Networks* (RANs) and the core network as well as between core networks (i.e., inter-machine facilities).

Backhaul, while not treating the packet data it is transporting, however, can use different protocols, and the type used depends on the type of circuit:

- What type of traffic is it carrying?
- Availability of existing circuit types for transport
- Resiliency required
- Reliability required
- Cost

The method of transport or connectivity can involve connecting distributed radio base stations (eNodeB) and a centralized switching center or *Enhanced Packet Core* (EPC) network. More specifically, the backhaul method could involve the aggregation of traffic from several radio base stations either using the same radio access method or multiple radio access methods. Additionally, a distributed core network backhaul can also be used to provide a transport layer between various elements of a service provider's network, that is, between EPC networks or between a centralized *Policy Control and Rating Function* (PCRF) and a *Mobility Management Entity* (MME) cluster. Backhaul connectivity can also include the service provider's LAN/WAN network, which is used for support of the broadband network itself.

In short, backhaul designs can be simply pipe connections for data transport following a best-effort approach or have defined *service level targets* that need to be met. However, backhaul designs typically include the need for unique traffic and application deliveries requiring separation and unique treatment of each type of traffic that is transported.

13.1 Backhaul Design Requirements

When designing the backhaul for a BWN network, whether it is a single link or part of a larger backhaul design, the main functions that need to be considered during the design phase include

- Broadband network transport requirements
- Broadband network backhaul topology/architecture alternatives
- Existing infrastructure owned, leased, or available
- Service level targets (kpi) per link
- Future proofing
- Interconnection rules and restrictions (regulatory)
- Land use acquisition
- Disaster recovery
- Timeframes
- Infrastructure vendors
- Existing infrastructure platforms
- Planned infrastructure platforms
- Leased facility vendors
- Ordering intervals
- Previously ordered facilities (pending)
- Backhaul path analysis
- Construction
- CapEx budget
- Recurring costs (OpEx)

Of course, a constant theme, or rather review point, in any backhaul design process is the determination of how much of the backhaul you want or need to have that relies on off-net components.

Some of the connectivity and transport media requirements for backhaul design include

- Wireless ISP/WISP solutions
- Wireless Access Node backhaul
- Carrier Ethernet IP-native networks
- Carry mixed IP/TDM traffic
- IP traffic only (TCP/UDP)
- IP and Ethernet services (video, VoIP, gaming)
- Business-class Ethernet services
- SS7/SIGTRAN

- Frame relay
- Circuit switched (TDM) voice traffic

The inclusion of circuit switched components could involve the voice traffic from a PBX, which is not converted to VoIP.

13.2 Backhaul Circuit Types

There are numerous circuit types that can be used for backhaul. However, the circuit types can be broken down into either structured or unstructured. A structured circuit typically refers to a circuit that is formatted, whereas an unstructured circuit refers to a circuit that has no formatting.

The advantage of an unstructured circuit is that you can send any protocol over it; however, it requires that you have the proper equipment at both ends of the circuit path. Structured circuits can also be used for obtaining timing information if needed.

Often a data path that the backhaul transverses includes many different transport circuit types and sizes and can involve different leased line service providers for a single connection. I cannot overstress the need to understand the entire path the circuit travels to ensure that the proper resiliency, bandwidth, and latency are met.

Backhaul transport includes

- Copper (T1/E1/DS3)
- Fiber (OC1-OCx, STM1-STMx)
- Metro Ethernet
- Microwave (licensed and unlicensed)
- Satellite
- DOCSIS (cable)

Protocols used in backhaul include

- TDM circuit switched (T1/E1/DS3)
- ATM
- IP (TCP/UDP)
- PDH
- SONET/SDH
- MPLS
- Frame relay
- Proprietary protocols (vendor-specific)

13.3 T1/E1

T1s and E1s are used extensively in 2G, 2.5G, and also with 3G wireless services. A T1 or E1 can be unstructured or structured. Additionally, a T1 or E1 can be provided on copper or fiber facilities with copper being the more prevalent. However, it is also

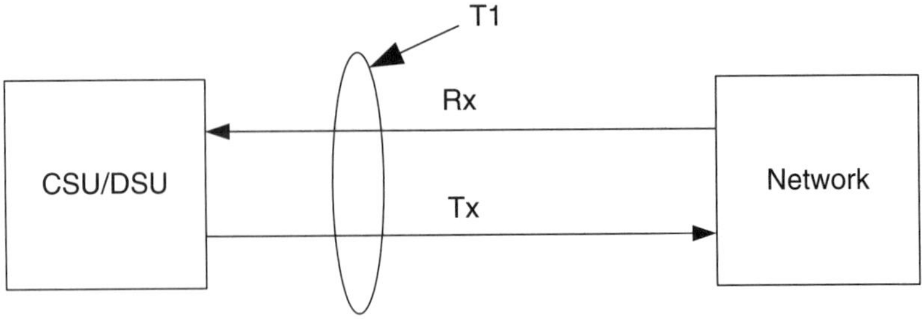

FIGURE 13.6 T1.

possible to concatenate several T1 or E1 circuits to increase the overall bandwidth needed.

T1 is also called a *Digital Signal 1* (DS1) and has a bandwidth of 1.544 Mbps in each direction. Unstructured T1s are a pipe that the *Wireless Broadband Networks* (WBN) provider can use in any protocol and are typically used for IP connectivity with 1.544 Mbps of available bandwidth. A structured T1 consists of 24 *Digital Signal 0* (DS0s) or time slots with each DS0 capable of 64 kbps. Additionally, a DS3 that is structured is made up of 28 T1s for a total bandwidth of 44.736 Mbps.

The T1 interface for the broadband network is via a CSU/DSU and is shown in Figure 13.6, which consists of four wires that are grouped into two pairs and use an RJ48 connection. Every reference for a T1 is from the network aspect, and for RX the RJ48 uses pins 1+2, and for TX pins 4+5 are used. However a T1 has a limited range which is typically 6000 ft. When the distance is greater than 6000 ft, a repeater will need to be used.

With a T1, a wireless service provider does not need to use all of the DS0s and can instead use a fraction of the available DS0s, which is referred to as a fractional T1. However, with the cost of T1s the use of fractional T1s is not as prevalent as it once was.

A structured T1, ITU-T Rec. G.711 (μ-law), is organized in frames and each frame includes 193 bits; each time slot, DS0, uses 8 bits. However there is one additional bit used, the F bit, which carries framing and supervisory information ($24 \times 8 + 1 = 193$).

There are several types of formats used for supervisory information and these are either

- Superframe (SF) (also called D4)—consists of 12 T1 frames
- Extended Superframe (ESF)—consists of 24 T1 frames

The SF format provides limited supervision capabilities while the ESF format provides improved supervision capabilities, which include online link performance monitoring through *Cyclic Redundancy Check* (CRC) and end-to-end supervision and control data link.

Structured T1s use either AMI or B8ZS where AMI has "ones" which are alternately transmitted as positive and negative pulses, whereas "zeros" are transmitted as a zero voltage level. The AMI format cannot transmit long strings of "zeros," because such strings do not carry timing information. Therefore, the AMI signal source must generate a signal with guaranteed minimum "ones" density which may reduce the data rate. B8ZS, however, is a zero-suppression method and provides clear channel capability, and the "ones" density requirement no longer restricts user data characteristics enabling a full 64 kbps per timeslot.

E1s are used primarily everywhere outside the United States. There are exceptions, like Japan and some other countries; however, E1s are used in most of the world.

An E1 has more bandwidth than a T1. An E1 has a bandwidth of 2.048 Mbps in each direction. E1 signal characteristics are specified in ITU-T Rec. G.-T Rec. G.703 and can be structured or unstructured.

- Structured—30 DS0s (30 × 64 kbps = 1.92 Mbps)
- Unstructured—2.048 Mbps (32 equivalent DS0s)

The DS0 timeslots for structured E1s can transport data or audio (voice) transmissions. Structured E1s use a different DS0 structure than that of a T1. For an E1 timeslot, 0 and 16 are used for signaling, meaning in a structured environment an E1 has 30 usable DS0s.

- Timeslot 0—reserved for Frame Alignment Signaling (FAS)
- Timeslot 16—Multiframe Alignment Signaling (MFAS) and Channel-Associated Signaling (CAS)

The data transferred over the E1 line is organized in frames using either 256N or 256S. 256N consists of 2 frames (one odd frame and one even frame). The 256N multiframe is generally used when timeslot 16 is to carry a user's payload. 256S also consists of 16 frames; however, 256S multiframe is generally used when timeslot 16 serves for the transmission of end-to-end signaling using CAS. Additionally, E1s use CRC-4 (ITU-T Rec. G706) for monitoring link quality. Additionally, an E3 consists of 16 E1s.

Both T1s and E1s can be configured as a *Primary Rate Interface* (PRI), which is based on *Integrated Services Digital Network* (ISDN). A PRI is typically used for interfacing with the PSTN/PTT to a PBX for *Direct Inward Dialing numbers* (DID). A PRI supports both voice as well as data at the same time and can dynamically adjust which DS0s will be used for data and voice usage.

PRIs consist of Bearer (B) channels and a D-Channel

- T1 (23 B + 1 D)
- E1 (30 B + 1 D)

Where the T1 PRI is capable of 1.472 Mbps data and the E1 PRI is capable of 1.920 Mbps data.

13.4 Fiber

The use of fiber optic facility is the preferred method of backhaul transport for WBNs. Fiber optic facilities have much more bandwidth capabilities than other circuit types like copper, microwave, or satellite connectivity.

Fiber optic circuits can transport all telecom protocols like

- TDM
- ATM
- IP

Signal Type			Mbps
OC-1	STS-1	STM-0	51.840
OC-3	STS-3	STM-1	155.520
OC-12	STS-12	STM-4	622.080
OC-24	STS-24	STM-8	1244.160
OC-48	STS-48	STM-16	2488.320
OC-192	STS-192	STM-64	9953.280
OC-768	STS-768	STM-256	39813.120

TABLE 13.1 Fiber

Fiber can be also be secured as *dark fiber* where the WBN operator has to provide the optics; this is used for unique protocols like *radio over fiber*.

Fiber optic bandwidths are categorized based on OC, STS, and STM levels as shown in Table 13.1.

However there are fiber optic circuits that can be transported. The two types of fiber optic cable are

- Single mode
- Multimode

Single mode (SM) uses a single optical frequency offered by a laser and offers higher bandwidths and is used for longer transmission paths.

Multimode (MM) typically uses LEDs for transmission/receiving (tx/rx) and is meant for short distance communication links, typically within a building, and has a lower bandwidth than single mode.

The choice of which fiber cable type to use is dependent upon the application that the backhaul needs to support. It is important to know whether the devices you are connecting have the correct interface cards for either single mode or multimode.

For example, SM has less attenuation than MM due to the characteristics of the fiber optic cables' properties. Specifically, SM fiber is used in many applications where data is sent at multifrequency *Wavelength-Division Multiplexing* (WDM).

However, MM typically uses LEDs while SM uses lasers for modulating the signal that is sent over the fiber optic cable. Table 13.2 shows the high-level differences between MM and SM. Please note that most frequencies used in MM and SM are typically different.

13.5 Microwave

Microwave is another method for providing backhaul capabilities to a WBN. It is a point-to-point RF connection that requires both line-of-sight and proper Fresnel zone clearance for the RF path between the sending and receiving microwave dishes.

Microwave links can consist of both licensed and licensed-exempt frequencies. There are a multitude of frequencies to pick from when selecting a microwave link. A simple rule of thumb, regardless of whether the microwave frequency is licensed or

Multimode	Single-Mode
850-nm and 1310-nm LEDs 850-nm lasers	1310 nm and 1550 nm 1260–1670 nm with DWDM
Lower installation cost	Higher installation cost
Lower system cost	Higher system cost
Higher loss, lower bandwidth	Lower loss, higher bandwidth
Distance up to 2 km	Distance up to 60 km+
Best for LAN, SAN, data centers, central offices	Best for WAN, MAN, access, campus

TABLE 13.2 MM and SM

unlicensed, is that the lower the frequency used the longer the range is between sender and receiver; however, the lower frequencies do not have the same bandwidth and transport capabilities as higher frequency microwave links. Therefore, if you need range you may need to compromise on bandwidth; however, if you need bandwidth then shorter distances should be used.

Some of the typical uses for microwave backhaul include

- Extend network reach, expand footprint
- Fail-safe redundancy to critical links
- Temporary deployment before wired solution is available
- Reduction in total cost of ownership (TCO)

Microwave system topology can include any or all of the following:

- Single link (aka point to point)
- Star
- Ring

Microwave link(s) may be constructed as a single hop of less than 2 km or as a backbone route (backhaul) that includes multiple hops spanning 15 to 20 km each.

Microwave radio links can be deployed as a last-mile access connection, aggregation or hub, or a combination of both.

Presently, digital microwave equipment is available with 10/100/1000BT Ethernet interfaces as well as SONET/SDH at various transmission rates depending on the frequency and channel bandwidths that are available for possible use.

The frequency bands used for digital microwave radio are recommended by *Consultative Committee for International Telegraphy and Telephony* (CCITT). Each band is defined for the frequency range, the channel spacing, and the bit rate.

- The lower-frequency bands are meant for longer distances while the higher-frequency bands are meant for shorter distances.
- The higher-frequency bands have more bandwidth capabilities than the lower-frequency bands

Regardless of whether the microwave link uses licensed or license-exempt frequencies, the microwave paths require *line-of-sight* and are often limited in use by building or terrain blockage or existing use of the licensed spectrum within urban areas.

When determining whether a microwave link or system meets the backhaul requirements, the following needs to be determined at the inception of the design process or consideration.

- Overall network architecture/topology plan
- Interface requirements (Ethernet/SDH/SONET)
- Link details (per link)
- Link capacity required
- Frequency of operation
- Polarization
- Antenna size requirement
- Antenna height
- Antenna azimuth
- Hop length
- Line-of-sight survey report for link
- Protection scheme (hot standby, loop, etc.)

Figure 13.7 is a diagram that illustrates the use of several microwave topologies in a single wireless network.

There are numerous books and vendor information that provide a host of design and implementation considerations.

13.6 Plesiochronous Digital Hierarchy (PDH)

Plesiochronous Digital Hierarchy (PDH) is primarily used for point-to-point applications involving backhaul. An advantage with using PDH is that it does not require network synchronization to function.

Microwave radio point-to-point links typically use PDH.

With PDH the basic transfer rate of the data is based on increments of 1.544/2.048 Mbps. A PDH connection typically involves a DS3/E3 or less.

- 34.368 Mbps (16 × E1)
- 44.736 Mbps (28 × T1)

There are several drawbacks with using PDH:

- Not easily scalable
- Used for non-optical interfaces
- Not an international standard

However, PDH can be more cost-effective for lower bandwidth requirements, and more important, PDH can be transported over SONET/SDH networks.

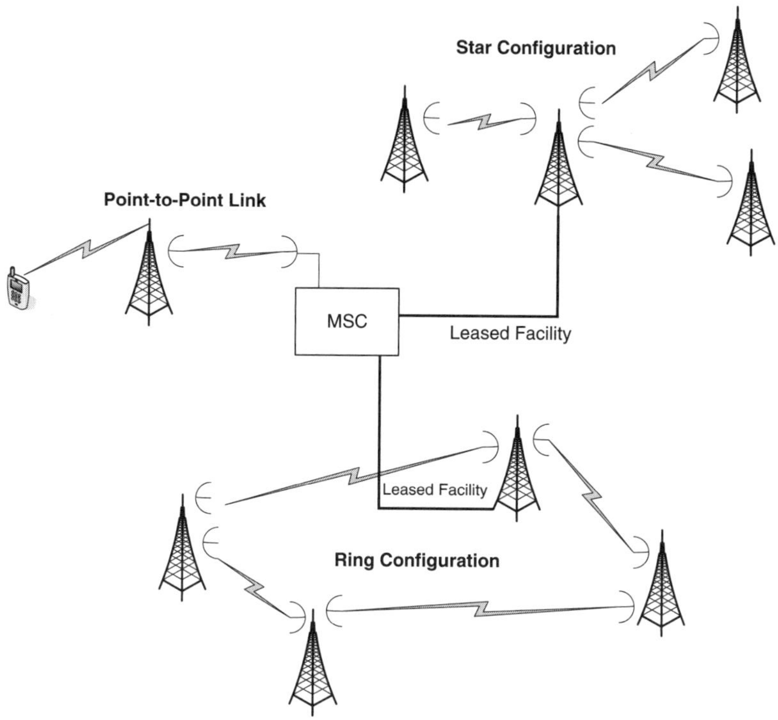

FIGURE 13.7 Microwave.

13.7 SONET/SDH

Synchronous Optical Network (SONET) and *Synchronous Digital Hierarchy* (SDH) are used extensively in backhaul for wireless service providers. SONET is an ANSI standard used in the United States, while SDH is an ITU standard.

SONET and SDH are similar but not the same, and the differences are important to know. Both SONET and SDH are linear networks and have

- Optical interfaces
- Capability of powerful management
- World standard digital formats
- Synchronous structure which is flexible
- Cost-effective and easy traffic cross-connection capacity and add and drop facility
- Reduced networking cost due to the transversal compatibility
- Forward and backward compatibility

Both SONET and SDH can support transversal capability where *Coarse WDM* (CWDM) and *Dense WDM* (DWDM) can be intermixed allowing for lower-cost components CWDM to be used with DWDM.

SONET and SDH have a transmission hierarchy that has many similarities.

- SONET—synchronous transport signal levels (STSs)
- SDH—synchronous transport modules (STMs)

The relationship between STS and STM is shown in Table 13.1.

Both SONET/SDH networks can be configured as linear networks where a typical network consists of a path, either fiber or microwave, with the use of *Add/Drop Multiplexers* (ADM) and Regenerators.

ADMs are used to allow traffic to enter and leave network. It's how you add and remove traffic from the SONET/SDH network. ADMs enable ring topologies to be established. There are several types of ADM.

- Terminal ADM (TM)—used to combine or divide PDH and SDH/SONET to or from higher bit rate STS/STM signals
- DXX—used to groom traffic and enable PDH tributaries to be mapped into VC/VTs in addition to switching VCs between STMs or VTs between STSs
- ADM—used to insert or extract PDH or SONET/SDH signals from higher order STS/STM signals

Regenerators, as their name indicates, are used to regenerate the clock signal as well as the amplitude and integrity of the signal that may have been attenuated or distorted due to dispersion.

The components of a SONET/SDH network consist of the following major parts as shown in Figure 13.8.

Figure 13.9 is a simplified SONET/SDH ring.

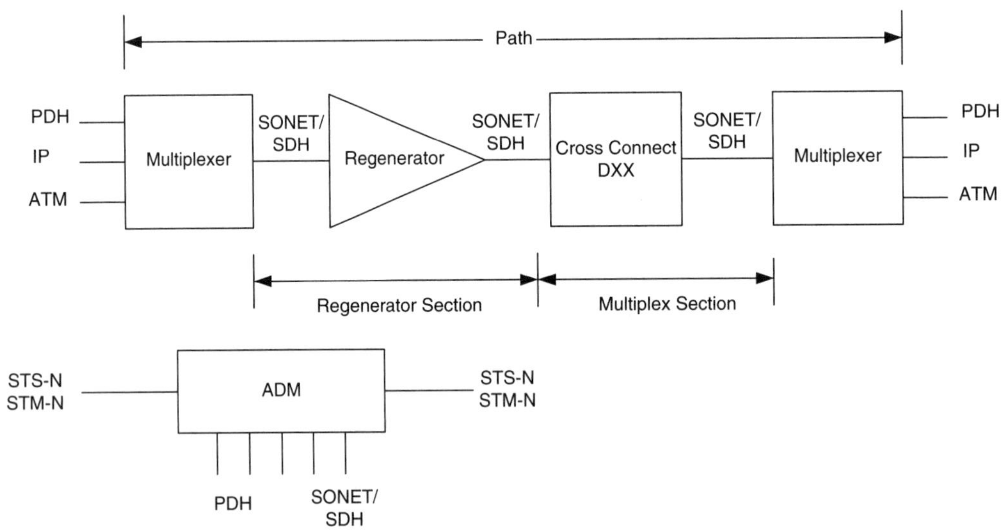

Figure 13.8 SONET/SDH.

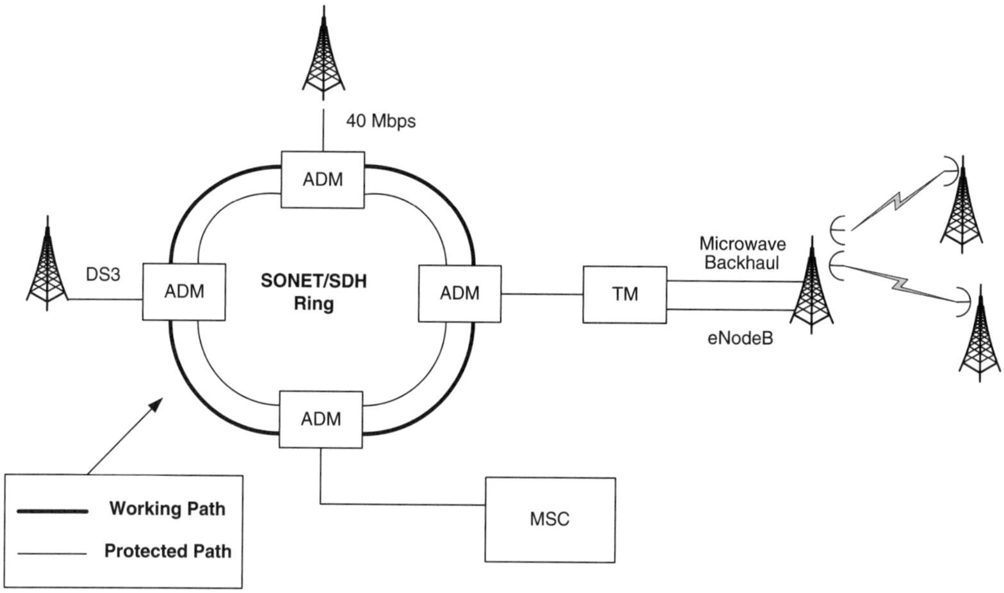

FIGURE 13.9 Simplified SONET/SDH ring.

Though SONET and SDH have a very similar structure, there are some differences between the two:

- Use different terms to describe layers.
- Multiplexing (MSOH) *Automatic Protection Systems* (APS) messages transmitted in the K1/K2 bytes are major differences.
- *Path Overhead* (POH) byte is different.

Table 13.3 helps refine some of the SONET/SDH differences.

SONET	SDH	Comments
Section Overhead	RSOH	Regeneration Section Overhead
Line Overhead	AU Pointer	Administrative Unit Pointer used for Tributary identification • Virtual Container (VC)-SDH • Virtual Tributary (VT)-SONET
Line Overhead	MSOH	Multiplexer Section Overhead
Path Overhead (POH)	Path Overhead (POH)	Path Overhead
Payload	Payload	

TABLE 13.3 SONET/SDH

With SONET/SDH, the use of regenerative and multiplexing information allows part of overhead to be coupled to network architecture. Some of the key attributes for SONET/SDH are listed next:

- SONET and SDH frame structure the same
  - Size: 270 columns with 9 rows
  - Frame repetition rate is 125 ms regardless of payload
  - Bit rate 155.52 Mbps
- Section overhead
  - Performance monitoring (STS-N signal)
  - Data communication channels
  - Framing
- Line overhead
  - Locating the payload in the frame
  - Multiplexing or concatenating signals
  - Performance monitoring
  - APS
  - Line maintenance
- Path overhead
  - Performance monitoring of payload
  - Signal label
  - Path status
  - Path trace

Figure 13.10 shows the SONET/SDH format.

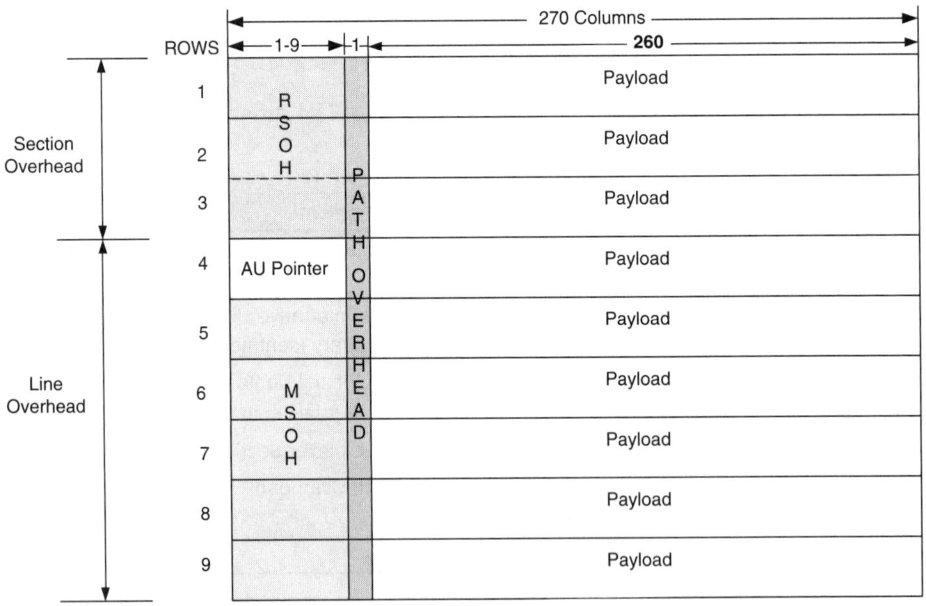

Figure 13.10 SONET/SDH format.

SONET		SDH	
Name	**Speed**	**Name**	**Speed**
VT-1.5	1.728 Mbit/s	VC-11	1.728 Mbit/s
VT-2	2.304 Mbit/s	VC-12	2.304 Mbit/s
VT-3	3.456 Mbit/s		
VT-6	6.912 Mbit/s	VC-2	6.912 Mbit/s
STS-1	50.112 Mbit/s	VC-3	48.960 Mbit/s
STS-3	150.336 Mbit/s	VC-4	150.336 Mbit/s

Table 13.4 VT and VC Relationships

With both SONET and SDH the actual transmission takes place serially where the left-most byte in the top row is transmitted, then the second byte in the first row, and so on. The process of mapping the various signals to the network is called mapping. However the two diverge here in that the SONET and SDH mapping is different where the mapping involves linking tributary signals with payload. SONET uses a *Virtual Tributary* (VT-n) and SDH uses a *Virtual Container* (VC). The VT and VC relationships are shown in Table 13.4. VTs and VCs are not the same.

One of the key attributes of SONET and SDH is the protection methods built in the architecture to ensure high reliability and availability for the SONET/SDH network. Specifically, SONET/SDH rings are *self-healing* and basically are designed to automatically recover from link failures.

Self-healing is achieved through the APS protocol and is meant to take less than 50 msec to restore the services. However there is another self-healing method, that is, *Linear Automatic Protection Switching* (LAPS).

LPS uses four fibers: two working fibers (one in each direction) and two protection fibers. The switching is based on the line state and can be unidirectional (each direction switching independently) or bidirectional.

Additionally, there are two types of protection schemes used in SONET/SDH architectures and they are *link* and *ring protection schemes*.

Link protection scheme is a 1:1 or 1:N, shown in Figure 13.11.

- One to one (1:1)
  - 1:1 basic; however, doubles network
  - 100% redundancy
- One to many (1:N)
  - Used where one protected platform or circuit is used for every N paths or platforms

Ring protection schemes, however, are different in that they can use path, line, span, or ring switching or a combination of them:

- Path switching
  - Restores the traffic on the paths affected by a link failure using a different path.
  - The path is an end-to-end connection on the ring for that service.

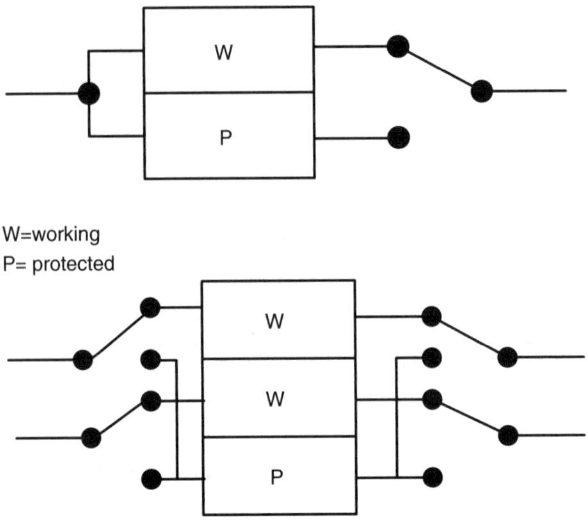

Figure 13.11 (a) 1:1, (b) N:1.

- Line switching
  - Restores all the traffic that passes through a failed link by rerouting all the traffic through a different path.
- Span switching
  - If a working fiber fails, the working traffic will be transferred over its protection ring.
- Ring switching
  - If the working and protection fibers are part of the same bundle of fibers when a bundle is cut, traffic is then switched to protection fibers. SONET and SDH have a limited number of architectures defined.

Every SONET/SDH connection on the optical physical layer uses two optical fibers regardless of the transmission speed (2Tx, 2Rx).

A SONET/SDH ring is either unidirectional or bidirectional. If the SONET ring is unidirectional, it is referred to as a *Unidirectional Path-Switched Ring* (UPSR) and if it is a bidirectional ring, it is referred to as a *Bidirectional Line-Switched Ring* (BLSR).

In a unidirectional ring, signals are transmitted in both directions of the ring and have two redundant (path-level) copies of protected traffic sent in either direction around the ring, as shown in Figure 13.12, and is called a 2F-UPSR or collector ring. With UPSR, a selector at the egress node determines which copy has the highest quality, and uses that copy and is usually near the edge of a large network.

The SDH equivalent of UPSR is *Subnet Connection Protection* (SNCP). However SNCP does not impose a ring topology and it is also used for mesh topologies.

If the SONET ring is a bidirectional ring, then the signals are transmitted in one direction of the ring. There are two types of BLSRs:

- Two-fiber BLSR (2F-BLSR)
- Four-fiber BLSR (4F-BLSR)

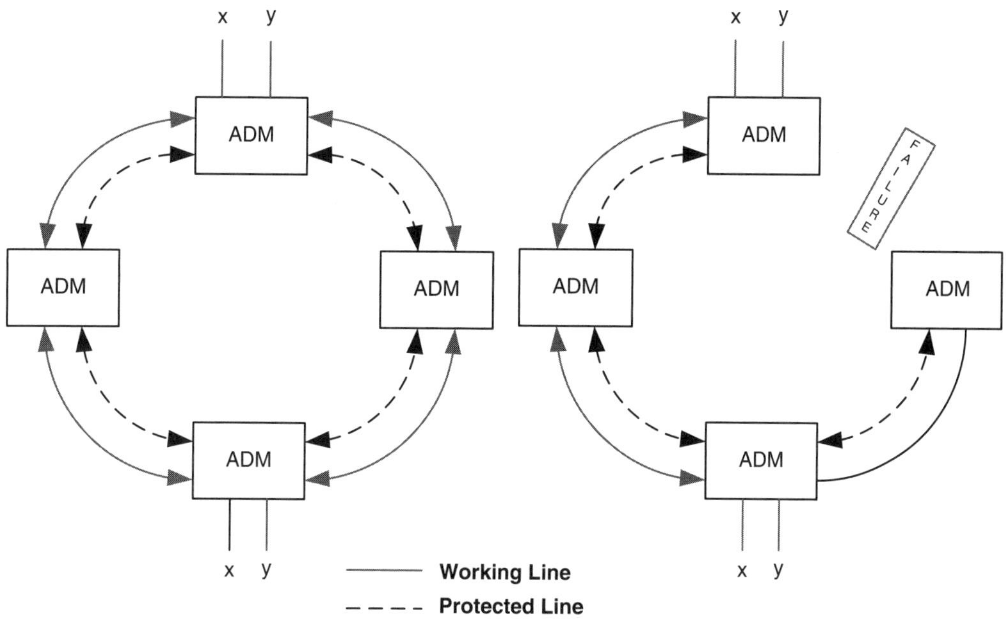

FIGURE 13.12 UPSR.

BLSR is at the line layer and does not send redundant copies from ingress to egress. Additionally, the BLSR ring nodes adjacent to the failure reroute the traffic "the long way" around the ring on the protection fibers. In four-fiber ring (4F-BLSR), single node failures or multiple line failures can be supported.

BLSRs trade cost and complexity for bandwidth efficiency, as well as the ability to support "extra traffic" that can be preempted when a protection switching event occurs. BLSRs can operate within a metropolitan region or, often, will move traffic between municipalities. They are rarely deployed in collector rings, but often deployed in inter-office rings.

The SDH equivalent of BLSR is called *Multiplex Section-Shared Protection Ring* (MS-SPRING).

In summary, the BLSR attributes are

- (2F-BLSR)
  - Used in metro core rings.
  - The ring provides line switching.
  - If working fiber fails, then the traffic that goes over the working ring will be automatically switched to the protection part of ring.
- (4F-BLSR)
  - Two working rings and two protection rings.
  - The two working rings transmit in opposite directions, and each is protected by a protection ring that transmits in the same direction.
  - The advantage of this four-fiber ring is that it can suffer multiple failures and still function.
  - Deployed in regional and national rings.

Figure 13.13 is an example of a 2F-BLSR.

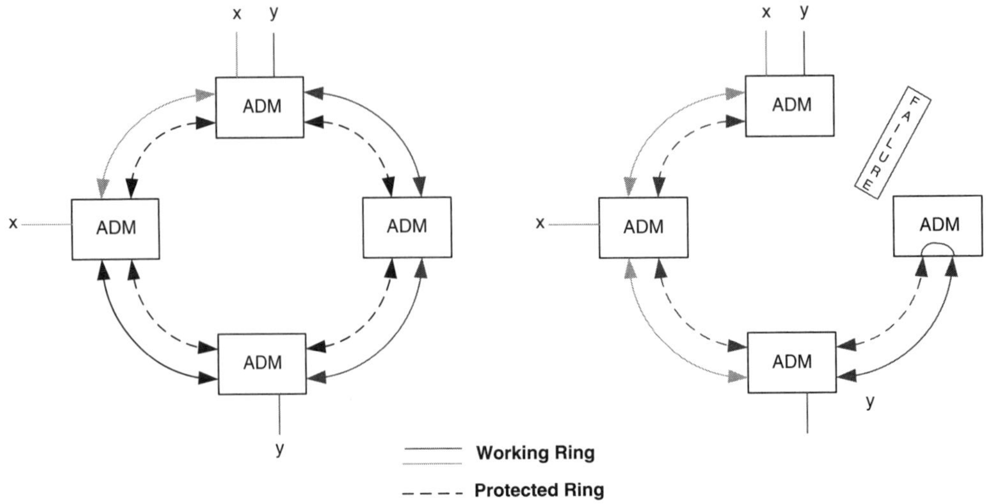

Figure 13.13 BLSR.

13.8 MPLS

Multiprotocol Label Switching (MPLS) is a unifying technology for various backhaul methods. MPLS, however, does not replace IP, it improves it. MPLS works with both IPv4 and IPv6 and is a method of improving IP performance through applying a label to IP packets at the edge router by combing layer 2 performance and layer 3 traffic management.

MPLS enables IP which is a connectionless per-hop protocol to become a connection-oriented network.

The MPLS label information include

- Destination
- Precedence
- *Virtual Private Network* (VPN) membership
- *Quality-of-Service* (QoS) information from RSVP
- The route for the packet, as chosen by traffic engineering (TE)

The basic elements in a label-switching network are

- Edge label switch routers
- Label switches
- *Label Distribution Protocol* (LDP)

LDP is used along with standard network layer routing protocols to distribute label information between devices in a label-switched network. LDP improves the efficiency of IP traffic and allows for carrier-grade capabilities to be implemented with MPLS. MPLS adds scalability, resiliency, network management, traffic engineering, and multiple service support.

13.9 WDM/DWDM

Wavelength-Division Multiplexing (WDM) is a method that enables multiple channels to be transmitted on the same fiber using different wavelengths of light. In short, WDM enables an existing fiber to transport more bandwidth and minimize the amount of fiber optic cable that needs to be deployed.

WDM uses a *Waveband Coupling System* to combine optical signals of different wavelengths on an existing fiber optic cable (span) as depicted in Figure 13.14.

WDM technology is implemented as

- Narrowband—doubles the fiber capacity and is implemented using 1533-nm and 1577-nm wavelengths
- Wideband—like narrowband, wideband doubles the fiber capacity; however, it uses a 1310-nm wavelength and combines it with another wavelength between 1528 and 1560 nm
- Dense—increases fiber capacity by implementing multiple wavelengths at the same time; has the capability to transport up to 32/64/128 discrete channels (wavelengths) requiring tight control over laser due to channel density

Figure 13.15 is an example depicting DWDM. Specifically, with DWDM, more capacity is possible with a single-mode fiber due to the amount of discrete wavelengths utilized. The table that is shown in Figure 13.15 is useful in that it identifies the laser frequency ranges used.

13.10 MetroEthernet (MEN)

MetroEthernet (MEN) is being used more and more by broadband service providers. In essence, MEN is an Ethernet technology in a metropolitan area that connects nodes to *Wide Area Network* (WAN). A MEN is an all-IP network using a fiber backbone which a

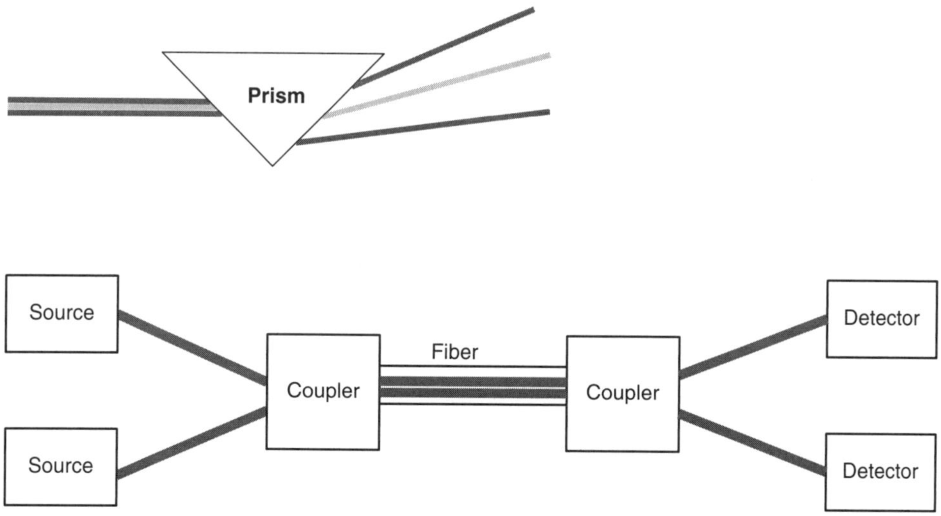

FIGURE 13.14 WDM.

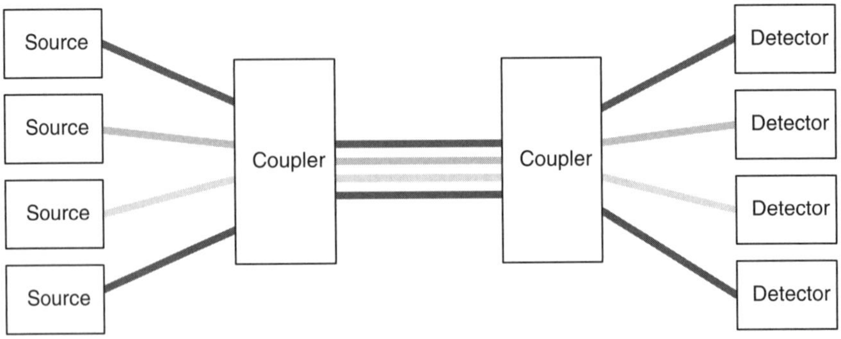

Band Name	Wavelengths	Description
O-band	1260 – 1360 nm	PON upstream
E-band	1360 – 1460 nm	Water peak band
S-band	1460 – 1530 nm	PON downstream
C-band	1530 – 1565 nm	Lowest attenuation, DWDM band
L-band	1565 – 1625 nm	Low attenuation, expanded DWDM band
U-band	1625 – 1675 nm	

FIGURE 13.15 DWDM.

broadband service provider leases facilities from. However, connectivity to MEN requires being on fiber ring and the connection is a *User to Network Interface* (UNI) connection for the broadband service provider.

There are two types of MEN connections

1. Ethernet Line (E-Line) point-to-point service
   - Ethernet Virtual Vircuit (VPN)
2. Ethernet LAN (E-LAN) multipoint–to-multipoint service
   - Multiple UNI connections
   - Appears as a LAN to UNI

MEN Links are primarily point to point; however, they can also be point to multipoint and consist of a layer 2 and layer 3 switches/routers.

Bandwidth and an all-IP pipe are the key advantages for using MEN. Typical speeds available from a MEN network are 10 Mbps, 100 Mbps, 1 Gbps, and 10 Gbps.

Because 4G wireless architectures are a flat architecture, they can easily take advantage of MEN where it provides connectivity between the eNodeB and the EPC for an LTE network.

MEN has typical bandwidth profiles that are provided as part of the leased circuit requirements. Some of the key bandwidth profiles include

- Committed Information Rate (CIR)
- Committed Burst Rate (CBR)

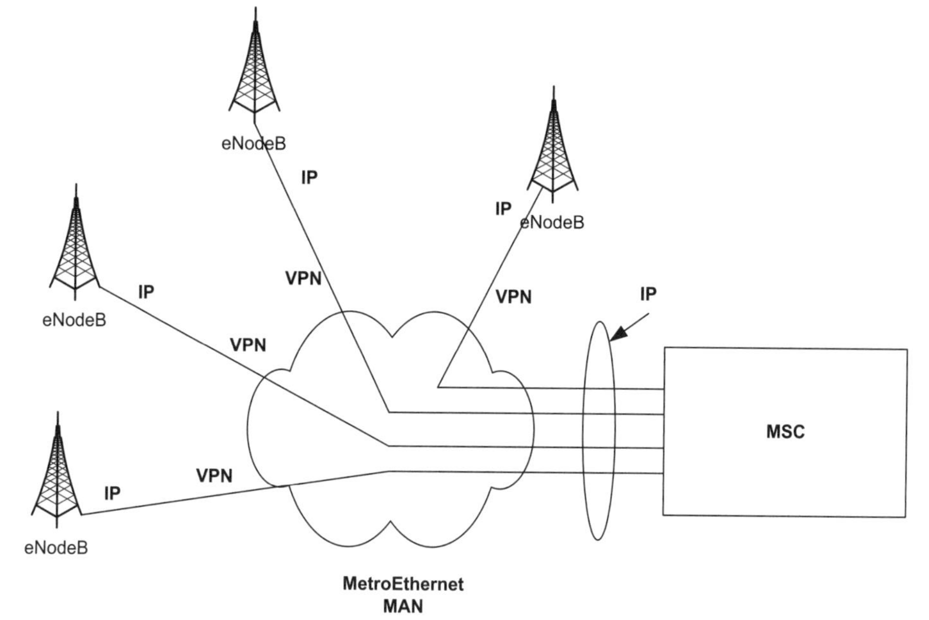

FIGURE 13.16 MEN.

- Excess Information Rate (EIR)
- Excess Burst Size (EBS)

Additionally, MEN affords *Class of Service Identifiers,* meaning a wireless mobile operator can offer or obtain different levels of performance per class of service offered or used.

Figure 13.16 is a simple depiction of an MEN configuration, where each eNodeB is connected via a VPN connection to the MSC, or rather the EPC.

However, MEN availability is not ubiquitous requiring the need for fiber build out where not present. Therefore, if MEN is not available, then an alternative backhaul method needs to be used.

13.11 Backhaul Topology

Figure 13.17 attempts to bring into perspective the diversity of backhaul types used in a typical wireless mobility network.

13.12 Backhaul Design

The design of a network topology that supports the transmission needs must consider the following information:

- Network size (number and location of base station sites)
- Capacity and traffic growth between each network element

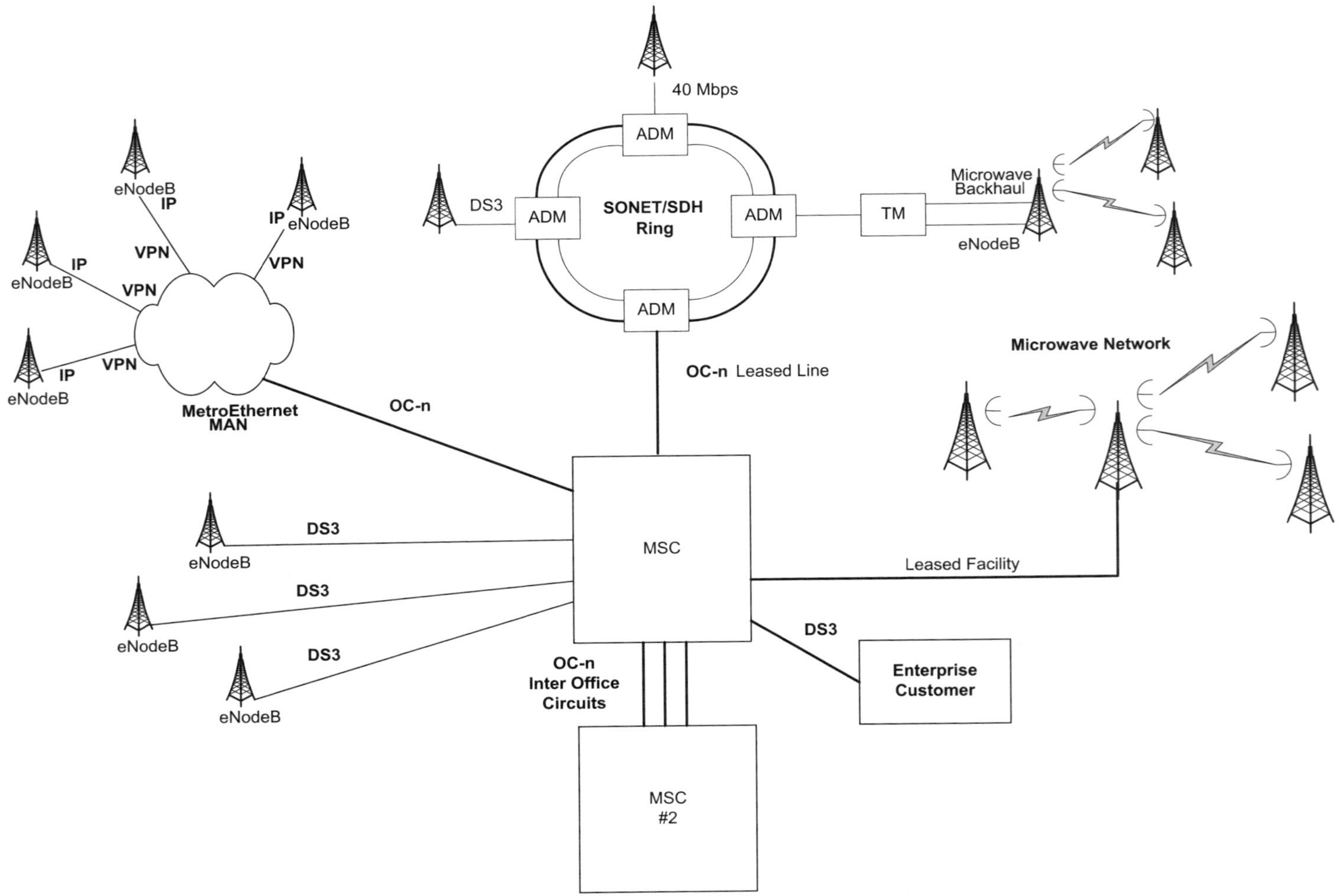

FIGURE 13.17 Backhaul topology.

- Interface and signaling requirements
- Preferred locations for effective interconnection to PSTN, Internet
- Availability of existing fiber optic capacity; cost and reliability
- Availability of microwave spectrum
- Availability of leased lines and related cost, reliability issues
- Availability of public and private right-of-way permits to construct or extend a fiber network
- Cost and budget constraints
- Requirements of design scalability and flexibility
- Multi-vendor environment issues related to problem isolation
- Space and power requirements

References

Stryland Bass, *Fiber Optic Handbook*, McGraw-Hill, New York, 2002.
Doug Bush, "Understanding Sonet VTs: A Tutorial," Xelic 1/6/2005.
Communication Developers Handbook, March 2007 NComm, Inc., Salem, NH.
Luyuan Fang, Doug Hunt, "MPLS in Mobile Backhaul," MR-234, Issue 2, May 2010.
"Introduction to MPLS," Cisco, April 2000.
Håkon Lønsethagen, Multilink Network Architecture, MARCH Workshop, Sept 21, 2011, Telenor, Fornebu, Telenor Group, Industrial Development.
"LTE L2 Backhaul Networks," NMC Consulting Group, 2011.
"LTE Backhaul: Meeting Operator Requirements," RAD Data Communications Ltd., 2011.
"MPLS Environment," RAD Data Communication, November 2007.
"Multi-Channel SDH Mapper Core," Aliathon Ltd., March 2004.
Torsten Musiol,"LTE Transport," Nokia Siemens, 2009.
Eitan Schwartz, "Mobile Backhaul Challenges and Opportunities Transporting Ethernet over SDH/SONET Networks," RAD Data Communications Ltd., 2008.
"SDH/SONET Environment," RAD Data Communication, September 2004.
"SONET Telecommunications Standard Primer," Tektronix, August 2001.
Clint Smith, *LMDS: Local Mutipoint Distribution Service,* McGraw-Hill, New York, 2000.
www.broadband-forum.org
www.3gpp.org.
www.cisco.com.
www.dragonwaveinc.com.
www.FCC.gov.
www.flukenetworks.com/FiberHandbook.
www.itu.int
www.Metroethernetforum.org.
www.rad.com.
www.telecomengineering.com.
www.wikipedia.org.

CHAPTER 14

Antenna System Selection

This chapter will briefly discuss some of the more important issues associated with an antenna system regarding *Third-Generation* (3G) and *Fourth-Generation* (4G) applications. More specifically, this chapter will provide the necessary information for selection of the antenna type to use for base stations whether for a macro-, micro-, or picocell. An important aspect is that the method used for selecting the antenna type is similar for all the technology platforms. There are, of course, some differences related to the different design issues associated with a *First-Generation* (1G), *Second-Generation* (2G), 2.5G, 3G, and 4G system. The key difference in antenna design lies in the desire to keep the systems either separate or unified depending on the underlying technology platform.

The antenna system for any radio communication platform is one of the most critical and least understood parts of the system. The antenna system is the interface between the radio system and the external environment. The antenna system can consist of a single antenna at the base station and one at the mobile or receiving station. Primarily, the antenna is used by the base station and the mobile unit for establishing and maintaining the communications link.

A great many types of antennas are available, and they all perform specific functions depending on the application at hand. Antenna types include collinear, log periodic, folded dipole, and yagi, to mention a few. Coupled with the type of antenna is the notion of an active or passive antenna. An *active* antenna usually has some level of electronics associated with it to enhance its performance. A *passive* antenna is more of the classic type in which no electronics are associated with its use, and it simply consists entirely of passive elements.

Along with the type of antenna, there is the relative pattern of the antenna, indicating in what direction the energy emitted or received by it will be directed. There are two primary classifications of antennas in terms of directivity: omni and directional. *Omni* antennas are used when the desire is to obtain a 360-degree radiation pattern. *Directional* antennas are used when a more refined pattern is desired. The directional pattern usually is needed to facilitate system growth through frequency reuse or to shape the system's contour.

The choice of which antenna to use will have a direct impact on the performance of either the cell or the overall network. The radio engineer is concerned primarily with the design phase of the base-station antenna because this is the fixed location, and there is some degree of control over the performance criteria that the engineer can exert at that location.

The correct antenna for the design can overcome coverage problems or other issues that are apparent. The antenna chosen for the application must take into account a multitude of design issues. Some of these issues involve the antenna's gain, its antenna pattern, the interface or matching with the transmitter, the receiver used for the site, the bandwidth and frequency range over which the signals must be sent, the antenna's power-handling capabilities, its *Intermodulation Distortion* (IMD) and *Passive Intermodulation* (PIM) performance. Ultimately, the antenna you use for a network needs to match the system's *Radiofrequency* (RF) design objectives.

14.1 Base-Station Antennas

A number of antennas can be used at a base station. However, the specifics of what makes up a base-station antenna or antenna system are determined by the design objectives for the site coupled with real-world installation issues. For 3G and 4G radio systems, as well as the 2.5G radio systems, most, if not all, of the antenna design decisions are determined by the type of base station at which the antenna will be employed. For instance, the antenna system for a macrocell most likely will be different from that used for a microcell and definitely different from that for a picocell.

Base-station antennas are either omnidirectional, referred to as *omni,* or directional antennas. The antenna selected for the application should be one that meets the following major points as a minimum:

- Elevation and azimuth patterns meet requirements.
- The antenna exhibits the proper gain.
- The antenna is available from common stock and company inventory.
- The antenna can be mounted properly at location; that is, it can be physically mounted at the desired location.
- The antenna will not adversely affect the tower, wind, and ice loading for the installation.
- Negative visual impact has been minimized in the design and selection phases.
- The antenna meets the required performance specifications.

This section will restrict itself to collinear, log periodic, folded dipole, yagi, and microstrip antennas with respect to a passive configuration; that is, no active electronics are located in the antenna system itself. Of the antenna classifications mentioned, two are more common for use in 1G and 2G communication systems for base stations and will be used for 2.5G and 3G as well: collinear and log periodic antennas. 3G and now 4G systems are also utilizing multiband antennas and some very interesting combining techniques to reduce the antenna real estate and the amount of coax cable runs at a given site.

14.2 Performance Criteria

The performance or performance criteria for an antenna are not restricted to its gain characteristics and physical attributes, that is, maintenance. With the introduction of 4G platforms, the performance criteria associated with new or existing antennas need to be reviewed because more emphasis is being placed on using existing or future antenna real estate at a communications site. When evaluating or defining the performance criteria for antennas, many parameters must be taken into account. The parameters that define the performance of an antenna can be referred to as the *Figures of Merit* (FOM) that apply to any antenna that is selected for use in a communications system:

- Antenna pattern
- Main lobe
- Sidelobe suppression
- Input impedance
- Radiation efficiency
- Horizontal beamwidth
- Vertical beamwidth
- Directivity
- Gain
- Antenna polarization
- Antenna bandwidth
- Front-to-back ratio
- Power dissipation
- Connector type
- Intermodulation suppression (PIM)
- Construction
- Cost

Just because an antenna is performing or appears to be performing properly in a 2G system does not mean that alteration of the existing antenna system is not necessary with the introduction of a 2.5G, 3G, or 4G platform. The alteration could involve replacement or addition of more antennas in order to meet the design and performance criteria of the new system.

There are many parameters and FOM that characterize the performance of an antenna system. The following is a partial coverage of the FOM for an antenna that should be quantified by the manufacturer of the antennas you are using. The tradeoffs that need to be made when selecting an antenna involve all the FOM issues discussed in the following paragraphs.

The antenna pattern, of course, is one of the key criteria the design engineer uses for directing the radio energy either into the desired area or for keeping it out of another area. The antenna pattern typically is portrayed by a graphic representation of the elevation and azimuth patterns.

The antenna pattern chosen should match the coverage requirements for the base station. For example, if the plan is to use a directional antenna for a particular sector of a cell site, 120 degrees, then choosing an antenna pattern that covers 360 degrees in azimuth would be incorrect. Care also must be taken in looking for electrical downtilt, which may or may not be referenced in the literature.

The sidelobes are important to consider because they can create potential problems with interference. Ideally, there would be no sidelobes for the antenna pattern. For downtilting, the sidelobes are important because they can create secondary interference.

The radiation efficiency of an antenna frequently is not referenced, but it should be considered in that it is the ratio of total power radiated by the antenna to the net power accepted from the transmitter by the antenna. The equation is as follows:

$$e = \text{power radiated}/(\text{power radiated} - \text{power lost})$$

The antenna would be 100 percent efficient if the power lost in the antenna were zero. This number indicates how much energy is lost in the antenna itself, assuming an ideal match with the feedline and the input impedance. Using the efficiency equation, if the antenna absorbed 50 percent of the available power, then it would have only 50 percent of the power for radiating, and thus the effective gain of the antenna would be reduced.

The beamwidth of the antenna, either elevation or azimuth, is important to consider. The beamwidth is the angular separation between two directions in which radiation of interest is identical. The half-power point for the beamwidth is usually the angular separation where there is a 3-dB reduction off the main lobe. Why this is important is that the wider the beamwidth, the lower is the gain of the antenna normally. A simple rule of thumb is for every doubling of the number of elements associated with an antenna, a gain of 3 dB is realized. However, this gain comes at the expense of beamwidth. The beamwidth reduction for a 3-dB increase in gain is about half the initial beamwidth, so if an antenna has a 12-degree beamwidth and has an increase in gain of 3 dB, then its beamwidth now is 6 degrees.

The gain of any antenna is a very important FOM. The gain is the ratio of the radiation intensity in a given direction to that of an isotropically radiated signal. The equation for antenna gain is as follows:

$$G \text{ maximum radiation intensity from antenna}/\text{maximum radiation from an isotropic antenna}$$

The gain of the antenna also can be described as

$$G = e \times G(D)$$

If the antenna were without loss, $e = 1$, then $G = G(D)$.

Polarization is important to note for an antenna because wireless mobility systems use vertical polarization, with some exceptions, when the use of X-pole antenna is in play.

The bandwidth is a critical performance criterion to examine because the bandwidth defines the operating range of the frequencies for the antenna. The *Standing Wave Ratio* (SWR) is usually how this is represented besides the frequency range over which it is constant. A typical bandwidth that is referenced is the 1:1.5 SWR for the band of interest. Antennas are now being manufactured that exceed this, having an SWR value of 1:1.2 at the band edges.

The antenna's bandwidth must be selected with extreme care to account not only for current but also for future configuration options with the same cell site. For example, an antenna selected for use as the receive antenna at a cell site also should operate with the same performance in the transmit band and vice versa. The rationale behind this dual-purpose use is that if the transmit antenna fails, a receive antenna can be switched internally in the cell for use as a transmit antenna.

The front-to-back ratio indicates how much energy is directed in the exact opposite direction of the main lobe of the antenna. The front-to-back ratio is a loosely defined term. IEEE Standard 145-1983 references the front-to-back ratio as the ratio of maximum directivity of an antenna to its directivity in a specified rearward direction. Front-to-back ratio is applicable only to directional antennas because with an omni antenna there is no rearward direction.

Many manufacturers reference high front-to-back ratios, but care must be taken in knowing just how the number was computed. In addition, if installation is, say, on a building, and the antenna will be mounted on a wall, then the front-to-back ratio is not an important FOM. However, if the antenna is mounted so that there are no obstructions between it and the reusing cell, then the front-to-back ratio can be important. Specifically, in the latter case, the front-to-back ratio should be at least the *Carrier-to-Interference* (C/I) ratio level required for operation in the system.

The power dissipation needs to be looked at when integrating a new platform. Power dissipation is a measure of the total power the antenna can accept at its input terminals. This is important to note because receive antennas may not need to handle much power, but the transmit antenna might have to handle 1500 W of peak power. The antenna chosen should be able to handle the maximum envisioned power load without damage.

The amount of intermodulation that the antenna will introduce to the network in the presence of strong signals, as referenced by the manufacturer, needs to be considered in antenna selection. The intermodulation that is referenced should be checked against how the test was run. For instance, some manufacturers reference the Intermodulation Distortion (IMD) to two tones, although some reference it to three or multiple tones. The point here is that the overall signal level at which the IMD is generated needs to be known, in addition to how many tones were used, their frequency of operation, bandwidth, and of course, at which power levels they caused the IMD level.

The connector type used in an antenna system affects the performance of the radio site, as does the feedline and, of course, the antenna jumper. A good connector does not contribute to IMD in the temperature and environment in which the antenna will operate. Two common types of connectors are used: the N-type and the 7/16 DIN. The N-type is easier to install, but the center pin can move under temperature variations, and there is less cross-sectional area for contact as compared with the 7/16 DIN. The 7/16 DIN connector has a larger cross-sectional area for contact and therefore has a higher power-handling capability and lower IMD contributor, making it a preferred connector.

PIM performance is IMD and is typically defined as the unwanted mixing of two or more signals in a nonlinear junction within a wireless or RF system creating harmonic signals that are not desirable. These undesired signals from PIM are of most concern if they fall in your own receive band and of course if they fall into another wireless operator's frequency band.

PIM, however, can and is generated by many components in an RF antenna system and the concern is the power levels that are created. In short, just because you have PIM does not mean it is causing a problem since its signal level may be so low that while the product is present it has no impact, or rather any measurable impact on the communication system.

PIM, in short, is not an IMD product as a result of other transmitters at the site, though it could be. However PIM is primarily a self-induced issue which is exasperated when using duplexers/triplexers. Additionally, the more components you place in the path from the transmitter to the antenna increases the potential for PIM.

The antenna construction attributes associated with its physical dimensions, mounting requirements, materials used, wind loading, connectors, and color constitute this FOM.

For instance, one of the items that needs to be factored into the construction FOM is the use of materials, whether the elements are soldered together or bolted. In addition, the types of metals used in the antenna and the associated hardware need to be evaluated with respect to the environment in which the antenna will be deployed. For instance, if you install an antenna near the ocean or an aircon unit that uses saltwater for cooling, then it will be imperative that the material chosen not corrode in the presence of saltwater.

How much the antenna costs is a critical FOM. No matter how well an antenna performs in a system, the cost associated with that antenna will need to be factored into the decision. For example, if the antenna chosen meets or exceeds the design requirements for the system but costs twice as much as another antenna that meets the requirements, the choice would seem obvious: Pick the antenna that meets the requirements at the lower cost.

Another example of cost implications involves selecting a new antenna type to be deployed in a network. The spares and stocking issues need to be factored into the antenna selection process. If the RF department designs every site's antenna requirements too uniquely, then it is possible to have a plethora of antenna types deployed in the network, leading to a multitude of additional stocking issues for replacements. Therefore, it is important to select a specific number of antennas that meet most, if not all, of the design requirements for a system and use only those antennas.

14.3 Diversity

Diversity, as it applies to an antenna system, refers to a system used in wireless communication as a method for comparing signal fading in the environment. Diversity gain is the gain in the signal that is experienced as compared to what it would be if fading would have taken place in the event that a diversity technique was not used. In the case of a two-branch diversity system, if the received signal into both antennas is not of an equal strength, then there cannot be any diversity gain. This is an interesting point considering that most link-budget calculations incorporate diversity gain as a positive attribute. The only way diversity gain can be incorporated into a link budget is if a fade margin is included in the link budget, and the diversity scheme that is chosen attempts to improve or reduce the fade margin.

Several types of diversity need to be accounted for in both legacy systems and 2.5G and 3G platforms. When discussing diversity, a radio engineer usually focuses on the receive path, the uplink from the mobile unit to the base station. With the introduction of 2.5G and 3G platforms, transmit diversity has been introduced, but is implemented in such a fashion that the subscriber does not need a second antenna.

Multiple-Input, Multiple-Output (MIMO) has been introduced with 3G and is being used extensively in 4G. However MIMO has been used in wireless communications for quite some time now. MIMO is a key attribute to WiFi performance with 802.11n. Additionally, MIMO has been used in CDMA with diversity transmit in diversity receive.

The type of antenna diversity used can be augmented with another type of diversity that is accomplished at the radio level:

- Spatial
- Horizontal
- Vertical
- Polarization

- Frequency
- Time
- Angle

Most 2G, 2.5G, 3G , and 4G systems use two antennas separated by a physical distance that is horizontal. In many cases the antennas are duplexed, facilitating transmit and receive functions from the same antenna. Some 2.5G platforms such as *Integrated Digital Enhanced Network* or *Integrated Dispatch Enhanced Network* (iDEN) use a three-branch diversity receive scheme, but they are the exception, and the usual method is to deploy only two antennas per sector for diversity reception.

The spacing associated with antennas located in the same sector normally is a design requirement stipulated by RF engineering. Diversity spacing is a physical separation between the receive antennas that is needed to ensure that the proper fade-margin protection is designed into the system. As mentioned earlier, horizontal space diversity is the most common type of diversity scheme used in wireless communications systems.

The following is a brief rule of thumb to determine the horizontal diversity requirements for a site (Figure 14.1):

$$n = h/d = 11$$

where h is height (ft) and d is the distance between antennas (ft). This equation was derived for cellular systems operating in the 800-MHz band but has been applied successfully to other wireless bands within the 1800- and 1900-MHz range and also will work for *Advanced Wireless Services* (AWS) deployments.

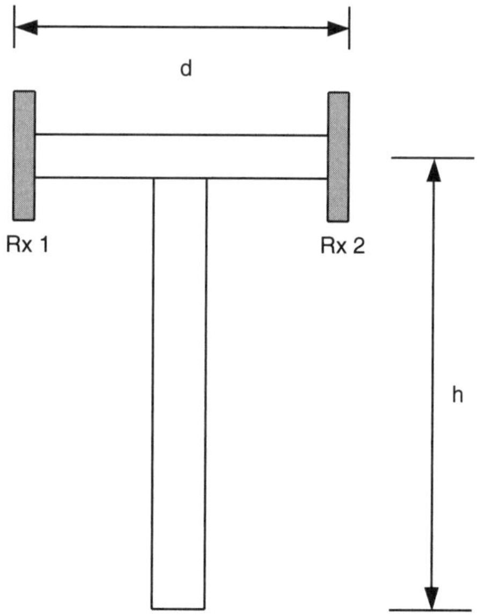

FIGURE 14.1 Two-branch diversity spacing.

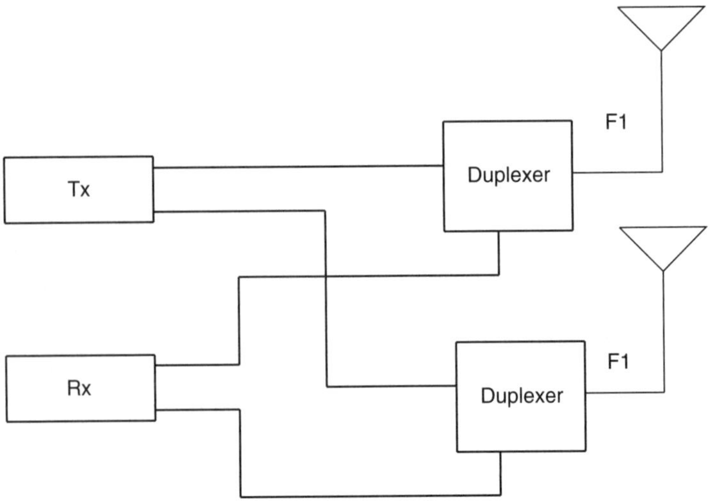

FIGURE 14.2 Sector STD transmit diversity scheme.

With the introduction of *Long-Term Evolution* (LTE), the use of MIMO technology is being used extensively where two antennas are installed for a 2 × 2 MIMO configuration, or possibly four antennas if space, OEM support, and physical installation issues permits, allowing for a 4 × 4 configuration. The issue of having multiple technologies sharing the same antenna system is also more prevalent as wireless carriers deploy more technologies often trifurcating their network in the process. CDMA2000, UMTS, TD-CDMA, and TD-SCDMA, introduced the use for transmit diversity needs to be factored into the antenna design. Two different transmit diversity schemes are possible with these platforms. Such use of two different transmit diversity schemes is driven by the practical issue of antenna installation. The two transmit diversity schemes are *Space Transmit Diversity* (STD) and *Orthogonal Transmit Diversity* (OTD). The preferred transmit diversity scheme, when implemented, is STD.

What follows are simplified diagrams of STD and OTD transmit diversity schemes. Figure 14.2 shows the STD transmit diversity scheme for a single channel when two antennas are available on a sector. The two antennas also could be separate ports on an X-pole antenna. The important issue is that when integrating a second carrier, either more antennas need to be added or additional *transmit* (Tx) combing losses will ensue.

Figure 14.3 shows the recommended configuration for the rest of the network that involves using OTD transmit diversity. In examining the differences between Figures 14.2 and 14.3, one immediate observation is that a second carrier is introduced with the same number of physical antennas.

Another immediate observation with the use of transmit diversity for the new radio platforms is the issue of what happened to the legacy systems. This will be covered shortly.

14.4 Cross-Pole Antennas

The use of cross-pole or cross-polarized antennas has proved to be very beneficial in wireless mobility applications when installation restrictions prevail. Two types of cross-pole antennas have been used, one of which can be used for diversity

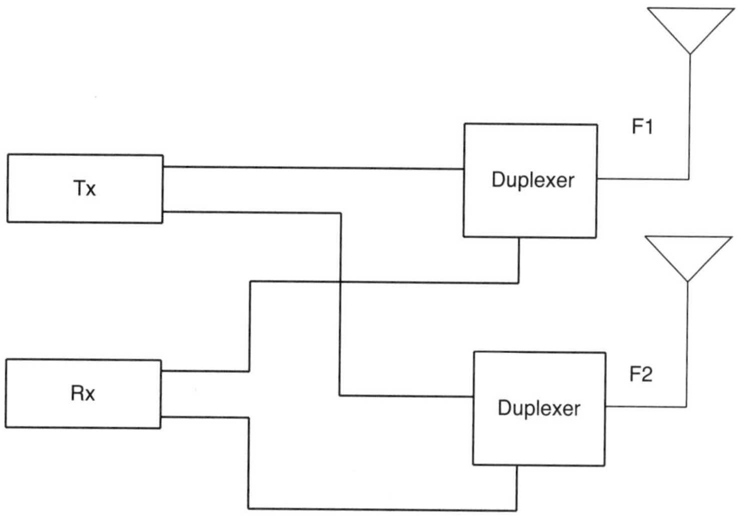

FIGURE 14.3 Sector OTD transmit diversity scheme.

reception, and the other of which is meant to facilitate either two transmit antennas in one housing or two different technology platforms that cannot share the same antenna for performance reasons.

The use of cross-pole antennas, while having advantages, has technical disadvantages depending on the application. Typically, a cross-pole antenna consists of two unique antennas that occupy the same general antenna housing, that is, radome. There are two main variants to the cross-pole antenna. One variant has a vertical antenna array and a separate horizontal antenna array. The other main variant uses two antennas where each has a 45-degree polarization, one at +45 degrees and the other at –45 degrees referenced to vertical. Figure 14.4 provides a visual representation of the two cross-pole variants. The cross-pole antennas in the figure consist of two separate antenna arrays that are oriented so that both arrays have an angular separation of 90 degrees.

Depending on which trade journal, antenna manufacturer, or technical paper you read regarding the advantages of using cross-pole antennas, a few things always need to be considered. The first item to consider is that a cross-pole antenna has a lower correlation between branches than ordinary spatial diversity. On the surface, this would appear to be the ticket; however, for diversity, you want two antennas to be correlated not 100 percent but 70 percent. Therefore, the lack of correlation does not provide as good a protection from fades as spatial diversity.

The other item to consider is that the use of cross-pole antennas does not reduce the number of cables and feeders that need to be installed for a communications site. Referring to Figure 14.4, each antenna array has its own feed, meaning that the installation advantage is appearance and some wind loading reduction. The wind loading reduction is not 50 percent owing to the physical housing requirements associated with two antenna arrays.

If you use a cross-pole antenna the 45-degree variant is preferred for several reasons. The first reason addresses the correlation issue and the fact that elements oriented

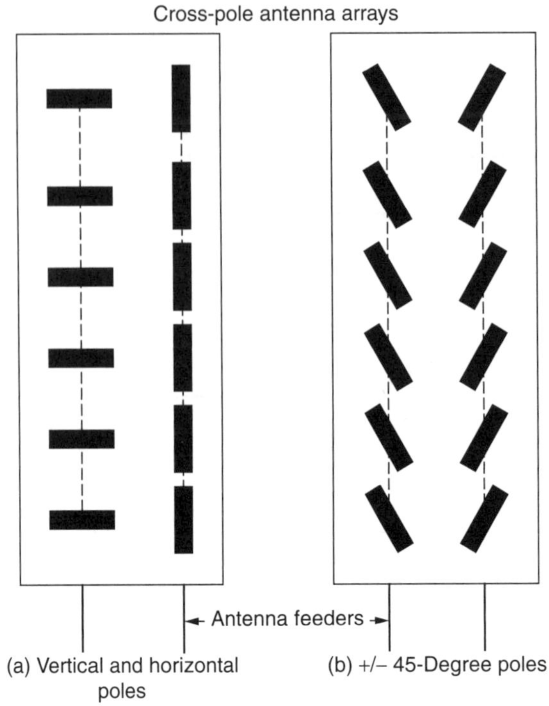

FIGURE 14.4 Cross-pole antennas.

45 degrees from vertical will provide the best correlation coefficient possible, assuming that diversity is a requirement with only one antenna.

The second reason addresses possibly having more than one technology type colocated at the same wireless facility, that is, UMTS/*High-Speed Packet Access* (HSPA) and *Global System for Mobile Communications* (GSM), to mention one obvious one. Use of the 45-degree cross-pole antenna would afford the possibility of having an UMTS/HSPA system on pole 1 and a GSM system on pole 2. If there were a second cross-pole antenna, then spatial diversity also would be possible using each technology on a separate antenna array.

When looking at the antenna array in Figure 14.4, the rationale for using the 45-degree antenna should be apparent; that is, which system goes on the horizontal leg if the 45-degree antenna is not used? Obviously, if more than one technology is used by the operator, the possibility exists that they will be in different frequency bands. When this occurs, the typical mode is to use one cross-pole antenna for technology *A* and another separate cross-pole antenna for technology *B*, again using a 45-degree antenna for each cross-pole antenna.

In addition, some dual-band antennas are being used, and in this situation the use of a 45-degree cross-pole antenna is a viable option.

Keep in mind that since the polarization for wireless mobility is vertical, spatial diversity using vertically polarized antennas is still the preferred method for aperture gain as well as cross-correlation of the diversity signals, thereby providing the best fade margin.

14.5 Dual-Band Antenna

When multiple frequency bands are used by the wireless operator and antenna real estate is restricted significantly, the use of dual-band antennas often can be pursued. As the name implies, dual-band antennas allow for the same antenna, that is, radome, to support two different frequency bands.

Figure 14.5 is an illustration of a dual-band antenna shared by 850- and 1900-MHz systems. This figure shows that the radome in fact contains two antennas and not just one. As with the cross-pole antenna discussed previously, the number of feedlines is not reduced, and at this time, the wind loading actually is increased owing to the increased size of the radome itself.

It is possible with a cross-band adaptor to reduce the number of feedlines, but as with all multipurpose devices, compromises must be made. Specifically, the design criteria for one system, say, GSM at 850 MHz and UMTS at 1900 MHz are not the same. Therefore, when using a dual-band antenna system, performance may be sacrificed to overcome installation circumstances.

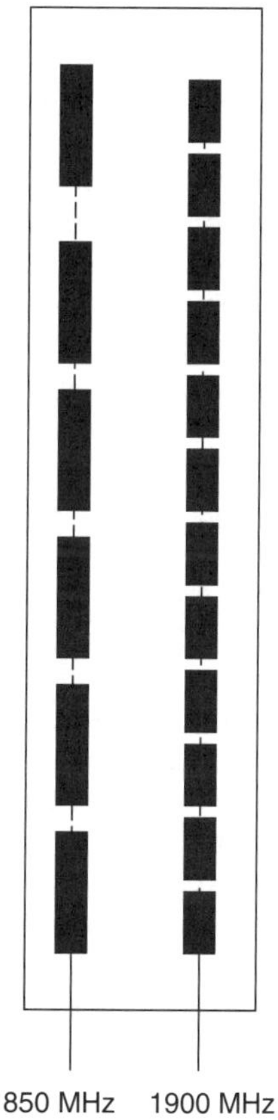

FIGURE 14.5 Dual-band antennas.

Additionally, dual-band antennas can be tailored for each band by employing a separate downtilt to each of the bands. For instance, the 850 band could have a 3-degree downtilt while the 1900 band might have 0 degrees, or no downtilt.

The introduction of LTE can also lead to the desire to have a triband antenna. However, with any design involving multiple-frequency bands the potential for performance degradations and necessary trade-offs are always present. The *kiss principle* is always the preferred method since technology requirements constantly change regardless of your future planning vision and knowledge. And, having an antenna system design which is overly complex will directly impact your ability to introduce technology changes, improvements, or modifications needed in the future, besides the obvious impact on performance-related activities and maintenance.

14.6 Intelligent Antennas

Intelligent antenna systems are being introduced to commercial wireless communications systems. The concepts and implementation of intelligent antenna systems have been used in other industries for some time, primarily the military.

Intelligent antenna systems can be configured for either receive-only or full-duplex operations. The configuration of an intelligent antenna can be arranged as either an omni or sector cell site depending on the application at hand.

With LTE, WiMAX, CDMA2000, WCDMA, TD-CDMA, and TD-SCDMA, the use of intelligent antenna systems is supported directly, unlike 1G and 2G systems, with the use of auxiliary and dedicated pilot channels.

Intelligent antennas were promoted initially as providing an increase to the *Signal-to-Noise* (S/N) *ratio* of a sector by reducing the amount of noise and interference and possibly increasing the serving signal in the same process. All the technologies referenced are based on the principle that narrower radiation beam patterns will provide increased gain and can be directed toward the subscriber and at the same time offer less gain to interfering signals that will arrive at an off-axis angle owing to the reduced beam width.

Intelligent antennas are now promoted as being able not only to improve the S/N ratio of a sector or system but also to more uniformly balance traffic between sectors and cells and improve the system performance by reducing soft handoffs, as one example. Figure 14.6 illustrates three types of intelligent antenna systems, each with positive and negative attributes.

All these illustrations can be either receive-only or full-duplex. The difference between the receive-only and the full-duplex systems involves the number of antennas and the potential number of transmitting elements in the cell site itself. The beam-switching antenna arrangement is the simplest to implement. It normally involves four standard antennas of narrow azimuth beam width, 30 degrees for a 120-degree sector, and based on the receive signal received, the appropriate antenna will be selected by the base-station controller for use in the receive path.

The multiple-beam array involves using an antenna matrix to accomplish the beam switching. The beam-steering array, however, uses phase shifting to direct the beam toward the subscriber unit. However, the direction chosen by the system for directing the beam will affect the entire sector. Normally, amplifiers for transmit and receive are located in conjunction with the antenna itself. In addition, the phase shifters are located

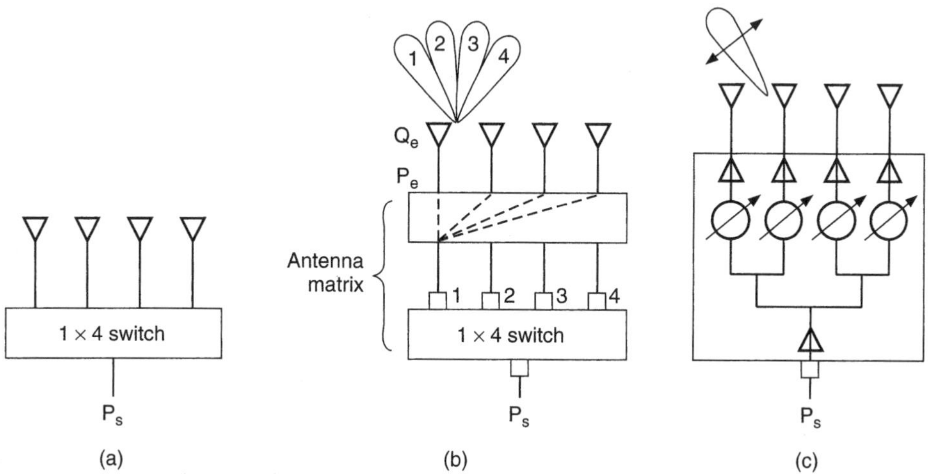

FIGURE 14.6 Intelligent antenna systems: (*a*) switched antennas; (*b*) multiple-beam array; (*c*) steered-beam array.

directly behind each antenna element. The objective of placing the electronics in the mast head is to maximize the receive sensitivity and exploit the maximum transmit power for the site.

14.7 MIMO

Multiple-Input, Multiple Output (MIMO) antenna systems were a new entrant in wireless and 3G networks. However with 4G the use of MIMO is essential for the performance required for these advanced wireless mobile systems. MIMO is also called *spatial multiplexing*. MIMO uses receive (Rx) and transmit (Tx) diversity. For a system to be defined as MIMO it has to have several transmit paths and not just multiple receive paths.

MIMO, in short, uses multiple antennas in both the uplink and the downlink paths. What is unique is that MIMO is used not only at the cell site but also at the subscriber device. However multiple antennas have been used for years in radio systems to combat fading through the use of diversity antennas both for receive and transmit RF paths.

With some 3G and primarily 4G, the use of MIMO is meant to improve the overall signal quality for end devices through improved reception and lower-power requirements, thus extending battery life. The use of MIMO techniques is an advance to decorrelate the radio paths in either the downlink or the uplink direction. The decorrelation allows for reduced power for mobile transmission owing to improved uplink performance coupled with the cell site's diversity (MIMO) antenna system. It is important to note that MIMO can have 2, 3, 4, or *n* number of inputs and outputs.

MIMO systems increase the *Signal-to-Noise* (S/N) *ratio* by transmitting data on multiple antennas using the same RF channel which is technically called *spatial multiplexing*. With MIMO the data bit stream is multiplexed to multiple transmitters without changing the symbol rate of each independent transmitter, increasing the data throughput without increasing the channel bandwidth. MIMO takes advantage of the fact that the RF signals emanating and received by different antennas experience different fading conditions resulting in overall S/N ratio improvements enabling higher throughputs.

Beam forming, however, is not MIMO. While both are similar in diversity transmission, beam forming is typically a physical antenna beam directed toward specific user equipment (UE).

MIMO can and does have many different possible configurations. The following are a few of the typical MIMO configurations (Tx × Rx):

- 2 × 2
- 2 × 4
- 4 × 4
- 8 × 4

Please note that a 1 × 2 configuration, which is typical for 3G/2G and lower, is not a MIMO configuration.

Beyond four antennas the practicality of using more antennas becomes nonpractical; however new antenna techniques are making the possibility of an 8 × 4 MIMO more prevalent for intense bandwidth requirements. The use of 8 transmit antennas enables the ability for LTE to achieve 1 Gbps in the downlink path.

Within MIMO, you have Single-User (SU) and Multiple-User (MU) MIMO. SU MIMO is where the specific data stream is transmitted to only one subscriber or UE at a given time. MU MIMO is where the data stream is transmitted to multiple subscribers or UEs at the same time.

There are several different configurations for MIMO. Figure 14.7 is a brief depiction of a 4 × 4 MIMO configuration for a base station and a 2 × 4 configuration for a subscriber unit.

14.8 dBi and dBd

All too often the calculated value that you have is not in the right scale required for the form or questioner. Therefore, it is necessary to convert from either dBi to dBd or from dBd to dBi. To convert a value in dBi to the equivalent dBd value, the following equation is used:

$$\text{dBd} = \text{dBi} - 2.14$$

Therefore, Table 14.1 can be used to reinforce the conversion process because it shows the calculated values along with the nearest approximate value found. To convert a value in dBd to the equivalent dBi value, the following equation is used:

$$\text{dBi} = 2.14 + \text{dBd}$$

Therefore, Table 14.2 can be used to reinforce the conversion process because it shows the calculated values along with the nearest approximate value found.

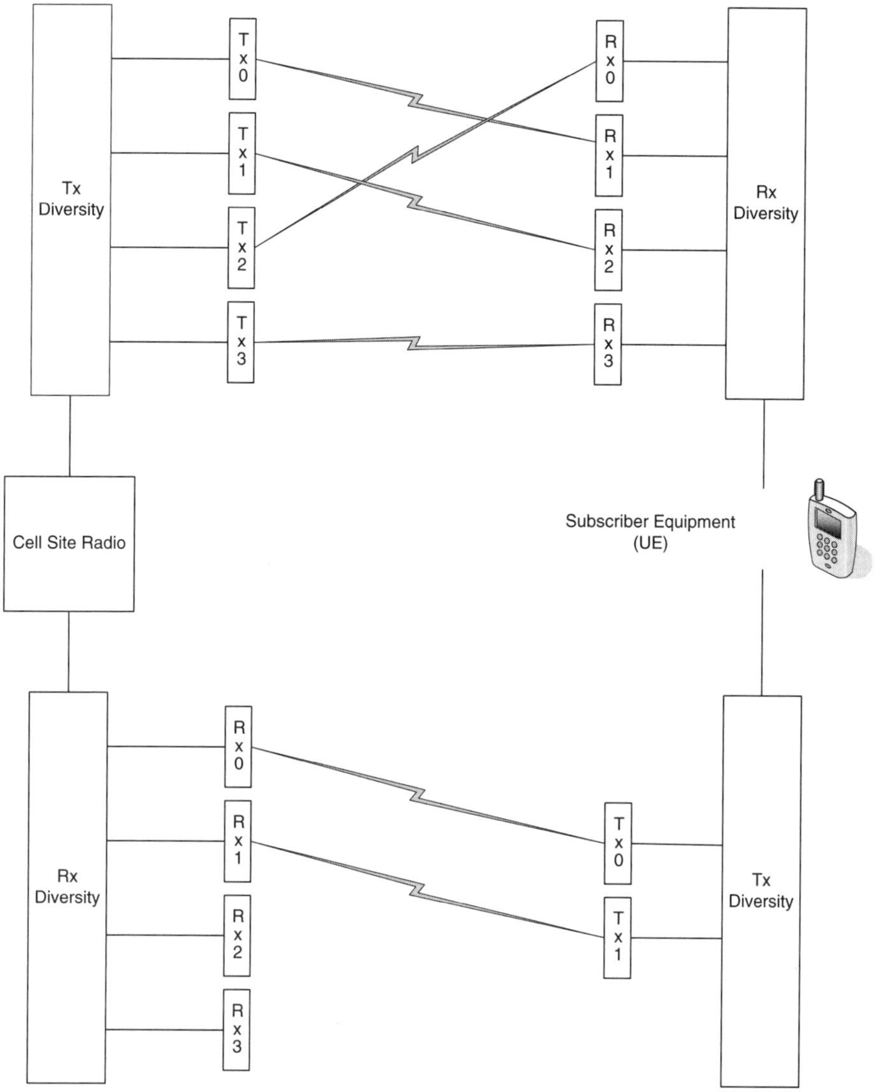

FIGURE 14.7 MIMO.

dBi	dBd
5	2.86 (3 dBd)
10	7.86 (8 dBd)
12	9.86 (10 dBd)
14	11.86 (12 dBd)
18	15.86 (16 dBd)
21	18.86 (19 dBd)

TABLE 14.1 dBi to dBd

dBd	dBi
3	5.14
10	12.14
12	14.14
14	16.14
18	20.14
21	23.14

TABLE 14.2 dBd to dBi

References

American Radio Relay League, *The ARRL Antenna Handbook,* 14th ed., American Radio Relay League, Newington, CT, 1984.

Carr, J. J., *Practical Antenna Handbook,* Tab Books, McGraw-Hill, Blue Ridge Summit, PA, 1989.

Fink, Beaty, *Standard Handbook for Electrical Engineers,* 13th ed., McGraw-Hill, New York, 1995.

Fink, Donald, and Donald Christiansen, *Electronics Engineers Handbook,* 3d ed., McGraw-Hill, New York, 1989.

Jakes, W. C., *Microwave Mobile Communications,* IEEE Press, New York, 1974.

Johnson, R. C., and H. Jasik, *Antenna Engineering Handbook,* 2d ed., McGraw-Hill, New York, 1984.

Kaufman, M., and A. H. Seidman, *Handbook of Electronics Calculations,* 2d ed., McGraw-Hill, New York, 1988.

Miller, Nathan, *Desktop Encyclopedia of Telecommunications,* McGraw-Hill, New York, 1998.

Mouly, Pautet, *The GSM System for Mobile Communications,* 1992.

Sams, *Reference Data for Radio Engineers,* 6th ed., 1983. Howard W. SAMS & Co, Inc., New York, 1983.

Schwartz, Bennett, and Stein, *Communication Systems and Technologies,* IEEE Press, New York, 1996.

Smith, Clint, *Practical Cellular and PCS Design,* McGraw-Hill, New York, 1997.

Smith, Clint, *Wireless Telecom FAQ,* McGraw-Hill, New York, 2000.

Steele, Raymond, *Mobile Radio Communications,* IEEE Press, New York, 1992.

Stimson, G.W., "Introduction to Airborne Radar," Hughes Aircraft Company, Inc., El Segundo, CA, 1983. 3GPP TR 37.976: Technical Specification Group Radio Access Network; Measurement of Radiated Performance for Multiple-Input, Multiple-Output (MIMO) and Multi-Antenna Reception for High-Speed Packet Access (HSPA) and LTE Terminals (Release 11), March 2012.

3GPP2 C.S0008-0: Multi-Carrier Specification for Spread Spectrum Systems on GSM MAP (MC-MAP) (Lower Layers Air Interface), June 9, 2000.

TIA/EIA IS-2000-1: Introduction to cdma2000 Standards for Spread Spectrum Systems, June 9, 2000.

TIA/EIA IS-2000-2: Physical Layer Standard for cdma2000 Spread Spectrum Systems, September 12, 2000.

TIA/EIA IS-2000-3: Medium Access Control (MAC) Standard for cdma2000 Spread Spectrum Systems, September 12, 2000.

TIA/EIA IS-2000-4: Signaling Link Access Control (LAC) Specification for cdma2000 Spread Spectrum Systems, August 12, 2000.

TIA/EIA-98-C: Recommended Minimum Performance Standards for Dual-Mode Spread Spectrum Mobile Stations (Revision of TIA/EIA-98-B), November 1999.

Webb, Hanzo, *Modern Amplitude Modulations,* IEEE Press, New York, 1994.

White, Duff, "Electromagnetic Interference and Compatibility," Interference Control Technologies, Inc., Gainesville, GA, 1972.

CHAPTER 15

UMTS System Design

In previous chapters we described the *Universal Mobile Telecommunications Service* (UMTS) from a pure-technology perspective. In this chapter we aim to address the design criteria and methodologies that apply to deploying UMTS technology in a real network. Numerous interrelated considerations must be addressed in the design and deployment of such a network. While some of these considerations are common to any wireless network design, a number are specific to the technology in question. Regardless, because of the multiple issues involved, it is very important that a well-understood methodology is in place so that the network design can proceed from the initial establishment of requirements to the final deployment of the network.

15.1 Network Design Principles

Figure 15.1 shows the overall network design and deployment process at a very high level. To begin with, we must specify a number of criteria regarding the set of services that we wish to provide and the estimated demand for those services. We must establish exactly where we wish to offer those services and any limiting factors that might constrain our ability to meet all objectives—such as spectrum limitations.

Based on the established input requirements, a number of network design activities take place. These can be broken into two main areas—*Radiofrequency* (RF) network design and core network design. Of course, within each of these areas is a myriad of individual design efforts.

Once designs are established, implementation is undertaken. This involves RF network implementation, core network implementation, integration, and optimization. Quite often during the implementation phase, one finds that it is not possible or optimal to deploy the system exactly as designed, in which case the design itself needs to be modified. There are many reasons why designs might need to be changed—such as an inability to acquire a Node B site in the exact location desired, coverage or quality problems discovered during integration or optimization, and so on.

Finally, statistics and measurements generated during performance of the network should be fed back to those who generated the design inputs as well as to those who were responsible for system design. This enables design modifications and expanded coverage, capacity, or service demand to be based on real experience.

15.2 RF Coverage Analysis

Looking again at Figure 15.1, we now delve into a little more detail in each of the areas of concern. We start with the input requirements, specifically the RF coverage

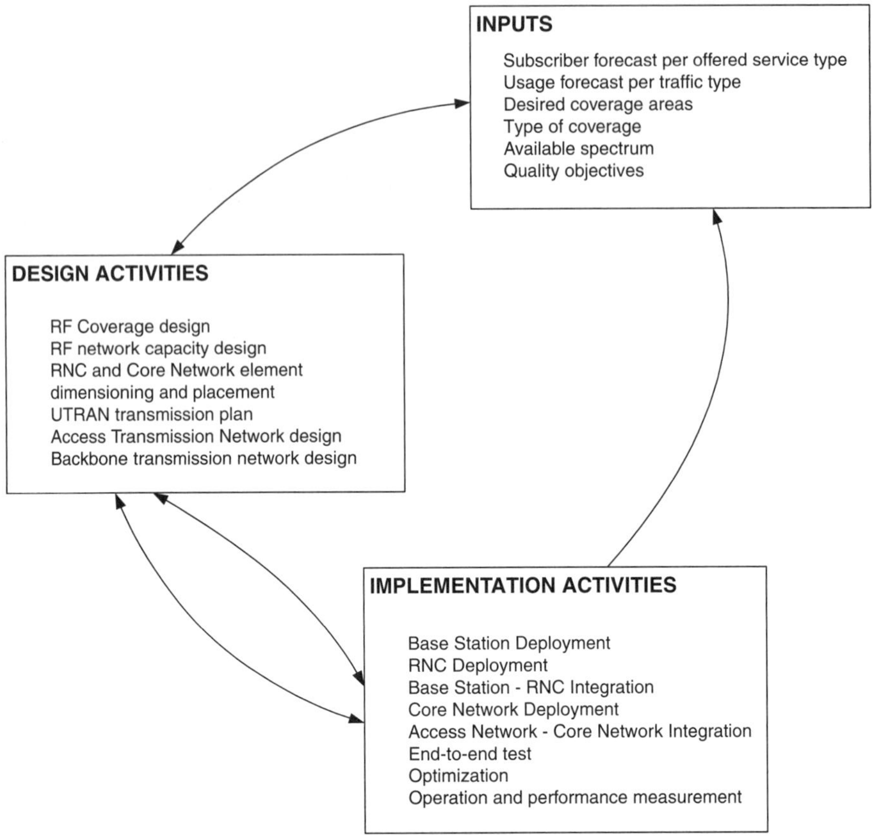

FIGURE 15.1 Design and deployment process (high level).

requirements, and consider the issues related to designing an RF network to meet those coverage requirements.

As described in Chapter 11, a good deal of detail should be specified regarding where coverage is to be provided and the type of coverage to be provided in those areas. It is not sufficient simply to state that we wish to provide coverage in a given market. Often it is necessary and desirable to provide coverage only in certain areas—such as commercial areas, areas with significant population density, and major highways. Therefore, we must obtain a good understanding of the market to be covered, which will require a great deal of map-based information specifying population densities; what areas are urban, suburban, and rural; what areas are primarily commercial, residential, industrial, and parkland; and so on. Figure 15.2 provides a simple example of a population-density map. An understanding of these factors is important for two reasons: First, we wish to make sure that we provide sufficient capacity in those areas where we expect the greatest traffic. Second, the type of environment will have a direct impact on propagation modeling, where we consider such issues as modeling correction factors and in-building penetration losses.

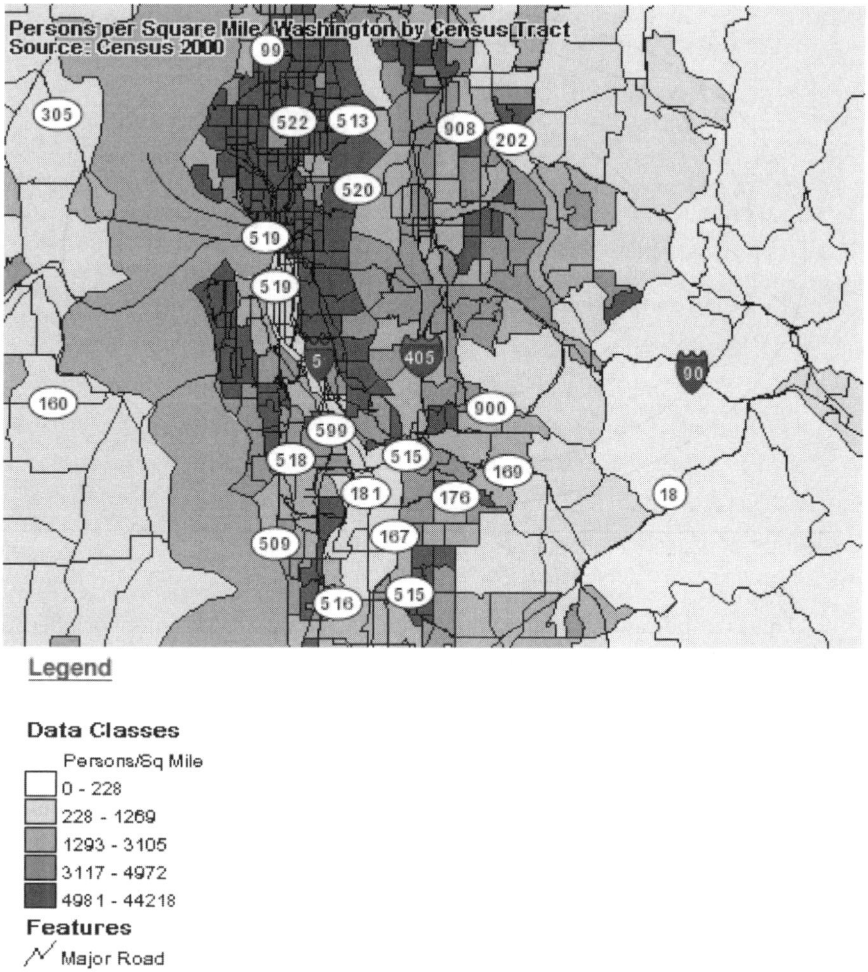

FIGURE 15.2 Example of a population-density map.

Depending on the type of environment, we may wish to provide different levels of coverage. For example, in urban and suburban areas, we may wish to provide in-building coverage. On highways, however, we will be interested only in in-vehicle coverage. In other areas, such as parkland, we likely will want to provide only outdoor coverage. In systems such as the *Global System for Mobile* (GSM) communications, a good understanding of these criteria may well be sufficient to start the design process. With UMTS, however, a further consideration is required—what types of services should be available in a given area. For example, in a given area, should a subscriber have access to data rates of up to, say, 480 kbps, or is a lower rate sufficient, or is speech-only service acceptable? These issues are important because, as shown in Figure 15.3, the effective footprint of a cell is influenced by the data rate to be supported. The higher the throughput, the smaller is the effective cell radius.

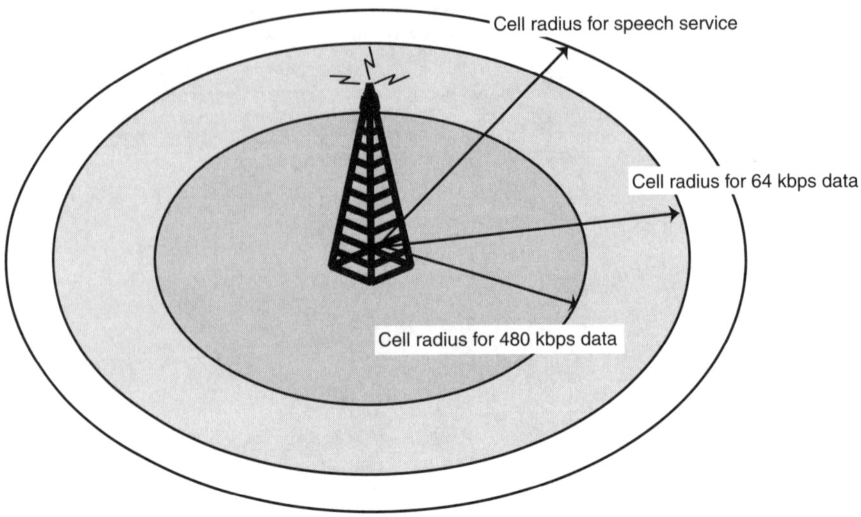

Figure 15.3 Relative cell footprints for different user data rates.

This is important to understand when designing a UMTS network because the RF environment has a direct influence on the user's data throughput. As modulation schemes used for *High-Speed Downlink Packet Access* (HSDPA) and *High-Speed Uplink Packet Access* (HSUPA) are employed, the effective radius for the site is related directly to its throughput for both the uplink and downlink directions.

Once we have a solid understanding of the coverage requirements, then we can use that information for the preparation of an initial RF coverage plan. Before generating that plan, however, another critical input is required—link budgets.

15.2.1 Link Budgets

A *link budget* is a calculation of the amount of power received at a given receiver based on the output power from a given transmitter. The link budget accounts for all the gains and losses that a radio wave experiences along the path from transmitter to receiver. For a given transmitter power, we determine the maximum path loss that the signal can experience and still be recoverable at the receiver. Given that the base station must be able to "hear" the mobile unit and the mobile unit must be able to "hear" the base station, we need to perform the calculation in both directions—from mobile to base station and from base station to mobile. We determine the maximum allowable path loss in each direction, and the lesser of the two corresponds to the *coverage limit* for the cell and service in question. For example, if the maximum allowable path loss in the uplink is, say, 130 dB and the maximum allowable path loss in the downlink is, say, 135 dB, then we should not exceed 130 dB, and we are said to be *uplink-limited.*

The link budget needs to include a margin (i.e., a buffer) to enable fading of the signal. In other words, we design the system such that service still will be supported even in the case of a significant signal fade. The greater the fade margin, the greater is the reliability of the service. Moreover, because a *Wideband Code-Division Multiple Access* (WCDMA) system is interference-limited, we also need to include an interference margin. As described later in this chapter, the size of that margin depends on the load.

As mentioned earlier, the effective cell coverage depends on the service to be provided. One reason for this is the fact that the higher the spreading factor (corresponding to a lower data rate), the higher is the processing gain, and the lower the spreading factor, the lower is the processing gain. Because the processing gain is one of the gains that needs to be included in a link budget, it follows that the lower the processing gain, the lower is the maximum allowable path loss and the smaller is the effective radius of the cell.

From a pure radio propagation point of view, we generally will find that coverage is uplink-limited, if for no other reason than that the output power of the base station is far greater than that of the mobile unit. As we shall see, however, cell loading also affects coverage, so the cell load also must be considered in coverage analysis.

Tables 15.1, 15.2, and 15.3 provide example uplink link budgets for three WCDMA services—speech service at 12.2 kbps outdoors, data service at 128 kbps indoors, and data service at 384 kbps indoors. In practice, the speeds desired are fundamental to the relationship between throughput and coverage via the link budget.

In Table 15.1, we have a link budget that would apply to outdoor (nonvehicular) speech service at 12.2 kbps. The user device has a nominal power output of 0.125 W (21 dBm). Thus it is likely to be a power class 4 device (maximum power of 21 dBm, ± 2 dB) or a power class 3 device (maximum power of 24 dBm, +1/–3 dB). We assume that there is no antenna gain for the device, and we assume 3 dB of body loss because the device is likely to be close to the user, and the signal will have to pass through the user.

At the receiving side, we assume a receiver noise figure of 5 dB and an interference margin of 4 dB. The interference margin accounts for the fact that there will be interference at the base station caused by multiple users. The greater the number of users, the greater is the interference, and the greater is the required interference margin. Also, at the receiving side, we specify the processing gain and the E_b/N_o required for the service. As we will describe shortly, the required E_b/N_o can vary according to the service in question.

If we include a typical antenna gain value for the base-station antenna and typical losses for cables and connectors, then simple addition gives the maximum path loss. In reality, however, we need to add some additional factors to account for real-world situations.

First, we need to add a fast-fading margin. This is a buffer to enable the mobile unit to adjust power according to closed-loop power control. If the link budget in Table 15.1 were prepared for a mobile unit moving at a fast speed (e.g., 60 mph), then closed-loop power control would be unlikely to be fast enough to change the transmitted power in response to the rapid changes in pass loss as the mobile unit moves. Thus, for a high-speed vehicular service, one would set the fast fading margin to zero.

We also need to add a log-normal fading margin with a value that is determined by the desired cell area (or cell edge) coverage reliability. The higher the desired coverage reliability, the higher is the log-normal fading margin. Finally, we can add a gain that results from soft handover. Basically, if a subscriber is being covered by more than one cell and is in a soft-handover situation, then the signal from the handset is being received by two base stations (or perhaps by two cells at the same base-station site). This is equivalent to an extra level of receiver diversity and offers a similar gain.

In Table 15.2, we have a link budget that would apply to an indoor data service at 128 kbps. In this case, the service is assumed to be provided by a base station located outside the building in question. The user device has a nominal power output of 0.25 W (24 dBm). Thus it is likely to be a power class 3 device (maximum power of 24 dBm, +1/–3 dB) or a power class 2 device (maximum power of 27 dBm, ±1/–3 dB). We assume that there

Transmitter (mobile)		
Mobile TX power (dBm)	21	
Antenna gain (dBi)	0	
Body loss (dB)	3.0	
EIRP (dBm)	18	Equivalent isotropic radiated power
Receiver (base station)		
Thermal noise density (dBm/Hz)	174.0	Thermal noise floor[1]
Receiver noise figure (dB)	5.0	Equipment/vendor-dependent
Receiver noise power (dBm), calculated for 3.84 Mcps	103.2	Thermal noise density, receiver noise figure 10 log (3.84 × 10^6)
Interference margin (dB)	4	Cell load-dependent
Total noise interference (dBm)	99.2	
Processing gain (dB)	25.0	10 log (3,840,000/12,200)
Required E_b/N_o (dB)	4	Service-dependent
Effective receiver sensitivity (dBm)	120.2	Total noise interference minus processing gain E_b/N_o
Base-station antenna gain (dBi)	18	
Base-station feeder and connector losses (dB)	2	
Fast fading margin (dB)	4	Enables room for closed-loop power control
Log-normal fade margin (dB)	7.5	Enables for greater cell-area reliability (note 2)
Building penetration loss (dB)	0	
Soft handover gain (dB)	2	
Maximum allowable path loss (dB)	144.7	

[1]Thermal noise density = kT, where k = Boltzmann's constant and T = temperature in Kelvin. T is usually assumed to be 293 K.

[2]The log-normal fade margin is a design value that depends on the required level of signal reliability over the cell area; 7.5 dB corresponds to 93.4 percent coverage probability.

Table 15.1 Example Link Budget for 12.2-kbps Speech, Outdoor Pedestrian Service

is no antenna gain for the device. We further assume that unlike the case for a speech service, the device is less likely to be very close to the user (i.e., not against the user's head). Therefore, we do not allow for any body loss.

At the receiving side, many of the parameters are the same as for the example in Table 15.1. The required E_b/N_o in this case, however, is 2 dB, and the processing gain is lower (owing to the higher data rate). The other margins, gains, and losses are the same as for Table 15.1, with the exception of the building-penetration loss, which we assume to be 15 dB. This figure is highly dependent on the area to be covered. In a dense urban environment, for example, the building-penetration loss could be significantly higher.

Transmitter (mobile)		
Mobile TX power (dBm)	24	
Antenna gain (dBi)	0	
Body loss (dB)	0	
EIRP (dBm)	24	Equivalent isotropic radiated power
Receiver (base station)		
Thermal noise density (dBm/Hz)	174.0	Thermal noise floor[1]
Receiver noise figure (dB)	5.0	
Receiver noise power (dBm), calculated for 3.84 Mcps	103.2	Thermal noise density receiver noise figure 10 log (3.84×10^6)
Interference margin (dB)	4	Cell load-dependent
Total noise interference (dBm)	99.2	
Processing gain (dB)	14.8	10 log (3,840,000/128,000)
Required E_b/N_o (dB)	2	Service-dependent
Effective receiver sensitivity (dBm)	112.0	Total noise interference minus processing gain E_b/N_o
Base-station antenna gain (dBi)	18	
Base-station feeder and connector losses (dB)	2	
Fast fading margin (dB)	4	Enables room for closed-loop power control
Log-normal fade margin (dB)	7.5	Enables for greater cell-area reliability (note 2)
Building penetration loss (dB)	15	Typical value for suburban building
Soft handover gain (dB)	2	
Maximum allowable path loss (dB)	127.5	

[1]Thermal noise density = kT, where k = Boltzmann's constant and T = temperature in Kelvin. T is usually assumed to be 293 K.

[2]The log-normal fade margin is a design value that depends on the required level of signal reliability over the cell area; 7.5 dB corresponds to 93.4 percent coverage probability.

Table 15.2 Example Link Budget for 128-kbps Data, Indoor Service

In Table 15.3, we have a link budget that would apply to an indoor data service at 384 kbps. We assume that the service is to be provided by a base station located outside the building in question. The user device has a nominal power output of 0.25 W (24 dBm). Given that this is likely to be a specialized data device with an external antenna, we assume an antenna gain of 2 dBi for the device. We also assume that there is no body loss.

At the receiving side, many of the parameters are the same as for the example in Table 15.2. The required E_b/N_o in this case, however, is 1 dB, and the processing gain is lower (owing to the higher data rate). The other margins, gains, and losses are the same as for Table 15.2. We assume the same building-penetration loss as in Table 15.2 because we are assuming that the service is provided from a base station outside the building.

Transmitter (mobile)		
Mobile TX power (dBm)	24	
Antenna gain (dBi)	2	
Body loss (dB)	0	
EIRP (dBm)	26	Equivalent isotropic radiated power
Receiver (base station)		
Thermal noise density (dBm/Hz)	174.0	Thermal noise floor[1]
Receiver noise figure (dB)	5.0	
Receiver noise power (dBm), calculated for 3.84 Mcps	103.2	Thermal noise density receiver noise figure 10 log (3.84×10^6)
Interference margin (dB)	4	Cell load-dependent
Total noise interference (dBm)	99.2	
Processing gain (dB)	10.0	10 log (3,840,000/384,000)
Required E_b/N_o (dB)	1	Service-dependent
Effective receiver sensitivity (dBm)	108.2	Total noise interference minus processing gain E_b/N_o
Base station antenna gain (dBi)	18	
Base station feeder and connector losses (dB)	2	
Fast fading margin (dB)	4	Enables room for closed-loop power control
Log normal fade margin (dB)	7.5	Enables for greater cell-area reliability (note 2)
Building penetration loss (dB)	15	Typical value for suburban building
Soft handover gain (dB)	2	
Maximum allowable path loss (dB)	125.7	

[1]Thermal noise density = kT, where k = Boltzmann's constant and T = temperature in Kelvin. T is usually assumed to be 293 K.

[2]The log-normal fade margin is a design value that depends on the required level of signal reliability over the cell area; 7.5 dB corresponds to 93.4 percent coverage probability.

TABLE 15.3 Example Link Budget for 384-kbps Data, Indoor Service

If we were to assume an in-building base station, then the penetration loss would be much lower—just enough to accommodate for losses in internal walls within the building.

In reviewing the three example link budgets, we see that the maximum allowed path loss decreases as the required data rate increases as expected. Thus the higher the data rate to be offered over a given area, the greater is the required density of base stations.

There is not an exact apples-to-apples comparison between the different services in our example link budgets because we have made different assumptions regarding mobile output power, antenna gains, and building-penetration losses.

If, for comparison purposes, we were to assume that these quantities were the same in each scenario, then we would have a clear picture of how the cell coverage reduces as the data rate increases (all other things being equal). The reduction in cell coverage is due to the reduced processing gain. It is noticeable, however, that while there is a reduced processing gain for higher data rates, this is somewhat counterbalanced by a lower E_b/N_o requirement for higher data rates.

The required E_b/N_o depends on many factors, including the mobile speed, data rate, and multipath profile. Why should the E_b/N_o decrease as the data rate increases? The answer is the fact that higher bit rates mean greater power output from the mobile. There is greater output power for both the *Dedicated Physical Control Channel* (DPCCH) and the *Dedicated Physical Data Channel* (DPDCH) as the data rate increases. The pilot symbols on the DPCCH are used for channel estimation and received *Signal-to-Interface Ratio* (SIR) estimation. As the DPCCH power increases, the better is the channel estimation, which means that a lower E_b/N_o can be accommodated. Of course, as the data rate increases, the DPDCH power also increases and, in fact, the relative power of the DPCCH versus the DPDCH decreases. In other words, as the data rate increases, a greater proportion of the total power is allocated to DPDCH rather than DPCCH. But the fact that the overall power increases with increasing data rate means that the total DPCCH power increases (albeit not as much as the DPDCH power). It is the absolute DPCCH power that is important in channel estimation, and because the absolute DPCCH power increases, the required E_b/N_o decreases.

The link-budget variables are a critical element and input into the design process when deploying UMTS and HSDPA and/or HSUPA. There obviously is a lot of room for manipulation, which has to be understood properly to avoid having a spread sheet design that cannot be implemented in practice.

15.3 RF Capacity Analysis

Based on link budgets and using an appropriate propagation model as described in Chapter 11, we can construct an initial RF coverage plan. Typically, this is done using a software-based planning tool. This will only be an initial plan, however. The next step requires that we validate the plan to ensure that it will support the expected load. Recall that the link budget includes an interference margin that is based on the loading expected on the cell. The greater the expected load, the greater the interference margin needs to be. Suppose, for example, that we perform an initial coverage analysis based on a nominal interference margin of, say, 3 dB, equivalent to approximately a 50 percent cell loading. Therefore, we must validate the initial coverage-based plan to ensure, based on the coverage provided and the expected traffic forecast in the covered area, that the interference margin chosen will be sufficient to support the expected load. Table 15.4 shows the required interference margins as a function of uplink cell load.

The reason for the interference margin is to account for the interference that will be caused by other users. That interference is effectively additional noise over and above thermal noise. In other words, the greater the cell load, the greater is the noise, and we need to include a greater margin to account for that noise. This increase in noise is known as the *noise rise,* and the margin we include in the link budget matches the noise rise generated by the expected cell load.

From Table 15.4 we can see that the noise rise tends toward infinity as the cell load tends toward 100 percent. In other words, 100 percent cell load is not achievable.

Uplink Cell Load	Required Interference Margin
0%	0 dB
10%	0.46 dB
20%	1 dB
50%	3 dB
75%	6 dB
90%	10 dB
95%	13 dB
99%	20 dB

Table 15.4 Required Interference Margins as a Function of Uplink Cell Load

Moreover, the greater the cell load, the greater is the noise rise, and the smaller is the effective cell coverage area.

We cannot achieve 100 percent cell load, but we can readily achieve a cell load of, say, 60 percent. We must, of course, be able to translate this percentage into some measure of subscriber usage—such as total number of subscribers for a given service or total throughput. This will allow us to verify whether the cell coverage we expect (assuming a particular interference margin) will be sufficient to support the offered load. Imagine, for example, that we have a nominal cell plan based on a link budget for a particular service (such as 128-kbps data) and with a particular interference margin (such as 4 dB or about 60 percent uplink load). This plan will mean that a given cell has a particular footprint. We then consider this footprint and determine whether the load expected within this footprint will be less than the loading for which the plan was designed in the first place. Clearly, this is an iterative process. If we find that the plan does not support the expected load in some areas, then we need to modify the plan, perhaps by the addition of extra base stations.

In order to determine whether a given cell can accommodate the expected load, we need to quantify that load. As described in Chapter 11, we first should determine the expected demand in the busy hour so that we can make sure that we design the system to accommodate peak demand. The peak demand needs to be specified for the various services we wish to offer—both voice and data at various rates. We then determine the cell capacity and check to make sure that it can support the expected demand. The expected demand should be specified for both the uplink and the downlink, and the cell capacity calculation also should be performed for both directions. This is of particular importance because UMTS services can be asymmetric.

15.3.1 Calculating Uplink Cell Load

The load placed on a cell is described in terms of *load factor*, which is some fraction of the maximum theoretical load. In other words, a load factor of 0.5 equates to 50 percent cell loading. The load placed on the cell can be viewed as the sum of the loads generated by all users of the cell. Alternatively, the *total load factor* is the sum of the load factors contributed by each user, as shown in Equation 15.1:

$$\text{Load factor} = \sum_{j=1}^{N} L_j \tag{15.1}$$

where L_j is the load factor of a single user j, and we assume N users in the cell. L_j is simply the fraction of the total power at the base station that user j generates. Thus

$$L_j = S_j / S_{\text{total}} \tag{15.2}$$

Alternatively,

$$S_j = L_j \times S_{\text{total}} \tag{15.3}$$

The signal power S_j for a given user needs to be such that the E_b/N_o requirement is met for the service that the user wishes to obtain. Moreover, E_b/N_o is a function of the total interference in the cell and can be expressed as follows:

$$(E_b/N_o)_j = \text{processing gain} \times (S_j/I) \tag{15.4}$$

where I is the total interference and is equal to the total received power at the base station minus S_j (the signal power from user j). This can be rewritten as

$$(E_b/N_o)_j = (C/a_j \times R_j)[S_j/(S_{\text{total}} - S_j)] \tag{15.5}$$

where C is the chip rate, a_j is the activity factor (such as about 65 percent for voice, including DPCCH overhead, and 100 percent for data), R_j is the user data rate, and S_{total} is the total received signal power at the base station. If we then solve for S_j, we get

$$S_j = S_{\text{total}}/\{1 + C/[a_j \times R_j(E_b/N_o)_j]\} \tag{15.6}$$

Substituting from Equation (15.2), we get

$$L_j = 1/\{1 + [C/(a_j \times R_j)(E_b/N_o)_j]\} \tag{15.7}$$

Using Equation (15.1), we now get

$$\text{Load factor} = \sum_{j=1}^{N} L_j = \sum_{j=1}^{N} 1/[1+(C/(a_j \times R_j)(E_b/N_o)_j)] \tag{15.8}$$

In addition to the interference generated by users on the local cell, there also will be interference caused by transmissions from users in nearby cells. If we define the quantity i to the ratio of nearby cell interference to the interference in our own cell, that is,

$$i = \text{nearby cell interference}/\text{local cell interference} \tag{15.9}$$

then the total load factor for the local cell is

$$\text{Load factor} = (1+i) \times \sum_{j=1}^{N} L_j = (1+i) \times \sum_{j=1}^{N} 1/[1+(C/(a_j \times R_j)(E_b/N_o)_j)] \tag{15.10}$$

As the load factor approaches unity, the cell has reached its maximum capacity.

15.3.1.1 Example Uplink Cell Loading for Voice Service

In this example, we calculate cell loading as a function of the number of users, assuming that all users are using standard voice service.

Assumptions

$a_j = 0.65$

$R_j = 12.2$ kbps

$E_b/N_o = -4$ dB (= 2.512) for all users (because all users are voice-only in this example)

$i = -$ 50% (i.e., of the total interference at the base station, one-third is being received from other cells)

Using Equation (15.10), we calculate the uplink load factor for a single user, that is,

$$\text{Load factor for one voice user} = 0.00774 = 0.774\%$$

Thus, for a load factor of 50 percent, we can accommodate approximately 65 simultaneous voice users. For a load factor of 60 percent, we can accommodate approximately 76 simultaneous voice users, and so on. The number of users as a function of load factor (and required interference margin) is shown in Figure 15.4.

Given that the required interference margin (i.e., noise rise) means a smaller allowable path loss, it is clear that the cell footprint reduces as the number of users increases. If we consider the link budget shown in Table 15.1 and consider the required interference margin as a function of the number of users, the allowable path loss (which determines the cell size) is as shown in Figure 15.5. We also should note that the maximum path loss shown does not consider building or vehicle penetration losses, which would need to be subtracted from the figures shown.

The calculation we performed previously provided the loading in terms of number of users. We could easily present the loading results in terms of kilobits per second. In fact, if we consider that there will be both data and voice usage, then presenting the information in terms of kilobits per second can be useful because that term will apply both to voice and data.

15.3.1.2 Example Uplink Cell Loading for Data Service

In this example, rather than showing the number of users, we show the total throughput (in kilobits per second) for a given cell loading. If we look again at Equation (15.10), we note that it can be simplified if we assume that all users have the same data rate. The equation then is as follows:

$$\text{Load factor} = N \times (1 + i)/(1 + \{C/[a \times R(E_b/N_o)]\}) \tag{15.11}$$

If we note that the term $C/[a \times R(E_b/N_o)]$ is far greater than 1 for most services, then we can simplify the equation further to

$$\text{Load factor} = N \times (1 + i)/\{C/[a \times R(E_b/N_o)]\} \tag{15.12}$$

To get the total throughput (rate times number of users), we rearrange to get

$$\text{Throughput} = R \times N = \text{load factor } C/[(E_b/N_o) \times (1 + i)] \tag{15.13}$$

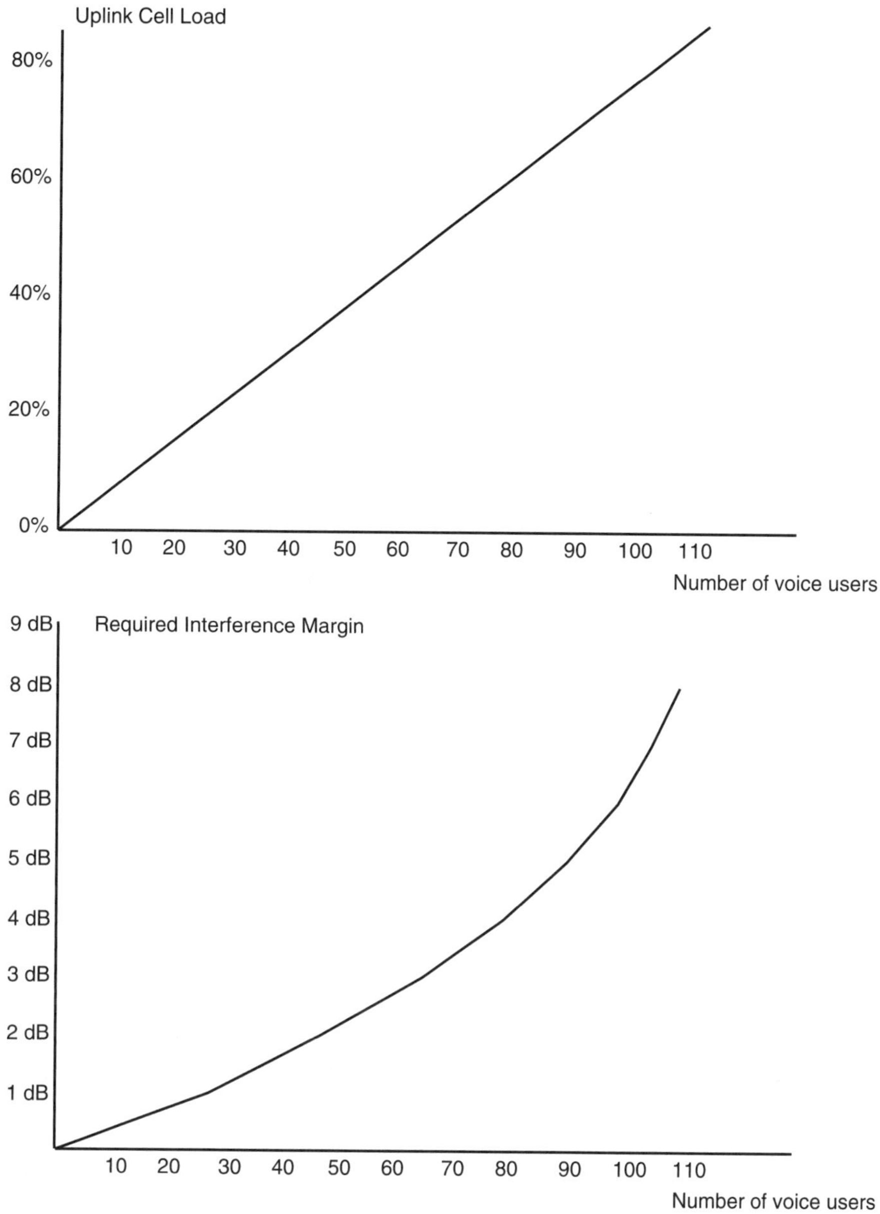

FIGURE 15.4 Example uplink cell loading and interference as a function of number of users.

Assumptions

$a = 1.0$

$E_b/N_o = 1$ dB (= 1.259)

$i = -50\%$ (i.e., of the total interference at the base station, one-third is being received from other cells).

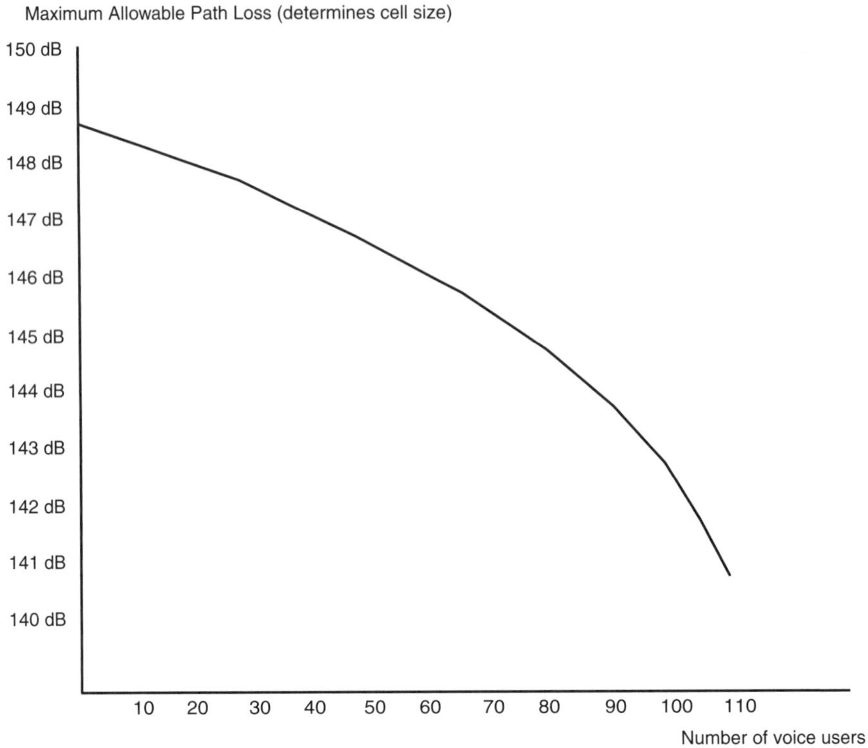

FIGURE 15.5 Example allowable uplink path loss as a function of number of users.

Thus, for a load factor of 50 percent, we have a total throughput of 1106 kbps. For a load factor of 60 percent, we have a total throughput of 1220 kbps.

15.3.2 Downlink Cell Load

In the downlink, the determination of cell loading uses the same basic approach as for the uplink. The same approach is applicable because the ability of a given mobile unit to recover a signal that is destined for that mobile depends on how many other signals are being sent to other mobiles in the cell. In other words, for a given user *j*, the signals that are being sent from the base station to other users are simply interference. The more such signals, the greater is the interference. As is the case for the uplink, the effect of the interference depends on the E_b/N_o requirement at the mobile unit. Interference also is caused by downlink common channels and other base stations. In the case of interference from other base stations, the amount of interference depends on the individual user's location. A user who is close to the serving base station is less likely to experience as much interference from neighboring cells as a user who is near the border between cells.

Finally, we need to factor in orthogonality. In the downlink, for a given scrambling code, transmissions to different users are sent using different channelization codes, which are chosen such that the codes are orthogonal. If the transmission from the base station to a single user arrives over multiple paths, however, and the delay spread across those paths is sufficiently large, the mobile unit will directly recover only a part of the signal from the base station. The other part of the signal, which arrives over a long delay path, will be seen as interference. This phenomenon needs to be accounted for in our calculation of downlink loading.

Because the same methodology for downlink load factor calculation applies in the uplink, Equation (15.10) still applies, but with some modifications to account for orthogonality and the fact that interference from neighboring cells is different for each user. Thus the equation for downlink load factor becomes

$$\begin{aligned}\text{Load factor} &= \sum_{j=1}^{N} L_j \times (1-\alpha_j + i_j) \\ &= \sum_{j=1}^{N} (1-\alpha_j + i_j)/[1+(C/a_j \times R_j(E_b/N_o)_j)] \qquad (15.14)\end{aligned}$$

where a_j is the orthogonality factor related to user j, and i_j is the interference from neighboring cells experienced by user j.

As in the downlink, for most services, the term $C/[a_j \times R_j(E_b/N_o)_j]$ is far greater than 1, which means that the equation can be simplified. Moreover, it is not realistic to determine the orthogonality factor for each mobile unit in the cell because this will depend on the exact user location and multipath profile. Nor is it realistic to determine the intercell interference experienced by each user because that also will depend on the user's exact location. Thus we need to consider average values of orthogonality a and intercell interference i. A typical value for a is 0.4, and a typical value for i is 0.5.

Including these considerations, the load equation becomes

$$\begin{aligned}\text{Load factor} = (1-\alpha+i) \times \sum_{j=1}^{N} L_j &= (1-\alpha+i) \\ &\times \sum_{j=1}^{N} 1/[C/(a_j \times R_j(E_b/N_o)_j)] \qquad (15.15)\end{aligned}$$

15.3.2.1 Example Downlink Cell Loading for Voice Service

In this example, we calculate cell loading as a function of the number of users, assuming that all users are using standard voice service.

Assumptions

$a_j = +0.65$ for all users

$R_j = 12.2$ kbps for all users

$E_b/N_o = -4$ dB (= 2.512) for all users (because all users are voice-only in this example)

$a = 0.4$

$i = 0.5$

Because of the fact that all users in this example have the same characteristics, Equation (15.15) becomes

$$\text{Load factor} = N \times (1 - a + i)/\{C/[a \times R(E_b/N_o)]\} \qquad (15.16)$$

Uplink Cell Load	Required Interference Margin
0%	0 dB
10%	0.46 dB
20%	1 dB
50%	3 dB
75%	6 dB
90%	10 dB
95%	13 dB
99%	20 dB

TABLE 15.5 Required Interference Margins as a Function of Uplink Cell Load

Using the previous assumptions, the load factor for one user is

$$(1 - 0.4 + 0.5)/[3{,}840{,}000/(0.65 \times 12{,}200 \times 2.512)] = 0.0057 = 0.57\%$$

Thus, for a downlink load factor of 50 percent, we can accommodate approximately 88 simultaneous voice users. For a load factor of 60 percent, we can accommodate approximately 105 simultaneous voice users, and so on.

As is the case for the uplink, the downlink link budget needs to include an interference margin equivalent to the noise rise. The required interference margin is a function of the cell load factor, and the same figures as in Table 15.4 apply. In other words, for a 50 percent load factor, we need a 3-dB interference margin in the downlink.

Assume, for example, a downlink link budget, as shown in Table 15.5, where there is a base-station transmitter output power of 10 W.

This link budget does not show an interference margin. Such a margin must be included, however. The exact value of the interference will equate to the noise rise, which increases with increasing cell load—that is, throughput. Using the example assumptions outlined previously, Figure 15.6 shows the cell load as a function of the number of users and also the noise rise/required interference margin as a function of the number of users. Figure 15.7 shows the allowable downlink path loss as a function of the number of users. If we compare Figure 15.7 with Figure 15.5, we can determine whether the system is uplink-limited or downlink-limited for a given number of voice users.

The foregoing examples show how cell loading, in terms of numbers of voice users, can affect uplink and downlink coverage. Using voice service is a convenient example to show how the calculations can be performed. In reality, however, we can expect a significant mix of services—with some subscribers using voice services and some subscribers using data services of one kind or another. Thus the calculations should be performed individually for each type of service.

While, for a service such as voice, the coverage is likely to be uplink-limited rather than downlink-limited, the same might not apply for data service. With UMTS, data services can be asymmetric—that is, different data rates in the uplink compared with the downlink. Moreover, for many data services (such as Web browsing), we will find that the downlink data rate is far greater than the uplink data rate. Consequently,

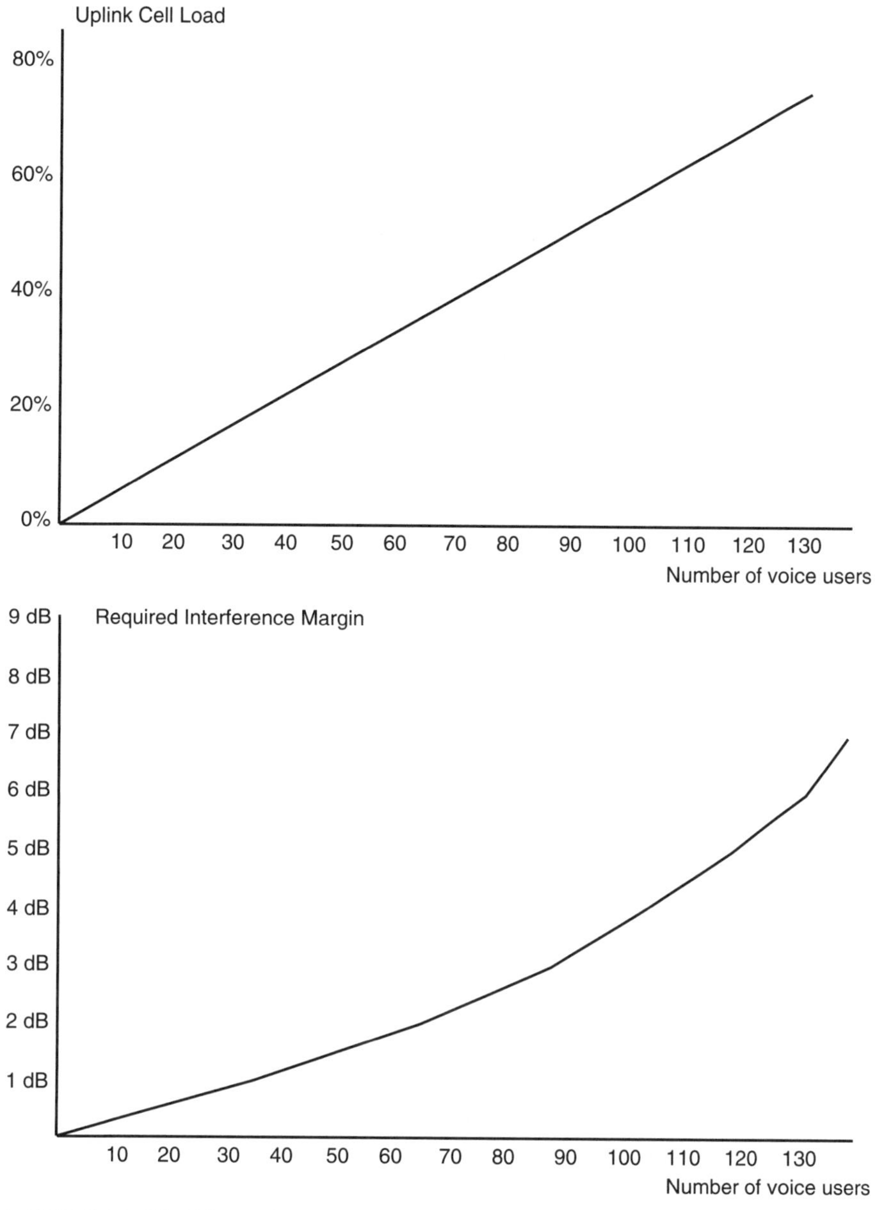

FIGURE 15.6 Example downlink cell loading and interference as a function of number of users.

the effect of interference in the downlink may well be greater than in the uplink, which means that the downlink load may become the limiting factor.

If we find that we are downlink-limited, then we may be able to increase the base-station output power and/or add an additional RF carrier subject to spectrum availability. As mentioned in Chapter 3, however, the addition of a second carrier will mean that compressed mode must be used (where the MS can tune to other carriers for potential

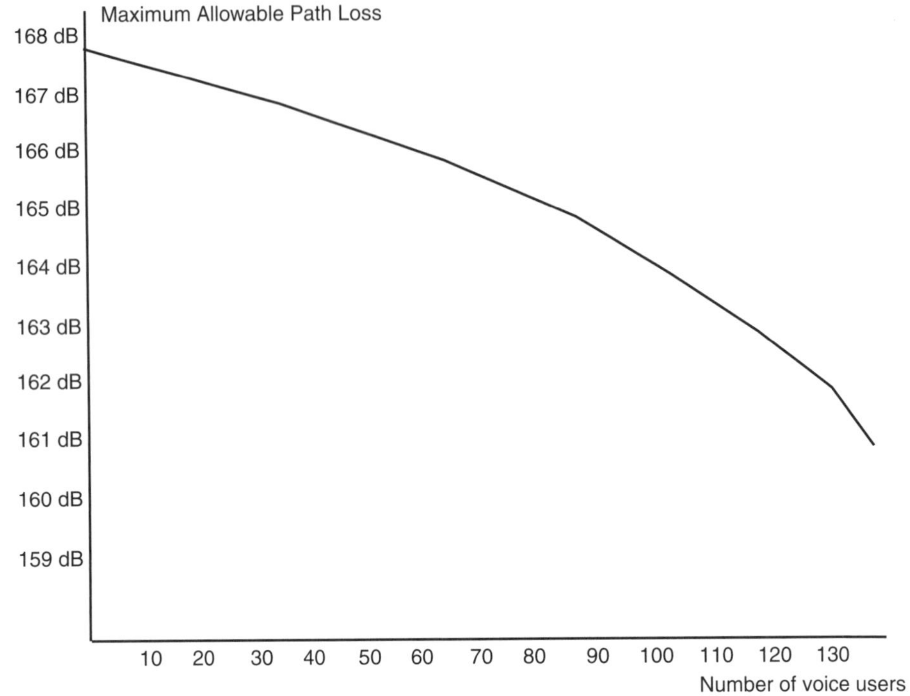

FIGURE 15.7 Example allowable downlink path loss as a function of number of users.

hard handover). Compressed mode means an aggregate lower throughput per carrier so that although a second carrier does provide significant additional capacity, it does not mean a capacity increase of 100 percent.

Another downlink-limiting factor for a single carrier base station is the availability of downlink channelization (spreading) codes. Recall from Chapter 3 that channelization codes are chosen from a code tree. Recall also that the use of a particular channelization code can preempt the use of other channelization codes on the same branch of the code tree. For example, consider the channelization code $C_{ch,4,0}$. This code is simply the sequence 1, 1, 1, 1 repeated over and over. Consider the channelization code $C_{ch,8,0}$. This is simply the sequence 1, 1, 1, 1, 1, 1, 1, 1 repeated over and over. Clearly, if the base station is using either of these codes in a transmission to a particular mobile, then it cannot use the other code (or any other code that is a series of all 1s) in transmission to any other mobile. One way to overcome this limitation, however, is for the base station to use multiple scrambling codes. A given cell can use up to 16 downlink scrambling codes.

15.3.3 Load Sharing

As described in the preceding discussions, intercell interference plays a role in the capacity of a given cell. In both the uplink and downlink, higher interference from nearby cells means a lower capacity and possibly a smaller footprint in the cell of interest. Conversely, lower interference from nearby cells means that the cell of interest can

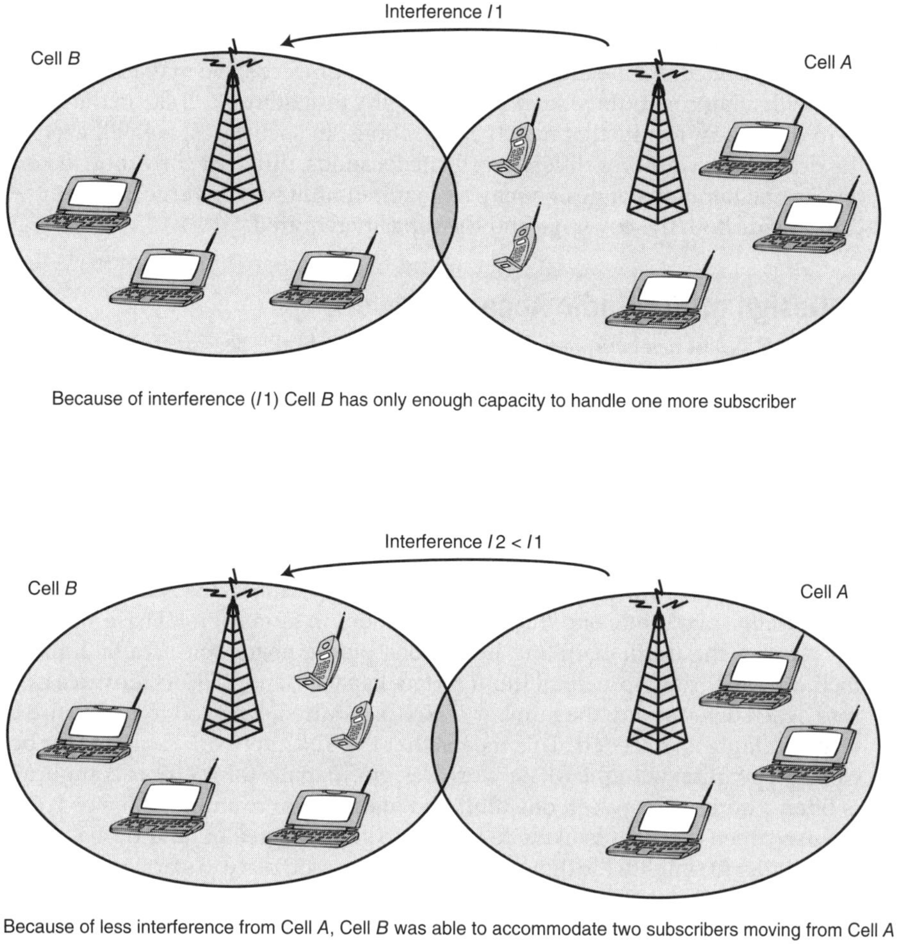

Figure 15.8 Example of load sharing.

have higher capacity and a larger footprint. This means that one cell can effectively "borrow" capacity from one or more nearby cells that are less loaded.

Consider Figure 15.8, for example. Some subscribers move from cell *A* to cell *B*. Thus cell *A* becomes less loaded, and cell *B* becomes more loaded. If cell *B* were already heavily loaded, then the existing intercell interference could mean that it might be able to accommodate more users in cell *B*. However, the fact that cell *A* now has fewer users means that it is generating less intercell interference in cell *B*. Thus it may well be possible to accommodate the additional load on cell *B*. This example shows that the capacity of a cell is not static but rather varies with the load on nearby cells.

The foregoing discussions regarding uplink and downlink capacity and their effect on coverage emphasize the fact that coverage and capacity are interrelated. Because we need to develop an RF design that supports both coverage and capacity requirements, and because capacity affects coverage, development of the RF design is an iterative process. We start with an initial coverage-based design, and we check that design against the expected demand. We then modify the design to allow for additional capacity where needed. As the implementation phase proceeds, we may find that we need to deal with other constraints, such as the inability to acquire a cell site in the ideal location or to derive test results that match expectations. In such cases, we will need to change the design to account for different cell-site locations, different correction factors, and so on. Several iterations of design may be required until we converge to a point where we can provide both the coverage and the capacity required.

15.4 Design of the Radio Access Network

Once an RF plan has been developed, the next step in the design effort is to design a network that will connect the various base stations to their *Radio Network Controllers* (RNCs). This means that we must determine the number of RNCs required, we must determine a suitable placement for the RNCs, and we must design a transmission network from RNCs to the various base stations and between RNCs (for inter-RNC soft handover).

For many GSM *Base Station Controllers* (BSCs), the main capacity limitations are in the numbers of base stations, cells, or transceivers that can be supported. In some cases, there are limits in terms of Erlangs, but such capacity limits are rarely encountered in real networks. With UMTS, however, the capacity of most RNCs is more tightly linked to the traffic mix. While one still finds limitations in terms of total base stations, cells, or RF carriers, the traffic-handling limitations play a major role. Traffic limits typically include total throughput, total Iub-interface capacity, and BHCAs for voice calls. Therefore, when determining the number of RNCs required, we need to make sure that none of these limits is exceeded. This means that the RNC network design must be done in close cooperation with the RF network design. To make things more complicated, there is often a trade-off between one limit and another. For example, if fewer voice Erlangs are used, then the RNC is likely to be able to support greater data-traffic demand.

In order to simplify the dimensioning effort, a good place to start is with dimensioning of the Iub interface. The total Iub-interface capacity is likely to be the most constraining factor. Moreover, the Iub interface is common for voice and packet data. Once we have determined the Iub capacity demand from each of the base stations, we can sum that capacity and determine the minimum number of RNCs needed. In practice, we should add an additional 25 to 35 percent to the RNC capacity that we have determined for three reasons. First, when allocating base stations to RNCs, we need to consider location areas/routing areas. It is common to assign registration areas such that they align with RNC boundaries. This means that we need some flexibility in how we allocate base stations to RNCs. Second, intra-RNC soft handover is preferable to inter-RNC soft handover because it helps to minimize Iub transport requirements, and it helps to minimize the total switching demand on the RNC. Thus we would like to define RNC boundaries such that they do not align with high-traffic areas. This also means that we need flexibility in how base stations are assigned to RNCs. Third, we never want to find ourselves in a situation where the addition of one or two extra sites (or even RF carriers) would require the addition of a new RNC. In other words, we need to leave some room for growth.

15.4.1 Iub-Interface Dimensioning

The physical interface to a base station will be such that the Iub capacity from a given base station has some discrete value. For example, a single T1 offers 1.5 Mbps.

Typically in North America, we will find that a UMTS base station has some number of T1, DS-3, or OC-3 interfaces. However, while determining that a particular base station needs one or two T1s is important, we need to determine the total Iub load at the RNC. We will not arrive at that total simply by summing the total Iub capacity available at each base station. Imagine, for example, that 100 base stations each have an Iub bandwidth demand of 1.7 Mbps. We could configure each such base station with, say, two T1s, equivalent to about 3 Mbps. However, the total load at the RNC still would be 170 Mbps, not 300 Mbps.

As described earlier in this chapter, the RF design is performed in accordance with both the coverage and capacity demand that we expect. Consequently, information will be available as to the traffic (in kilobits per second) to be carried on the Iub interface from each base station. Unlike other parts of the network, however, the RF design is unlikely to have a very long build-ahead included. While some build-ahead should be factored into the design, a build-ahead of 9 or 12 months is not pragmatic. This is so because the RF network usually represents the greatest component of the total network cost. A large build-ahead could mean a drastic increase in capital expenditure far in advance of when the capacity is needed. If, however, there is a large build-ahead, then we can simply size the Iub interface based on the expected throughput (including the build-ahead) and with the addition of perhaps 40 percent for overhead. While this approach is less than scientific, the inclusion of a long build-ahead will mean that the interface will have sufficient capacity for some time in the future. During that time, we have the opportunity to observe the increase in demand and make more accurate predictions of future interface capacity needs. If there is only a small build-ahead (such as 3 to 6 months), then we need to be more discerning in our determination of Iub capacity.

To determine the actual Iub capacity required, we need to add a certain amount of overhead to the user throughput. This overhead needs to allow for burstiness of traffic, signaling load, and *Operation and Maintenance* (O&M) load. Moreover, we need to add ATM overhead because all the user traffic, signaling, and O&M loading is carried in ATM cells.

The amount of burstiness will depend on the mix of traffic. If only voice service is to be offered, then we can assume zero burstiness. On the other hand, an all-data service could require an overhead of up to 40 percent. An allowance of 25 percent would be typical. In addition, we can assume that for a given throughput, an extra 10 percent will be required for signaling. We also can assume that we need an additional 10 percent for O&M load. To each of these, we then must add ATM overhead, which will vary according to the service. To begin with, the cell structure of ATM means that there are 5 octets of overhead for every 48 octets of payload. This alone means an overhead of 10.4 percent. In addition, as described in Chapter 3, we have *ATM Adaptation Layers* (AALs), which also consume bandwidth. Each AAL consumes some number of octets in each ATM cell, in addition to the 5 octets of the ATM header. For AAL2, 3 of the 48 payload octets are consumed by AAL2 information. Thus, for AAL2, the total ATM overhead is approximately 18 percent. For AAL5, 4 of the 48 payload octets may be consumed, meaning that the total overhead is approximately 20 percent. For signaling, the *Service-Specific Connection-Oriented Protocol* (SSCOP) and *Service-Specific Coordination Function* (SSCF), as described in Chapter 3, reside on top of AAL5 and generate even more overhead. In order to make calculations straightforward, however, the SSCOP and SSCF overhead should be included as part of the total signaling overhead.

Based on the foregoing, the total required Iub bandwidth is given by

$$\begin{aligned}\text{Iub bandwidth} = -&\text{expected user traffic} \times (1 + \text{burstiness}) \times \\ &(1 + \text{signaling overhead} + \text{O\&M overhead}) \times \\ &(1 + \text{ATM overhead})\end{aligned} \tag{15.17}$$

If we take typical examples as described previously, this equation becomes

$$\text{Iub bandwidth} = -\text{expected user traffic} \times (1 + 0.25) \times (1 + 0.1 + 0.1) \times (1 + 0.2)$$
$$\text{Iub bandwidth} = \text{expected user traffic} \times 1.8$$

Thus, because of signaling, O&M, and ATM overhead, the Iub interface should be sized to a bandwidth that is almost twice that of the actual raw user traffic. Of course, the user traffic is likely to be asymmetric, and we are likely to find that the downlink traffic is greater than the uplink traffic. The actual Iub transmission facilities, however, will be symmetric. In other words, if there is 2 Mbps capacity on one direction, there is also 2 Mbps in the other direction. Therefore, when dimensioning the Iub, we need only to consider the user traffic in one direction—the direction of greater demand. This usually will be the downlink direction.

15.4.2 Determining the Number of RNCs

As mentioned previously, the capacity of an RNC typically is limited by some or all of the following factors:

- Total Erlangs
- Total BHCAs
- Total Iub-interface capacity (Mbps)
- Total Iur-interface capacity (Mbps)
- Total Iu-interface capacity (Mbps)
- Total switching capacity (Mbps)
- Total number of controlled base stations
- Total number of RF carriers

In most cases, one will find that the Iub-interface capacity is likely to be the limiting factor. For example, a typical Iub limit for an RNC is between 150 and 200 Mbps. The same RNC might well have a limit of 500 or more RF carriers (i.e., cells if only one carrier per cell). Given that we might expect a cell to support 500 kbps to 1 Mbps, it is clear that the number of RNCs is likely to be driven by the total Iub-interface bandwidth than by the other factors. Of course, once we determine the number of RNCs based on the Iub bandwidth required, we need to validate that no other RNC dimensioning limits have been exceeded. If they have been exceeded, then additional RNC capacity needs to be added according to the most constraining factor. That, however, would be an uncommon situation.

15.4.3 Designing the UTRAN Transmission Network

Once we have determined the number of RNCs required, based on Iub bandwidth requirements, we need to develop a homing plan that specifies which base stations are to be controlled by which RNCs. This will define the RNC borders. Analysis of the cells at or near the borders then will allow us to estimate the amount of inter-RNC handover traffic we can expect so that we can determine the Iur connections required and the bandwidth needed for those connections.

The exact number of inter-RNC handovers will depend on the RF environment near the RNC borders. A reasonable approach, however, is to assume that 50 percent of traffic in border cells is being served by two base stations on different RNCs. This would be a conservative estimate that would allow for additional inter-RNC soft handovers involving cells that are not defined on the border. Imagine, for example, a user near the top of a tall building. That user might be served by a cell that is not on the border between RNCs, but because of the user's location, he or she also might be able to hear and be heard by a Node B on another RNC. Of course, the exact number of soft handovers to be allowed in the network will be specified as datafill within the RNCs. However, at the point in the design effort, where Node B homing and transmission network design is being performed, that datafill may not yet be defined.

Given that the Iur acts in many ways as a conduit for Iub traffic from a mobile unit to its controlling RNC, the basic assumptions for determining the Iub bandwidth can be applied to determining the Iur bandwidth. For example, if we assume that the Iub bandwidth needs to be approximately twice the user throughput, then the Iur bandwidth should be close to twice the user throughput for that portion of the traffic that is in inter-RNC soft handover.

Now that we have established the Node-B-to-RNC homing plan and we know the Iub and Iur interface requirements, we need to design a transport network to support all the necessary connections between Node Bs and RNCs, between RNCs, and between RNCs and *Service GPRS Support Nodes* (SGSNs) and *Mobile Switching Centers* (MSCs). Given that all the interfaces in question are ATM interfaces, we are effectively talking about designing an ATM network.

In the example of Figure 15.9, we have three RNCs, each controlling a number of Node Bs and with each RNC connected to each of the other two RNCs. All three RNCs are connected back to a single SGSN and a single MSC. In this example, all three RNCs are in different locations, and all are remote from the MSC and SGSN. This is a somewhat unrealistic situation, but we use this example in order to show complexity and how that complexity could be managed. The figure shows the logical connections between the various nodes. To implement each of those interfaces individually, however, would be impractical. Rather, one would like to implement a cost-effective transport arrangement that will support each of the logical interfaces. One way to do this could be through the use of a ring, arrangement, as also shown in the figure. Basically, each of the locations in question would become nodes on the ring, which might be an OC-12 or perhaps an OC-48 ring, or even have a higher capacity depending on demand.

In many cases, however, the distances between nodes could mean that the cost of such a ring could be prohibitive. In such a case, one might want to employ one of the configurations shown in Figure 15.10. In the first case, we deploy a separate ATM switching layer that takes care of switching the various ATM paths between the various nodes. By deploying such a layer, we can reduce the overall transmission cost. Of course, there is a capital cost that must be paid, plus the operational cost of deploying new equipment. In the second configuration of Figure 15.10, we use one of the RNCs as an ATM switch. Back at the MSC site, we may have the possibility to use an SGSN or an RNC at that site as an ATM switch. This option is possible for some equipment vendors because an RNC is fundamentally an ATM switch with additional UMTS-specific functionality. It is not

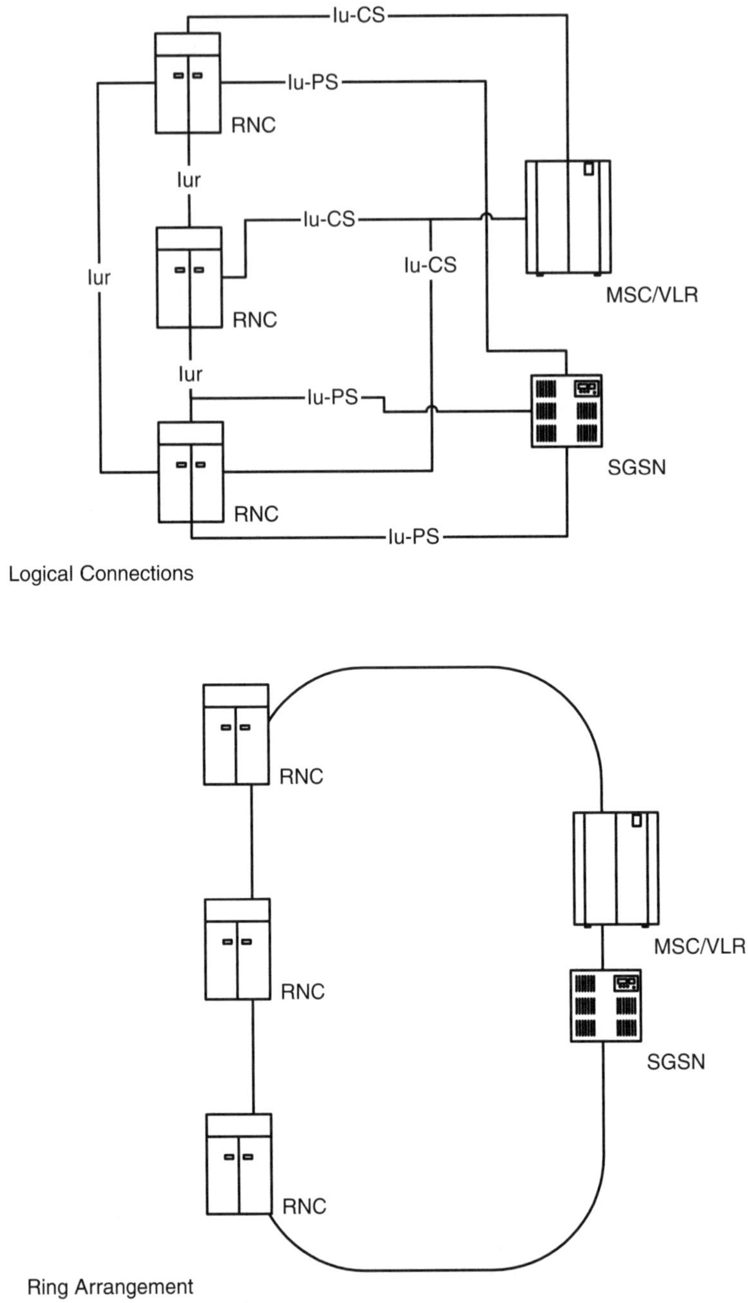

FIGURE 15.9 RNC connectivity—logical connections and possible ring transport.

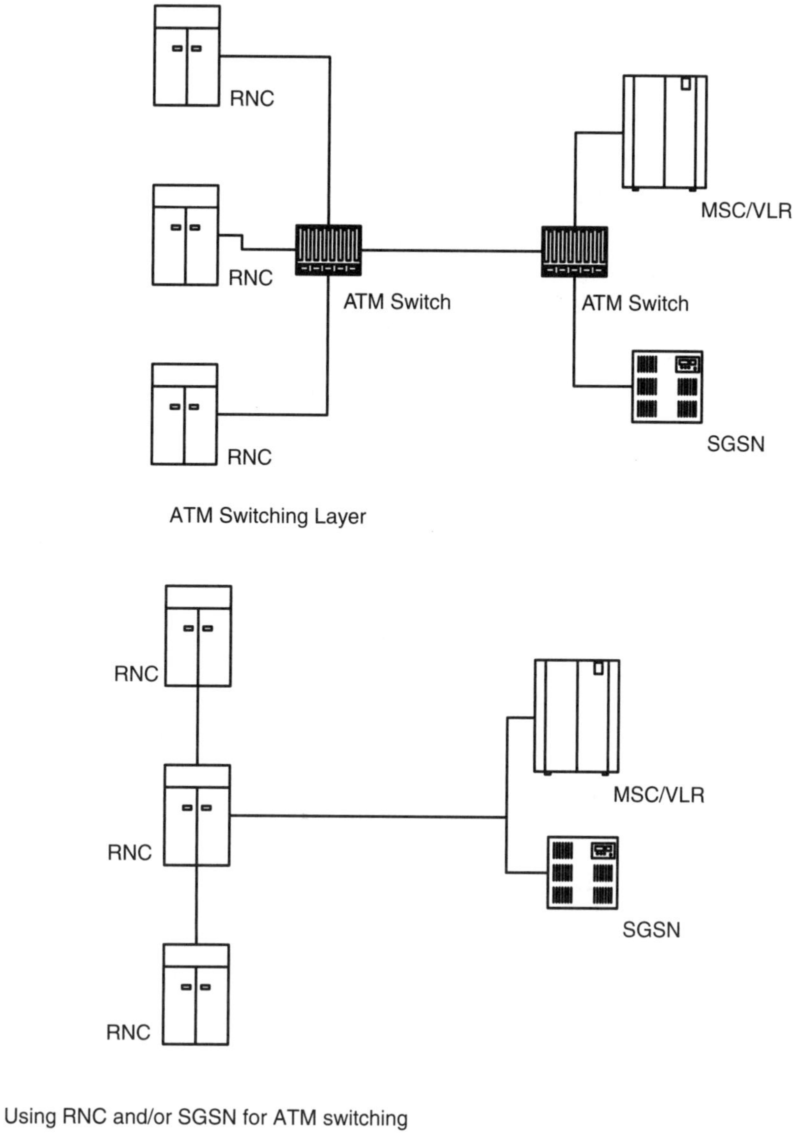

FIGURE 15.10 Separate ATM switching layer or use of an RNC and/or an SGSN for ATM switching.

uncommon to find that the total switching capacity of an RNC is several gigabits per second, whereas the Iub-interface capacity may be limited to perhaps 200 Mbps. Thus we are likely to find that the RNC can switch more ATM traffic than would be required of it as a pure RNC. We can take advantage of this extra switching capacity and reduce overall transmission cost without having to deploy a separate ATM switching network.

The design and cost of the *UMTS Terrestrial Radio Access Network* (UTRAN) transmission network is interwoven with the placement of the RNCs. There may be multiple options for placement of RNCs. We may choose to place all RNCs at the MSC location, all remotely, or some mix of remote and local RNCs. The placement of the RNCs will be related to the capacity of an RNC, the cost of the RNC, the availability of suitable locations, and the cost of transmission. The final solution must aim for a network topology that strikes a balance between capital cost, operational cost, and network reliability.

15.5 UMTS Overlaid on GSM

Some network operators will deploy a greenfield UMTS network. For many, however, UMTS will be deployed alongside an existing GSM network as an overlay network requiring new spectrum or spectrum segmentation. These operators will wish to reuse the components of the GSM network to the greatest extent possible. There is a desire to reuse everything from cell-site locations to MSCs, SGSNs, and *Home Location Registers* (HLRs). Because of the fact that the core network of UMTS is essentially the same core network as is used for GSM/GPRS, there is a significant opportunity to reuse existing equipment. For example, a GSM MSC can be upgraded to support both GSM and UMTS simultaneously. Similarly, SGSNs and *Gateway GPRS Support Nodes* (GGSNs) can be upgraded to support both UMTS and *General Packet Radio Service* (GPRS) simultaneously.

In the radio access network, there is also some opportunity for reuse. For most vendors, it will not be possible to upgrade a BSC to function as a BSC and an RNC simultaneously. For base stations, however, several vendors support both GSM and UMTS within the same base-station cabinet. In such a situation, it is possible for the GSM and UMTS transceivers to use the same antennas. Even if a given vendor does not support both UMTS and GSM transceivers within the same cabinet, or if the UMTS and GSM systems are provided by different vendors, there still may be an opportunity to colocate a UMTS base-station cabinet with a GSM base-station cabinet. This can reduce site-acquisition costs and some construction costs.

It is important to remember that UMTS and GSM have different RF design criteria. For GSM systems operating in the same frequency band, that is, 800, 900, 1800, or 1900 MHz, as will the new UMTS cell, the footprint for both UMTS and GSM cells will be very similar. In fact, for a cell loading factor of up to approximately 65 percent, the footprint of a UMTS cell for voice service is slightly greater than the equivalent footprint of a GSM cell. However, if the desire is to have UMTS operate in a different frequency band, say, 1900 MHz for UMTS and 800 MHz for GSM, the difference in frequencies is such that the GSM signal propagates a great deal farther, which means that the coverage of a UMTS cell will be less than that of a GSM cell. Thus, when deploying UMTS1800/1900 over an existing GSM900/800 network, extra cell sites will be required for UMTS. In urban areas, the number of extra UMTS cell sites is likely to be quite limited because the GSM sites will have been deployed in a more dense arrangement for capacity reasons rather than just for coverage reasons. In rural/highway areas, however, there will need to be many more UMTS1800/1900 sites than GSM900/800 sites simply because of less attenuation for the lower-frequency GSM900/800 signal.

In the case where a GSM base station and a UMTS base station are colocated or even share the same cabinet, then they also can share the transmission facilities back toward the BSC and RNC. Figure 15.11 shows how this can be done. In this example, a UMTS cabinet is colocated with an existing GSM cabinet. The GSM cabinet already has a T1 connection back to the BSC. Given that a GSM base station requires between two and three DS0s per transceiver, it is quite possible that the T1 is not fully used. In fact, less than half the T1 might be used, as would be the case for, say, a three- transceiver GSM BTS.

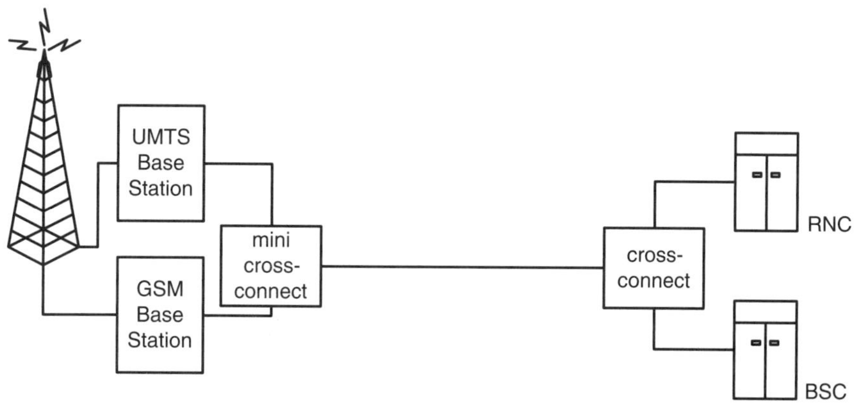

Figure 15.11 Sharing Iub and Abis transport.

Provided that the expected Iub will consume less than the remaining bandwidth, then we can use that fractional T1 capacity for the Iub interface. In other words, we can carry ATM on a fractional T1. Back at the BSC/RNC location, we need to have a cross-connect that can perform DS0-level grooming. This cross-connect strips out that part of the T1 that is used by the GSM BTS and sends it to the BSC. The part of the T1 that carries the Iub interface is sent to the RNC. Of course, we also need a mini-cross-connect at the base-station site.. For many GSM base stations, such one-card devices are available because it is not uncommon in many countries to daisy-chain GSM base stations in order to reduce transmission cost.

Of course, we are likely to find that this type of transport sharing will be possible only for sites that expect relatively low demand. If we are co-siting a UMTS base station with a GSM base station in an urban area, for example, we may find that the GSM base station is already consuming more than half of a T1 (as would be the case for a six-transceiver base station). We are likely to find that the UMTS base station also will require more than half of a T1, particularly when we consider the overhead that the Iub interface needs to include. In this case, we have little choice but to increase the transport bandwidth to the site.

Figures 15.12 and 15.13 show several spectrum-allocation schemes for UMTS systems being overlaid on top of an existing GSM network; if a third or fourth RAN technology exists in the system, then the spectrum obviously will be further subdivided, or "*x*-furcated." For these very simple examples, it has to be assumed that there is an ability to migrate existing legacy GSM users into a smaller portion of the spectrum, freeing up spectrum. However, if you have only 5 MHz of spectrum for a market and it is being used for GSM presently, then deploying UMTS is not an option unless you pursue a temporary *Mobile Virtual Network Operator* (MVNO) arrangement.

In Figure 15.12 it is assumed that only the A cellular band is available, and the launch of UMTS is shown. In Figure 15.13, a 5-MHz PCS band is available along with the existing 850-MHz band. The operator has the choice in this situation of deploying UMTS in the 850- or 1900-MHz band. Each band selection has a benefit and trade-off. The 850-MHz band would require fewer UMTS cell sites to provide initial coverage than the 1900-MHz band. However, the 850-MHz band usually has other legacy systems,

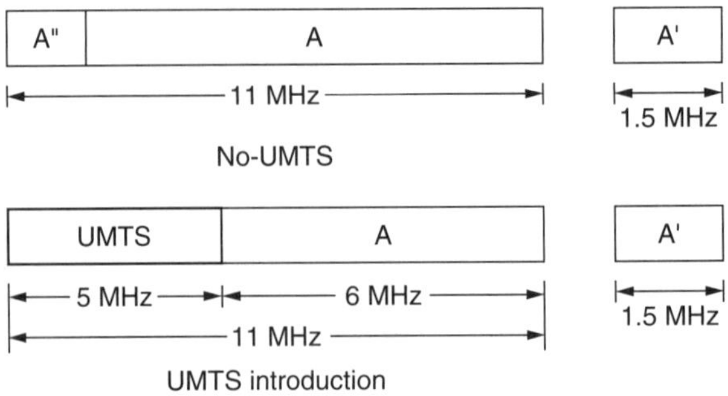

Figure 15.12 UMTS spectrum-deployment schemes (850-MHz cellular only).

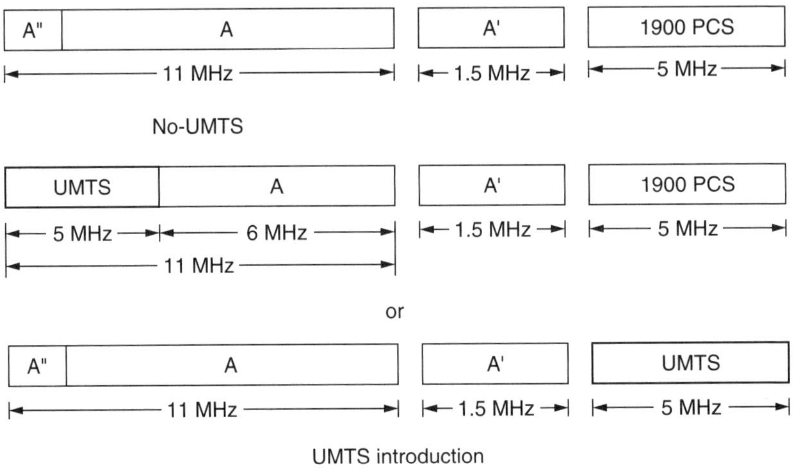

Figure 15.13 UMTS spectrum-deployment schemes (850 and 1900 MHz).

and clearing the required spectrum may prove more challenging than deploying in the 1900-MHz band.

Additionally, you need to determine if you want to have a 1:1 or *N*:1 design for UMTS. Using an *N*:1 technique, UMTS can be deployed in the core, or high-data-usage areas, of the network in a limited rollout. Depending on the extent of the rollout, a guard zone like that used in IS-95 networks also could be established. The trade-off with the limited rollout for UMTS is the difference in RAN technologies making the limited rollout an island deployment requiring dual-band/technology handsets.

A 1:1 deployment not only has the advantage of matching the coverage of the existing system but also of having a more robust service area from which to offer the service. Obviously, covering the rural areas with UMTS may not be practical initially, but a wide-scale deployment for initial service offering is essential when other broadband wireless mobility access systems are available in the same market area.

15.6 HSPA

High-Speed Packet Access (HSPA) and enhanced HSPA (HSPA+) are a major enhancement to UMTS as discussed in Chapter 3 and as a result requires some important deployment considerations. HSPA, unlike UMTS, is designed specifically to support multimedia services like streaming video, interactive gaming, telemetrics, and, of course, faster downloads and uploads for high-resolution images and large files.

Recapping HSPA also has improved spectral efficiency over UMTS; however, it can be deployed using a similar RF footprint like WCDMA using a (5 × 5)-MHz RF channel. HSPA is also backward-compatible with 3GPP Release 99.

However, HSPA requires some modifications to the WCDMA network in order to be implemented, and the elements in the WCDMA network requiring some modification are shown in Figure 15.14 and are the Node B, RNC, and SGSN.

The primary change to a UMTS system in the deployment of HSPA involves the changes to the Node B as well as the RNC and, of course, the SGSN. With the deployment of HSPA and the support of multimedia and higher bandwidth capabilities the backhaul connection between the Node B and the core network needs to also be increased, otherwise the increased throughput capability cannot be achieved. Although the enhancement to the backhaul, Iub, may seem obvious, several deployments of HSPA have been hampered due to the lack of increased backhaul support.

As part of the HSPA design considerations, the following factors need to be included. However the following list is not all-inclusive and is meant as a general guide or reminder to factor into the design process.

- Coverage changes
- RF performance differences
- Node B modifications

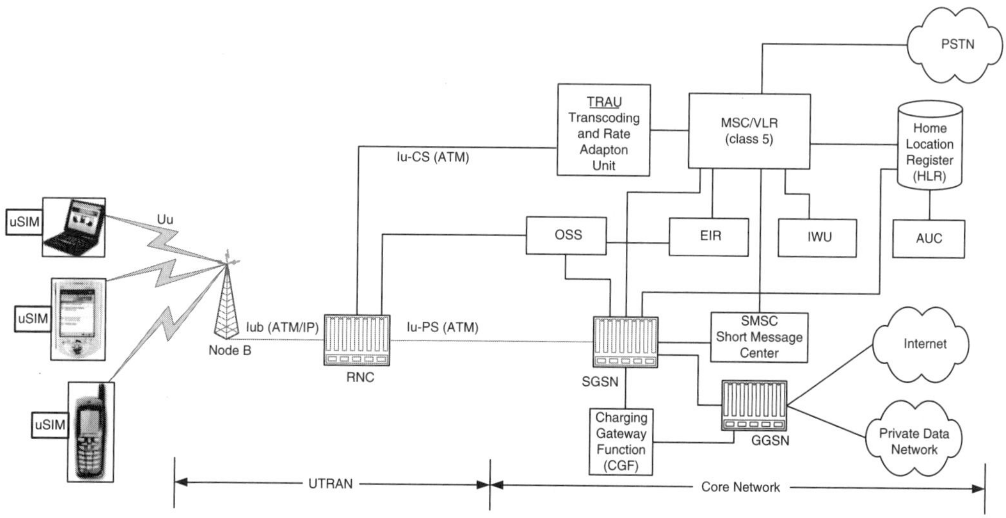

Figure 15.14 HSPA high-level architecture.

- Backhaul
- RNC upgrade
- SGSN upgrade
- UE support
- RF spectrum to deploy in
- Multicarrier considerations (intraband or interband)

With HSPA+, the ability to support higher-modulation format also places increased pressure on the RF environment as well as backhaul needs. The UEs in the network if deploying HSPA or HSPA+ need to also be able to support the new format and of course the higher-modulation schemes.

For example, HSPA+ supports a 2 × 2 downlink MIMO enabling the Node B to send parallel data streams to the UE in order to improve the overall downlink throughput.

With HSPA and HPSA+, the radio at the Node B needs to be enhanced to support the new transport signaling and setup interfaces. Because logistics always play a part in the deployment of new radio systems, the notion of either deploying HSPA or HSPA+ in limited areas for initial service offering and then expanding the coverage area is the preferred method.

However with the increased modulation format, the range of the Node B for coverage will decrease. It is recommended that you deploy HSPA initially as a 1:1 overlay and then deploy additional Node Bs, that are HSPA-capable based on where data usage is the highest.

With HSPA the fundamental dimensioning process is basically the same as in UMTS; however, the overall capacity dimensioning needs to factor in the use of an RF carrier as a shared resource for both HSPA and UMTS. Regarding the capacity estimates for HSPA, it is recommended that you initially utilize an average throughput level expected instead of designing for peak levels. However if data usage is known for an area then adjustments for that part of the network should be made since you would potentially have to immediately add capacity to support the usage requirements. However the advantage of using average throughput levels in the initial planning is associated with the power levels needed for the HSPA.

In HSPA the *Radio Network Controller* (RNC) dimensioning is dependent upon the Node Bs that are connected to it. The RNC dimensions are based on not only the overall throughput requirements, but also on the number of Node Bs that are associated with the RNC. Additionally, the RNC dimensioning needs to account for the connectivity for the Iub, Iur, and Iu-CS interfaces.

With HSPA the introduction of carrier aggregation needs to be accounted for in the design process. Specifically, depending on where HSPA is deployed needs to factor in both current RF spectrum inventory and potentially future RF spectrum inventory.

Therefore HSPA can be deployed as a single-carrier or multiple-carrier deployment. However physics plays an important role in the decision of how to deploy multiple carriers with HSPA. Specifically, if you want to deploy two 5 × 5 RF channels using HSPA for carrier aggregation, the choice is either intraband or interband allocation. With intraband allocation the deployment is more straightforward in that the same antenna system can be used; however, you need to potentially refarm the spectrum since in all likelihood it is already being used with a legacy system.

If you are deploying in an interband method, then the higher frequency selected for the multicarrier deployment will determine the RF coverage of the Node B resulting in the potential impacts to the existing subscribers on the legacy network components using the lower-frequency band.

References

"Release 7 HSPA+ For Mobile Broadband Evolution," Qualcomm, 2007.

Smith, Gervelis, *Wireless Network Performance Handbook*, McGraw-Hill, New York, 2003.

3GPP TR 25.863: Universal Terrestrial Radio Access (UTRA); Uplink Transmit Diversity for High-Speed Packet Access (HSPA) (Release 2011).

3GPP TR 25.942: RF System Scenarios.

3GPP TS 23.002: Network Architecture (Release 1999).

3GPP TS 23.101: General UMTS Architecture (Release 1999).

3GPP TS 25.101: UE Radio Transmission and Reception (FDD).

3GPP TS 25.104: UTRA (BS) FDD; Radio Transmission and Reception. 3GPP TS 25.317: Universal Terrestrial Radio Access (UTRA); High-Speed Packet Access (HSPA); Requirements on User Equipments (UEs) Supporting a Release-Independent Frequency Band Combination (Release 2011).

3GPP TS 25.401: UTRAN Overall Description. 3GPP TS 25.401: UTRAN Overall Description (Release 2012).

www.3gpp.org.

"Understand HSPA: High-Speed Packet Access For UMTS," Qualcomm, November 2006. www.umtsworld.org.

CHAPTER 16

CDMA2000 System Design

The system design associated with *Code-Division Multiple Access* (CDMA2000) systems has multiple factors that are interwoven with each other. The CDMA2000 designer must account for the inclusion of packet-data services not only into the radio and fixed network access system but also into the legacy systems, variants to 1xRTT, and *Evolution Data Optimized* (EVDO). This chapter will attempt to quantify some of the more salient aspects with CDMA2000 system design, looking at several key scenarios or issues that need to be addressed when considering or expanding CDMA2000-compatible infrastructure within a wireless system. Throughout this chapter, CDMA2000 will refer to IS-2000 and IS-856 systems.

The topics covered in this chapter are

- Design criteria
- Traffic assumptions
- Link budgets
- Deployment issues
- Network node dimensioning

for the following three general types of systems:

- CDMA2000 1xRTT (greenfield)
- EVDO (IS-856) (greenfield)
- EVDO overlay onto CDMA2000-1xRTT

It will be assumed throughout this chapter that migration of legacy systems has been done successfully because it is not the intention of this chapter to cover the design aspects of the various legacy wireless systems. The reference section at the end of this chapter, however, has several excellent sources addressing legacy system design guidance and examples.

With this said, the key factor that needs to be addressed—but often one of the most difficult—is, What do you want to do? It is a simple question but one that has profound implications with (all too often) no real answer. The usual questions are more marketing and business questions that are intertwined with technical issues that exist in a wireless system. Therefore, through the examples listed next, it is hoped that the technical issues associated with decisions about service and network deployments can be better weighed, enabling a better implementation of this exciting technology platform.

16.1 Design Methodology

The methodology for network and *Radiofrequency* (RF) design for CDMA2000 needs to be established at the beginning of the process. Establishment of the methodology used for the formulation of the report is essential in the beginning stages to ensure that the proper baseline assumptions are used to facilitate flexibility in design and implementation. Flexibility is needed in the design and implementation to address many of the future issues that are really unknown and therefore cannot be foreseen properly at the onset of the design process.

Some of the issues that need to be identified at the beginning of study are

- Time frames for the growth plan
- Subscriber growth projections
- Services offered with take rates
- Design criteria
- Baseline system numbers for building on the growth study
- Construction expectations
- Legacy and future technology systems

The time frames for the growth plan must be determined at the beginning. The time frames will define what the baseline, foundation, and future look of the plan will present. Therefore, data from the baseline month or time frame associated with an existing system is critical for generating the plan because the wrong baseline will alter the outcome of the report.

The amount of time the plan projection is to take into account is also critical for the analysis. The decision to project 1 year, 2 years, 5, or even 10 years has a dramatic effect on the final outcome and accuracy of the forecast. In addition to the projection time frame, it is important to establish the granularity of the reporting period—monthly, quarterly, biannually, yearly—or some perturbation of them all.

The particular marketing plans also need to be factored into the report. The marketing department's plans are the leading element in any network and RF growth study. The basic input parameters to the network and RF growth plan provided by the marketing department are listed in the following:

- Projected subscriber growth for the system over the time frame
- Projected milli-Erlangs per subscriber expected at discrete time intervals
- Projected megabits per second (Mbps) per subscriber expected at discrete time intervals
- The subscriber voice usage over the time frame
- The dilution rates for legacy subscriber equipment over the time period
- The types of subscriber equipment used in the network and percent distribution of CPE projections
- Special promotion plans over the time frame of the report, such as free Internet access
- The projected amount of mobile data users over the time frame of the study

- Data services offered over the time period, with the take rates for each service type and treatment

There are a multitude of other items needed from the marketing department for determining network and RF growth. However, if you obtain the information on these basic nine topics listed previously from the marketing department, it will be enough to adequately start the RF, fixed network, voice, and packet design.

The design, whether it is for new or expansion systems, needs to factor in the following key elements along with the expected traffic-loading forecast:

- Spectrum available for use
- Spectrum required and methods for achieving the required bandwidth
- Flexibility to meet the ever-changing market conditions
- Cost-effective use of the existing and future capital infrastructure
- Standardized systems that enable backward- as well as forward-compatibility with other networks and data platforms
- *Quality of Service* (QoS)/*Grade of Service* (GoS) for each type of service offering
- Coverage requirements for either new coverage or enhancements to existing coverage

With the previous said, the next step is to establish some guidelines that are specific to CDMA2000 systems.

16.2 Deployment Guidelines

The deployment of CDMA2000 can and does have different faces presented to the designer depending on the situation the designer is trying to solve. If the design is for a new system (greenfield) the deployment is driven by coverage and then by capacity. On the other hand, if the design is for integrating CDMA2000 into an existing system, such as IS-95, then the design is more focused on capacity and possibly inclusion of packet-data services. Alternatively, if the operator is deploying EVDO into an existing 1xRTT network, which is an overlay, then coverage and capacity are some of the immediate considerations.

Because CDMA2000-1xRTT and EVDO occupy the same bandwidth as IS-95A/IS-95B, this obviously facilitates the introduction of this platform into that type of system. CDMA2000 can be deployed as a distinct carrier or shared carrier with IS-95 systems, leading to many possibilities that a design engineer can possibly utilize to achieve the desired design requirements. Some of the design options for integrating CDMA2000-1xRTT into an existing IS-95 system are shown in Table 16.1, with 1xRTT being the logical introduction or upgrade and with EVDO being deployed as an overlay.

For cellular, the F1 is the primary channel for the hunt, although for *Personal Communications Services* (PCS), it is the first channel in the channel-selection sequence provided by the operator. The CDMA2000 channel-assignment scheme shown earlier is revisited in Table 16.2 for cellular systems and Table 16.3 for U.S. PCS systems. The asterisks in both tables represent the preferred *Third-Generation* (3G) F1 carrier recommendation. Presently, 3X configurations, also called *Multicarrier* (MC) as defined in IMT2000 are not being pursued in the standard's bodies at this time. However, with

Option	Method	Advantages	Disadvantages
1	Deploy 1xRTT across all F1 channels.	Maximizes service footprint for packet-data services. Seamless 1xRTT service.	Method is capital-intensive and depends on the penetration of CDMA2000 handsets, which limits full use of services.
2	Deploy 1xRTT on any CDMA channel other than F1.	Focused on high-capacity locations.	Limited service area for high-speed packet data. Hard handoff with IS-95.
3	Deploy 1xRTT in own spectrum.	Provides additional capacity without affecting IS-95.	Spectrum-clearing requirements. Possible reallocation of traffic.
4	Deploy 1xRTT on all CDMA channels.	Full potential for packet and voice services realized.	Extremely capital-intensive, except when only one CDMA channel is operational.

TABLE 16.1 CDMA2000-1xRTT Deployment Schemes

Cellular System Carrier	Sequence	A	B
1	F1	283	384
2	F2	242*	425*
3	F3	201	466
4	F4	160	507
5	F5	119	548
6	F6	78	589
7	F7	37	630
8	F8 (Not advised)	691	777

TABLE 16.2 Cellular CDMA2000-1x Carrier-Assignment Scheme

PCS System Carrier	A	B	C	D	E	F
1	25	425	925	325	725	825
2	50	450	950	350*	750*	850*
3	75*	475*	975*	375	775	875
4	100	500	1000	NA	NA	NA
5	125	525	1025	NA	NA	NA
6	150*	550*	1050*	NA	NA	NA
7	175	575	1075	NA	NA	NA
8	200	600	1100	NA	NA	NA
9	225*	625*	1125*	NA	NA	NA
10	250	650	1150	NA	NA	NA
11	275	675	1175	NA	NA	NA

TABLE 16.3 PCS CDMA2000-1x Carrier-Assignment Scheme

regards to EVDO Rev B the use of multicarriers is essential to achieve the desired throughput envisioned. Therefore, it is important to factor in the possible inclusion or exclusion of EVDO in the initial system design and, of course, the relative location of the particular 1x carriers envisioned so that it facilitates a logical expansion and migration with future releases.

When introducing CDMA2000 into an existing IS-95 system, there is a need to upgrade different network elements depending on the radio infrastructure supplier utilized by the operator. The upgrade from IS-95 to CDMA2000 requires new network elements, namely, the *Packet Data Service Node* (PDSN) and *Authentication Authorization Accounting* (AAA). However, regardless of which radio vendor is chosen or used, all the existing CDMA *Base Transceiver Stations* (BTSs) need some type of modification or upgrade. Additionally, upgrading a base station from 1xRTT to EVDO involves platform changes at both the BTS and *Base Station Controller* (BSC). With CDMA2000-1xRTT, owing to enhancements in modulation schemes, as well as vocoders, a net voice capacity gain of 1.5 times in the reverse link and 2 times in the forward link is seen over that of 8-kbps *Enhanced Variable Rate CODEC* (EVRC). Not only is the 1xRTT platform meant for improvements in overall voice system capacity, but as mentioned many times, the introduction of packet data is the driving force for CDMA2000 to be deployed in an existing network because 1xRTT is able to support both packet- and circuit-switched services. The introduction of EVDO into a CDMA2000 network greatly facilitates the delivery of packet-data services.

Because 1xRTT and EVDO support packet data, the estimation of packet data is essential. The average packet-data user is expected to use the service for the following services:

- E-mail—65 percent
- Web browsing—30 percent
- Extension of company network (LAN)—27 percent
- Address book/calendar functions—27 percent

The use of EVDO over 1xRTT has many advantages previously covered in Chapter 11. The proliferation of 1xRTT devices in comparison to EVDO, and the fact that 1xRTT supports circuit-switched voice services have operators deploying 1xRTT for voice switched and EVDO for data services.

The expected migration path either for a new CDMA2000 system, for an upgrade from IS-95, or for one system that chooses to bifurcate its network is to first deploy CDMA2000-1xRTT and then to overlay an EVDO system on top of it.

The 1xEVDV versions of CDMA2000 are no longer under development at this time. CDMA2000-1xEV-DO is a data-only service that has a wide deployment, whereas 1xEV-DV is a data and voice offering that allows for higher throughput for data services while sharing resources for voice services. EVDO can be overlaid on top of a CDMA2000-1xRTT system; it was designed specifically to enable an EVDO system to be also overlaid on existing IS-95 systems.

Regarding EVDV 3X systems, they were envisioned to be overlay systems. In 3X, the forward link breaks up the data into three carriers, each of which is spread 1.2288 Mcps, hence the term *Multicarrier* (MC). The reverse link in 3X uses three aggregated 1x carriers that have a combined carrier spread of 3.6864 Mcps.

	Downlink	Uplink
Rev 0	2.4 Mbps	144 kbps
Rev A	3.1 Mbps	1.8 Mbps
Rev B	73.5 Mbps	27 Mbps
Rev C	129 Mbps	75.6 Mbps

TABLE 16.4 EVDO

16.2.1 1xRTT

This is the initial deployment for CDMA2000 involving a single carrier and is typically referred to as *CDMA2000 phase 1*. The 1xRTT system introduces the use of packet- data services for wireless operators. The 1xRTT system utilizes an SR1 and will transport both voice and packet data over the same physical resources.

The data rates for 1xRTT are a maximum of 153 kbps for the downlink and 64 kbps for the uplink. The specification allows for higher data rates, but the aforementioned values are what has been deployed in the field.

16.2.2 1xEV-DO

1xEV-DO is the terminology used to describe non-real-time, high-packet-data services that will be offered on an SR1 channel transporting packet data only, hence the name *DO*. The objective behind deploying a 1xEV-DO service is to enable a higher number of users of the system to utilize packet-data services. Separating voice users from data users onto two carriers, 1xRTT for voice and EVDO for data, results in higher data rates for users as well as a higher throughput per carrier.

The data rates envisioned for 1xEV-DO are listed in Table 16.4.

16.2.3 1xEVDV

1xEVDV is the last evolution expected for a CDMA2000-1xRTT platform. The 1xEVDV system will enable both voice and packet-data services to share the same resource FA, similar to 1xEV systems, but they have the packet-data throughput associated with 1xEVDO systems.

16.3 System Traffic Estimation

Traffic estimation for a CDMA2000 system depends directly on the type and quality of services that will be offered and how they will be transported. The traffic estimation process involves not only the radio link but also the other fixed facilities that make up the network.

The process and methodology for conducting system traffic engineering for a CDMA2000 system with 1xRTT and/or EVDO involves determining the amount of physical and logical resources that need to be in place at different points and nodes within the network to support current and future traffic. Determination of existing traffic loads is rather more straightforward in that you have existing information from which to make decisions. For future forecasts, the level of uncertainty grows exponentially the further the forecast or planning takes you into the future. However, many elements in the network require long lead times, ranging from 3 weeks to over 1 year to implement. Obviously, the goal of traffic engineering is to design the network and its subcomponents not only to meet the design criteria, which should be driven by technical, marketing, and sales information, but also to do so in such a way that the design is achieved in a cost-effective manner. It is not uncommon to have conflicting objectives within a design, that is, to ensure that the customers have the highest QoS/GoS for both voice and packet data but yet have a limited amount of capital by which to achieve this goal. Therefore, it is important to define the design objective at the onset of the design process and have some interim decision points where the design can be reviewed and altered, if required, either by increasing the capital budget, revisiting the forecast input, or altering the QoS/GoS expectations.

Because there are potentially a number of variations to the circuit- and packet-switched services offered, it is important to recognize the commonalties that can be drawn on. Additionally, if a 1xRTT and EVDO network is deployed, then the type of traffic and which system it will reside on need to be determined. Remember that when outside the EVDO service area, a subscriber device can migrate down to a 1xRTT network but cannot migrate back.

Several methods can be used for calculating or estimating network traffic.

It is essential to note that there are several key points within the network where the traffic engineering calculations need to be applied:

- BTS-to-subscriber terminal
- BTS to BSC
- BSC-to-packet network
- BSC-to-voice network

There are several situations and an unknown level of perturbations that can occur in the estimation of traffic for a system. In an ideal world, the traffic forecast would be projected by integrating the marketing plan with the business plan, coupled with the products that should be integral to both the marketing and business plans. However, reality is much harsher, and usually, very little information is obtainable by the technical team from which to dimension a network. Therefore, the following is meant to help steer system planners in determining their traffic-transport forecast.

Two key concepts about packet data: (1) initially, packet-data traffic is low owing to the penetration rate (the higher data speed is a result of the data not being as time-sensitive as voice), and (2) packet-data services are an enabler for more services offered by the operator.

The forecast would be much more simplified if the system were operational because there would be real traffic information as well as a minimal set of products. The forecast, or growth, could be extrapolated from the business plan or simplified marketing plans, which would specify a desired growth level.

The equation to follow for an existing system would be

$$\text{Total traffic} = \text{existing traffic} + \text{new traffic expected}$$

The new traffic expected could be a simple multiplication of the existing traffic load.

If the system is new or a new service/platform is to be introduced, then the traffic needs to be distributed in a weighted proportion to each of the markets being designed for the system or homogeneously distributed for a given market or submarket.

Symbol	Service Type	Transport Method
S	Voice	Circuit switch
SM	Short message	Packet
SD	Switched data	Circuit switch
MMM	Medium multimedia	Packet
HMM	High multimedia	Packet
HIMM	High interactive multimedia	Packet

TABLE 16.5 Circuit and Packet Data

The forecasting for voice traffic and packet services, as discussed in Chapter 11, is well-documented and will get only superficial treatment here. However, the real issue with traffic dimensioning is the ability to forecast both the circuit-switched and new packet services related to 1xRTT and EVDO that will be used by customers of the wireless operator.

The ultimate question that the designer must answer is, "How do you plan on supporting the traffic with the prescribed services?"

There are numerous types of services available for both circuit-switched and packet data, and some generalizations need to be made in order to have a chance at arriving at the conclusions necessary for input into the design phase.

For a new or existing system, the issue of where to begin is always the hardest part. However, one of the key parameters that you need to determine from marketing and/or sales is the penetration rate; the take rate for each of the service types offered in 1xRTT and/or EVDO can be categorized into the segments defined in Table 16.5. Please note that voice traffic also can be *Multimedia* (MM) or *High Multimedia* (HMM) with *Voice over IP* (VoIP) for EVDO.

Traffic estimation for both circuit and packet services can be achieved by several methods, such as a general approach where a standard percent is used for, say, packet services. Or, you could base the amount of packet-data subscribers on the number of handsets expected to be purchased for resale in the market.

Regardless, the first step in any traffic study is to determine the population density for a given market; in the case of an existing system, the population density and primary penetration rates are already built into the system owing to known loading issues. However, especially for new services such as packet data, the process of determining the population density for a given area, followed by the multiplication of this by the penetration rate, will help greatly in determining the expected traffic load on which to base the system design.

For the traffic forecast, the following marketing attributes need to be defined to the best level of granularity and as clearly as possible:

- Population density
- Penetration rate
- Coverage area
- Quality of service (QoS)/service-level agreement (SLA)

With the introduction of packet data with CDMA2000-1xRTT and EVDO, the traffic modeling for packet-switched data involves the interaction of the following items:

- Number of packet bursts per packet session
- Size of packets
- Arrival time of packet burst within a packet session
- Arrival times for different packet sessions

Packet usage is evolving with improvements in RANs and devices for wireless mobility systems on a mass-market basis. The issue of where, when, and how much do you dimension a system for packet data always will be a debate between marketing and technical teams. However, in light of the fact that packet-data usage is still in its infancy, there is little guidance from which to go forth and design the network. However, ITU-R M.1390, which gives a methodology for calculating the spectrum requirements for IMT-2000, has some guidelines for data dimensioning, and the following tables are extracted from that specification. The values in the tables should be used as a guide to establish packet loading for dimensioning when market-specific data are not available for many reasons.

It is important to note that all the circuit-switched services defined previously in Table 16.5 are either symmetric or asymmetric. Of the services listed previously, only MMM and HMM are asymmetric; the rest are symmetric service offerings.

16.4 Radio Elements

Because CDMA2000 is a radio access platform, it stands to reason that the driving force for dimensioning the network to meet customer demands is to ensure that the radio system is dimensioned accordingly. The radio elements, which include the base radius, BTS, channel elements, and BSCs, directly influence the circuit-switched and packet-data network requirements.

A few key elements are associated with radio dimensioning for a CDMA2000 system whether it is 1xRTT or EVDO. These key elements are

- Spectrum
- Channel-assignment scheme
- Site configuration (antennas)
- Channel elements (1xRTT)
- Link budget

The spectrum requirements for a CDMA2000-1xRTT or EVDO network, of course, depend directly on the number of radio channels required to meet the current or expected demand. In addition, the channel-assignment scheme that is used will have a direct impact on the expandability of 1xRTT and or EVDO in the future, as well as optimal spectrum management of the existing system.

16.4.1 Antenna Configurations

The configurations for the CDMA2000 sites can and do take advantage of many of the IS-95 lessons learned through the deployment phases. Taking a simplistic view of CDMA2000 antenna requirements, a total of two receive antennas (or paths) are needed

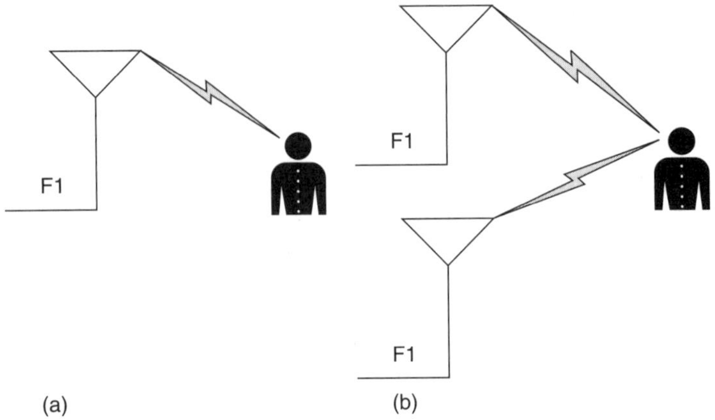

FIGURE 16.1 CDMA2000 Tx configurations: (*a*) single antenna; (*b*) Tx diversity (STD).

per sector, as was the case with IS-95 systems. Figure 16.1 illustrates the requirement for a single CDMA2000 1xRTT or EVDO *transmission* (Tx) channel.

Figure 16.1 addresses two issues with 1xRTT deployments: to utilize or not to utilize transmit diversity. The figure shows that for a CDMA2000-1xRTT a single Tx antenna is needed for a CDMA2000-1xRTT carrier; however, the figure also shows that two antennas are needed for the *Single Transmit Diversity* (STD) method. The transmit diversity scheme has technical advantages that can be exploited by the system operator, but at the cost of deploying or using a second antenna for Tx diversity. Normally, this is not an issue with a CDMA carrier because there are usually two antennas or a cross-pole is used, and the duplexers provide the dual path. Where the rub comes is when a second carrier is deployed, and unless the link budget shows the splitting loss that can be accommodated, more antennas will need to be added to the system or sector.

Tx diversity also can be deployed with a single antenna used for transmit following the *Orthogonal Transmit Diversity* (OTD) method.

16.4.2 BTS

The BTS controls the interface between the CDMA2000 network and the subscriber unit. The BTS controls many aspects of the system that are directly related to network performance. Some of the items the BTS controls are the multiple carriers that operate from the site, the forward power (allocated for traffic, overhead, and soft handoffs), and of course, assignment of the Walsh codes.

With CDMA2000 systems, the use of multiple carriers per sector, as with IS-95 systems, is possible. Therefore, when a new voice or packet session is initiated, the BTS must decide how to best assign the subscriber unit to meet the demands of the services being delivered. The BTS in the decision process not only examines the service requested but also must consider the radio configuration and the subscriber type, as well and, of course, whether the service requested is voice or packet. Therefore, the resources the BTS has to draw on can be both physically and logically limited, depending on the particular situation involved.

The following is a brief summary of some of the physical and logical resources the BTS must allocate when assigning resources to a subscriber:

- *Forward Channel* (FCH) Number of physical resources available
- *FCH forward power* Power already allocated and that which is available
- *Walsh codes* required and those available
- *Total FCHs* used in that sector

For 1xRTT systems, the physical resources the BTS draws on also involve management of the channel elements that are required for both voice and packet-data services. Although discussed in more detail later, handoffs are accepted or rejected on the basis of available power only.

Integral to the resource-assignment scheme is Walsh code management, covered below in more detail. For 1xRTT, EVDO, and EVDV, a total of 128 Walsh codes can be drawn on. However, the Walsh codes are expanded to a total of 256 in later releases initially associated with 3X.

For CDMA2000-1xRTT, the voice and data distribution is handled by parameters that are set by the operator and involve

- *Data resources* Percent of available resources, which includes FCH and SCH
- *FCH resources* Percent of data resources
- *Voice resources* Percent of total available resources

These are best described by a brief example to help clarify the issue of resource allocation shown in Table 16.6. Obviously, allocation of data/FCH resources directly controls the number of simultaneous data users on a particular sector or cell site. With Table 16.6 it is assumed that there are only 64 resources to draw upon.

16.4.3 Channel Element (CE) Dimensioning

The CE dimensioning for 1xRTT obviously will be based on the requirements for both voice and data services. The total number of CEs required will be the summation of both the fundamental and supplemental CEs defined for voice and data services.

The new CE that is being offered by all the major vendors is compatible with the existing IS-95 system and can be substituted directly for an existing CE. However, as discussed later, the full replacement of all CEs is not practical; based on the deployment options, active IS-95 legacy systems should be left in place.

For simplification, a CE is required for each

- Voice cell
- Leg of the soft handoff

Topic	Percentage	Resources
Total resources		64
Voice resources	70%	44
Data resources	30%	20
FCH resources	40%	8

TABLE 16.6 Carrier Resource-Allocation Example

- Overhead channel
- Data call

The dimensioning of the CEs is done in increments of 32 or 64 for CDMA2000-1x RTT-capable CEs. Because CEs typically come in 32/64/96/128 cards, this leads to the fact that if 20 CEs are required, a 32-CE card is acquired. Although, in the same sense, if 33 CEs are needed, a choice needs to be made to either underequip or obtain another CE card to bring the count to 64 when only 33 are needed:

- 32 CEs for 13 to 17 percent of CDMA2000-1xRTT full capacity
- 64 CEs for 29 to 44 percent of CDMA2000-1xRTT full capacity

CDMA2000-1xRTT full capacity is derived based on a fixed environment and availability of 128 Walsh codes. Obviously, the percentages shown previously depend on the mix of voice and data traffic within the system, as well as their mutual interference.

A rule of thumb to follow is that for sites requiring fewer than 40 CEs, a 32-element card should be used, but for sites requiring more than 40 CEs, then the 64-element CE card is expected to be used.

Larger CE cards are now available, and pooling of CEs between sectors (same FA), is now standard, enabling a more efficient use of CE resources.

16.4.4 Packet-Data Services (RF Environment)

Packet-data services have different implications when introduced into a radio environment as compared with a fixed-network environment. More specifically for the radio link, how packet data are handled depends on whether the radio link is sharing its resources with voice services or is a data-only sector use. The data-only possibility with CDMA2000 1xEVDO packet-data services is permitted only on the new channel and is deployed as an overlay. However, regardless of this issue, when involved with the wireless link, data services are still a best effort. In addition, signaling traffic has higher priority than voice traffic, but voice and circuit-switched services have a higher priority than packet data.

Regarding packet-data resource dimensioning, it is important to remember that the packet session is considered active when data are being transferred. During this process, a dedicated FCH and/or DCCH for traffic signaling and power control exists between the mobile unit and the network. In addition, the high-speed supplemental channel can be used for large data transfers. An important issue that needs to be considered in the allocation of system resources is that while the session is active, channel elements, as well as Walsh codes, are consumed by the subscriber and are system-independent if data actually are being transferred.

The packet session is considered dormant when no data is being transferred, but a PPP link is maintained between the *Packet Data Server Network* (PDSN) and the subscriber. It is important also to note that no system resources are consumed relative to channel elements or Walsh codes while the packet session is considered dormant in both 1xRTT and EVDO.

16.5 Fixed Network Design Requirements

The introduction of packet-data services for a CDMA2000 system requires a focus not only on the radio environment but also on all the supporting elements that make up the wireless system. Therefore, the following are the major elements that need to be factored into the design of a CDMA2000 system:

- Mobile Switching Center (MSC)
- Base-Station Controller (BSC)
- Base Transceiver Station (BTS)
- Packet Data Service Node (PDSN)

In reviewing this list, several nodes or elements directly associated with radio elements are listed. The reason the BSC and BTS are listed in the fixed network design requirements is the simple fact that connectivity needs to be established between the BTS and the BSC, whether it is via landline services or a microwave link. The BSCs are listed not only because they route packet and voice traffic, which requires a certain link dimensioning, but also because the BSCs can be local or remote to the MSC depending on the network configuration deployed.

The fixed network design includes not only element dimensioning but also dimensioning of the links that connect the various nodes or elements of the wireless system. Some of the connectivity requirements involve the following elements. It is important to note that all elements require some level of connectivity whether it is from the *Digital Cross Connect* (DXX), also referred to as *DACS*, to the MSC or between the voice-mail platform and the switch. However, the list that follows involves elements that usually require an external group with which to interface, whereas the internal nodes are more controllable.

- Link between BTS and BSC (usually a leased line)
- Link between BSC and MSC (if remote)
- BTS/BSC concentration method
- Connectivity of the PDSN
- Router (for internal)
- Router (for external)
- AAA
- HA
- Service Creation Server (SCS) and other servers
- Interconnection to public and/or private data networks

Note as a general practice that the routers used for the packet-data network for 1xRTT and EVDO applications should not be used for other functions, such as company *Local-Area Network* (LAN) work.

16.5.1 PDSN

The PDSN needs to have connectivity with the following major nodes:

- CDMA radio network
- Either a public data network, private data network, or both
- AAA server

- Dynamic Host Configuration Protocol (DHCP) server
- Service-creation platform that contains the configuration, policy, profile, subprovisioning, and monitoring capability

The PDSN usually is connected to the packet-data network via an OC3/STM1 or 100baseT connection. The choice of which bandwidth to use is determined not only by the proximity of the PDSN to the BSC but also by traffic requirements. In summary, the PDSN design is based on many factors, including the following:

- Number of BSC locations
- Access type supported (simple IP, mobile IP, and so on)
- Connectivity between nodes
- Network performance requirements
- Routing areas

16.5.2 Packet Zone

For network layout, several zones can be assigned within a PDSN network. These zones are referred to as *packet zones* and should be distributed via the same deployment and logical assignment method used for assigning BSCs. In other words, every BTS connected to a particular BSC should have the same packet zone assigned to it. However, it is possible to have several BSCs residing in the same packet zone, but it is recommended that a separate packet zone be assigned to every BSC.

16.5.3 Design Utilization Rates

The facility utilization goal for the network should be 70 percent of capacity for the line rate over the time period desired. The time period that should be used is a 9-month sliding window that needs to be briefly revisited on a monthly performance report and then during a quarterly design review following the design review guideline process.

The facility will need to be expanded once it is understood that the 70 percent utilization level will be exhausted within 9 months, with continued growth showing 100 percent within 18 months.

For the packet-switched network, the following should be used as the general guidelines:

- *Processor occupancy* 70 percent
- *Switching platform* SCR (25 percent), PCR, (90 percent)
- *Port capacity* Design for 70 percent port utilization (growth projection for additional ports based on 9-month forecast)

Naturally, buffer dimensioning, as well as throughput, is also an important design and dimensioning requirement.

16.5.4 IP Addressing

The issue of *Internet Protocol* (IP) addressing is important to a CDMA2000 system design. The introduction of Simple IP and Mobile IP with and without a *Virtual Private Network* (VPN) requires the use of multiple IP addresses for successful transport of the packet services envisioned. It is therefore imperative that the IP addresses used for the network be approached from the initial design phase to ensure a uniform growth that is logical and easy to maintain over the life cycle of the system.

Not only does the introduction of packet data require an IP address scheme for the mobility portion of the system, but each of the new platforms introduced also needs to have its own IP address or range of IP addresses. Some of the platforms requiring IP addresses involve

- PDSN
- FA
- HA
- Routers

Some of these new devices require the use of private addresses as well as some public addresses. However, because the range of perturbations for IP address schemes is so vast and requires a specific look at how the existing network is set up, factoring into the mix the desires for the future, a generic discussion on IP address schemes will follow.

Use of the IPv4 format is shown in the following discussion. IPv6 or IPng is the next generation; it enables for QoS functionality to be incorporated into the IP offering. However, this discussion will focus on IPv4 because it is the protocol used today and has legacy transparency for IPv6.

Every device that wants to communicate using IP needs to have an IP address associated with it. The addresses used for IP communication have the following general format:

Network number Host number

Network prefix Host number

There are, of course, public and private IP addresses. The public IP addresses enable devices to communicate using the Internet, whereas private addresses are used for communication in a LAN/WAN intranet environment. The CDMA2000 system will use both public and private addresses. However, the bulk of the IP addresses will be private in nature and, depending on the service offering, will be dynamically allocated or static in nature.

Table 16.7 represents the valid range of public IP addresses that can be used. The private addresses will not be recognized on the public Internet system, and this is why they are used. Also, it will be necessary to reuse private addresses within sections of the network, profound as this may sound. Because the packet system is segregated based on the PDSN, each PDSN can be assigned the same range of IP addresses. Additionally, based on the port involved with the PDSN, the system can be segregated into localized nodes, and the segregation enables the reuse of private IP addresses, ensuring a large supply of a seemingly limited resource.

Network Address Class	Range
A (/8 prefix)	1.xxx.xxx.xxx through 126.xxx.xxx.xxx
B (/16 prefix)	128.0.xxx.xxx through 191.255.xxx.xxx
C (/24 prefix)	192.0.0.xxx through 223.255.255.xxx

TABLE 16.7 Public IP Addresses

Private Network Address	Range
10/8 prefix	10.0.0.0 through 10.255.255.255
172.16/16 prefix	172.16.0.0 through 172.31.255.255
192.168/16 prefix	192.168.0.0 through 192.168.255.255

Table 16.8 Private IP Addresses

Mask	Effective Subnets	Effective Hosts
255.255.255.192	2	62
255.255.255.224	6	30
255.255.255.240	14	14
255.255.255.248	30	6
255.255.255.252	62	2

Table 16.9 Subnets

The private addresses that should be used are shown in Table 16.8.

To facilitate the use of IP addressing, the use of a subnet further helps to refine the addressing by extending the effective range of the IP address itself. The various subnets are defined in Table 16.9. The IP address and its subnet directly affect the number of subnets that can exist, and from those subnets, the number of hosts that can also can be assigned to that subnet is determined.

It is important to note that the IP addresses assigned to a particular subnet include not only the host IP addresses but also the network and broadcast addresses. For example, the 255.255.255.252 subnet that has two hosts requires a total of four IP addresses to be allocated to the subnet: two for the hosts, one for the network, and the other for the broadcast address. Obviously, as the number of hosts increases with a valid subnet range, the more efficient the use of IP addresses becomes. For instance, the 255.255.255.192 subnet enables for 62 hosts and utilizes a total of 64 IP addresses.

Therefore, you might say, Why not use the 255.255.255.192 subnet for everything? However, this would not be efficient either, so an IP-address plan needs to be worked out in advance because it is extremely difficult to change once the system is being or has been implemented.

Just what is the procedure for defining the IP addresses and their associated subnets? The following rules apply when developing the IP plan for the system; the same rules are used for any LAN or *Internet Service Provider* (ISP) that is designed. There are four basic questions that help to define the requirements:

1. How many subnets are needed presently?
2. How many are needed in the future?
3. How many hosts are on the largest subnet presently?
4. How many hosts are on the largest subnet in the future?

You might be wondering why the use of multiple hosts should be factored into the design phase for CDMA2000. The reason is that it is possible to have several terminals for a fixed application using a single CDMA2000 subscriber unit or fixed unit.

Therefore, using the previous methods, an IP plan can be formulated for the wireless company's packet-data platforms. It is important to note that the IP plan should factor into the design not only the end customers' needs but also the wireless operators' needs.

Specifically, the CDMA2000 operators' needs will involve IP addresses for the following platforms at a minimum. The platforms requiring IP addresses are constantly growing as more and more functionality for the devices is done through *Simple Network Management Protocol* (SNMP).

- Base stations
- Radio elements
- Microwave point to point
- Subscriber units
- Routers
- ATM switches
- Work stations
- Servers (e.g., AAA, HA, FA, and PDSN)

The list can and will grow when you tally up all the devices within the network both from a hardware and network management aspect. Many of the devices listed previously require multiple IP addresses in order to ensure their functionality of providing connectivity from point *A* to point *B*. It is extremely important that the plan follow a logical method. Some CDMA2000 network equipment may also may require an IP plan that incorporates the entire system and not just pieces.

A suggested methodology is to:

- List out all the major components that are, will be, or could be used in the network over a 5- to 10-year period.
- Determine the maximum number of these devices that could be added to the system over 5 to 10 years.
- Determine the maximum number of packet-data users per BSC.
- Determine the maximum number of packet-data users per PDSN.
- Determine the maximum number of Mobile IP users with and without a VPN.
- Determine the maximum number of Simple IP users with and without a VPN.

The reason for the focus on the amount of Simple- and Mobile-IP users lies in the fact that these devices will have the greatest demand for IP addresses due to their sheer volume in the network.

Naturally, each wireless system is unique and will require implementation of a different IP address scheme to be implemented. However, the concept presented has been beneficial and should prove useful. If you need more information on IP addressing schemes, an excellent source for such information is available on the Web at www.cisco.com.

16.6 Traffic Model (1xRTT)

The capacity for a CDMA2000-1xRTT cell site is determined through the interaction of several parameters and is driven by the radio access portion of the system, provided that the fixed network has the requisite number of modules for each platform. The parameters for determining the traffic load at a CDMA2000-1xRTT site are similar to those used for an IS-95 system with the exception that CDMA2000-1xRTT introduces packet data and the inclusion of 128/256 Walsh code, to mention a few of the previously covered issues.

As with IS-95 base stations, the use of CE cards is essential for the handling of traffic, whether it is for voice or data. The desired result from traffic engineering for a CDMA2000-1xRTT base station is to be able to determine the number of CE cards required to support the expected traffic. Another factor that fits into the traffic calculations for the site involves system noise. There is a simple relationship between system noise and the capacity of the cell site. Typically, the load of the cell-site design is somewhere in the vicinity of 40 to 50 percent of the pole capacity, with a maximum of 75 percent.

The next major element in determining the capacity of a CDMA cell is the soft and softer handoff factor. Because CDMA2000-1xRTT, like IS-95, relies on soft and softer handoffs as part of the fundamental network design for the network, this must also be factored into the usable capacity of the site. The reason for factoring soft and softer handoffs into capacity is that if 35 percent of the calls are in a soft handoff mode, then this will require more CE elements to be installed at the neighboring cell sites to keep the capacity at the desired levels.

The pole capacity for CDMA is the theoretical maximum number of simultaneous users that can coexist on a single CDMA carrier. However, at the pole, the system will become unstable, and therefore, operating at less than 100 percent of the pole capacity is the desired method of operation. Typically, the design calls for 50 percent of the pole capacity for the site.

However, because soft handoffs are an integral part of CDMA, they also need to be included in the calculation for capacity. In addition, for each traffic channel that is assigned for the site, a corresponding piece of hardware is needed at the cell site.

The actual traffic channels for a cell site are determined using the following equation:

Actual traffic channels = (effective traffic channels + soft handoff channels)

The maximum capacity for a CDMA cell site should be 75 percent of the theoretical limit. Unlike IS-95 systems, which are power-limited, the CDMA2000 system is Walsh code–limited.

16.6.1 Walsh Codes

Reiterating, the utilization of Walsh codes has a direct impact upon the radio network's ability to carry and transport the various services. With the introduction of CDMA2000, there are several alterations to the use of Walsh codes that were discussed previously, although only briefly.

With CDMA2000, the Walsh codes now have variable lengths that range from 4 to a total of 256, which is an expansion over IS-95 systems, which had only 64 codes. The one effect of utilizing variable-length Walsh codes is that if a shorter Walsh code is being used, then it precludes the use of the longer Walsh codes that are derived from it. For instance, if Walsh code 2 is used, then it precludes the use of all the Walsh codes in the code tree that were derived from it.

Table 16.10 helps in establishing the relationship between which Walsh code length is associated with a particular data rate. However, how does this play into the use of determining the radio network?

Walsh Code Tree								
	RC	256	128	64	32	16	8	4
SR1	1	Na	Na	9.6	Na	Na	Na	Na
	2	Na	Na	14.4	Na	Na	Na	Na
	3	Na	Na	9.6	19.2	38.4	76.8	153.6
	4	Na	9.6	19.2	38.4	76.8	153.6	307.2
	5	Na	Na	14.4	28.8	57.6	115.2	230.4
SR3	6		9.6	19.2	38.4	76.8	153.6	307.2
	7	9.6	19.2	38.4	76.8	153.6	307.2	614.4
	8		14.4	28.8	57.6	115.2	230.4	460.8
	9	14.4	28.8	57.6	115.2	230.4	460.8	1036.8

TABLE 16.10 Walsh Codes

For an SR1 and RC1, there are a maximum number of users that have individual Walsh codes equating to 64, a familiar number from IS-95A. However, if we were to have a Release 3–capable base radio with an SR1, phase 1 CDMA2000, and we had a total of 12 RC1 and RC2 mobile units in that sector, then this would allow for only three data users at 153.6 kbps, or 6 at 76.8 kbps, 13 at 38.4 kbps, 26 at 19.2 kbps, or 104 at 9.6 kbps. Obviously, the negotiated mobile data rate complicates the determination for the total throughput of traffic levels. The real issue behind this is that the type of data that will be enabled to be transported over the network has a direct impact on the available users. If, for example, the need were for high-speed data for interactive video with a Release 3–capable mobile unit, 384 kbps of bandwidth would not be feasible.

At times, it is best to see the preceding example in a visual format in order to better understand the relationship between the short and long Walsh codes. Table 16.11 shows the Walsh tree for 4 to 256 Walsh codes and their relative relationship with one another; the relationship is illustrated in Table 16.12.

Table 16.12 is an illustration of the interaction of a Walsh code and that of its higher or lower branches. The Walsh codes that are consumed are depicted in the shaded area. Also, the use of Walsh codes for the various channels that are associated with CDMA2000 are not included here because they too draw upon the same Walsh code pool. For ease of illustration, they were left out for the example.

In the example shown in Table 16.12, the use of Walsh code 48, which is CDMA2000-capable, and is set up for low-speed packet data and voice applications, precludes the use of high-speed packet data based on utilizing this set of Walsh codes, thereby effectively reducing the site's data-handling capability by 25 percent with the use of a single voice call. Alternatively, the use of a single high-speed data session using Walsh code 1 eliminates from possible use a total of 64/32 Walsh codes. Now, there is a difference between both examples: The first is that the data session will end sooner, at least in concept, than the voice call, thereby replenishing the Walsh code pool.

Another very important issue regarding the Walsh code pool and 3X is that with the 3X channels, the same Walsh code is used for all three carriers associated with the 3X platform.

Therefore, based on the expected traffic mix that is anticipated for the system, the choice of how to deploy the services relative to the carriers is important. If there is a

256	128	64	32	16	8	4
0	0	0	0	0	0	0
128						
64	64					
192						
32	32	32				
160						
96	96					
224						
16	16	16	16			
144						
80	80					
208						
48	48	48				
176						
112	112					
240						
8	8	8	8	8		
136						
72	72					
200						
40	40	40				
168						
104	104					
232						
24	24	24	24			
152						
88	88					
216						
56	56	56				
184						
120	120					
248						
4	4	4	4	4	4	
132						
68	68					
196						
36	36	36				
164						

TABLE 16.11 Walsh Code Tree

256	128	64	32	16	8	4
100	100					
228						
20	20	20	20			
148						
84	84					
212						
52	52	52				
180						
116	116					
244						
12	12	12	12	12		
140						
76	76					
204						
44	44	44				
172						
108	108					
236						
28	28	28	28			
156						
92	92					
220						
60	60	60				
188						
124	124					
252						
1	1	1	1	1	1	1
129	1					
65	65					
193						
33	33	33				
161						
97	97					
225						
17	17	17	17			
145						
81	81					

Table 16.11 Walsh Code Tree (*Continued*)

256	128	64	32	16	8	4
209					5	
49	49	49				
177						
113	113					
241						
9	9	9	9	9		
137						
73	73					
201						
41	41	41				
169						
105	105					
233						
25	25	25	25			
153						
89	89					
217						
57	57	57				
185						
121	121					
249						
5	5	5	5	5		
133						
69	69					
197						
37	37	37				
165						
101	101					
229						
21	21	21	21			
149						
85	85					
213						
53	53	53				
181						
117	117					
245						
13	13	13	13	13		

TABLE 16.11 Walsh Code Tree (*Continued*)

256	128	64	32	16	8	4
141						
77	77					
205						
45	45	45				
173						
109	109					
237						
29	29	29	29			
157						
93	93					
221						
61	61	61				
189						
125	125					
253						
2	2	2	2	2	2	2
130						
66	66					
194						
34	34	34				
162						
98	98					
226						
18	18	18	18			
146						
82	82					
210						
50	50	50				
178						
114	114					
242						
10	10	10	10	10		
138						
74	74					
202						
42	42	42				
170						

Table 16.11 Walsh Code Tree (*Continued*)

256	128	64	32	16	8	4
106	106					
234						
26	26	26	26			
154						
90	90			10		
218						
58	58	58				
186						
122	122					
250						
6	6	6	6	6	6	
134						
70	70					
198						
38	38	38				
166						
102	102					
230						
22	22	22	22			
150						
86	86					
214						
54	54	54				
182						
118	118					
246						
14	14	14	14	14		
142						
78	78					
206						
46	46	46				
174						
110	110					
238						
30	30	30	30			
158						
94	94					
222						

TABLE 16.11 Walsh Code Tree (*Continued*)

256	128	64	32	16	8	4
62	62	62				
190						
126	126					
254						
3	3	3	3	3	3	3
131						
67	67					
195						
35	35	35				
163						
99	99					
227						
19	19	19	19			
147						
83	83					
211						
51	51	51				
179						
115	115					
243						
11	11	11	11	11		
139						
75	75					
203						
43	43	43				
171						
107	107					
235						
27	27	27	27			
155						
91	91					
219						
59	59	59				
187						
123	123					
251						
7	7	7	7	7	7	

TABLE 16.11 Walsh Code Tree (*Continued*)

256	128	64	32	16	8	4
135						
71	71					
199						
39	39	39				
167						
103	103					
231						
23	23	23	23			
151						
87	87					
215						
55	55	55				
183						
119	119					
247						
15	15	15	15	15		
143						
79	79					
207						
47	47	47				
175						
111	111					
239						
31	31	31	31			
159						
95	95					
223						
63	63	63				
191						
127	127					
255						

TABLE 16.11 Walsh Code Tree (*Continued*)

256	128	64	32	16	8	4
0	0	0	0	NA	NA	NA
128						
64	64					
192						
32	32	32				
160						
96	96					
224						
16	16	16	NA			
144						
80	80					
208						
48	NA		NA			
176						
112	112					
240						
8	8	8	8	8		
136						
72	72					
200						
40	40	40				
168						
104	104					
232						
24	24	24				
152						
88	88					
216						
56	56	56				
184						
120	120					
248						
4	4	4	4	4		
132						
68	68					
196						
36	36	36				

Table 16.12 Walsh Code Pool Usage Example (*Continued*)

256	128	64	32	16	8	4
164						
100	100					
228						
20	20	20	20			
148						
84	84					
212						
52	52	52				
180						
116	116					
244						
12	12	12	12	12		
140						
76	76					
204						
44	44	44				
172						
108	108					
236						
28	28	28	28			
156						
92	92					
220						
60	60	60				
188						
124	124					
252						
NA	NA	NA	NA	NA	NA	1
NA						
NA	NA					
NA						
NA	NA	NA				
NA						
NA	NA					
NA						

TABLE 16.12 Walsh Code Pool Usage Example (*Continued*)

256	128	64	32	16	8	4
NA	NA	NA	NA			
NA						
NA	NA					
NA						
NA	NA	NA				
NA						
NA	NA					
NA						
NA	NA	NA	NA	NA		
NA						
NA	NA					
NA						
NA	NA	NA				
NA						
NA	NA					
NA						
NA	NA	NA	NA			
NA						
NA	NA					
NA						
NA	NA	NA				
NA						
NA	NA					
NA						
NA	NA	NA	NA	NA	NA	
NA						
NA	NA					
NA						
NA	NA	NA				
NA						
NA	NA					
NA						
NA	NA	NA	NA			
NA						
NA	NA					

TABLE 16.12 Walsh Code Pool Usage Example (*Continued*)

256	128	64	32	16	8	4
NA						
NA	NA	NA				
NA						
NA	NA					
NA						
NA	NA	NA	NA	NA		
NA						
NA	NA					
NA						
NA	NA	NA				
NA						
NA	NA					
NA						
NA	NA	NA	NA			
NA						
NA	NA					
NA						
NA	NA	NA				
NA						
NA	NA					
NA						

TABLE 16.12 Walsh Code Pool Usage Example (*Continued*)

50/50 mix between packet and voice traffic and the packet usage is 70 kbps or higher, then it is advisable to use a separate channel for packet data only when deploying CDMA2000, thereby preserving the imbedded voice platforms and, of course, throughput. The second carrier should be EVDO but it also could be a 1xRTT carrier.

16.6.2 Packet-Data Rates

The next part of the puzzle when performing the design aspect is to review the relationship between the data rates and the other components that are affected by the choice of data and their requisite selected speed.

Looking at Table 16.13, one is drawn to the conclusion or suspicion that there must be some other factor involved with system capacity other than the Walsh codes as described earlier. As suspected, with the differing data rates, there is a corresponding alteration to the link budget and pole capacity found in the processing-gain determination aspect. Naturally, as data rate increases, the processing gain is reduced because the overall spreading rate remains constant. In Table 16.13, the asterisks refer to the fact that there is now a reverse pilot involved with those configurations of CDMA2000.

Forward			
RC	**SR**	**Data Rates**	**Characteristics**
1	1	1200, 2400, 4800, 9600	$R = 1/2$
2	1	1800, 3600, 7200, 14400	$R = 1/2$
3	1	1500, 2700, 4800, 9600, 38,400, 76,800, 153,600	$R = 1/4$
4	1	1500, 2700, 4800, 9600, 38,400, 76,800, 153,600, 307,200	$R = 1/2$
5	1	1800, 3600, 7200, 14,400, 28,800, 57,600, 115,200, 230,400	$R = 1/4$
6	3	1500, 2700, 4800, 9600, 38,400, 76,800, 153,600, 307,200	$R = 1/6$
7	3	1500, 2700, 4800, 9600, 38,400, 76,800, 153,600, 307,200, 614,400	$R = 1/3$
8	3	1800, 3600, 7200, 14,400, 28,800, 57,600, 115,200, 230,400, 460,000	$R = 1/4$ (20 ms) $R = 1/3$ (5 ms)
9	3	1800, 3600, 7200, 14,400, 28,800, 57,600, 115,200, 230,400, 460,800, 1,036,800	$R = 1/2$ (20 ms) $R = 1/3$ (5 ms)
Reverse			
RC	**SR**	**Data Rates**	**Characteristics**
1	1	1200, 2400, 4800, 9600	$R = 1/3$
2	1	1800, 3600, 7200, 14,400	$R = 1/2$
3*	1	1200, 1350, 1500, 2400, 2700, 4800, 9600, 19,200, 38,400, 76,800, 153,600, 307,200	$R = 1/4$ $R = 1/2$ for 307,200
4*	1	1800, 3600, 7200, 14,400, 28,800, 57,600, 115,200, 230,400	$R =1/4$
5*	3	1200, 1350, 1500, 2400, 2700, 4800, 9600, 19,200, 38400, 76,800, 153,600, 307,200, 614,400	$R = 1/4$ $R = 1/2$ for 307,200 and 614,400
6*	3	1800, 3600, 7200, 14,400, 28,800, 57,600, 115,200, 230,400, 460,800, 1,036,800	$R = 1/4$ $R = 1/2$ for 1.036,800

Table 16.13 Packet-Data Rates

Table 16.14 shows the relationship between the data rates, defined in kilobits per second (kbps), and the processing gain. It is important to note that the SR and RC are also involved with the decisions, hence their inclusion in the table.

What follows next is an example of how to determine the relative number of users that can be accommodated by a single CDMA2000 (1xRTT) channel is calculated as follows:

$$N = [(W/R)/[\alpha(E_b/N_o)(1 + \beta)] + 1$$

Reverse Link											
RC1, kbps	**PG**	**RC2, kbps**	**PG**	**RC3, kbps**	**PG**	**RC4, kbps**	**PG**	**RC5, kbps**	**PG**	**RC6, kbps**	**PG**
9.6	128	14.4	85.33	9.6	128	14.4	85.33	9.6	384	14.4	256
				19.2	64	28.1	42.67	19.2	192	28.1	128
				38.4	32	57.6	21.33	38.4	96	57.6	64
				76.8	16	115.2	10.67	76.8	48	115.2	32
				153.6	8	230.4	5.33	153.6	24	230.4	16
				307.2	4			307.2	12	460.8	8
								614.4	6	1036.8	4
Forward Link											
RC1, kbps	**PG**	**RC2, kbps**	**PG**	**RC3, kbps**	**PG**	**RC4, kbps**	**PG**	**RC5, kbps**	**PG**	**RC6, kbps**	**PG**
9.6	128	14.4	85.33	9.6	128	9.6	128	14.4	85.33	9.6	384
				19.2	64	19.2	64	28.1	42.67	19.2	192
				38.4	32	38.4	32	57.6	21.33	38.4	96
				76.8	16	76.8	16	115.2	10.67	76.8	48
						153.6	8	230.4	5.33	153.6	24
						307.2	4			307.2	12
RC7, kbps	**PG**	**RC8, kbps**	**PG**	**RC9, kbps**	**PG**						
9.6	384	14.4	256	14.4	256						
19.2	192	28.1	128	28.1	128						
38.4	96	57.6	64	57.6	64						
76.8	48	115.2	32	115.2	32						
153.6	24	230.4	16	230.4	16						
307.2	12	460.8	8	460.8	8						
614.4	6			1036.8	4						

TABLE 16.14 Data Rate and Processing Gain Interaction

where

W/R = Process gain

α = Activity factor = 0.479 for voice and 1.0 for data (generally)

$E_b/N_o = 7$

β = 0.6 (omni) and 0.85 (sector)

Examples:

(a) RC = 2 and SR = 1,
$W/R = 85.33$, $\alpha = 0.479$, $E_b/N_o = 7$, and $\beta = 0.85$ (sector). Thus

$$N = (85.33)/[(0.479)(7)(1.85)] + 1 = 14.756$$

Now if $\alpha = 1.0$, then

$$N = (85.33)/[(1)(7)(1.85)] + 1 = 7.58$$

(b) RC = 3, SR = 1,
Data rate = 76.8 kbps. Therefore, $W/R = 16$, $\alpha = 1.0$, and $\beta = 0.85$ (sector). Thus

$$N = (16)/[(1)(7)(1.85)] + 1 = 2.235$$

16.7 Handoffs

CDMA2000 systems utilize several types of handoffs for both voice and packet data. The types of handoffs involve soft, softer, and hard handoffs. The difference between the types is dependent upon what one is trying to accomplish and, of course, if the system is 1xRTT, EVDO, or both (with EVDO, soft and softer handoffs are not utilized). The process for having a call or packet session in handoff for soft, softer, or hard is the same as that used for IS-95 or for a 1xRTT system. When a packet session is in progress and the subscriber exits the PDSN coverage area, the result is a termination of the packet session. There are several user-adjustable (user being the carrier) parameters that help the handoff process take place. The parameters that need to be determined involve the values to add or remove a pilot channel from the active list, and the search window sizes. There are several values that determine when to add or remove a pilot from consideration. In addition, the size of the search window cannot be too small or too large.

When introducing CDMA2000 into an existing IS-95 system or overlaying an EVDO system on top of a 1xRTT network, the choice of how to set up the neighbor list and search windows should mirror the existing system except where there is a transition zone.

16.7.1 Search Window

There are several search windows in CDMA2000 for both 1xRTT and EVDO, and they are the same as those used for IS-95, facilitating integration and compatibility. As with IS-95 systems, each of the search windows has its own role in the process, and it is not uncommon to have different search window sizes for each of the windows for a particular cell site. Additionally, the search window for each site needs to be set based on actual system conditions. The *search window* is defined as the amount of time, in terms of chips, that the CDMA subscriber's receiver will hunt for a pilot channel. There is a slight difference in how the receiver hunts for pilots, depending on its type.

The search windows that needed to be determined for CDMA involve the following sets:

- Active
- Neighbor
- Remaining

The method for determining the search window sizes for CDMA2000, 1xRTT, and EVDO, is the same as that done for IS-95 and is covered in Appendix B.

16.7.2 Soft Handoffs (1xRTT)

Soft handoffs are an integral part of CDMA in both IS-95 and 1xRTT systems. The determination of which pilot channels will be used in the soft handoff process has a direct impact on the quality of the voice call or packet-data session, as well as on the capacity for the system. Therefore, setting the soft handoff parameters is a key element in the system design for CDMA2000.

The parameters associated with soft handoffs involve the determination of which pilot channels are in the active, candidate, neighbor, and remaining sets. The list of neighbor pilots is sent to the subscriber unit when it acquires the cell site or is assigned a traffic channel.

A brief description of each type of pilot is the same as that used for IS-95 systems and is discussed in Appendix B; however, it is repeated here for clarity.

The *active set* is the set of pilots associated with the forward traffic channels assigned to the subscriber unit. The active set can contain more than one pilot because a total of three carriers, each with its own pilot, could be involved in a soft handoff process.

The *candidate set* includes the pilots that the subscriber unit has reported are of sufficient signal strength to be used. The subscriber unit also promotes the neighbor set and remaining sets of pilots that meet the criteria to the candidate set.

The *neighbor set* is a list of the pilots that are not currently on the active or candidate pilot list. The neighbor set is identified by the base station via the neighbor list and neighbor list update messages.

The *remaining set* is the set of all possible pilots in the system that potentially could be used by the subscriber unit. However, the remaining set pilots that the subscriber unit looks for must be a multiple of the Pilot_INC.

An example of the interaction among active, candidate, neighbor, and remaining sets is shown in Figure 16.2 and the associated description that accompanies the figure.

Several issues need to be addressed regarding soft handoffs with 1xRTT. The issues that need to be factored in are the different radio configurations among all the base stations involved with the soft handoff process. More specifically, the radio configurations involved must be the same. In addition, radio resources must be available for use by the mobile unit during soft handoff with all involved base stations. The resources available could possibly involve excluding the subscriber unit soft handoff with a target cell owing to the lack of resources available.

16.7.3 EVDO

The handoff process for EVDO is very similar to that used with 1xRTT with the exception that there is no soft handoff. With EVDO, the pilots of all the surrounding cells are monitored. With EVDO, there is an active, candidate, and of course, neighbor set.

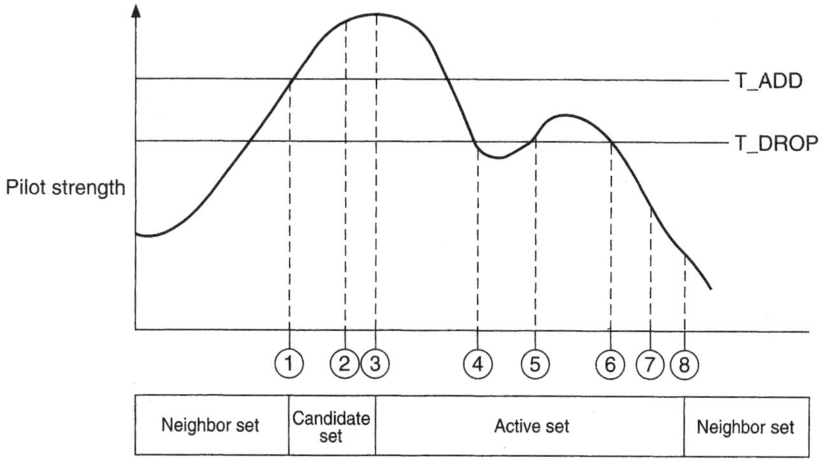

FIGURE 16.2 The pilot elevation and demotion process.

The active set is the pilot, only one, associated with the forward traffic channels assigned to the subscriber unit. The active set consists of only one pilot because there is no soft handoff process.

The candidate set consists of the pilots that the subscriber unit has reported are of sufficient signal strength to be used. The subscriber unit also promotes the neighbor set and remaining set pilots that meet the criteria to the candidate set.

The neighbor set is a list of the pilots that are not currently on the active or candidate pilot lists. The neighbor set is identified by the base station via the neighbor list and neighbor list update messages.

The remaining set is the set of all possible pilots in the system that potentially can be used by the subscriber unit. However, the remaining set pilots that the subscriber unit looks for must be a multiple of the Pilot_INC.

During the course of the packet session, a subscriber unit can switch between base stations or sectors as the RF condition dictates (besides capacity). The chief difference is that the handoff is a break-and-then-make instead of the soft handoff method usually associated with CDMA networks.

16.8 PN Offset Assignment

The assignment of the *Pseudorandom Number* (PN) offset for each CDMA2000 channel and/or sector utilizes the same rules that were and are used for IS-95 systems. In CDMA2000-1xRTT and EVDO, just as with IS-95 systems, the forward pilot channel carries no data, but it is used by the subscriber unit to acquire the system and assist in the process of soft handoffs, synchronization, and channel estimation. A separate forward pilot channel is transmitted for each sector of the cell site. The forward pilot channel is uniquely identified by its PN offset, or rather, PN short code that is used. The reverse pilot channel introduced in CDMA2000, however, does not utilize the PN offset.

The PN sequence has some 32,768 chips that, when divided by 64, result in a total of 512 possible PN codes that are available for potential use. The fact that there are

Sector	PN Code
Alpha	3*P*N-2P
Beta	3*P*N
Gamma	3*P*N-P
Omni	3*P*N

N = reusing PN cell, and *P* = PN code increment.

Table 16.15 PN Reuse Sequence

512 potential PN short codes to pick from almost ensures that there will be no problems associated with the assignment of these PN codes. However, there are some simple rules that must be followed in order to ensure that no problems are encountered with the selection of the PN codes for the cell and its surrounding cell sites. It is suggested that a reuse pattern be established for allocating the PN codes. The rationale behind establishment of a reuse pattern lies in the fact that it will facilitate the operation of the network for maintenance and growth.

Table 16.15 shows what can be used for establishing the PN codes for any cell site in the network. The method that should be used is to determine whether you wish to have a 4, 7, 9, 19, and so on, reuse pattern for the PN codes.

The suggested PN reuse pattern is an $N = 19$ pattern for a new CDMA2000 system. If you are overlaying the CDMA system onto a cellular system, an $N = 14$ pattern should be used when the analog system utilizes an $N = 7$ voice channel reuse pattern, and if a PN code scheme has been established for the sector or site, then the same PN code should be used for that sector/cell.

Figure 16.3 is an example of an $N = 19$ PN code reuse pattern. Please note that not all the codes have been utilized in the $N = 19$ pattern. The remaining codes should be left in reserve for use when there is a PN code problem that arises. In addition, a suggest PN_INC value of 6 is also recommended for use.

The PN short code used by the pilot is an increment of 64 from the other PN codes by which an offset value is defined. The Pilot_INC is the value that is used to determine the number of chips, or rather phase shift, one pilot has versus another pilot. The method that is used for calculating the PN offset is shown in Figure B.34 of Appendix B and applies to CDMA2000 as well as IS-95 systems.

Pilot_INC is valid from the range of 0 to 15. Pilot_INC is the PN sequence offset index and is a multiple of 64 chips. The subscriber unit uses the Pilot_INC to determine which are the valid pilots to be scanned. The method for calculating the Pilot _INC is the same as that used for IS-95 systems and is a function of the distance between reusing sites.

16.9 Link Budget

The link-budget process, as defined in previous sections of this book, is essential for the establishment of a valid RF design to take place. The link budget helps to define the cell-site spacing. The cell-site spacing is determined by the link budget using a signal level that is exceeded 50 percent of the time.

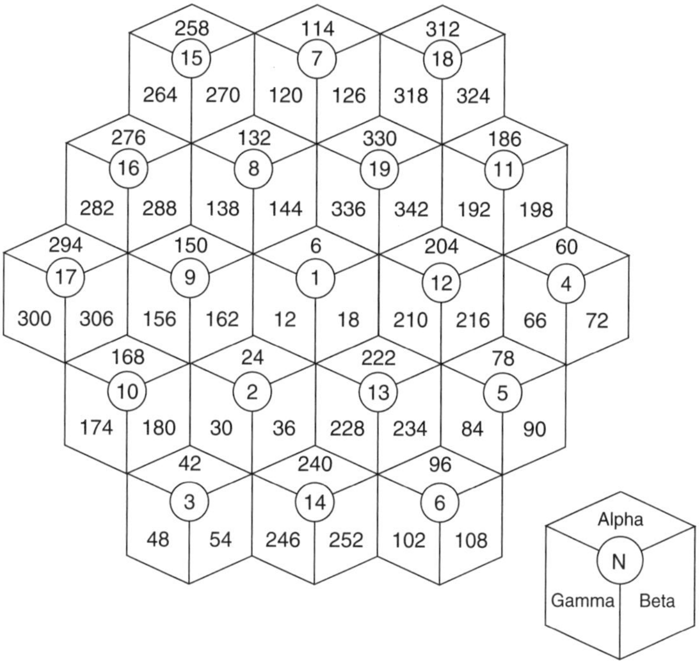

FIGURE 16.3 PN reuse pattern.

There are two links that need to be determined in the establishment of a link budget: forward and reverse. The forward and reverse links utilize different coding and modulation formats in both 1xRTT and EVDO systems. The first step in the link-budget process is to determine the forward link maximum path loss before the reverse link's maximum path losses. The link budget is defined in Appendix B.

16.9.1 1xRTT

CDMA2000-1xRTT has a better link budget than IS-95A/IS-95B at the same traffic loading, therefore offering a higher overall capacity at the same traffic load owing to vocoder improvements as well as utilizing a coherent demodulation for the reverse link. However, for the link budget that will be used for the design, the link budget parameters primarily associated with IS-95 are utilized owing to the prevalence of the RC1 and RC2 subscriber units in the market.

Regarding packet-data services, owing to the improved modulation and coding scheme (resulting in a lower target E_b/N_o), the 38.4-kbps packet-data rate for CDMA2000-1xRTT has approximately the same link budget as the IS-95 13-kbps voice vocoder, but at higher data rates, the service coverage will shrink owing to a variety of factors that include process gain as well as power allocation. With 1xRTT, voice is given a priority, and therefore, data petitions for all available remaining power. Therefore for the design effort put forth, a lower data rate of 38.4 kbps was used per packet-data subscriber in the link-budget calculations, but 70 kbps was used for subscriber packet throughput. The disparity is allowed for ease of discussion.

As stated previously, the link-budget calculations utilized directly influence the performance of the CDMA system because they are used to determine power setting and capacity limits for the network. Proper selection of the variables that make up the link budget is a very obvious issue owing to its impact on a successful design.

Tables 16.16 and 16.17 represent the link budgets for a CDMA2000 system. Obviously, the issue of differing data rates and subscriber and base radio configurations make the possible combinations daunting. However, the basic principles that make up these link-budget tables presented in Tables 16.16 and 16.17 can be modified with different process gains as well as a different spreading rate for the uplink path.

16.9.2 EVDO

The link budget for EVDO consists of different components from that of an IS-95 or 1xRTT system. However, EVDO was designed from the onset to be overlaid onto

		Value	Comment
Subscriber terminal	Tx power	23 dBm	Maximum power per traffic channel
	Cable loss	2 dB	
	Antenna gain	0 dBd	
	Tx power per traffic channel	**21 dBm**	
External factors	Fade margin	5 dB	Log Normal
	Penetration loss		(Street/vehicle/building)
	External Losses	**15 dB**	
Base station	Rx Antenna Gain	15 dBd	(Approximately 17.25 dBi)
	Tower-top amp net gain	0	
	Jumper and connector loss	0.25 dB	
	Feedline loss	1 dB	
	Lightning arrestor loss	0.25	
	Jumper and connector loss	0.25	
	Duplexer loss	0.5	
	Receive configuration loss	0	
	Handoff gain	4 dB	
	Rx diversity gain	0 dB	
	Rx noise figure	5 dB	
	Receiver interference margin	3.4 dB	55% pole
	Receiver noise density	174 dBm/Hz	
	Information rate	41.58 dB	14.4
	Rx sensitivity	124.02 dBm	
E_b/N_o	$\boldsymbol{E_b/N_o}$	**7.00 dB**	
	Maximum path loss	123.02 dB	

TABLE 16.16 Reverse Link Budget

		Value	Comment
Tx power distribution	Tx PA power	39.0 dBm	8 W
	Pilot channel power	30.8 dBm	15.0% of maximum power per channel
	Sync channel power	20.8 dBm	10.0% of pilot power
	Paging channel power	26.2 dBm	35.1% of pilot power
	Traffic channel power	**38.0 dBm**	78.2% of maximum power per channel
	Number of mobile units per carrier	13	
	Soft/softer handoff traffic	13	1.85 overhead factor
	Maximum number of active traffic channels	26	
	Average traffic channel power	**23.8 dBm**	26 total traffic channels
	Voice activity factor	0.479	Voice = 0.479; data =1.0
	Peak traffic channel power	27.0 dBm	Average traffic channel power/voice activity factor
Base Station	Traffic channel Tx power	27.0 dBm	
	Duplexer loss	0.5 dB	
	Jumper and connector loss	0.25 dB	
	Lightning arrestor loss	0.25 dB	
	Feedline loss	1 dB	
	Tower-top amp loss	0 dB	
	Antenna gain	15 dBd	
	Net base-station Tx power	**39.8 dBm**	10-W ERP per traffic channel (voice)
	Total base-station Tx power	51.8 dBm	151-W ERP per carrier
Environmental	Fade margin	5 dB	Log normal
	Penetration loss	10 dB	(Street/vehicle/building)
	Cell overlap	3 dB	
	External losses	**18 dB**	
Subscriber	Antenna gain	0 dBd	
	Cable loss	2 dB	
	Rx noise figure	10 dB	
	Receiver noise density	174 dBm/Hz	
	Information rate	60.90 dB	1238 kbps
	Rx sensitivity	101.1 dBm	
Subscriber traffic channel RSSI	Base Tx	39.8	
	Environmental loss	18	
	Maximum path loss	123.02	Obtained from uplink path analysis
	RSSI at subscriber antenna	**–101.22**	

Table 16.17 Forward Link Budget

an existing 1xRTT system with the intention of having the same RF coverage in both the uplink and the downlink. In other words, under the same RF channel conditions and environment, the coverage should be the same, facilitating an overlay network approach. Of course, for equality in coverage, the EVDO system needs to operate in the same frequency band as 1xRTT, as well as have the same antenna characteristics.

Some of the important items:

- EVDO is TDM and has a constant power output for the cell site.
- Of the total power for the site, 15.3 percent typically is allocated for the control channel, whereas the rest is allocated for the traffic time slots. Additionally, the downlink path is not shared and is dedicated for a single user when transferring packets, and thus the total power is allocated for one user.
- When comparing the link budgets of EVDO and 1xRTT, they are similar for the same data rate, as it should be. Obviously, as the data rate increases with EVDO, pressure is applied to the link budget, as indicated in Table 16.18, where higher data rates require a good RF environment. The values listed in Table 16.18 are for 1 percent *Packet Error Rate* (PER) and can and will be relaxed for a 2 percent PER if the design and market dictate.

The EVDO sample link budget indicates a slightly better forward link than reverse link. The schedule and rate adaptation are not included here, leading the real design to have coverage bands defined through computer simulation. However, as with all designs, the inputs to the design need to be understood. Table 16.19 represents the reverse link budgets for a CDMA2000 EVDO system. Table 16.20 therefore represents the forward link budget. Obviously, the issue of differing data rates and subscriber and base radio configurations makes the possible combinations daunting. The basic principles that underlie the link budgets presented in these tables can be modified with different data rates to estimate the range of the site or bands of coverage for each data rate offered, enabling a robust packet-data design.

Data Rate (kbps)	SINR (dB), 2% Packet Error Rate 1%
38.4	–12.5
76.8	–9.5
153.6	–6.5
307.2	–4.0
614.4	–1.0
921.6	1.3
1228.2	3.0
1843.2	7.2
2457.6	9.5

Table 16.18 SINR and Data Rate

		Value	Comment
Subscriber terminal	Tx power	23 dBm	
	Cable loss	2 dB	
	Antenna gain	0 dBd	
	Tx power	21 dBm	
External factors	Fade margin	5 dB	Log normal
	Penetration loss	10 dB	(Street/vehicle/building)
	External losses	15 dB	
Base station	Rx antenna gain	15 dBd	(approximately 17.25 dBi)
	Tower-top amp net gain	0	
	Jumper and connector loss	0.25 dB	
	Feedline loss	1 dB	
	Lightning arrestor loss	0.25	
	Duplexer loss	0.5	
	Receive configuration loss	0	
	Select handoff gain	4.1 dB	
	Rx diversity gain	0 dB	
	Rx noise figure	5 dB	
	Receiver interference margin	3.4 dB	
	Receiver noise density	174 dBm/Hz	
	Information rate	41.58 dB	14.4
	Rx sensitivity	124.17 dBm	
E_b/N_o	**E_b/N_o**	**7.00 dB**	
	Maximum path loss	**130.17 dB**	

Table 16.19 EVDO Reverse Link Budget

16.10 Sample Basic Designs

Over the next few pages, three basic designs will be covered. The designs will be rudimentary in nature because the concept of what has to be done needs to be stressed, not a particular design for a particular market that will not be relevant for any other system.

- CDMA2000-1xRTT (greenfield)
- EVDO (greenfield)
- CDMA2000-1xRTT, with EVDO overlay

The traffic estimate for all designs will be the same fundamentally, with a few variants that are relative to the access platform being deployed. However, a key element in the traffic-forecast method is the overbooking of data services, as well as the issue of the volume of CDMA2000-ready subscriber units. One method of determining the number of available subscribers who will be CDMA2000-ready is to obtain an estimate of subscriber handsets that will be procured by the company over the next 6 to 12 months.

		Value	Comment
Tx power distribution	Tx PA power	39.0 dBm	8 W
	Pilot channel power	30.8 dBm	15.0% of maximum power per channel
	Sync channel power	20.8 dBm	10.0% of pilot power
	Paging channel power	26.2 dBm	35.1% of pilot power
	Serving time	14.3%	
Data channel power	38.0 dBm	79.9% of maximum power	
Base station	Traffic channel Tx power	38.0 dBm	
	Duplexer loss	0.5 dB	
	Jumper and connector loss	0.25 dB	
	Lightning arrestor loss	0.25 dB	
	Feedline loss	1 dB	
	Jumper and connector loss	0.25 dB	
	Tower-top amp loss	0 dB	
	Antenna gain	15 dBd	
	Total base-station Tx power	50.8 dBm	151 W ERP per carrier
Environmental	Fade margin	5 dB	Log normal
	Penetration loss	10 dB	(Street/vehicle/building)
	Cell overlap	3 dB	
	External losses	**18 dB**	
Subscriber	Antenna gain	0 dBd	
	Cable loss	2 dB	
	Rx noise figure	10 dB	
	Receiver noise density	174 dBm/Hz	
	Information rate (I_{or}/N_o)	60.90 dB	1238 kbps
Rx sensitivity	101.1 dBm		
Subscriber	Base Tx	50.8	
	Environmental loss	18	
	Maximum path loss	130.17	Obtained from uplink path analysis
	RSSI at subscriber antenna	**–98.17 dBm**	

Table 16.20 EVDO Forward Link Budget

16.10.1 CDMA2000-1xRTT

The following is a brief design example that is relevant to a new CDMA2000-1xRTT system being deployed as a greenfield situation. The design example focuses on the issues that are more relevant to the internal network and does not factor into the mix any possible networking and coordination issues with adjacent systems.

Because this is a new CDMA2000 system, the concerns of legacy equipment are not relevant, and it will be assumed that only CDMA2000-capable handsets are used by the system. However, in real life, the issue of roaming mobile units in the system that are legacy (IS-95) will need to be factored into the design, even though product life cycles should have resolved this issue.

The initial design calls for coverage of a selected area within the network. The first step in this case is to determine the desired traffic load for both circuit switched as well as packet data. Using the traffic-loading numbers, Tables 16.21 and 16.22 show the expected traffic load for a total of 50,000 people who sales and marketing expect will use the system. Because the actual throughput is undefined owing to the lack of actual traffic data from the network, the design will encompass all the possible traffic loads.

Naturally, if packet-data services of only 70 kbps will be offered, then some of the services included in the example can be eliminated. Table 16.21 shows the expected load on the overall system in Erlangs and megabits per second. Erlangs are related to circuit-switched data, whereas packet data is reflected in megabits per second. However, it should be noted that for 1xRTT VoIP on the RAN is not used, and therefore, voice traffic is represented as circuit-switched traffic only. In previous comments, if only an estimate from marketing is available regarding packet-data usage given in a percentage of voice usage, then the estimation should be done using an Erlang C model.

Table 16.23 is a summary of the calculations derived for the system traffic load. However, some additional information is contained in the table, which relates to geographic areas associated with each type of traffic. For the purposes of simplifying this example, the areas will be considered to be adjacent to each other. However, in real life, the areas will be intertwined.

The next step is to determine the number of sites required to support the expected load. An assumption needs to be made at this time: All the CDMA2000-1xRTT sites will be sector sites, three sectors per cell. In addition, it is assumed that for this design, a total of 8.2 Erlangs per sector can be supported for circuit switch per sector, which is derived from a 2 percent GoS using Erlang B with 14 trunk members. The packet throughput is based on 2.35 trunk members at 76.8 kbps. Both the packet- and circuit-switched traffic-handling capacities are very conservative and are driven by the link budget and process gain used.

$$\text{Cell voice Erlangs} = -8.2 \text{ Erlangs/sector} \times 2.64 \text{ (sector gain)}$$
$$= 21.648 \text{ Erlangs per cell}$$

$$\text{Packet throughput} = -2.35 \times 76.8 \text{ kbps/sector} \times 2.64 = 453.15 \text{ kbps per cell}$$

$$N_{\text{circuit switched}} = -\text{estimated traffic/cell capacity} = 21.648/680.15$$
$$= 32 \text{ cells total for the system}$$

Building									
			Activity Factor		Call Seconds		Penetration	Population	Net Users
	BHCA	Call Duration (s)	UL	DL	UL	DL	35%	50000	
S	0.9	180	0.5	0.5	81	81	73%	17500	12775
SM	0.006	3	1	1	0.018	0.02	40%	17500	7000
SD	0.2	156	1	1	31.2	31.2	13%	17500	2275
MMM	0.5	3000	0.15	0.0029	225	4.28	15%	17500	2625
HMM	0.15	3000	0.15	0.0029	67.5	1.28	15%	17500	2625
HIMM	0.1	120	1	1	12	12	25%	17500	4375
Pedestrian									
			Activity Factor		Call Seconds		Penetration	Population	Net Users
	BHCA	Call Duration (s)	UL	DL	UL	DL	15%	50000	
S	0.8	120	0.5	0.5	48	48	73%	7500	5475
SM	0.03	3	1	1	0.09	0.09	40%	7500	3000
SD	0.2	156	1	1	31.2	31.2	13%	7500	975
MMM	0.4	3000	0.15	0.0029	180	3.42	15%	7500	1125
HMM	0.06	3000	0.15	0.0029	27	0.51	15%	7500	1125
HIMM	0.05	120	1	1	6	6	25%	7500	1875
Vehicular									
			Activity Factor		Call Seconds		Penetration	Population	Net Users
	BHCA	Call Duration (s)	UL	DL	UL	DL	70%	50000	
S	0.4	120	0.5	0.5	24	24	73%	35000	25550
SM	0.02	3	1	1	0.06	0.06	40%	35000	14000
SD	0.02	156	1	1	3.12	3.12	13%	35000	4550
MMM	0.008	3000	0.15	0.0029	3.6	0.07	15%	35000	5250
HMM	0.008	3000	0.15	0.0029	3.6	0.07	15%	35000	5250
HIMM	0.008	120	1	1	0.96	0.96	25%	35000	8750

Note: Expect to sell 50,000 handsets that are CDMA2000-capable.

TABLE 16.21 CDMA2000-1xRTT Greenfield Traffic Forecast

Net User Bit Rates DL kbps	UL kbps	Building	Circuit Switch Usage UL	DL		Packet Service Usage UL	DL
16	16		1034775	1034775			
14	14					315	126
64	64		70980	70980			
384	64					3937500	11221.875
2000	128					1181250	3366.5625
128	128					210000	52500
		Total BHCS	1105755.0	1105755.0	**Total BHPS**	5329065.0	67214.4
		Erlang	307.2	307.2	**Mbps**	53.29	0.67
Net User Bit Rates DL kbps	**UL kbps**	**Pedestrian**	**Circuit Switch Usage UL**	**DL**		**Packet Service Usage UL**	**DL**
16	16		262800	262800			
14	14					675	270
64	64		30420	30420			
384	64					1350000	3847.5
2000	128					202500	577.125
128	128					45000	11250
		Total BHCS	293220	293220	**Total BHPS**	1598175.0	15944.6
		Erlang	81.45	81.45	**Mbps**	15.98	0.16
Net User Bit Rates DL kbps	**UL kbps**	**Vehicular**	**Circuit Switch Usage UL**	**DL**		**Packet Service Usage UL**	**DL**
16	16		613200	613200			
14	14					2100	840
64	64		14196	14196			
384	64					126000	359.1
2000	128					126000	359.1
128	128					33600	8400
		Total BHCS	627396	627396	**Total BHPS**	287700	9958.2
		Erlang	174.28	174.28	**Mbps**	2.88	0.10
		System Totals					
		Erlang	562.88	562.88	**Mbps**	72.15	0.93

TABLE 16.22 CDMA2000-1xRTT Traffic Forecast (Erlangs and Megabits per Second)

	Region	Area (km²)	Erlangs	Erlang/km	Mbps	Mbps/km
Building	1	100	307.2	3.0715417	53.29	0.5329065
Pedestrian	2	900	81.45	0.0905	15.98	0.0177575
Vehicular	3	4000	174.28	0.0435692	2.88	0.0007193
Unserved	4	6000				
Total		10000	680.15		72.15	

TABLE 16.23 Traffic Loading Summary Table

Packet data = (estimated traffic/overbooking)/cell capacity

= (72.15 Mbps/10)/453.15 kbps = 16 total for the system

The next step is to determine the radius for the site(s) involved with each area. In this example, the same path loss will be used because it is assumed the same morphology is used for all three areas. From the link budget, PL_{max} = 140 dB. Therefore, radius $r = 140 = 132 + 38 \log r$. The results for coverage sites are shown in Table 16.24.

Obviously, from the example, the system is coverage-limited and not capacity-limited. However, in looking briefly at the traffic data, the treatment of one section of the system, in building, needs a higher throughput than the vehicular areas, which is obvious. Therefore, the deployment recommendation is to have two carriers for the building area, where F2 is not to initially exceed 50 percent for data (50/50 voice) and F1 is configured for 25 percent data. Figure 16.4 represents possible channel deployment schemes that apply to a PCS system operating with 15 MHz of duplexed spectrum. EVDO is included but is really left up to the traffic mix, as well as true availability for the technology. Additionally, EVDV was included to facilitate the concept of future migration.

Now, the next issue is what do you do with this wonderful information. Well, you need to lay out a rough system topology from which you can begin to determine if it is valid to centralize or decentralize the BSCs or have intermediate nodes in the network. Typically, for a system having 1100 km^2, it would be expected to have several MSCs or concentration nodes to reduce the leased-line costs and improve on interconnection transport fees.

It is recommended that the core of the network consisting of the building environment use two CDMA2000 carriers, whereas the pedestrian and vehicular zones use only one carrier. A hard handoff will need to take place between the F2 and F1 zones. However, it is recommended in a situation such as this that the BTS F1 carriers process

	Region	Area (km²)	Coverage
Building	1	100	12
Pedestrian	2	900	60
Vehicular	3	4000	96
Unserved	4	6000	
Total		11000	168

TABLE 16.24 System Sites

PCS 15 MHz												
GB	F7-DO	F5-DO	F6-DO	F8-DO	F1-1X	F2-1X	F3-1X	F4-1X	F10 1X	F9-1X	F11 1X	GB
GB	F7-DO	F5-DO	F6-DO	F8-DO	F1-DV	F2-DV	F3-DV	F4-DV	F10 DV	F9-DV	F11 DV	GB
5 MHz					5 MHz				5 MHz			

FIGURE 16.4 1 × channel allocation scheme.

primarily voice traffic, whereas F2 is more a data-only situation. As mentioned earlier, this configuration can be supported via software and user-definable parameters.

Figure 16.5 shows a sample FA assignment for the greenfield deployment, assuming that traffic density is centered at the core of the network.

Figure 16.6 takes the design one step further in that the various pipe sizes are estimated for the initial concept. From this figure it is apparent that it would be advantageous to locate BSC 1 with the MSC, provided that the MSC is located near a tandem. The other BSCs, however, owing to their initial traffic load, should be considered to be remotely located, provided that the operational and support issues can be met. In addition, the BSCs will have, on average, 15 sites connected to them for the design example, with the exception of the core, where a total of 12 BTSs are associated with the BSCs.

The facilities between the BTSs and BSCs are assumed to be unstructured TDM because this is a more readily available circuit type. The connectivity to the off-net data networks assumes an 80/20 mix of public versus private networks. The assumption used is that 100 percent of the packet traffic is off-net.

The PN offset assignment scheme that was presented earlier in this chapter should be used for the system design following an $N = 19$ reuse pattern for the PN offsets.

Obviously, there are more issues involved when designing a CDMA2000 system, but the preceding material should help in setting up the thought processes to achieve the desired goal of supporting the customer requirements for service delivery and transport.

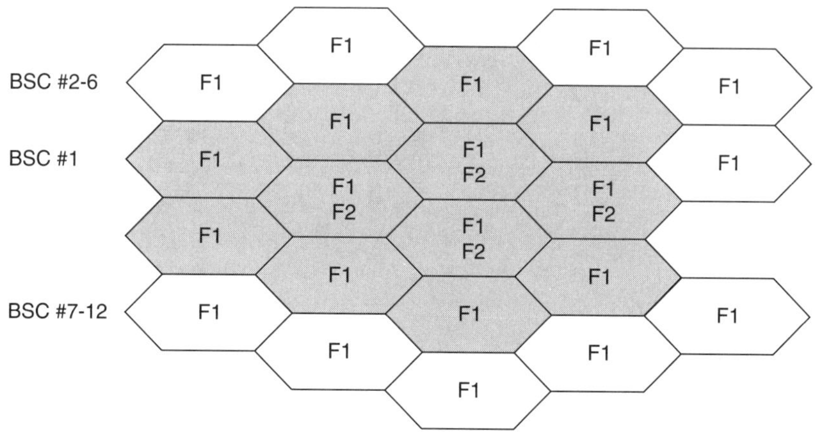

FIGURE 16.5 CDMA2000-1xRTT carrier deployment scheme.

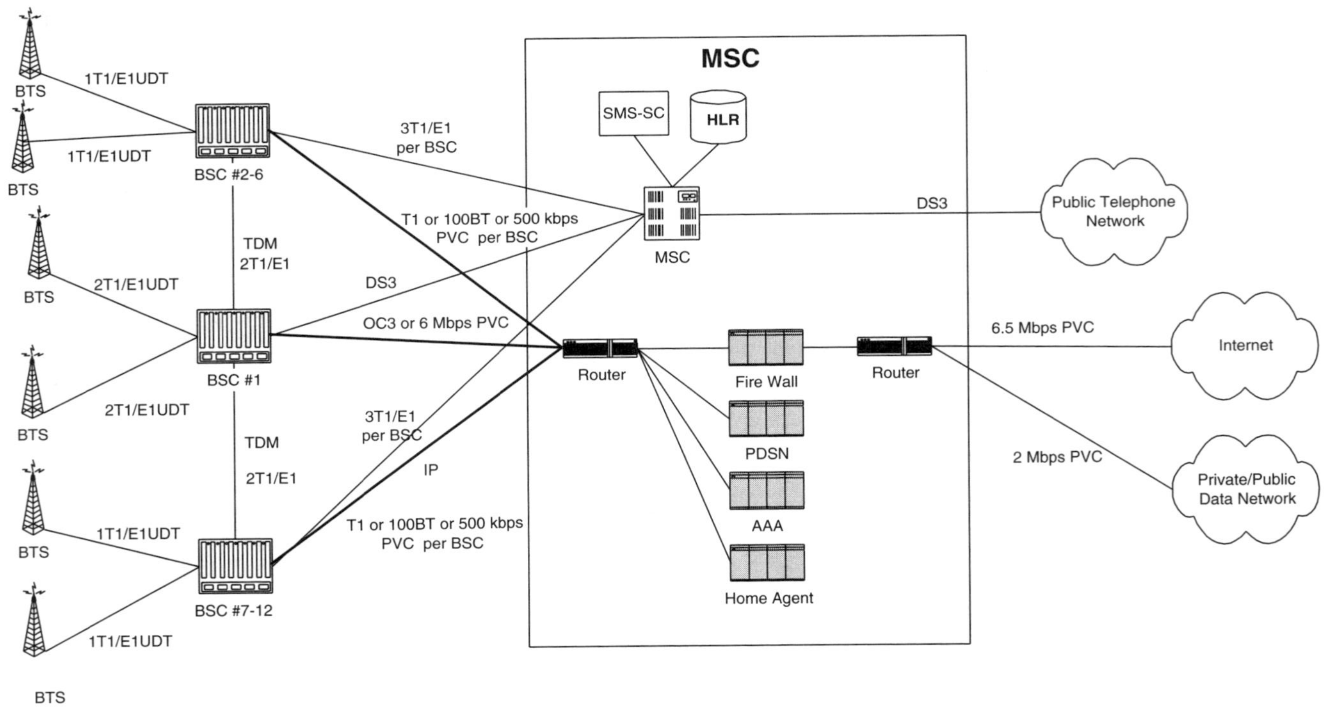

FIGURE 16.6 Sample CDMA2000-1xRTT system configuration.

16.10.2 EVDO (IS-856)

The following is a brief design example that is relevant for a new EVDO (IS-856) system being deployed as a greenfield situation. The design example focuses on the issues that are more relevant to the internal network and does not factor into the mix any possible networking and coordination issues with adjacent systems. Because this is a new system, the concerns of legacy equipment are not relevant, and it will be assumed that only EVDO-capable handsets are used by the system. However, it is important to remember that IS-856–capable devices can use CDMA2000-1xRTT networks, but a 1xRTT device cannot use an IS-856 network. Therefore, roaming mobiles that are not EVDO-capable cannot be factored into the design.

For this design, IS-856 channel types will be available for deployment. The specific release, 0, A, B, etc., will not be differentiated for this design concept. However, the relative differences and capabilities are important among the releases; this was discussed previously and will not be repeated here.

The initial design calls for coverage of a selected area within the network. The first step in this case is to determine the desired traffic load for both circuit-switched and packet data. Use the traffic-loading numbers presented in Tables 16.21 and 16.22. Because the actual throughput is undefined owing to a lack of actual traffic data from the network, the design will encompass all possible traffic loads.

Naturally, if packet-data services of only 70 kbps will be offered, then the issue of deploying EVDO should be reevaluated because this by itself would not justify deployment of a new EVDO network.

Table 16.22 showed the expected load on the overall system in Erlangs and megabits per second. However, EVDO is a packet system. Therefore, voice services are VoIP, and a conversion from circuit-switched to VoIP needs to take place and depends entirely on the vocoder selected for use.

The initial Erlang calculation is based on the premise of a DS0 at 64 kbps. With VoIP, less bandwidth is required, but when it is being passed, a higher throughput is needed in terms of data speed. For planning purposes,

$$\text{VoIP (bandwidth), kbps} = (\text{Erlangs} \times 0.45) \times 28 \text{ kbps}$$

It is important to remember that Erlangs are relative for circuit switched data, whereas Mbps are for packet data. In previous comments, if only an estimate from marketing is available regarding packet-data usage given in a percentage of voice usage, then the estimation should be done using an Erlang C model.

Table 16.25 is a summary of the calculations derived for the system traffic load. However, some additional information is contained in the table, and this is the relative geographic areas associated with each type of traffic. For the purposes of simplifying this example, the areas will be considered to be adjacent to each other. However, in real life, the areas will be intertwined.

The next step is to determine the number of sites required to support the expected load. An assumption needs to be made at this time: All the EVDO sites will be sector sites, three sectors per cell. In addition, it is assumed that for this design, a total of 2 Mbps per sector can be supported; we picked this number just to start.

$$\text{Cell voice Erlangs} = 8.2 \text{ Erlangs/sector} \times 2.64 \text{ (sector gain)} = 21.648 \text{ Erlangs per cell}$$

$$\text{Data capability} = 2000 \text{ kbps/sector} \times 2.64 = 5280 \text{ kbps per cell}$$

	Region	Area (km^2)	Erlangs	Mbps (VoIP)	Mbps/km (VoIP)	Mbps (IP)	Mbps/km (IP)
Building	1	100	307.2	387.32	3.87324	53.29	0.5329065
Pedestrian	2	900	81.45	102.627	1.1403	15.98	0.0177575
Vehicular	3	4000	174.28	2.19592	0.054898	2.88	0.0007193
Unserved	4	6000	0	0	0	0	0
Total		10000	680.15	8.568		72.15	

TABLE 16.25 EVDO Traffic Loading Summary Table

$$\text{Total IP bandwidth} = \text{VoIP} + \text{data/overbooking} = 8.568 + 72.15 = 80.718 \text{ Mbps}$$

$$80.718 \text{ Mbps}/5.28 \text{ Mbps/cell} = 15.29 \text{ cells, or 16 cell sites for EVDO}$$

Note that no overbooking for data was included here; alternatively, the overbooking factor could be in the overall data demand estimation calculation.

The next step is to determine the radius for the site(s) involved with each area. In this example, the same path loss will be used because it is assumed the same morphology is used for all three areas. However, this is difficult to estimate because we have not defined the data rate on which the design for the site will be based. Therefore, keeping with the path loss from the link-budget example for EVDO, the radius for the site can be found using the same method used for 1xRTT in Table 16.24.

Obviously, from the example, the system is coverage-limited and not capacity-limited for any of the areas, remembering that the link budgets for EVDO and 1xRTT are similar. However, EVDO with VoIP has additional capacity. Therefore, the deployment recommendation is to have one carrier deployed.

Figure 16.7 shows possible channel-deployment schemes that apply to a PCS system operating with 15 MHz of duplexed spectrum. EVDV also was added to the figure for future expansion, primarily for future spectrum allocation.

The next issue is that you need to lay out a rough system topology. Typically, for a system having 1100 km^2, it would be expected to have several MSCs or concentration nodes to reduce the leased-line costs and improve on interconnection transport fees. In our simple example, though, only 1 MSC is needed. Figure 16.8 shows an FA deployment scheme for a single-carrier EVDO network.

The various pipe sizes were estimated for the initial concept and are shown in Figure 16.9. The BSCs will have, on average, 15 sites connected to them for the

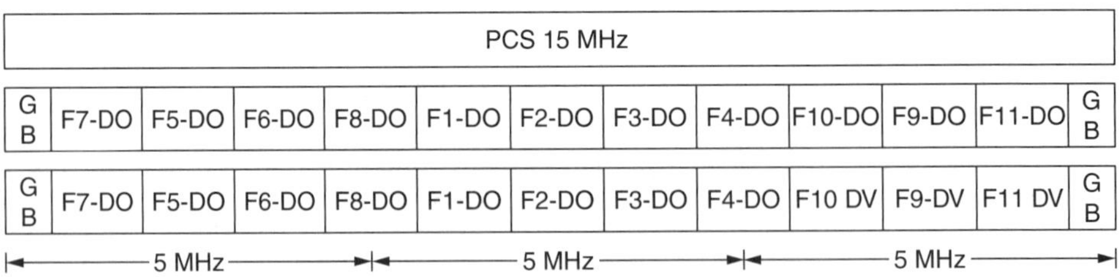

FIGURE 16.7 EVDO channel scheme.

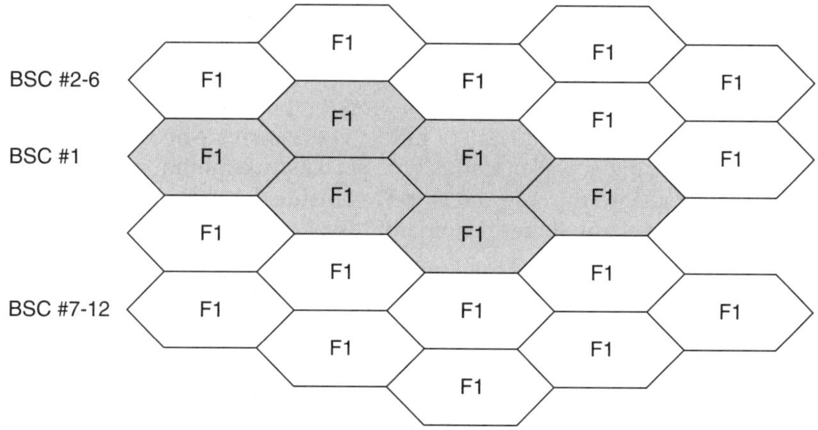

Figure 16.8 EVDO FA deployment.

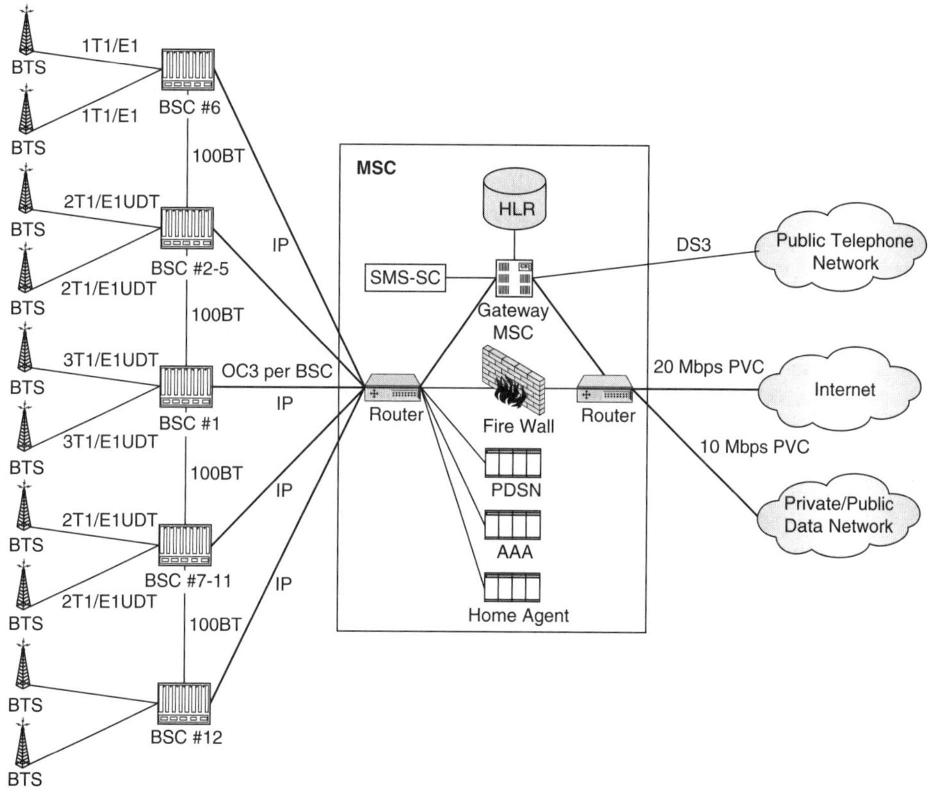

Figure 16.9 EVDO network configuration (sample).

design example, with the exception of the core, where a total of 12 BTSs are associated with the BSCs.

The facilities between the BTSs and the BSCs are assumed to be unstructured TDM because this is a more readily available circuit type. The connectivity to the off-net data networks assumes an 80/20 mix of public versus private networks. The assumption is that 100 percent of the packet traffic is off-net. Lastly, a minimum configuration involves three T1s per site for the core and two T1s outside the core.

As with all designs, there are numerous issues that need to be factored into the design, and this example only begins to cover some of those topics. However, this material should help in setting up the thought process to achieve the desired goal of supporting the customer requirements for service delivery and transport.

16.10.3 CDMA2000-1xRTT with EVDO Overlay

An all-too-common wireless situation is that operators need to address the issue of how to integrate EVDO into their CDMA2000-1xRTT network. Many operators have devised their own methods for implementing EVDO into an existing CDMA2000 network. However, not all of them have implemented EVDO. Therefore, this section will attempt to bring to light many of the issues associated with integrating an EVDO system into a CDMA2000-1xRTT system.

EVDO and CDMA2000 carriers have fundamental differences whose nuances were discussed previously. However, the primary concept is that EVDO is a packet network that is designed to be an overlay network on top of or in parallel with an existing CDMA2000-1xRTT network. Therefore, EVDO, when deployed in a CDMA2000-1xRTT network, will require an individual FA that is dedicated to the EVDO network. This will create some pressure on existing 1xRTT capacity owing to spectrum and capacity reallocation.

Additionally, EVDO's link budget is similar to that of 1xRTT for the same data rate. Therefore, the coverage footprint initially should be the same, assuming that the same data rates are used. However, it is strongly recommended that EVDO be deployed with its own antenna system because optimization will be different from that with CDMA2000-1xRTT. The mobile units, or terminal devices, in an EVDO system will tune periodically to a CDMA2000 channel when it is active, and when they are idle, they will monitor both EVDO (IS-856) and CDMA2000-1xRTT paging channels, allowing the terminal to resume high-speed packet data and also to receive incoming voice and *Short Message Service* (SMS) messages.

When deploying an EVDO overlay network, it is important to note that guard bands are not needed between EVDO and 1xRTT. The items in the network requiring upgrades and/or enhancements for EVDO are

- *BTS* The power amplifier and chip need to be set to support EVDO.
- *BSC* Software upgrades are needed to support the new interface.
- *PDSN* Software upgrades are needed to support QoS.
- *BTS/BSC* Facility enhancement (separate circuits) are needed.

These are in addition to any other enhancements, such as VoIP gateways, soft switches, or other IMS-related platforms that enable the migration and evolution to a full IP-based system.

There are several ways to implement an EVDO overlay network:

- 1:1 deployment
- *N*:1 deployment
- Targeted areas initially

The issue at hand is to marry the technical and implementation side with the marketing and budgeting process. Obviously, based on the size of the network, a 1:1 implementation may not be practical from both a monetary and an implementation timeline perspective.

An *N*:1 deployment is not really practical if it is not a targeted approach, which is covered next. The *N*:1 drawback is that the link budgets are relatively similar between 1xRTT and EVDO at lower data rates, but at higher data rates, the effective coverage area of the site is reduced owing to RF channel conditions. Therefore, using an *N*:1 deployment scheme implies coverage gaps and possibly a lackluster service introduction.

The targeted approach for deployment of an EVDO network is strongly suggested. EVDO should be deployed in high-traffic areas such as airports and business campus areas. This will facilitate a more rapid service offering and allow for gradual or graceful deployment following traffic usage. Traffic usage can be tracked because EVDO-capable devices still can use 1xRTT for data delivery when they are outside the EVDO coverage area. Figure 16.10 is an example of an initial core deployment where EVDO is added to the 1xRTT service offering. The figure only shows a single 1xRTT channel, but obviously, many systems have deployed more than a single 1xRTT channel.

Now, using the same traffic data as in the preceding two examples for greenfield deployments of 1xRTT and EVDO, the system is, of course, coverage-limited. With EVDO, the use of VoIP now is possible, but the migration to VoIP will not take place overnight and initially will be lightly loaded. Additionally, the amount of preceding EVDO-capable devices in the network can be considered initially not to factor into traffic loading for the new EVDO network.

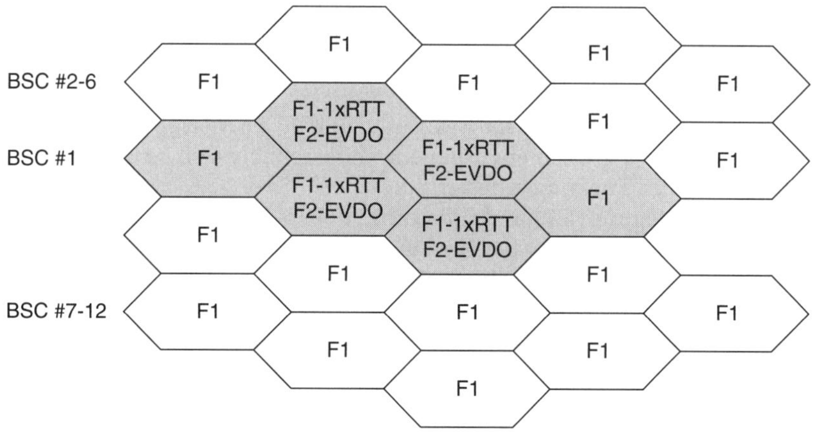

Figure 16.10 EVDO overlay FA deployment scheme, F1-1xRTT and F2-EVDO.

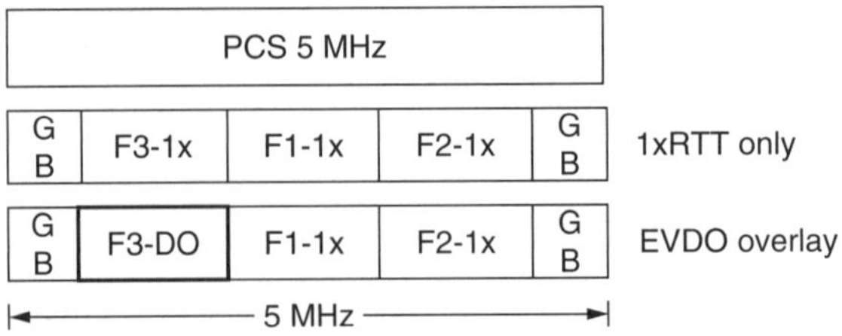

FIGURE 16.11 EVDO overlay FA deployment scheme, PCS.

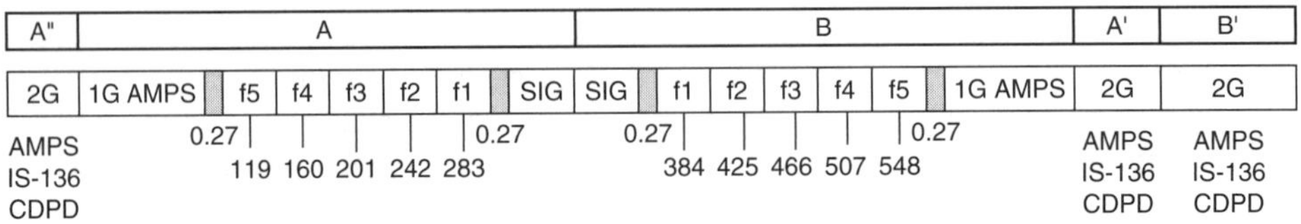

FIGURE 16.12 EVDO overlay FA deployment scheme, cellular.

Figure 16.11 presents an example of an EVDO overlay approach where 1xRTT is currently deployed with a 5-MHz PCS channel allocation, and a 1xRTT channel needs to be displaced in order to facilitate the introduction of EVDO. With migration of a 1xRTT channel, the need potentially to build more capacity may be required. However, for this example, the traffic loading also does not require the use of multiple FAs for 1xRTT, and therefore, the issue of spectrum migration for this example is not relevant.

Alternatively, Figure 16.12 indicates the possible deployment of EVDO into the cellular band. Looking at the figure, EVDO could be placed into service as an overlay as F5 or be allocated for use in the expanded-spectrum portion of both the A- and B-block systems. The figure also includes other legacy systems as part of the channel scheme.

A very crude diagram of an EVDO overlay network is shown in Figure 16.13. A few items need to be mentioned, and the first is that the BTS and BSC need to be upgraded to support EVDO. Additionally, the PDSN needs to have the correct software loaded to support EVDO's enhancements such as QoS. Also, the facilities connecting the BTSs to the BSCs need to be separate for EVDO and 1xRTT. There is the chance for facility concentration, but in a growth environment, this probably will be prohibitive in terms of operations and engineering redesigns.

Looking at Figure 16.13 and comparing it with Figure 16.6, the one item that leaps out is the size of the facilities between the BTSs and BSCs. The facilities for EVDO result in the addition of two T1s, unstructured, for use with EVDO.

The PN offset assignment scheme presented earlier in this chapter should be used for system design following an $N = 19$ reuse pattern for the PN offsets. Just as with the design example for a new CDMA2000-1xRTT or EVDO system, there is a plethora of issues not covered in this example. However, the preceding material should help in setting up the thought processes to achieve the desired goal of supporting the customer requirements for service delivery and transport.

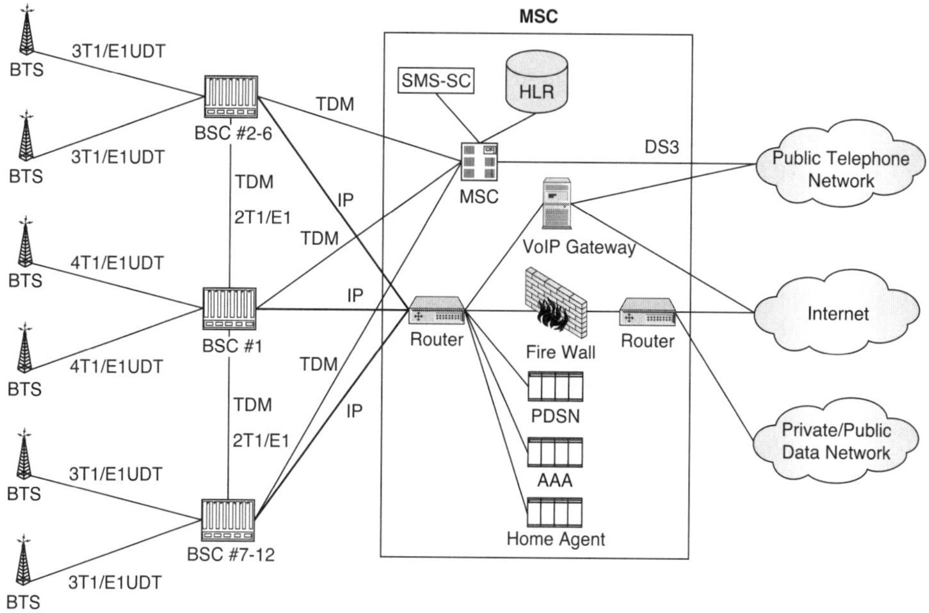

FIGURE 16.13 EVDO overlay network sample configuration.

References

Barron, Tim, "Wireless Links for PCS and Cellular Networks," *Cellular Integration*, September 1995, pp. 20–23.

Bates, Gregory, "*Voice and Data Communications Handbook*," McGraw-Hill, New York, 1998.

Carr, J. J., "*Practical Antenna Handbook*," Tab Books, McGraw-Hill, Blue Ridge Summit, PA, 1989.

DeRose, James E., "*The Wireless Data Handbook*," Quantum Publishing, Mendocino, CA, 1994.

Dixon, Robert C., "*Spread Spectrum Systems*," 2d ed., Wiley, New York, 1984.

Jakes, W.C., "*Microwave Mobile Communications*," IEEE Press, New York, 1974.

Lynch, Dick, "Developing a Cellular/PCS National Seamless Network," *Cellular Integration*, September 1995, pp. 24–26. Kasera, Sumit, and Nishit Narang, "3G Mobile Networks," McGraw-Hill, 2005.

Molisch, Andreas F., "*Wideband Wireless Digital Communications*," Prentice-Hall, Englewood Cliffs, NJ, 2001.

Morrow, Robert, "*Wireless Network CoExistance*," McGraw-Hill, New York, 2004.

Qualcomm, "An Overview of the Application of Code Division Multiple Access (CDMA) to Digital Cellular Systems and Personal Cellular Networks," Qualcomm, San Diego, CA, May 21, 1992. Rosenberg, Adam, and Sid Kemp, "CDMA Capacity and Quality Optimization," McGraw-Hill, New York, 2003.

Salter, Avril, "W-CDMA Trial & Error," *Wireless Review*, November 1, 1999, p. 58.

Shank, Keith, "A Time to Converge," *Wireless Review*, August 1, 1999, p. 26.

Siddiqui, Shahid, "Roaming in Wireless Networks," McGraw-Hill, New York, 2006.

Smith, Clint, "*Practical Cellular and PCS Design*," McGraw-Hill, New York, 1997.

Smith, Clint, "*Wireless Telecom FAQ*," McGraw-Hill, New York, 2000.

Smith, Gervelis, *Cellular System Design and Optimization,* McGraw-Hill, New York, 1996.

Smith, Meyer, *3G Wireless with WiMAX and WiFi,* McGraw-Hill, New York, 2004.

Stimson, G.W., "*Introduction to Airborne Radar,*" Hughes Aircraft Company, El Segundo, CA, 1983.

3GPP2 C.S0008-0: "Multicarrier Specification for Spread Spectrum Systems on GSM MAP (MC-MAP) (Lower Layers Air Interface)," June 9, 2000.

TIA/EIA-98-C: "Recommended Minimum Performance Standards for Dual-Mode Spread Spectrum Mobile Stations (Revision of TIA/EIA-98-B)," November 1999.

TIA/EIA IS-2000-1: "Introduction to cdma2000 Standards for Spread Spectrum Systems," June 9, 2000.

TIA/EIR IS-2000-2: "Physical Layer Standard for cdma2000 Spread Spectrum Systems," September 12, 2000.

TIA/EIA IS-2000-3: Medium Access Control (MAC) Standard for cdma2000 Spread Spectrum Systems, September 12, 2000.

Webb, William, "CDMA for WLL," *Mobile Communications International,* January 1999, p. 1.

Willenegger, Serge, "cdma2000 Physical Layer: An Overview," Qualcomm 5775, San Diego, CA.

www.3gpp2.org

CHAPTER 17

TD-CDMA and TD-SCDMA System Design

TD-CDMA and TD-SCDMA are part of the *International Mobile Telephony 2000* (IMT-2000) group of specifications using *Time-Division Duplex* (TDD) as their radio access method. The system design associated with a TDD system, whether it is *Time Division–Code Division Multiple Access* (TD-CDMA) or *Time Division–Synchronous Code-Division Multiple Access* (TD-SCDMA), has some unique characteristics that differentiate it from a typical *Frequency-Division Duplex* (FDD) design that is normally associated with wireless mobility.

Some of the unique characteristics of a TDD design aspect are as follows:

- TDD has no need for paired frequencies; it uses the same frequency for uplink and downlink transmission.
- TDD is suitable for asymmetric uplink and downlink transmission rates, especially for *Internet Protocol* (IP)–type data services.
- The TDD system's major attribute is its spectral efficiency with asymmetric traffic such as IP.

TD-CDMA and TD-SCDMA use different TDD techniques for their *Radio Access Network* (RAN), but both use the same core network as *Wideband Code-Division Multiple Access* (WCDMA), enabling both technologies to share a common core network with WCDMA and/or *Global System for Mobile* (GSM) communications. TD-CDMA has been deployed in only a few markets and is primarily a wireless data solution, whereas TD-SCDMA is deployed commercially primarily in China and has an evolution path like TD-CDMA to *Long-Term Evolution Time-Division Duplex* (LTE-TDD).

Table 17.1 highlights some of the similarities and differences between TD-CDMA and TD-SCDMA.

The rest of this chapter will discuss the overall design aspects expected for TD-CDMA and TD-SCDMA networks. Because the core network is the same for both TDD RAN systems, this is not be repeated in any detail (because it was covered previously in Chapters 5 and 6). However, the RAN portions have different *Radiofrequency* (RF) characteristics and dimensioning requirements, necessitating separate sections for this coverage.

One of the profound questions that has been echoed with each design consideration and then often forgotten is, What is your objective with the business plan, and how can the system design help to achieve the overall financial plan? Therefore, the initial questions posed are more marketing and business decisions that are intertwined with the

	TD-CDMA	TD-SCDMA
Multiple access	HCR TDD	LCR TDD
Handoff	Hard	Hard
Modulation	QPSK	QPSK/8PSK
Bandwidth	5 MHz	1.6 MHz
Chip rate	3.84 Mcps	1.28 Mcps
Time slots per frame	15	7
Spreading Factor	1,2,4,8,16	1,2,4,8,16
Receiver	Joint Detection Rake (mobile)	Joint Detection Rake (mobile)

Table 17.1 TD-CDMA and TD-SCDMA Comparison

technical platforms that exist for a wireless system. Thus, through a presentation of some high-level system design examples, this chapter hopes to assist you in better weighing the technical issues associated with decisions surrounding service and network deployments, enabling better implementation of these exciting technology platforms.

17.1 Design Methodology

The methodology for the network and RF design of a TDD system needs to be established at the beginning of the process. Establishing a solid methodology is essential in the beginning stages to ensure that the proper baseline assumptions are used to facilitate flexibility in design and implementation. Flexibility is needed in the design and implementation in order to address the many future issues that are really unknown and therefore cannot be foreseen properly at the onset of the design process.

Some high-level questions for a TDD system designer involve the desired services, if this is a greenfield deployment, and whether the TDD system needs to be integrated into an existing network as an overlay or is simply a replacement. Numerous issues need to be identified at the beginning of the design process, and the following are some of the more important technical issues:

- Time frames for the growth plan
- Subscriber growth projections
- Services offered and take rates for each
- Design criteria for RAN and core
- Construction expectations
- Legacy and future technology systems
- *Capital Costs* (CapEx) and *Operations Costs* (OpEx) objectives

It is essential to determine the time frame for the growth plan at the beginning. The time frame will define what the baseline, foundation, and future look of the the plan will present. Therefore, establishing the baseline month or time frame associated with the data used for generating the plan is critical because the wrong baseline will alter the outcome of the report.

The amount of time the plan projection is to take into account is also critical for the analysis. The decision to project 1, 2, 5, or even 10 years has a dramatic effect on the final outcome and accuracy of the forecast. In addition to the projection time frame, it is important to establish the granularity of the reporting period—monthly, quarterly, biannually, yearly, or some perturbation of them all.

The particular marketing plans also need to be factored into the report. The marketing department's plans are the leading element in any network and RF growth study. The input parameters to the network and RF growth plan provided by the marketing department are as follows:

- Projected subscriber growth for the system over the time frame
- Projected kilobits or megabits per subscriber expected at discrete time intervals
- Dilution rates for legacy subscriber equipment over the time period, specifically the migration from legacy systems, if still applicable, to TDD
- Types of subscriber equipment used in the network and percent distribution of CPE projections
- Special promotion plans over the time frame of the report, such as free Internet access
- The projected number of mobile data users over the time frame of the study
- Data services offered over the time period and the take rates for each service type and treatment

There are a multitude of other items needed from the marketing department for determining network and RF growth. However, if you obtain information on these basic eight topics from the marketing department, it will be enough to start the RF, fixed network, voice, and packet design.

The design, whether it is for new or expansion systems, needs to factor in the following key elements along with the expected traffic-loading forecast:

- Spectrum available
- Spectrum required
- Options and feasibility of achieving the required bandwidth needs
- Flexibility to meet ever-changing market conditions
- Cost-effective use of the existing and future capital infrastructure
- Standardized systems that enable backward- as well as forward-compatibility with other networks and data platforms
- *Quality of Service* (QoS)/*Grade of Service* (GoS) for each type of service offering
- Coverage requirements, either new or enhancements to existing coverage

With this said, the next step is to establish some guidelines that are specific to TD-CDMA and TD-SCDMA systems. There are numerous guidelines that a designer or design team must factor into the process. However, the guidelines can be boiled down to two major areas:

- Deployment
- Dimensioning

Additionally, the migration plan from TD-SCDMA or TD-CDMA to LTE-TDD needs to be considered in the up-front design process.

17.2 Deployment Guidelines

The deployment of a TDD system presents different challenges to the designer depending on design objectives and the configuration of any existing infrastructure. If the design is for a new system (greenfield), the deployment is driven by coverage and then by capacity. Alternatively, if the design is for integrating TD-CDMA or TD-SCDMA into an existing system, such as a GSM or CDMA system, then it is more focused on capacity and possibly inclusion of packet-data services as the new service offering.

TD-CDMA and TD-SCDMA require different radio bandwidths that need to be included in the spectrum plan. As an overlay network, either TDD system can be deployed as a separate and unique system, providing many possibilities for a design engineer to use in achieving the desired design requirements. Some of the design options for integrating TD-SCDMA or TD-CDMA into an existing legacy system are

- Greenfield
- Overlay
- Migration

There are obviously several variations to the three main areas listed. For example, a greenfield deployment can consist of a new operator deploying the system using new spectrum. Another example of a greenfield deployment would be the situation where the new system is built adjacent to an existing system but is run autonomously.

The overlay approach is the most probable scenario, in which the TDD network is implemented into an existing wireless network that most likely contains GSM and/or *Universal Mobile Telecommunications Service* (UMTS); in addition, both also could have WiFi, WiMax, and/or CDMA2000 as well. Additionally, the overlay approach will involve either new or dedicated spectrum or spectrum clearing.

For an overlay approach, the use of new or dedicated spectrum is the preferred method because it facilitates implementation of the new RAN platform into the network without potentially causing any adverse effects for the existing customer base. Spectrum clearing, on the other hand, may be required when new spectrum is not available and the existing services use the entire license allocation. Spectrum clearing typically results in the need to build more legacy cell sites and supporting infrastructure because of the reduced spectrum of operation resulting in the need to build capacity-relief sites, further driving up operations costs for an operator.

Additionally, when introducing either TD-CDMA or TD-SCDMA into an existing GSM and/or UMTS system, there is a need to upgrade different network elements depending on the radio infrastructure supplier that is used by the operator. Obviously, a new RAN will need to be implemented, and this involves new equipment for the Node B and *Base Station Controller* (BSC) to support the unique TDD network. The *Serving General Packet Radio Service (GPRS) Support Node* (SGSN) may need to have software upgrades to support multiple RAN interfaces.

However, several key elements that need to be resolved at the onset of the design process, besides harmonization with the business plan, include

- Establishment of design criteria
- Traffic estimation
- Spectrum management
- Implementation (civil work)
- Interconnect
- Optimization
- Sustainment (operations, maintenance, and refresh)

17.3 System Traffic Estimation

Traffic estimation for a TDD system or, in fact, any communications platform depends directly on the type and quality of services that will be offered and how they will be transported. The process and methodology for conducting system traffic engineering for a TDD system involve determining the amount of physical and logical resources that need to be in place at different points and nodes within the network to support the current and future traffic. The determination of existing traffic loads is rather more straightforward in that you have existing information on which to base decisions, that is, the legacy network. For future forecasts, the level of uncertainty grows exponentially the further the forecast or planning takes you into the future. However, many elements in the network require long lead times, ranging from 3 weeks to over 1 year to implement. Obviously, the goal of traffic engineering is to design the network and its subcomponents not only to meet the design criteria, which should be driven by technical, marketing, and sales information, but also to do so in a cost-effective manner. It is not uncommon to have conflicting objectives within a design, that is, to ensure that the customers have the highest QoS/GoS for both voice and packet data but yet have a limited amount of capital by which to achieve this goal. Operations costs also need to be factored into the design process given the long-term impact they will have on maintaining a healthy business. Therefore, it is important at the onset of the design process to have some interim decision points where the design process can be reviewed and altered, if required, either by increasing the capital budget, revisiting the forecast input, or altering the QoS/GoS expectations.

The variations to the circuit- and packet-switched services offered may be vast, but there are some commonalities that can be drawn on. Additionally, if a TDD system is deployed with a GSM/UMTS network, then the type of traffic and which system it will reside on need to be differentiated. It is important to remember that TD-CDMA and UMTS have the ability, depending on the handset employed, to potentially hand off. Additionally, TD-SCDMA has the ability to hand off with UMTS as a blind handoff. TD-CDMA and TD-SCDMA do not have that ability to hand off to legacy systems like GSM, UMTS, or CDMA, except as a potential blind handoff.

There are several methods that can be used for calculating the required or estimated traffic for the network and RAN, and these were covered in Chapters 5 and 6. Regarding the various network nodes, the dimensioning for each is vendor-specific; that is, a BSC from one vendor may be able to support 32 Node Bs, whereas another vendor's product may be able to support 40. Therefore, the specific node-dimensioning guidelines need to be obtained from each vendor. However, the vendors will not provide your traffic requirements, only the equipment-dimensioning criteria.

It is essential that the desired occupancy percentage be established, and it should not exceed 70 percent for any node. This level is meant to handle peak loads, provide ample time for capacity augmentation, and also help to facilitate network reconfigurations and expansions. Keeping in mind that not all nodes have a 70 percent design goal (SS7/CC7 usually has 40 percent), each node must be evaluated on its own merits.

With this said, there are several key points within the network where traffic engineering calculations need to be applied:

- Node B to subscriber terminal
- Node B to *Radio Network Controller* (RNC)
- RNC to packet network
- RNC to voice network

There are several situations and an unknown number of perturbations that can occur in the estimation of traffic for a system. In an ideal world, the traffic forecast would be projected by integrating the marketing plan with the business plan, coupled with the products that should be integral to both the marketing and business plans. However, reality is much harsher, and usually very little information from which to dimension a network is obtainable by the technical team. Therefore, the following is meant to help steer new system planners in determining their traffic-transport forecast.

Because TDD supports asymmetric data, the estimation of packet data associated with the data services offered, as well as *Voice over IP* (VoIP), that will transverse the system is essential. The average packet-data user is expected to use the following services:

- E-mail: 65 percent
- Web browsing: 30 percent
- Extension of company network (LAN): 27 percent
- Address book/calendar functions: 27 percent

Two key concepts about packet data:

1. Initially, packet-data traffic is low owing to the penetration rate. The higher data speed is a result of the data not being as time-sensitive as voice.
2. Packet-data services are an enabler for more services offered by the operator.

However, with the wide adoption of 3G Smartphone terminals, the demand for data has seen a strong demand and deploying a TD-CDMA or TD-SCDMA network will need to anticipate this type of heavy data usage. The forecast, or growth, should be extrapolated from the business plan or simplified marketing plan, which would specify a desired or expected growth level.

The equation to follow for an existing system would be

$$\text{Total traffic} = \text{existing traffic} + \text{new traffic expected}$$

Symbol	Service Type	Transport Method
S	Voice	Circuit switch
SM	Short message	Packet
SD	Switched data	Circuit switch
MMM	Medium multimedia	Packet
HMM	High multimedia	Packet
HIMM	High interactive multimedia	Packet

Table 17.2 Circuit and Packet Data

The new traffic expected could be a simple multiplication of the existing traffic load.

If the system is new, or a new service/platform is introduced, then the traffic needs to be distributed in a weighted proportion to each of the markets being designed for the system or homogeneously for a given market or submarket. The forecasting for voice traffic and packet services was discussed in Chapter 11 and is well documented; thus, it will get only superficial treatment here. However, the real issue with traffic dimensioning lies in the ability to forecast both circuit-switched and new packet-data services that will be used by customers of the wireless operator related to TDD.

The ultimate question that the designer must answer is, "How do you plan on supporting the traffic with the prescribed services?"

Numerous types of services are available for both circuit switched and packet data, so some generalizations are needed in order to have a chance at arriving at the necessary inputs for the design phase. For a new or existing system, the issue of where to begin is always the hardest part. However, one of the key parameters you need to obtain from marketing and/or sales is the penetration rate or take rate for each of the service types offered, which can be grouped into the segments defined in Table 17.2. Please note that voice traffic also can be *Multimedia* (MM) or *High Multimedia* (HMM) with VoIP.

It is important to note that all the circuit-switched services in Table 17.2 are either symmetric or asymmetric. Of the services listed previously, only MM and HMM are asymmetric; the rest are symmetric service offerings, and this has a direct impact on the amount of *Time Slots* (TSs) that will be available to the customer base.

Traffic estimation for both circuit-switched and packet-switched services can be achieved by several methods, such as a general approach where a standard percentage is used for, say, packet services. Or, you could base the amount of packet-data subscribers on the number of handsets expected to be purchased for resale in the market.

Regardless, the first step in any traffic study is to determine the population density for a given market; in the case of an existing system, the population density and primary penetration rates are already built into the system because of known loading issues. However, especially for new services, such as packet data, the process of determining the population density for a given area, followed by the multiplication of that number by the penetration rate, will greatly help in determining the expected traffic load from which to design the system.

For the traffic forecast, the following marketing attributes need to be defined to the best level of granularity and clarity possible:

- Population density
- Penetration rate
- Coverage area
- QoS/*Service-Level Agreement* (SLA)

With the introduction of packet data with TD-CDMA or TD-SCDMA, traffic modeling for packet-switched data involves interaction of the following items:

- Number of packet bursts per packet session
- Size of packets
- Arrival time of a packet burst within a packet session
- Arrival times for different packet sessions

17.4 Radio Elements

Both TD-CDMA and TD-SCDMA are radio access platforms, and the RAN needs to be dimensioned properly to meet customer demand. The radio elements for both TDD systems include the antenna system, Node B, and the RNCs, and they have a direct influence on the circuit-switched and packet-data network requirements.

There are a few key elements that are associated with the radio dimensioning of a TDD system, whether it is TD-CDMA or TD-SCDMA. These key elements include

- Spectrum
- Channel-assignment scheme
- Site configuration (antennas)
- Link budget

17.4.1 Spectrum

The spectrum requirements for TD-CDMA and TD-SCDMA are different: TD-CDMA requires 5 MHz, whereas TD-SCDMA requires 1.6 MHz. Depending on the deployment method used, spectrum segmentation may be required, as well as the use of a guard band and possibly a guard zone.

When deploying the second channel of a TDD system, there are a number of methods that can be used. Typically, a second or third channel will be required because of capacity needs. However, the second or additional channels can be used for specific services, allowing for differentiation or segmentation of service offerings. The method for selecting the next channel depends on numerous factors.

If the spectrum that the TDD system is being deployed in is meant for TDD initially, then the channel-assignment scheme is more or less sequential, starting with a middle-band frequency and then using the next channel on either side and then working outward from that point until the available spectrum has been exhausted. If the spectrum that the TDD system is deployed in is an FDD allocation, then the decision needs to be based partly on the available spectrum and channels cleared or available. For instance, if only one FDD channel is cleared, then the initial allocation should be what would be the base *transmission* (Tx) channel for the FDD system. The second TDD channel would follow to be the other part of the paired-channel set.

The use of a guard band will be necessary for the deployment of a TDD system in an FDD network. The guard band is used to protect the mobile units that are operating on the FDD system from being interfered with by the TDD base stations and mobile units.

A guard zone will be necessary if the spectrum is not entirely cleared. More specifically, the TDD system is deployed in a limited fashion either by design or through implementation constraints. The guard zone will be necessary to create an interference-free area where neither the TDD nor the FDD system is allowed to use spectrum. Obviously, this is not a good long-term solution because the spectrum cannot be used in that area.

17.4.2 Antenna Configurations

The site configurations for TD-CDMA and TD-SCDMA systems should be considered with care from the onset. Taking a simplistic view of both TDD architectures, the antenna requirements dictate a minimum of two antennas per sector. When implementing a TDD system into a legacy network, it is strongly advised that the antenna system for the TDD system be independent of that of the legacy network. The rationale behind this lies in the fact that the design criteria are different, and the sector or site is differentiated by code not by a *Time-Division Multiplexing* (TDM) carrier.

If the legacy system is GSM, the antenna system relies on a D/R ratio that is largely influenced by the orientation or grid of the system. TDD systems, however, do not have the restrictions that a TDM system has, and therefore, the antennas can be oriented to achieve the maximum coverage and capacity for a given area and therefore are not subject to a rigid or semirigid reuse grid.

Additionally, when a TDD system is implemented in conjunction with a WCDMA network, again, the desire is to have the antenna system separated. The reason for the separation lies in transmit-combining issues for power in addition to orientation flexibility.

Obviously, the antenna real estate for many locations is quite limited, and compromises may have to be made. However, compromises should not become the normal method of operation or desired result for the antenna system.

17.4.3 Node B

Node B controls the interface between the TD-CDMA or TD-SCDMA network and the subscriber unit. Node B controls many aspects of the system that are directly related to the performance of the network. Some of the items Node B controls are the multiple carriers that operate from the site, the forward power (allocated for traffic, overhead, and soft handoffs), and the assignment of the codes.

With all TDD systems, the use of multiple carriers per sector is possible. Therefore, when a new voice or packet session is initiated, Node B must decide how best to assign the subscriber unit to meet the services being delivered. Node B in the decision process not only examines the service requested but also must consider the radio configuration and subscriber type, as well as, of course, whether the service requested is voice or packet. Therefore, the resources Node B has to draw on can be both physically and logically limited depending on the particular situation.

A key concept with TDD systems is that soft and softer handoffs typically associated with CDMA2000 and WCDMA systems are not used in TD-CDMA and TD-SCDMA systems, freeing up additional resources for traffic handling.

The following is a brief summary of some of the physical and logical resources Node B must allocate when assigning resources to a subscriber:

- Number of physical resources available
- Power already allocated and that which is available
- Codes required and those available to support the services
- Total TSs used in that sector

The capacity for the TDD Node B needs to be looked at from two primary aspects. First, the traffic-handling capability of the site must be examined, where the cell or sector is designed to support a traffic load for voice of, say, a 2 percent GoS. The second issue is the threshold or occupancy level that the site is allowed to achieve prior to having a relief plan become implemented. The relief plan would involve offloading the capacity either through additional hardware or implementation of another site. The occupancy percentage should be 70 to 75 percent when this triggers action. The specific percentage needs to be evaluated in the market to determine a realistic time frame for remediation; that is, if it take 9 months to order a circuit and you have a 40 percent growth rate, then your threshold might be 30 percent.

17.5 Fixed Network Design Requirements

Traditional voice services and their dimensioning requirements have been well documented and have been the primary dimensioning and growth driver for wireless mobile systems since their inception. However, with the introduction of packet-data services, this requires a focus not only on the radio environment for either of the TDD systems but also on all the supporting elements that make up the wireless system. Therefore, the following are the major elements that need to be factored into the design of a TD-CDMA or TD-SCDMA system:

- MSC (Media Gateway, IMS)
- RNC
- Node B
- SGSN

In reviewing this list, several nodes or elements directly associated with radio elements are listed. The reason that Node B and RNC are listed in the fixed network design requirements lies in the simple concept that connectivity needs to be established between Node B and the RNC, whether it is via landline services or a microwave link. The RNCs are listed not only because they route packet and voice traffic, which requires a certain link dimensioning, but also because they can be local or remote to the MSC depending on the ultimate network configuration deployed.

The fixed network design includes not only element dimensioning but also dimensioning of the links that connect the various nodes or elements in order to establish a wireless system. Some of the connectivity requirements involve the following elements. It is important to note that all elements require some level of connectivity, whether it is from the *Digital Cross Connect* (DXX), also referred to as *DACS,* to the MSC or between the voice-mail platform and the switch. However, the list that follows involves elements that usually require an external group to interface with, while the internal nodes are more controllable.

- Link between Node B and RNC (usually a leased line)
- Link between RNC and MSC (if remote)
- Node B/RNC concentration method
- Connectivity of the SGSN
- Router (for internal)
- Router (for external)
- *Session Creation Server* (SCS) and other servers
- Interconnection to the public and/or private data networks

Note that as a general practice, the routers used for the packet-data network for SGSN applications should not be used for other functions, such as company LAN work.

The SGSN needs to have connectivity with the following major nodes:

- TD-CDMA or TD-SCDMA RNC
- Legacy GSM system (UMTS or CDMA)
- Either a public data network, private data network, or both
- HSS/HLR
- GGSN
- Service creation platform that contains the configuration, policy, profile, subprovisioning, and monitoring capability

The SGSN is usually connected to the packet-data network via an OC3/STM1 or 100baseT connection. The choice of which bandwidth to use is determined not only by the proximity of the SGSN to the RNC, but also by traffic requirements. In summary, the SGSN design is based on many factors, including the following basic issues:

- Number of RNC locations
- Access type supported (Simple IP, Mobile IP, etc.)
- Connectivity between nodes
- Network performance requirements
- Number of active and idle sessions supported

It is important to remember that the SGSN may have to interface with legacy systems, which also have to be factored into the dimensioning requirements not only for traffic volume but also for port availability.

17.5.1 Utilization Rates

The utilization goal for the network should be 70 percent of capacity for the line rate over the time period desired. However, each major node in the system needs to be evaluated using vendor and operation experience, as well as financial objectives, to guide the decision process.

For example, the time period that should be used is a 9-month sliding window that needs to be revisited briefly on a monthly performance report and then during a quarterly design review (following the design review guidelines process). The facility will need to be expanded once it is understood that the 70 percent utilization level will be exhausted within 9 months, with continued growth showing an exhausting 100 percent within 18 months.

For the packet-switched network, the following should be used as the general guidelines:

- Processor occupancy: 70 percent
- Switching platform: SCR (25 percent); PCR (90 percent)
- Port capacity: Design for 70 percent port utilization (growth projection for additional ports based on 9-month forecast)

Naturally, buffer dimensioning and throughput are also important design and dimensioning requirements.

17.5.2 IP Addressing

The issue of IP addressing is important to a TD-CDMA and TD-SCDMA system. The introduction of Simple IP and Mobile IP with and without a *Virtual Private Network* (VPN) requires the use of multiple IP addresses for successful transport of the packet services envisioned to be offered. It is therefore imperative that the IP addresses used for the network be approached from the initial design phase to ensure a uniform growth that is logical and easy to maintain over the life cycle of the system.

Not only does the introduction of packet data require an IP address scheme for the mobility portion of the system, each of the new platforms introduced needs to have its own IP address or range of IP addresses. Some of the platforms requiring IP addresses involve

- SGSN
- Routers
- Subscriber units

Some of these new devices require the use of private addresses as well as some public addresses. However, every device that wants to communicate using IP needs to have an IP address associated with it. There are, of course, public and private IP addresses. The public IP addresses enable devices to communicate using the Internet, whereas private addresses are used for communications in a *Local-Area Network* (LAN)/*Wide-Area Network* (WAN) intranet environment. The TD-CDMA and TD-SCDMA systems will use both public and private addresses. However, the bulk of the IP addresses will be private in nature and, depending on the service offering, will be dynamically allocated or static in nature.

The private addresses will not be recognized on the public Internet system, and this is why they are used. Also, it will be necessary to reuse private addresses within sections of the network, profound as this may sound. Because the packet system is segregated based on the SGSN, it can be assigned the same range of IP addresses. Additionally, based on the port involved with the SGSN, the system can be segregated into localized nodes, and the segregation enables for the reuse of private IP addresses, ensuring a large supply of a seemingly limited resource.

To facilitate the use of IP addressing, a subnet further helps to refine the addressing by extending the effective range of the IP address itself. The IP address and its subnet directly affect the number of subnets that can exist and, from those subnets, the number of hosts that also can be assigned to that subnet.

Just what is the procedure for defining the IP addresses and its associated subnet? The following rules apply when developing the IP plan for the system; the same rules are used for any LAN or *Internet Service Provider* (ISP) designed. There are four basic questions that help to define the requirements:

1. How many subnets are needed presently?
2. How many are needed in the future?
3. How many hosts are on the largest subnet presently?
4. How many hosts are needed on the largest subnet in the future?

You might be wondering why the use of multiple hosts should be factored into the design phase for TD-CDMA and TD-SCDMA. This is so because it is possible to have several terminals for a fixed application using a single TDD subscriber unit or fixed unit.

Therefore, using the previous methods, an IP plan can be formulated for the wireless company's packet-data platforms. It is important to note that the IP plan should factor into the design not only the end customers' needs but also the wireless operators' needs.

Specifically, TD-CDMA/TD-SCDMA operators' needs will involve IP addresses for the following platforms at a minimum. The platforms requiring IP addresses are constantly growing as more and more functionality for the devices is done through *Simple Network Management Protocol* (SNMP).

- Base stations
- Radio elements
- Microwave point to point
- Subscriber units and their capability
- Routers
- ATM switches
- Workstations
- Servers (AuC, SGSN, others)

The list will grow when you tally up all the devices within the network from both a hardware and a network management aspect. Many of the devices just listed require multiple IP addresses to ensure their functionality of providing connectivity from point *A* to point *B*. It is extremely important that the plan follow a logical method.

A suggested methodology is to

- List all the major components that are, will be, or could be used in the network over a 5- to 10-year period.
- Determine the maximum number of these devices that could be added to the system over 5 to 10 years.

- Determine the maximum number of packet-data users per RNC.
- Determine the maximum number of packet-data users per SGSN.
- Determine the maximum number of Mobile IP users with and without a *Virtual Private Network* (VPN).
- Determine the maximum number of Simple IP users with and without a VPN.

The reason for the focus on the amount of Simple and Mobile IP users lies in the fact that these devices will have the greatest demand for IP addresses owing their sheer volume in the network.

Naturally, each wireless system is unique and will require implementation of a different IP address scheme. However, the concept presented has been beneficial and should prove useful. If more information is sought on IP address schemes, an excellent source for information is the Web site www.cisco.com.

17.6 Sample Basic Designs

Several implementation suggestions were put forth for TD-CDMA and TD-SCDMA in Chapters 5 and 6, respectively, and will not be repeated here. The implementation examples in those chapters focused on RAN deployment schemes, with various spectrum allocations being available. The concepts presented are rudimentary in nature because the concept of what needs to be done is stressed, not a particular design for a particular market that will not be relevant for any other system.

Typically, an operator will implement his or her TD-SCDMA or TD-CDMA system as an overlay approach. There are several ways to implement an overlay network, provided that the spectrum is available for use. Often it is spectrum availability, land-use acquisition, and budget constraints that ultimately determine how a system will be rolled out. However, for this discussion, three general approaches for an overlay network will be discussed briefly, and they are

- 1:1 deployment
- *N*:1 deployment
- Targeted areas initially

For a 1:1 deployment, the obvious advantage is coverage parity with the legacy system, enabling a ubiquitous service offering. The 1:1 approach, however, has the longest lead time of any of the overlay approaches. More specifically, besides being financially inhibitive, the 1:1 approach also typically has land-use acquisition issues such as leases, antenna and equipment space, power, HVAC, and permitting issues that require time and do not lend themselves to quick implementations.

The *N*:1 or rather 1:*N* approach has the advantage of deploying the new equipment at selective sites to provide the best coverage and service-offering footprint possible. The advantage with this approach is that it is quicker to market and does not require as much up-front capital and operating money. The negative associated with this approach is the potential for less than optimal service performance owing to coverage and capacity constraints.

The middle-ground approach to both the 1:1 and *N*:1 approach is the targeted-area method. This process involves deploying the new service in selected areas of the network, typically the core of high-traffic areas such as airports, where data services are in high demand. The advantage to this method is that time to market and the service offerings can be designed to provide proper coverage and capacity. The negative aspect with this approach is that it is a limited offering and available only in the targeted area.

The issue at hand with any overlay or system implementation is to marry the technical and implementation aspects with the marketing and budget process with the end goal of maximizing the customer experience while obtaining the desired financial benefits.

References

Bates, Gregory, *Voice and Data Communications Handbook,* McGraw-Hill, New York, 1998.

Chen, Xinhua and Liu Hu, "Technology Advantage of TD-SCDMA," *TDIA,* February 2005.

Huang, Yin-Fu and Tsung-Yi Chiu, "Radio Resource Management for a Mobile Network with TD-CDMA," IEEE 6th CAS Symposium on Emerging Technologies: Mobile and Wireless Communications, Shanghai, China, May 31–June 2, 2004.

Rosenberg, Adam, and Sid Kemp, *CDMA Capacity and Quality Optimization,* McGraw-Hill, New York, 2003.

Smith, Clint, *Practical Cellular and PCS Design,* McGraw-Hill, New York, 1997.

Smith, Gervelis, *Cellular System Design and Optimization,* McGraw-Hill, New York, 1996.

Smith, Gervelis, *Wireless Network Performance Handbook,* McGraw-Hill, New York, 2003.

Sumit Kasera, and Nishit Narang, *3G Mobile Networks,* McGraw-Hill, New York, 2005.

TS C002 V3.0.0 (1999-10): Services Provided by the Physical Layer.

TS C403 V1.2.0 (1999-10): RF Parameters in Support of Radio Resource Management.

www.3GPP.org

www.ccsa.org.cn

www.tdscdma-forum.org.

www.umtstdd.org.

CHAPTER 18

Long-Term Evolution System Design

There are multiple considerations needed when designing an LTE system. Typically, an LTE system design falls into a few general categories of either an overlay or greenfield implementation with the vast majority falling into the overlay scenario. When designing an LTE system, the inclusion or interaction with legacy wireless networks should be considered as part of the up-front design process. The decision to either make the system interoperable with legacy devices or establish it as a stand-alone network is directly dependent upon your business plan.

Whether you are designing a greenfield, overlay, or expansion of the LTE, there are several concepts that need to be stressed. The first concept is that for the design to have any relevance there is a requirement for external inputs from various departments within the company. Specifically, the need to define what services along with their volume and treatment requirements need to be defined at the onset of the design process. Without critical marketing, sales, customer care, and of course, legal/regulatory guidance, the technical design will in all likelihood not support the desired services the company will seek to sell.

It is recommended that a three-step approach takes place beginning with the design concept, or rather, *High-Level Design* (HD) which then feeds a *Preliminary Design* (PD). The PD is where most, if not all, the interactions take place. This PD is then followed by a *Critical Design* (CD), which is where you lock the design in for implementation.

It is recommended that each of the functional disciplines within the company needs to be involved with the overall high-level design review but after that they can be broken down by functional group. There would of course be several layers within each of the efforts involving a PD and a CD. Regardless of the amount of sub-layers within the PD, there would also be a *final design review* meeting where the compilation of all the different functional groups is combined into one unified package which has both Network and RF components. Additionally, the network components should also include the transmission planning aspects and not just the infrastructure requirements.

It is important to note that while operations and implementation are not included in the design process by name, they are involved with the system planning. The engineering departments are responsible for the generation of the technical design and will need to coordinate their information with operations and implementation to ensure that the planned design is possible and that specific network requirements from Operations are not excluded by mistake. Additionally, if your organization has the performance group incorporated into the Operations division or department, then they will need to be directly involved with the RF planning aspects.

This chapter will cover many of the aspects needed when formulating an LTE system design. However, the use of computer models for simulating traffic loads has to be done in order to determine what is realistically possible based on your current inventory of site locations, time to launch, and expansion plans and services that you plan to offer either in the short term or at some practical point in the future.

18.1 Design Methodology

The design methodology for the LTE system whether it is an overlay, greenfield deployment, or expansion needs to be established at the beginning of the process. The establishment of the design methodology utilized is essential in the beginning to ensure the proper baseline assumptions are defined, documented, and communicated from the inception. Included with the baseline assumptions are marketing plans, budget constraints, implementation considerations, and of course, the services that will be offered. Knowing these items at the beginning will greatly improve the design process.

Therefore, the design methodology that is followed is essential in establishing a good system design, which correlates with the business plans of the wireless company. One key element mentioned previously for the overall design process involves both a look at what services need to be supported, where they will be supported, and of course, how it will be supported. Following the age-old what, why, when, and how of any decision process, answering those four fundamental questions will determine the methodology for the design.

- What—defines what you are trying to accomplish with the design
- Why—clarifies the issue of why the platform and/or services need to be introduced
- When—defines a time frame to follow
- How—clarifies the concept of how this will be realized

Since LTE is a high-speed packet network is your plan to offer only a best effort approach or tiered service offering. More specifically, if the plan is to offer high-speed packet data services for fixed applications only and medium-speed packet services for mobility, then the methodology for implementing the packet data services will have a direct impact on the RF system design. Another issue is a decision regarding the legacy network, if it exists, and how to leverage the existing assets or reclaim some of the assets for the LTE network to utilize.

Regarding the implementation, a decision needs to be made as to how it will be introduced. Two possible alternatives involve a direct overlay on the existing system for a 1:1 overlay or proceed with only introducing the new LTE system with a limited footprint, say in core of the network.

Therefore, the methodology chosen determines the fundamental direction of the design itself. The methodology obviously should not be left to the purview of the technical services group but needs to have direct involvement with senior management from marketing, sales, customer service, new technology, operations, implementation, network engineering, and of course, the RF engineering group.

For a successful design to take place, reviews and interim steps in the process should be implemented. The interim steps, that is, not leaping to the final design, are driven by the very fact that you will require to obtain projection information from other departments that are critical to the growth assumptions. Therefore, it is strongly advised that the three-step process of *High-Level Design Review* (HDR) be followed by a *Preliminary Design Review* (PDR) and then the *Critical Design Review* (CDR).

The general flow of the design process is important to stress because it works regardless of the system's age, technology used, and organization. Often problems arise in the design due to shortcuts taking place that are typically driven by either time constraints, lack of technical discipline by management, or both.

Some methodology issues that need to be identified at the beginning of study:

1. Time frames for the design to be based on, that is, whether it is a 1-year or multiyear projection
2. Subscriber demographic information, that is, geographic locations where they are clustered
3. Services offered or planned and if there are any legacy issues to be concerned with
4. Subscriber usage projection (current and forecasted by month and/or quarter) by service type
5. Design criteria (technology-specific issues)
6. Baseline system numbers for building on the growth study
7. Construction expectations (ideal and with land-use entitlement issues factored in)
8. Maximum and minimum offloading for cell sites
9. Utilization factors for CPU and network equipment including transmission facilities
10. Resiliency requirements for transmission, network, and RF (i.e., how many 9s)
11. 911 Phase 2 Implementation or 112 Requirements
12. Advanced platforms and features and time frames
13. Staffing requirements (technical staff loading, training)
14. Technical support equipment (software and hardware)
15. Budget constraints

At each stage of the design process, the department heads need to sign off on the design process phases, where relevant. For example, in the HD phase the head of marketing needs to sign off on their objectives, after review by peers, saying this is what they want the technical services group to implement. For the PD the technical services department head signs off on the design phase while the CD is signed off by corporate.

The amount of time the report projection is to take into account is also critical for the analysis. The decision to project 1 year, 2 years, 5 years, or even 10 years has a dramatic effect on the final outcome. In addition to the projection time frame, it is important to establish the granularity of the reporting period—monthly, quarterly, biannually, yearly—or some perturbation of them all.

The particular marketing plans also need to be factored into the report itself. The marketing department's plans are the leading element in any network and RF growth study. The basic input parameters to the technical design provided by the marketing department are listed next.

1. The projected subscriber growth for the system over the time frame for the design
2. The projected Mbps or kbps per subscriber expected at discrete time intervals for the study
3. Types of subscriber equipment used in the network and percent distribution of CPE projections
4. Special promotion plans over the time frame of the study
5. *Applications* (aps) that will be directly supported by the company

There are a multitude of other items needed from the marketing department for determining design. However, if you obtain the aforementioned basic information, it will be enough to adequately start the design.

The establishment of the design criteria used for the report is another key criteria when putting together the plan. It is recommended that the design criteria used for the study be signed off by the Director of Engineering to ensure that nothing is missed.

The items in the RF and network growth plan design criteria need to include the following as a minimum.

1. Marketing input
2. RF spectrum available for the growth plan
3. Types of services needed to be supported with QoS, latency requirements
4. Coverage requirements
5. Identification of coverage sites
6. BH peak fixed data traffic, 10-day high average for month (or best guess)
7. BH peak mobile data traffic, 10-day high average for month (or best guess)
8. Infrastructure equipment constraints
9. Network topology
10. MME and SGW design criteria including clustering
11. PGW design criteria
12. Resiliency for radio, fixed network, and transmission
13. Facility concentration, that is, remote hubs, relay sites
14. WiFi and femto cell use
15. Interconnection data and VoIP delivery requirements
16. 911 Phase 2 implementation
17. CALEA/lawful intercept requirements
18. New technology and service considerations

19. Cell-site configurations used for new cells
20. Baseline system numbers
21. Backhaul design

The above list should be used as the foundation that can be built upon for establishing the design criteria.

The baseline system numbers should be listed in the design criteria for the growth plan. In particular, the decision to say use June data instead of July data has a significant impact in the plan as well as using peak-versus-an-average value for traffic. It is also important to note that Fixed and Mobile IP traffic will in all likelihood have different BHs.

The budget constraints must be folded into the design itself. Failure to incorporate budget constraints into the plan will only make the plan unrealistic and squander everyone's time needlessly.

18.2 Spectrum

What spectrum to utilize for an LTE deployment may or may not be an option with the initial system deployment. However, as a good design practice, it is best to utilize the lowest frequency block in your spectrum portfolio. Specifically, if you have bandwidth available for the LTE deployment in both the 700 MHz and AWS bands, it would be best to deploy LTE for a macro system in the 700 MHz band first and use the AWS band for *Relay Nodes* (RN).

The standard LTE frequency bands, as of this moment, are called out in Figures 7.3 and 7.4. However, Figure 18.1 shows the spectrum allocation for the various spectrum blocks allocated for wireless mobility in the United States.

It is important to note that when selecting the spectrum for allocating the deployment in LTE that the associated UE device is also capable for operation in the requisite band or bands.

Because LTE can be deployed as either an FDD or a TDD network, there are numerous combinations possible; however, there are practical limits to the spectrum allocation for LTE. Some of the limitations involve not only the UE but the antennas used for the eNodeB and the associated receivers and transmitter along with any unique antenna system combining process that usually has a spectrum bandwidth component to factor into the decision.

As mentioned earlier, it is very possible to deploy LTE in both an FDD and a TDD configuration where different LTE frequency bands or even the same frequency band are allocated for FDD and TDD use. However, the typical LTE deployment utilizes an FDD configuration as a 10 × 10-paired channel assignment. Additionally, when LTE is deployed in the WiMAX frequency band, it will likely be TDD deployment though it is possible to have some of it operate in an FDD mode as well.

18.2.1 Carrier Aggregation

Many wireless carriers have multiple frequency bands that they have in their spectrum portfolio. LTE capacity is directly related to the spectrum bandwidth. The most straightforward way to increase a carrier's capacity is to add more bandwidth so it can operate with more than a 10 × 10 LTE channel. The use of carrier aggregation is possible if the spectrum is available for use. With carrier aggregation, a wireless carrier can increase its

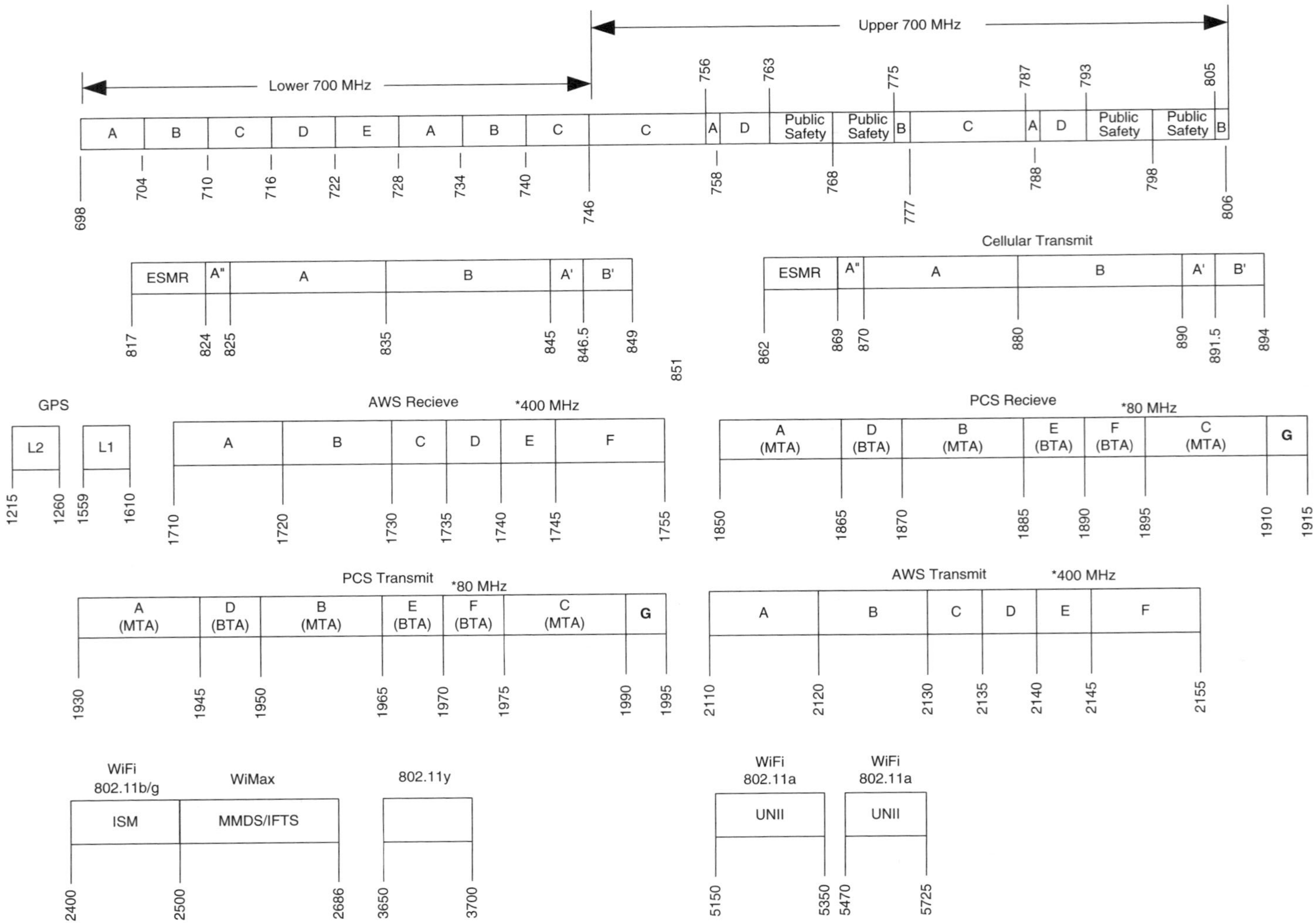

FIGURE 18.1 U.S. wireless mobility spectrum allocation (partial).

LTE	1.4 × 1.4	3 × 3	5 × 5	10 × 10
GSM	5	10	17	34
UMTS	NA	NA	1	2
EVDO	2	3	4	8

Table 18.1 Refarming

overall capacity through increasing the bandwidth of the LTE service by aggregating a 10 × 10 channel with a 5 × 5 or another 10 × 10.

Carrier aggregation affords the wireless carrier the ability to utilize frequency bands, say the PCS E block with a 700 MHz C block. However, when using carrier aggregation, the higher frequency spectrum allocation is the design-limiting factor in the coverage for carrier aggregation due to the increased propagation losses.

18.2.2 Refarming

As LTE capacity requirements increase, the need for more spectrum will place demands on the wireless carrier to refarm its existing spectrum holdings for use with LTE. Table 18.1 depicts the amount of carriers needed to be refarmed based on the LTE bandwidth requirements.

If you are refarming GSM spectrum for LTE use, you will have to reclaim for a 10 × 10 LTE channel, a total of 34 GSM channels which are contiguous. Because this might not be possible, the use of lower bandwidth LTE channels may be more practical with the knowledge that throughput may be compromised. I mention "may be compromised" only because it depends on the services offered and the amount of active UEs on a particular site.

With refarming, the use of carrier aggregation is attractive because you can aggregate small blocks of contiguous channels.

18.2.3 Relay Node (RN)

The use of RNs requires the use of RF spectrum for connecting to the UE and also to the *Donner eNodeB* (DeNB). Allocation of spectrum for use with an RN needs to be thought out in advance. There are several methods for RN spectrum allocation. However, a Type 1 configuration is the preferred method provided the spectrum is available to be used. It is advisable that the higher frequency block be assigned for RN to UE communication; (Uu) operates on the higher spectrum allocation while the (Un) operates on the lower spectrum. This configuration allows the RN to be placed as far away from the DeNB as possible, that is at the cell-edge boundary of the DeNB.

18.3 Coverage Requirements

Coverage, as in all wireless communication systems, is essential for a good communication link. LTE in no exception in that there is a direct relationship between coverage and performance of the LTE system.

As suspected, the range or coverage area for the LTE eNodeB has many variables which impact its overall performance. However, as with all wireless technologies the farther you are away from the eNodeB the lower the radio link quality will be regardless of how well the design is and how many eNodeBs are deployed. Additionally, the link quality has a direct impact on the throughput of the data connection assuming that there are no other bottlenecks for bandwidth in the LTE ecosystem.

Therefore, the first step in the LTE RF design process is to determine all the coverage requirements needed for the network after you know the services you need to support.

The LTE RF design is typically bounded by the cell-edge performance. Cell-edge performance in LTE relates to the overall throughput of the cell and its effective coverage area. The scheduler in the eNodeB, however, greatly enhances the LTE system for improving cell-edge performance. However, cell-edge performance is directly related to the uplink, UE to eNodeB, interference during loaded conditions.

Therefore, as one would expect, as the cell range increases, the performance due to cell-edge conditions causes the eNodeB not to perform as best as it can. At the cell edge the UE uplink causes interference to other eNodeBs due to its range from the serving eNodeB and therefore increases the noise floor of other eNodeBs reducing the SINR and, therefore, performance.

Since LTE sites are frame synchronized, it is not possible to avoid inter-site interference altogether. Therefore, achieving a balance between cell-site coverage and improved cell-edge performance is essential.

The cell-edge performance optimization is handled through the use of computer simulation tools. However, there are several key criteria that are required to determine cell-edge performance and overall coverage of an LTE system. The two criteria are traffic and the RF link budget.

18.4 Link Budget

The link budget for LTE systems is different than that used for GSM, CDMA, EVDO, WCDMA, and HSPA. A representative radio system is shown in Figure 18.2. However, there are similarities in which it is a wireless network requiring a radio transmitter and receiver at both the eNodeB and the CPE (UE). The link budget parameters are essential in determining the overall coverage and cell-edge performance of an LTE eNodeB and system in total. There are many parameters which can and do go into a link budget.

Uplink:

- UL cell-edge bite rate
- Cell-edge bit rate per resource block
- # resource blocks
- SINR required per RB
- RB sensitivity
- Uplink interference margin
- Uplink link budget (path loss and distance)

Downlink:

- Bandwidth
- Power per RB
- UE sensitivity

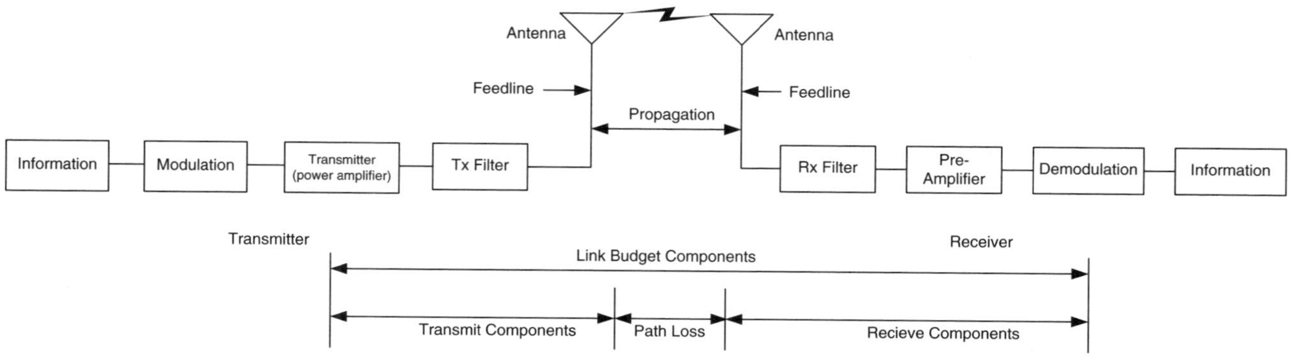

FIGURE 18.2 Radio.

- Max path loss (uplink)
- System load on downlink
- Cell-edge bit rate on DL

Path loss is an essential element in the link-budget determination. The link budget typically is the same for both uplink and downlink provided the frequency separation between the uplink and the downlink is not too large which may require a separate path-loss calculation for the uplink and the downlink paths.

Path loss by itself is a function of distance and in free space is a 20-dB-per-decade loss factor. However, in the wireless mobility free space path loss is only a portion of the path loss calculation. The details of path loss calculations are covered in Chapter 11. The equation utilized for the path loss determination needs to be applicable to the frequency band used, as well as properly defining the correct propagation parameters associated with the morphology.

However, path loss is only a portion of the link-budget calculations. Therefore, the following is a typical link-budget calculation for LTE in the downlink direction.

$$\text{DL path loss} = P_{tx} - S_{ue} - B_{IDL} - FM_{DLF} - L_{BL} - L_{CPL} - L_{BPL} + G_a - L_f - OH_{cc} - P_{DLf} \tag{18.1}$$

The definitions of each of the downlink parameters are defined in Table 18.2.

The following is the uplink link-budget equation for LTE.

$$\text{UDL path loss} = P_{RB} - S_{eNB} - B_{IUL} - FM_{UL} - L_{BL} - L_{CPL} - L_{BPL} + G_a - L_f - OH_{cc} - P_{ULf} \tag{18.2}$$

The uplink parameters are in Table 18.3.

A quick review of the downlink and uplink equations has several parameters being the same as expected since these are primarily environmental conditions that impact the RF path. In some simulations several of the uplink and downlink parameters are included in the path loss equations; however, they are broken out here.

The amount of power per RB is essential in LTE. The DL power budget is such that the RBs can have different power levels while the uplink has a consistent power level during the transmission.

Downlink: 40 dBm (1Tx)

- 6 RB = 35 dB/RB
- 25 RB = 29 dB/RB
- 50 RB = 26 dB/RB

If UE puts out 23 dBm

- 9 RB = 13.5 dBm/RB
- 4 RB = 17 dBm/RB
- 1 RB = 23 dBm

A simple link-budget calculation is shown next in Table 18.4, which takes a little liberty since it is meant to show that the objective is to have a balanced equation between uplink and downlink paths for a good LTE radio design.

Due to the relationship between cell-edge performance, interference, and path loss, the use of computer simulation tools is needed to account for all the perturbations.

P_{tx}	Transmit Power at Antenna Input per RB (dBm)	Power Is Shared Between All RBs, There Is More Bandwidth Used, Is Less Power per RB
S_{ue}	UE Sensitivity (dBm) $S_{ue} = N_t + N_f + 10\log(B_w) + SINR_{CB}$	N_t Thermal noise = 174dBm/Hz N_f Noise figure (dB) 6 dB – 11 dB B_w Bandwidth per RB (180kHz) $SINR_{CB}$ SINR at cell border (dB)
B_{IDL}	Downlink Interference margin or noise floor (dB) $B_{IDL} = 1 + (P_{tx} * SL_{DL} * F_c)/(N_{RB} * L_{SA})$	SL_D Downlink system load (dB) F_c Cell-edge interference factor (other or own) (dB) N_{RB} Noise floor per RB $N_{RB} = N_t + N_f + 10\log(B_w)$ L_{SA} Maximum signal attenuation at cell edge (dB)
FM_{DL}	Fading margin downlink (dB)	
L_{BL}	Body loss (dB)	
L_{CPL}	Car or vehicle penetration loss (dB)	
L_{BPL}	Building penetration loss (dB)	
G_{eNBa}	Gain of eNB antenna (dBi)	
G_{UEa}	Gain of UE antenna (dBi)	
L_j	Antenna jumper loss (dB)	
OH_{cc}	Control channel overhead	PBCH, PDCCH, and PHICH (10-25% or 0.4 dB-1.0 dB)
P_{DLf}	Downlink path loss as a function of distance and frequency (dB)	

TABLE 18.2 Downlink Parameters

An important aspect is that LTE designs are cell-edge limited. The UL SINR becomes Tx-power limited at the cell edge.

However, when you compare the LTE link budget with WCDMA and EVDO, their effective coverage areas are about the same although their capacities are different than LTE. Regardless, the important point with this comment is that LTE overall link budget is similar to WCDMA and EVDO and therefore deploying LTE in 1:1 with existing cell sites is viable.

18.5 Network Topology

LTE can be implemented as a stand-alone system or as an overlay system. Typically, the LTE network is implemented as a single frequency 10-MHz FDD network that is overlaid onto an existing 3G network.

Figure 18.3 shows a high-level diagram of an LTE network with signaling links included for reference. If more detail is desired, 3GPP TS 23.401 and 3GPP TS 23.402 should

P_{tx}	Transmit power at antenna input per RB (dBm)	Power Is Shared Between All RBs, There Is More Bandwidth Used, Is Less Power per RB
S_{eNB}	eNB Sensitivity (dBm) $S_{eNB} = N_t + N_f + 10 \log (B_w) + SINR_{CB}$	N_t Thermal noise = 174 dBm/Hz N_f Noise figure (dB) 6 dB–11 dB B_w Bandwidth per RB (180 kHz) $SINR_{CB}$ SINR at cell border (dB)
B_{IUL}	Uplink Interference margin or noise floor (dB) $B_{IUL} = (N_{RB} + F)/ N_{RB}$ $= 1/(1 - (SL_{UL} \times F \times SINR_{UL}))$	SL_U Uplink System Load (dB) F Interference Factor caused by other eNBs (dB) N_{RB} Noise floor per RB $N_{RB} = N_t + N_f + 10\log (B_w)$ $SINR_{UL}$ Uplink SINR target (dB)
FM_{UL}	Fading margin uplink (dB)	
L_{BL}	Body loss (dB)	
L_{CPL}	Car or vehicle penetration loss (dB)	
L_{BPL}	Building penetration loss (dB)	
G_{eNBa}	Gain of eNB antenna (dBi)	
G_{UEa}	Gain of UE antenna (dBi)	
L_j	Antenna jumper loss (dB)	
	Coax combining losses (dB)	
OH_{cc}	Control Channel Overhead	(10-25% or 0.4 dB-1.0 dB)
P_{ULf}	Uplink path loss as a function of distance and frequency (dB)	

TABLE 18.3 Uplink Parameters

provide sufficient information or direct you to the appropriate specification for the information desired.

In Figure 18.3 depicting the LTE architecture, several new signaling references have been added and they are briefly described next.

S1: It provides access for the radio resources in the eNodeB and used for the transport of user plane and control plane traffic

- S1-U provides user plan traffic to and from the SGW the eNodeB is connected to
- S1-MME provides for the control traffic to and from the MME or MME cluster the eNodeB is connected to

S2a: It provides the user plane with its related control and mobility support between a trusted non-3GPP IP access and the PGW (HSGW-PGW)

S2b: It provides the user plane with related control and mobility support between ePGW and the PGW

S3: It enables user and bearer information exchange for inter 3GPP access system mobility in idle and/or active state

10-MHz FDD	DL (2Tx)	UP (1Rx)
Max power output	40 W (46 dBm)	23 dBm
RB	50	9
Power per RB	29	13.5 dBm
Thermal noise	–174	–174
Noise figure	7 dB (UE)	2.5 dB (eNB)
User bit rate	5 Mbps	500 kbps
SNR	–0.4 dB	–2.9 dB
Ga eNB	12	12
Ga UE	0	0
Sensitivity	–114.9 dBm (Ue)	–121.0 dBm (eNB)
Penetration loss (total)	15 dB	15 dB
Fading margin	8,4	8.4
Interference margin	11.5 dB	2.1 dB
Max. path loss (dB)	139	139

TABLE 18.4 Link Budget

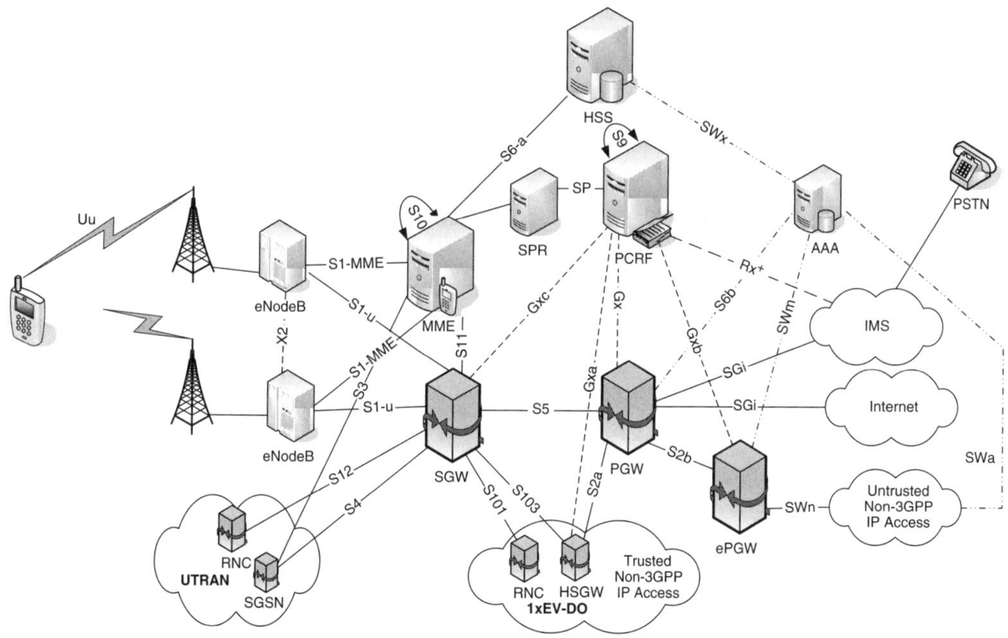

FIGURE 18.3 LTE architecture diagram.

S4: It provides the user plane with related control and mobility support between SGSN/GGSN and the PGW

S5: It provides the user plane with related control and mobility support between SGW and PGW

S6: It enables the transfer of subscription and authentication data for authenticating/authorizing user access to the evolved system (AAA interface)

SGi: It is the reference point between the LTE network and other 3GPP and non-3GPP networks like the Internet or an IMS ring

Table 18.5 has a more complete listing of many of the LTE interfaces and their related primary 3GPP TS specification shown in Figure 18.3.

18.5.1 Self-Organizing Network (SON)

One of the key attributes for LTE is its ability to utilize *Self-Organizing Network* (SON) functionality.

Effectively SON enables the LTE network to self-configure based on the addition or subtraction of eNodeB and other resources in the network itself. For example, SON can make automatic neighbor relations evaluations based on the UE reports and adjust the neighbor lists accordingly.

One very important feature with SON is its ability to perform load balancing in an LTE network where an eNodeB is overloaded, the traffic can be shed, or spread to other eNodeBs. With SON, coverage can also be dynamically adjusted based on the traffic loading as well as facilitating a more optimum network resource utilization with improved performance.

SON also enables the eNodeBs to become self-aware and minimize the initial configuration settings required to be populated manually into the system.

Name	Interface	3GPP TS	Name	Interface	3GPP TS
S1-MME	eNB-MME	36.413	S101	MME-EVDO RNC	29.276
S1-U	eNB- SGW	36.414	S103	SGW-HSGW	29.276
S2a	HSGW-PGW	23.402/275	X2	eNB-eNB	36.423/424
S2b	PGW-ePGW	23.402/275	SG i	PGW-Internet	29.061
S3	MME-SGSN	29.274	Sn	HSS-PCRF	29.328
S4	SGW-SGSN	29.274	Sp	PCRF-SPR	29.203
S6a	HSS-MME	29.272	SWa	AAA-Untrusted	29.234
S6b	AAA-PGW	29.273	SWd	AAA-3PGG	29.234
S11	MME-SGW	29.274	SWm	3GPPAAA-ePGW	29.234
S12	RNC-SGW	29.281	SWn	ePGW-untrusted	29.234
			SWx	HSS-3GPP AAA	29.273

Table 18.5 Interface specs

18.5.2 Tracking Areas

Tracking Area (TA) designs are important in an LTE network. TAs are used for service area definition, paging, and location updates. TAs have a similar function as a *Routing Area* (RA) which is in WCDMA and CDMA. In short, a TA can consist of a single eNodeB or multiple eNodeBs. In practice, a TA usually is configured with multiple eNodeBs that have some commonality in terms of traffic handling and is defined as a *Tracking Area Code* (TAC).

Designing TAs in an LTE network involves a few decisions. However, the end result is to determine which eNodeBs are associated with a particular TA that produces the least amount of location updates for a UE and reduces the overall paging and location signaling needed to track and communicate with the UE.

For example, a small allocation of eNodeBs to a TA may require frequent TA updates if the majority of the traffic is mobile. TA updates, if done too frequently, increase the MME processor load, and, more importantly, it drains the UE battery due to the amount of updates that take place. Also, as an interesting comment, when the TA update is taking place and if the TA area is small, the UE will be unable to acknowledge a page if one is directed toward it when it is undergoing the update process.

However, the flip side to a small TA is to make it as large as possible. This has a negative impact of increasing the paging load on the MME and eNodeB. Depending on the paging load that occurs due to the TA size under signaling, congestion can occur in both the MME and eNodeB. Therefore, the amount of eNodeBs associated with a TA is dependent upon traffic patterns and paging.

Figure 18.4 is an example where one TA is assigned for the whole network. TAs can be set up in an LTE network as either one massive TAC for the entire system or with multiple TACs in a network as shown in Figure 18.5.

Because the TA in LTE is largely driven by RF coverage, the potential for ping-ponging between TAs can occur. Therefore, a TA list is used to eleviate this. A TA list can have several TACs associated with it. In fact, as shown in Figure 18.6, the TA list for TAC 1 has common TAs that are part of the TA list for TAC 3.

Additionally, in an LTE system there will be borders between different TAs and the TA lists. There are a few practical guidelines to follow when picking the location of the TA and TA list borders.

- High mobile traffic should not be on a TA or TA list border, instead it should be ideally located in low mobility or low-traffic areas
- TA and specifically TA lists should be configured to minimize the amount of updates that are required

For example, a bridge like the George Washington Bridge in NYC should not have a separate TA or TA list for each side of the bridge. A better approach would be to have one TA list for the bridge and the eNodeBs, which feed the bridge from both the sides.

When generating a TA and TA list, the following are the high-level capacity triggers needed when designing TA. The particular capacity values need to be extracted from the OEM vendor.

- Paging capacity of the MME
- Paging capacity of the RBS
- Paging intensity per subscriber during busy hour

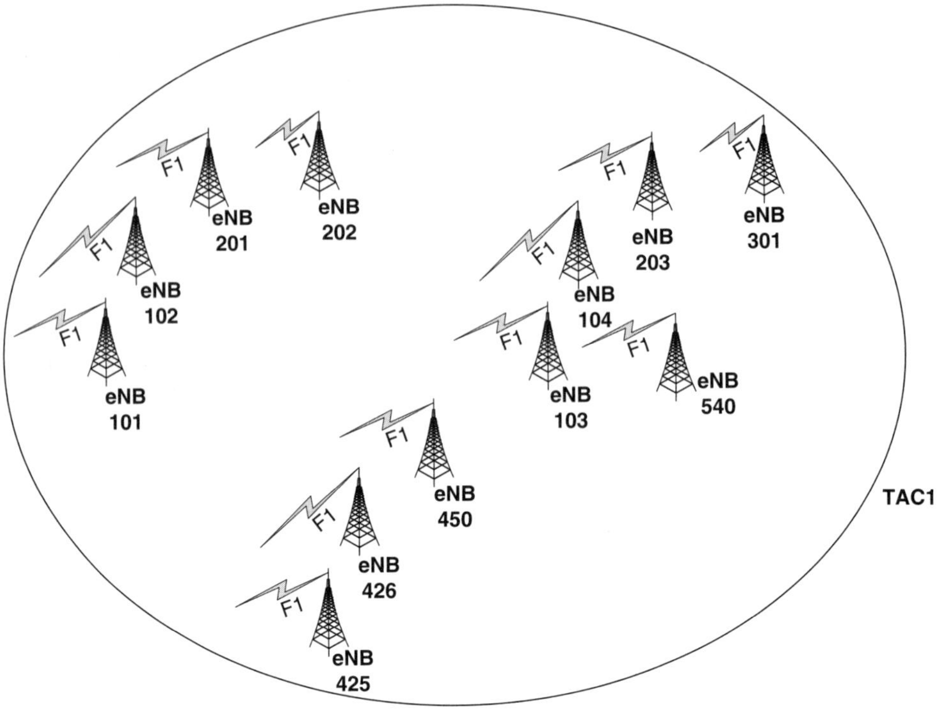

Figure 18.4 Single TA network.

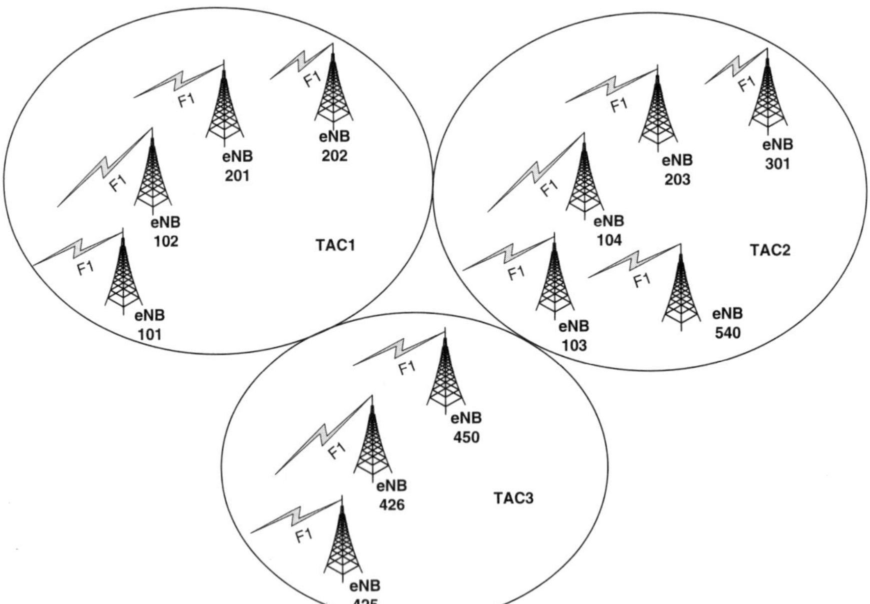

Figure 18.5 Nonoverlapping TAs.

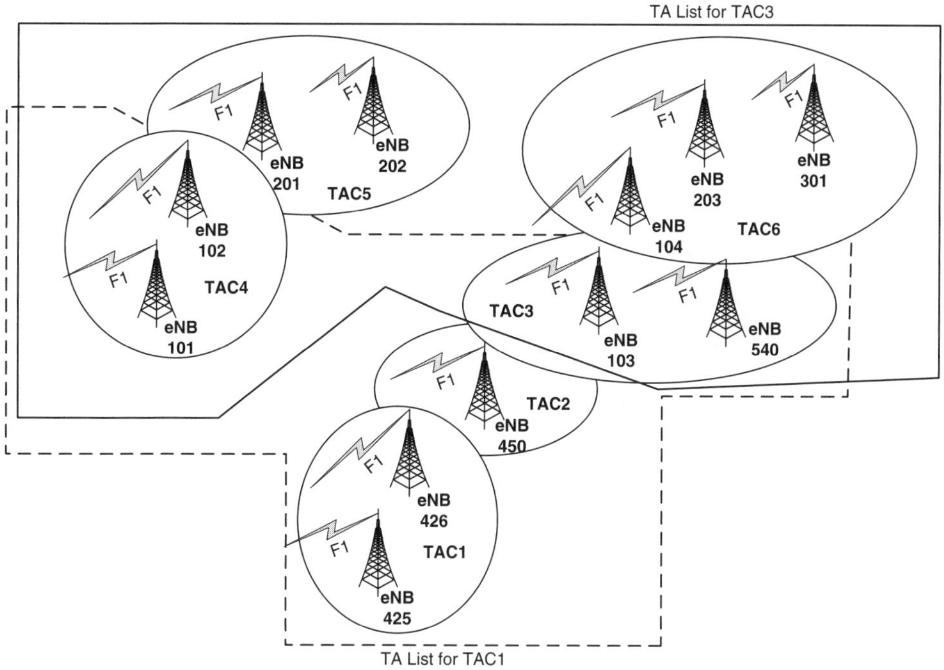

FIGURE 18.6 TA list with multiple TAs.

- Number of simultaneously attached users to the MME during busy hour
- Average number of subscribers per eNodeB during busy hour

18.5.3 Radio Resources Parameters

There are numerous radio parameters for LTE; however, the three most important radio parameters after traffic dimensioning are:

1. Physical Cell Identity (PCI) allocation,
2. Physical Random Access Channel (PRACH) parameter planning,
3. Uplink reference signal sequence planning.

Obviously, there are numerous other types of air interface parameters that can be tuned in LTE, such as DL power, UL power control, and handover thresholds to mention a few of the obvious ones.

18.5.3.1 PCI Planning

The *Physical Cell Identity* (PCI) allocation takes on a similar role that scrambling codes did in CDMA and WCDMA systems. PCI is encoded in the physical layer synchronization signal transmission and is used by the UE for obtaining neighbor cell handover measurement reports. However, there are only 504 PCI codes available for possible use and this means in larger networks that there will be reuse of the PCI codes.

A reuse factor greater than 15 should be sufficient for the PCI code assignments with a possibility of approaching 24 as the reuse factor. The rationale behind a reuse factor is to introduce some order to the PCI planning process. Additionally, with a 15- or 24-reuse pattern, not all the PCI codes are consumed and are available for local resolution of PCI pollution problems.

The PCI also serves as a resource allocator parameter for downlink and uplink *Reference Signals* (RSs).

18.5.3.2 DL Reference Signal

The downlink reference signal or symbol called the LTE pilot signal is allocated in a time and frequency. In time domain the RSs are always transmitted in the same OFDM symbol. However, in the frequency domain each eNodeB has a different frequency shift determined by the modulo-3 of the PCI itself. The reason for this is to ensure that reference signals of different eNodeBs do not overlap in frequency, which results in less interference on UE channel estimation.

Determining the PCI reuse pattern desired can result in the RS overlapping with the control and data resource elements of the neighboring eNodeBs. Therefore, a design choice must be made between RS-to-RS interference and RS-to-PDSCH/PDCCH-interference. If neither can be avoided, the RS to PDSCH/PDCCH-interference is more acceptable since it will not directly interfere with cell search and handover measurements.

18.5.3.3 UL Reference Signal

LTE *Physical Uplink Shared Data Channel* (PUSCH) carries *Demodulation Reference Signal* (DMRS). Optionally, also *Sounding Reference Signal* (SRS) is transmitted in the uplink. The uplink DMRSs are constructed from Zadoff-Chu sequences that are divided into 30 groups. Therefore, for a given number of PRBs allocated in the uplink, there are 30 different base sequences that can be used as the RS.

The planning requirement is that the neighboring cells should be allocated in different base sequences to ensure that inter-cell interference is minimized. One recommended method is to introduce a reuse factor of 30 for the base sequences.

18.6 System Traffic Estimation

The process and methodology for conducting system traffic engineering, that is, determining the amount of physical and logical resources that need to be in place at different points and nodes within the network to support the current and future traffic. The determination of existing traffic loads is rather more straightforward in that you have existing information from which to make decisions upon. For future forecasts, the farther the forecast or planning takes you into the future, the level of uncertainty grows exponentially. However, many elements in the network require long lead times, ranging from 3 weeks to over 1 year to implement. Obviously, the goal of traffic engineering is to design the network and its subcomponents not only to meet the design criteria, which should be driven by both technical, marketing, and sales, but also be done so that it is achieved in a cost-effective manner. It is not uncommon to have conflicting objectives within a design, that is, to ensure that the customers have the highest QoS but yet have a limited amount of capital from which to achieve this goal. Therefore, it is important to define at the onset of the design process, and have some interim decision points where the design process can be reviewed and altered, if required, either by increasing the capital budget, revisiting the forecast input, or altering the QoS expectations.

There are several methods which can be used for calculating the required or estimated traffic for the network. I cannot provide an infamous "F1" button approach for traffic engineering or forecasting. However, what I can provide is a guide from which you can, when applying engineering skills, craft the traffic projections to meet your individual market requirements.

It is essential to note that there are several key points within the network where the traffic engineering calculations need to be applied.

- eNodeB to User Equipment (UE)
- eNodeB to eNodeB (X2)
- eNodeB to MME
- eNodeB to SGW
- SGW to PGW
- PGW to IMS
- PGW to private and public data network (Internet)

There are several situations and an unknown level of perturbations, which can occur in the estimation of traffic for a system. In an ideal world, the traffic forecast would be projected by integrating the marketing plan with the business plan and coupled with the products, which should be integral to both the marketing and business plan. However, reality is much harsher and usually very little information is obtainable by the technical team from which to dimension a network with.

The key LTE RF target growth parameters can be distilled into five categories:

1. Bandwidth available (5, 10, 15, 20 MHz)
2. Power (20, 40, 60 W)
3. Throughput DL/UL (cell throughput)
4. PRB (RB utilization %)
5. Connected users (# active sessions)

The key network growth parameters can also be distilled into five general categories:

1. Throughput DL/UL
2. # Active sessions
3. # Bearers
4. Handover rates (intra and Inter MME/EPC)
5. # Transactions

18.6.1 Radio Data Traffic Projection

Mobile data traffic volume is directly dependent upon the type and quality of services that will be offered and how they will be transported. The traffic estimation process involves not only the radio link but also the other fixed facilities that comprise the network. Additionally, the inclusion of 802.11 and WiFi, services which utilize the unlicensed bands, commonly referred to as *license exempt*, complicates the situation for estimating wireless data usage since it is an off-net situation.

Mobile data traffic, however, is still a contention-based service, and therefore, the issue of how much blocking, that is, denied access due to radio facilities, will continue to prove to be difficult. The difficulty lies in the nature of mobile data in that it is packet based and not circuit switched, meaning that you can and should overbook the service.

Therefore, the forecasting of the Mobile IP traffic would be much more simplified if the usage patterns were understood better. But in light of this issue the forecast, or growth, could be extrapolated from the business plan or simplified marketing plans, which would specify a specific growth level desired.

The equation to follow for an existing system would be

$$\text{Total traffic} = \text{existing traffic} + \text{new traffic expected}$$

The new traffic expected could be a simple multiplication of the existing traffic load. For instance, if the traffic is 25 Mbps and the plan is to increase the traffic by 25 percent over the next year, then the traffic forecast for the 1-year forecast would be altered by increasing the current traffic load at each node by that amount and determining the requisite amount of logical and physical elements needed in addition to any load sharing that might be achievable.

The ultimate question that the designer must answer is, "How do you plan on supporting the traffic with their prescribed services?"

Since there are numerous types of services available, some generalizations need to be done in order to have a chance at arriving at some conclusions necessary for input into the design phase. Therefore, the symbols defined in Table 18.6 will be used to define the different classifications of transport services required.

For a new or existing LTE system, the issue of where to begin is always the hardest part. However, one of the key parameters you need to obtain from marketing and/or sales is the penetration rate, or take rate, for each of the service types offered. This can be achieved via several methods, a general approach where a standard percentage, %, is used for each of the packet services.

In review of Table 18.6, the inclusion of voice services is included. The inclusion of voice services in this is driven by the fact that VoLTE could be in play.

However, ITU-R M.1390 has some guidelines for dimensioning, and the tables below are extracted from that specification. The values in the tables should be used as a guide to establish packet loading for dimensioning when market-specific data are not available for a numerous amount of reasons.

Table 18.7 has projected net user bit rates by service type. It is important to note that all the service types defined are either symmetric or asymmetric. Of the services in Table 18.7 only MMM and HMM are asymmetric; the rest are symmetric service offerings

The next step is to determine the traffic forecast by user service type. The method for achieving this value is determined by the equation for each of the service types and locations defined, building/pedestrian/vehicular (Table 18.8) and the activity factor coming from Table 18.9.

$$\text{Traffic/user} = \text{BHCA} \times \text{duration} \times \text{activity factor}$$

Symbol	Service Type
S	Voice
SM	Short message
MMM	Medium multimedia
HMM	High multimedia
HIMM	High interactive multimedia

TABLE 18.6 Packet Data Service Types

	Net User Bit Rate	
Service Type	Downlink (kbps)	Uplink (kbps)
S	16	16
SM	14	14
MMM	384	64
HMM	2000	128
HIMM	128	128

TABLE 18.7 Net User Bit Rate

	Call Duration (sec)		
Service Type	Building	Pedestrian	Vehicular
S	180	120	120
SM	3	3	3
SD	156	156	156
MMM	3000	3000	3000
HMM	3000	3000	3000
HIMM	120	120	120

TABLE 18.8 Call and Session Duration

	Activity Factor	
Service Type	Downlink	Uplink
S	0.5	0.5
SM	1	1
SD	1	1
MMM	0.015	0.00285
HMM	0.015	0.00285
HIMM	1	1

TABLE 18.9 Activity Factor

MMM/HMM			
Type	**Description**	**Downlink**	**Uplink**
NPCPS	# Packet calls per session	5	5
NPPPC	# Packets per packet call	25	25
NBPP	# Bytes per packet	480	90
PCIT	Packet call inter arrival time	120	120
PIT	Packet inter arrival time	0.01	0.01

TABLE 18.10 Packet Data

Now the next question is to define the next set of variables, which need to be established to help dimension the rest of the packet network. For symmetric services, the dimensioning is straightforward, as well as straight as it can get. However, for asymmetric service, a few more details are required that are used for the selection and performance of the PDSN that are included in Table 18.10

$$\text{Transmission time(s)} = \text{NPCPS} \times \text{NPPPC} \times \text{NBPP} \times 8 \text{ bits per byte}/1024 \text{ kbps}$$

$$\text{Total session time} = \text{packet transmission} + [(\text{PCIT} \times (\text{NPCPS} - 1)] + [\text{PIT} \times (\text{NPPPC} - 1)]$$

Activity factor = packet transmission time/total session time

Going through the above is essential. However, equating it to the S1 and X2 links associated with an eNodeB may be more difficult. Therefore, the following brief example may provide some assistance.

If the traffic on the eNodeB is 4:1, or rather 80 percent downlink and 20 percent uplink and there are 1000 subscribers per eNodeB that need to be supported with a 10-MHz FDD channel offering, then the S1 and X2 link sizes can be determined as follows.

Additionally, let's assume that each subscriber during the BH consumes 25 kbps of data combined for uplink and downlink.

S1 and X2 Dimensioning

$$\text{BH data traffic/sub} = 25 \text{ kbps}$$

$$\text{Traffic ratio 80 percent DL 20 percent UL (i.e., 4:1)}$$

$$\text{\# subs/eNB} = 1000$$

The downlink data rate is then determined by

$$\begin{aligned} \text{DL} &= \text{BH traffic/sub} \times \text{traffic ratio} \\ &= 25 \text{ kbps} \times 0.8 = 20 \text{ kbps} \\ \text{MAC layer signaling} &= 1.37 \times 20 \text{ kbps} = 27.4 \text{ kbps} \end{aligned}$$

$$\text{Peak traffic rate} = 1.2$$

Therefore,

$$27.4 \text{ kbps} \times 1.2 = 32.88 \text{ kbps/sub}$$

and since there are 1000 subscribers per eNodeB, this is how much data traffic needs to be supported by the

$$32.88\ \text{kbps/sub} \times 1000\ \text{subs} = \mathbf{32.88\ Mbps}$$

The uplink data rate is then determined by

$$\text{UL} = \text{BH traffic/sub} \times \text{traffic ratio}$$
$$\text{BH traffic/sub} \times \text{traffic ratio} = 25\ \text{kbps/sub} \times 0.2 = 5\ \text{kbps/sub}$$
$$\text{MAC layer signaling} \times \text{peak traffic level} = 5\ \text{kbps/sub} \times 1.37 \times 1.2 \times 1000 = \mathbf{8.22\ Mbps}$$

However, for the S1 link, for the S1-U traffic, and for the uplink and downlink, traffic data need to be combined. Therefore, the link for supporting user traffic needs to be

$$\text{DL} + \text{UL or } 32.88\ \text{Mbps} + 8.22\ \text{Mbps} = 41\ \text{Mbps}$$

For the control-plane calculation, 2 percent is used for the rule for determining control plane traffic; therefore,

$$41\ \text{Mbps} \times 0.02 = -0.82\ \text{Mbps}$$

The S1 interface which carries both the S1U and S1-MME needs to be

$$41\ \text{Mbps} + 0.82\ \text{Mbps} = \mathbf{41.82\ Mbps}$$

S1 bandwidth requirements can also be done by using

$$\text{S1} = (1.002) \times \text{user plane traffic}$$

Now, the X2 link size is related to the S1 link size and this is typically 3 percent; however, I have seen values of 4 percent used. Therefore,

$$\text{X2} = (0.03) \times 41.82\ \text{Mbps} = 1.26\ \text{Mbps}$$

So, in brief summary, the S1 link needs to support 41.82 Mbps and the X2 link needs to support 1.26 Mbps. If the X2 is backhauled to the EPC, then link size needs to include the X2 link to every eNodeB that the eNodeB has an X2 link to.

18.7 Fixed Network Design Requirements

The fixed network portion of the planning phase obviously is primarily packet based; however, the inclusion of VoLTE introduces an element of circuit-switched components in the IMS section of the fixed network when interfacing with the PSTN. However, if the VoLTE is sent to a third-party provider for later treatment, the inclusion of VoLTE is more in the support of the particular service focusing on the QoS aspects as well as on the incremental aspects of increased packet loads it brings.

For the planning purposes, the integration of the RF design has always been an integral part of the fixed network design and will continue to be the situation. The fixed network design can proceed in parallel with the RF design process since many of the same marketing inputs serve both interests. However, the final configuration associated with the number of eNodeBs, their locations, and the other LTE node concentration considerations are dependent upon the RF group.

With that said there are certain commonalities that are applicable for both the voice and data services which involve utilization rates. The facility utilization goal for the network should be 70 percent of capacity for the line rate over the time period desired. The time period that should be used is a 9-month sliding window, which needs to be briefly revisited on a monthly performance report and then during a quarterly design review, following the design review guideline process.

The facility will need to be expanded once it is understood that the 70 percent utilization level will be exhausted within 9 months with continued growth showing exhausting, 100 percent, within 18 months. The reason for the large lead time lies in the implementation cycle, that is, characteristic of the fixed network environment. If your market significantly deviates from the suggested lead time recommendation, you should make the requisite changes.

18.7.1 MME

MMEs provide the control-plane messaging with the eNodeB. An MME is typically connected with multiple eNodeBs. One important concept is that several MMEs can be clustered or pooled together.

An MME cluster will have a single-group identifier; however, each of the MMEs in the cluster will have its own identity since it is involved with the control-plane signaling.

- Enables geographic redundancy
- Increases overall capacity

Another key advantage with the cluster of the MMEs is that an MME can be removed or added to the cluster pool without adversely affecting the service delivery and eliminates a single point of failure.

The pooling or clustering of MMEs enables inter-MME *Tracking Area Updates* (TAUs) to now be handled as intra-MME TAUs reducing the signaling load on the MME itself.

The MME design parameters also need to include the overall process (CPU) loading on the MME by setting a % utilization. Naturally, in an initial system design these values are estimated based on initial projections, however, in a post-turn on system data is available from which to establish a more realistic value based on real world data for your network.

The MME design parameters that follow are not all inclusive; however, they should cover most of the bases and the values are all in subscriber/hour interval. It is important to remember that you should not exceed the recommended utilization levels for any of these design parameters defined by the OEM vendor you are using.

- MME design parameters (subscribers/hour)
- Initial UE attach/detach
- Idle-active transactions
- Active to idle transactions
- Bearer channel activation/activation per session
- # bearers per session

- Session setup/teardown
- # Sessions
- Ingress paging
- Egress paging
- Intra-MME S1 handover SGW relocation
- Intra-MME S2 handovers without SGW relocation
- X2 handovers
- Inter-MME handovers
- # Intra-MME TAUs
- # Inter-MME TAUs

Additionally, the use of MME clustering can and will help reduce hot spot loading where one MME has substantially higher traffic than other MMEs.

18.7.2 Gateway

There are several gateways in an LTE network. The two primary gateways are the *Service Gateway* (SGW) and the *Packet-Data Network Gateway* (PGW). There can be multiple SGWs and PGWs that a subscriber utilizes based on the service that is being used. Additionally, PGWs and SGWs can be clustered together for traffic efficiency and resiliency.

18.7.2.1 Service Gateway (SGW)

Service gateways, SGWs, should be placed as close to the edge of the network as possible to support bandwidth-intensive applications. An SGW handles all the user traffic to and from an eNodeB with the packet-data networks. Additionally, an SGW can and is connected to multiple eNodeBs at the same time.

One key design factor for SGWs is that SGWs can be clustered and referred to as a serving area. The advantage of SGW clustering is that the MME has a pool of SGWs to select, which is meant to reduce the amount of inter-SGW handovers that can take place. In addition, the eNodeB's S1U connection can be linked to multiple SGWs in the pool.

However, a key advantage with the cluster of the SGWs is that an SGW can be removed or added to the cluster pool without adversely affecting the service delivery.

18.7.2.2 Packet-Data Network Gateway (PGW)

The PGW is used to send and receive data to and from off-net data networks. The PGW typically is connected to multiple SGWs. Additionally, there can be multiple PGWs in the LTE network with each PGW being responsible for a particular service delivery type.

The PGW is the interface for service delivery and reception and, of course, treatment between an LTE core and other network elements like an IMS platform or the Internet, to mention a few possible PGW destinations.

18.7.2.3 Design Parameters

The SGW and PGW design parameters that follow are not all inclusive; however, they should cover most of the bases. It is important to remember that you should not exceed the recommended utilization levels for any of these design parameters defined by the OEM vendor you are using.

The following are the key design parameters:

- # Simultaneous active subs
- Percentage of subscribers sending and receiving data
- DL average packet size (Mbps)
- UL average packet size (Mbps)
- # Active sessions (subs/hour)
- # Active bearers (subs/hour)
- # Bearers set up and torn down (sub/hour)
- # Idle-active and active-idle transactions (sub/hour)
- # Intra-SGW handovers (sub/hour)
- # Inter-SGW handovers (sub/hour)
- # Inter systems handovers (sub/hour)
- # Subs using IP4 (# and %)/hour
- # Subs using IP6 (# and %)/hour
- # Gx transactions (connections/hour)
- # Gy transactions (connections/hour)

18.8 Implementation

The system design process needs to factor into it how you intend on implementing the LTE network. For example, will the LTE system be a stand-alone network or will it need to interface with existing legacy wireless networks? If you do have a legacy system, is it possible to collocate the LTE equipment at the same locations as the legacy cell sites and, coupled with that, though, is there any reuse of the antenna system desired or will the LTE antenna system be entirely new?

Additionally, will the system design include the use of RN, femto, and other HetNet elements?

Some other preimplementation questions, but not all, are:

- What services will be supported?
- Are there sufficient CPEs deployed that can take advantage of the LTE system or expansion in the geographic area?
- What spectrum resources are or will be available for use?
- When does the service have to be operational?

The other question that needs to be resolved is "Where LTE will be deployed and will it be in a phased approach or provide system-wide coverage on the first day?"

18.8.1 Interconnection with UMTS and GSM

If you have a UMTS or possibly a GSM system as part of the your legacy network, then you need to consider whether you indeed want to interconnect LTE with the rest of the network. Or, do you want to focus more on deploying and populating the LTE system and migrate users from the UMTS/GSM system to the LTE network to refarm the spectrum for use by LTE?

However, if the desire is to interconnect UMTS/GSM with LTE, it is possible to achieve this under current 3GPP specifications. LTE interconnection with UMTS/GSM is needed for existing operators when launching LTE as an overlay to handle coverage gaps. Therefore, using of UMTS/GSM with LTE is a viable deployment scheme for providing service when there is a limited LTE footprint or gaps exist in an LTE deployment.

The *LTE Enhanced Packet Core* (EPC) needs to be physically connected with the UMTS/GSM core network via the GGSN/SGSN as shown in Figure 18.3. Additionally, the UE needs to have its USIM provisioned to enable it to operate in a multiple RAN environment.

Traffic steering, or preference selection, needs to also take place and this is done through cell reselection where the UE is instructed to use the LTE network, or it can be instructed to use the UMTS/GSM network depending on the conditions or criteria you desire. The cell reselection is handled by an *Inter Radio Access Technology* (IRAT), and this message is sent on SIB6 and SIB7 in LTE.

Handovers of active sessions are a little different in that the LTE anchor, MME, and PGW are connected to the UMTS/GSM SGSN. The handover that does occur will be a hard handover and the credentials need to follow in order for the session to remain intact.

18.8.2 Interconnection with CDMA/EVDO Networks

If you have an EVDO system as part of the your legacy network, then you need to consider whether you indeed want to interconnect LTE with the rest of the network. Or, do you want to focus more on deploying and populating the LTE system and migrate users from the EVDO system to the LTE network to refarm the spectrum for use by LTE.

However, if the desire is to interconnect EVDO with LTE, it is possible to achieve this under current 3GPP specifications. LTE interconnection with EVDO is needed for existing operators when launching LTE as an overlay to handle coverage gaps. Therefore, using EVDO with LTE is a viable deployment scheme for providing service when there is a limited LTE footprint or gaps exist in an LTE deployment.

The LTE EPC needs to be physically connected with the EVDO core network via the HSGW and RNC as shown in Figure 18.3. Additionally, the UE needs to have its USIM provisioned to enable it to operate in a multiple RAN environment.

Traffic steering, or preference selection, needs to also take place and this is done through cell reselection where the UE is instructed to use the LTE network, or it can be instructed to use the EVDO network depending on the conditions or criteria you desire. The cell reselection is handled by an IRAT, and this message is sent on SIB8 8 in LTE.

Handovers of active sessions is a little different in that the LTE anchor, SGW, and PGW, are connected to the RNC and HSGW as depicted in Figure 18.3. The handover that does occur will be a hard handover and the credentials need to follow in order for the session to remain intact.

18.8.3 Interconnection with WiMAX Networks

If you have a WiMAX system as part of your legacy network, then you need to consider whether you indeed want to interconnect LTE with the rest of the network. Or, do you want to focus more on deploying and populating the LTE system and migrate what users you have from the WiMAX system to the LTE network to refarm the spectrum for use by LTE.

What is interesting about the WiMAX integration or interconnection is that it operates as a TDD system and the implementation of LTE should be a TDD configuration in order to best fit into the existing frequency and channel scheme.

However, unlike WiMAX, which needs an $N = 3$ reuse pattern, an LTE TDD system only needs one channel for system-wide deployment.

It is recommended that the channel or channels selected for use with TDD are aligned with the CPE capabilities that are available. Ideally, it would be best for the LTE system to use the lowest portion of the WiMAX band for deploying LTE. However, the CPE and UE equipment may perform better at the higher end of the frequency band for WiMAX.

Furthermore, it would be beneficial to make sure that the LTE channel used is cleared from use with WiMAX to avoid any misconfigurations and deployment issues that may arise.

Additionally, as in previous recommendations for deployments, the reuse of existing assets, that is, cell sites, needs to be heavily relied upon except in those instances where the lease cost or other constraining issues limit the sites' short- and long-term viability.

The LTE EPC needs to be physically connected with the WiMAX CSN core network via the S2 and S103 links as shown in Figure 18.3. Also, the UE needs to have its USIM provisioned to enable it to operate in a multiple RAN environment. This is assuming that the UE is capable of both LTE and WiMAX operation.

Traffic steering, or preference selection, needs to also take place and this is done through cell reselection where the UE is instructed to use the LTE network, or it can be instructed to use the WiMAX network depending on the conditions or criteria you desire. The cell reselection is handled by an IRAT, and this message is sent on SIB 7 in LTE.

18.9 Backhaul Considerations

A separate chapter has been dedicated to backhaul. However, stating the obvious, the backhaul critical used will directly determine if you can or are meeting the desired LTE data rates with the proper latency. This said, it is important to remember that LTE is a packet system and because it is more robust and has more traffic-handling capability, the backhaul links for LTE are much greater than those used for EVDO and HSPA.

A typical backhaul link that connects the eNodeB to the core network via a VPN or dedicated circuit is 40 Mbps with many approaching 100 Mbps. Fiber, the preferred connection level for backhaul with LTE, however, depending on the existing cable plant involved in concatenating DS1s and DS3s may be required.

When collocating LTE with HSPA/EVDO, the existing backhaul that the legacy site uses normally does not have the head room for LTE to utilize.

In more urban environments the use of Metro Ethernet connections is the preferred method of connectivity followed by microwave.

18.10 Deployment

The deployment of LTE into an existing network should be done as a 1:1 configuration with existing sites. The link budget for LTE in a 10×10 configuration has similar overall performance expectations as those with EVDO and UMTS.

When deploying LTE into a system having LTE available across the entire network on day one is not realistic. Therefore, a phased approach needs to be utilized. In the

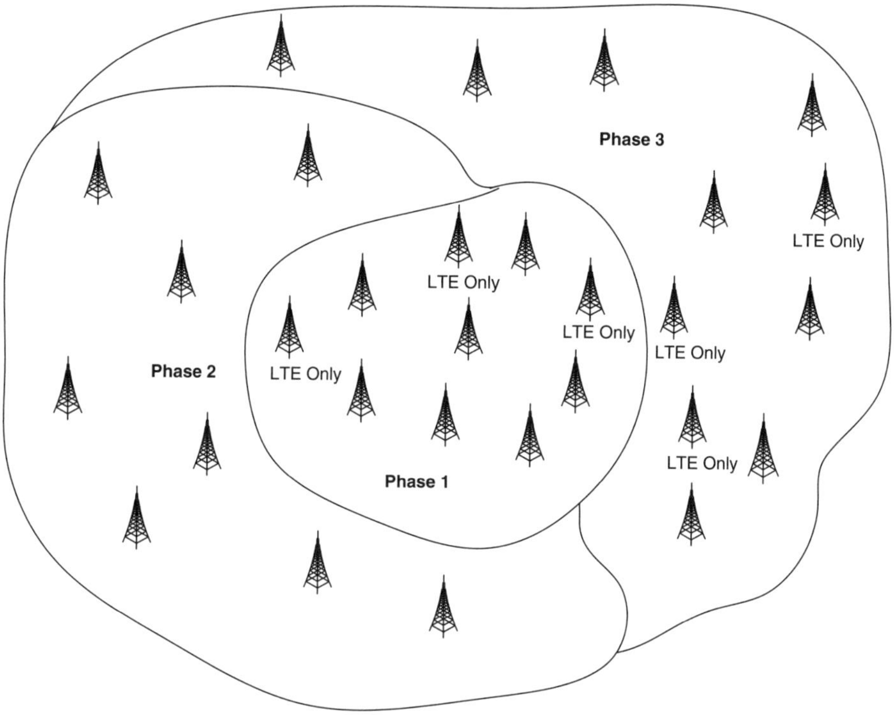

FIGURE 18.7 Overlay.

past, the phased approach involved having the limited coverage and service area matching the dense part of the wireless network and for LTE.

Figure 18.7 depicts a three-phase implementation sequence where phase 1 is the core of the network and phases 2 and 3 are outside of the core. In both phases, 1 and 2, additional LTE-only sites are defined in addition to having LTE being collocated with existing sites. Phase 3 involves only collocating LTE into existing sites.

The use of small cells, WiFi hot spots, as well as RN, is not included for brevity. However, the use of small cells for coverage in buildings and local hot spots on the street makes it a great deployment tool. WiFi hot spots, of course, continue to play a valuable role in a wireless system in that the WiFi systems offload the existing carriers' networks with another service provider's network, usually at no cost to the carrier.

There are numerous perturbations for implementing LTE either into an existing UMTS, EVDO, or WiMAX system. I will only attempt to give some high-level concepts for each. However, you need to ensure that your implementation plan matches the company's business plan and objectives.

Figure 18.8 depicts just a few possible scenarios for deploying LTE into an existing wireless network. Of course, having new spectrum and no legacy systems coupled with ample capital and a defined customer base would be the ideal situation. However, in more cases than not you have existing wireless assets and the asset types are not the same in each market you may operate in. Some of the asset differences could easily be the spectrum portfolio or it can be the OEM used or you will be using and any legacy construction constraints to mention a few items.

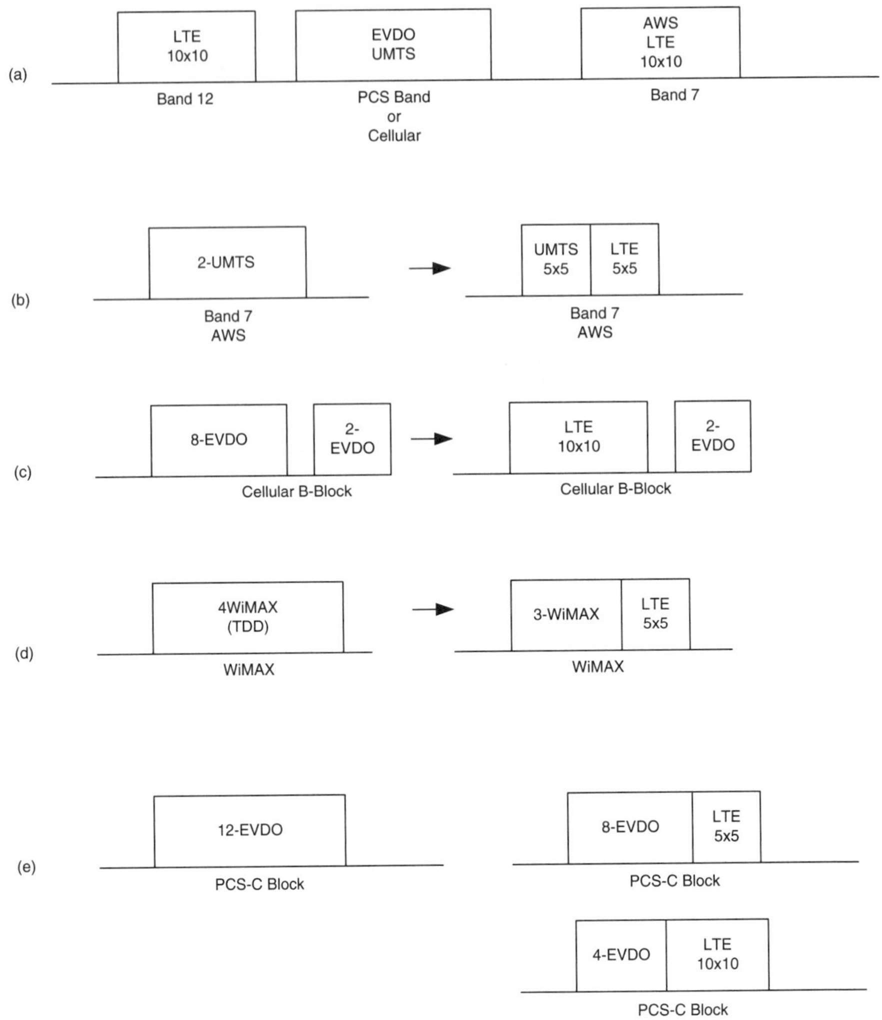

FIGURE 18.8 LTE deployments.

Figure 18.8a shows the situation where an operator has three bands of spectrum with an EVDO legacy system. It is advisable that in the scenario highlighted in Figure 18.8a that the lower band be used for LTE initially and that the AWS band be used for secondary LTE performance either as a Uu link for RNs if used, or combined with channel aggregation for a small footprint cell site.

Figure 18.8b shows the situation where you might have the AWS spectrum and are fully using it for UMTS. However, to deploy LTE, you need to remove one of the UMTS channels from service-making room for LTE. The LTE channel should start off as a 5×5 until all the subscribers are migrated to it, then the channel can be configured through software for a 10×10, assuming the OEM hardware is capable. The issue that could happen in this migration scenario is that the legacy UMTS system will need to be expanded upon or you will need to enter into an agreement with your competition to off-load subscribers until the LTE system is in place. Alternatively, you can flash cut the LTE system when you turn off one of the UMTS channels provided you have populated the subscriber base with enough dual mode devices to make this happen; ideally you need to have over 75 percent of the subscriber-based dual mode capable for a flash cut to be successful.

Figure 18.8(c) shows the situation where the cellular B block is currently being used by EVDO. To deploy LTE into the cellular B block, a 10 × 10 LTE channel is used to displace some EVDO channels. The B prime block is left initially with EVDO; however, it is possible to have most of the B prime operate as a 1.4-MHz LTE channel and utilize channel aggregation with Rel 10.

Figure 18.8(d) shows the situation where you want to deploy LTE into a WiMAX band. In this scenario a 5-MHz TDD channel is shown; however, it could easily be a 10-MHz TDD channel that is deployed for LTE depending on the WiMAX spectrum portfolio in the given market.

Figure 18.8(e) shows a situation where the wireless operator is utilizing the PCS C block and wishes to utilize LTE. In this situation it is proposed that initially a 5 × 5 LTE channel be used to displace or use only 1/3 of the band. Once the subscriber population and LTE buildout has progressed, then through software it is proposed to configure the sites for a 10 × 10 LTE channel.

18.11 Antenna Configurations

The overlay or codeployment of an LTE system with legacy 2G/3G wireless networks presents several unique problems primarily with the issue of antenna real estate. The following scenarios or sample configurations are meant to highlight some possible antenna deployment options. What is important to note is that only a 2 × 2 MIM for LTE is used in these examples.

Figure 18.9 (a) through (d) where the new LTE systems is connected to antennas in a variety of fashions; however, in all cases the use of a coax cable from the equipment room to the antenna mounting location is used.

However, in Figures 18 (e) and (f) the new LTE system has a remote RF head which is connected to the equipment room via a pair of dark fiber cables.

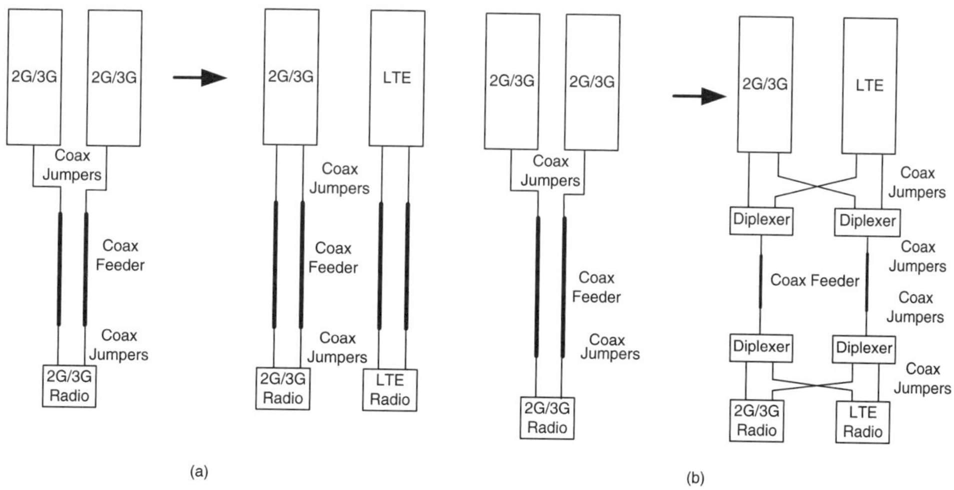

FIGURE 18.9 LTE System Antenna Configurations.

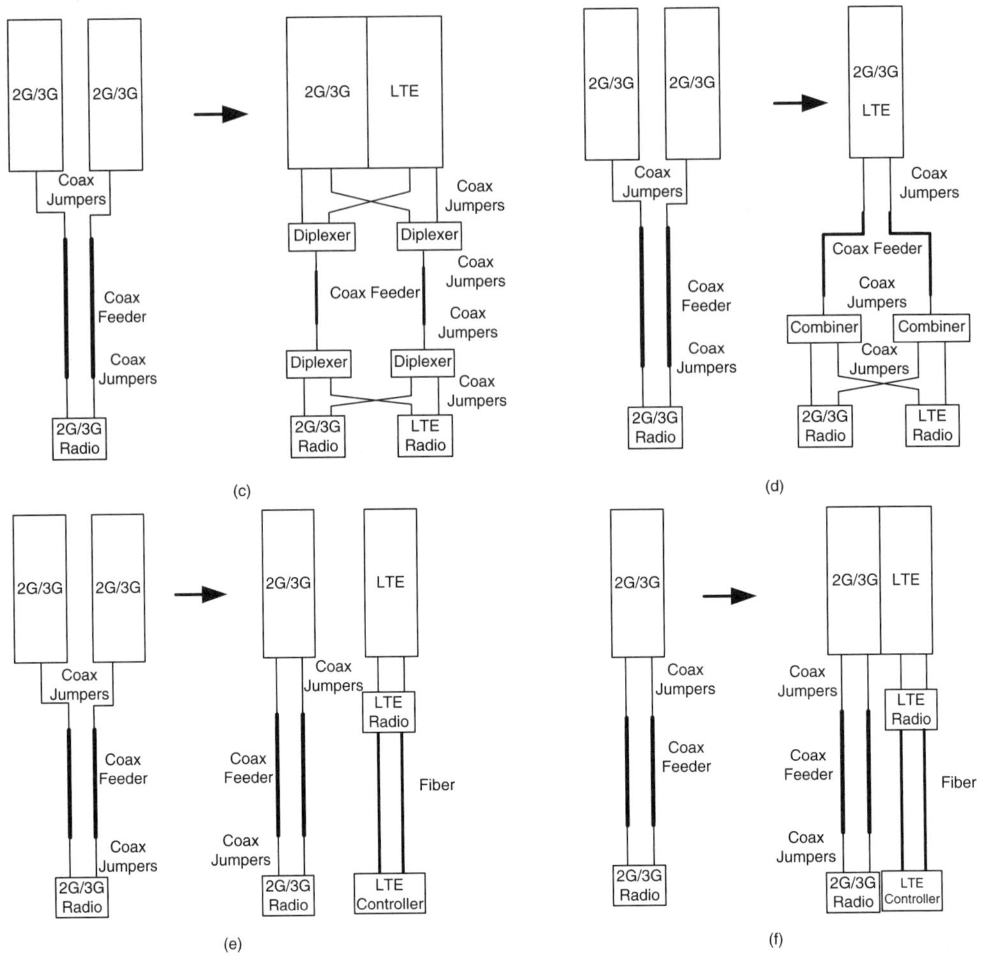

FIGURE 18.9 *(Continued).*

References

TS 36.211 Evolved Universal Terrestrial Radio Access (E-UTRA); Physical channels and modulation

TS 36.212 Evolved Universal Terrestrial Radio Access (E-UTRA); Multiplexing and channel coding

TS 36.213 Evolved Universal Terrestrial Radio Access (E-UTRA); Physical layer procedures

TS 36.300 Evolved Universal Terrestrial Radio Access (E-UTRA) and Evolved Universal Terrestrial Radio Access Network (E-UTRAN); Overall description; Stage 2

TS 36.321 Evolved Universal Terrestrial Radio Access (E-UTRA); Medium Access Control (MAC) protocol specification

TS 36.331 Evolved Universal Terrestrial Radio Access (E-UTRA); Radio Resource Control (RRC); Protocol specification

TS 36.413 Evolved Universal Terrestrial Radio Access Network (E-UTRAN); S1 Application Protocol (S1AP)
TS 36.423 Evolved Universal Terrestrial Radio Access Network (E-UTRAN); X2 Application Protocol (X2AP)
TR 36.806 Evolved Universal Terrestrial Radio Access (E-UTRA); Relay architectures for E-UTRA (LTE-advanced)
TR 36.808 Evolved Universal Terrestrial Radio Access (E-UTRA); Carrier aggregation; Base Station (BS) radio transmission and reception
TR 36.814 Evolved Universal Terrestrial Radio Access (E-UTRA); Further advancements for E-UTRA physical layer aspects
TR 36.815 Further Advancements for E-UTRA; LTE-advanced feasibility studies in RAN WG4
TR 36.817 Evolved Universal Terrestrial Radio Access (E-UTRA); Uplink multiple antenna transmission; Base Station (BS) radio transmission and reception
TR 36.826 Evolved Universal Terrestrial Radio Access (E-UTRA); Relay radio transmission and reception
TR 36.871 Evolved Universal Terrestrial Radio Access (E-UTRA); Downlink Multiple Input Multiple Output (MIMO) enhancement for LTE-advanced
TR 36.912 Feasibility Study for Further Advancements for E-UTRA (LTE-Advanced)
TR 36.913 Requirements for Further Advancements for Evolved Universal Terrestrial Radio Access (E-UTRA) (LTE-Advanced)
TS 36.211 Evolved Universal Terrestrial Radio Access (E-UTRA); Physical channels and modulation
TS 36.212 Evolved Universal Terrestrial Radio Access (E-UTRA); Multiplexing and channel coding
TS 36.213 Evolved Universal Terrestrial Radio Access (E-UTRA); Physical layer procedures
TS 36.216 Evolved Universal Terrestrial Radio Access (E-UTRA); Physical layer for relaying operation
TS 36.300 Evolved Universal Terrestrial Radio Access (E-UTRA) and Evolved Universal Terrestrial Radio Access Network (E-UTRAN); Overall description; Stage 2
www.3GPP.org

CHAPTER 19

WiMAX System Design

19.1 Design Methodology

The process used for a WiMAX system design involves many aspects which draw upon both microwave point to point system and cellular/PCS system designs. It is important to note that the system design process is extremely fluid and is equally complex when addressing system growth issues. The design process can be extremely simplistic or elaborate.

There are numerous design considerations that need to be factored into the possible use and deployment of an 802.16 system. The decisions relate to the economical deployment and operation of the system.

However, the fundamental issue that needs to be defined for the deployment of an 802.16 wireless system or any wireless system is

- What are you trying to accomplish (i.e., the objective derived from the business plan)?
- Why are you doing it?
- When does it need to be operational (i.e., the schedule)?
- How is this process going to be done?

Answering the four basic items above will go a long way in helping define and deliver a system. For example, is the design for mobile cellular network or is it for a fixed service either a subscriber station in a residential location, or is it for a point-to-point backhaul link. It is important to note that the design with WiMAX can involve one or all of the above, where a system design involves mobility, point-to-point links as well as fixed stations depending on the service offerings that the company wants to offer.

19.2 Coverage

For any wireless technology to function properly, it requires a good communication link between the sending and receiving devices. Coverage requirements vary depending on which 802.16 protocol is used in the application. 802.16 requires line of sight while 802.16e and 802.16m do not require *Line-of-Sight* (LOS) and are *Non-Line-of-Sight* (NLOS) access protocols.

With mobile, WiMAX coverage is still king in that in order to have good performance to maximize the user's experience there is a need to have the best DL and UL paths possible. Assuming the DL and UL paths are balanced then coverage for both is defined by:

- Height above ground of the sending and receiving antennas
- Terrain (mountains, valleys, etc.)
- Morphology (i.e., buildings, trees, etc.)
- Antenna gain and ERP
- Modulation format
- Link reliability (ITU rain zone)
- Polarization (PtP)
- Throughput requirements

Ultimately, the question comes about to ask what the range of the base station will be. The answer for a PtP link or 802.16 is based on path loss that is dictated by free-space loss. The 802.16e and 802.16m coverage, however, is more a function of propagation path loss that is in the order of $1/r^4$ or 40 $\log_{10}$.

19.3 Link Budget

The link budget is one of the key technical parameters used for the design process. It is absolutely essential in order to properly set up and configure the RF planning tool used to predict system coverage and performance. Obviously all the components in the design process are important; however, the link budget directly determines the range and deployment pattern used for a given system. An important aspect is that there can be several link budgets in any system based on the operating frequency, spectrum allocated, link reliability, physical components for the system, differences in up- and downstream modulation methods or protocols, and rain fade issues to mention some of the elements that need to be considered.

The objective behind the link budget is to determine the path length, or rather the size of the cell sites needed for the network design.

As with all wireless technologies, the available throughput is a function of several key elements.

- Radio access protocol
- Bandwidth available
- Radio link quality
- Backhaul

The radio access protocol is a determining factor in that you should apply the correct protocol to the application at hand because no single protocol is completely universal. Bandwidth availability will determine how much capacity, and hence the services, both individual and aggregated can be supported by the radio link.

The radio link quality will directly alter the throughput availability. As with all wireless access systems, the lower the radio link quality the lower the throughput allowed or possible. The radio link quality is a function of coverage and interference coupled with modulation format used in an allotted bandwidth and spectrum. Figure 19.1 is an

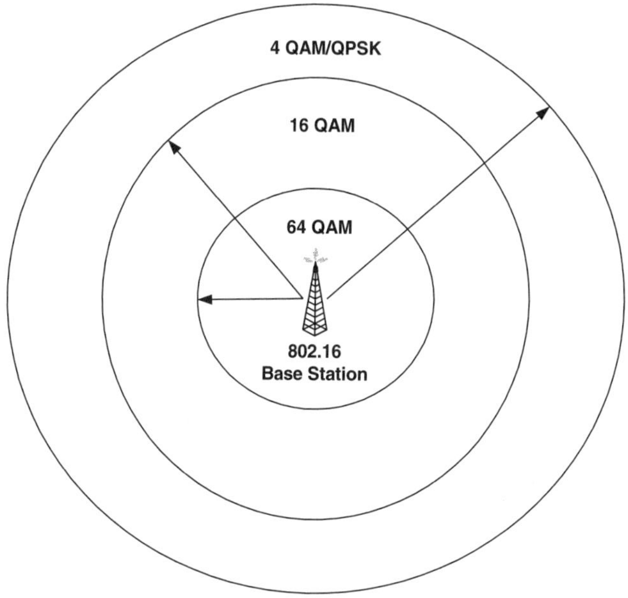

Figure 19.1 Range and modulation scheme.

attempt to depict conceptually the relationship between modulation scheme and the radius or rather coverage of an 802.16 site. Figure 19.1 illustrates that as the site tries to cover more geographic area, a tradeoff is made in terms of throughput allowed due to the lower modulation scheme employed.

Therefore, a proper link budget needs to account for each traffic type, or rather data rate that can be used in the system that is basic to the formulation of the RF plan. The link budget combines all the gains and losses in the system and will demonstrate the maximum path loss that the system can withstand at the edge of the WiMAX cell under the assumed conditions of data rate and reliability.

The WiMAX forum has published a typical link budget which can be used as a starting point to the development of the specific link budget that can be then adjusted based on the infrastructure vendor that is chosen to use either for the network or a portion of the network itself.

19.4 RF Propagation Analysis

Accurate prediction of path loss is the foundation of any wireless network plan and is usually done with computer simulation tools. Based on detailed terrain and overlay clutter databases at varying spatial resolutions (from 30 m down to 1 m), this module provides the capability to analyze the radio environment at both a macro-cellular and microcellular level.

Propagation models selected for the computer simulation enable accurate coverage analysis for *Line-of-Sight* (LOS) and *Non-Line-of-Sight* (NLOS) situations for both indoor and outdoor environments. The propagation model accuracy used for simulations depending on the specific details related to the network configuration, network topology, equipment choices, terrain, and morphology.

19.5 Parameters

WiMAX, like all other wireless mobility access systems, has a multitude of parameters. The parameter planning needed for WiMAX is similar to that of any other wireless system. The following are a few of the parameters needed for planning a mobile WiMAX network.

- BSID
- Preamble index
- Neighbor cell parameters

BSID

In a WiMAX network every carrier has a BSID. The BSID, however, needs to be provisioned such that the AS-Base Station and the AS-Gateway can communicate with each other. A BSID consists of an Operator ID as well as a Sector ID to help uniquely identify it. Therefore, the BSID = Operator ID + Sector ID. Each Sector ID should be uniquely identified so that each sector can be uniquely identified.

Preamble Index

The preamble in the WiMAX frame, discussed in Chapter 8, is also used to help uniquely identify a particular sector. Each preamble has an index, ID cell, and segment.

The preamble index should be arranged into multiplexing groups whose size is determined by the network scale and should be selected via a network planning tool. Specifically, all cell sites using different preambles are grouped as one multiplexing cluster. When the preambles are reused creating another multiplexing cluster, the two clusters need to have physical separation. For example, since there are 96 preamble indexes (0-95), a three-sector cell site facilitating an $N = 3$ pattern would then have a maximum multiplexing cluster of 32 sites. However, reality dictates that problems will arise in real-world implementations. Therefore, some preamble index codes should be held in reserve to be implemented for resolving interference issues.

Neighbor Lists

Neighbor cell planning in WiMAX is very similar to that used in other wireless mobility systems. For instance, adjacent cells to the serving cell should be included in the neighbor list and should be mutually assigned. However, in more densely populated areas where cell sites are spaced closer, additional tiers of neighbors will need to be added based on the mobility of subscriber and their ability to pass by an adjacent site before a handover can be initiated.

19.6 Configurations

There are multiple network configurations possible with an 802.16 wireless access system. The configuration that the network takes on is dependent upon the services offered and/or transported, resiliency, growth, and flexibility.

Some of the services that an 802.16 system can support are listed below for quick reference. It should be noted that the services listed are not all inclusive of what can be delivered or will be delivered. The only exception is that the service offered cannot have a bandwidth requirement greater than what the radio transport layer can support.

- LAN/WAN (VPN)
- T1/E1 replacement (clear and channelized)

- Fraction T1/E1 (clear and channelized)
- Frame relay
- Voice telephony (POTS and enhanced services)
- Video conferencing
- Internet connectivity
- Web services (e-mail, hosting, virtual ISP, etc.)
- E-commerce
- VoIP
- FaxIP
- Long distance and international telephony
- ISDN (BRI and PRI)

For 802.16a and 802.16e, the services center around IP and do not include CBR-type traffic as 802.16 does. Regardless, the host of services and perturbations to those just listed above makes an impressive portfolio to offer. Of course, the necessary platforms and connectivity for the network need to be in place in order to ensure that these services can and will be able to be offered and effectively delivered.

802.16 systems, like all other networks, can have both on-net and off-net traffic considerations as depicted in Figure 19.2. Ideally, the traffic should be all on-net but when the system initially goes on line, most if not all, the traffic goes off-net and the use of the PTT or another CLEC will be required to facilitate the delivery of the service almost exclusively.

There are several general configurations for 802.16 wireless access systems as they apply to wireless mobile systems. The wireless mobile systems that can take advantage of the robustness of 802.16 fall into the 2.5G and 3G network platforms that utilize a packet network for delivering services.

- Mobility
- Backhaul
  - 3G/4G cellular backhaul
  - WiFi hot spot backhaul
  - Mobile WiMAX backhaul
- DSL/fiber alternative
- 3G/4G alternative route/resiliency
- Integration with 3G/4G networks

The following are a few of the many network configurations that are possible with the use of 802.16-specified equipment.

19.6.1 WiMAX Mobility

Figure 19.3 depicts a high-level view of a mobile WiMAX network involving one core network, CSN, with several ASN base stations not all of which are connected to the same AS-Gateway. Additionally, a relay node is also shown in Figure 19.3 to illustrate the potential use of relay nodes to extend the coverage of a WiMAX network.

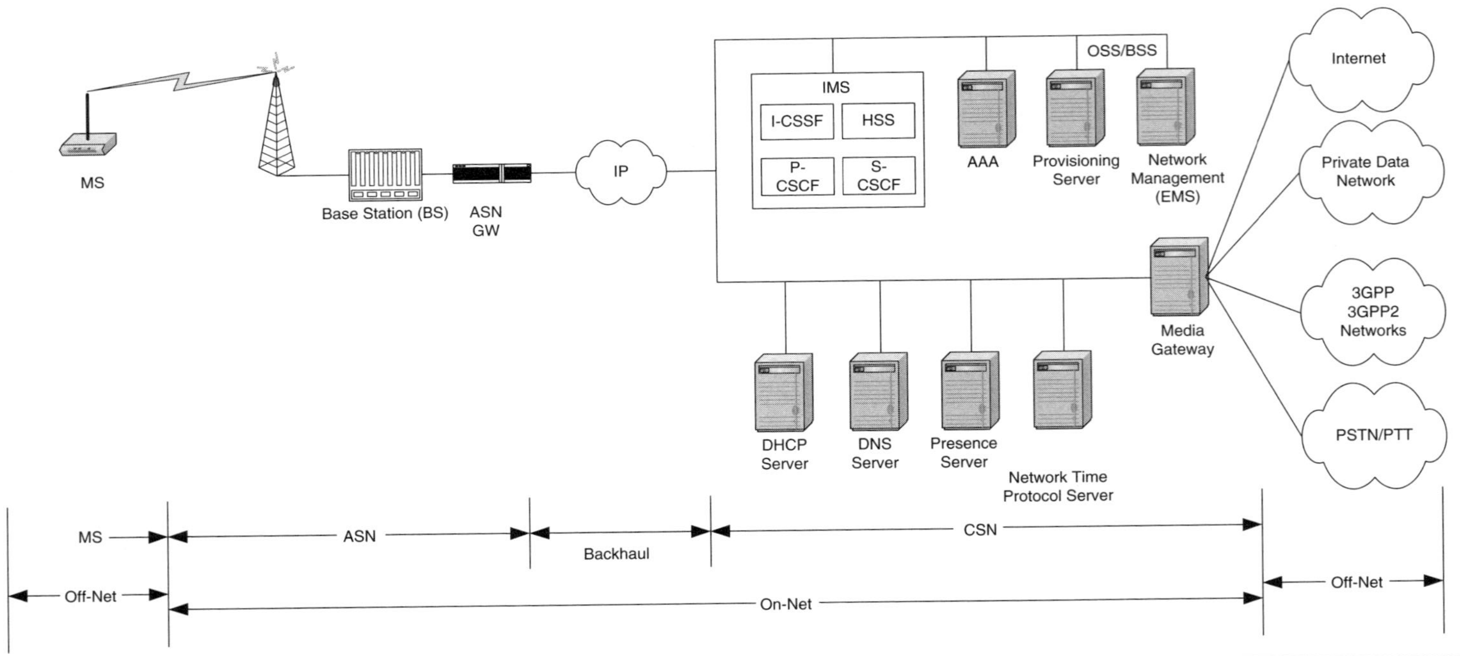

FIGURE 19.2 On-net and off-net.

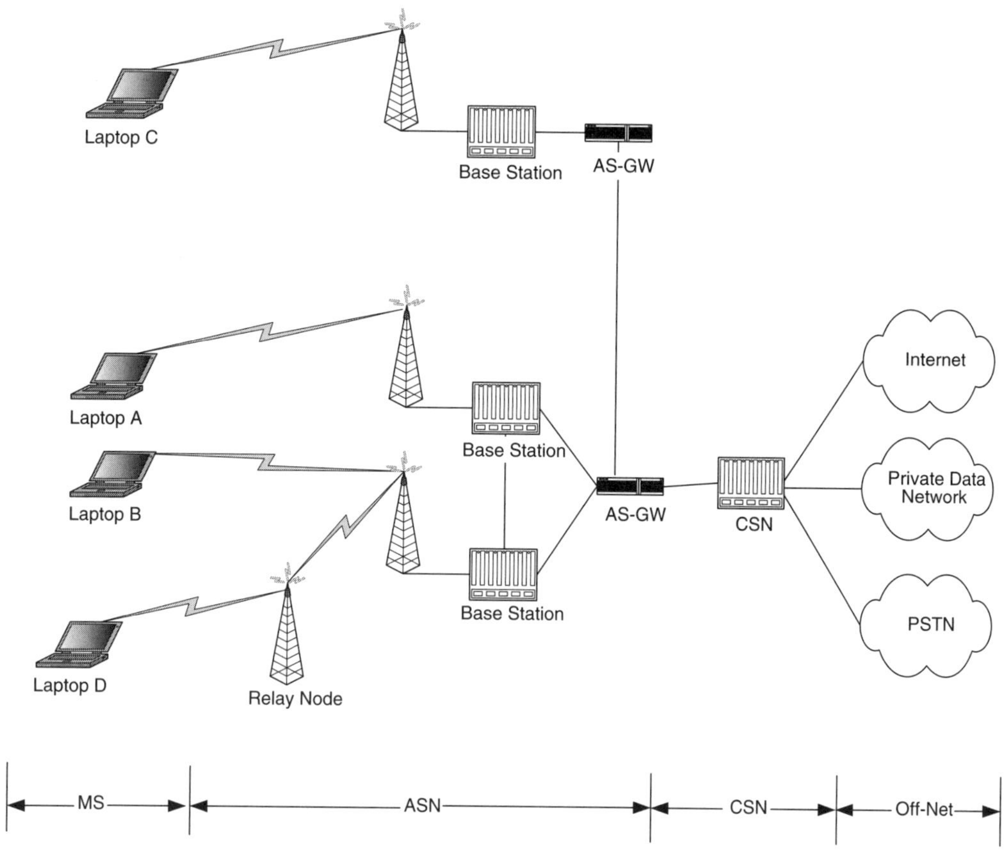

FIGURE 19.3 Mobile WiMAX.

The frequency plan for a WiMAX mobile network should be an $N = 3$ reuse pattern. However, sometimes spectrum is not available to leverage an $N = 3$, or additional carriers are needed to be deployed in order to handle high-density traffic areas without the need to build a new WiMAX cell site.

To facilitate the frequency planning aspect, the use of *Fractional Frequency Reuse* (FFR) is utilized, where all the subchannels for the RF carrier are available close to the site or in a particular zone. Outside of that zone a smaller set of the subchannels is used, which is allocated so that they are not as heavily used in the adjacent cell site so as to minimize the amount of mutual interference it will generate.

19.6.2 Cellular Backhaul

The use of 802.16 for backhaul can take on the configuration as shown in Figure 19.4. The configuration in Figure 19.4 can also be a star configuration having multiple BTSs connected to a few 802.16 hub sites that act as traffic concentrators.

WiMAX point-to-point links can consist of both licensed and licensed-exempt frequencies. There are a multitude of frequencies to pick from; however, a simple rule is that the longer range you need to connect the cell site to another cell site, or hub location, the lower the potential throughput will be.

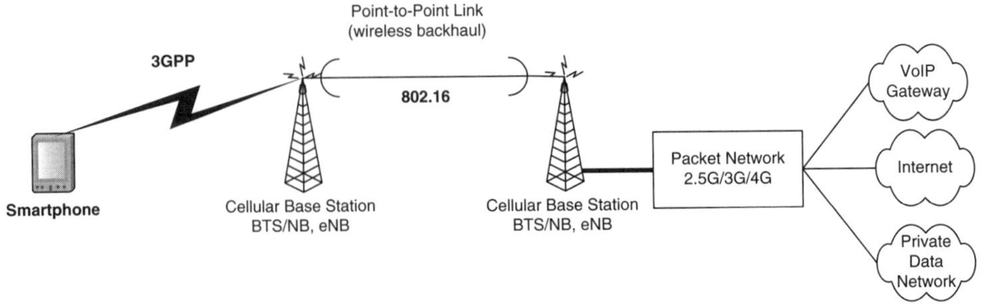

FIGURE 19.4 Cellular backhaul.

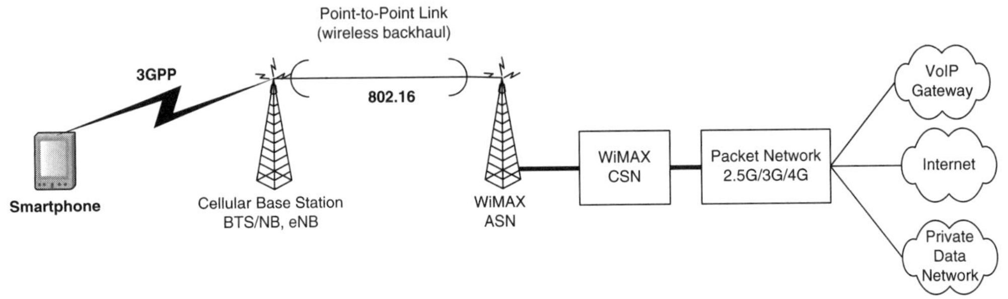

FIGURE 19.5 WiMAX backhaul.

Selecting licensed, unlicensed, or lightly licensed spectrum to use is largely driven by the spectrum portfolio you have at your disposal. If you have WiMAX, frequencies using licensed channels for backhaul will likely be the correct option. However, if you do not have spectrum to leverage, then your options are primarily unlicensed. The lightly licensed band, whose details can be found at www.fcc.gov is good for more rural applications; however, it may not be desirable in an urban or more dense area or along borders with Canada or Mexico in the United States.

Figure 19.5 is another example of cellular backhaul with WiMax except the transport is done through the WiMAX network providing the backhaul connectivity is between the cellular base station and their core network.

WiMAX can also be the backhaul for mobile WiMAX or act as a backhaul for WiFi hot spots or even serve as both as shown in Figure 19.6.

Historically, 802.16 by itself for wireless backhaul needs to have the economics of the transport method reviewed in detail. However, when Metro Ethernet facilities are not available, or the cost structure of those circuits has not been lowered due to competition, the possibility of using 802.16 for backhaul becomes more compelling.

19.6.3 Fixed Last-Mile Broadband

WiMAX can be and is used for equipment in the network, which can be used for residential, small offices, and campus environments where the mobility to the 3G/4G network is not needed. Instead, the service offered is the ability to provide packet-data network services and have traditional voice services offered via legacy devices.

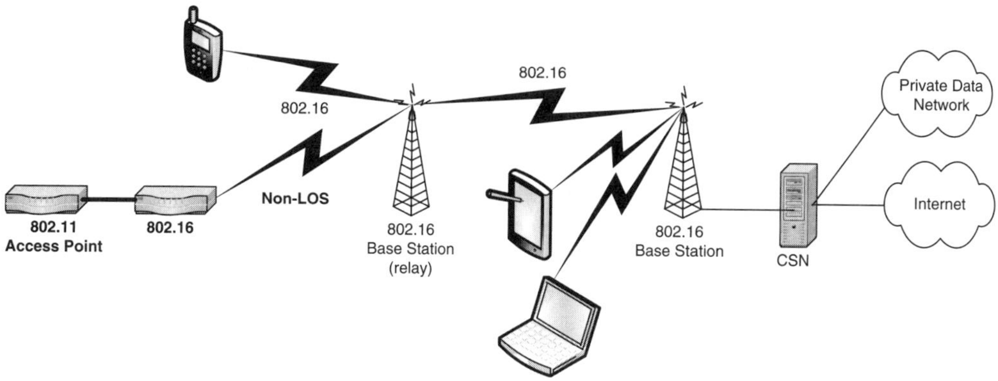

FIGURE 19.6 Wifi and WiMAX backhaul.

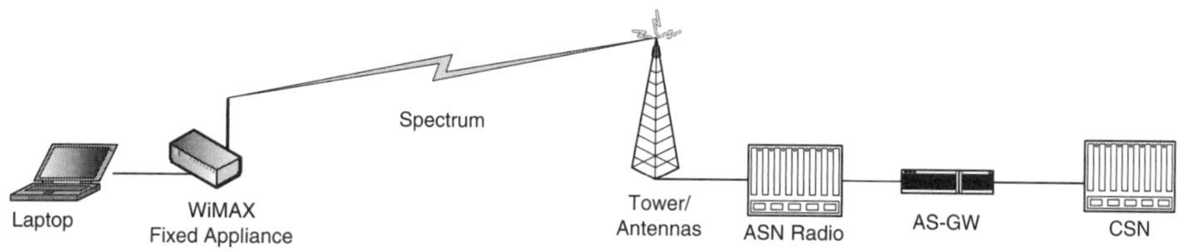

FIGURE 19.7 WiMAX fixed appliance.

Figure 19.7 illustrates a subscriber accessing the WiMAX network that can be any variant of WiMAX or just another use for mobile WiMAX.

19.6.4 Wireless Packet Network Offload

In addition, the use of 802.16 can be used to augment the resiliency of a 3G/4G wireless network for critical paths or sites as depicted in Figure 19.8. Typically, a radio site is connected to a single central office, and even though it may be on a fiber ring, the last 1000 feet is usually a single point of failure. Therefore, 802.16 can be used to provide not only backhaul but alternative routes for data traffic to and from various cell sites and other off-net facilities.

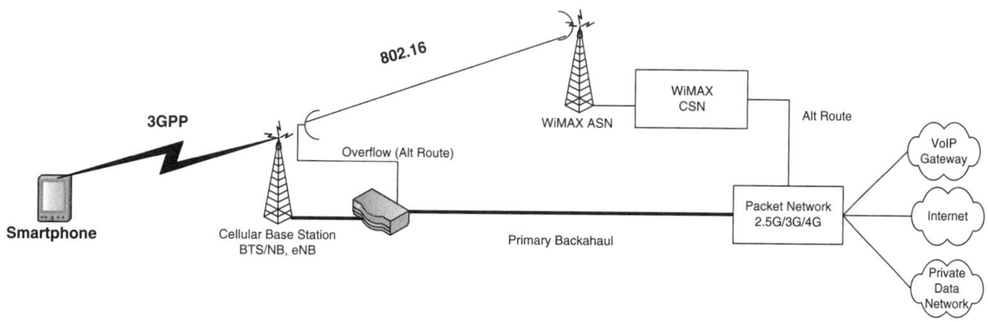

FIGURE 19.8 Packet offload.

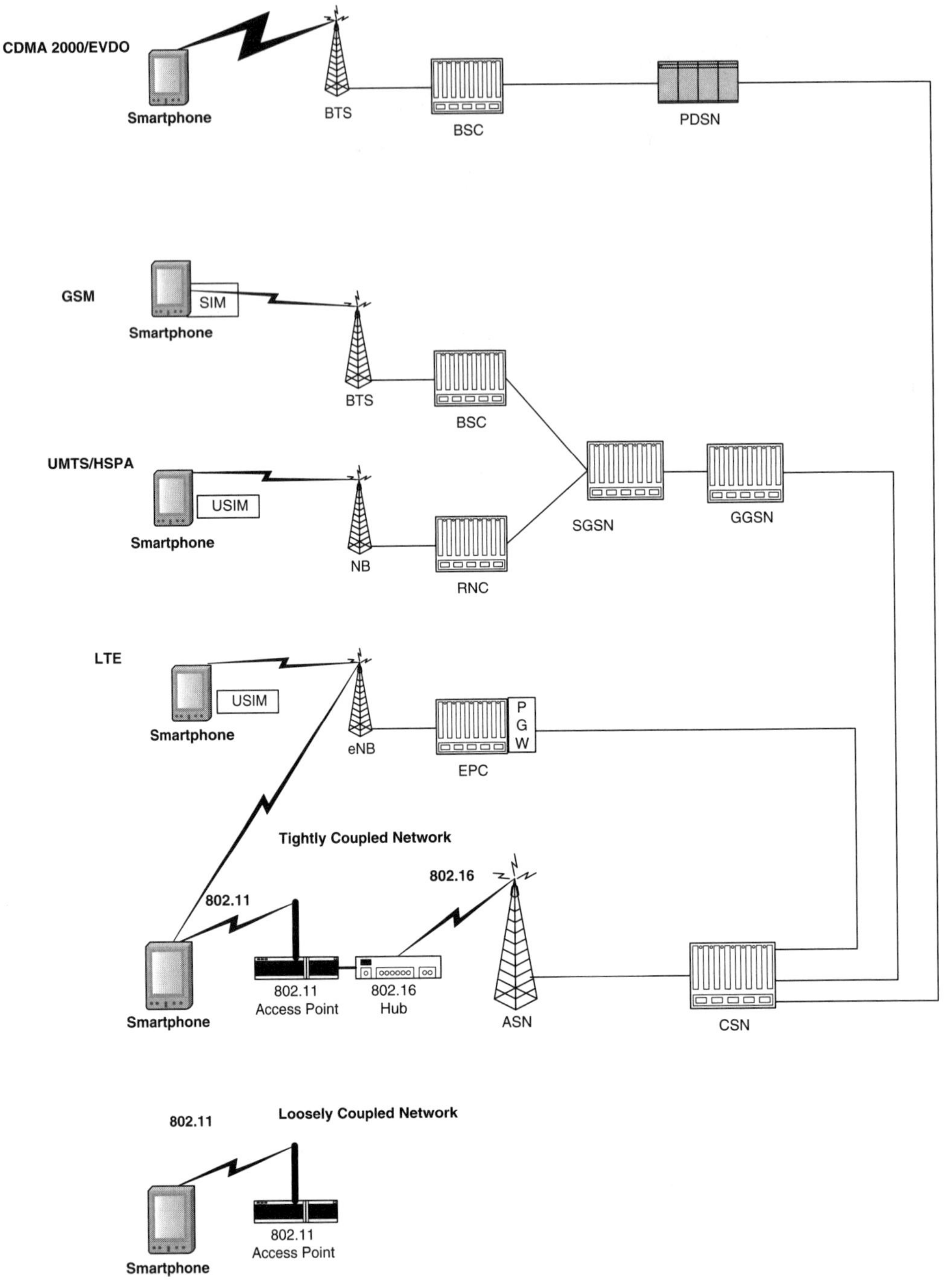

FIGURE 19.9 WiMAX facilitating WiFi integration into 3G/4G.

19.6.5 Integration with 3G/4G and 802.11 Mobility

An obvious question would begin to arise regarding the use of 802.11 and 802.16 and that is why not just use 802.11 instead and keep one type of infrastructure. The introduction of 802.16/16d and 16e and recently 16m into a mobile wireless network can run counterintuitive to the simplification process.

802.16 can help reduce churn by enticing enterprise customer and fleets with higher bandwidth, VoIP integration, and remote access services. In addition, the advantage that a wireless mobile operator has for using 802.16 in their system would be the freeing up of 3G/4G radio spectrum for use by other customers that are truly mobile and do not require the use of packet data services.

802.16 also has the advantage of providing a more cost-effective link for connecting several BTSs, or more importantly, several 802.11 access nodes back to the core packet network. In short, 802.16 advantages over an 802.11 system involve:

- Multiple QoS, not just contention based
- Higher data rates
- Longer distances (access node range)

A key reason for integrating 802.16 into a 3G/43G network is the ability to transport the IP traffic generated on the 802.11 WiFi networks as shown in Figure 19.9

However, WiMAX can directly integrate with 3G and 4G networks as part of the interoperability standards. Primarily, this is possible due to the fact that 3G, and especially 4G wireless networks, are packet networks as is WiMAX.

Figure 19.10 depicts a 4G network, LTE, having an integrated approach with WiMAX. The handset, also in figure 19.10, is dual mode and most likely multifrequency, where it can access both an LTE (4G) and WiMAX network. Therefore, Figure 19.10 shows an edge device using two different wireless access technologies to receive and send information.

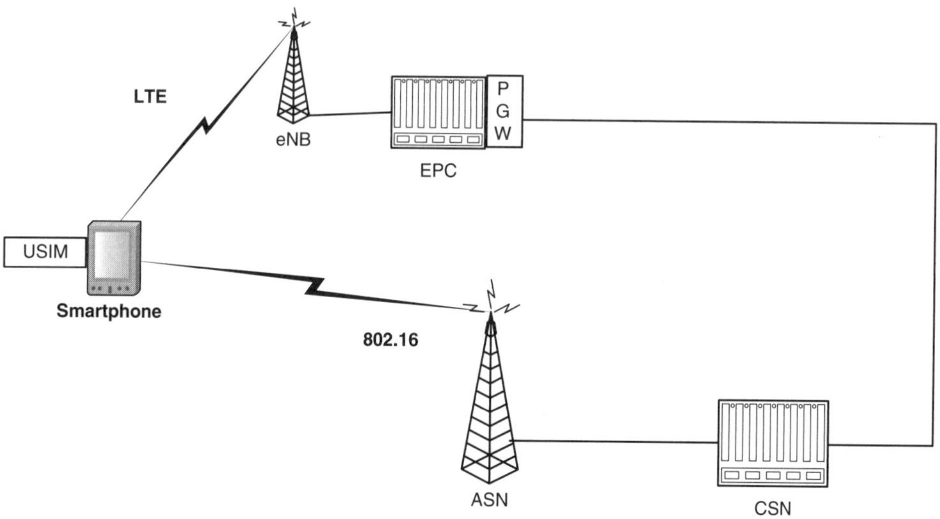

FIGURE 19.10 LTE and WiMAX Interoperability.

References

Chang, Dean, and Subir Varma, "WiMax Rolls Interop Guidelines for 802.16a," Electronic Engineering Times, May 16, 2003.

Eklund, Carl, Rodger Marks, and Kenneth Stanwood, "IEEE Standard 802.16: A Technical Overview of the WirelessMAN Air Interface for Broadband Wireless Access," IEEE Communications Magazine, June 2002, pp. 98-107.

IEEE Standard 802.16: A Technical Overview of the Wireless MAN Air Interface for Broadband Wireless Access. (Carl Eklund, Rodger Marks, Kenneth Stanwood) IEEE Communications Magazine, June 2002, pp. 98-107.

IEEE 802.16-2001, IEEE Standard for Local and Metropolitan Area Networks Part 16: Air Interface for Fixed Broadband Wireless Access Systems, April 8, 2002.

IEEE 802.16-2004, IEEE Standard for Local and Metropolitan Area Networks, Air Interface for Fixed Broadband Wireless Access Systems, October 2004.

IEEE 802.16 and WiMAX , Intel , Broadband Access for Everyone, July 8th, 2003.

IEEE Std 802.16™-2012, IEEE Standard for Air Interface for Broadband Wireless Access Systems.

IEEE Std 802.16.1b™-2012, IEEE Standard for WirelessMAN-Advanced Air Interface for Broadband Wireless Access Systems Amendment 1: Enhancements to Support Machine-to-Machine Applications.

IEEE Std 802.16.1™-2012, IEEE Standard for WirelessMAN-Advanced Air Interface for Broadband Wireless Access Systems.

Marks, Roger B., Kenneth L. Stanwood, Stanley Wang, "IEEE Standard 802.16: A Technical Overview of the WirelessMAN™ Air Interface for Broadband Wireless Access," IEEE Communications Magazine, June 2002, pp. 98-107.

Ohrtman, Frank, *WiMAX: Handbook Building 802.16 Wireless Networks*, McGraw-Hill, NY, 2005.

Sassan, Ahmadi, *Mobile WiMAX: A Systems Approach to Understanding IEEE 802.16m Radio Access Technology*, Academic Press, NY, 2011.

Smith, LMDS, McGraw-Hill, 2000.

Smith, Meyer, *WiMAX and WiFi*, McGraw-Hill, 2005.

Wirbel, Loring, "LMDS, MMDS Race for Low-Cost Implementation," Electronic Engineering Times, November 29, 1999, p. 87.

www.wimaxforum.org.

www.ieee802.org.

CHAPTER 20

Communication Sites

The communication site, usually referred to as the *base station* or *cell site*, is a critical component in any wireless system. A communication site is a physical location where there is radio equipment that is intended for either receiving, transmitting, or both. With the advent of multiple-technology platforms and the need to collocate wireless services on a single structure, normally more than one technology and service operator are located at any one site. For the design engineer involved in designing either a greenfield or a collocation service, an almost infinite number of different types of communication site configurations and perturbations can be considered.

This chapter will cover some of the more salient issues associated with a wireless communication site and the implications that should be considered in installing 2.5G, 3G, and *Forth-Generation* (4G) equipment. Therefore, the focus of attention will be directed toward the *Radiofrequency* (RF) engineer and the issues associated with the design phase. The particulars of operation and construction that are an integral part of communication site design will not be covered here because they are beyond the scope of this book.

20.1 Communication-Site Types

There are numerous types of communication sites that make up the *First-Generation* (1G), *Second-Generation* (2G), 2.5G, 3G, current *Fourth-Generation* (4G), and future *Fifth-Generation* (5G) configurations associated with wireless mobility systems. There are also a plethora of other communication sites that the design engineer may encounter in the design process, such as existing mobility systems like LTE, CDMA, GSM, UMTS, WiMAX, *Specialized Mobile Radio* (SMR), *Enhanced Specialized Mobile Radio* (ESMR), paging, broadcast, FM, AM, microwave backhaul, and so on. Each of these different types of wireless sites, depending on proximity, may need to be included in the design phase.

The usual collocation considerations are

- Antenna placement
- Frequency of operation of adjacent channels and co-channels (adjacent markets)
- Intermodulation—third- and fifth-order *Intermodulation Distortion* (IMD) products along with spectral regrowth
- Site maintenance obstructions—window washing equipment, sand blasting, and so on

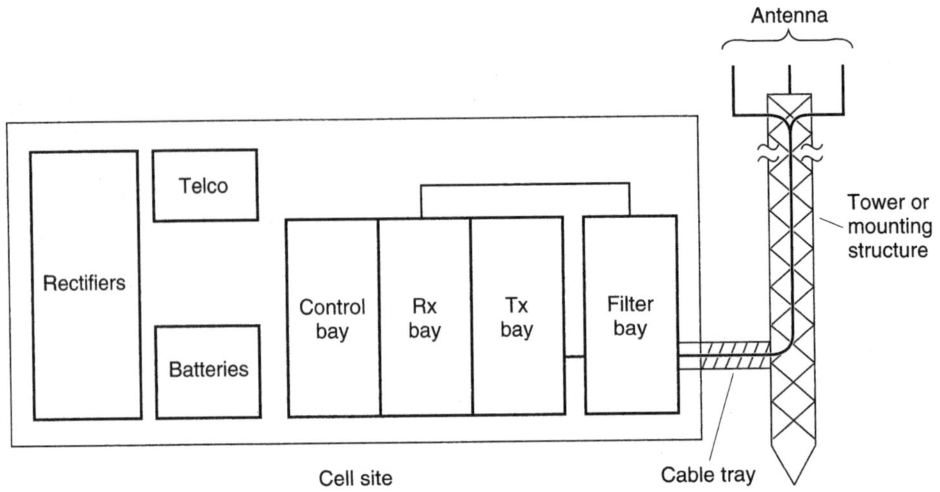

Figure 20.1 Communication site.

The most common types of sites that would be considered for a 2.5G and 3G implementation are

- Macro
- Omni
- Sector
- Micro
- Pico

The definitions of what macro-, micro-, and picocells are really depend on the service area the base station will cover. For instance, if a site is to cover 25 square miles, it is considered a *macrocell site*. However, if the site is to cover 0.25 mile, it is usually referred to as a *microcell*, whereas a site that is meant to cover a meeting room is often referred to as a *picocell*. Because there is no specification that defines the service area and the name for the particular communication site, the definitions of what constitutes a macro-, micro-, and picocell will remain somewhat vague.

A typical cell site or, rather, communication site consists of the following components, as shown in Figure 20.1. The piece components are the same whether it is for a macro-, micro-, or picocell site. The chief difference lies in the form factors that affect the overall capacity carrying capability for the site and, of course, power.

20.1.1 Macrocell Site

A macrocell site is what most people have come to know or expect to see for a communication site for all forms of mobile communications systems. With the advent of *Personal Communications Services* (PCS), the need for macrocells was portrayed initially as being an item of the past. However, the need to provide coverage to compete with existing wireless operators made the use of macrocell sites a necessity.

Multiple configurations are associated with each type of technology platform chosen for a communications system. For instance, *Advanced Mobile Phone System* (AMPS), *Total Access Communications System* (TACS), *Global System for Mobile* (GSM) communications, *Code-Division Multiple Access* (CDMA), *North America Digital Cellular* (NADC) (IS-136), *Integrated Dispatch Enhanced Network/Integrated Digital Enhanced Network* (iDEN) UMTS, and *Long Term Evolution* (LTE), to mention a few, all can be configured either as an omni-, bidirectional, or three-sector cell depending on the application at hand.

For the design engineer, the decision to use a macrocell is driven by multiple factors. However, there are a few different perturbations with regard to cell sites that lead to interesting designs. As is often the case in real life, in a city, the number of greenfield locations is not large, and in fact, the desire to use an existing communication site is receiving much pressure. There are a multitude of reasons why an existing communication site should be used and also why it should not be used.

The reasons for using an existing communication site lead to the issue of community affairs in that there is a strong public awareness of mobile phone systems and the need to limit the number of towers or any new communication sites that are located in a community.

There are several types of macrocell sites that a design engineer considers for possible use depending on the design objectives. The types of macrocell sites can be classified as either omnidirectional or directional sites. The omnicell sites have a coverage pattern of 360 degrees in the horizontal plane, whereas directional sites usually consist of three sectors, each covering 120 degrees of the horizontal plane.

20.1.2 Omnidirectional Cell Sites

The omnidirectional cell site is used typically in a low-capacity area of the network where the system is noise-limited and not interference-limited. The omnicell typically is used to cover uniformly in all directions (360 degrees). Under ideal conditions, the omnicell site would have a circular pattern when there are no obstructions and the coverage is purely line-of-sight.

With an omnidirectional site, there are several methods that can be used for antenna installations. The first is a simple installation on a monopole, as shown in Figure 20.2. The transmit antenna is highest on the structure, with the receive antennas located under the platform, which was the standard installation for 1G and typically is a

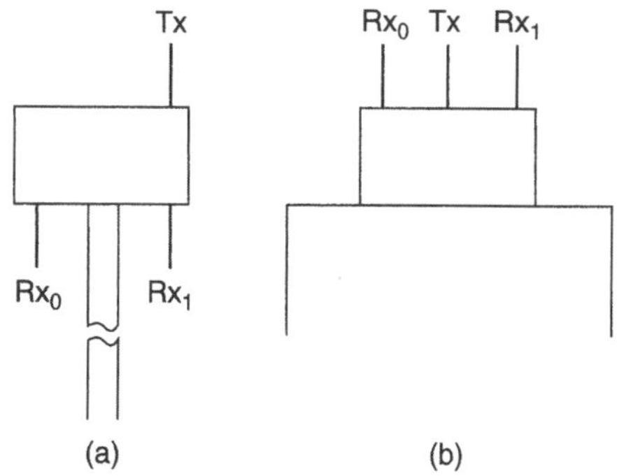

Figure 20.2 Monopole.

hold-over from that design era. The distance between the receive antennas usually is determined for maximizing spatial diversity so that the mobile signals arriving at both receive antennas are somewhat decorrelated, enabling a diversity gain or, rather, fade-margin protection.

There are, of course, other variants to omnicell antenna installations, and they involve installing on a building and when the number of antennas is limited. Figure 20.3 is an example of an antenna installation that occurs on a building. Please note that the location of the antenna needs to meet the required setback rules. If the setback rules cannot be adhered to, then it is possible to install the antennas near the edge of the roof; however, they may become visible to the public at this point, the landlord may not wish this type of installation to take place, or the local ordinances may prohibit this from occurring.

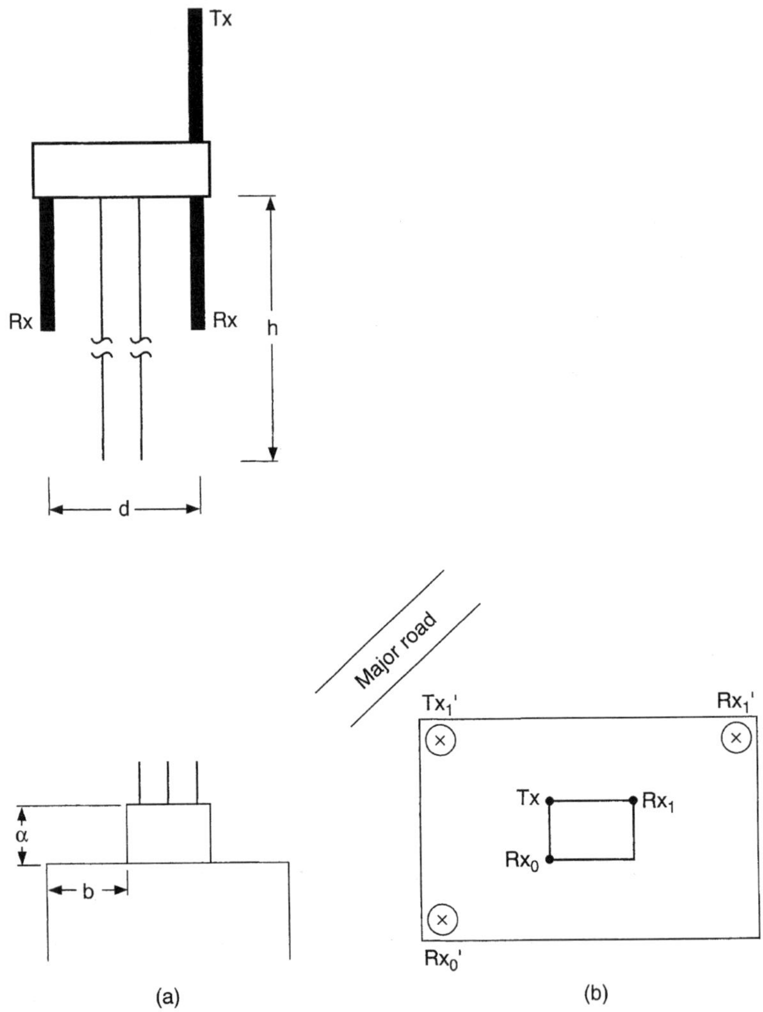

FIGURE 20.3 Existing rooftop installation.

Please note that placement of the receive antennas on the rooftop should be such that if there is only one major road in the area for the cell to cover, then the horizontal diversity placement for the antennas should be maximized in the direction toward the road. Lastly, if the primary location is not achievable for mounting the antennas, then moving them to a lower level is possible. However, based on the penthouse size, significant blockage may occur in one direction, and this needs to be factored into the design process.

20.1.3 Directional Cell Site

The directional cell site employing three sectors is one of the most popular cell-site configurations used in the wireless industry, next to the omnicell. The three-sector cell is one that has sectors that cover 120 degrees each; thus, having three sectors makes a full circle.

There are a multitude of combinations for transmit and receive that can be used for establishing a three-sector cell site. However, the following example is the basic configuration that is used and involves three antennas per sector, as shown in Figure 20.4.

The configuration shown in the figure has a single transmit and two receive antennas per sector. Naturally, the number of receive and transmit antennas can change depending on the technology implemented for 2G, 2.5G, or 3G. For example, iDEN 2G systems typically involve three antennas per sector, but all three usually are duplexed to keep the antenna count down while at the same time enabling for three-branch diversity reception.

When designing antenna placements for a site, it is strongly recommended that future configurations be considered at the onset of the design process. For instance, implementing 3G services may require the use of a separate *Transmission* (Tx) antenna and possibly *Receive* (Rx) antennas. While 4G has placed a premium on the antenna real estate for a cell site.

Please keep in mind that with *Multiple Input Multiple Output* (MIMO), the potential for three or four or more antennas for use in both Tx and Rx is becoming more common. Additionally, "smart" and active antenna systems also can be deployed, and the configuration and requirements need to be factored into the initial design as well.

20.1.4 Microcells

Microcells are prevalent in wireless mobility systems as the operators strive to reduce the geographic area each cell site covers, thus facilitating more reuse in the network. Microcells are also deployed to provide coverage in buildings, subway systems, and

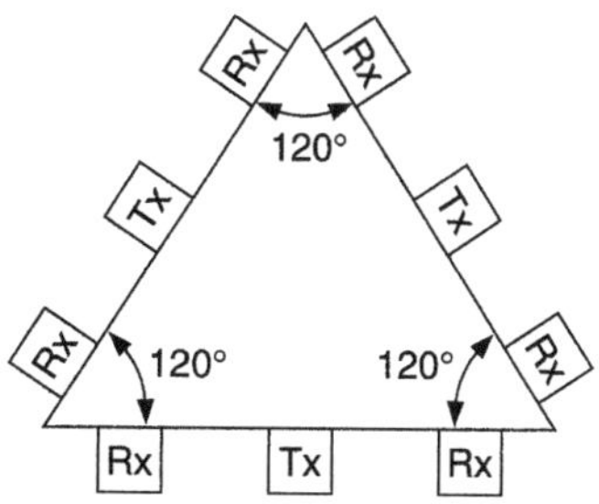

FIGURE 20.4 Directional cell site.

tunnels and to resolve unique coverage problems. The technology platforms that tend to be referenced as *microcells* involve any communications system that is less than ½ km in radius. Typically 4 to 10 microcells make up the footprint that a macrocell site might be able to cover.

There are currently several types of technology platforms that fall into the general categorization of microcells:

- Fiber-fed microcells
- T1/E1 microcells
- Microwave microcells
- High-power ReRad cells
- Low-power ReRad cells
- Bidirectional amplifiers

The choice of which technology platform to use is driven by a variety of factors that are unique to the specific situation. One driving factor for the technology platform chosen is the application that is being engineered for capacity, coverage, or wireless *Private Branch Exchange* (PBX). Another important factor in the technology platform decision is the configuration options available at that location for providing radio capacity. A third factor in the decision about which technology to use is the overall cost of the solution for the network.

A microcell typically uses an omni antenna for transmission and reception. The microcell also has less Tx power and lower-gain receive antennas than does a macrocell site. In addition, a microcell site typically exhibits a lower elevation than a macrocell site, helping to contain its coverage area and leading to selected trouble-spot resolutions. The use of an omni antenna for microcells facilitates a smaller physical appearance, leading to installations in more difficult land-use areas.

An example of a microcell is shown in Figure 20.5. The form factor of the radio hardware is not shown and is assumed to be insignificant or located in the utility box next to the traffic light shown.

20.1.5 Picocell Sites

The use of picocell sites by wireless operators is driven by the desire to provide very targeted coverage and capacity to a given area or application. The picocell has a very small service area, where several picocells can cover the same area as a microcell.

The picocell is a spot-coverage and low-capacity site as compared with a macrocell site. Picocell sites typically have a single omni antenna, as do microcells. However, the power and thus the coverage of the picocell are less than those of a microcell. An example of a potential picocell site is shown in Figure 20.6.

20.1.6 Femto Cell

Femto cells are a new entrant to the coverage portfolio and require little or no engineering from the designers aspect whether RF or core network. Femto cells can provide coverage for a very localized coverage area, which could be an SOHO, venue application, or home network, to mention a few possibilities.

The femto cell configuration is shown in Figure 20.7 where the femto cell is connected to a DSL or cable modem which then allows the femto cell to connect to a femto

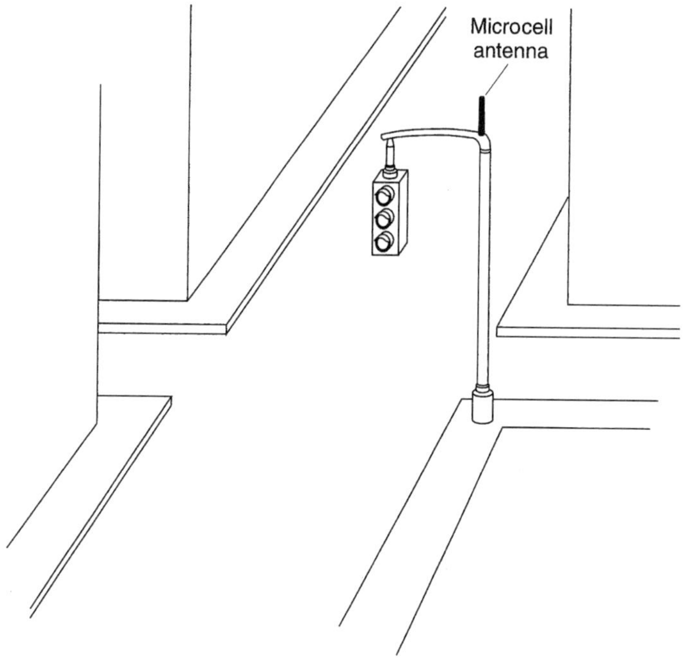

Figure 20.5 Microcell.

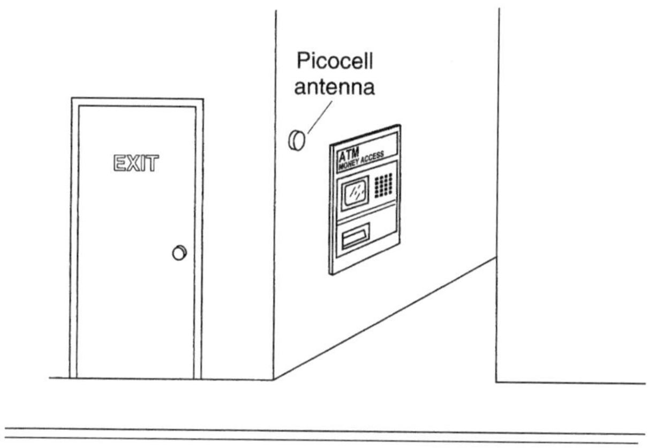

Figure 20.6 Picocell.

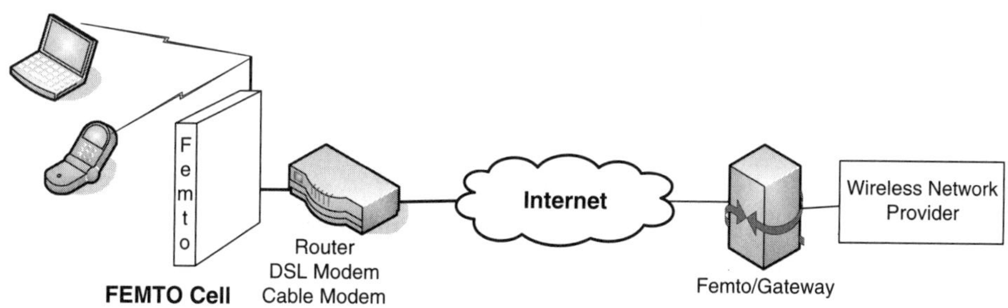

Figure 20.7 Femto cell.

gateway located at the wireless network provider's system. What is most interesting about the use of a femto cell is that the backhaul is typically provided by another provider and many times may be a competitor to the wireless operator.

The femto cell, however, has limited coverage due to its transmit power. Yet, the coverage for a femto cell is similar to that experienced in building with WiFi. However, unlike WiFi the femto cell uses licensed spectrum. The femto cell also typically has a unique ID that informs the subscriber device that it is roaming; however, it is roaming on the home network just using a femto cell.

Some other key attributes of a femto cell involve its ability to facilitate hand-ins and handouts of the femto cell with the macrocellular network. However, the handovers are hard handovers.

With all the many attributes of a femto cell, its primary limitation, besides coverage, is the overall capacity it can truly support.

Additionally, a femto cell is technology dependent for operation and is found in UMTS, CDMA, and LTE; however, it is not available with GSM.

20.2 Installation

The installation of a cell needs to accommodate a number of issues that have different elements of importance within them. Some of the issues that need to be factored into the design are the physical placement of the antennas themselves. The physical placement of the antennas depends on some factors that may or may not be in the design engineer's control.

20.2.1 Cable Runs

Some of the physical installation issues that need to be factored into the design involve cable runs from the antenna system, each leg, to the base-station equipment. Although this may seem to be an obvious point, often there are situations where the desired routing of the cables is not practical, making the real installation length much longer than desired. The additional cable run length, when installation reality is factored in, may have made the site undesirable; however, if this situation occurs too far downstream from the construction process, it is too late to reject the site or make the appropriate design alterations to correct the situation.

20.2.2 Antenna Mounting

Obviously, the mounting of the antennas must be done with extreme care. The following is a brief checklist to ensure that antenna mounting concerns are checked prior to acceptance from a cell site:

- What are the number and type of antennas to be installed?
- What is the maximum cable run allowed?
- Identify and rank obstructions that would alter the desired coverage.
- Make sure that Rx antenna spacing is adequate and that diversity requirements are met.
- Make sure that isolation requirements meet with those of other services.

- Make sure that antenna *Above-Ground-Level* (AGL) requirements are met.
- Make sure that antenna-mounting parameters are met.
- Make sure that intermodulation analysis is completed.
- Make sure that path clearance analysis is verified (if applicable).

This list is just preliminary and can be altered easily based on the situation at hand. However, the list should be modified to meet your particular system design requirements.

When installing on a tower, the physical spacing or offset from the tower must be selected based on the tower's structure, which either enhances or does not alter the antenna pattern desired. In addition to the pattern issue, care must be taken to ensure that there are no degradations caused to the system because of unwanted energy from adjacent systems. It is suggested that an interference analysis be conducted for every site to ensure that proper isolation requirements are met.

When installing on an existing building, the following few items should be considered in the design phase.

20.2.3 Diversity Spacing

The diversity spacing for the receive antennas needs to ensure that the proper fade-margin protection is designed into the system. Diversity spacing is meant to achieve some decorrelation between the mobile received signal. Numerous studies have been conducted on diversity reception and the system performance improvements associated with proper implementation of a receive diversity scheme. The diversity scheme typically is achieved through horizontally placed antennas that then are fed to the radio receiver at the base station.

The base-station receiver typically would use either maximum-ratio combing or select diversity as the method of achieving system performance improvements. However, for the diversity reception, the antennas for mobile communications for 1G, 2G, 2.5G, 3G, and 4G involve horizontal diversity spacing. The initial objective would be to place the receive antennas so that they are as decorrelated as possible. However, there is a practical limit: The spacing between the receive antennas when the feedline length between the antennas becomes such that either the feedline loss exceeds the diversity advantage or the signals are completely decorrelated as to eliminate any diversity combining gain possible.

For a micro- or picocell site, the use of diversity reception is usually a forgone conclusion owing to the antenna configuration—one omni antenna. However, when looking at a macrocell or even a microcell with multiple antennas for receive, the question arises as to what spacing is needed between the antennas. The following equation should be used for a two-branch receive system:

$$\text{Diversity spacing (ft)} = \{[[\text{AGL of antenna (ft)}/11] \times (835/f_o)]\}$$

where f_o is the center receive frequency in megahertz.

20.2.4 Roof Mounting

When installing antennas on an existing roof or penthouse, consideration must be given to how high the antenna must be with respect to the roof surface. Obviously, the ideal location is to place the antenna right at the roof edge. However, placing the antenna at

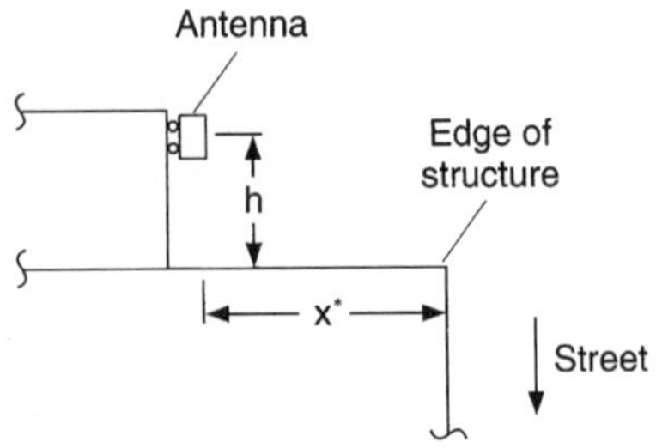

FIGURE 20.8 Roof mounting.
*x = 5 *h

the roof edge may not be viable owing to aesthetics, local ordinances, or practical mounting issues. When the antenna cannot be placed at the edge of the roof, a relationship between the distance from the edge of the roof and antenna height exists and must be accounted for. This relationship between antenna height and the roof edge of a building is depicted in Figure 20.8.

The preceding example assumes that there are no additional obstructions between the antenna and the roof edge. If there are obstructions between the antenna and the roof edge, then additional height may be needed. Examples of additional obstructions involve HVAC units and window-cleaning apparatus. Please remember that if there is a desire to implement severe downtilt into the design either at present or in the future, then the height requirements above the rooftop may need to be increased.

20.2.5 Wall Mounting

For many building installations, it may not be possible to install antennas above the penthouse or other features of the building. Often it is necessary to install antennas on the penthouse or a water tank. When installing antennas onto an existing structure, rarely has the building architect factored in the effects of antenna installation at the onset of the building design. Therefore, as shown in Figure 20.9, the building walls may be sufficient for one orientation of the system but rarely all three for a three-sector configuration.

Therefore, it is necessary to determine what the offset from the wall of the building structure needs to be. Figure 20.10 illustrates the wall-mounting offset required to ensure proper orientation for each sector.

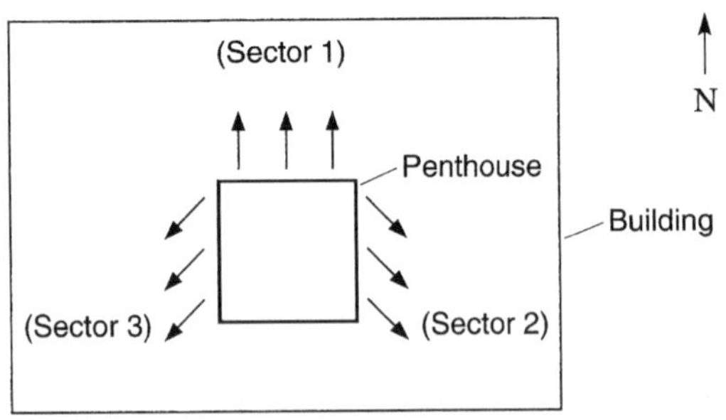

FIGURE 20.9 Three-sector building configuration.

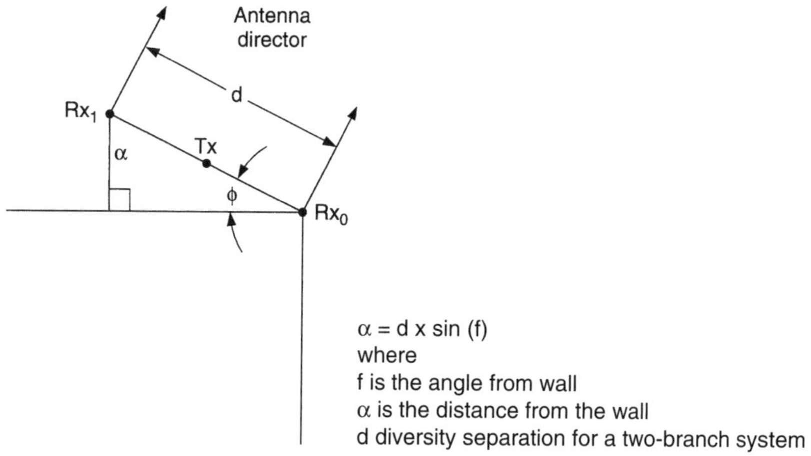

FIGURE 20.10 Antenna offset mounting.

Obviously, common sense must enter into the situation here when the inclusion of the offset brackets makes the site a metal monster. Tradeoff can be made in the design when the orientation for each sector is within the design-tolerance limits for sector orientation. The design tolerance should be within + or −5 degrees for a three-sector cell.

Lastly, the wall-mounting offset must meet the setback requirements for both antennas and local ordinances.

20.3 Towers

Numerous types of towers can and do exist in wireless networks. However, three basic types of towers are most common: self-supporting, guy wire, and monopole. The general configuration for each of the towers is shown in Figures 20.11 through 20.13.

The cheapest to construct is the guy-wire tower, followed by the monopole and then the self-supporting tower. Each has its advantages and disadvantages. The guy-wire tower requires a large amount of room for its guy wires and is shown in Figure 20.12. This can be either relaxed or increased depending on loading and height issues.

The self-supporting tower will enable multiple carriers to entertain operation at the facility, whereas the monopole also will accommodate multiple users, although not as many as a similar-height self-supporting tower.

20.4 Stealth

The proliferation of wireless sites has also seen an increase in the use of camouflage or stealth technology. The ability to camouflage or reduce the visual impact of a wireless site is essential to many communities. Several types of stealth technologies are used, with Figure 20.14 showing three common techniques.

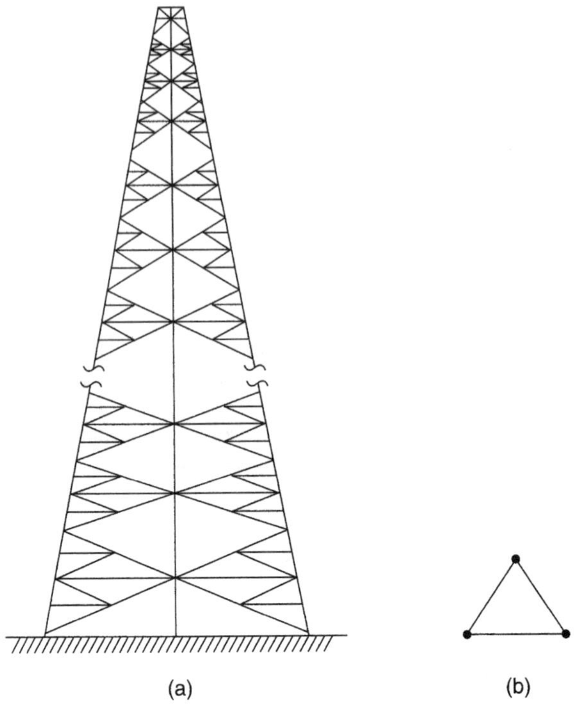

FIGURE 20.11 Self-supporting tower: (a) side view; (b) top-down view.

20.5 In-Building and Tunnel Systems

Wireless systems have numerous in-building applications, including improving coverage for a convention center or large client, disaster recovery, and a wireless PBX, to mention a few. With the advent of better transport for data services, the possibility exists that 3G will find more uses for in-building systems.

Designing an interior wireless system to support in-building service, subway, or tunnel application can and should be a rewarding situation. The issues encountered in interior communications system applications are unique and similar to those associated with a macrocell system.

There has been a lot of focus on in-building applications over the years, with surges in effort taking place. The fundamental problem that keeps arising is the inherent cost of the facility. The cost of an in-building system typically far outweighs any potential for increased wireless revenue for the operator. The cable and antenna installation, whether distributed or leaky feeder antennas, are the real driving costs of these installations. The inclusion of 802.11g wireless *Local-Area Networks* (LANs) has proved exceptional for configuration and installation ease but has contributed little in the way of increased revenues for wireless companies.

One method that has been pursued to offset the cost for inbuilding systems is the multicarrier approach. The multicarrier approach involves having a host system that enables several of the wireless operators to share the same antenna distribution system

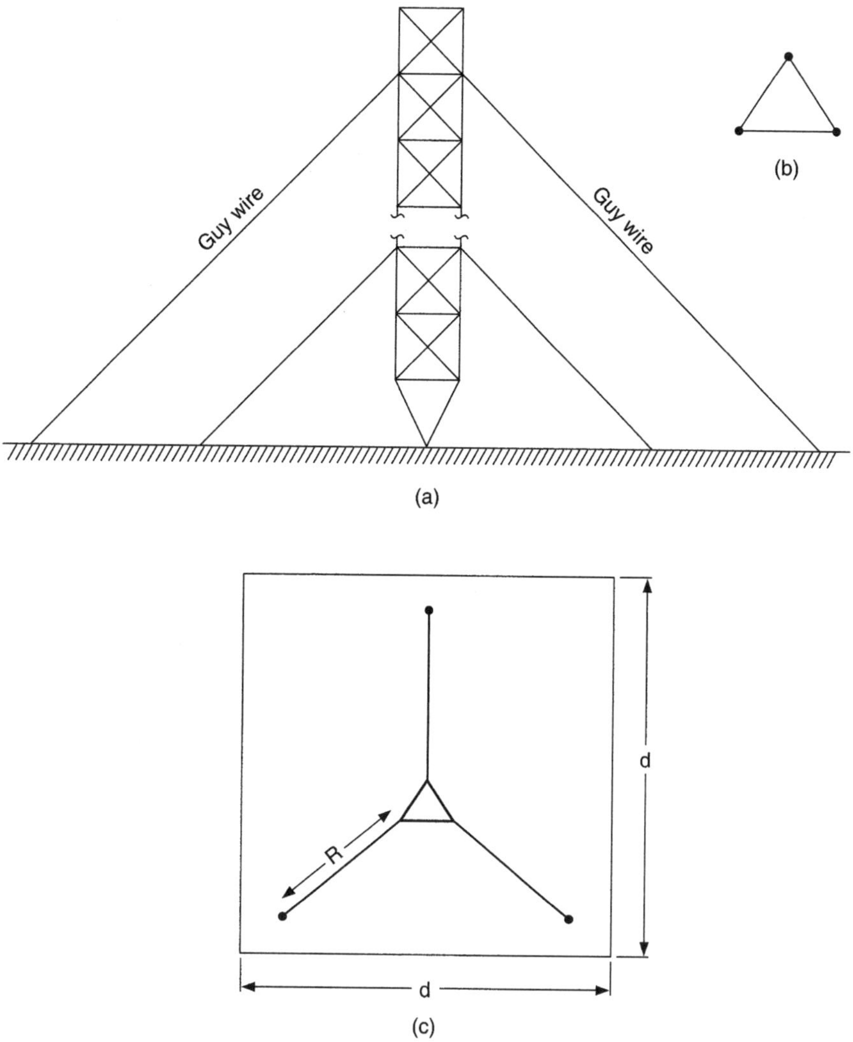

Figure 20.12 Guy-wire tower: (*a*) side view; (*b*) top-down view; (*c*) footprint of tower.

or raceway for their own antennas. The advantage with the multicarrier approach is that the cost of the installation is spread over several carriers, thus reducing the largest cost component of the installation.

Regardless of the application and the solutions sought, we have yet to get involved with any interior communications system that requires fewer than three visits to resolve all the issues that consistently arise. Typically, the in-building design approach has been the "surround and drown" method, which should be evident in the link-budget calculations associated with urban environments. However, the propagation of the RF energy takes on unique characteristics in an interior or confined application as compared with an outdoor environment.

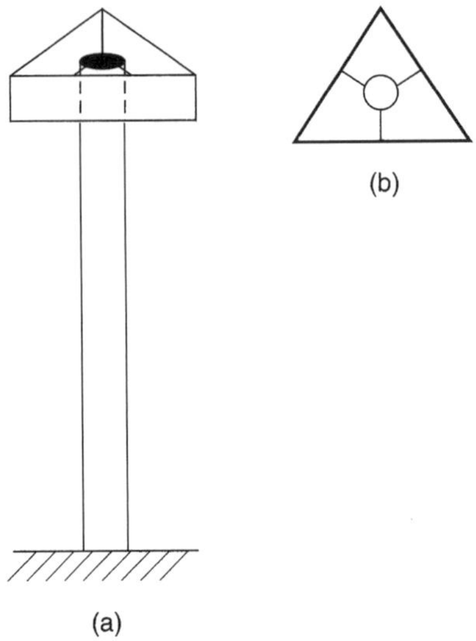

Figure 20.13 Monopole: (*a*) side view; (*b*) top-down view.

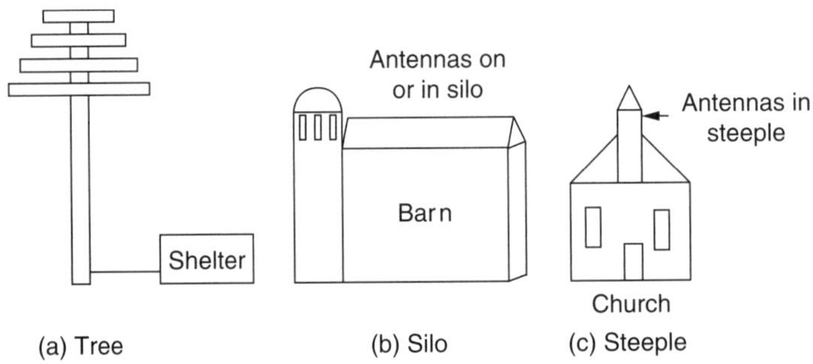

Figure 20.14 Stealth techniques.

The primary difference in propagation characteristics for interior versus outdoors is the fading, shadowing, and interference. The fading situation for interior situations results in a deeper fade that is spatially closer than that normally encountered in an exterior application. Shadowing is also quite different owing to the lower antenna heights; excessive losses through floors, walls, and cubicles; and vehicle blockage, as in the case of a tunnel. The shadowing effects severely limit the effective coverage area to almost *Line-of-Sight* (LOS) for wireless communications whether in the cellular or PCS band. The interference issue with an interior communications system actually can be a benefit because it is primarily noise driven and not co-channel interference, assuming that no reuse is involved in the interior applications (obviously, this is not a CDMA issue). The reasoning behind this interference comment is that interior communications systems are primarily noise driven owing to the attenuation experienced by external cell sites as they transverse into buildings and various structures. Now, this is not to say that there cannot be IMD or out-of-band emissions problems that disrupt the communications system.

There are some unique considerations that must be taken into account regarding interior communications system designs. Some of these considerations include

- Base-to-subscriber-unit power
- Subscriber-unit-to-base power
- Link budget
- Coverage area
- Antenna system type and placement
- Frequency planning

The base-to-subscriber-unit power needs to be considered carefully to ensure that the desired coverage is met, deep fades are mitigated in the area of concern, the amplifier is not being overdriven or potentially underdriven, and mobile overload does not take place. The desired coverage that the interior system is to provide might require several transmitters because of the limited output power available from the units themselves. For example, if the desired coverage area required 1 W ERP to provide the desired result, a 10-W amplifier would not be able to perform the task if you needed to deliver a total of 40 channels to that location, meaning that only 25 mW of power per channel were really available. The power limitation can and often does make the limiting path in the communications system for an in-building system the forward link.

The forward-link power problem is complicated further by the fact that portable and potentially mobile units will be operating in very close proximity to the interior systems antenna. If the forward energy is not set properly, a subscriber unit easily could go into gain compression, causing the radio to be desensitized.

The subscriber-unit-to-base power also needs to be factored into the interior design. If the power windows, that is, dynamic power control, are not set properly, then imbalances could exist in the talk-out to talk-back path. Usually the reverse link in any interior system is not the limiting factor, but the subscriber-unit-to-base path should be set so that there is a balanced path between the talk-out and talk-back paths.

Most interior systems have the ability to use diversity receive, but do not use it for a variety of reasons. The primary reason for not using diversity receive in an interior system is the need to place two distinct antenna systems in the same area, which is rarely possible owing to installation and cost restrictions.

The link budget for the communications system needs to be calculated in advance to ensure that both the forward and reverse links are set properly. The link-budget analysis plays a very important role in determining where to place the antenna system, distributed or leaky feeder, and the number of micro-/picocell systems required to meet the coverage-area requirement. The link budget associated with an in-building/tunnel system is for all intents and purposes an LOS model. The simple rule is that if you can see the antenna, you have coverage. Interior fading and attenuation are very severe, and rounding a corner in an office usually will result in loss of signal, which will deteriorate the call or make it terminate prematurely.

Several methods are available for wireless operators to ensure interior coverage:

- Macro system ("surround and drown")
- Micro-/picocells
- Reradiators

Please note that frequency planning for an interior system needs to be coordinated with the external cellular network. Coordination is needed because most interior communications systems are designed to facilitate handoffs with the macro system. If the in-building system is using a microcell with its own dedicated channels, then it is imperative that the interior system be integrated into the macro network.

Reiterating, the concepts for in-building or tunnel coverage are very similar in that they both rely on LOS and not multipath to ensure a reliable communications link. This is a fundamental change from macro-system design, which, by default, relies on the use of multipath to ensure the communications link.

20.5.1 Antenna System

The antenna system selected for an interior application is directly related to the uniformity of the coverage and quality of the system. An antenna system with no diversity primarily provides LOS coverage to most areas in the defined coverage area. Based on link-budget requirements, the antenna system can be either passive or active. The antenna system for an in-building system may take on the role of having passive and active components in different parts of the system to satisfy design requirements. Figure 20.15 shows a *Distributed Antenna System* (DAS) that can be augmented by additional active components if the design warrants it.

Typically, a passive antenna system is made up of a single or distributed antenna system or uses a leaky coaxial system. The in-building system shown in Figure 20.15 uses a distributed antenna for delivering the service. A leaky coaxial system also could be deployed within the same building to provide coverage for the elevator in the building.

The advantage a leaky coaxial system has over a distributed antenna is that it provides more uniform coverage to the same area. However, a leaky coaxial system does not lend itself to an aesthetic installation in a building. The use of a distributed antenna system for providing coverage in an in-building system makes the system stealthy. However, for providing coverage for elevators, the only method we have found to be successful is the leaky feeder owing to the "metal coffin" the elevator makes when the doors are closed.

If the antenna system requires the use of active devices in the communication path, the level of complexity increases. The complexity increases for active devices because they require ac or dc power and introduce another failure point in the communications system. However, the use of active devices in an in-building system ultimately can make the system work in a more cost-effective fashion. The most common active device in an in-building antenna system is a bidirectional amplifier.

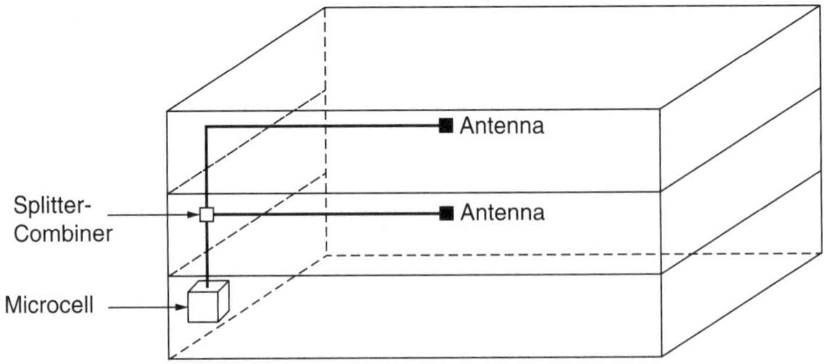

Figure 20.15 Active in-building system.

20.5.2 In-Building Application

For covering a simple room or meeting areas within a hotel, technology has advanced considerably providing a host of potential solutions allowing the design engineer to better tailor the solution to the particular application. A traditional method for providing in-building coverage would involve the use of a bidirectional system, whether permanent or temporary, may prove most viable. Figure 20.16 is an example of an in-building bidirectional system. Please note the reliance on the external system and the need to provide isolation between the external directional antenna and the DAS antenna shown in Figure 20.16. Preventing feedback typically requires 70 dB of port-to-port isolation, which can be achieved by antenna placement or use of an attenuator for minor adjustments.

However, the introduction of femto cells enables a more tailored approach for limited coverage areas for in-building applications. A typical femto cell configuration would involve placing the femto cell in a meeting room and having it connect to a DSL or cable modem for backhaul.

The disadvantage of using a femto cell is that it typically is configured for a single wireless carrier and to enable multiple carriers the need for a femto cell for each wireless carrier is then required.

However, the use of femto cells enables the coverage of a subscriber's apartment similar to the coverage expected with WiFi for range. This use of femto cells enables better control of the subscriber equipment, especially in high-rise conditions where the origination and session management are now handled by the femto cell instead of a macro/micro cell that maybe several cell tiers away.

Femto cells can also be deployed for enterprise conditions and done so with minimal effort.

20.5.3 Tunnel Applications

Tunnel systems are unique, and many papers have been written on the topic of tunnel coverage. What we have found is that trying to provide coverage for a tunnel using external illumination works up to the first bend in the tunnel. Therefore, unless it is a small tunnel or short underpass, reliable coverage can be provided only by some form of cell site or if reamplication of the macro system is accomplished.

For tunnel coverage, we have used distributed antenna systems, using bidirectional amplifiers, and the leaky feeder approach. Regarding the bidirectional amplifier method,

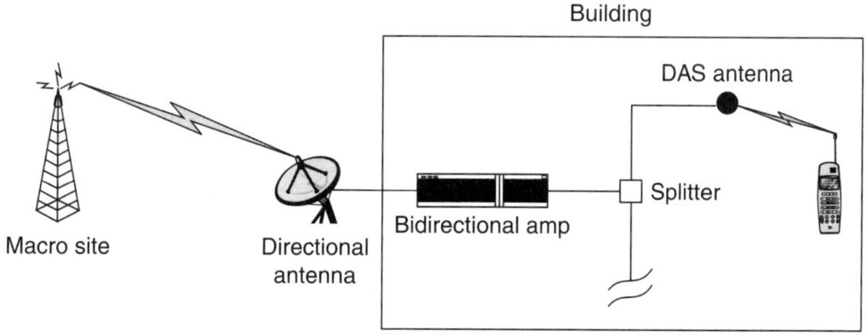

Figure 20.16 Bidirectional antenna system.

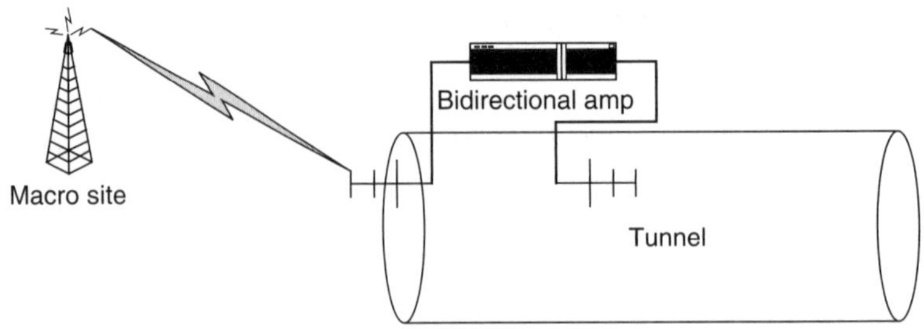

Figure 20.17 Tunnel coverage using the rerad approach.

an example is shown in Figure 20.17, which could be done with a micro- or picocell. The antennas should be yagi antennas, and they should be installed after the first bend or as deep in the tunnel as possible.

The leaky feeder approach provides more uniform coverage and accounts for bends. Location of the leaky feeder has been on the top of one of the sidewalls, placing the feeder against the roof, with the appropriate standoffs provided by the leaky feeder vendor. An example of a leaky feeder installed for a tunnel is shown in Figure 20.18, where a microcell is used as the antenna system feeder instead of relying on a reradiation technique.

At both ends of the leaky feeder, a 50-ohm antenna is installed, and this provides both a load for balancing and a method for providing better overlap with the macro system. Additionally, if the tunnel is excessively long, which you can determine by evaluating the link budget, two microcells or bidirectional amplifiers can be used. The middle point of the tunnel has a termination on both ends to allow for a handoff zone (Figure 20.19).

Please note that a leaky feeder installation is effectively a single antenna with non-diversity. This is not an issue because in most cases LOS is maintained. However, installing a second leaky feeder for diversity will not produce any realizable benefit and will escalate the installation costs needlessly.

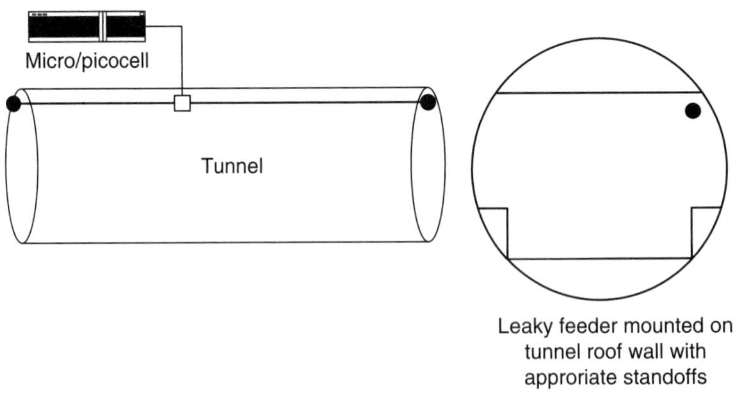

Figure 20.18 Leaky feeder system for a tunnel.

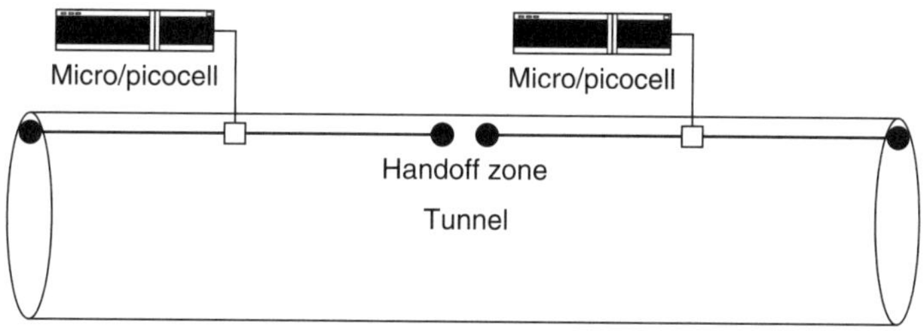

Figure 20.19 Long tunnel application.

Anyone who has ventured into a tunnel system knows some of the unique installation concerns. One of the issues is access for maintenance. This is why we are strong proponents of placing the equipment near the entrance of the tunnel and having the RF distribution system remain passive.

Lastly, there is the question of what to do when several operators want to use the same leaky feeder. The obvious issue is a combining technique for the forward energy and a receiver multiplex system for the receive. However, this will prove problematic because someone has to maintain the Rx multiplexer if it is active, as most are. Therefore, cross-band coupling is recommended because it allows all the operators to use a duplexer for Tx and Rx of the signal. However, you must pay attention to the operating band of the coupler port, the losses of the combing system (which can get excessive), and the overall power being delivered to the unit. Once all the losses are understood, you need to redo the link budget to ensure that you still can have reliable communications.

20.5.4 Planning

The issues associated with designing an in-building or tunnel system for a wireless system are listed below. The time durations that accompany each of the steps are not included because they depend directly on the size of the system and the time-to-market requirements.

- Project kick-off meeting
- Antenna mount installation
- Equipment installation
- UPS system installation
- Equipment rack installation
- Equipment installation
- Microcell installation
- Cable installation to micro-/picocell
- Radiax cable and/or distributed antenna system installation
- RF engineering project design kick-off meeting
- Establishment of responsibility centers
- Performance criteria
- Review system designs

- Micro-/picocell system review
- Macrocell system review
- Coverage requirements
- Handover
- Cable design
- Performance criteria
- Tunnel operators system design review
- System requirements
- Design criteria defined
- Link budget
- Intermodulation
- Noise levels
- Filter requirements
- Microcell system review
- Radiax system review
- RF plumbing design
- Path analysis
- Antenna system
- Antenna selection
- Lightning protection
- Tx combing
- Frequency plan
- Translations
- Parameter adjustments
- Design review
- Establishment of equipment list and ordering
- Acceptance test plans
- Criteria establishment
- Generate
- Test plan
- Issue ATP
- Conduct tests
- Facilities
- Fiber testing
- Power acceptance test
- ATP signoff

20.6 Intermodulation

Intermodulation is the mixing of two or more signals that produce a third or fourth frequency that is undesired. All radio communication sites produce intermodulation no matter how good the design is. However, the fact that intermodulation products are produced does not mean that there is a problem.

Just what is intermodulation, and how does one go about calculating an intermodulation product? Various intermodulation products are shown in the following examples for reference. The values used are simplistic in nature so as to facilitate the examples. In each of the examples, $A = 880$ MHz, $B = 45$ MHz, $C = 931$ MHz, and D is the intermodulation product. The following example does not represent all the perturbations possible.

Second order: $A + B = D$ (925 MHz)

$A - B = D$ (835 MHz)

Third order: $A + 2B = D$ (970 MHz)

$A - 2B = D$ (790 MHz)

$A + B + C = D$ (1856 MHz)

$A - B + C = D$ (1766 MHz)

Fifth order: $2A - 2B - C = D$ (739 MHz)

The various products that make up the mixing equation determine the order of the potential intermodulation. All too often when you conduct an intermodulation study for a cell site, a number of potential problems are identified in the report. The key concept to remember is that the intermodulation report you are most likely looking at does not take into account power, modulation, or physical separation between the source and the victim, to mention a few. Therefore, the intermodulation report should be used as a prerequisite for any site visit so that you have some potential candidates to investigate.

Intermodulation also can be caused by your own equipment through bad connectors and antennas or faulty grounding systems. However, most intermodulation problems are a result of a problem in the antenna system for the site and well within the control of the operator to fix.

Just how you go about isolating an intermodulation problem is part art and part science. We prefer the scientific approach because it is consistent and methodical in nature.

The biggest step is identifying the actual problem; the rest of the steps will fall in line. Therefore, the following procedure is recommended for intermodulation-site investigations.

20.6.1 IM Check Procedure

1. Determine if there are any collocated transmitters at this facility.
2. Collect information on each of the following transmitters:
    - Antenna types
    - Emission type
    - Transmit power

  - Location of antennas
  - Operator of equipment
  - FCC license number

3. Conduct an intermodulation study report looking for hits in your own band or in another band based on the nature of the problem.
4. Allocate sufficient time to review the report.
5. Determine if there is a potential problem.
6. Formulate a hypothesis for the cause of the problem, and engineer a solution.

Based on the actual problem encountered, the resolution can take on many forms:

- Is the problem identified feasible?
- Can the problem be resolved through isolation alone?
- Is the problem receiver overload–related?

If the intermodulation product is caused by the frequency assignment at the cell site, then it will be necessary to alter the frequency plan for the site, but first you remove the offending channels from service. If the intermodulation problem is due to receiver overload, the situation can be resolved by placing a notch filter in the receive path if it is caused by a discrete frequency. If the overload is caused by cellular mobiles, using a notch filter will not resolve the situation. Instead, mobile overload can be resolved by placing an attenuation in the receive path prior to the first preamp, effectively reducing the sensitivity of the base-station receiver.

20.6.2 Passive Intermodulation (PIM)

Passive Intermodulation (PIM) is IMD and is typically defined as the unwanted mixing of two or more signals in a non-linear junction within a wireless or RF system, creating additional undesirable harmonic signals. These unwanted IMD signals are of most concern if they fall in your own receive band, and of course, if they fall into another wireless operator's frequency band.

PIM, however, can and is generated by many components in an RF antenna system and the concern is the power levels that are created. PIM, in short, is not an IMD product as a result of other transmitters at the site, though it could be. However, PIM is primarily a self-induced issue, which is exasperated when using duplexers/triplexers. Additionally, the more components you place in the path from the transmitter to the antenna increases the potential for PIM.

Reducing PIM has always been a design attribute for any good RF component manufacturer. However, the reason why PIM has received as much focus in wireless mobility in recent years is because of the increased use of multiple radio access techniques and its associated combining techniques, and of course, the increased amount of data usage being experienced.

PIM, as earlier mentioned, is IMD. However, PIM is typically self-generated by the wireless RF antenna system itself. Typically, PIM IMD is found at connector junctions due to poor connector construction or water intrusion.

PIM is tested by using the classic two-tone test where you select two different frequencies and their IMD product falls into the receive band.

$$\text{PIM frequency} = 2A - B$$

The power level used by both frequencies, A and B, is typically 20 W each. The reason for the power level is the need to create the nonlinear effect at junctions.

As a note, please make sure that the PIM frequency product is in your authorized frequency band if you are testing the antenna system itself since it radiates into the system as opposed to transmitting into an RF load.

Now since PIM is prevalent in all wireless communication sites, the question that comes to light is: Is it really affecting my communications? This is best determined by the RF level of the IMD product itself and typically should be below –100 dBm or better and is, of course, dependent upon the age of the site since many older sites utilized N connectors and not 7/16 DIN.

Figure 20.20 is an illustration of a basic RF cabling connection where the radio cabinet is connected to an antenna with the use of a feedline and two coax jumpers.

However, Figure 20.21 illustrates the increase in the amount of connectors, coax jumpers, and additional components in the RF path when bringing together multiple

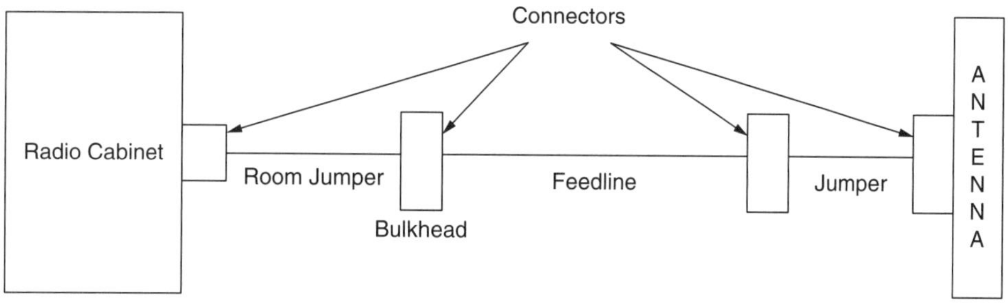

Figure 20.20 RF cabling.

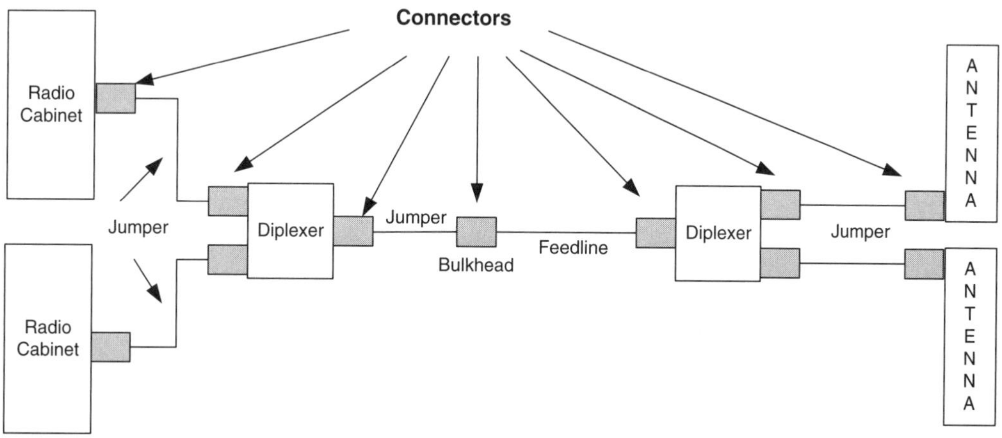

Figure 20.21 Multiple technology combining.

technologies and or frequency bands with a shared antenna feed. The antennas shown in Figure 20.21 could be housed in the same radome; however, they were separated here for illustrative purposes.

What is important about Figure 20.21 is the possibility of increasing PIM and other performance issues with this combining technique. However, sometimes lease considerations as well as tower loading and antenna space become an issue. With every connector or junction present, the possibility of harmful PIM or a combined PIM effect caused by multiple connectors increases.

20.7 Collocation

Many wireless mobility operators approach collocation requirements largely driven by rules of thumb that may or may not be entirely applicable considering the original application. Use of rules of thumb for defining the isolation requirements can make communication sites less than optimal or potentially undesirable from an RF, permitting, or construction aspect. With the added pressure to use existing structures for wireless communications, there is a strong need to establish mutually acceptable collocation guidelines for all the operators in a given market.

Besides physically ensuring that there is space and that there are no structural problems for the installation, the following issues need to be addressed:

- Isolation
- Intermodulation
- Grounding
- EMF compliance

Antenna collocation requirements stem from the need for isolation of one service provider's transmitter (source) from another service provider's receiver (victim). However, when collocating antennas for your own service, and when working with another service provider, the need to maximize performance for the site from both perspectives is mandatory.

Other sections of this book address intermodulation and EMF compliance, and these topics will not be repeated. However, grounding is an entire discipline in itself, and each vendor and wireless operator has his or her own particular requirements for grounding. The common issues, however, revolve around ground loops that result in voltage differences, which, in turn, induce current to flow between the loops, raising the noise floor as well as compromising grounding-system integrity. The specific requirements for grounding at a tower or building application should be obtained from your infrastructure vendor.

Antenna collocation guidelines are centered around the issue of required isolation between the source and the victim. For the isolation study, several critical assumptions are made that form the basis of this guideline. The critical assumptions are listed below for reference and will be used throughout this chapter:

1. The equipment used by both the source and the victim must meet or exceed the *Federal Communications Commission* (FCC) requirements and must be fundamentally well designed.

2. The grounding system at each of the sites must be adequate.
3. There are no intermodulation products caused by mixing of frequencies.
4. Receiver blocking will not be the driving issue for collocation isolation requirements because receiver desensitization will take place before blocking occurs.
5. The isolation requirements should be based on a rise in the usable noise floor by 0.5 dB, which places the source of out-of-band interference 10 dB below the victim's usable receiver sensitivity.
6. Relaxation of the isolation requirements should take place if the margin between the usable noise floor and the operating signal for the service has enough headroom to relax the minimal threshold.
7. All radio transmitters emit low-level emissions outside their intended channel and band.
8. The isolation requirements should be compiled from input from the operators and manufacturers involved with this report.

20.7.1 Isolation Requirements

The isolation requirements for each of the operators depend on a multitude of variables. These requirements are best defined by referencing the antenna input port of the equipment for the base station. This normalization then can be used for further refining the requirements.

Some of the items that need to be factored into the isolation requirements evolve around the receiver and performance requirements for each operator. The isolation requirements for each of the services can be defined either at the 1-dB compression point for out-of-band energy or at the level of inband energy that begins to degrade the receiver's sensitivity.

For collocation situations with other wireless operators, the recommended method for determining the isolation requirements is based on the impact to the receiver's sensitivity. The starting point for determining receiver desensitization would be first to establish what the effective sensitivity of the receiver is and then determine what level of receiver desensitization would be acceptable in a collocation environment. For example, if the operating signal is –80 dBm and receiver sensitivity is –115 dBm with an E_b/N_o requirement of 20 dB, then there is almost 15 dB of headroom still available for collocation situations. Specifically, in this case, the isolation requirements could be relaxed by 15 dB to enable collocation to take place.

Figure 20.22 depicts the inband interference issues caused by a collocated transmitter in another band. For this example, the source is a cellular operator using AMPS technology, and the victim is a PCS-1900 service provider in the PCS A-block.

The opposite scenario is depicted in Figure 20.23, where the source is now the PCS-1900 system, and the victim is a cellular operator.

Figures 20.22 and 20.23 attempt to highlight the fact that isolation requirements apply in both directions. However, the amount of isolation needed depends directly on the technology chosen, the operational service issues, and the physical separation of the antennas.

It is possible, of course, to improve the isolation requirements by further filtering the transmitter of the source to remove more out-of-band energy, making collocation

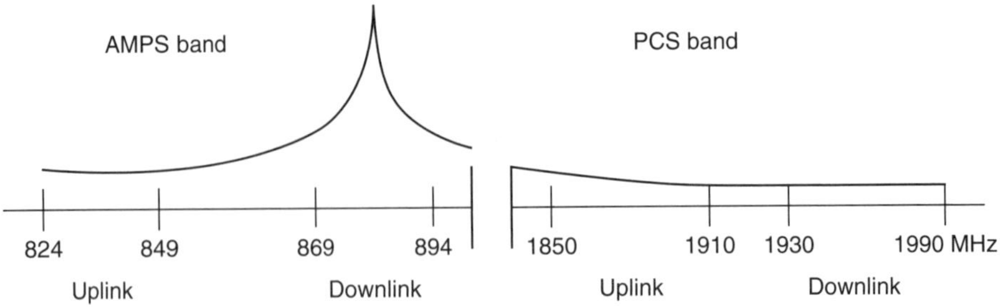

FIGURE 20.22 AMPS to PCS inband interference depiction.

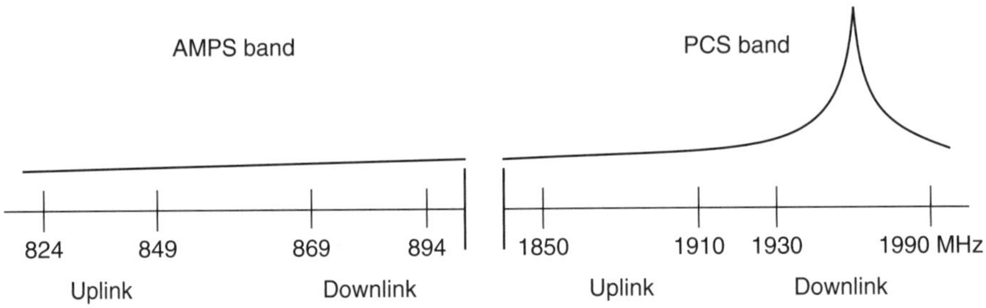

FIGURE 20.23 PCS to AMPS inband interference depiction.

less onerous. However, the inclusion of additional filtering at the transmitter comes with a cost in terms of insertion loss, to mention but one of the drawbacks.

20.7.2 Calculating Needed Isolation

The method used for calculating the required isolation is both straightforward and extremely complicated. The complication for determining isolation requirements is driven by the fact that not all the issues are known for any location. Also, there are different infrastructure vintages and models available for an operator to use at any location. The infrastructure used for location *A* may not be the same as for location *B*.

It is strongly suggested that you include in your initial site design the isolation requirements for both your system and those of other operators so as to avoid redesigns that will need to be done early in the project. The method that should be used to determine the isolation required for inband interference is as follows:

1. Determine if the site is a collocation site (if it is not, then the following is not applicable).
2. Determine who the other service providers are at the facility (including in the future).
3. Determine the optimal antenna locations for your installation.
4. Compare the desired locations with the ones already allocated.

5. Reposition the antennas for your service if the location already has been allocated for another.
6. Determine the required isolation needed for your system from each of the other operators.
7. Determine the required isolation for each of the other operators from your system.
8. Reconfigure the antenna system if there is an isolation violation in parts 6 and 7, and redo the calculations.
9. Generate a lease exhibit drawing showing the antenna locations chosen.
10. Submit for review to the other operators involved in your antenna placement scheme.
11. Allow 5 business days for feedback on the design.
12. Give operators initial drawing for approval; if no feedback occurs, use an escalation procedure.
13. If there is a discrepancy in the design, use reasonable efforts to resolve the issue to all the parties' mutual satisfaction.
14. Generate a revised lease exhibit drawing, and resubmit it to the operators for their records.

The method for calculating the out-of-band interference allowed in any system is given here.

$$\text{ES (effective receiver sensitivity, in} -\text{dBm}) = \text{TN} + \text{BW} + \text{NF}$$

$$\text{Y (maximum allowable in-band signal permitted, in} -\text{dBm}) = \text{ES} + 10 \text{ dB}$$

where TN = thermal noise flow = − 174 dBm/Hz
BW = bandwidth = 10log B
B = bandwidth of signal in Hz
NF = front-end noise figure (includes cable loss and preamplifier NF), in dB

20.7.3 Isolation Requirements

Now that we have determined how much isolation is required, the next step is to determine just how this can be achieved. Determination of how much isolation is afforded owing to the physical separation of the antennas themselves is indicated below. There are several components that must be physically separated, and they involve

- Free-space loss
- Antenna patterns of source and victim at the victim frequency
- Cable and connector losses
- Physical isolation
- Horizontal isolation
- Vertical isolation

- Slant isolation
- The transmit filtering of out-of-band energy from the source
- The antenna "filtering" at the victim frequency band

20.7.4 Free Space

For a given spatial separation, the application of the free-space path loss is considered to be a reasonable assumption. The free-space path loss (in decibels) is shown below for both the 800- and 1900-MHz bands. It is important to note that the free-space equation used for the isolation requirements needs to be at the victim's frequency.

$$L = 20 \log f_{\text{MHz}} + 20 \log l_{\text{m}} - 31.02 \qquad \text{(cellular/ESMR)}$$

$$L = 20 \log f_{\text{MHz}} + 20 \log l_{\text{m}} - 38.02 \qquad \text{(PCS)}$$

where f_{MHz} is the frequency in megahertz, and l_{m} is the distance in feet. Also, 10 ft is equal to 3.048 m.

The free-space calculation assumes that the antennas, both source and victim, are in the far field when, in fact, most of the time they are actually in the near field. This is a valid assumption when trying to determine the expected field strength in a given area. The free-space path loss, however, will be less, that is, conservative, or equal to the actual value at the site. OST 65 and antenna theory books indicate why this is a valid assumption when trying to estimate field strength. Additionally, an α of 2 is used for the slope, and this is a standard free-space loss model that again will be applicable to the situation.

The frequency of operation that is chosen for the free-space calculation should involve the lower portion of the receive band you are using as the victim. The rationale behind this method is the fact that the propagation characteristics are better at lower frequencies, and when comparing inband interference, the use of a transmitter frequency is not appropriate unless it is an on-channel hit.

20.7.5 Antenna Patterns

A critical part of the isolation determination has to do with the antenna patterns for both the source and the victim. The particular pattern used for both the source and the victim needs to be evaluated at the victim's frequency band of operation.

The antenna pattern used will have to factor into it the full gain of the antenna at the victim's receive frequency band because a far-field assumption is being made. Again, this is a valid assumption for modeling estimates because the actual gain for both the source and the victim will vary both positive and negative until the far-field region is achieved. Therefore, for the modeling aspects, the maximum gain obtainable at any one point is determined by the antenna pattern itself and is entered into the equation.

The antenna gain, however, can be reduced from maximum depending on the angle of attack from the source and victim antennas. Both antenna patterns will need to be corrected for the situation at hand.

It is assumed that antenna 1 is for the existing radio system (source) and that antenna 2 is for the collocating equipment (victim). The combined antenna gain in decibels thus is

$$G = G_{\text{max},1} - G_1(\theta_1, \phi_1) + G_{\text{max},2} - G_2(\theta_2, \phi_2)$$

where $G_{\text{max},1}$ and $G_{\text{max},2}$ are the maximum gains of the respective antennas. The antenna gain at a particular azimuth θ elevation ϕ is $G(\theta, \phi)$.

20.7.6 Vertical Separation

Vertically separating the source and victim antennas affords the greatest amount of isolation. The use of vertical separation may not be practical for certain applications. An example of what consitutes vertical separation is given in Figure 20.24. Please note that the vertical separation is defined from the base on the top antenna to the top of the bottom antenna.

The method that is used for determining the isolation requirements is defined in the following two equations:

$$I_v = L - G \tag{20.1}$$

$$I_v = 28 + 40 \log (Sv/\lambda) \qquad \textit{Note: } I_v \max = 70 \text{ dB} \tag{20.2}$$

Equation (20.1) is used when the antenna information is available, and Equation (20.2) is used when no information is available about the other service provider's antenna system.

An example for determining vertical isolation is shown below, in which the source is an AMPS system and the victim is a PCS-1900 (GSM) system. For this example, several assumptions are made.

Antenna System

AMPS: ALP9214 (14 dBd)

GSM: DB980H90-M (15 dBd)

Therefore, when vertically separated, the antenna factors are

$$G = 14 - 25 + 15 - 20 = -16 \text{ dB}$$

Similarly, at 1865 MHz and 10 ft of separation, the path loss is

$$L = 65.41 + 20 - 38.02 = 47.39 \text{ dB}$$

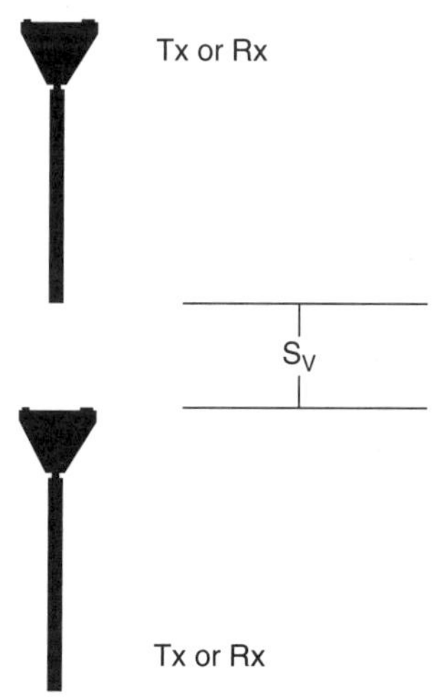

Figure 20.24 Vertical separation.

The total isolation is

$$I = 47.39 + 16 = 63.39$$

Since the isolation required is 30 dB, a vertical separation of 10 ft gives an additional *isolation margin of* 33.39 dB (63.39 – 30 dB).

This is even without factoring in the cable losses for the source and the victim and the source antenna frequency attenuation. Therefore, the physical separation requirements are limited to practical installation requirements, that is, avoid touching the antennas, etc.

20.7.7 Horizontal Separation

Horizontal separation of antennas is not by itself a practical method of obtaining large amounts of isolation through physical separation. The method for determining the isolation is shown in the two equations below. It is important to note that the horizontal separation equation can be used when the particulars about the other antenna system are known.

If the particulars of the antenna system are known, use the following:

$$H = L - G$$

If the particulars of the antenna system are not known, then use the following:

$$I_H = 22 + 20 \log (S_H/\lambda) - (G_T + G_R)$$

Note: $S_H/\lambda > 10$.

The distances between the antennas used for horizontal separation are shown in Figure 20.25. The reference points for both the source and victim antennas are the base or middle of the antenna itself. For the preceding equations, it is assumed that both the source and the victim antennas are on the same horizontal plane and that no variation for slight changes in antenna height owing to gain and frequency issues needs to be factored into the equation.

Using the same methodology as for vertical separation, the following example is meant to illustrate horizontal isolation. For this example, two antenna systems are pointing straight at each other and are at the same height.

$$G = 14 + 15 = 29 \text{ dB}$$

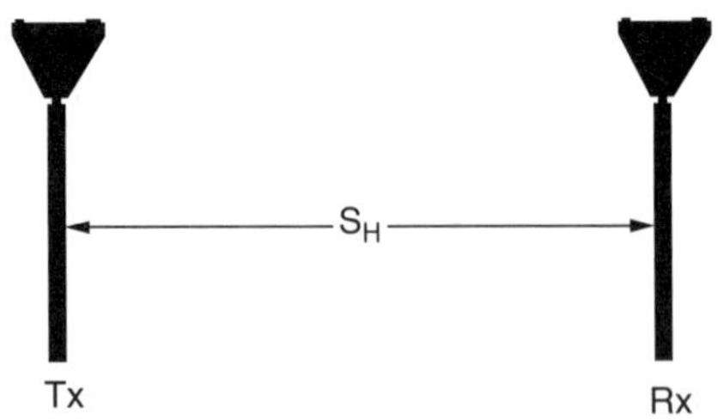

FIGURE 20.25 Horizontal separation.

Similarly, at 1865 MHz with 20 ft of horizontal separation, the path loss is

$$L = 65.41 + 26.02 - 38.02 = 53.41 \text{ dB}$$

and the total isolation is

$$I = 53.41 - 29 = 24.41 \text{ dB}$$

Thus, with antenna systems directly pointing at each, which is a possibility for adjacent building applications, this required isolation of 30 dB is not met for this application by close to 6 dB (30 – 24.41).

However, the frequency correction for the antenna system used by the source at the victim's frequency band was not factored into the preceding calculation, nor were the cable or connector losses. Assuming the following additional items, a refinement to the isolation situation can be made.

Victim cable and connector loss: 2 dB

Source antenna correction: 10 dB

$$\begin{aligned} I_{adj} &= I + C_L + SAC \\ &= 24.41 + 2 + 10 \\ &= 36.41 \text{ dB} \end{aligned}$$

Note: The source antenna correction factor needs to be provided by the manufacturer of the antenna in question. However, if this is not available, a minimum of 10 dB can be assumed when comparing frequency bands that are separated significantly.

The isolation provided by cables, connectors, and the source antenna provides additional isolation needed for this example. However, if, say, the distance were reduced to 10 ft, then the isolation requirements with all the correction factors added would just meet the requirements with no safety margin. If the isolation required for reliable communications cannot be achieved with the proposed antenna configuration, the following are some steps that can be taken to try to achieve the required solation.

The desired solution is to

1. Add vertical separation isolation.
2. Change the orientation of the antenna systems between the source and the victim.
3. Add additional transmit filtering to the source transmit path to reduce out-of-band energy emissions.

If the directional antennas are mounted side by side, then the antenna pattern attenuation can be included to reflect the situation, thus improving the isolation greatly. This type of installation is shown in Figure 20.26.

It is important to note that the separation between the mounting configurations is shown as I_{sx}, and this information needs to be delivered to the construction team associated with the implementation of the site itself.

20.7.8 Slant Separation

As is often the case in wireless installations, the antennas for the source and the victim are neither purely horizontally nor vertically separated. When the exact parameters for both the source and victim antennas are known, then corrections for horizontal and

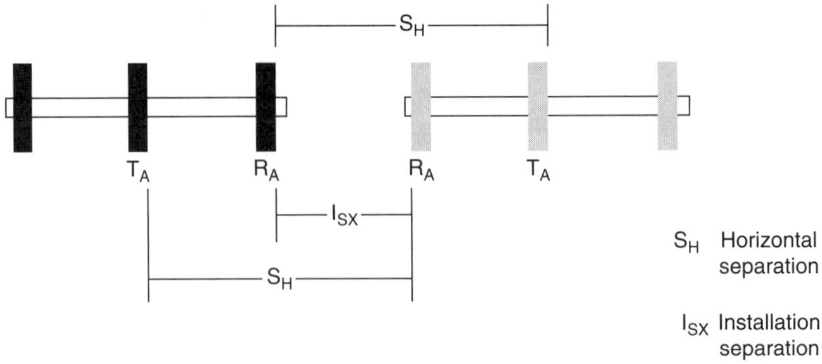

FIGURE 20.26 Horizontal separation installation guideline.

vertical issues are accounted for in the antenna pattern factors. The distance that is used for free-space calculations is the shortest distance between the source and the victim, as depicted in Figure 20.27.

However, when the particulars are not known for the location of the other antenna system, the following equation can be used:

$$I = (I_V + I_H) \times \theta/90 + I_H \text{ (dB)}$$

where θ = angle of 0 to 90 degrees and where

$I_V = 28 + 40 \log (Sv/\lambda)$

$I_H = 22 + 20 \log (S_H/\lambda) - (G_T + G_R)$

$I_{\text{Vmax}} = 70$ dB, $S_H/\lambda > 10$

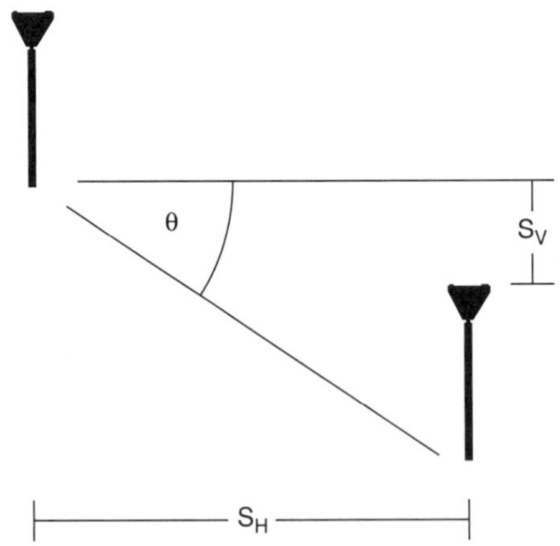

FIGURE 20.27 Slant angle.

20.7.9 Collocation Guidelines

The following is a brief overview of the collocation guidelines that you should use for your market. We have included some of the more common installation types encountered that apply to collocation. The installation types that are covered herein are not all-inclusive but serve as the starting point.

However, as in all cases with collocation, the one true rule of thumb that always needs to be applied is "fair and equitable treatment for all parties involved." We cannot overstress this important point, especially with the need to mutually coexist on structures.

20.7.9.1 Roof Top

Roof-top installation is prevalent in urban and dense suburban areas. There are several key elements that need to be factored into antenna design for a roof top. When designing an antenna system for a roof top, the potential for the same site to be used by another wireless operator is high on the probability list. Obviously, this is not the pressing issue when you are the first tenant to locate on a roof. However, when another operator is already at the facility, care must be exercised to ensure that your installation does not have a negative impact on the performance of the existing system.

Therefore, the items to address are

- Antenna mounting
- Blockage
- Adjacent buildings

Figure 20.28 shows the situation where one operator is currently located at the building, depicted by the black squares. The new operator is depicted by the gray squares, and this is the location that is desired for the antennas. At this time, the critical assumption is that the isolation requirements have been met for both the current and new operator.

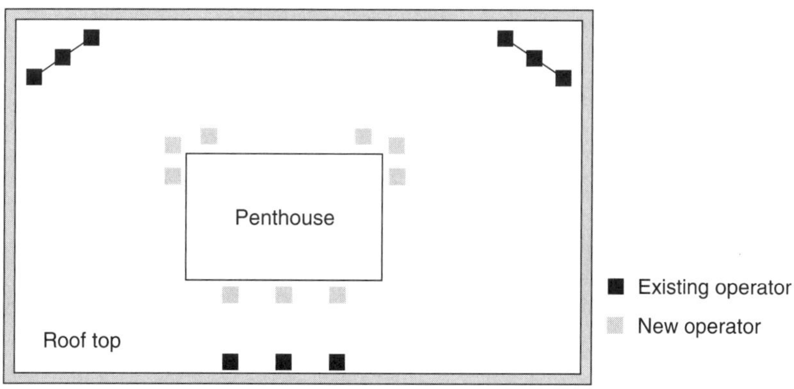

FIGURE 20.28 Collocation roof layout.

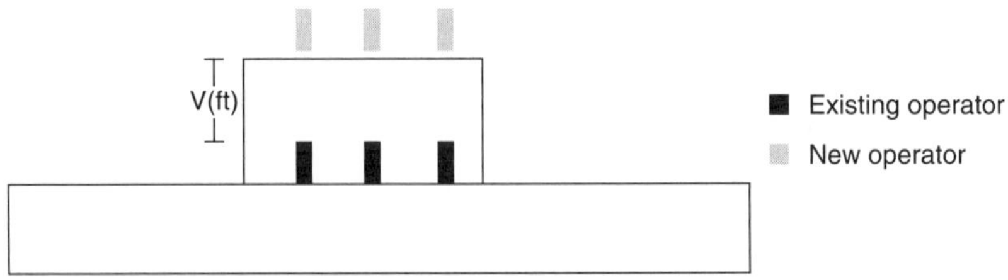

FIGURE 20.29 Collocation vertical.

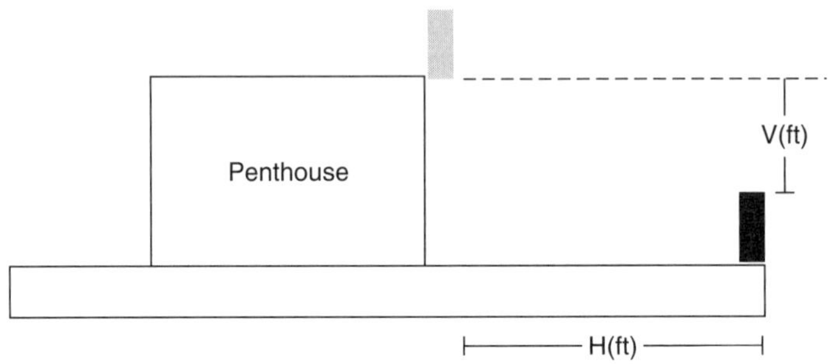

FIGURE 20.30 Collocation vertical depiction.

Figure 20.29 is a side view of the same situation. In this case, the antennas of the new service are located in such a manner that they will not have an impact on the existing operator and at the same time will maximize the performance requirements for the new operator.

Figure 20.30 shows that the first Fresnel zone is not violated by the new operator and that all the obstructions have been cleared. It is important to note that the presence of HVAC equipment can block an antenna's clearance from the roof.

A relatively good figure of merit for antenna placement to determine if the obstructions are cleared involves the following equation:

$$V = \text{distance in from roof edge and/or nearest obstruction}/5 \text{ (ft)}$$

20.7.9.2 Water Tank

The antenna installations for water tanks are rather unique in that the installation requirements are driven by the type of tank involved. However, in most situations, the ability to collocate on the same structure is very possible and is extremely viable from a land-use acquisition position.

Figure 20.31 shows a possible scenario where three operators are using an existing water tank on which to place their antenna systems. The figure depicts vertical separation as the method used for collocation. However, based on the situation, horizontal separation is achievable as long as the isolation requirements are met.

Figure 20.32 represents another scenario for a water tank installation. For this installation, all the carriers use horizontal separation as the main method for achieving the required isolation.

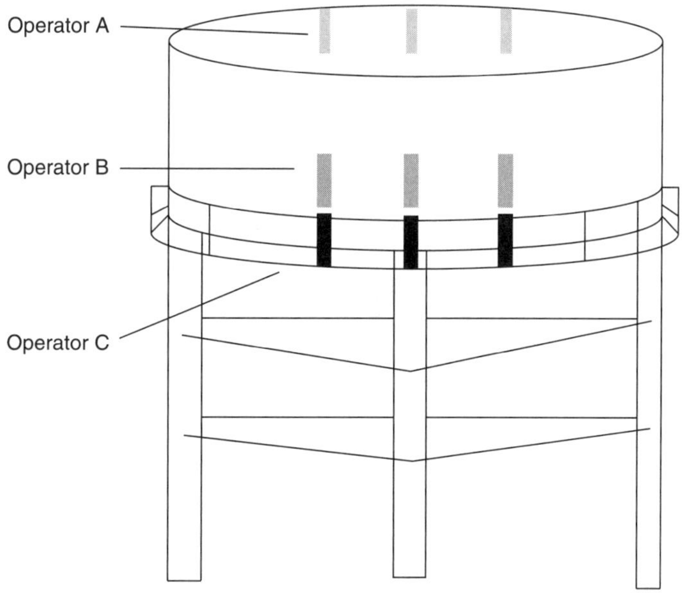

Figure 20.31 Water tank vertical install.

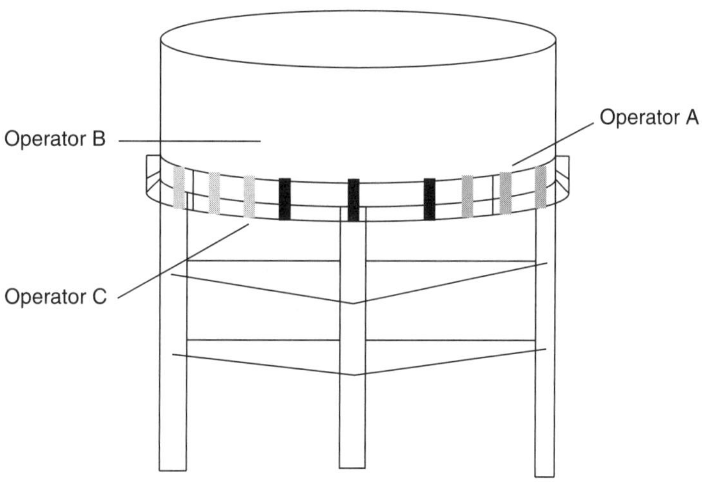

Figure 20.32 Water tank horizontal install.

20.7.9.3 Monopole

The use of a monopole for collocation is yet another option. The antenna mounting situation, of course, is different for each installation encountered. For example, some monopoles have a larger top hat than needed for existing operators, and with the required isolation being met, collocation on the same top hat can be achieved.

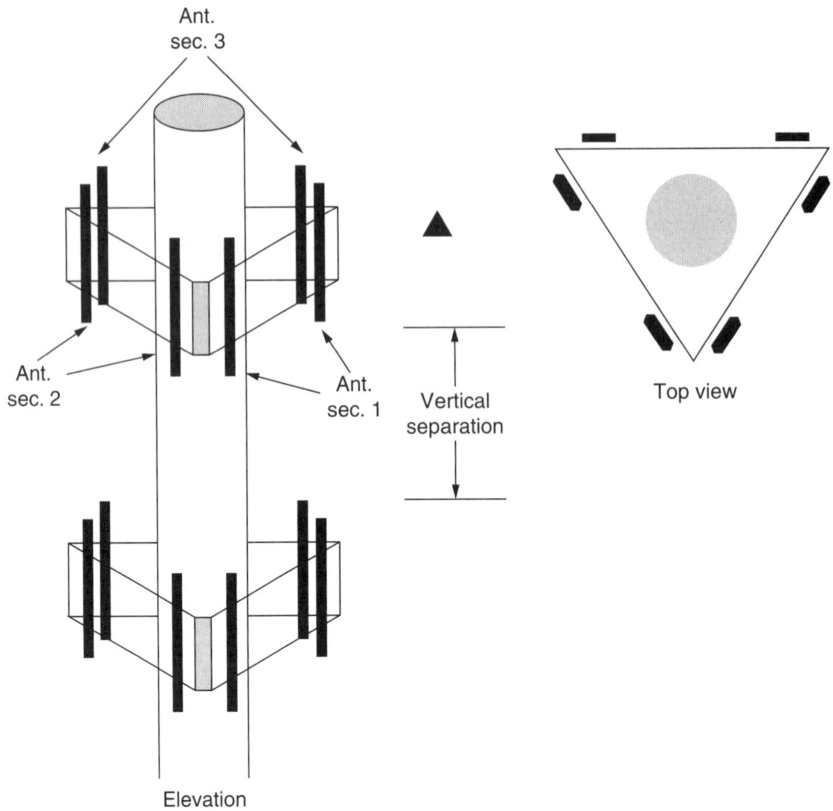

FIGURE 20.33 Monopole vertical collocation.

Figure 20.33 shows an installation where a monopole with a collar is used for collocating another service. The physical distance (vertical) needed from one operator to the next is determined by the isolation requirements defined in the preceding section. For this part, it is assumed that the monopole is of sufficient structural strength to facilitate placement of an additional service provider.

Figure 20.34 shows another collocation scenario where two carriers are located on the same top hat using horizontal separation.

20.7.9.4 Tower

The use of a tower for collocation is yet another option. The antenna mounting situation, of course, is different for each installation encountered. Figure 20.35 shows an installation involving a tower in which the new service provider is located below the existing operator. This situation is assumed for illustrated purposes; the physical location will depend on available space and structural loading of the tower itself. However, the physical distance (vertical) needed from one operator to the next is determined by the isolation requirements defined in the preceding section.

Next, Figure 20.36 shows another collocation situation where two antennas are mounted at the same elevation and therefore use horizontal separation to achieve the necessary isolation.

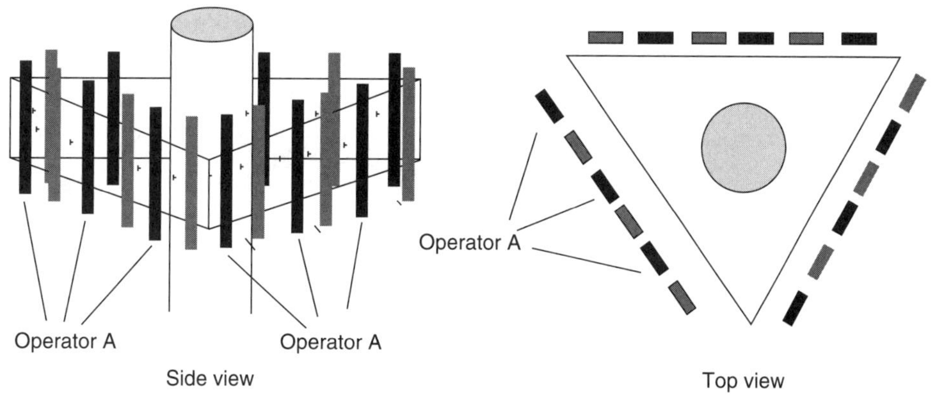

Figure 20.34 Monopole horizontal collocation.

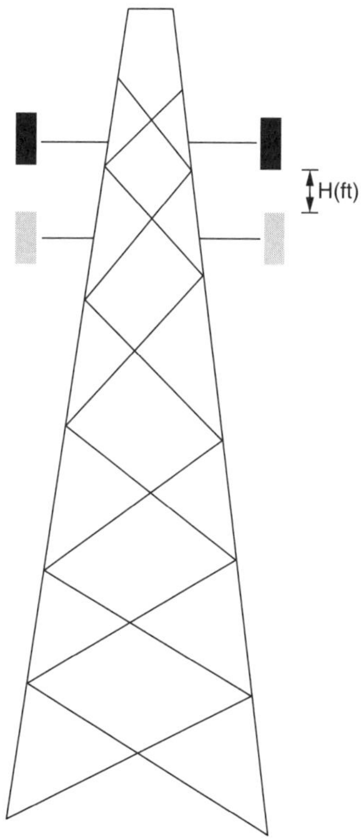

Figure 20.35 Tower vertical collocation.

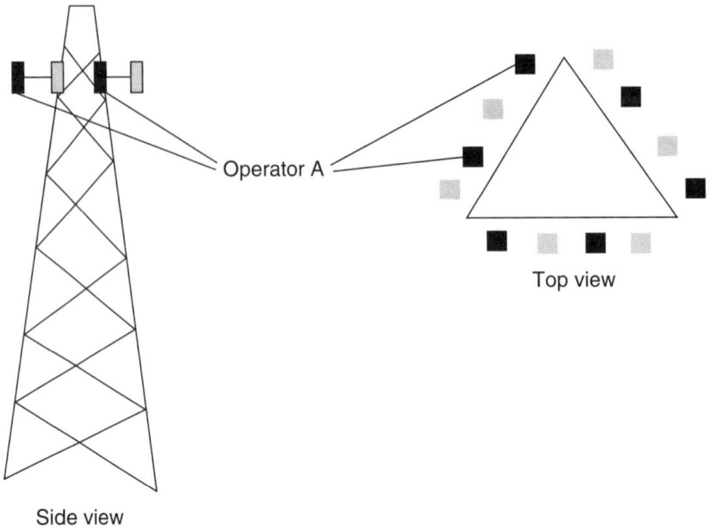

Figure 20.36 Tower horizontal collocation.

20.7.10 Collocation Rules

The following are the general rules that should be used when a collocation situation arises:

- Fair and equitable treatment for all parties involved.
- Cooperation between PCS/cellular and ESMR service providers for technical considerations when collocating on a facility is essential in minimizing interference issues and maximizing performance of the site.
- Unwanted emissions, that is, those that fall out of band, should be kept to a minimal level. The use of additional transmit filtering may be warranted in collocation situations.
- Relaxation of the isolation requirements can take place when the operating power level of the system is substantially greater than the effective noise floor.

Note: With RF emission regulations, typically the "last one on" is responsible for mitigating any violations. However, the guidelines presented will help mitigate any problems that can arise and facilitate the use of the communication site by multiple operators.

20.8 Path Clearance

There are many different types of point-to-point communications systems that can be utilized in wireless mobility for backhaul. However, radio path clearance is an essential criterion for any point-to-point communications system and especially with fixed wireless solutions like a microwave backhaul link and reradiators (bidirectional amplifiers).

The path clearance analysis needs to be performed for every RF point-to-point communication link in the network. The path clearance analysis is usually done with the aid of a simple computer model and terrain data. However, it is essential to understand the principles for path clearance analysis. Additionally, for link reliability, rain fade zone data may also be used but only after assurance that path clearance has been achieved.

Therefore, the first step in determining if the point-to-point path is appropriate for the communication link is to determine the Fresnel zone.

The Fresnel zone is shown in Figure 20.37. There are effectively an infinite amount of Fresnel zones for any communication link. The Fresnel zone is a function of the frequency of operation for the communication link. The primary energy of the propagation wave is contained within the first Fresnel zone. The Fresnel zone is important for the path clearance analysis because it determines the effect of the wave bending on the path above Earth and the reflections caused by Earth's surface. The odd-numbered Fresnel zones will reinforce the direct wave while the even Fresnel zones will cancel.

In a point-to-point communications system, it is desirable to have at least a 0.6 first Fresnel zone clearance to achieve path attenuation approaching free-space loss between the two antennas. The clearance criteria applies to all sides of the radio beam, not just the top and bottom portions represented by the drawing in Figure 20.38.

Environmental effects on the propagation path have a direct influence on the point-to-point communications system. The environmental effects altering the communications system are foliage, atmospheric moisture, terrain, and antenna height of the transmitter and receiver.

The *K* factor utilized for point-to-point radio communication is 1.333 or 4/3 of Earth's radius. The *K* factor ties in the relationship between Earth's curvature and the atmospheric conditions that can bend the electromagnetic wave.

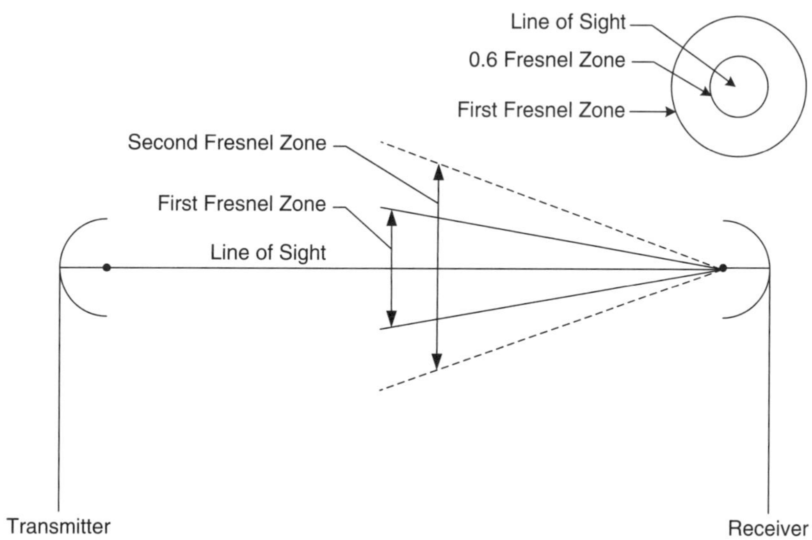

FIGURE 20.37 Fresnel zone.

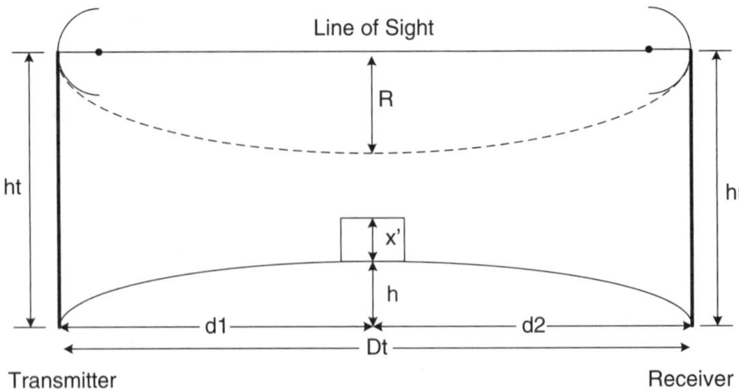

FIGURE 20.38 Path clearance.

A quick example of determining the path clearance required for a point-to-point communication site is shown next utilizing the variables in Figure 20.38.

First Fresnel zone:

$$R = 72\sqrt{((\mathrm{d1} \times \mathrm{d2})/(\mathrm{Dt} \times f))}$$

Earth curvature:

$$h = (\mathrm{d1} \times \mathrm{d2})/(1.5 \times K)$$

with f = frequency in GHz,
d1, d1, Dt in miles
x, h, ht, and hr in feet

Setting d1 = 1.6 mi, d2 = 2.1 mi, Dt = 3.7 mi, and f = 0.88 GHz or 880 MHz for this example.

$$R = 72\sqrt{((1.6 \times 2.1)/(3.7 \times 0.88))} = 73.14 \text{ ft}$$

$$R' = (0.6);\ R' = 43.884 \text{ ft}$$

Where R' is 0.6 of the first fresnel zone.

$$H = (1.6 \times 2.1)/(1.5 \times 4/3) = 1.68 \text{ ft}$$

For this example, the transmitter and receiver will have the same ASML and AGL.

Earth's curvature = 1.68
R' (0.6 Fresnel zone) = 43.88
Obstruction height (x') = 100.00

145.56 ft

Therefore, the minimum height for path clearance for both the transmitter and receiver is 145.56 ft.

20.9 Tower-Top Amplifiers

The use of *Tower-Top Amplifiers (TTA)* has many applications for use in wireless communications. The use of TTA was initially utilized in two-way communications and with the advent of 3G and 4G has found its way into wireless mobile communication systems.

The primary purpose of the TTA is to improve the receivers sensitivity through elimination of the feedline loss component. The TTA effectively establishes the noise figure and ultimate receiver sensitivity for the communication site. Care, however, must be exercised in ensuring the installation of the TTA does not cause more problems than it was meant to solve. Specifically, the TTA net gain must be such that it does not overdrive the front end of the cell site, therefore, reducing the real dynamic range of the site and also its sensitivity.

In addition, the filtering used for the TTA needs to be studied since too wide a front end for the TTA can in fact cause it to go into compression when collocated with other transmitters; even the site's own transmitter.

Figure 20.39(a) is a depiction of a passive receive path for a wireless mobile communications system. Figure 20.39(b) shows the inclusion of a TTA for the receive path along with a variable attenuator.

The TTA net gain should be set so that it effectively removes the line loss component for the system and nothing more. This adjustment is done with the variable attenuator shown in Figure 20.39(b), which is called a pad. There have been false reports issued indicating that an overall increase in gain from the TTA nets a similar gain in the cell's receive path. The latter is not true and usually comes about as a result of incorrect analysis being performed.

An example of how to set the net gain for a TTA situation is presented here. The simple application here is where the site's coaxial cable run is excessive and it is desired to improve the site's performance by installation of a TTA.

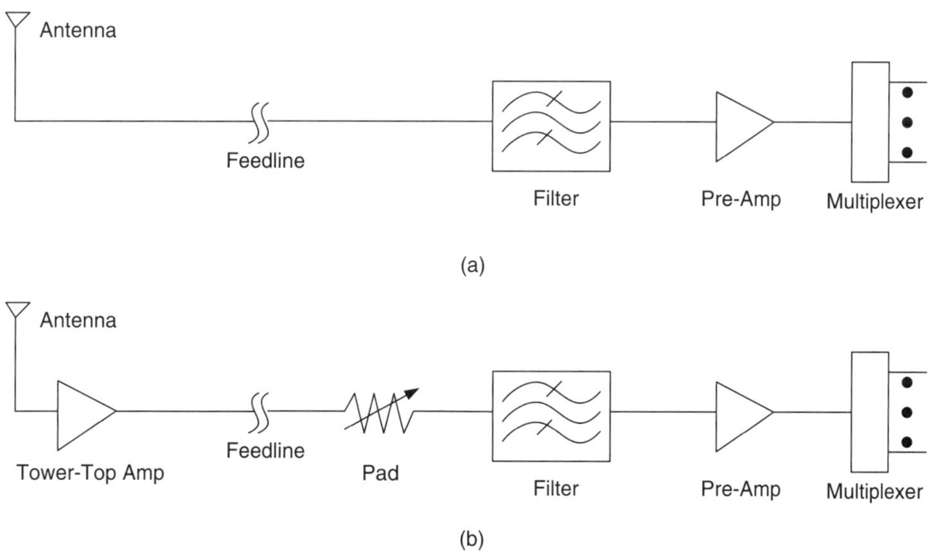

Figure 20.39 Tower-top amplifiers.

Example TTA Calculation
Design parameters:
Feedline for site = 250 ft
Attenuation 7/8 in at 2000 MHz = 1.97 dB/100 ft
Feedline loss = (250/100) × 1.97 = 4.925 dB = 5 dB (reality)
TTA gain for linear operation = 15 dB

Therefore,
TTA gain = 15 dB
Feedline loss = 5 dB
----
10 dB pads required

Now assuming that you can only use ½" cable which has a 3.45 dB/100-ft attenuation

$$\text{Feedline loss} = (250/100) \times 3.45 = 8.625 \text{ (approx 9 dB)}$$

Therefore,

TTA Gain = 15 dB
Feedline loss = 9 dB

6-dB pad (could also use a 7-dB pad)

It is interesting to note that with the use of a TTA, it is possible to relax the installation requirements for the site either in terms of overall cable length or changing the cable size itself. For example, as shown in Figure 20.39, a smaller cable size was picked, therefore, reducing the cost and potential tower loading at the site. However, the relaxation of the installation requirements comes at the expense of increased operational costs and additional points for system failure.

20.10 Multiple Technology Configuration Considerations

With the seeming ever-increasing proliferations of wireless access methods being deployed at this time by wireless operators increased pressure is being applied to antenna designs to support not only multiple access technologies but also multiple frequency bands.

When multiple radio access technologies are needed to be supported and they are going to be using the same frequency band, the operator has limited choices. The choices the wireless operator has is to either add additional antennas and feedlines, or have a different radio access technologies share some or all of the same RF components. The preferred design method is to have a separate antenna system for each radio access technology since performance requirements are always different. Using one combined antenna system usually means compromising either one or both networks.

In the event that you do need to share antenna systems, Figure 20.40(a) to (f) illustrate several combining techniques that could be used. In all cases, receive multicoupler is not used. I have used LTE as the example for the 4G platform being introduced.

In Figure 20.40(c) and (f) a multiband antenna is illustrated. Additionally, in Figure 20.40(e) and (f), a remote LTE head is shown indicating active electronics are placed close to the antenna itself.

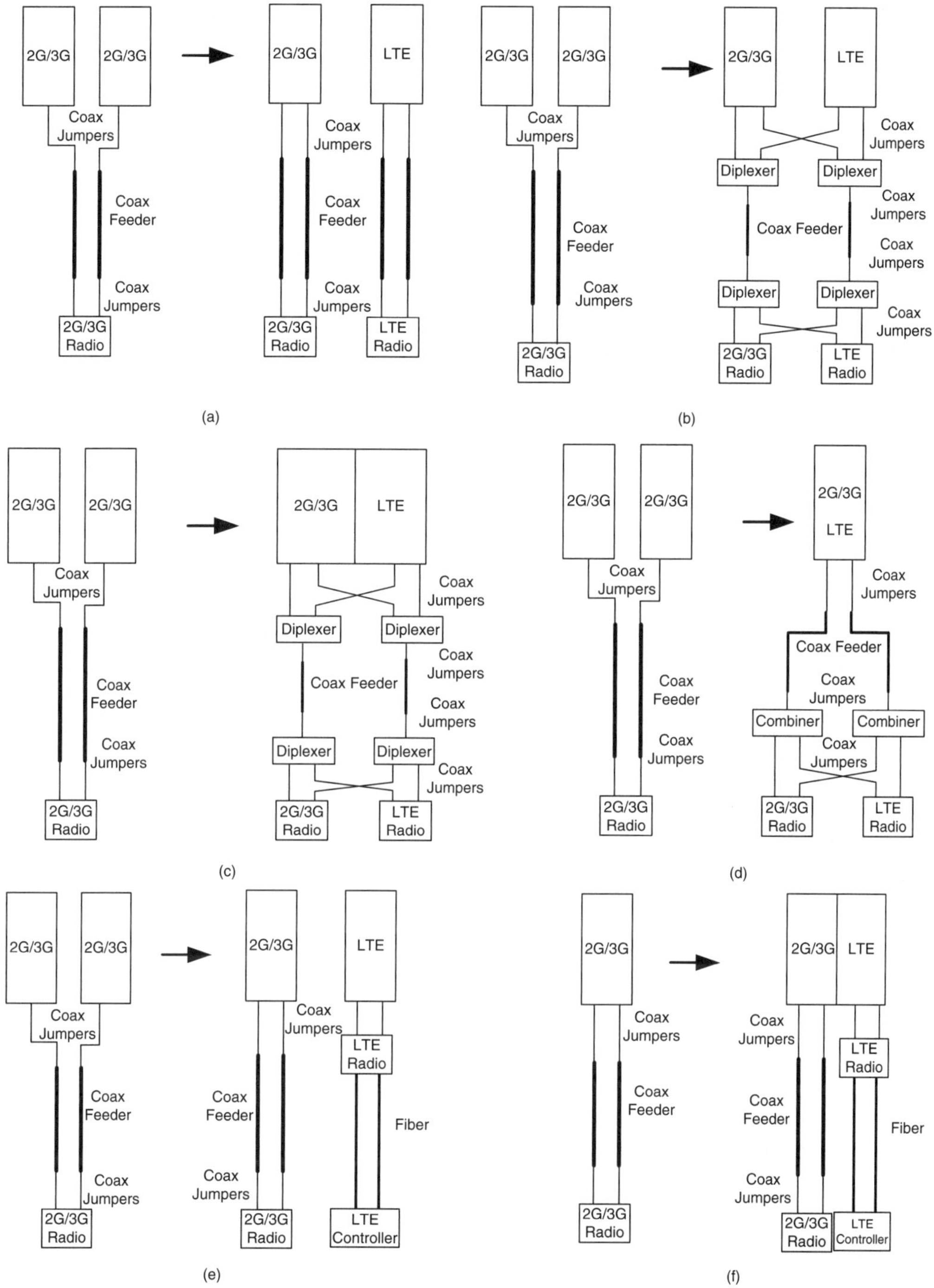

FIGURE 20.40 Antenna system combining possibilities. (a) LTE added with separate antenna system and coax; (b) LTE added with no additional coax; (c) Multiple antennas no additional coax; (d) Multiband antenna system; (e) LTE added with separate antenna system and coax and Remote LTE head, and (f) LTE added with Multiband antenna system and Remote LTE head.

It is also recommended that any active electronics are not included with the antenna itself. This recommendation is driven by the issue of operational considerations for accessing the critical equipment when it possibly goes into failure. Opening connectors and replacing equipment often does not correspond to favorable weather conditions.

Additionally, the use of LNAs should also be revisited since the supposed improvement with noise figure is often offset in urban and suburban environments with increased out-of-band energy amplification, usually the competitor's system, due to the wide front-end filters on an LNA.

20.11 Communication-Site Checklist

Table 20.1 is a brief summary of the major items that need to be checked prior to or during the commissioning of a communication site. This checklist is generic and should be tailored to your particular application; that is, add or remove parts where applicable. However, the checklist is an excellent first step in ensuring that everything is accounted for prior to the communication site going commercial.

Topic	Received	Open
Site-location issues		
1. 24-hour access		
2. Parking		
3. Direction to site		
4. Keys issued		
5. Entry/access restrictions		
6. Elevator operation hours		
7. Copy of lease		
8. Copy of building permits		
9. Obtain lien releases		
10. Certificate of occupancy		
Utilities		
1. Separate meter installed		
2. Auxiliary power (generator)		
3. Rectifiers installed and balanced		
4. Batteries installed		
5. Batteries charged		
6. Safety gear installed		
7. Fan/venting supplied		

Table 20.1 Cell-Site Checklist

Topic	Received	Open
Facilities		
1. Copper or fiber		
2. Power for fiber hookup (if applicable)		
3. POTS lines for operations		
4. Number of facilities identified by engineering		
5. Spans shacked and baked		
HVAC		
1. Installation completed		
2. HVAC tested		
3. HVAC system accepted		
Antenna system		
1. FAA requirements met		
2. Antennas mounted correctly		
3. Antenna azimuth checked		
4. Antenna plumbness check		
5. Antenna inclination verified		
6. SWR check of antenna system		
7. SWR record given to operations and engineering		
8. Feedline connections sealed		
9. Feedline grounds completed		
Operations		
1. User alarms defined		
Engineering		
1. Site parameters defined		
2. Interference check completed		
3. Installation MOP generated		
4. FCC requirements document filled out		
5. Drive test complete		
6. Optimization complete		
7. Performance package completed		
Radio infrastructure		
1. Bays installed		
2. Equipment installed according to plans		
3. Rx and Tx filters tested		
4. Radio equipment ATP'd		
5. Tx output measured and correct		
6. Grounding complete		

Table 20.1 Cell-Site Checklist (*Continued*)

References

Smith, Clint, *Practical Cellular and PCS Design,* McGraw-Hill, New York, 1997.

Smith, Clint, *Wireless Telecom FAQ,* McGraw-Hill, New York, 2000.

Smith, Gervelis, *Cellular System Design and Optimization,* McGraw-Hill, New York, 1996.

Smith, Gervelis, *Wireless Network Performance Handbook,* McGraw-Hill, New York, 2003.

TIA/EIA IS-2000-1: Introduction to cdma2000 Standards for Spread Spectrum Systems, June 9, 2000.

CHAPTER 21

5G and Beyond

Right now, *Third-Generation* (3G) and *Fourth-Generation* (4G) are two important topics that carriers are wrestling with. However, *Fifth-Generation* (5G) and even *Sixth-Generation* (6G) comments are appearing in presentations. The fundamental question that is driving the wireless industry is, "Where is this all headed?"

The answer in a simple word is "convergence." However, the simple word has many different variations and meanings. Convergence at a high level is the intersection of data and circuit-switched services where the user no longer has to make the decision as to which transport method they need for the application. Carried further, it also is the merging of fixed and mobile phones, tablets, *Personal Digital Assistants* (PDAs), laptops, and even wearable broadband devices so that the consumer can either have one device to use. Therefore, the goal is to enable end user devices to perform with the same user experience regardless of the location or situation they are in.

The choice of whether a single or multiple devices should be used should be left to the consumer to decide. However, the industry needs to provide devices that are configurable to the requirements of the customer. This is required because every customer has different needs. Also, these needs change over time and certain applications or features may not be needed on a regular basis. This single platform, or chameleon platform, is not far from reality.

The single platform is the envisioned X-Car where the subscriber purchases a *Software Definable Radio* (SDR) platform module and selects which application modules they want to use. The subscriber, through a user-friendly *Graphical User Interface* (GUI), instructs the SDR platform what services or functions are desired. This can be augmented through a personal module key that contains all the requisite authentication material.

Currently, the fixed and wireless telecommunications industry is vying for the same market space, where data and access to the Internet and/or intranet are provided by a single source, which provides telephony services, video, mobility, and Internet/intranet access.

The fundamental protocol that is now being used is IP. However, there are numerous access methods and schemes which while using IP still require a proxy system whether it is LTE, WiMAX, WiFi, CDMA, HSPA, DOCSIS, or any other method to bridge the gap between the access stratum and the application network.

This chapter will provide some additional information and thoughts about a variety of technical topics that may or may not pertain to the future direction of the wireless mobility industry as it progresses toward convergence.

The topics to be covered over the next few pages include

- Technology path
- Wireless Broadband Edge Device (WBED)

- Software-Defined Radio (SDR)
- Applications
- Advanced Broadband Wireless Access (ABWA)
- Ultrawideband (UWB)
- Unlicensed Wireless Access (UWA)
- Fixed Wireless Access (FWA)
- Mobile Broadband Wireless Access (MBWA)
- Multimedia (Mobile TV) technologies
- Mobile Virtual Network Operators (MVNO)
- First responders
- Business requirements

For more detailed and in-depth discussions of these topics, the reader is referred to the reference section at the end of this chapter as a good starting point.

21.1 Technology Path

The technology path that one needs to choose can be daunting. However, if the goal is to offer services that are ubiquitous as well as a host of value-added services, then an all-IP platform should be the desired outcome. For new entrants to wireless, the choices are easier owing to the availability of new platforms and the lack of a legacy system on which to build. However, as indicated earlier, existing operators have to wrestle with the issue of just how to get to that all-IP end state in an economical fashion.

Figure 21.1 shows the general path for technology migration or evolution. It is important to note that the use of 4G platforms in concert with *Wireless Fidelity* (WiFi) is part of the vision leading to the all-IP platform.

The need for having WiFi as part of the service offering allows for local broadband access on unlicensed spectrum that does not use the mobile carrier's spectrum, freeing up radio resources for other services. Obviously, Mobile IP and location-based services have a large play in the implementation of this approach.

21.2 Smartphones and Wireless Broadband Edge Device (WBED)

The future of wireless and its ability to deliver broadband services at the edge of the network to the end user depends not only on the various elements and components that make up the network but also on the edge device that are used by subscribers. The edge device is the window into the world and applications for the end user, and it is important that the device not only supports the vast array of advanced services but also be designed ergonomically.

Smartphones have become the dominant device subscribers use to access the wireless broadband networks and are no longer a mobile phone but a personal appliance. A Smartphone is in essence a *Wireless Broadband Edge Device* (WBED). However, there is no single definition or configuration that defines what a WBED is or consists of. A WBED can be either a mobile or fixed wireless access device. Some examples of WBED devices besides Smartphones are data aircards, USB devices, tablets, and femto cells.

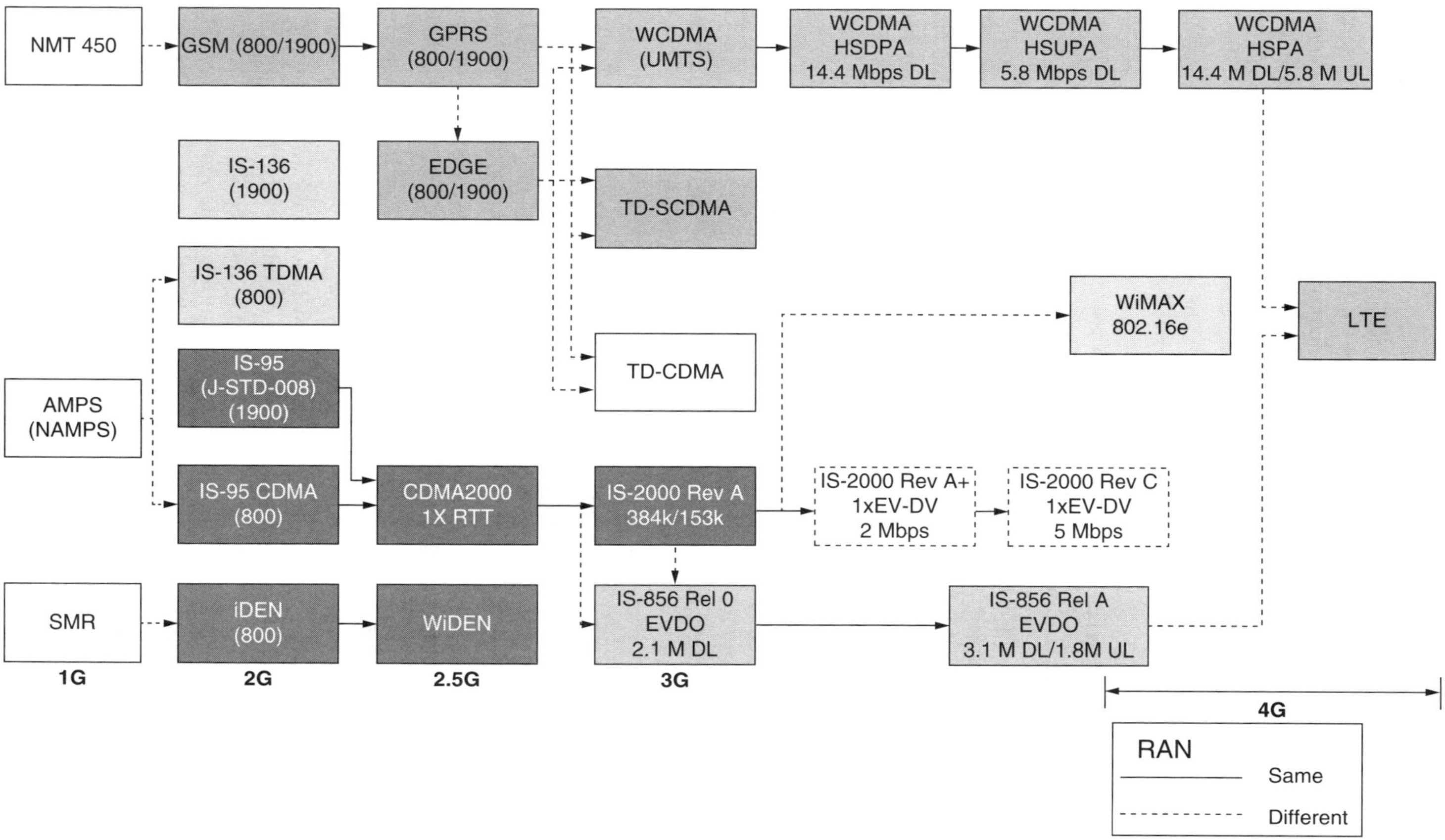

FIGURE 21.1 Technology migration path.

Many WBED devices have attributes that combine a mobile phone with an Internet appliance.

At the heart of a WBED is the *Operating System* (OS). However, there is not one OS that is used by all the broadband devices. The multiple of OS platforms has segmented the broadband market since applications run on one OS may or may not work on another OS or at the least have to be converted. Additionally, within the OS platform itself, there are different processors and software versions that need to be understood and supported.

Examples of some of the OSs that are in play right now are

- Google—Android
- Apple—iOS
- RIM—Blackberry
- Microsoft—Windows Mobile

WBEDs are typically capable of heterogeneous network operations enabling them to utilize multiple radio access systems operating on disperse frequency bands. In order to accomplish this massive task, the current technique is to enable the WBED via hardware solution where there is a separate radio, transmitter, and receiver for each frequency band of operation as well as the radio access system. Therefore, a typical WBED has multiple radios integrated into its platform that pose some interesting challenges for the device manufacturers due to size, heat, battery power, and capabilities, let alone keeping the device to a small form factor.

Figure 21.2 is a block-level depiction of a WBED of a multiple band handset that has WiFi and Bluetooth capabilities. Typical edge device has multiple radios integrated into its platform. The current technique for a WBED is to implement a hardware solution where a separate radio for Tx and Rx is used for each RAN and frequency band of operation.

However, when different wireless mobility radio access systems are needed to be supported by the WBED, the complexity increases as expected. Therefore, from a wireless operator's perspective, it is extremely important to understand and define the capabilities in the different WBEDs that you intend on offering to your customers.

Figure 21.2 is a brief diagram of a WBED device.

21.3 Software-Defined Radio (SDR)

SDRs have been available for many years but have seen most of their use focused on military applications. The commercial world has long sought an SDR, but the cost, until recently, has been the inhibiting factor for its potential use.

Presently, there are some practical limits to know how many different radio access systems and frequency bands possible to be placed on one device. Network operators typically have multiple *Radiofrequency* (RF) bands used for broadband service delivery. Additionally, RF bands as well as radio access technology changes are a constant for wireless network operators. Therefore, an SDR would enable one device to be able to communicate with all heterogeneous networks.

An SDR would enable a wireless broadband device to support a vast array of RAN technologies. In fact, an SDR could be used to provide one device to the subscriber that would interface with any of the multitude of RAN protocols such as LTE, WiMAX, HSPA, UMTS, EVDO, TD-SCDMA, GSM, CDMA2000, etc., making it RAN-agnostic.

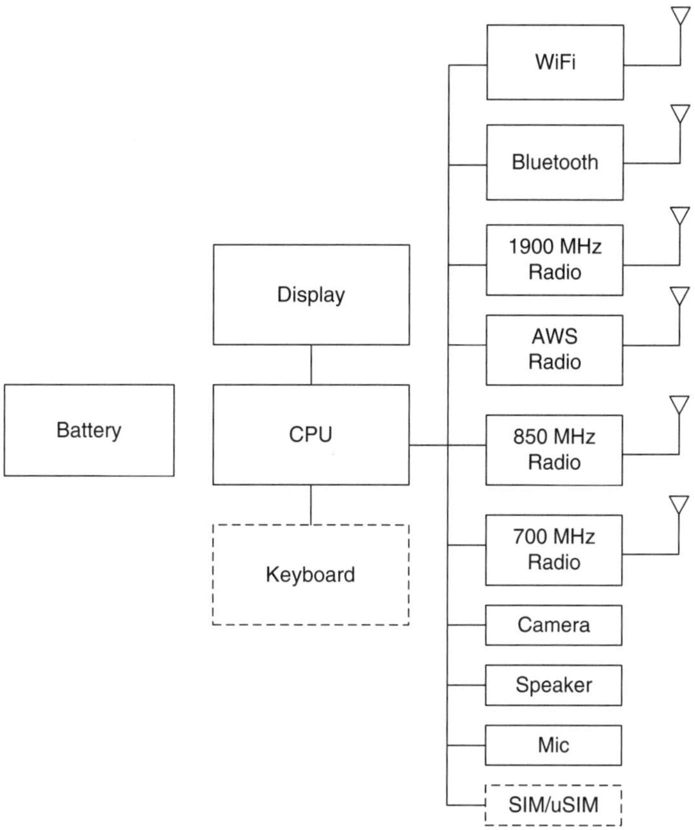

FIGURE 21.2 WBED architecture.

The SDR would be able to support multimodes, modulation schemes, as well as support the interface to many services. For instance, an SDR could be used by a wireless operator who has several RAN technologies in diverse markets facilitating roaming from one market to the other without concern about the underlying RAN in each market.

An SDR is capable of covering the wide range of spectrum used for mobility 400 MHz to 2.4 GHz and beyond. The SDR will also be able to support 802.11b/g access methods further enhancing the mobile data environment.

The use of an SDR enables the possibility of a single platform for all mobile devices. The SDR would enable a subscriber to purchase one device that allows them to configure it to meet their specific user requirements. This method would enable a subscriber to tailor their specific data and mobile requirements to match their specific needs instead of making numerous compromises.

However, the SDR, due to its wide receiver bandwidth, will be more susceptible to interference and possible range reductions. But the ubiquity and overall functionality of the device has exceptional potential to enhancing the customers overall experience with fixed and mobile data.

SDRs presently have been deployed commercially for cell sites and access nodes. To enable an SDR to function, the modulation format, that is, the waveform, needs to be programmed in the SDR.

As a rule of thumb, SDR is chosen when more than three technologies are deployed since the added cost and complexity with deployment of a single or dual RAN technology usually will not warrant the need for an SDR solution.

SDR is a radio that can change its RF characteristics via software. However, SDR, while an exceptional technology, has limits. Specifically, an SDR cannot replace the need for separate power amplifier and filters for each frequency band of operation. However, this limitation can be overcome depending on the spectrum and RAN portfolio of the wireless operator.

SDR can be seen for both the edge device as well as the access platform the edge device interfaces with.

For example, an SDR approach can greatly simplify the amount of hardware that exists on an edge device leading to more possible applications ranging from body worn devices with a small form factor or increased battery area to extend the time the WBED can operate without being recharged.

Figure 21.3 is an example of a WBED using SDR. You can contrast the complexity of the Figure 21.2 with that of the same notional device shown in Figure 21.3.

Taking a similar approach, SDR has great potential for fixed infrastructure deployments where the desire is to future-proof the platform.

Figure 21.4 is an example of a tri-technology/quad frequency band cell site following a conventional approach.

However, Figure 21.5 is the same as tri-technology/quad frequency band cell site utilizing SDR technology.

21.4 Millimeter Wave Backhaul

The need to improve the backhaul throughput as well as reduce operating costs has lead to the introduction of millimeter wave technology into the wireless mobility network. The millimeter technology covers a host of functions and can operate in a multitude of frequency bands; however, the bands of most interest for wireless operators fall into the 70-GHz to 95-GHz band.

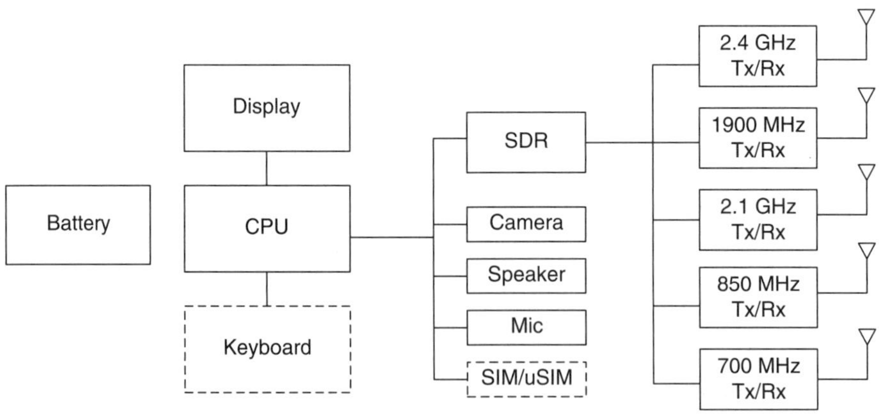

Figure 21.3 Tri-technology/quad frequency band SDR.

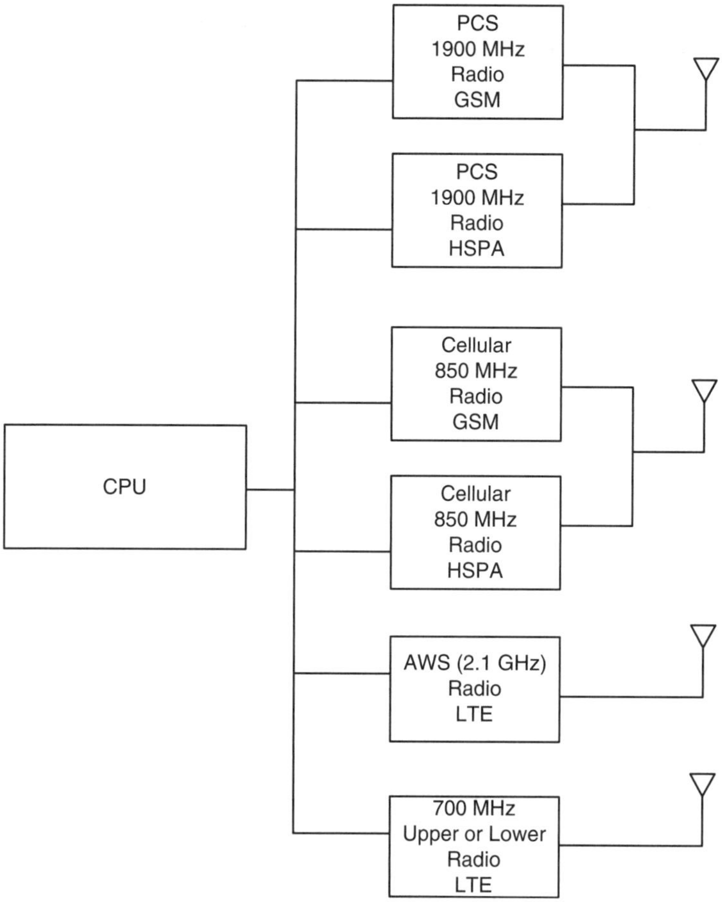

FIGURE 21.4 A tri-technology/quad frequency band cell site convention approach.

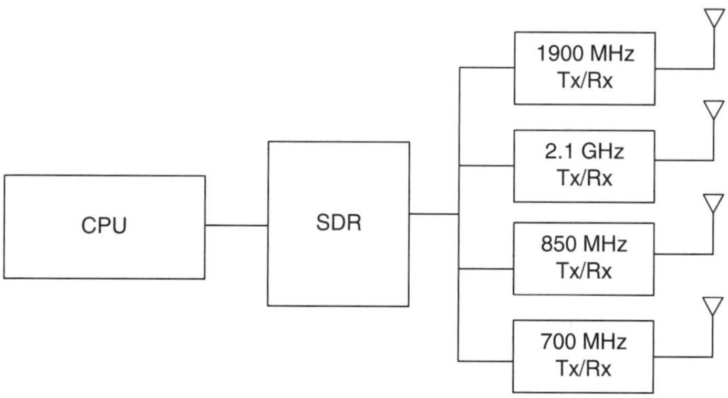

FIGURE 21.5 Tri-technology/quad frequency band cell site SDR approach.

However, millimeter backhaul is meant for very short distances.. Therefore, due to its limited range obtaining licenses for the use of millimeter wave backhaul is not envisioned to be a hindrance.

A key attribute with millimeter wave backhaul is that it is an enabler for HetNet deployments since they are a layered system involving both point-to-point and mesh configurations. Millimeter wave can be used to provide the necessary bandwidth for transporting the backhaul traffic from multiple APs to an aggregation point.

Millimeter wave can also be used for edge device access in mobility for delivering extreme bandwidth-intense bearer traffic.

21.5 Applications

Applications are a key driver in enabling the use of broadband edge devices. Presently, there are many applications available for use by WBED and the number keeps growing. Some applications are free, others are based on subscription.

Depending on the device's operating system, applications can be obtained from multiple sources.

- Android—Android Market
- RIM—RIM Market
- Apple—Apple Store

As expected, applications are operating systems as well as device specific. For example, an App used for Apple's OS platform cannot run on a device using RIM, Microsoft, or Android operating system. Additionally, the current broadband edge devices physical platform may not support the application based on the vintage of the device as well as its hardware configuration, that is, forward and rear-facing cameras.

Wireless operators also offer applications that are unique to their system providing value-added features. Additionally, all WBED devices support Web browsing enabling the broadband device to access Web content and, more important, Web browsing in independent operating systems.

Additionally, many broadband devices also can act as a WiFi hot spot.

21.5.1 Cloud Computing

Broadband edge devices enable cloud computing. In essence, cloud computing is network centric and location independent. The WBED devices access cloud services via the Web. A key advantage with utilizing the cloud is that when an edge device uses cloud resources, it reduces the platform and resource requirements for running those applications.

Additionally, depending upon the application used, the edge device can dynamically increase or decrease its requirements for network resources. Cloud computing enables applications to be made available quicker since it is not resident on the edge device.

21.5.2 Sensor Integration

A key area for future growth and applications in broadband edge devices involves the evolution and utilization of multiple sensors that can be used on a device to enhance the user experience. Some sensors that exist or can exist on a WBED include

- GPS
- Accelerometer

- Compass
- 2 and 3 axis gyros
- Pressure
- RFID

For example, GPS is used for location determination as well as E911. Accelerometers are used for orienting the display based on the orientation of the device to the user. A compass can be used for bearing information and reading the magnetic field. Gyros are being installed in many devices either with 2 or 3 axis facilitating movement tracking and improvement location determination. Pressure sensors are used to assist in the determination of elevation of the device. *Radiofrequency Identification* (RFID) sensors can be used to help identify the WBED without the need to use a wireless network. And lastly, for sensor integration is the possibility of biosensors either to assist in life services or use in detecting bio or radiological material.

21.5.3 PTT

Push to Talk (PTT) is included here not because it is a new feature or concept, but because it offers a glimpse into a strong push to have *Land Mobile Radio* (LMR) capabilities as standard offerings in wireless networks. Presently, PTT and cellular are associated with *Integrated Digital/Dispatch Enhanced Network* (iDEN), which was a successful service and platform offering.

However, with the success of iDEN, many operators are vying for the position of offering PTT on their networks. PTT can be used not only for commercial and private enterprises but also for public safety, fostering interoperability between public safety networks and commercial cellular networks, and enabling an extensive expansion of coverage and capacity.

21.6 Services

Services, services, services. The objective of convergence is to provide more services that will be purchased and used by customers. Technology is nice, but if the technology deployed does not fill a need in the market, then it will fail, no matter how good it is, either to be profitable or be replaced with something else that meets the demand.

The onslaught of 2.5G/3G and now 4G devices introduces a multitude of possibilities, all centered around the IP. The use of IP has been predominantly centered on fixed locations; however, now it has entered the any-place any-time realm.

The services offered are partly dependent upon the RAN but more on the device itself. This is a fundamental shift in service offering from the past in that the RAN was the driving force in mobility for service. But with the proliferation of IP and the ever-maturing RAN for mobility, the next real quest for service offering and enhancements is device driven.

In order for the mobile devices to achieve real convergence it will still be necessary to achieve the following with regards to the handset or user device.

1. More stable software for the devices, no BSOD.
2. Simplify choices and eliminate the configuration headaches.
3. Truly eliminate the interoperability issues between networks.

4. Improve data security by protecting the information in the device as well as the network.
5. Improve the technical support skill for data integration.
6. Battery life improvements, as more is demanded from a device the battery life diminishes.

In addition, some of the enhancements to the user's experience will involve actively informing the user via their mobile device that the call dropped or the session was terminated prematurely. The user also needs to be informed that the transaction was completed or not completed.

As mentioned above, the desire of any operator is to offer and support real applications, which provide some sticky content that helps retain the customer. The specific services that can and will be offered by any mobile wireless operator are of course market- and customer-segment dependent.

21.7 Advanced Broadband Wireless Access

Numerous broadband wireless access technologies are either available today or will be available in the near future. Some of the notable broadband wireless access platforms include LTE, WiMAX, and WiFi, previously covered in other chapters. However, this section will address several other access technologies that have not been discussed previously.

21.7.1 Ultrawideband (UWB)

UWB is an emerging wireless access method that has been talked about for many years. UWB falls under the 802 suite of standards, namely, 802.15, and will operate in the unlicensed bands. UWB is a *Personal Area Network* (PAN) technology that is meant to improve the throughput capability now offered by Bluetooth. Presently, Bluetooth can support PAN data rates of about 1 Mbps, but UWB has the possibility of offering rates exceeding Bluetooth's data rates.

UWB has some unique advantages owing to its unique propagation properties. More specifically, owing to the short time duration of the signal, it can support extremely high data rates and multiple users in a small area (PAN). The short duration of the signal and the resulting modulation and demodulation scheme make it very immune, at short range, to multipath interference, which is a significant problem in both mobile and in-building situations.

The specific access method that will be used with UWB is undefined at this moment, with the *Institute of Electrical and Electronic Engineers* (IEEE) and *International Telecommunications Union* (ITU) apparently supporting two different standards for the new and improved PAN. The potential for UWB is vast, however, and the range of applications is still being put together. At the moment, UWB devices are not available commercially.

21.7.2 Unlicensed Wireless Access (UWA)

UWA is used to describe a host of radio access methods that employ unlicensed spectrum in concert with the licensed spectrum. UWA is now being referenced as *Generic Network Access* (GNA).

UWA effectively enables a 3G/4G mobile device to seamlessly roam between the commercial network and a WiFi network. The ability to roam seamlessly requires Mobile IP and interoperability between the carrier's network and the WiFi network.

Obviously, an IMS platform can and does play heavily into this type of arrangement. However, at the heart of UWA and WiFi's integration with 3G/4G networks is the fundamental proliferation of 802.11, which has seen widespread acceptance. WiFi equipment cost for the end device as well as for the infrastructure *Access Points* (APs) is relatively low, unlike its 3G/4G counterparts. The inexpensiveness of WiFi tends to imply that the service is easy to deploy and affordable. However, as always, the real issue regarding ease of deployment and affordability is that it depends on what the objective or service offering really is. Presently, the primary services are e-mail and Internet/intranet access.

21.7.3 FOMA and iMODE

Freedom of Mobile Multimedia Access (FOMA) is the brand name for the 3G offering under NTT DoCoMo. It is mentioned here along with other wireless access methods because iMODE is commonly interchanged with the issue of broadband access. FOMA uses WCDMA or UMTS as the primary transport mechanism. A unique function of FOMA is the iMODE.

iMODE is an enabler for broadband services. iMODE is a Web browser based on Netscape that uses *compact HyperText Machine Language* (cHTML). iMODE was introduced initially as its own platform and is different from *Wireless Application Protocol* (WAP). However, WAP2.0 will be able to process iMODE services.

The advantage of cHTML is that it is designed for use on a handset and has many of the same attributes as HTML, allowing for easier service delivery and integration as compared with WAP.

21.7.4 WiBro

Wireless Broadband (WiBro) is a wireless broadband access method developed and currently deployed in Korea and operating in the 2.3-GHz band. WiBro is similar to WiMAX but is a proprietary technology. WiBro has been standardized and falls under the 802.16e grouping.

WiBro has the advantage of being uniquely defined, enabling it to be more integrated from inception with 3G wireless devices for seamless roaming. Some of the technical attributes of WiBro include

- 18.4 Mbps in the downlink
- 6.1 Mbps in the uplink
- Bandwidth 9 MHz
- Access: *Time-Division Duplex* (TDD)
- Modulation: OFDM

21.7.5 FWA

Fixed Wireless Access (FWA) is another name that is used to describe UWA/GNA. FWA is the convergence of WiFi with 3G/4G mobility networks, enabling the use of unlicensed spectrum to interoperate harmoniously with licensed spectrum.

The primary facilitator for FWA is an IMS platform, which is IP- and SIP-based.

21.8 Multimedia (Mobile TV)

For some time now there has been a push in the industry to offer broadcast services on a mobile device enabling a separate radio link to be used for delivery of multimedia content to the subscriber. The broadcast services utilize a separate frequency than that used by the mobile license operator. However, to take advantage of the broadcast media, the subscriber equipment needs to be able to receive the information on whatever frequency band the service is being sent at. This is in addition to the need to have the subscriber pay for the service.

There are multiple ways that the mobile can receive broadcast TV signals. The primary method being pursued involves a separate channel for broadcast operating usually in the *very high frequency* (VHF) range. The other method utilizes broadcast over the Internet where the subscriber device using 3G is able to receive the signal. However, this can be rather expensive from the subscriber and operator's point of view since the RAN and network resources are being used on a peer-to-peer basis when a multicast method can be better employed. Hence, the reason for the pursuit of Mobile TV with the use of a separate radio channel that supports the broadband signal required for properly sending high-resolution full-motion video the end user's device.

Three promising but competing technologies exist for Mobile TV.

21.8.1 MediaFLO

MediaFLO was a unique mobile broadband approach using multiple frequency bands at the same time. Although a promising technology, it went dark near the end of 2011. Regardless, MediaFLO uses broadcast techniques to send high-resolution video to multiple users. The MediaFLO method uses a multiband CDMA/OFMA handset that subscribes to the MediaFLO service.

The user has regular cellular/PCS service in the United States using a *Code-Division Multiple Access* (CDMA) carrier. The MediaFLO information is sent via several broadcast base stations operating in the 700-MHz band. The subscriber unit decodes and displays the MediaFLO signal and acknowledges the MediaFLO information, when needed, using commercial cellular systems. This is unique in that the MediaFLO information is sent on one frequency band, and the acknowledgment message is sent via the wireless mobility operator, facilitating high-resolution broadband services to be delivered to the subscriber, and thereby freeing up valuable resources on the commercial cellular systems for other traffic that is not as bandwidth intensive.

21.8.2 T-DMB

Terrestrial Digital Multimedia Broadcasting (T-DMB) is one of the several broadcast transmission methods being pursued to deliver multimedia services to the subscriber's handset. T-DMB is the terrestrial portion of the *Digital Multimedia Broadcasting* (DMB) methods, one complementary method involved with *Satellite Digital Multimedia Broadcasting* (SDMB).

T-DMB and the DMB suite of standards are derived from the *Digital Audio Broadcasting* (DAB) standard, which has been highly successful. T-DMB has been developed under EU147. T-DMB is made for transmissions on VHF and UHF RF bands and can utilize WiFi/WiMAX Access Points as well. T-DMB is presently being targeted toward GSM/UMTS markets but could be used in conjunction with other 3G RAN technologies.

T-DMB requires a 1.1712-MHz radio channel for the broadcast and uses the GSM/UMTS network for uplink communication when required.

T-DMB's technical discussion is beyond the scope of this book. There are several URL links provided in the reference section, which can help provide more information.

21.8.3 DVB-H Digital

Digital Video Broadcasting–Hybrid/Handhelds (DVB-H) is a recent addition to the DVB transmission standards Terrestrial DVB-T and Satellite DVB-S. DVB-H is meant for mobile terminals, hence the term handset or hybrid. Because DVB-H was designed for the mobile, special care was taken into consideration for battery life. Therefore, the technology was developed with the intention of it being used by a subscriber's handheld device that could access the DVB-H system either from a separate radio channel or via an IP broadband link. The IP broadband link could be from the 3G wireless network or from a WiFi/WiMAX Access Point. DVB-H is being targeted for GSM and UMTS markets but could easily be applied to a CDMA, TD-CDMA, or TD-SCDMA market.

DVB-H uses a broadcast channel that is either 5-, 6-, 7-, or 8-MHz wide, which enables this service to offer high data rates and multiple streaming video channels. DVB-H is competing with or will coexist with T-DMB in the same marketplace as the preferred broadcast service.

DVB-H's technical discussion is beyond the scope of this book. There are several URL links provided in the reference section, which can help provide more information.

21.9 MVNO

Mobile Virtual Network Operator (MVNO) has become a viable method for many companies to offer mobile wireless services without owning and physically operating the wireless infrastructure and has been garnishing a lot of attention. MVNOs are based on a business arrangement made between a reseller and a wireless operator for use of its spectrum or system. In essence, the reseller purchases bulk minutes from a wireless operator for its exclusive use under a predefined service management agreement that includes a *Service-Level Agreement* (SLA). An MVNO is a mobile operator that does not own wireless spectrum or have its own network infrastructure. Because it does not own spectrum or the infrastructure, it resells the services of other, network-based mobile operators or purchases bulk minutes of use for sale to its own customers.

However, MVNOs are not pure resellers of a service. Instead, MVNOs have an SLA with the wireless operator, making it possible for the MVNO to participate in network service rollouts, infrastructure investments, and overall network buildouts in order to improve its service offerings.

There have been and continue to be many uses of an MVNO. Some agreements for MVNOs can and do involve leasing spectrum for use from the license holder while at the same time using the license holder's infrastructure to deliver the service. This is important and a permitted use of spectrum in the United States. As consolidations continue, the ability to offer niche services falls to MVNOs. MVNOs differentiate themselves through content, price, and use of specialized *Customer Premise Equipment* (CPE) or WBED.

21.10 First Responders

The proliferation of commercial cellular devices into the first-responder and general population enables a host of possible service options. In the United States, an LTE nationwide network is currently under deployment for all first responders. LTE was chosen as the technology making it possible for first responders to take advantage of technology leaps that have occurred over the last 10 years.

Public safety operates in what is referred to as Band 14 for LTE.

Commercial cellular operations are presently an integral part of the first-responder communications net. But commercial cellular is the primary method during normal communications and LMR during emergency.

With the advent of priority access and deployable cellular systems that operate independent of the commercial network during disasters, commercial cellular communications for first responders is a viable method for enabling broadband data, both downlink and uplink.

21.11 Business Requirements

At the heart of the convergence decision lies the business considerations that influence the deployment and implementation decisions directly for a company. For example, what are the services that will be or can be offered? This is a simple question, but often it is extremely difficult to answer. The services selected should match the market that is being sought, which is obvious. However, just how will the selected services be distributed, that is, selective or global? Will the service offering be only selectively targeted to islands of interest (niche), or will the services be advertised for the entire market? Both these approaches bring with them technical and financial issues.

Regardless, the primary objective with any wireless company, or any company, is not to explore technology but to make money. The governing method for determining the money made is driven by the following equation.

$$\text{Gross profit} = \text{gross revenue} - \text{gross expenses}$$

and

$$\text{Gross profit} > 0$$

While this may seem primitive in nature, the fundamental issue is that this concept is not fully understood. As it has been witnessed many times in the wireless industry, technology platforms and systems have been converted based on the quest for technology instead of answering the three questions:

- Who is the customer?
- What do they really want?
- What will they be willing to pay for this service?

Failure to answer these questions can and often does lead to the deployment of technology for technology's sake. More specifically, if you have an *Average Revenue per User* (APRU) = US\$50, but the service costs US\$51 to deliver including capital expenditures, then its viability is questionable.

Ultimately, the decision comes down to vision: What is your vision for the future? You must have a firm understanding of where the company is today and what will it become when it grows up. Again, another profound question but one that needs to be resolved at the beginning of any major shift in network architecture or direction. The marketing department should be at the center of helping to provide the vision. The vision for the company and how it envisions where the revenue will come from is the billion-dollar question that has to be answered.

The four basic steps are the same as those that govern marketing, and they are referred to as the four *Ps*: product, place, promotion, and of course, price.

- *Product.* This involves defining the services offered, installation, warranty, product lines, packaging, and branding, to mention the most important.
- *Place.* This refers to what the marketing objectives are, that is, penetration, the channels used for achieving the penetration (direct and indirect), market exposure and competition, distribution methods (direct and indirect), VAR partnerships, distribution of add-on services and products, and sales locations like direct and indirect stores.
- *Promotion.* This basically involves the bundling or unbundling method, direct and indirect sales force (i.e., number, training, incentives), advertising (i.e., media type, copy trust, trade shows, press, ad types, agency, if used). Also included are the sales promotion, involving commissions, customer retention programs, discounts, and VAR products.
- *Price.* This involves, of course, the objective, which can be to ensure profitability for every product offering or have a loss leader in the mix: flexibility in terms of pricing for volume purchases or bundling techniques, product life cycle pricing, local market pricing differentiation, and sales allowances for indirect sales or CPE replacement and/or subsidization.

All four Ps center around the important issue of enticing the customer to want and perceive value with the product, use the service, and continue to use the service and possibly utilize additional services either offered now or some time in the future.

From the inception, the marketing department needs to feed many groups within the company. The sales department obviously needs to know what the vision and product offerings are that will be offered for initial sale. The technical community needs information related to traffic or capacity planning to ensure that the factory is in place for the sales force to sell while minimizing the capital and operating expenses. The customer care, billing, and provisioning groups need to know the market and services so that billing can be issued and of course the customer serviced in as high a quality as economically possible based on the services offered.

Marketing information is not static but is dynamic in that market conditions always change and the company must respond to those changes; the response could be not to act. Therefore, the marketing department needs to interact with the other functioning entities within the company to ensure continued success and refinements take place.

From the technical communities' aspect, the marketing information will drive the infrastructure layout and design. The RF and network design aspects, while related, are not the same but utilize the same fundamental information to derive their answers and designs.

The overriding point is that convergence is not only a technology issue but also an approach that the wireless companies need to adhere to in their internal workings as well.

References

Chang, Dean, and Subir Varma, "WiMax Rolls Interop Guidelines for 802.16a," Electronic Engineering Times, May 16, 2003.

Eklund, Carl, Roger Marks, and Kenneth Stanwood, "IEEE Standard 802.16: A Technical Overview of the Wireless MAN Air Interface for Broadband Wireless Access," IEEE Communications Magazine, June 2002, pp. 98–107.

Fettweis, Gerhard, "A 5G Wireless Communication Vision," Microwave Journal, p. 24, Dec 2012.

"Future Mobile Handsets 8th Edition Worldwide Market Analysis & Strategic Outlook, 2006 -2011," INFORMA, 2006.

Hosking, Rodger, "Software Defined Radio Handbook, 9th edition," Pentek Inc, New Jersey, 2011.

IEEE 802.16-2001: IEEE Standard for Local and Metropolitan Area Networks, Part 16: Air Interface for Fixed Broadband Wireless Access Systems, April 8, 2002.

IEEE Std 802.16.2™-2004: Coexistence of Fixed Broadband Wireless Access Systems.

Koukoulidis, Vassilios, and Mehul Shah, "The IP Multimedia Domain in Wireless Networks: Concepts, Architecture, Protocols and Applications," Proceedings of the IEEE Sixth International Symposium on Multimedia Software Engineering (ISMSE '04), 2004.

Location Services (LCS); LCS Application Protocol (LCS-AP) between the Mobile Management Entity (MME) and Evolved Serving Mobile Location Centre (E-SMLC); SLs interface (Release 9).

"Mobile Application Platforms and Operating Systems: Technology Trends and Challenges 3rd Edition," INFORMA, 2006.

O'Keefe, Sue, "TDD vs FDD: The Next Hurdle," Telecommunications, December 1999, p. 40.

Roessler, A., and M. Kottkamp, LTE-Advanced (3GPP Rel.11) Technology Introduction White Paper 7.2013 - 1MA232_1E, Rhode& Schwarz.

Roos, Anders, Magnus Artman, and Stephen Dutnall, "Critical Issues for Roaming in 3G," IEEE Wireless Communications, February 2003.

Sheppart, Steven, *IMS Crash Course*, McGraw-Hill, New York, 2006.

Shneydermanm, Alex and Alessio Casati, *Fixed Mobile Convergence*, McGraw-Hill, New York, 2008.

Smith, Clint, *LMDS*, McGraw-Hill, New York, 2000.

Smith, Clint, *Practical Cellular and PCS Design*, McGraw-Hill, New York, 1997.

Smith, Meyer, *3G Wireless with WiFi and WiMax*, McGraw-Hill, New York, 2004.

3GPP TS 29.171 V9.2.0 (2010-09) Technical Specification Group Core Network and Terminals;

www.android.com.

www.dvb.org.

www.ieee802.org.

www.microsoft.com.

www.t-dmb.org.

www.wikipedia.org.

www.wimaxforum.org.

www.wirelessman.org.

Index

A

D

E

F

G

H

I

J

K

L

M

N

O

P

S

T

U

V

X